Microbial Nutraceuticals

Microbial Nutraceuticals

Products and Processes

Edited by Sudhir Pratap Singh and Santosh Kumar Upadhyay

WILEY

This edition first published 2025

Registered Offices
John Wiley & Sons, Inc., 111 River Street, Hoboken, NJ 07030, USA
John Wiley & Sons Ltd, New Era House, 8 Oldlands Way, Bognor Regis, West Sussex, PO22 9NQ, UK

For details of our global editorial offices, customer services, and more information about Wiley products visit us at www.wiley.com.

The manufacturer's authorized representative according to the EU General Product Safety Regulation is Wiley-VCH GmbH, Boschstr. 12, 69469 Weinheim, Germany, e-mail: Product_Safety@wiley.com.

Wiley also publishes its books in a variety of electronic formats and by print-on-demand. Some content that appears in standard print versions of this book may not be available in other formats.

Library of Congress Cataloging-in-Publication Data applied for

Hardback: 9781394241507

Cover Design: Wiley
Cover Image: © Per Bengtsson/Shutterstock

Set in 9.5/12.5pt STIXTwoText by Straive, Pondicherry, India

Contents

List of Contributors

Amin Abbasi
Department of Food Science and Technology
National Nutrition and Food Technology
Research Institute, Faculty of Nutrition
Science and Food Technology
Shahid Beheshti University of Medical Sciences
19395-4741, Tehran, Iran

Neeraj Agarwal
Department of Biotechnology
Noida Institute of Engineering & Technology
Greater Noida, 201310
Uttar Pradesh, India

Cristobal Noe Aguilar
Bioprocesses Research Group
Food Research Department, School of Chemistry
Universidad Autonoma de Coahuila
Unidad Saltillo, 25280
Coahuila, Mexico

Shayma Thyab Gddoa Al-Sahlany
Department of Food Science
College of Agriculture, University of Basrah
Basra City, 61004, Iraq

Hend Altaib
Department of Research and Development
Bacterial Vaccines Section
Middle East for Veterinary Vaccines (MEVAC)
Company, El-Salhya
El-Gdeda, 44813, EL-Sharkya
Egypt

Yassien Badr
Department of Infectious Diseases and
Epidemics, Faculty of Veterinary Medicine
Damanhour University, Damanhour
El-Beheira, 22511, Egypt

Mamoni Banerjee
Bio-Research Laboratory
Rajendra Mishra School of Engineering
Entrepreneurship
Indian Institute of Technology Kharagpur
Kharagpur, West Bengal, 721302, India

Sara Bazzaz
Department of Food Science and Technology
National Nutrition and Food Technology
Research Institute, Faculty of Nutrition
Science and Food Technology
Shahid Beheshti University of Medical Sciences
19395-4741, Tehran, Iran

Susana Calderón-Toledo
Laboratorio de Biología Molecular
Facultad de Farmacia y Bioquímica
Universidad Nacional, Mayor de San Marcos
Lima, Perú

Ailton Cesar Lemes
School of Chemistry, Department of
Biochemical Engineering
Federal University of Rio de Janeiro (UFRJ)
Rio de Janeiro, 21941-909
Brazil

Prakram Singh Chauhan
DBT-IOC Centre for Advanced
Bioenergy Research, R&D Centre
Indian Oil Corporation Ltd. Sector-13,
Faridabad, Haryana
121007, India

Poonam Choudhary
Biomanufacturing Process Lab
Department of Biosciences and Bioengineering
Indian Institute of Technology Roorkee
Roorkee, 247667
Uttarakhand, India

Tripti Dadheech
Department of Chemistry, Queen's University
Kingston, Ontario, K7L 3N6
Canada

Mariana Buranelo Egea
Goiano Federal Institute of Education
Science and Technology, Campus Rio Verde
Goiás, Rio Verde, 75901-970
Brazil

Mahmoud A. M. El-Nouby
Department of Pesticide Chemistry and
Technology, Faculty of Agriculture
Alexandria University, Alexandria
21545, El-Shatby, Egypt

Unit of Analysis and Scientific Services
(Central Lab), Faculty of Agriculture
Alexandria University, Alexandria
21545, El-Shatby, Egypt

R. Elangovan
Department of Biochemical Engineering and
Biotechnology, Indian Institute of Technology
Delhi, Hauz-Khas, New-Delhi
110016, India

Josemar Gonçalves de Oliveira Filho
Brazilian Agricultural Research Corporation
Embrapa Instrumentation, São Carlos
13561-206, São Paulo, Brazil

Atiyeh Ghafouri Ghotbabad
Department of Food Science and Technology
National Nutrition and Food Technology
Research Institute, Faculty of Nutrition
Science and Food Technology
Shahid Beheshti University of Medical Sciences
19395-4741, Tehran, Iran

Neha Goel
Biomanufacturing Process Lab
Department of Biosciences and Bioengineering
Indian Institute of Technology Roorkee
Roorkee, 247667, Uttarakhand, India

Mónica L. Chávez González
Bioprocesses Research Group
Food Research Department
School of Chemistry, Universidad Autonoma
de Coahuila, Unidad Saltillo, 25280
Coahuila, Mexico

Hedayat Hosseini
Department of Food Science and Technology
National Nutrition and Food Technology
Research Institute, Faculty of Nutrition
Science and Food Technology
Shahid Beheshti University of Medical Sciences
19395-4741, Tehran, Iran

Avijeet S. Jaswal
Department of Biochemical Engineering and
Biotechnology, Indian Institute of Technology
Delhi, Hauz-Khas, New-Delhi
110016, India

Chayanika Kala
Department of Pathology
G.S.V.M. Medical College
Kanpur, 208002, Uttar Pradesh
India

Sanjay Kala
Department of General Surgery
G.S.V.M. Medical College
Kanpur, 208002, Uttar Pradesh
India

Shweta Kamboj
Department of Biotechnology, Chandigarh College of Technology, Chandigarh Group of Colleges, Landran, Mohali, Punjab
140301, India

Maninder Jeet Kaur
Department of Biotechnology, Chandigarh College of Technology, Chandigarh Group of Colleges, Landran, Mohali, Punjab
140301, India

Mohit Kumar
Department of Life Sciences
Sharda School of Bioscience & Technology
Sharda University, Greater Noida, 201310
Uttar Pradesh, India

Narendra Kumar
Department of Biotechnology and Bioengineering, Institute of Advanced Research
Gandhinagar, 382426, Gujarat, India

Rajni Kumari
Microbial Catalysis and Process Engineering Laboratory, Department of Microbiology
School of Life Sciences, Central University of Rajasthan Bandarsindri, Kishangarh, Ajmer
Rajasthan, 305817, India

Leili Aghebati Maleki
Immunology Research Center
Tabriz University of Medical Sciences
19294-4753, Tabriz, Iran

Parmis Mirzaei
Department of Clinical Nutrition & Dietetics
Faculty of Nutrition Science and Food Technology, Shahid Beheshti University of Medical Sciences, 19395-4741
Tehran, Iran

Santosh Kumar Mishra
Department of Life Sciences
Sharda School of Bioscience & Technology
Sharda University, Greater Noida, 201310
Uttar Pradesh, India

Saroj Mishra
Department of Biochemical Engineering and Biotechnology, Indian Institute of Technology Delhi, Hauz-Khas, New-Delhi
110016, India

Monika
Microbial Catalysis and Process Engineering Laboratory, Department of Microbiology
School of Life Sciences, Central University of Rajasthan Bandarsindri, Kishangarh, Ajmer
Rajasthan, 305817, India

Alaa Kareem Niamah
Department of Food Science
College of Agriculture, University of Basrah
Basra City, 61004, Iraq

Nupur
Department of Life Sciences
Sharda School of Bioscience & Technology
Sharda University, Greater Noida, 201310
Uttar Pradesh, India

Parvin Orojzade
Department of Food Sciences and Technology
Faculty of Nutrition & Food Sciences
Tabriz University of Medical Sciences
19294-4753, Tabriz, Iran

Nidhi Pareek
Microbial Catalysis and Process Engineering Laboratory, Department of Microbiology
School of Life Sciences, Central University of Rajasthan Bandarsindri, Kishangarh, Ajmer
Rajasthan, 305817, India

Ami R. Patel
Division of Dairy Microbiology
Mansinhbhai Institute of Dairy and Food Technology-MIDFT
Kamdhenu University, Mehsana
384002, Gujarat, India

Satya Narayan Patel
Department of Food Science and Technology
Texas A&M University, College Station, Texas
77843, United States

Suresh D. Pillai
Department of Food Science and Technology
Texas A&M University, College Station, Texas
77843, United States

Hadi Pourjafar
Dietary Supplements and Probiotic Research
Center, Alborz University of Medical
Sciences, 19395-4963, Karaj, Iran

Pawan Prabhakar
Bio-Research Laboratory
Rajendra Mishra School of Engineering
Entrepreneurship
Indian Institute of Technology Kharagpur
Kharagpur, West Bengal, 721302, India

School of Business, Woxsen University
Hyderabad, Telangana, 502345, India

Amit Kumar Rai
Food and Nutritional Biotechnology Division
National Agri-Food and Biomanufacturing
Institute, BRIC-NABI, Sector-81
(Knowledge city)
SAS Nagar, Mohali, Panjab
140306, India

Govindan Rajamohan
Council of Scientific Industrial Research-
Institute of Microbial Technology, Chandigarh
160036, India

Ranju Kumari Rathour
Department of Biotechnology
Chandigarh College of Technology
Chandigarh Group of Colleges, Landran
Mohali, Punjab, 140301, India

Sahar Sabahi
Department of Nutrition
School of Allied Medical Sciences
Ahvaz Jundishapur University of Medical
Sciences, 19294-3342
Ahvaz, Iran

César Salcedo-Okuma
Laboratorio de Biología Molecular
Facultad de Farmacia y Bioquímica
Universidad Nacional, Mayor de San Marcos
Lima, Perú

José Sandoval-Cortés
Bioprocesses Research Group
Food Research Department
School of Chemistry, Universidad Autonoma
de Coahuila, Unidad Saltillo
25280, Coahuila, Mexico

Arunika Saxena
Biology Division
DC Model School Group of Institution
Sector –9, Faridabad, Haryana
121006, India

Deepak Sharma
Department of Biotechnology
Chandigarh College of Technology
Chandigarh Group of Colleges
Landran, Mohali, Punjab
140301, India

Nitish Sharma
Department of Genetics
Waksman institute of Microbiology
Rutgers University (A State University of New
Jersey), New Brunswick, NJ
08901, USA

Sweety Sharma
Indian Institute of Science Education and
Research Mohali, SAS Nagar
Punjab, 140306, India

Elham Sheykhsaran
Department of Medical Bacteriology and
Virology, Tabriz University of Medical
Sciences, 19294-4753
Tabriz, Iran

Rakesh Mohan Shukla
Processing and Food Engineering Division
College of Agricultural Engineering and Technology (CoAE&T)
Sher-e-Kashmir University of Agricultural Sciences and Technology of Kashmir
Srinagar, 190025, Jammu and Kashmir
India

Aditi Singh
Department of Life Sciences
Sharda School of Bioscience & Technology
Sharda University, Greater Noida, 201310
Uttar Pradesh, India

Amit Kumar Singh
Department of Post Harvest Technology
College of Horticulture
Banda University of Agriculture and Technology
Banda, 210001, Uttar Pradesh
India

Anurag Singh
Department of General Surgery
G.S.V.M. Medical College
Kanpur, 208002, Uttar Pradesh
India

Ashish Kumar Singh
Center of Innovative and Applied Bioprocessing
BRIC-NABI, Knowledge City
SAS Nagar Mohali, Punjab, 140306
India

Ashutosh Kumar Singh
Department of Biotechnology
Panjab University, Chandigarh, 160014
India

Balvinder Singh
Council of Scientific Industrial Research-Institute of Microbial Technology, Chandigarh
160036, India

Manish Kumar Singh
Department of Vegetable Science, College of Horticulture, Banda University of Agriculture and Technology, Banda, 210001, Uttar Pradesh
India

Shashank Singh
Multi-Disciplinary Research Unit
G.S.V.M. Medical College
Kanpur, 208002, Uttar Pradesh
India

Smita Singh
Department of Allied Health Sciences
Chitkara School of Health Sciences
Chitkara University, Shambu Kalan, Punjab
140401, India

Sudhir Pratap Singh
Department of Industrial Biotechnology
Gujrat Biotechnology University, GIFT City
Gandhinagar, Gujarat, 382355
India

Vijaya Bharathi Srinivasan
Council of Scientific Industrial Research-Institute of Microbial Technology
Chandigarh, 160036
India

Prem Prakash Srivastav
Agricultural and Food Engineering Department Indian Institute of Technology Kharagpur
Kharagpur, 721302, West Bengal
India

Santosh Kumar Upadhyay
Department of Botany, Panjab University
Chandigarh, Sector 14, 160014
India

Gemilang Lara Utama
Faculty of Agro-Industrial Technology, Universitas Padjadjaran
Sumedang, West Java, 45363
Indonesia

Center for Environment and Sustainability Science, Universitas Padjadjaran, Bandung
Jawa Barat, 40132
Indonesia

Deepak Kumar Verma
Agricultural and Food Engineering Department
Indian Institute of Technology Kharagpur
Kharagpur, West Bengal, 721302
India

V Vivekanand
Centre for Energy and Environment
Malaviya National Institute of Technology
Jaipur, Rajasthan, 302017
India

Alexandra Wallimann
CK-CARE,
Christine Kühne-Center for Allergy Research and Education, 7265, Davos Wolfgang
Graubünden, Switzerland

Amparo Iris Zavaleta
Laboratorio de Biología Molecular
Facultad de Farmacia y Bioquímica
Universidad Nacional, Mayor de San Marcos
Lima, Perú

Jyoti Yadav
Department of Industrial Biotechnology
Gujarat Biotechnology University
GIFT City, Gandhinagar, Gujarat
382355, India

Hamideh Fathi Zavoshti
Department of Food Hygiene and Aquatics
Faculty of Veterinary Medicine
Tabriz University, 19294-2253
Tabriz, Iran

Mohammadreza Ziavand
Department of Food Sciences and Technology
Faculty of Nutrition & Food Sciences
Tabriz University of Medical Sciences
19294-4753, Tabriz, Iran

About the Editors

Dr. Santosh Kumar Upadhyay is currently working as an assistant professor of Botany at the Panjab University, Chandigarh, India. He has been working in the field of Plant Biotechnology for more than 16 years. His present research focuses in the area of functional genomics for stress tolerance. He is involved in the bio-prospecting and characterization of various important proteins from plant biodiversity, and defense and stress signaling genes in bread wheat. His research group has developed a pool of vital genes and long noncoding RNAs associated with abiotic and biotic stress tolerance and signaling in plants. He has also established the method for genome editing in bread wheat using CRISPR-Cas system and developed a tool SSinder for CRISPR target site prediction. His research contribution led to the publication of more than 80 research papers in leading journals of international repute. Further, there are more than five national and international patents, 50 book chapters, and 16 books in his credit. In recognition of his substantial research record, he has been awarded NAAS Young scientist award (2017–2018) and NAAS-Associate (2018) from the National Academy of Agricultural Sciences, India, INSA-Associate Fellow (2023), and INSA Medal for Young Scientist (2013) from the Indian National Science Academy, India, NASI-Young Scientist Platinum Jubilee Award (2012) from the National Academy of Sciences, India, and Altech Young Scientist Award (2011). He has also been the recipient of the prestigious DST-INSPIRE Faculty Fellowship (2012) and SERB-Early Career Research Award (2016) from the Ministry of Science and Technology, Government of India. Dr. Upadhyay also serves as a member of the editorial board and reviewer of several peer-reviewed international journals.

Prof. Sudhir Pratap Singh is currently Professor at the Department of Industrial Biotechnology, Gujarat Biotechnology University (DST, Govt. of Gujarat), Gandhinagar, Gujarat, India. He has been working in the area of molecular biology and biotechnology for more than a decade. Currently, his primary focus of research is gene mining, biocatalyst engineering, and characterization of novel enzyme variants to produce high-value functional biomolecules from low-cost feedstock. His team generated metagenomic data resources from diverse ecological niches and ethnic food samples and characterized novel genes encoding enzyme variants with desirable catalytic properties for the biosynthesis of functional sugar molecules such as D-allulose, turanose, trehalose, and trehalulose. Further, his group achieved enzymatic production of prebiotic molecules, such as fructooligosaccharides, glucooligosaccharides, 4-galactosyl-Kojibiose, xylooligosaccharides, levan, dextran, type III resistant starch, etc. Prof. Singh has published more than 80 research articles (SCI) and 10 books (edited). Further, he has 12 patents (granted) to his credit as an inventor. Prof. Singh has been conferred TATA Innovation Fellowship-2024 (DBT, Govt. of India), Indian National Science Academy (INSA) Associate Fellowship-2024, Professor SB Chincholkar Award-2022 (BRSI), Young Scientist Award-2016/17 (IBA), SBS-MKU Genomics Award-2016/17 (BRSI), and

Professor Hira Lal Chakravarty Award-2017/18 (ISCA, DST, Govt. of India). His team was awarded Gandhian Young Technological Innovation Award (SRISTI) in 2019. He is a life member of the National Academy of Sciences, India (NASI). Further, Prof. Singh is the Fellow of the National Academy of Agricultural Sciences (FNAAS) and the Fellow of the International Society of Energy, Environment, and Sustainability (FISEES).

Preface

Nowadays, the pursuit of well-being has been extended beyond traditional medicine, driving a mounting interest in nutraceuticals—bioactive compounds that deliver both therapeutic and nutritional benefits. These nutraceuticals are pivotal in boosting immunity, nurturing health, and mitigating several chronic diseases like diabetes, obesity, and cardiovascular diseases. As the research on these compounds continued, the urgent need for efficient, innovative, and sustainable production approaches has become more critical. So, microbial biotechnology has emerged as a powerful tool for the large-scale production of these nutraceuticals. The incredible ability of microorganisms to alter essential nutrients into complex bioactive molecules makes it a versatile platform for the production of nutraceuticals. The microbial community can produce several health-promoting molecules such as vitamins, amino acids, peptides, rare sugars, and polyphenols. The efficient and eco-friendly microbial approach is a potential alternative for producing nutraceuticals over traditional methods. This book compiles the current developments in microbial biotechnology to produce different therapeutic molecules. The successive chapters highlight the significant development in microbial engineering, which has metabolically and genetically reformed cells to act as efficient cell factories. These bioengineered microbial cells have the promising potential to produce several bioactive molecules of enormous health potential. This book covers various topics, from microbial production of peptides, lipids, short-chain fatty acids, and polyphenols to prebiotics, postbiotics, synbiotics, and human milk oligosaccharides. The emerging roles of microalgae as a sustainable source of nutraceutical production and the potential applications thereof have been illustrated in the book. Due to the advent of new health challenges rising globally, it will be requisite to produce bioactive therapeutics molecules effectively and sustainably to boost community wellness and longevity. By exploiting microorganisms' potential, researchers can make meaningful progress toward confronting global health and enhancing individual wellness.

Prof. Sudhir Pratap Singh
Department of Industrial Biotechnology
Gujarat Biotechnology University (DST, Govt. of Gujarat),
Gandhinagar, Gujarat, India

Dr. Santosh Kumar Upadhyay
Department of Botany
Panjab University, Chandigarh, India

1

Microbial Nutraceuticals: An Overview

Ashish Kumar Singh[1], *Santosh Kumar Upadhyay*[2], *and Sudhir Pratap Singh*[3]

[1] *Center of Innovative and Applied Bioprocessing, BRIC-NABI, Knowledge City, SAS Nagar, Mohali, Punjab, 140306, India*
[2] *Department of Botany, Panjab University, Chandigarh, Sector 14, 160014, India*
[3] *Department of Industrial Biotechnology, Gujrat Biotechnology University, GIFT City, Gandhinagar, Gujarat, 382355, India*

1.1 Introduction

1.1.1 Overview of Microbial Nutraceuticals

Food and medicine are the two fundamental requirements in the quest for a healthy life. Food is a source of nourishment and vital in nurturing the body, delivering essential nutrients for development, growth, and well-being (Helal et al. 2019; Maurya et al. 2020). Conversely, a medicine derived from natural sources or chemically synthesized addresses only specific types of diseases or physiological anomalies in the body (Maurya et al. 2020). Although diverse in their purpose, these two elements merge into the concept of nutraceuticals. Products that seamlessly blend the benefits of medicine and food are known as nutraceuticals (Maurya et al. 2020). Even though the term nutraceuticals seems new, the concept finds its roots in the ancient civilizations of Indian, Roman, Sumerian, and Chinese cultures. These ancient civilizations incorporated nutraceuticals into their traditional medicine practices, which are recognized nowadays as alternative or complementary medicine systems (Singh et al. 2018). They know that good nutrition is vital for maintaining health and preventing disease, embodying the concept of "Prevention is better than cure (Cardenas 2013; Maurya et al. 2020)."

Nutraceuticals are an important category of molecules known for their long-term health benefits. Nutraceuticals can help in the reduction of diseases related to aging, *viz.*, inflammation, arthritis, depression, cancer, diabetes, and gastrointestinal and cardiovascular diseases (CVDs) (Wang et al. 2016; Yuan and Alper 2019; Chhabra et al. 2021). Traditionally, these nutraceutical molecules, such as polysaccharides, phytochemicals, carotenoids, amino acids, vitamins, pigments, and polyphenols, are sourced and isolated from animals, plants, and microbes (Das et al. 2012; Singh and Upadhyay 2021, 2023; Upadhyay and Singh 2021, 2023, 2024; Bisht et al. 2024). However, direct extraction methods are restricted to accessibility and cost of raw material, low natural content, quality control issues, and purity of these nutraceuticals. Although chemical synthesis can be an alternative source, it is suitable only for producing simple compounds and not for complex molecules like oligosaccharides (OSs), polyphenols, and pigments (Wang et al. 2016).

Microbial Nutraceuticals: Products and Processes, First Edition. Edited by Sudhir Pratap Singh and Santosh Kumar Upadhyay.

Microbial-based metabolic engineering has become a promising method to overcome these challenges and has gained significant attention for producing value-added nutraceuticals. After understanding the genetic flexibility of microbes and the metabolic pathway of natural products, scientists can create microbial hosts for large-scale production of nutraceuticals using simple carbon sources. These microbial cells are used for the production of different nutraceuticals like amino acids, bioactive peptides, short-chain fatty acids (SCFAs), polyphenols, OSs, functional rare sugars, polymeric nutraceuticals, pigments, and specialized nutraceuticals. A few specific examples have been briefly covered in this chapter, and the following chapters will explore these topics in detail.

1.2 Microbial Production of Nutrients

1.2.1 Microbial Amino Acid and Peptide Production

Today, food is known for its nutritional value and biologically active compounds that promote health and well-being. As the awareness of the effect of diet on health grows, people choose natural products rich in minerals, vitamins, and bioactive peptides (Zaky et al. 2022). Due to their physicochemical functions and role as essential dietary nutrients, proteins in food are gaining more attention. These physicochemical functions of food are due to the presence of peptide sequences in protein, which become active after enzymatic digestion during food processing or microbial fermentation (Chakrabarti et al. 2018).

Bioactive peptides are defined as sequences within proteins that confer health benefits beyond essential nutrition (Chakrabarti et al. 2018). These bioactive peptides can offer various health benefits, including lowering blood pressure, fighting infections, preventing blood clots, boosting the immune system, acting as antioxidants, and aiding in mineral absorption (Sarker 2022; Zaky et al. 2022). Various plant and animal food proteins are used to extract bioactive peptides. Familiar animal protein sources include meat, milk (casein and whey), and eggs. Plant sources typically used are oats, soy, pulses, wheat, flaxseed, and hemp seed. Marine proteins from sources like salmon, fish, sea urchins, etc., are also used (Chakrabarti et al. 2018; Tachie et al. 2023). Bioactive peptides can be synthesized through three primary methods: (i) *via* microorganisms, primarily through fermentation processes; (ii) through the action of gastrointestinal digestive enzymes such as pepsin, trypsin, and chymotrypsin; and (iii) using *in vitro* techniques, which involve treatment with proteolytic enzymes, exposure to high temperatures, processed food by-products, and extreme pH conditions (Mahgoub et al. 2023).

Amino acids, recognized as the fundamental units of proteins and peptides, are essential components in biological systems, often referred to as the building blocks of life. Notably, amino acids contain approximately 16% nitrogen, differentiating them from fats and carbohydrates (Mahmood 2014; Liu et al. 2018; Ding et al. 2024; Madhushan et al. 2024). Amino acids are classified into two groups: essential amino acids, which the human body does not synthesize, and non-essential amino acids, which are synthesized by the human body (Mahmood 2014). The amino acids have great demand and application in several industries like food, feed, cosmetics, and pharmaceuticals, due to their critical role as primary metabolites in protein constituents within cells (Mahmood 2014; Wendisch 2014; Hirasawa and Shimizu 2016; Liu et al. 2018). A significant percentage of the amino acid market, nearly two-thirds, is dominated by L-glutamic acid and its salts (Wendisch 2014). These compounds, particularly monosodium glutamate, are extensively utilized as flavor enhancers in the food industry (Hirasawa and Shimizu 2016).

Additionally, incorporating amino acids into animal feed can enhance animal growth and improve meat quality (Liu et al. 2018). Historically, it was believed that microbial cells could not produce amino acids. However, in 1956, this concept was challenged by Japanese scientists, Dr Shukuo Kinosita and Dr Shigezo Udaka, after discovering *Corynebacterium glutamicum*, which makes a substantial amount of glutamic acid (Hirasawa and Shimizu 2016). Consequently, many amino acids have been produced using microbial cells, leading to significant growth in the amino acid market (Hirasawa and Shimizu 2016; Sanchez et al. 2018). Also, current advancements in metabolic engineering are used for the development of enhanced microbial cell design for the production of amino acids and peptides by adding a system biology concept along with innovative technologies such as synthetic biology, evolutionary engineering, and single-cell analysis (Hirasawa and Shimizu 2016; Liu et al. 2018; Sanchez et al. 2018; Wang et al. 2022; Zaky et al. 2022; Tachie et al. 2023). The subsequent chapter in this book discusses the production and processing of bioactive peptides and amino acids through enzymatic hydrolysis and microbial fermentation. It explores and covers conventional purification and identification methods, highlights their potential health benefits, and examines possible applications of these amino acids and bioactive peptides.

1.2.2 Dietary Short-Chain Fatty Acid Production

Microbes are found throughout the body, such as on the skin, in saliva, in the oral and vaginal mucosa, and in conjunctiva. Still, most of them are found in the gastrointestinal tract. The gut microbiota consists of trillions of microorganisms, including viruses, bacteria, archaea, and eukaryotes (Tang et al. 2019; Fusco et al. 2023; Facchin et al. 2024). Microbes produce many metabolites that benefit human health, particularly those from gut microbes, which are crucial for interactions between the intestinal microbiota and the host. These metabolites include polyamines, *viz.*, putrescine, spermidine, and spermine, secondary bile acids, indoles, succinate, and SCFAs (Gill et al. 2018; Lavelle and Sokol 2020; Krautkramer et al. 2021). SCFAs are among the most frequently studied small-molecule metabolites produced by the fermentation of dietary fiber by gut microbes (Krautkramer et al. 2021). The gut microbiota produces a different type of SCFAs, namely butyrate, acetate, and propionate, as critical metabolites through the fermentation of different dietary fibers such as polysaccharides, resistant starch, and simple sugars (Schönfeld and Wojtczak 2016; Nogal et al. 2021). These SCFAs are very crucial for human health because they influence different physiological processes like regulating glucose, lipid, and cholesterol metabolism, along with modulating immune response and maintaining gut barrier integrity by preventing the entry of harmful substances into the bloodstream (Morrison and Preston 2016; Schönfeld and Wojtczak 2016; Wegh et al. 2019; Lavelle and Sokol 2020; Nogal et al. 2021).

1.3 Oligosaccharide Production

1.3.1 Prebiotic Oligosaccharide Molecule Production in Microbial Cells

Nowadays, worldwide consumers are progressively conscious of the quality and safety of food (Patel and Goyal 2011). As health consciousness increases among consumers, the demand for specific dietary carbohydrates such as OSs and resistant starch also increases due to these products' several eminent health benefits (Patel and Goyal 2011; Meyer et al. 2015). Functional OSs have been commercially utilized since the 1980s as low-calorie bulking agents, particularly gaining significant acceptance in pharmaceuticals and food industries due to their physiological benefits (Patel and Goyal 2011; Nguyen and Haltrich 2013; Meyer et al. 2015; Cezar et al. 2017). In Japan, the concept of functional food was introduced in 1991, which led to the classification of OS as "food for specified health," and current global research also supported the invasion of non-digestible oligosaccharides (NDOs) in the same category (Patel and Goyal 2011; Bujna et al. 2022). OS with the degree of polymerization DP3 to DP10 plays a critical role in biological processes, and they mainly stimulate the beneficial colon microorganisms (Meyer et al. 2015; Cezar et al. 2017; Davani-Davari et al. 2019; Bujna et al. 2022). Over the current era, the significance of prebiotic OSs has grown due to several health benefits, such as enhanced calcium absorption, prevention of colon cancer, increased immune defense system, and reduced allergy risk (Belorkar and Gupta 2016; Bujna et al. 2022).

1.3.2 Microbial Transformation and Bio-production of High-Value Rare Functional Sugars: Sources, Methods, and Safety Aspects

Worldwide, the rate of noncommunicable diseases such as hyperlipidemia, diabetes, and obesity has increased significantly due to the overconsumption of high-fat food and simple sugars (Li et al. 2017; Smith et al. 2022). Therefore, food manufacturers are gradually reformulating the products with polyols, high-intensity sweeteners, and dietary fibers instead of different simple sugars like fructose, sucrose, and glucose. However, alternative sweeteners have been linked to potential health issues, including appetite dysregulation and glucose intolerance (Smith et al. 2022). So, low-calorie and non-calorie sugars have come in demand for food production (Sharma et al. 2024). Rare sugars, in particular, have become increasingly popular as they provide sweetness without the high-calorie content (Singh et al. 2023). Rare sugars, defined by the International Society of Rare Sugars as "monosaccharides and their derivatives that hardly exist in nature," have received limited attention historically. These sugars, present in small quantities in foods such as honey, certain fruits, vegetables, and grains, are gaining interest due to recent advancements in their commercial-scale biosynthesis (Ahmed et al. 2022; Smith et al. 2022; Patel et al. 2023). Rare sugars are low-energy monosaccharides that exhibit sweetness comparable to sucrose. Specifically, D-psicose or D-allulose and D-tagatose have been considered safe for consumption and are already incorporated into various products such as biscuits, chocolates, jams, protein bars, soft drinks, and commercial sweetener blends across Europe, Asia, and the United States (Mijailovic et al. 2021; Smith et al. 2022). Rare sugars can be produced using both chemical and biological methods. Chemical synthesis of rare sugars is hindered by challenges such as low yield, the need for extensive purification processes, and the generation of toxic by-products, which can harm the environment (Sharma et al. 2024). Microbial production of rare sugars is particularly significant due to its potential for higher efficiency and reduced environmental impact compared to chemical methods (Patel et al. 2023; Singh et al. 2023).

Milk consumption is high worldwide, with India reaching 81 million tons in 2021. Milk production is projected to increase by 1.6% annually from 2022 to 2029 (Gambelli 2017). This rise in production leads to more dairy waste, primarily whey, which contains lactose (45–52 g/L) and contributes to environmental pollution due to high biological oxygen demand (BOD) and chemical oxygen demand (COD) levels (Zhang et al. 2022). Directly using whey is difficult due to sweetening properties and low solubility of lactose, but it can be transformed into valuable rare sugars through microbial fermentation or enzymatic reactions. The microbial production of rare sugars is environmentally friendly, sustainable, and cost-effective compared to chemical synthesis. Different rare sugars, such as lactulose, tagatose, and epilactose, have several applications in the pharmaceuticals, agriculture, and cosmetics industries (Roy et al. 2018; Vera et al. 2022; Moss et al. 2024). These rage sugars have health benefits in conditions like diabetes, obesity, and chronic constipation (Lu et al. 2008; Vera et al. 2022; Milasing et al. 2024; Moss et al. 2024).

1.3.3 Microbial Production of High-Value Polyphenolics

The plant produces different types of natural compounds in their metabolic pathways, known as phytochemicals. These phytochemicals are a rich source of bioactive substances with several applications in biotechnology and medicine (Chouhan et al. 2017; Delmulle et al. 2018). Since ancient times, humans have used plants for flavors, medicine, colorants, and fragrances due to their unique structural and functional groups (Chouhan et al. 2017; Shah et al. 2019; Rainha et al. 2020). More than 200,000 plant-based natural products have nutritional and pharmacological importance because of their lower side effects than synthetic compounds (Chouhan et al. 2017; Sharma et al. 2019).

Despite food production and hygiene developments in the current era, microbial contamination still causes severe loss of food, health, and the economy. Different traditional and synthetic food preservatives, such as benzoates and nitrates, are commonly used as food preservatives, but they pose severe health risks like cancer and liver damage. Consumers are increasingly favoring natural antimicrobials due to their growing demand for healthier options. Among the several natural antimicrobials, polyphenols have gained more attention due to their antimicrobial properties and health benefits, and they are now used as an emerging natural food preservative (Ofosu et al. 2020). Different natural phytochemicals that have been synthesized or modified by chemical methods possess several disadvantages, such as toxicity and high cost of production, making them impractical for their output (Stephanopoulos et al. 1998). So, microbial biosynthesis with advanced synthetic biology techniques enhances safe and cost-effective microbial biosynthetic production (Chouhan et al. 2017).

1.3.4 Specialized Carbohydrate Production

Human milk is dominated by lactose-based carbohydrates known as human milk oligosaccharides (HMOs) (5–20 g/L), which play a crucial role in an infant's well-being and development (Kulinich and Liu 2016; Yu et al. 2018; Zeuner and Meyer 2020; Zhu et al. 2022). On the other hand, bovine milk used for infant formula lacks the HMO-like structure. So, research is growing on producing HMOs in infant formula similar to human milk with the same health benefits (Zeuner and Meyer 2020). To date, more than 200 different types of HMOs have been identified in human milk, of which about 100 were studied (Kulinich and Liu 2016; Zeuner and Meyer 2020; Zhu et al. 2022). Natural HMOs are composed of five monosaccharides: α-L-fucose (Fuc), β-D-galactose (Gal), β-D-*N*-acetylglucosamine (GlcNAc), β-D-glucose (Glc), and α-D-*N*-acetylneuraminic acid (Neu5Ac), always with lactose at the reducing end (Zeuner and Meyer 2020; Zhu et al. 2022). The most common are the fucosylated HMOs, which act as decoy receptors in the infant's gut, preventing pathogens from adhering to the gut lining and supporting immune system development (Kulinich and Liu 2016; Zeuner and Meyer 2020; Zhu et al. 2022). HMOs are the third largest group of bioactive molecules in human milk, with colostrum containing 20–25 g/L of HMOs, decreasing to 5–20 g/L as milk production matures (Kulinich and Liu 2016; You et al. 2024). The composition of these OSs varies among women and changes with lactation stage, diet, and other biological factors (Kulinich and Liu 2016).

2′-Fucosyllactose (2′-FL) is one of the most abundant (30%) HMOs in human milk, which is approved as generally recognized as safe (GRAS) by the United States FDA (Liu et al. 2020; You et al. 2024). The safety of 2′-FL for infants is also recognized by the European Food Safety Authority (EFSA) Panel on Nutrition, Novel Foods, and Food Allergens (You et al. 2024). 2′-FL is added to premium infant formula, which supports diverse intestinal flora, precisely *Bifidobacteria*, reduces neurodegradation in stroke, prevents pathogen adhesion to the gut, and protects infant intestinal epithelial cells from apoptosis (Yu et al. 2018; Chen et al. 2023; Kuntz et al. 2024; You et al. 2024). Nowadays, biosynthesis of 2′-FL is an active research area due to its application in pharmaceuticals and additives. Also, extraction of 2′-FL directly from human milk is not cost-effective, so research is going on its production from metabolically engineered microorganisms (Hollands et al. 2019; Chen et al. 2023; Zhang et al. 2023; Moya-Gonzálvez et al. 2024; You et al. 2024). So, subsequent chapters in this book provide brief insight into microbial production of different HMOs, their application, and health benefits.

1.3.5 Polymeric Nutraceuticals

Microbial polymers or biopolymers are naturally derived polymers produced by microorganisms such as yeast, molds, and bacteria (Ghosh et al. 2021). Microorganisms produce several types of polymers, such as polysaccharides, polyesters, polyphosphates, and polyamides, *via* enzymatic methods using building blocks such as amino acids, hydroxyl fatty acids, and sugars (Wang et al. 2016; Ghosh et al. 2021; Paliwal et al. 2021). These biopolymers are of great interest due to their production potential, therapeutic or nutraceutical applications, and biodegradability (Nicolescu et al. 2023). Nutraceuticals, distinct from functional foods and pharmaceuticals, encompass structurally and functionally diverse bioactive compounds offering long-term medicinal or physiological benefits (Wang et al. 2016). These nutraceuticals are sourced from plants, animals, microorganisms, and marine life; these compounds are used to promote health and prevent diseases associated with aging, such as oxidative stress, depression, inflammation, arthritis, osteoporosis, gastrointestinal diseases, CVDs, diabetes, and cancer (Paliwal et al. 2021). However, conventional nutraceutical production faces challenges like limited raw material availability, high costs, and issues with quality control. While chemical synthesis offers an alternative, it is often unsuitable for complex biochemicals (Wang et al. 2016; Paliwal et al. 2021). The promising solution to enhance the production of nutraceutical biopolymers is the metabolic engineering of microbes. Recent developments in the field of genetic engineering significantly enhanced the production of microbial nutraceuticals such as prebiotics, amino acids, phytochemicals, and polysaccharides (Wang et al. 2016; Ghosh et al. 2021; Paliwal et al. 2021).

1.4 Advanced Nutraceutical Products and Processes

1.4.1 Functional Nutraceutical Products

Due to increased awareness about healthy lifestyles and balanced diets, the market for healthy foods has increased significantly (Salminen et al. 2021; Egea et al. 2023; Thangaraj and Kim 2024). Consumers are interested in foods that meet nutritional needs and promote well-being, prompting the development of prebiotics, probiotics, and synbiotics (Pandey et al. 2015; Palai et al. 2020; Salminen et al. 2021; Khursheed et al. 2022). Prebiotics and probiotics have become prominent functional foods over the past three decades (Khursheed et al. 2022). Still, Gibson and Roberfroid (1995) introduced the term "synbiotics," which describes the combined or synergistic effects of prebiotics and probiotics (Zepeda-Hernández et al. 2021; Khursheed et al. 2022). Synbiotics are products that precisely help in the growth and development of probiotic strains, and in synbiotics, probiotics and prebiotics combine on a single matrix, which boosts intestinal growth and the immune system (Markowiak and Ślizewska 2017; Upasana 2022; Campaniello et al. 2023; Singh et al. 2023; Thangaraj and Kim 2024). So, research on the effects of synbiotics on several diseases, such as CVDs, inflammatory bowel diseases, diabetes, and cancer, is carried out (Markowiak and Ślizewska 2017; Khursheed et al. 2022; Singh et al. 2023).

Postbiotics are nonviable metabolites or soluble components produced by live bacteria or released after bacterial lysis, such as peptides, cell-surface proteins, enzymes, peptidoglycan-derived muropeptides, organic acids, teichoic acids, and polysaccharides (Thangaraj and Kim 2024). When administered sufficiently, these compounds exert biological effects on hosts, promoting health and well-being (Mishra et al. 2024; Thangaraj and Kim 2024). The International Scientific Association for Probiotic and Prebiotics defines postbiotics as "a preparation of inanimate microorganisms and their components that confers a health benefit on the host" (Salminen et al. 2021; Aggarwal et al. 2022; Liang and Xing 2023). These postbiotic substances are typically recovered *in situ* from producer strains found naturally in various fermented foods (Salminen et al. 2021; Mishra et al. 2024; Thangaraj and Kim 2024). They are produced by bacterial and fungal species like *Faecalibacterium*, *Lactobacillus*, *Eubacterium*, *Bifidobacterium*, *Saccharomyces*, and *Streptococcus*, which are commonly found in fermented foods like pickled vegetables, kombucha, yogurt, and sauerkraut (Aggarwal et al. 2022; Thorakkattu et al. 2022; Mishra et al. 2024; Thangaraj and Kim 2024). Unlike probiotics, postbiotics do not contain live microbes, thus reducing associated risks. They offer numerous health benefits, including improved immunological and digestive health, and possess antioxidant, antihypertensive, anti-obesity, anti-inflammatory, and immune-modulating properties (Thorakkattu et al. 2022; Liang and Xing 2023; Mishra et al. 2024; Thangaraj and Kim 2024). The subsequent chapter presents the multifaceted roles of postbiotics and synbiotic products with a nutraceutical impact. It highlights the selection criteria for effective synbiotics, different types, mechanisms of action, and their benefits for human health.

Pigments, which are colored substances, can be produced synthetically or naturally (Sundarajan and Ramasamy 2024). Synthetic pigments derived from chemical processes have detrimental effects on the environment and human health, including oncogenicity, teratogenicity, and toxicity (Chandi and Gill 2011; Kirti et al. 2014; Lyu et al. 2022; Sundarajan and Ramasamy 2024). Due to these detrimental effects, industries have prioritized producing natural pigments as eco-friendly alternatives (Sundarajan and Ramasamy 2024). Microbial pigments are especially promising due to their sustainability and high economic value. Among these, carotenoids stand out for their cost-effectiveness and high production rates, making them valuable for various industrial applications (Lyu et al. 2022; Barreto et al. 2023). Ancient civilizations used natural pigments from plants, insects, and minerals to dye textiles, color foods, paint, and perform religious ceremonies. The discovery of the first synthetic organic dye by William Henry Perkin in 1856 marked the beginning of the commercial synthetic pigment industry, which, despite its economic benefits, has posed significant environmental and health risks (Lyu et al. 2022).

Carotenoids, known for their yellow, orange, and red colors, are lipid-soluble pigments produced by plants, bacteria, algae, yeast, and fungi (Chandi and Gill 2011; Mata-Gómez et al. 2014; Joshi et al. 2023; Sundarajan and Ramasamy 2024). These pigments, categorized as hydrocarbons (carotenes) or oxygenated derivatives (xanthophylls), are extensively used in food, feed, cosmetics, and pharmaceuticals (Kirti et al. 2014; Mata-Gómez et al. 2014; Martínez-Cámara et al. 2021; Sundarajan and Ramasamy 2024). Carotenoids are vital for their antioxidant properties and as precursors to vitamin A, but humans must obtain them through diet or supplements (Joshi et al. 2023). There are over 1204 known carotenoids, with their synthesis regulated by environmental factors like pH, temperature, salinity, dissolved oxygen, and light (Martínez-Cámara et al. 2021; Joshi et al. 2023). In 2019, the global carotenoid market was valued at $1.44 billion and is projected to reach $1.84 billion by 2027. Despite this demand, 80–90% of commercially available carotenoids are produced synthetically, which generates toxic waste and raises health concerns (Mendes-Silva et al. 2020; Joshi et al. 2023; Ochoa-Viñals et al. 2024; Sundarajan and Ramasamy 2024). Only a tiny fraction is currently derived from microbial sources. However, microbial fermentation is gaining interest due to its lower space requirements and independence from seasonal conditions. Advances in biotechnology and bioprocessing are further enhancing the efficiency of microbial carotenoid production (Chandi and Gill 2011; Mendes-Silva et al. 2020; Martínez-Cámara et al. 2021; Joshi et al. 2023; Sundarajan and Ramasamy 2024). The next chapter talks about important strategies and new developments in improving the production of carotenoids. It also shows how microbial biotechnology could change things to meet the growing need for natural pigments in many industrial settings.

1.4.2 Specialized Nutrient Molecules

Several commonly used industrial products and chemicals highly rely on fossil fuels, which is unsustainable and environmentally harmful (Sundarajan and Ramasamy 2024). Natural gas and crude oil refining into valuable products is also highly energy-intensive, producing greenhouse gases, especially CO_2 (Karayannis et al. 2023). The global chemical industry was also responsible for 2.6 billion tons of CO_2 emission in 2019, and to alleviate these environmental effects, there is a growing focus on reforming the existing processes as well as developing new methods to use renewable raw materials (Karthikeyan et al. 2020; Karaba et al. 2021; Restrepo et al. 2021; Glowka and Krawczyk 2023). So, the research is assessing the potential of renewable resources such as saccharides, lignin, and algae for the production of their derivatives like 1,2-propanediol, ethanol, and glycerol (1,2,3-propanetriol) (Sun et al. 2018; Cao et al. 2019; Lv et al. 2020; Glowka and Krawczyk 2023). These renewable resources are being explored as sustainable substitutes for fuel and petrochemicals (Glowka and Krawczyk 2023). The initiative of these alternatives is due to environmental concerns and a finite supply of fossil fuels (Cao et al. 2019; Restrepo et al. 2021). New methods for making biofuels and biochemicals are being explored, and biobased industries are increasing due to lower environmental effects, safe processes, and diverse biomass resources (Tao et al. 2021).

1,2-Propanediol (1,2-PDO) is a versatile and highly demanding substance in several industries like pharmaceuticals, building materials, and chemicals (Saxena et al. 2010; Nonaka et al. 2021; Glowka and Krawczyk 2023). It is a monomer for producing biofuels, cosmetics, detergents, polyester resins, antifreeze agents, and food products (Saxena et al. 2009; Nonaka et al. 2021; Tao et al. 2021). Low-yield microbial production of 1,2-PDO is not suitable for industrial production. Therefore, developing new biological processes, genetic engineering, and novel strains has considerably enhanced the commercial availability of 1,2-PDO.

A new health approach highlights the link between diet, nutrition, and health, enhancing the use of functional food, nutraceuticals, and dietary supplements (Chhabra et al. 2021; Chopra et al. 2022). These products have high demand and are safe due to their natural origin (Chhabra et al. 2021; Sut and Dall'Acqua 2023). Nutraceuticals and supplements significantly prevent CVD, which include prebiotics, probiotics, phytochemicals, fibers, and polyunsaturated fats that are available in the form of tablets, capsules, powders, or supplemented in foods such as yogurt or soup (Johnston et al. 2017; Korcz et al. 2018; Soliman 2019; Sommella et al. 2019; Chhabra et al. 2021; Chopra et al. 2022; Sut and Dall'Acqua 2023). Globally, CVD causes severe illness and is the leading cause of death that affects about 523 million people (Cicero et al. 2017; Sosnowska et al. 2017; Hung et al. 2022; Liu et al. 2023). CVD and other risk factors such as hypertension, dyslipidemia, and diabetes mellitus are significantly affected by dietary habits (Cicero et al. 2021; Sut and Dall'Acqua 2023; Ge et al. 2024; Russo et al. 2024). Among these, hyperlipidemia (HLP) is characterized by elevated cholesterol, low-density lipoprotein, triglycerides in plasma, and decreased high-density lipoprotein caused by genetic or environmental factors (Attardo et al. 2022). So, diet is crucial in preventing and treating HLP (Cicero et al. 2021; Al-Okbi et al. 2023). Hypercholesterolemia, a significant risk factor for CVD, is prevalent globally, especially in low- and middle-income countries (Ge et al. 2024). The following chapter in this book focuses on the microbial production of dietary nutraceuticals used for lowering lipid levels.

Vitamins and minerals are distinct micronutrients necessary in trace amounts for survival and maintaining health (Acevedo-Rocha et al. 2019). Humans and animals cannot synthesize vitamins independently, necessitating their intake through diet or supplements. In contrast, plants and microorganisms naturally produce these essential organic compounds, which primarily function as coenzymes in metabolic reactions (Survase et al. 2006; Acevedo-Rocha et al. 2019; Wang et al. 2021). The 13 vitamins are categorized as either fat-soluble (A, D, E, K) or water-soluble (C, B group) (Acevedo-Rocha et al. 2019). Pathological conditions, malnutrition, poor diet, high levels of exercise, pregnancy, stress, and drug abuse can increase the demand for vitamins. Additionally, food processing and preservation methods can reduce vitamin content due to their sensitivity to light, heat, oxygen, and pH (Vandamme and Revuelta 2016; Acevedo-Rocha et al. 2019). Consequently, the production of vitamins for various industries, including feed, food, cosmetics, chemicals, and pharmaceuticals, has become crucial (Survase et al. 2006). Animals, unable to produce sufficient vitamins, require these micronutrients for proper growth and health. Vitamins are made either through chemical synthesis or fermentative processes (Survase et al. 2006; Acevedo-Rocha et al. 2019; Wang et al. 2021). Beyond their nutritional roles, vitamins are increasingly used as food and feed additives, therapeutic agents, health aids, and technical aids. Today, many processed foods, feeds, pharmaceuticals, cosmetics, and chemicals contain added vitamins or vitamin-related compounds to meet consumers' nutritional and health needs (Survase et al. 2006; Acevedo-Rocha et al. 2019). Vitamins are currently produced through chemical synthesis and biotechnology approaches (Survase et al. 2006). Chemical synthesis often uses non-renewable chemicals and generates hazardous waste, raising sustainability and economic concerns (Survase et al. 2006; Acevedo-Rocha et al. 2019). Bioprocesses, however, can positively impact both the economy and the environment. Advances in understanding vitamin biosynthesis and regulation in microbes, along with new metabolic engineering tools, make it feasible to engineer high-yield vitamin-producing strains (Survase et al. 2006; Eggersdorfer et al. 2012; Calvillo et al. 2022; Cho et al. 2022). Industrial-scale production requires efficient and cost-effective purification methods (Acevedo-Rocha et al. 2019; Wang et al. 2021). Techniques such as genetic engineering, mutation screening, medium optimization, and biocatalyst conversion are being used to develop more sustainable vitamin synthesis processes (Survase et al. 2006; Acevedo-Rocha et al. 2019; Wang et al. 2021; Zhang et al. 2021; Cho et al. 2022; Nielsen et al. 2022).

Nowadays, several factors like lifestyles, work pressure, geography, diet, and antibiotic use influence major health concern problems, i.e. depression and mood disorders (Devi et al. 2023).γ-Aminobutyric acid (GABA) is a postbiotic that gained attention due to its potential health benefits as an antidepression, antihypertensive, immune-enhancer, and antidiabetic (Cui et al. 2020; Cha et al. 2023; Devi et al. 2023; Braga et al. 2024; Cataldo et al. 2024; Zhu et al. 2024). GABA has also helped to treat stress, insomnia, anxiety, and other psychological disorders, as well as regulate blood pressure, control hormone secretion, and improve liver and kidney function (Devi et al. 2023; Milon et al. 2024). The therapeutic application of GABA also extended up to several neurological illnesses like Alzheimer's disease, Huntington's chorea, and Parkinson's disease, and it was reported that it also increases the growth hormone level and protein synthesis in the brain (Milon et al. 2024; Zhu et al. 2024). Also, by inhibiting cancer and promoting cancer cell death, GABA becomes a promising bioactive molecule in several food and pharmaceutical industries (Li et al. 2019; Sahab et al. 2020; Galli et al. 2022; Milon et al. 2024). GABA production from plants and animals is challenging due to concentration constraints and obscure mechanisms, whereas microbial production is safe and eco-friendly, and different factors are easily controllable (Cui et al. 2020; Sahab et al. 2020; Li et al. 2023). Therefore, current research primarily focuses on GABA production using microbial bioresources. So, the successive chapter in this book highlights the importance, application, and production of GABA using innovative methods like metabolic engineering, fermentation, and whole-cell conversion.

1.5 Safety and Regulatory Aspects

Widespread nutrient deficiency leads to several health issues, and due to enhanced health and quality of life, people are more conscious about nutritional needs (Cui et al. 2020; Sahab et al. 2020; Li et al. 2023). Also, dietary supplements do not fulfil the increasing demands of nutraceuticals, leading to a significant increase in industrial extraction, synthesis, and production (Liu et al. 2017). For decades, several traditional methods of natural extraction and chemical synthesis of nutraceuticals have been used, but these methods depend on the petrochemicals and consume high-energy (Wang et al. 2016; Liu et al. 2017). However, the natural extraction of nutraceuticals is declining rapidly due to overuse, climate change, and environmental pollution (Wang et al. 2016). So, a more sustainable alternative for the production of nutraceuticals is through microbial synthesis. Microbial biosynthesis and high-yield nutraceutical production use abundant, low-cost, renewable agricultural and industrial waste residues (Liu et al. 2017; Yuan and Alper 2019; Paliwal et al. 2021; Janifer Raj et al. 2023). Several microbial strains that produce these molecules have been designated as GRAS (Liu et al. 2017). GRAS is a crucial aspect of US regulatory standards under the Federal Food, Drug, and Cosmetic Act (FDCA), established by the Food Additives Amendment of 1958. According to this, substances that are expected to become part of food or impact the properties of food are classified as food additives. These food additives require prior approval from the US FDA unless considered in the GRAS category by an expert committee (Sewalt et al. 2016; Liu et al. 2017). It is crucial to use the GRAS strain for the microbial production of nutraceutical products like probiotic lactic acid bacteria, and *Propionibacterium* bacteria have several health benefits and are extensively used in dairy products (Hugenholtz et al. 2002; Burdock and Carabin 2004; Sewalt et al. 2016; Liu et al. 2017). Another eukaryotic model organism, *Saccharomyces cerevisiae,* widely used as a starter culture for producing wine and pastries, is extensively used in molecular and synthetic biology research (Ostergaard et al. 2000).

Similarly, *Yarrowia lipolytica* is also valuable for producing unsaturated and organic fatty acids (Xue et al. 2013). Most GRAS strains are in the research stage due to their low-yield production, which limits their industrial application. Therefore, we are gradually applying synthetic biology and metabolic engineering to explore these GRAS strains for the industrial production of nutraceuticals. The subsequent chapter in this book describes promising GRAS microbial strain studies for the production of different nutraceuticals.

1.6 Alternative Sources

The global population is expected to reach 9.5 billion by 2050, raising significant concerns about food security and malnutrition. Over 10.8% of the world's population does not have enough food

(Kaur et al. 2023). To address these challenges, food production must double (Udayan et al. 2023). This requirement burdens the agriculture system to rotate crops, increase cultivation area, and implement new technologies to enhance production (Naik et al. 2024). As the demand for energy and food increases rapidly, limited resources create pressure on productivity and the supply chain. This makes it essential to develop alternative food sources to address the global food crisis (Mehariya et al. 2021; Naik et al. 2024). Microalgae are a promising solution because they produce high-quality digestible lipids and proteins (Galasso et al. 2019; Mehariya et al. 2021; Kaur et al. 2023). These organisms efficiently produce essential nutrients and different bioactive molecules, which are particularly beneficial for low-income countries (Mehariya et al. 2021). Algae are a diverse group of autotrophic organisms that proliferate, utilize light energy efficiently, fix atmospheric CO_2, and produce more biomass per acre as compared to vascular plants (Galasso et al. 2019; Kratzer and Murkovic 2021; Bhagea et al. 2022). Historically, algae are used as food and to treat different diseases due to the accumulation of various molecules such as lipids, amino acids, carbohydrates, pigments, and carotenoids. These compounds make the application of algae in food, feed, and health products, with growing use as nutraceuticals (Mehariya et al. 2021; Bhagea et al. 2022; Kaur et al. 2023; Garg et al. 2024). Microalgae's protein content exceeds meat, yeast, soy, dried skimmed milk, and peanuts, requiring significantly less extraction of water than animal proteins (Kaur et al. 2023; Garg et al. 2024). Types of microalgae like *Nostoc*, *Botryococcus*, *Anabaena*, *Chlamydomonas*, *Scenedesmus*, *Synechococcus*, *Parietochloris*, and *Porphyridium* are notable for producing essential vitamins (A, B1, B2, B3, B6, B9, B12, C, D, E, H) and concentrating crucial elements (potassium, zinc, iodine, selenium, iron, manganese, copper, phosphorus, sodium, nitrogen, magnesium, cobalt, molybdenum, sulfur, calcium) (Bishop and Zubeck 2012; Galasso et al. 2019; Mobin et al. 2019; Kratzer and Murkovic 2021; Mehariya et al. 2021; Ferreira de Oliveira and Bragotto 2022) Additionally, they are rich in essential amino acids and omega-3 and omega-6 fatty acids (Bishop and Zubeck 2012). Thus, algae represent a sustainable and efficient resource for addressing global nutritional needs. The successive chapter explores microalgal nutraceuticals, their health benefits, and their large-scale production methods. It covers bioprocessing techniques, commercialization challenges, and the economic viability of these products.

Acknowledgements

Authors acknowledge the Gujarat Biotechnology University (DST, Govt. of Gujarat) and Panjab University for all support. AKS acknowledges ICMR-RA fellowship.

References

Acevedo-Rocha, C.G., Gronenberg, L.S., Mack, M. et al. (2019). Microbial cell factories for the sustainable manufacturing of B vitamins. *Current Opinion in Biotechnology* 56: 18–29. https://doi.org/10.1016/j.copbio.2018.07.006.

Aggarwal, S., Sabharwal, V., Kaushik, P. et al. (2022). Postbiotics: from emerging concept to application. *Frontiers in Sustainable Food Systems* 6: https://doi.org/10.3389/fsufs.2022.887642.

Ahmed, A., Khan, T.A., Ramdath, D.D. et al. (2022). Rare sugars and their health effects in humans: a systematic review and narrative synthesis of the evidence from human trials. *Nutrition Reviews* 80 (2): 255–270. https://doi.org/10.1093/nutrit/nuab012.

Al-Okbi, S.Y., Ali, O., Aly, A.S. et al. (2023). Management of metabolic syndrome by nutraceuticals prepared from chitosan and ferulic acid with or without beta-sitosterol and their nanoforms. *Scientific Reports* 13 (1): 1–15. https://doi.org/10.1038/s41598-023-38837-9.

Alam, M.Z. (2015). Microbial amino acids production. In: *Microbial Biotechnology: Progress and Trends*, 1e (ed. F.D. Harzevili and C. Hongzhang), 187–212. CRC Press https://doi.org/10.1201/b17587-10.

Attardo, S., Musumeci, O., Velardo, D., and Toscano, A. (2022). Statins neuromuscular adverse effects. *International Journal of Molecular Sciences* 23: 8364. https://doi.org/10.3390/ijms23158364.

Barreto, J.V.d.O., Casanova, L.M., Junior, A.N. et al. (2023). Microbial pigments: major groups and industrial applications. *Microorganisms* 11 (12): https://doi.org/10.3390/microorganisms11122920.

Belorkar, S.A. and Gupta, A.K. (2016). Oligosaccharides: a boon from nature's desk. *AMB Express* 6 (1): https://doi.org/10.1186/s13568-016-0253-5.

Bhagea, R., Hossen, A.M., Ruhee, D. et al. (2022). Microalgae as sources of green bioactives for health-enhancing food supplements and nutraceuticals: a review of literature. *American Journal of Biopharmacy and Pharmaceutical Sciences* 2 (10): 10. https://doi.org/10.25259/ajbps_6_2022.

Bishop, W. and Zubeck, H.M. (2012). Evaluation of microalgae for use as nutraceuticals and nutritional supplements. *Journal of Nutrition & Food Sciences* 02 (05): https://doi.org/10.4172/2155-9600.1000147.

Bisht, D., Bhatt, V., Arya, R.K.K. et al. (2024). Plants Based Nutraceuticals: An Overview. *Herbal Nutraceuticals*. Wiley; Portico. https://doi.org/10.1002/9781394241576.ch1.

Braga, J.D., Thongngam, M., and Kumrungsee, T. (2024). Gamma-aminobutyric acid as a potential postbiotic mediator in the gut–brain axis. *npj Science of Food* 8 (1): https://doi.org/10.1038/s41538-024-00253-2.

Bujna, E., Styevko, G., Laskawy, P. et al. (2022). Synthesis of oligosaccharides with prebiotic potential by crude enzyme preparation from *Bifidobacterium*. *Food Chemistry* 367 (July 2021): 130696. https://doi.org/10.1016/j.foodchem.2021.130696.

Burdock, G.A. and Carabin, I.G. (2004). Generally recognized as safe (GRAS): history and description. *Toxicology Letters* 150 (1): 3–18. https://doi.org/10.1016/j.toxlet.2003.07.004.

Calvillo, A., Pellicer, T., Carnicer, M., and Planas, A. (2022). Bioprocess strategies for Vitamin B12 production by microbial fermentation and its market applications. *Bioengineering* 9 (8): 1–24. https://doi.org/10.3390/bioengineering9080365.

Campaniello, D., Bevilacqua, A., Speranza, B. et al. (2023). A narrative review on the use of probiotics in several diseases. Evidence and perspectives. *Frontiers in Nutrition* 10 (July): 1–12. https://doi.org/10.3389/fnut.2023.1209238.

Cao, Y., Chen, S.S., Zhang, S. et al. (2019). Advances in lignin valorization towards bio-based chemicals and fuels: lignin biorefinery. *Bioresource Technology* 291 (July): https://doi.org/10.1016/j.biortech.2019.121878.

Cardenas, D. (2013). Let not thy food be confused with thy medicine: the hippocratic misquotation. *E-SPEN Journal* 8 (6): e260–e262. https://doi.org/10.1016/j.clnme.2013.10.002.

Cataldo, P.G., Urquiza Martínez, M.P., Villena, J. et al. (2024). Comprehensive characterization of γ-aminobutyric acid (GABA) production by *Levilactobacillus brevis* CRL 2013: insights from physiology, genomics, and proteomics. *Frontiers in Microbiology* 15 (June): 1–13. https://doi.org/10.3389/fmicb.2024.1408624.

Cezar, M., Mano, R., Neri-numa, I.A. et al. (2017). Oligosaccharide biotechnology: an approach of prebiotic revolution on the industry. *Applied Microbiology and Biotechnology* 102 (1): 17–37.

Cha, X., Ding, J., Ba, W. et al. (2023). High production of γ-aminobutyric acid by activating the *xyl* operon of *Lactobacillus brevis*. *ACS Omega* 8 (8): 8101–8109. https://doi.org/10.1021/acsomega.2c08272.

Chakrabarti, S., Guha, S., and Majumder, K. (2018). Food-derived bioactive peptides in human health: challenges and opportunities. *Nutrients* 10 (11): 1–17. https://doi.org/10.3390/nu10111738.

Chandi, G.K. and Gill, B.S. (2011). Production and characterization of microbial carotenoids as an alternative to synthetic colors: a review. *International Journal of Food Properties* 14 (3): 503–513. https://doi.org/10.1080/10942910903256956.

Chen, Y., Zhu, Y., Wang, H. et al. (2023). De novo biosynthesis of 2′-fucosyllactose in a metabolically engineered *Escherichia coli* using a novel α1,2-fucosyltransferase from *Azospirillum lipoferum*. *Bioresource Technology* 374 (1800): 128818. https://doi.org/10.1016/j.biortech.2023.128818.

Chhabra, S., Bakshi, A., and Kaur, R. (2021). Role of nutraceuticals in health promotion and disease prevention: a review in the Indian context. *Current Topics in Nutraceutical Research* 19 (2): 139–145. https://doi.org/10.37290/ctnr2641-452X.19:139-145.

Cho, J.S., Kim, G.B., Eun, H. et al. (2022). Designing microbial cell factories for the production of chemicals. *JACS Au* 2 (8): 1781–1799. https://doi.org/10.1021/jacsau.2c00344.

Chopra, A.S., Lordan, R., Horbańczuk, O.K. et al. (2022). The current use and evolving landscape of nutraceuticals. *Pharmacological Research* 175: 106001. https://doi.org/10.1016/j.phrs.2021.106001.

Chouhan, S., Sharma, K., Zha, J. et al. (2017). Recent advances in the recombinant biosynthesis of polyphenols. *Frontiers in Microbiology* 8 (Nov): 1–16. https://doi.org/10.3389/fmicb.2017.02259.

Cicero, A.F.G., Bove, M., Cincione, R.I. et al. (2021). Effect of combined lipid-lowering and antioxidant nutraceutical on plasma lipids, endothelial function, and estimated cardiovascular disease risk in moderately hypercholesterolemic patients: a double-blind, placebo-controlled randomized clinical trial. *Archives of Medical Science – Atherosclerotic Diseases* 6 (1): 145–151. https://doi.org/10.5114/amsad.2021.107843.

Cicero, A.F.G., Colletti, A., Bajraktari, G. et al. (2017). Lipid-lowering nutraceuticals in clinical practice: position paper from an International Lipid Expert Panel. *Nutrition Reviews* 75 (9): 731–767. https://doi.org/10.1093/nutrit/nux047.

Cui, Y., Miao, K., Niyaphorn, S., and Qu, X. (2020). Production of gamma-aminobutyric acid from lactic acid bacteria: a systematic review. *International Journal of Molecular Sciences* 21 (3): 1–21. https://doi.org/10.3390/ijms21030995.

Das, L., Bhaumik, E., Raychaudhuri, U., and Chakraborty, R. (2012). Role of nutraceuticals in human health. *Journal of Food Science and Technology* 49: 173–183. https://doi.org/10.1007/s13197-011-0269-4.

Davani-Davari, D., Negahdaripour, M., Karimzadeh, I. et al. (2019). Prebiotics: definition, types, sources, mechanisms, and clinical applications. *Food* 8 (3): 1–27. https://doi.org/10.3390/foods8030092.

Delmulle, T., De Maeseneire, S.L., and De Mey, M. (2018). Challenges in the microbial production of flavonoids. *Phytochemistry Reviews* 17 (2): 229–247. https://doi.org/10.1007/s11101-017-9515-3.

Devi, P.B., Rajapuram, D.R., Jayamanohar, J. et al. (2023). Gamma-aminobutyric acid (GABA) production by potential probiotic strains of indigenous fermented foods origin and RSM based production optimization. *Lwt* 176 (January): 114511. https://doi.org/10.1016/j.lwt.2023.114511.

Ding, M., Song, S., and Li, X. (2024). A perspective on renewable production of amino acids from biomass through the chemocatalytic method. *Green Chemistry* 26: 4468–4476. https://doi.org/10.1039/D4GC00846D.

Egea, M.B., Oliveira Filho, J.G.d., and Lemes, A.C. (2023). Investigating the efficacy of *Saccharomyces boulardii* in metabolic syndrome treatment: a narrative review of what is known so far. *International Journal of Molecular Sciences* 24 (15): https://doi.org/10.3390/ijms241512015.

Eggersdorfer, M., Laudert, D., Létinois, U. et al. (2012). One hundred years of vitamins - a success story of the natural sciences. *Angewandte Chemie – International Edition* 51 (52): 12960–12990. https://doi.org/10.1002/anie.201205886.

Facchin, S., Bertin, L., Bonazzi, E. et al. (2024). Short-chain fatty acids and human health: from metabolic pathways to current therapeutic implications. *Lifestyles* 14 (5): 559. https://doi.org/10.3390/life14050559.

Ferreira de Oliveira, A.P. and Bragotto, A.P.A. (2022). Microalgae-based products: food and public health. *Future Foods* 6 (December 2021): https://doi.org/10.1016/j.fufo.2022.100157.

Fusco, W., Lorenzo, M.B., Cintoni, M. et al. (2023). Short-chain fatty-acid-producing bacteria: key components of the human gut microbiota. *Nutrients* 15 (9): https://doi.org/10.3390/nu15092211.

Galasso, C., Gentile, A., Orefice, I. et al. (2019). Microalgal derivatives as potential nutraceutical and food supplements for human health: a focus on cancer prevention and interception. *Nutrients* 11 (6): 1–22. https://doi.org/10.3390/nu11061226.

Galli, V., Venturi, M., Mari, E. et al. (2022). Gamma-aminobutyric acid (GABA) production in fermented milk by lactic acid bacteria isolated from spontaneous raw milk fermentation. *International Dairy Journal* 127: 105284. https://doi.org/10.1016/j.idairyj.2021.105284.

Gambelli, L. (2017). Milk and its sugar-lactose: a picture of evaluation methodologies. *Beverages* 3 (3): 35. https://doi.org/10.3390/beverages3030035.

Garg, D., Bala, S., and Phutela, U.G. (2024). Microalgal nutraceuticals: unravelling solution to food-health nexus. *Phytochemistry Reviews* 23 (2): 321–348. https://doi.org/10.1007/s11101-023-09888-7.

Ge, Q., Yan, Y., Luo, Y. et al. (2024). Dietary supplements: clinical cholesterol-lowering efficacy and potential mechanisms of action. *International Journal of Food Sciences and Nutrition* 75 (4): 349–368. https://doi.org/10.1080/09637486.2024.2342301.

Ghosh, S., Lahiri, D., Nag, M. et al. (2021). Bacterial biopolymer: its role in pathogenesis to effective biomaterials. *Polymers* 13 (8): 1–28. https://doi.org/10.3390/polym13081242.

Gill, P.A., van Zelm, M.C., Muir, J.G., and Gibson, P.R. (2018). Review article: short chain fatty acids as potential therapeutic agents in human gastrointestinal and inflammatory disorders. *Alimentary Pharmacology and Therapeutics* 48 (1): 15–34. https://doi.org/10.1111/apt.14689.

Glowka, M. and Krawczyk, T. (2023). New trends and perspectives in production of 1,2-propanediol. *ACS Sustainable Chemistry & Engineering* 11: 7274–7287. https://doi.org/10.1021/acssuschemeng.3c1018.

Helal, N.A., Eassa, H.A., Amer, A.M. et al. (2019). Nutraceuticals' novel formulations: the good, the bad, the unknown and patents involved. *Recent Patents on Drug Delivery & Formulation* 13 (2): https://doi.org/10.2174/1872211313666190503112040.

Hirasawa, T. and Shimizu, H. (2016). Recent advances in amino acid production by microbial cells. *Current Opinion in Biotechnology* 42: 133–146. https://doi.org/10.1016/j.copbio.2016.04.017.

Hollands, K., Baron, C.M., Gibson, K.J. et al. (2019). Engineering two species of yeast as cell factories for 2′-fucosyllactose. *Metabolic Engineering* 52 (December 2018): 232–242. https://doi.org/10.1016/j.ymben.2018.12.005.

Hugenholtz, J., Hunik, J., Santos, H., and Smid, E. (2002). Nutraceutical production by propionibacteria. *Le Lait* 82: 103–112. https://doi.org/10.1051/lait:2001009.

Hung, P.M., Thanh, V.H., Van Sy, H. et al. (2022). Adherence to hypertension and dyslipidemia treatment and its implication on control of cardiovascular disease in Vietnam: a semi-systematic review. *Medicine* 101: e32137. https://doi.org/10.1097/MD.0000000000032137.

Janifer Raj, X., Ilaiyaraja, N., Pal Murugan, M. et al. (2023). Biotechnological production of nutraceuticals and their delivery in different foods forms. *Journal of Food and Agriculture Research* 3 (2): 169–185.

Johnston, T.P., Korolenko, T.A., Pirro, M., and Sahebkar, A. (2017). Preventing cardiovascular heart disease: promising nutraceutical and non-nutraceutical treatments for cholesterol management. *Pharmacological Research* 120: 219–225. https://doi.org/10.1016/j.phrs.2017.04.008.

Joshi, K., Kumar, P., and Kataria, R. (2023). Microbial carotenoid production and their potential applications as antioxidants: a current update. *Process Biochemistry* 128 (November 2022): 190–205. https://doi.org/10.1016/j.procbio.2023.02.020.

Karaba, A., Patera, J., Ruskayova, P.D. et al. (2021). Experimental evaluation of hydrotreated vegetable oils as novel feedstocks for steam-cracking process. *Processes* 9 (9): 1–11. https://doi.org/10.3390/pr9091504.

Karayannis, D., Vasilakis, G., Charisteidis, I. et al. (2023). Screening of new industrially important bacterial strains for 1,3-propanediol, 2,3-butanediol and ethanol production through biodiesel-derived glycerol fermentations. *Microorganisms* 11 (6): https://doi.org/10.3390/microorganisms11061424.

Karthikeyan, S., Periyasamy, M., Prathima, A., and Ajai, M. (2020). Effect of biosolar fuels and S. Marginatum algae biofuel on equivalence ratio in diesel engine with EGR. *Materials Today Proceedings* 33: 3443–3448. https://doi.org/10.1016/j.matpr.2020.05.349.

Kaur, M., Bhatia, S., Gupta, U. et al. (2023). Microalgal bioactive metabolites as promising implements in nutraceuticals and pharmaceuticals: inspiring therapy for health benefits. *Phytochemistry Reviews* 22 (4): Springer Netherlands. https://doi.org/10.1007/s11101-022-09848-7.

Khursheed, R., Gulati, M., Wadhwa, S. et al. (2022). Multifaceted role of synbiotics as nutraceuticals, therapeutics and carrier for drug delivery. *Chemico-Biological Interactions* 368 (October): 110223. https://doi.org/10.1016/j.cbi.2022.110223.

Kirti, K., Amita, S., Priti, S. et al. (2014). Colorful world of microbes: carotenoids and their applications. *Advances in Biology* 2014: 1–13. https://doi.org/10.1155/2014/837891.

Korcz, E., Kerényi, Z., and Varga, L. (2018). Dietary fibers, prebiotics, and exopolysaccharides produced by lactic acid bacteria: potential health benefits with special regard to cholesterol-lowering effects. *Food and Function* 9 (6): 3057–3068. https://doi.org/10.1039/c8fo00118a.

Kratzer, R. and Murkovic, M. (2021). Food ingredients and nutraceuticals from microalgae: main product classes and biotechnological production. *Food* 10 (7): https://doi.org/10.3390/foods10071626.

Krautkramer, K.A., Fan, J., and Bäckhed, F. (2021). Gut microbial metabolites as multi-kingdom intermediates. *Nature Reviews Microbiology* 19 (2): 77–94. https://doi.org/10.1038/s41579-020-0438-4.

Kulinich, A. and Liu, L. (2016). Human milk oligosaccharides: the role in the fine-tuning of innate immune responses. *Carbohydrate Research* 432: 62–70. https://doi.org/10.1016/j.carres.2016.07.009.

Kuntz, S., Kunz, C., Borsch, C. et al. (2024). Influence of microbially fermented 2′-fucosyllactose on neuronal-like cell activity in an in vitro co-culture system. *Frontiers in Nutrition* 11 (February): 1–15. https://doi.org/10.3389/fnut.2024.1351433.

Lavelle, A. and Sokol, H. (2020). Gut microbiota-derived metabolites as key actors in inflammatory bowel disease. *Nature Reviews Gastroenterology and Hepatology* 17 (4): 223–237. https://doi.org/10.1038/s41575-019-0258-z.

Li, A., Cai, L., Chen, Z. et al. (2017). Recent advances in the synthesis of rare sugars using DHAP-dependent aldolases. *Carbohydrate Research* 452: 108–115. https://doi.org/10.1016/j.carres.2017.10.009.

Li, H., Li, B., Gao, L. et al. (2023). Gamma-aminobutyric acid (GABA) promotes characteristics of Levilactobacillus sp. LB-2. *Lwt* 184 (July): 115014. https://doi.org/10.1016/j.lwt.2023.115014.

Li, W., Wu, X., Yuan, X. et al. (2019). Rapid evaluation of γ-aminobutyric acid in foodstuffs by direct real-time mass spectrometry. *Food Chemistry* 277 (July 2018): 617–623. https://doi.org/10.1016/j.foodchem.2018.10.127.

Liang, B. and Xing, D. (2023). The current and future perspectives of postbiotics. *Probiotics and Antimicrobial Proteins* 15 (6): 1626–1643. https://doi.org/10.1007/s12602-023-10045-x.

Liu, L., Chen, Y., Chen, B. et al. (2023). Advances in research on marine-derived lipid-lowering active substances and their molecular mechanisms. *Nutrients* 15: 5118. https://doi.org/10.3390/nu15245118.

Liu, L., Guan, N., Li, J. et al. (2017). Development of GRAS strains for nutraceutical production using systems and synthetic biology approaches: advances and prospects. *Critical Reviews in Biotechnology* 37 (2): 139–150. https://doi.org/10.3109/07388551.2015.1121461.

Liu, Q., Yu, T., Campbell, K. et al. (2018). Modular pathway rewiring of yeast for amino acid production. In: *Methods in Enzymology*, 1e, vol. 608 (ed. N. Scrutton). Elsevier Inc. https://doi.org/10.1016/bs.mie.2018.06.009.

Liu, Y., Zhu, Y., Wang, H. et al. (2020). Strategies for enhancing microbial production of 2′-fucosyllactose, the most abundant human milk oligoschharide. *Journal of Agricultural and Food Chemistry* 70 (37): 11481–11499. https://doi.org/10.1021/acs.jafc.2c04539.

Lu, Y., Levin, G.V., and Donner, T.W. (2008). Tagatose, a new antidiabetic and obesity control drug. *Diabetes, Obesity and Metabolism* 10 (2): 109–134. https://doi.org/10.1111/j.1463-1326.2007.00799.x.

Lv, M., Xin, Q., Yin, D. et al. (2020). Magnetically recoverable bifunctional catalysts for the conversion of cellulose to 1,2-propylene glycol. *ACS Sustainable Chemistry and Engineering* 8 (9): 3617–3625. https://doi.org/10.1021/acssuschemeng.9b06264.

Lyu, X., Lyu, Y., Yu, H. et al. (2022). Biotechnological advances for improving natural pigment production: a state-of-the-art review. *Bioresources and Bioprocessing* 9 (1): https://doi.org/10.1186/s40643-022-00497-4.

Madhushan, K.W.A., Wickramasinghe, W.M.D.M., Herath, H.M.A.M. et al. (2024). Microbial production of amino acids and peptides. In: *Biorationals and Biopesticides: Pest Management*, 1e (ed. R. Kumar, M.S. de Oliveira, E.H.d.A. Andrade, et al.), 295–334. De Gruyter.

Mahgoub, S., Alagawany, M., Nader, M. et al. (2023). Recent development in bioactive peptides from plant and animal products and their impact on the human health. *Food Reviews International* 39 (1): 511–536. https://doi.org/10.1080/87559129.2021.1923027.

Markowiak, P. and Ślizewska, K. (2017). Effects of probiotics, prebiotics, and synbiotics on human health. *Nutrients* 9 (9): https://doi.org/10.3390/nu9091021.

Martínez-Cámara, S., Ibañez, A., Rubio, S. et al. (2021). Main carotenoids produced by microorganisms. *Encyclopedia* 1 (4): 1223–1245. https://doi.org/10.3390/encyclopedia1040093.

Mata-Gómez, L.C., Montañez, J.C., Méndez-Zavala, A., and Aguilar, C.N. (2014). Biotechnological production of carotenoids by yeasts: an overview. *Microbial Cell Factories* 13 (1): 1–11. https://doi.org/10.1186/1475-2859-13-12.

Maurya, A.P., Chauhan, J., Yadav, D.K. et al. (2020). Nutraceuticals and their impact on human health. In: *Preparation of Phytopharmaceuticals for the Management of Disorders: The Development of Nutraceuticals and Traditional Medicine*, 1e (ed. C. Egbuna, A.P. Mishra, and M.R. Goyal), 229–254. Elsevier Inc. https://doi.org/10.1016/B978-0-12-820284-5.00011-3.

Mehariya, S., Goswami, R.K., Karthikeysan, O.P., and Verma, P. (2021). Microalgae for high-value products: a way towards green nutraceutical and pharmaceutical compounds. *Chemosphere* 280: 130553. https://doi.org/10.1016/j.chemosphere.2021.130553.

Mendes-Silva, T.d.C.D., Andrade, R.F.d.S., Ootani, M.A. et al. (2020). Biotechnological potential of carotenoids produced by extremophilic microorganisms and application prospects for the cosmetics industry. *Advances in Microbiology* 10 (08): 397–410. https://doi.org/10.4236/aim.2020.108029.

Meyer, T.S.M., Miguel, A.S.M., Fernandez, D.E.R., and Ortiz, G.M.D. (2015). Biotechnological production of oligosccharides- applications in the food industry. In: *Food Production and Industry*, 1e (ed. A. Eissa), 1–48. IntechOpen Limited.

Mijailovic, N., Nesler, A., Perazzolli, M. et al. (2021). Rare sugars: recent advances and their potential role in sustainable crop protection. *Molecules* 26 (6): https://doi.org/10.3390/molecules26061720.

Milasing, N., Amornrattanachart, T., Khuwijitjaru, P., and Adachi, S. (2024). Isomerization of lactose to lactulose in an aqueous solution containing arginine. *International Food Research Journal* 31 (1): 80–86. https://doi.org/10.47836/ifrj.31.1.07.

Milon, R.B., Hu, P., Zhang, X. et al. (2024). Recent advances in the biosynthesis and industrial biotechnology of Gamma-amino butyric acid. *Bioresources and Bioprocessing* 11 (1): https://doi.org/10.1186/s40643-024-00747-7.

Mishra, B., Mishra, A.K., Mohanta, Y.K. et al. (2024). Postbiotics: the new horizons of microbial functional bioactive compounds in food preservation and security. *Food Production, Processing and Nutrition* 6 (1): https://doi.org/10.1186/s43014-023-00200-w.

Mobin, S.M.A., Chowdhury, H., and Alam, F. (2019). Commercially important bioproducts from microalgae and their current applications-a review. *Energy Procedia* 160 (2018): 752–760. https://doi.org/10.1016/j.egypro.2019.02.183.

Morrison, D.J. and Preston, T. (2016). Formation of short chain fatty acids by the gut microbiota and their impact on human metabolism. *Gut Microbes* 7 (3): 189–200. https://doi.org/10.1080/19490976.2015.1134082.

Moss, M.M., Taylor, B.J., Griffitts, J.S., and Kenealey, J.D. (2024). Simultaneous production of D-allulose and D-tagatose from lactose. *International Dairy Journal* 157: https://doi.org/10.1016/j.idairyj.2024.106022.

Moya-Gonzálvez, E.M., Zeuner, B., Thorhallsson, A.T. et al. (2024). Synthesis of fucosyllactose using α-L-fucosidases GH29 from infant gut microbial metagenome. *Applied Microbiology and Biotechnology* 108 (1): https://doi.org/10.1007/s00253-024-13178-3.

Naik, B., Mishra, R., Kumar, V. et al. (2024). Micro-algae: revolutionizing food production for a healthy and sustainable future. *Journal of Agriculture and Food Research* 15 (August 2023): 100939. https://doi.org/10.1016/j.jafr.2023.100939.

Nguyen, T.H. and Haltrich, D. (2013). Microbial production of prebiotic oligosaccharides. In: *Microbial Production of Food Ingredients, Enzymes and Nutraceuticals* (Issue i) (ed. B. McNeil, D. Archer, I. Giavasis, and L. Harvey). Woodhead Publishing Limited https://doi.org/10.1533/9780857093547.2.494.

Nicolescu, C.M., Bumbac, M., Buruleanu, C.L. et al. (2023). Biopolymers produced by lactic acid bacteria: characterization and food application. *Polymers* 15 (6): 1–48. https://doi.org/10.3390/polym15061539.

Nielsen, J., Tillegreen, C.B., and Petranovic, D. (2022). Innovation trends in industrial biotechnology. *Trends in Biotechnology* 40 (10): 1160–1172. https://doi.org/10.1016/j.tibtech.2022.03.007.

Nogal, A., Valdes, A.M., and Menni, C. (2021). The role of short-chain fatty acids in the interplay between gut microbiota and diet in cardio-metabolic health. *Gut Microbes* 13 (1): 1–24. https://doi.org/10.1080/19490976.2021.1897212.

Nonaka, D., Fujiwara, R., Hirata, Y. et al. (2021). Metabolic engineering of 1,2-propanediol production from cellobiose using beta-glucosidase-expressing E. coli. *Bioresource Technology* 329 (December 2020): https://doi.org/10.1016/j.biortech.2021.124858.

Ochoa-Viñals, N., Alonso-Estrada, D., Pacios-Michelena, S. et al. (2024). Current advances in carotenoid production by *Rhodotorula* sp. *Fermentation* 10 (4): 1–15. https://doi.org/10.3390/fermentation10040190.

Ofosu, F.K., Daliri, E.B.M., Elahi, F. et al. (2020). New insights on the use of polyphenols as natural preservatives and their emerging safety concerns. *Frontiers in Sustainable Food Systems* 4 (November): https://doi.org/10.3389/fsufs.2020.525810.

Ostergaard, S., Olsson, L., and Nielsen, J. (2000). Metabolic engineering of *Saccharomyces cerevisiae*. *Microbiology and Molecular Biology Reviews* 64: 34–50. https://doi.org/10.1128/MMBR.64.1.34-50.2000.

Palai, S., Derecho, C.M.P., Kesh, S.S. et al. (2020). Prebiotics, Probiotics, Synbiotics and Its Importance in the Management of Diseases. In: *Functional Foods and Nutraceuticals* (ed. C. Egbuna and D. Tupas), 173–196. Springer https://doi.org/10.1007/978-3-030-42319-3_10.

Paliwal, A., Verma, A., Nigam, A.K. et al. (2021). Microbial engineering and applications for the development of value-added products. In: *Microbial Technology for Sustainable Environment*, 1e (ed. P. Bhatt, S. Gangola, D. Udayanga, and G. Kumar), 203–215. Springer Nature Singapore Pte Ltd https://doi.org/10.1007/978-981-16-3840-4_7.

Pandey, K.R., Naik, S.R., and Vakil, B.V. (2015). Probiotics, prebiotics and synbiotics- a review. *Journal of Food Science and Technology* 52 (12): 7577–7587. https://doi.org/10.1007/s13197-015-1921-1.

Patel, S. and Goyal, A. (2011). Functional oligosaccharides: production, properties and applications. *World Journal of Microbiology and Biotechnology* 27 (5): 1119–1128. https://doi.org/10.1007/s11274-010-0558-5.

Patel, S.N., Sharma, S., Singh, A.K., and Singh, S.P. (2023). Potential microbial bioresources for functional sugar molecules. In: *Microbial Bioreactor for Industrial Molecules*, 1e (ed. S.K. Upadhyay and S.P. Singh), 211–236. Wiley.

Rainha, J., Gomes, D., Rodrigues, L.R., and Rodrigues, J.L. (2020). Synthetic biology approaches to engineer *Saccharomyces cerevisiae* towards the industrial production of valuable polyphenolic compounds. *Lifestyles* 10 (5): 1–21. https://doi.org/10.3390/life10050056.

Restrepo, J.B., Paternina-Arboleda, C.D., and Bula, A.J. (2021). 1,2—propanediol production from glycerol derived from biodiesel's production: technical and economic study. *Energies* 14 (16): 1–15. https://doi.org/10.3390/en14165081.

Roy, S., Chikkerur, J., Roy, S.C. et al. (2018). Tagatose as a potential nutraceutical: production, properties, biological roles, and applications. *Journal of Food Science* 83 (11): 2699–2709. https://doi.org/10.1111/1750-3841.14358.

Russo, V., Napolitano, N., Ascrizzi, A. et al. (2024). The lipid-lowering efficacy of a nutraceutical combination including leucoselect phytosome, red yeast rice, policosanol and folic acid in dyslipidaemia patients: real-world insights. *Pharmaceuticals* 17 (4): https://doi.org/10.3390/ph17040447.

Sahab, N.R.M., Subroto, E., Balia, R.L., and Utama, G.L. (2020). γ-Aminobutyric acid found in fermented foods and beverages: current trends. *Heliyon* 6 (11): e05526. https://doi.org/10.1016/j.heliyon.2020.e05526.

Salminen, S., Collado, M.C., Endo, A. et al. (2021). The international scientific association of probiotics and prebiotics (ISAPP) consensus statement on the definition and scope of postbiotics. *Nature Reviews Gastroenterology and Hepatology* 18 (9): 649–667. https://doi.org/10.1038/s41575-021-00440-6.

Sanchez, S., Rodríguez-Sanoja, R., Ramos, A., and Demain, A.L. (2018). Our microbes not only produce antibiotics, they also overproduce amino acids. *Journal of Antibiotics* 71 (1): 26–36. https://doi.org/10.1038/ja.2017.142.

Sarker, A. (2022). A review on the application of bioactive peptides as preservatives and functional ingredients in food model systems. *Journal of Food Processing and Preservation* 46 (8): 1–18. https://doi.org/10.1111/jfpp.16800.

Saxena, R.K., Anand, P., Saran, S., and Isar, J. (2009). Microbial production of 1,3-propanediol: recent developments and emerging opportunities. *Biotechnology Advances* 27 (6): 895–913. https://doi.org/10.1016/j.biotechadv.2009.07.003.

Saxena, R.K., Anand, P., Saran, S. et al. (2010). Microbial production and applications of 1,2-propanediol. *Indian Journal of Microbiology* 50 (1): 2–11. https://doi.org/10.1007/s12088-010-0017-x.

Schönfeld, P. and Wojtczak, L. (2016). Short- and medium-chain fatty acids in energy metabolism: the cellular perspective. *Journal of Lipid Research* 57 (6): 943–954. https://doi.org/10.1194/jlr.R067629.

Sewalt, V., Shanahan, D., Gregg, L. et al. (2016). The generally recognized as safe (GRAS) process for industrial microbial enzymes. *Industrial Biotechnology* 12 (5): 295–302. https://doi.org/10.1089/ind.2016.0011.

Shah, F.L.A., Ramzi, A.B., Baharum, S.N. et al. (2019). Recent advancement of engineering microbial hosts for the biotechnological production of flavonoids. *Molecular Biology Reports* 46 (6): 6647–6659. https://doi.org/10.1007/s11033-019-05066-1.

Sharma, S., Patel, S.N., and Singh, S.P. (2024). A novel thermotolerant l-rhamnose isomerase variant for biocatalytic conversion of d-allulose to d-allose. *Applied Microbiology and Biotechnology* 108 (1): https://doi.org/10.1007/s00253-024-13074-w.

Sharma, K., Zha, J., Chouhan, S. et al. (2019). Biosynthesis of polyphenols in recombinant microorganisms. *Recent Advances in Polyphenol Research* 6: 237–262. https://doi.org/10.1002/9781119427896.ch9.

Singh, S., Kumar, A., Kumar, R. et al. (2018). Nutraceutical: a new scope and opportunity of healthcare. *Pharma Tutor* 6 (8): 9–16. https://doi.org/10.29161/PT.v6.i8.2018.9.

Singh, A.K., Upadhyay, S.K., and Singh, S.P. (2023). Microbial bioreactor: an introduction. In: *Microbial Bioreactor for Industrial Molecules*, 1e (ed. S.K. Upadhyay and S.P. Singh), 1–17. Wiley.

Singh, S.P. and Upadhyay, S.K. (2021). *Bioprospecting of Microorganism-Based Industrial Molecules*. Wiley. https://doi.org/10.1002/9781119717317.

Singh, S.P. and Upadhyay, S.K. (2023). *Microbial Bioreactors for Industrial Molecules*. Wiley. https://doi.org/10.1002/9781119874096.

Smith, A., Avery, A., Ford, R. et al. (2022). Rare sugars: metabolic impacts and mechanisms of action: a scoping review. *British Journal of Nutrition* 128 (3): 389–406. https://doi.org/10.1017/S0007114521003524.

Soliman, G.A. (2019). Dietary fiber, atherosclerosis, and cardiovascular disease. *Nutrients* 11: 1155. https://doi.org/10.3390/nu11051155.

Sommella, E., Badolati, N., Riccio, G. et al. (2019). A boost in mitochondrial activity underpins the cholesterol-lowering effect of annurca apple polyphenols on hepatic cells. *Nutrients* 11 (1): 1–14. https://doi.org/10.3390/nu11010163.

Sosnowska, B., Penson, P., and Banach, M. (2017). The role of nutraceuticals in the prevention of cardiovascular disease. *Cardiovascular Diagnosis and Therapy* 7 (Suppl 1): S21–S31. https://doi.org/10.21037/cdt.2017.03.20.

Stephanopoulos, G., Aristodou, A., and Nielsen, J. (1998). *Metabolic Engineering*. Academic Press.

Sun, Z., Fridrich, B., De Santi, A. et al. (2018). Bright side of lignin depolymerization: toward new platform chemicals. *Chemical Reviews* 118 (2): 614–678. https://doi.org/10.1021/acs.chemrev.7b00588.

Sundarajan, P. and Ramasamy, S. (2024). Current perspective on industrial application of microbial carotenoid as an alternative to synthetic pigments. *Sustainable Chemistry and Pharmacy* 37: 101353. https://doi.org/10.1016/j.scp.2023.101353.

Survase, S.A., Bajaj, I.B., and Singhal, R.S. (2006). Biotechnological production of vitamins. *Food Technology and Biotechnology* 44 (3): 381–396.

Sut, S. and Dall'Acqua, S. (2023). Food-derived nutraceuticals for hypercholesterolemia management, mode of action and active ingredients. *Food Bioscience* 54 (June): 102866. https://doi.org/10.1016/j.fbio.2023.102866.

Tachie, C.Y.E., Onuh, J.O., and Aryee, A.N.A. (2023). Nutritional and potential health benefits of fermented food proteins. *Journal of the Science of Food and Agriculture* 104: 1223–1233.

Tang, Z.Z., Chen, G., Hong, Q. et al. (2019). Multi-omic analysis of the microbiome and metabolome in healthy subjects reveals microbiome-dependent relationships between diet and metabolites. *Frontiers in Genetics* 10 (May): https://doi.org/10.3389/fgene.2019.00454.

Tao, Y.M., Bu, C.Y., Zou, L.H. et al. (2021). A comprehensive review on microbial production of 1,2-propanediol: microorganisms, metabolic pathways, and metabolic engineering. *Biotechnology for Biofuels* 14 (1): 1–16. https://doi.org/10.1186/s13068-021-02067-w.

Thangaraj, S.G.P. and Kim, J.-H. (2024). Postbiotics: functional food materials and therapeutic agents for cancer, diabetes, and inflammatory diseases. *Food* 13 (89): 1–36.

Thorakkattu, P., Khanashyam, A.C., Shah, K. et al. (2022). Postbiotics: current trends in food and pharmaceutical industry. *Food* 11 (19): 1–29. https://doi.org/10.3390/foods11193094.

Udayan, A., Pandey, A.K., Sirohi, R. et al. (2023). Production of microalgae with high lipid content and their potential as sources of nutraceuticals. *Phytochemistry Reviews* 22 (4): 833–860. https://doi.org/10.1007/s11101-021-09784-y.

Upadhyay, S.K. and Singh, S.P. (2021). *Bioprospecting of Plant Biodiversity for Industrial Molecules*. Wiley. https://doi.org/10.1002/9781119718017.

Upadhyay, S.K. and Singh, S.P. (2023). *Plants as Bioreactors for Industrial Molecules*. Wiley. https://doi.org/10.1002/9781119875116.

Upadhyay, S.K. and Singh, S.P. (2024). *Herbal Nutraceuticals: Products and Processes*. Wiley. https://doi.org/10.1002/9781394241576.

Upasana (2022). Application of probiotics, prebiotics and synbiotics in maintaining gut health. *The Indian Journal of Nutrition and Dietetics* 59 (3): 388–397. https://doi.org/10.21048/ijnd.2022.59.3.29516.

Vandamme, E.J. and Revuelta, J.L. (2016). Vitamins, biopigments, antioxidants and related compounds: a historical, physiological and (bio)technological perspective. In: *Industrial Biotechnology of Vitamins, Biopigments, and Antioxidants* (ed. E.J. Vandamme and J.L. Revuelta), 1–14. Wiley-VCH Verlag GmbH & Co. KGaA.

Vera, C., Guerrero, C., and Illanes, A. (2022). Trends in lactose-derived bioactives: synthesis and purification. *Systems Microbiology and Biomanufacturing* 2 (3): 393–412. https://doi.org/10.1007/s43393-021-00068-2.

Wang, J., Guleria, S., Koffas, M.A.G., and Yan, Y. (2016). Microbial production of value-added nutraceuticals. *Current Opinion in Biotechnology* 37: 97–104. https://doi.org/10.1016/j.copbio.2015.11.003.

Wang, Y., Liu, L., Jin, Z., and Zhang, D. (2021). Microbial cell factories for green production of vitamins. *Frontiers in Bioengineering and Biotechnology* 9 (June): 1–15. https://doi.org/10.3389/fbioe.2021.661562.

Wang, J., Ma, W., Zhou, J., and Wang, X. (2022). Microbial chasis design and engineering for production of amino acids used in food industry. *Systems Microbiology and Biomanufacturing* 3: 28–48.

Wegh, C.A.M., Geerlings, S.Y., Knol, J. et al. (2019). Postbiotics and their potential applications in early life nutrition and beyond carrie. *International Journal of Molecular Sciences* 20: 4673. https://doi.org/10.3390/ijms20194673.

Wendisch, V.F. (2014). Microbial production of amino acids and derived chemicals: synthetic biology approaches to strain development. *Current Opinion in Biotechnology* 30: 51–58. https://doi.org/10.1016/j.copbio.2014.05.004.

Xue, Z., Sharpe, P., Hong, S. et al. (2013). Production of omega-3 eicosapentaenoic acid by metabolic engineering of *Yarrowia lipolytica*. *Nature Biotechnology* 31: 734–740. https://doi.org/10.1038/nbt.2622.

You, R., Wang, L., Hu, M., and Tao, Y. (2024). Efficient production of 2′-fucosyllactose from fructose through metabolically engineered recombinant *Escherichia coli*. *Microbial Cell Factories* 23 (1): 1–14. https://doi.org/10.1186/s12934-024-02312-5.

Yu, S., Liu, J.J., Yun, E.J. et al. (2018). Production of a human milk oligosaccharide 2′-fucosyllactose by metabolically engineered *Saccharomyces cerevisiae*. *Microbial Cell Factories* 17 (1): 1–10. https://doi.org/10.1186/s12934-018-0947-2.

Yuan, S.F. and Alper, H.S. (2019). Metabolic engineering of microbial cell factories for production of nutraceuticals. *Microbial Cell Factories* 18 (1): 1–11. https://doi.org/10.1186/s12934-019-1096-y.

Zaky, A.A., Simal-Gandara, J., Eun, J.B. et al. (2022). Bioactivities, applications, safety, and health benefits of bioactive peptides from food and by-products: a review. *Frontiers in Nutrition* 8 (January): 1–18. https://doi.org/10.3389/fnut.2021.815640.

Zepeda-Hernández, A., Garcia-Amezquita, L.E., Requena, T., and García-Cayuela, T. (2021). Probiotics, prebiotics, and synbiotics added to dairy products: uses and applications to manage type 2 diabetes. *Food Research International* 142 (October 2020): https://doi.org/10.1016/j.foodres.2021.110208.

Zeuner, B. and Meyer, A.S. (2020). Enzymatic transfucosylation for synthesis of human milk oligosaccharides. *Carbohydrate Research* 493 (May): 108029. https://doi.org/10.1016/j.carres.2020.108029.

Zhang, Z., Liu, L., Liu, C. et al. (2021). New aspects of microbial vitamin K2 production by expanding the product spectrum. *Microbial Cell Factories* 20 (1): 1–12. https://doi.org/10.1186/s12934-021-01574-7.

Zhang, G., Zabed, H.M., An, Y. et al. (2022). Biocatalytic conversion of a lactose-rich dairy waste into D-tagatose, D-arabitol and galactitol using sequential whole cell and fermentation technologies. *Bioresource Technology* 358 (May): 127422. https://doi.org/10.1016/j.biortech.2022.127422.

Zhang, Y., Zhang, X., Liu, H. et al. (2023). Efficient production of 2′-fucosyllactose in unconventional yeast *Yarrowia lipolytica*. *Synthetic and Systems Biotechnology* 8 (4): 716–723. https://doi.org/10.1016/j.synbio.2023.11.002.

Zhu, Y., Cao, H., Wang, H., and Mu, W. (2022). Biosynthesis of human milk oligosaccharides via metabolic engineering approaches: current advances and challenges. *Current Opinion in Biotechnology* 78: 102841. https://doi.org/10.1016/j.copbio.2022.102841.

Zhu, F., Hu, S., and Mei, L. (2024). Production and quality evaluation of a novel γ-aminobutyric acid-enriched yogurt. *Frontiers in Nutrition* 11 (May): 1–11. https://doi.org/10.3389/fnut.2024.1404743.

2

Microbial Cell Factories for the Production of Essential Amino Acids

Susana Calderón-Toledo, César Salcedo-Okuma, and Amparo Iris Zavaleta

Laboratorio de Biología Molecular, Facultad de Farmacia y Bioquímica, Universidad Nacional, Mayor de San Marcos, Lima, Perú

2.1 Introduction

The global amino acids market has been valued at USD 27,498 million by 2022 and is estimated to reach USD 55,428 million by 2031, with the production of L-glutamate as the main source of revenue, followed by L-lysine (Straits research 2023). This market has gained a high demand in the food, pharmaceutical, livestock, and cosmetics sectors (Sanchez et al. 2018) and production is currently dominated by companies such as Ajinomoto Co Inc, KYOWA HAKKO BIO Co LTD, Evonik Industries AG, AMINO GmbH, Taiwan Amino Acids Co Ltd, and Adissseo, among others (Straits research 2023). The first production of amino acids took place in Japan (1908) with the extraction of monosodium glutamate from acidic protein hydrolysates. Thereafter, other production methods such as chemical and enzymatic synthesis as well as fermentation were developed (Nakamori 2017). Currently, fermentation processes using bacterial strains are more widely used on an industrial scale because they are more cost-effective by requiring inexpensive carbon and nitrogen sources, such as sugarcane molasses and ammonia; in contrast, other methods such as enzymatic methods tend to be less cost-effective because they require synthetic substrates that tend to be more expensive (Kumagai 2014; Singh and Upadhyay 2023).

Different bacterial species possess biochemical routes for amino acid synthesis, however, *Corynebacterium glutamicum* presents, in its natural form and under optimized fermentation conditions, a high metabolic capacity to produce L-glutamate on an industrial scale (Leuchtenberger et al. 2005). In addition, *Corynebacterium* mutant strains synthesize L-lysine, L-valine, and L-histidine with high levels of productivity and yield. On the other hand, recombinant strains of *Escherichia coli* have been successfully used to produce L-threonine, L-tryptophan, L-isoleucine, L-aspartic acid, and L-alanine (Leuchtenberger et al. 2005; Liu et al. 2022b). This anabolic facility of microorganisms such as *C. glutamicum* and *E. coli* has encouraged the development of metabolic engineering and bioprocess techniques in order to optimize their industrial production; for example, modification of the culture medium using alternative substrates, the addition of biocatalysts, genetic mutations, and metabolic modeling, among others (Ivanov et al. 2013). Currently, the production of most amino acids is carried out with mutant strains of both *C. glutamicum* and

Microbial Nutraceuticals: Products and Processes, First Edition. Edited by Sudhir Pratap Singh and Santosh Kumar Upadhyay.

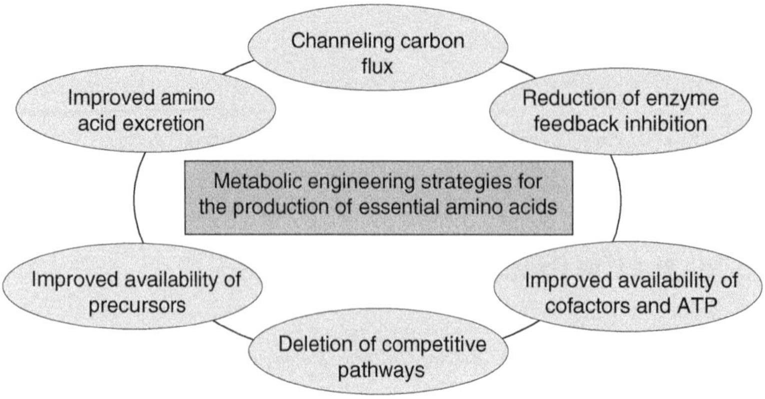

Figure 2.1 Main metabolic engineering strategies for the production of essential amino acids.

E. coli, to which different cultivation modalities have been applied in industrial bioreactors of 50–500 m^3 capacity (Hermann 2003).

Among the 20 amino acids, 9 are essential (L-leucine, L-isoleucine, L-threonine, L-methionine, L-phenylalanine, L-tryptophan, L-lysine, L-valine, and L-histidine) because they need to be acquired in the diet due to its inability to be synthesized in animals and humans. This chapter describes the biosynthetic routes used in the production of these amino acids, as well as the metabolic, protein, and bioprocess engineering strategies applied in recent years to improve their industrial production (Figure 2.1).

2.2 Essential Amino Acid Biosynthesis

2.2.1 Methionine

L-methionine is one of the amino acids that continue to be used in chemical synthesis due to the challenges involved in its biosynthesis. One of the alternatives has been the production by fermentation of the enantiomeric mixture and its subsequent conversion to L-methionine by enzymatic catalysis (Cai et al. 2023). This amino acid has a high demand as a food additive. For its biosynthesis, both *C. glutamicum* and *E. coli* use L-aspartate as a precursor in three steps corresponding to the synthesis of the four-carbon base structure O-succinyl-L-homoserine (OSH), followed by the assimilation of the sulfur group and then the addition of a carbon unit (Cai et al. 2023). Thus, glucose is transformed by glycolysis and the pentose phosphate pathway (PPP), then synthesized to L-aspartate, L-homoserine and subsequently acylated with succinyl-CoA. Later, sulfide assimilation in *C. glutamicum* can be performed by two pathways, transsulfuration and sulfhydrylation, whereas *E. coli* uses only transsulfuration with L-cysteine as a sulfide donor. Finally, a methyl of the C1 unit is transferred to homocysteine to generate L-methionine in both microorganisms (Ferla and Patrick 2014; Bastard et al. 2017). The conversion of homoserine to OSH is encoded by *metA*, sulfuration using cysteine by *metB,* and its conversion to L-homocysteine by *metC* (Gruzdev et al. 2023).

Methionine biosynthesis has complex regulatory mechanisms that limit its further accumulation, reasons for which have been traced as the removal of inhibition mechanisms and the repression of its genes, as well as the modification of the transport system to promote the efflux of L-methionine, most of these applied in *E. coli*. Among the strategies to control its metabolic flux, overexpression of MetC, the export protein YjeH, and $MetA^{fbr}$, involving in the latter the mutations

I296S-P298L-R27C to attenuate its feedback resistance, and blockade of the absorption transporter MetD, which led to the enhancement of methionine production to 1.93 g/L, have been performed. This was coupled with the deletion of *pykA* and *pykF* genes, which increased the synthesis of L-methionine precursors, evidenced by 2.51 g/L L-methionine (Li et al. 2023b). Additionally, overexpression of *gltA*, which encodes citrate synthase, enhanced L-methionine production from 2.38 to 2.73 g/L and cell growth, yielding up to 17.74 g/L L-methionine in 5 L reactor (Niu et al. 2023).

Another strategy has been to increase precursors such as L-homoserine by overexpressing *thrA*^{fbr} feedback resistant and *rhtA* and increasing Nicotinamide adenine dinucleotide phosphate (NADPH) generation and heterologous expression of Nicotinamide adenine dinucleotide (NADH) used by hydrogenases. This modified strain achieved 85.29 g/L of L-homoserine with a productivity of 1.78 g/L/h in 5 L bioreactor (Cai et al. 2022). Among the genes controlling the OSH production flow, the *metB* gene would be related to OSH lyase and its overexpression would lead to the decrease of L-methionine, while the overexpression of *metL* increases OSH production (Shen et al. 2023).

Likewise, an *E. coli* with the capacity to synthesize O-acetyl-L-homoserine (OAH), an alternative intermediate to OSH produced from homoserine (reaction encoded by *metX*) and converted to L-homocysteine, reaction encoded by *metY*, has been constructed. For that case, *metXbc/metA* from *Bacillus cereus*, responsible for homoserine acetylation, was introduced, *metL* was also overexpressed, which decreased lysine production, and *pykA* was deleted, which reduced carbon flux to pyruvate and increased OAH, resulting in 24.33 g/L OAH in fed-batch modality bioreactor (Li et al. 2023a). Also, the transsulfuration pathway of *E. coli* has been replaced by the direct sulfuration pathway of other bacteria, which involves replacing the three steps of transsulfuration with only two. Therefore, the deletion of *metA* and *metB* and their replacement by *metX* and *metY* from *Cyclobacterium marinum*, added to the deletion of the repressor gene *metJ* and the overexpression of YjeH, was applied, which allowed an increase in the intra- and extracellular levels of methionine (Gruzdev et al. 2023).

Furthermore, a limiting factor for L-methionine production is the intracellular abundance of L-serine, which is regulated by the *serB* and *serC* genes, and their overexpression would decrease L-methionine production (Shen et al. 2023). Likewise, blocking L-lysine synthesis, by deletion of *lysA*, has been applied, which improved L-methionine accumulation to 1.9 g/L (Huang et al. 2018a).

In recent years, optimization has focused on improving the production of OSH and OAH intermediates in *E. coli*, considering the bottleneck of L-methionine production and giving rise to the application of metabolic engineering techniques for the optimization of their synthesis pathways. Also, the synthesis of L-methionine has been improved through the deletion of genes responsible for the competitive synthesis of other amino acids, the redistribution of the carbon flux to improve the biosynthesis of intermediates, the replacement of synthesis pathways by shorter sequences, among other techniques, which should be applied together in order to lead to the improvement of the yield of the fermentation production of this amino acid.

2.2.2 Valine

L-valine biosynthesis has gained great importance due to its pharmacological application in the treatment of chronic liver diseases as well as dietary supplement (Holeček 2018; Lo et al. 2022). Various species, such as *C. glutamicum*, *Bacillus subtilis*, *Bacillus licheniformis*, *Saccharomyces cerevisiae,* and *E. coli*, have been evidenced to produce this amino acid (Gao et al. 2021). Its biosynthesis starts with pyruvate and requires four enzymes, acetohydroxy acid synthase (AHAS), acetohydroxy acid isomeroreductase (AHAIR), dihydroxy acid dehydratase (DHAD) and transaminase B (TA). AHAS catalyzes the synthesis of 2-acetolactate from two pyruvate molecules, encoded by *ilvB* and *ilvN* and regulated by environmental redox levels. AHAIR, encoded by *ilvC*, catalyzes the isomerization and reduction of 2-acetolactate releasing dihydroxy-isovalerate. DHAD, encoded by *ilvD*, dehydrates dihydroxy-isovalerate to form 2-ketoisovalerate; and TA, encoded by *ilvE*, transforms to L-valine (Wang et al. 2018). Whereby, carbon flux to valine biosynthesis was enhanced by overexpression of *ilvBN*, *ilvC*, *ilvD*, and *ilvE* genes and downregulation of *ilvA*, *alr*, and *leuB* genes, responsible for the synthesis of L-isoleucine, L-alanine, and L-leucine, respectively (Zhang et al. 2018).

Because pyruvate precursors, such as acetyl-CoA, acetate, oxaloacetate, and L-isoleucine, are also used in the synthesis of L-alanine, their use in competitive pathways with L-valine synthesis has been decreased. Thus, genes linked to the synthesis of acetyl-CoA and acetate from pyruvate, and oxaloacetate from phosphoenolpyruvate (PEP) have been blocked resulting in better L-valine synthesis (2.64 g/L) with decreased growth rate (Wang et al. 2018).

In *C. glutamicum*, mutations that ameliorated isocitrate dehydrogenase and activated glyoxylate shunting were performed to improve growth and intracellular levels of L-valine (Schwentner et al. 2018). Likewise, the application of irradiation improved L-valine synthesis by 21.47% and demonstrated the genetic stability of phenotypes (Han et al. 2020). *Brevibacterium flavum* has also been mutated by applying UV irradiation and binary ethyleneimine to obtain 30.1 g/L of L-valine, showing sufficient stability to be scaled up. Thus, the mutated strain produced 66.8 g/L L-valine after 64 hours of fermentation in a 120 m^3 reactor by fed-batch mode (Huang et al. 2018b).

Transcription factors such as PdhR, Crp, and Lrp have also been identified to promote L-valine production, with the first exerting the greatest effect, while the RpoS factor would exert a negative effect, whose alterations result in 17.4 g/L of L-valine (Hao et al. 2022). The introduction of constitutive strong promoters has been used to regulate the transcriptional levels of *C. glutamicum* genes to increase the availability of pyruvate for L-valine synthesis (Ma et al. 2018). Also, glucose uptake has been facilitated by introducing ppGpp 3-pyrophosphohydrolase and increased pyruvate by blocking GTP pyrophosphoinositide kinase (Hao et al. 2020). In *S. cerevisiae*, amino acid modifications at the level of the ilv6 regulatory subunit of AHAS have been applied, which allowed it to be less sensitive to feedback inhibition and increased intracellular L-valine content fourfold (Takpho et al. 2018).

Enzymes such as AHAS and AHAIR require Mg^{2+} and cofactors such as FAD and NADPH, which have led to increased flux to the PPP in order to generate more NADPH or conversion of the NADPH requirement to NADH (Bartek et al. 2010; Hasegawa et al. 2012). NADPH availability has been increased by integrating *edd* and *eda* genes, and NADPH depletion has been ameliorated by increasing NAD kinases, which increases L-valine production; however, deletion of *yahK*, encoding NADPH-dependent aldehyde reductase, evidenced no change in L-valine production (Hao et al. 2022). Another mechanism for enhancing NADPH availability is overexpression of the *pntAB* gene in *E. coli*, encoding pyridine nucleotide transhydrogenase, which enhances L-valine synthesis (Hao et al. 2020). The enhanced generation of NADPH and ATP has also been achieved by the upregulation of genes associated with the PPP and tricarboxylic acid pathway (Zhang et al. 2018). Also, deletion of genes encoding ATP-consuming enzymes has been identified to enhance L-valine synthesis, as was possible by deleting the *fecE* gene (Hao et al. 2022).

Additionally, L-valine efflux has been improved by introducing the protein complex BrnFE from *C. glutamicum* into *E. coli* (Hao et al. 2020). The application of multiple genetic modifications has achieved maximum L-valine production at 48 hours (92 g/L) with a productivity of 1.92 g/L/h in 5 L reactor by fed-batch mode (Hao et al. 2022). Therefore, the application of metabolic engineering techniques has allowed obtaining modified strains with high levels of L-valine biosynthesis, which combined with mathematical simulation tools could turn these strains into cell factories at the industrial level.

2.2.3 Tryptophan

L-tryptophan is an amino acid used both in the food industry and in the health industry due to its ability to improve sleep quality and treat anxiety (Kikuchi et al. 2021; Sutanto et al. 2022). Its biosynthesis begins with the condensation of PEP and erythrose 4-phosphate (E4P) to generate 3-deoxy D-arabino heptulosonate 7 phosphate (DAHP), which will be transformed to chorismate (CHA) through the shikimate pathway, and from this L-tryptophan, L-tyrosine, and L-phenylalanine are synthesized (Niu et al. 2019). In order to improve its production, genes related to the shikimate pathway, such as *aroE* and *aroK*, whose deficient expression would be a limiting factor for L-tryptophan synthesis, have been overexpressed (Liu et al. 2023a). Also, increased conversion of oxaloacetic acid to PEP and overexpression of tricarboxylic acid cycle and citric acid transport genes enhanced cell growth and tryptophan production by 40% (Du et al. 2019).

In order to increase the formation of erythrose 4-phosphate, the phosphoketolase coding genes of *Bifidobacterium adolescens* were introduced in *E. coli*, which increased the productivity to 10.8 g/L of tryptophan. In addition, the phosphotransferase system was replaced by a glucose-independent phosphoenolpyruvate transporter and the pyruvate kinase-encoding *pykF* gene was deleted, which improved the availability of PEP, achieving 41.7 g/L of tryptophan after 40 hours of fed-batch fermentation with decreased formation of the by-products such as acetate and lactate (Xiong et al. 2021).

Moreover, the accumulation of intermediates has been improved by the inactivation of *trpP*, *ptsH*, *pykF/pykA* and tryptophan attenuator, glucose utilization by modulation of *galP* and *glk*, and to reduce acetic acid accumulation, the *pta* gene was repressed, obtaining a conversion rate of 16.7% with a productivity of 1.6 g/L/h of tryptophan by fed-batch fermentation (Chen et al. 2018b). Also, the intracellular amounts of L-serine and L-glutamine have been increased, as their low amounts would be a limiting factor (Liu et al. 2023b). Therefore, L-tryptophan biosynthesis can be enhanced by introducing a variant of *serA* resistant to feedback inhibition; however, although anthranilate synthase adds an amino group from L-glutamine to CHA to produce anthranilate and L-glutamate, tryptophan production can be limited by high glutamine concentrations, indicating that high glutamine concentrations negatively affect anthranilate synthase activity (Chen and Zeng 2017). Likewise, control by inhibition of some enzymes such as indole glycerol phosphate synthase of the multifunctional enzyme TrpC mediated by anthranilate has caused genes to be introduced from *Aspergillus niger*, which are activated by anthranilate, producing an *E. coli* with improved growth stability and tryptophan production (Chen et al. 2018a).

To the improvement in the availability of serine and the intracellular levels of erythrose 4-phosphate and PEP, we have added the blockage of tryptophan degradation pathways and the improvement in its export, which increased the productivity to 52.1 g/L of tryptophan (Guo et al. 2022). Also, the *tnaB* gene, encoding aromatic amino acid permease, has been blocked, which prevents the reentry of this amino acid and increases its extracellular amount by 6.7% (Liu et al. 2023b).

Another technique applied was the blockade of the competitive pathways of other amino acid synthesis such as *tyrA* and *pheA* genes, encoding enzymes that synthesize L-tyrosine and L-phenylalanine from CHA, respectively, which induced an improvement in the synthesis of L-tryptophan by 23.84%. Also, the transcriptional factors YihL and TrpR that negatively regulate L-tryptophan synthesis were suppressed. On the other hand, L-tryptophan synthesis requires NADPH and ATP, so transhydrogenase (encoded by *pntAB*) and ATP-forming phosphoenolpyruvate carboxykinase (encoded by *pckA*) have been overexpressed, which would exert a positive effect on the production of L-tryptophan, and added to the regulation of its transcriptional factors and the inhibition of synthesis of other amino acids allows obtaining 42.50 g/L of L-tryptophan at 48 hours of fermentation (Tang et al. 2023).

2.2.4 Phenylalanine

L-phenylalanine has become widespread within the food additive industry due to its presence in the production of aspartame, as well as in the pharmaceutical industry for the synthesis of compounds such as fluorinated phenylalanine and polyphenylalanines for application as therapeutic agents, and drug transporters, among others (Humphries et al. 2008; Awad and Ayoup 2020; Meng and Wu 2021). This amino acid is also produced from CHA, after which prephenate and phenylpyruvate are synthesized using fused chorismate mutase. Subsequently, a transamination reaction of phenylpyruvate with an amino group donated by glutamate takes place, releasing L-phenylalanine as a product (Gu et al. 2023). Biosynthesis is regulated by TyrR and TrpR, transcriptional regulatory proteins of L-phenylalanine production, while aminotransferases are encoded by *tyrB*, *aspC*, and *ilvE* (Pittard and Yang 2008).

Strategies to enhance L-phenylalanine production include directing the carbon flux for its synthesis by overexpression of the *glf* gene, which enhances fast glucose utilization (Gu et al. 2023). Among the key intermediates for L-phenylalanine synthesis from glycerol, fructose 6-phosphate and PEP have been distinguished, the latter also limiting using glucose as a carbon source (Weiner et al. 2017). Another strategy to improve the supply of precursors such as PEP and erythrose 4-phosphate has been the mutation of the repressor/activator protein FruR, which modified the transcript levels of genes involved in central carbon metabolism and amino acid biosynthesis (Chen et al. 2022).

Also, modification of metabolic flux has been applied by deletion of *pykA*, *pykF*, *ppc*, *zwf*, and *adhE* genes, where the first three are linked to the consumption of PEP for the formation of other by-products, while *zwf* produces pyruvate from the ED pathway and *adhE* is involved in the NAD^+ regeneration for serine synthesis. This strain has been used in coculture with an auxotrophic phenylalanine-producing strain with *pheA* deletion, which led to the evolution of the initially designed strain in increasing its growth rate and phenylalanine production (Kawai et al. 2022). To improve growth and product rate in strains with multiple modifications, dynamic regulation using modified promoters in the *tyrP* promoter region has been applied to control the expression of the AroK enzyme, shikimate pathway kinase, which increases the transcriptional levels of this enzyme and improves L-phenylalanine production by up to 61.3 g/L, with an increase in glucose production yield by 1.22-fold compared to the native strain (Wu et al. 2019).

Multiple modular modifications have been applied such as the construction of a hybrid glucose transport system by integrating the *iolT2-ppgK* genes, which increased extracellular accumulation of L-phenylalanine by inactivating *aroP* and blocked the acetate and lactate synthesis pathways in *C. glutamicum* by deleting the *aceE* and *ldh* genes, achieving the production of 15.76 g/L of L-phenylalanine by fed-batch mode (Zhang et al. 2015). Regarding the costs associated with the fermentation medium, the pyruvate kinase A (*pykA*) gene has been deleted, after which it was transformed with pYHP, reaching 38.6 g/L production in 30 L bioreactor per fed-batch culture with a productivity of 1.07 g/L/h, without the supplementation of essential amino acids, which would lower its production cost (Bang et al. 2021).

These strategies of precursor increase, transcriptional regulation of enzymes associated with metabolic flux, feedback inhibition, and fermentation strategies, among others, have been applied for several years and have improved their production and productivity. However, there is still a need to improve their performance due to the toxicity of aromatic compounds at the cellular level (Liu et al. 2021).

2.2.5 Lysine

This amino acid, necessary for humans and mammals, is important for the prevention of cerebrovascular and cardiovascular diseases, so it is used in food additives and pharmaceuticals (Shimomura et al. 2014; Liu et al. 2022a). Therefore, genetic modifications have been made in *C. glutamicum* and *E. coli* in order to facilitate their growth and production of L-lysine. Its biosynthesis in *C. glutamicum* is known as the diaminopimelate (DAP) pathway and uses oxaloacetate, which is transformed to L-aspartate and subsequently to L-4-aspartyl phosphate, which through successive steps will be transformed to tetrahydrodipicolinate. This intermediate is catalyzed in four consecutive reactions, after which meso 2,6-diaminopimelate is obtained, which will be decarboxylated to lysine using diaminopimelate deacetylase (DAPDC) (Divya et al. 2022; Liu et al. 2022a). This bacterium possesses a LysE exporting system, whose expression is controlled by the transcriptional activator LysG. Likewise, it possesses an L-lysine reuptake system LysP, an L-lysine and L-alanine transporter protein, which maintains the intra- and extracellular balance of this amino acid (Liu et al. 2022a).

Oxaloacetate is one of the major precursors of L-lysine, which improves its availability by deletion of the *pck* gene, encoding PEP carboxykinase and contributes to L-lysine production; however, its increase by overexpression of pyruvate carboxylase is directed to the glycolysis pathway (Becker et al. 2011). Likewise, genes associated with sugar metabolism have been modified, such as the heteroexpression of fructokinase in *C. glutamicum*, which benefits the metabolism of sucrose and other sugar mixtures containing fructose in their composition and improves the production of L-lysine with these carbon sources. On the other hand, due to the low expression level of myoinositol permease, associated with carbohydrate reuptake, the promoters of the *iolT1* and *iolT2* coding genes have been replaced, which increased the activities of PEP carboxylase and pyruvate kinase, and consequently led to increased cell growth, L-lysine production, and sugar consumption rate. Glucokinase from *C. glutamicum* possesses a low level of glucose affinity, so GLK_{Cg} glucokinase was replaced by GLK_{Bs} from *B. subtilis*, having a high level of glucose affinity and being regulated by a strong promoter, which increased the rate of glucose consumption, cell growth, and L-lysine production (Xu et al. 2020).

However, the introduction of these enzymes increases ATP consumption, which can be reduced by the introduction of ADP-GlK/PFK from *Methanococcus maripaludis* that increases intracellular levels of NADH and ATP and enhances ATP synthesis *via* electron transport phosphorylation by introducing a strong promoter in the *ndh* gene (NADH dehydrogenase), leading to enhanced activities of glycolysis-related enzymes. Likewise, the SigH deletion increases the activity of F_0F_1-ATPase, which catalyzes ATP formation, and thus increases the carbon flux toward L-lysine biosynthesis, which reached 221.3 g/L of L-lysine with a productivity of 5.53 g/L/h (Xu et al. 2020). Also, ATP content can be increased using the regulation of the strong *trc* promoter for phosphoglycerate kinase and pyruvate kinase enzymes (Liu et al. 2023a).

The presence of NADPH is required for L-lysine biosynthesis, thus genes associated with the expression of PPP enzymes such as *zwf* and *gnd* have been overexpressed; however, this increase generates the underlying release of CO_2, which involves a loss of carbon (Wang et al. 2020b; Jojima et al. 2021). This resulted in the replacement of endogenous glyceraldehyde 3-phosphate dehydrogenase (encoded by *gapA*) with non-phosphorylated NADP-dependent glyceraldehyde 3-phosphate dehydrogenase (encoded by *gapN*) and promoter-mediated regulation of expression to optimize intracellular NADPH availability and achieving 80.9 g/L L-lysine, 75% more than with the native strain (Liu et al. 2023a). Also, the transcriptional regulator Cgl2680 has been deleted, which improved the intracellular NADPH/$NADP^+$ rate, and it is inferred that its presence would have a negative regulatory effect on the *zwf* gene. This leads to an increase in ATP, because NADPH is used as a hydrogen and electron donor to produce ATP. However, excessive amounts of NADPH can exert a cellular toxic effect, as it slows cell growth (Wang et al. 2020b).

Other mutation techniques include exposure to UV irradiation and protoplast fusion technology, resulting in production rates 7.2% higher than with the native strain and with growth stability (Li et al. 2023c). Also, the regulation of enzymes has been applied using promoters with different intensities, as was applied for the decrease in the regulation of *sucC*, encoding succinyl-CoA synthetase, which allows an increase in lysine production by 20% (Shang et al. 2018). The elimination of proteins that are not necessary for the production of L-lysine has been another optimization method in *C. glutamicum*, since it has the transmembrane transporters GluE, BrnE/BrnF and LysP, the first two are used for the secretion of other amino acids. Consequently, deletion of *gluE* increases the content of citrate and isocitrate (precursors of L-lysine), while deletion of the others affects the tricarboxylic acid cycle. However, the simultaneous deletion of more than two would not have as much effect on the increase in L-lysine as separately, since the three transporters would be involved in the metabolism of fatty acids and nucleotides (Xiao et al. 2020). Therefore, the improvement in the biosynthesis of this amino acid by global regulation requires dynamic monitoring of cellular metabolites in order to achieve the balance between L-lysine production and cell growth.

2.2.6 Leucine

L-leucine is used as a dietary supplement to regulate glucose homeostasis, cell regeneration, and protein synthesis (Pedroso et al. 2015). It stimulates insulin secretion and inhibits protein degradation, as well as serving as a nitrogen donor to glutamate (Myles et al. 2018). For its production, *E. coli* and *C. glutamicum* with different mutations have been designed and selected to improve their productivity due to high market demand. Its biosynthesis requires pyruvate as a precursor, which after being transformed by AHAS, AHAIR, and DHAD is converted into 2-ketoisovalerate, also an intermediate of L-valine and L-isoleucine, since they follow parallel routes. This intermediate will be transformed by α-isopropylmalate synthase (IPMS), β-isopropylmalate dehydratase (IPMD), and β-isopropylmalate dehydrogenase (IPMDH) to α-ketoisocaproate, which is converted to L-leucine by a transaminase (encoded by *ilvE*). This will be exported *via* a two-component system consisting of BrnF and BrnE in *C. glutamicum*, which is also responsible for the export of the other branched amino acids and L-methionine. In *E. coli*, leucine export is encoded by *leuE* and possesses a *brnFE*-like system known as *ygaZH*, which would be transcriptionally activated by Lrp (Wang et al. 2019a).

The enzymatic activity in its biosynthesis has been improved, by decreasing the feedback inactivation of IPMS and AHAS by replacing them with low-level inactivation enzymes and eliminating the premature termination of transcription of *leuA* (encoding IPMS) (Vogt et al. 2014; Luo et al. 2021; Ding et al. 2023; Guo et al. 2023). Increasing the transcriptional levels of *ilvB*, *ilvN*, and *ilvC* (encoding AHAS and AHAIR) and overexpressing the *leuA* gene improve carbon input for the specific production of L-leucine (Wang et al. 2020a). Also, the transcriptional regulator LtbR has been deleted, which represses the expression of *leuCD* and *leuB* when there are high levels of L-leucine, increasing IPMD activity and L-leucine production (Vogt et al. 2014).

In order to improve glucose reuptake, the transcriptional regulator IolR was eliminated, which de-represses genes associated with inositol metabolism and transport, leading to increased cell growth and the production of 2-ketisovalerate, which in excess is not fully used by IPMS, which can be improved by increasing the concentration of acetyl-CoA, the second substrate required for the reaction catalyzed by IPMS (Vogt et al. 2014). The deletion of *sugR* also improves the glucose reuptake pathway by phosphotransferase, causing an increase in the glucose consumption rate by 11% and the production of L-leucine. Acetyl-CoA can be increased by the introduction of acetyl-CoA synthetase and NAD^+-dependent deacetylase from *E. coli*, supplementing acetyl-CoA from acetate and producing 40.11 g/L of L-leucine by fed-batch culture (Wang et al. 2022). Also, acetyl-CoA supplementation can be achieved by the introduction of the non-oxidative glycolysis (NOG) pathway, which provides three acetyl-CoA molecules from a glucose molecule without CO_2 co-release, as well as the retargeting of acetylphosphate for the production of acetyl-CoA by overexpression of phosphotransferase (encoded by *pta*) and deletion of acetate kinases (encoded by *ackA*) (Ding et al. 2023).

2-Ketisovalerate is an intermediate that can be increased by the introduction of resistant to feedback inhibition acetolactate synthase, promoting L-leucine biosynthesis (Hao et al. 2023). On the other hand, pyruvate is an intermediate that improves its availability by overexpressing enzymes responsible for its synthesis, such as pyruvate kinase (encoded by *pyk*) and oxalacetate decarboxylase (encoded by *odx*) modified at the level of its promoters (Luo et al. 2021). However, it is also used in competitive pathways, so the elimination of the *ldhA*, *pflB,* and *poxB* genes directs its use to the synthesis of L-leucine (Hao et al. 2023). While the deletion of *pyc* (encoding pyruvate carboxylase) deregulates the pyruvate flux toward the synthesis of oxalacetate for entry into the tricarboxylic acid cycle, the pathway that can be decreased by dynamic regulation of genetic transcription levels (Luo et al. 2021; Ding et al. 2023; Hao et al. 2023; Wang et al. 2020a).

NADPH is required in L-leucine biosynthesis, therefore the *pntAB* gene has been overexpressed, and the specificity of AHAIR for NADPH has been modified, replacing it with NADH and introducing NADH-dependent glutamate dehydrogenase and leucine dehydrogenase, the latter being a replacement for transaminase, but with greater specificity for the L-leucine biosynthesis pathway than for L-valine (Wang et al. 2019b; Chen et al. 2022; Ding et al. 2023; Hao et al. 2023). Also, L-leucine efflux has been improved by overexpressing *leuE* and blocking reuptake by removal of *livK*, which improves L-leucine production to 63.29 g/L with a productivity of 2.64 g/L/h (Ding et al. 2023). Finally, the elimination of the reabsorption system of this amino acid encoded by *livJ* and the introduction of the *brnFE* genes, encoders of the export system, would have a positive effect on the production of L-leucine (Hao et al. 2023).

2.2.7 Threonine

Threonine is an amino acid of importance in the medicine and food industry due to its role in the immune system, growth, and food preservation (Duan et al. 2023). Its biosynthesis requires oxalacetate, which is transformed to aspartate by aspartate aminotransferase (encoded by *aspC*), after which aspartate kinase isoenzymes I, II, and III (encoded by *thrA*, *metL,* and *lysC*) catalyze its conversion to aspartyl phosphate, which will be transformed to aspartyl semialdehyde by aspartate semialdehyde dehydrogenase (encoded by *asd*), requiring NADPH for the latter two reactions. This intermediate is converted by aspartate kinase I and II to homoserine, which by homoserine kinase (encoded by *thrB*) is transformed to homoserine phosphate, which finally using threonine synthase (encoded by *thrC*) is converted to L-threonine. For export, *E. coli* uses the RhtA, RhtB, RhtC, and YecC proteins, while *C. glutamicum* has ThrE (Yin et al. 2014). However, L-threonine is a feedback inhibitor of homoserine kinase (ThrB), which has been reduced by the A20G mutation in its active site while maintaining its enzymatic activity (Petit et al. 2018).

The availability of PEP has been increased by deletion of *crr* and *ptsG* genes, encoding EIIAGlc and EIIBCGlc, respectively, which increases the production of this amino acid, but slows cell growth due to damage to the phosphotransferase system (Wang et al. 2021). Likewise, the expression levels of the *arcA*, *iclR*, *aspC*, *cpxR*, *gadE*, *fadR*, and *pykF* genes have been regulated according to the fermentation phases and the intracellular concentration of L-threonine. This is because *arcA* and *iclR* act in the repression of genes associated with the tricarboxylic acid cycle and glyoxylate bypass, and *aspC* and *pykF* are encoders of aspartate aminotransferase and pyruvate kinase, respectively, and *cpxR*, *gadE*, and *fadR* are regulators of fatty acid metabolism. Its regulation showed an increase of 49.4% in the production of L-threonine, with levels of 116.62 g/L in fed-batch culture (Zhao et al. 2020b). Thus, the deletion of *fadR* and *iclR* genes accelerates glucose consumption and improves L-threonine production but slightly inhibits cell growth, which can be countered by deleting FabR, a negative regulator of fatty acid synthesis. Likewise, in order to improve the conversion of acetate to acetyl-CoA, the *aceBA* operon has been overexpressed, as well as the carbon flux when inserting *ppc* and *aspC*, resulting in 103.89 g/L of L-threonine at 48 hours of fed-batch culture (Yang et al. 2019).

Glyoxylate bypass has been regulated by removing *iclR* and replacing the *aceBA* operon promoter, thereby improving glyoxylate bypass and the production of acetyl-CoA derivatives and allowing the accumulation of oxaloacetate. In addition, the overexpression of *thrA*, *thrB*, and *thrC* genes, the introduction of *asd*, and the regulation of *aspC*, associated with the biosynthesis, lead to the increase of L-threonine production by 70% (Zhao et al. 2018). These modifications, along with the deletions of *tdh*, *dapA*, and *metL* and the guarantee of sufficient L-glutamate for transamination, have proven to contribute better to the production of L-threonine (Hanke et al. 2023). As well as the application of promoters of different strengths that effectively regulate the expression of genes is associated with biosynthesis (Wei et al. 2018).

In order to improve its production, two strategies are applied, the first in which the overexpression of *pckA* and *pycA* and the introduction of *citA* improve glucose consumption and the tricarboxylic acid pathway for cell growth, and the second is the production phase in which overexpression of *aspC* and *gdhA* directs the carbon flux toward L-aspartate production. (Liu et al. 2019). Thus, the ideal metabolic analysis would lead to 44% of the carbon flux being converted into oxaloacetate and 56% into pyruvate, by which the reduction in the cycle of tricarboxylic acids improves the production of this amino acid (Yang et al. 2022). Also, a thermal exchange system has been applied along with the strong expression of pyruvate carboxylase to channel carbon for the production of oxalacetate and balance the availability of cofactor NADPH for L-threonine synthesis (Fang et al. 2020).

The protein transporters ProP and ProVWX are regulated by osmotic pressure, and they participate as osmoprotectors. Their deletion deregulates the intracellular biosynthesis of betaine and galactose transport, while the genes associated with glucose and galactose transport, as well as NADH dehydrogenase are overregulated, so that due to osmotic pressure changes the cells would increase the use of these carbon sources and the production of energy in order to maintain their physiological activity, leading to an increase in L-threonine production (Wang et al. 2021). NADH can be increased by overexpressing *fdh* and converted to NADPH by overexpressing *pntAB*, or its availability can be improved by inserting *ppnK* (Liu et al. 2019; Yang et al. 2019). On the other hand, the increase in the secretion of this amino acid is greater when *rhtC*, *yecC*, or *yecC-yecS-fliY* are overexpressed, and the deletion of *tdcC* prevents the reuptake of threonine (Fang et al. 2020, Xu et al. 2017). In order to decrease the production of coproducts such as L-lysine, *lysE* has been mutated, improving the production of L-threonine and eliminating the excretion of L-lysine, but with slower cell growth (Dong et al. 2016).

Also, *Halomonas bluephagenesis* has been constructed to produce this amino acid by deletion of bypasses, improvement of metabolic fluxes and transport, and reduction of its reuptake, reaching 33 g/L at laboratory scale with a productivity of 1.4 g/L/h (Du et al. 2020). Recently, an *E. coli* with more than two chromosomes has been designed with overregulation of replication, transcription, and translation of its genes, tolerance to low pH levels, resistance to acetate, and better growth when glucose is used as carbon source, reaching 160 g/L of L-threonine with a productivity of 1.66 g/L/h (Wang et al. 2023a). These designs have evidence to be suitable at the industrial level, which should be applied together to optimize the production of this amino acid.

2.2.8 Isoleucine

Isoleucine performs important functions at the neurological level by being used for the synthesis of glutamate, as well as within the immune system, for growth, in protein and fatty acid metabolism, and in the transport of glucose, which has increased its presence in the nutraceutical industry (Gu et al. 2019). Its synthesis requires oxalacetate as an intermediate, which is successively transformed into aspartate, aspartyl phosphate, aspartate semialdehyde, and homoserine. This intermediate is converted to homoserine phosphate, after which it is transformed to threonine, and then converted to 2-ketobutyrate using threonine dehydrogenase (encoded by *ilvA*). It is transformed parallelly to L-leucine and L-valine, by AHAS, AHAIR, DHAD, and transaminase of branched amino acids, after which it is exported by the BrnFE permease in *C. glutamicum* using a proton motif (YgaZH of *E. coli*) (Eggeling et al. 1997).

The *gnd*, *pgl*, and *fbp* genes associated with the PPP have been overexpressed, which redirects carbon flux and increases NADPH for L-isoleucine production, increasing L-isoleucine production by 25.1% (Ma et al. 2016). Mutations in AHAS, aspartate kinase, homoserine dehydrogenase, homoserine kinase, and threonine dehydrogenase have also been done in order to generate resistance to branched amino acids feedback inhibition (Wang 2019). Likewise, in *E. coli* the genes *thrABC*, *ilvA*, *ilvIH*, and *ygaZH* have been overexpressed, which increases the carbon flux for its biosynthesis and export, and the expression of *ilvCED* and *lrp* has been amplified, the latter being a regulator that activates the expression of *ilvIH* and *ygaZH* (Park et al. 2012).

The availability of intermediates such as pyruvate has been increased by deletion of *alaT* (aminotransferase encoder) and 2-ketobutyrate, by overexpressing the *thrABC* genes and increasing the transcription levels of *ilvA*, which led to its redistribution toward L-isoleucine production with lower co-products excretion (Wang et al. 2013; Zhang et al. 2021b). In order to increase NADPH availability, poly(P)/ATP-NAD kinase (encoded by *ppnK*), which uses ATP and poly(P) as a phosphoryl donor for NAD^+ and NADH, and glucose 6 dehydrogenase (encoded by *zwf*) have been overexpressed which improved NADPH supply and increased L-isoleucine production by 85.9%, showing a slight increase in the growth rate (Shi et al. 2013). Point mutations P57S and P117S of poly(P)/ATP-NAD kinase have been identified to increase its activity and NADPH concentration compared to the absence of these mutations (Shi et al. 2012). Also, overexpression of the *fusA* and *frr* genes, linked to the ribosome elongation factor G and recycling factor, evidenced an increase in

L-isoleucine levels, which along with the overexpression of *ilvA*, *ilvB*, *ilvN*, and *ppnK* produced 28.5 g/L of L-isoleucine at 72 hours of fed-batch culture (Zhao et al. 2015). Likewise, deletion of the *brnQ* gene, encoder of the L-isoleucine recapturing protein, increases the amount of L-isoleucine by 21.9% and accelerates cell growth.

Improved carbon flux also includes decreased resistance to branched amino acids inhibition, which along with overexpression of enzymes linked to intermediary biosynthesis have improved L-isoleucine production to about 32 g/L; however, combinations with other modifications leading to improvement in these levels should be considered (Zhang et al. 2021b).

2.2.9 Histidine

This amino acid is required for protein synthesis as well as for hyaluronic acid synthesis (along with other essential amino acids), which is involved in tissue repair and cell proliferation, and is used in the treatment of neurodegenerative diseases and metabolic syndrome (Moro et al. 2020). It has also been used in the biotechnology industry for protein purification, but its use in the formation of nanoparticles and as a drug transport system in bionanomedicine has recently been investigated (Ferrer-Miralles et al. 2011). Histidine has phosphorybosyl pyrophosphate (PRPP) and ATP as precursors, from which 10 biochemical reactions are carried out in *C. glutamicum*, involving nine enzymes (HisG, HisE, HisI, HisA, HisF, HisH, HisB, HisC, and HisN) along with histidinol dehydrogenase (HisD) (Kulis-Horn et al. 2014).

In order to improve feedback inhibition and attenuation of *his* operon transcription, HisG has been mutated and HisL regulated, respectively, and to increase the metabolic flux, promoters have been replaced or the biosynthesis operons have been overexpressed (Schwentner et al. 2019; Wu et al. 2020; Chai et al. 2023; Kim et al. 2024). Biosynthesis of this amino acid is associated with purine biosynthesis, as to achieve the cellular requirement of ATP, regeneration of ATP from 5-aminoimidazol-4 carboxamide ribonucleotide (AICAR) is key, and to achieve high carbon flux, the availability of 10-formyltetrahydrofolate (fTHF) as a cofactor of the purine pathway must be ensured (Schwentner et al. 2019). This coupled with the overexpression of genes linked to ATP regeneration has improved the flux of this nucleotide (Wu et al. 2020; Kim et al. 2024).

The HisD promoter, which regulates HisD and HisC, has been modified, thereby enhancing the expression levels of genes associated with the PPP and, therefore, the levels of NADPH (Kim et al. 2024). Additionally, the introduction of RocG improves the regeneration of NAD^+ and glutamate, which is necessary for the transamination reaction catalyzed by HisC (Wu et al. 2020). Redirection of carbon flux to the PPP and increased synthesis of the PRPP precursor have been achieved by blocking *pgi*, by overexpressing *zwf*, *prs*, and *gnd*, and by deleting *purR*, the negative regulator of *prs* (Wu et al. 2020; Chai et al. 2023; Kim et al. 2024). Finally, modified *E. coli* has been able to produce up to 66.5 g/L with a productivity of 1.5 g/L/h, being one of the highest, so the application of modular metabolic engineering would be effective for its optimization (Wu et al. 2020).

2.3 Fermentation Strategies

Metabolic engineering strategies coupled with bioprocess engineering techniques have improved production levels. Process optimization by analyzing different fermentation factors such as pH, temperature, incubation time, and agitation is an approach traditionally used to improve the yield and efficiency of fermentation processes (Zhang et al. 2021c). For example, in *B. cereus* isolated from Nigerian soils, L-valine production was optimized to 3.53 mg/mL by applying Luria-Bertani (LB) culture medium, pH 7.5, 40 °C for 58 hours incubation, and 125 rpm (Aliyu et al. 2022). The composition of the culture medium regulates gene expression, thus, supplementation of 9.8 g/L phosphate in the production of L-threonine improves cell growth and accumulation of this amino acid, with a glucose conversion rate of 45.5% (Yang et al. 2022).

In order to make the production of these amino acids profitable, lower cost substrates are used. Such is the case of whey as a carbon source, which increased L-valine production by 36% compared to glucose or lactose in modified *E. coli*. In addition, the Luedeking-Piret model is applied, which allows the simulation of L-valine biosynthesis considering parameters associated and not associated with growth (Carranza-Saavedra et al. 2021). The use of complex substrates requires pretreatments to facilitate the availability of sugars; however, they are low-cost sources of nutrients, for example, the use of pretreated beet molasses by combining tricalcium phosphate, sulfuric acid, and activated carbon is applied in the L-lysine synthesis (He et al. 2015). Also, κ-carrageenan hydrolysates have been used to release D-galactose in the production of L-hisitidine (Kim et al. 2024).

In the same way, the effect of oxygen on the cultures has been studied, since in the production of L-lysine the aeration system has been improved during the 5 L batch fermentation, which increases the level of dissolved oxygen and therefore its viability in the medium and late growth phase leading to levels of 185.3 g/L of L-lysine with a glucose conversion rate of 74.57% (Wang et al. 2023b). Likewise, in the production of L-valine, the culture has been applied in two phases, the first being the aerobic phase in which the increase in biomass was evidenced and then followed by a second phase with oxygen limitation, which evidenced a better yield of L-valine (0.41 g/g glucose) compared to the aerobic phase (0.33 g/g glucose) (Hao et al. 2020). In the production of L-tryptophan, the concentration of dissolved oxygen and pH was modulated, and thus, maintaining 20% of dissolved oxygen and pH 7 in the first 20 hours of fermentation followed by an increase to 30% and pH 6.5 for 20 hours leads to the production of 52.57 g/L of L-tryptophan. Similarly, high levels of dissolved oxygen produce increased glutamate, which has been negatively correlated with tryptophan production, while phosphofructokinase activity is inhibited at low pH levels, limiting cell growth, while high pH produced by the NH_4 release decreases energy efficiency and inhibits growth (Zhao et al. 2020a).

The continuous culture mode has advantages over the fed-batch by continuously renewing the culture medium and eliminating cellular toxic products, thus increasing the yield of L-leucine from glucose and decreasing the production of other amino acids (Zhang et al. 2021d). On the other hand, in order to improve extracellular L-leucine levels, ultrasonication has been applied during fermentation, which increases cell permeability and, therefore, the amount of L-leucine by 21.6% (Zhang et al. 2021a). For the production of L-tryptophan and L-phenylalanine, fermentation with immobilized cells has been applied, and it is achieved through the use of positively charged particles at the surface level, which contributes to cell adsorption and reusability by up to five times (Su et al. 2023).

2.4 Perspectives and Challenges

The great applicability and economic contribution have motivated the improvement in the production levels of essential amino acids in *C. glutamicum* and *E. coli*, which has led to their great potential for industrial application, especially in the pharmaceutical sector, which covers a large part of the market. In addition, the increased demand for products free of animal cruelty has driven the implementation of microbial modifications and bioprocess technology that improve the production of amino acids by fermentation. These microorganisms are generally considered as safe (GRAS), which has extended its use in the food industry. Among the most applied strategies are the improvement of precursors availability, the decrease in feedback inhibition, the channeling of carbon flux, the improvement of the excretory system, and the decrease of amino acid reuptake systems and competitive pathways, for which the regulation of transcription and translation, the dynamic regulation of enzymes, deletion of genes linked to competitive pathways, and the overexpression of operons associated with biosynthetic pathways, among others, have been applied, which together have improved the production and productivity levels of these amino acids. These strains have been used at laboratory scale using fed-batch culture, with which promising results have been obtained for their industrial application. However, successful industrialization requires the application of bioprocess engineering for scale-up using these cellular factories and further development to meet demanding quality standards that guarantee purity and safety, as well as efficient and sustainable production processes. Therefore, industrial production will be subject to raw materials requirements and processing costs for synthesis in a highly competitive market.

References

Aliyu, A., Nasiru, S., Mari, J.U. et al. (2022). Optimization of valine production using *Bacillus cereus* isolated from soil. *Scholars International Journal of Biochemistry* 5 (2): 37–49. https://doi.org/10.36348/sijb.2022.v05i02.002.

Awad, L.F. and Ayoup, M.S. (2020). Fluorinated phenylalanines: synthesis and pharmaceutical applications. *Beilstein Journal of Organic Chemistry* 16: 1022–1050. https://doi.org/10.3762/bjoc.16.91.

Bang, H.B., Choi, I.H., Jang, J.H. et al. (2021). Engineering of *Escherichia coli* for the economic production l-phenylalanine in large-scale bioreactor. *Biotechnology and Bioprocess Engineering* 26 (3): 468–475. https://doi.org/10.1007/s12257-020-0313-1.

Bartek, T., Blombach, B., Zönnchen, E. et al. (2010). Importance of NADPH supply for improved l-Valine formation in *Corynebacterium glutamicum*. *Biotechnology Progress* 26 (2): 361–371. https://doi.org/10.1002/btpr.345.

Bastard, K., Perret, A., Mariage, A. et al. (2017). Parallel evolution of non-homologous isofunctional enzymes in methionine biosynthesis. *Nature Chemical Biology* 13 (8): 858–866. https://doi.org/10.1038/nchembio.2397.

Becker, J., Zelder, O., Häfner, S. et al. (2011). From zero to hero-design-based systems metabolic engineering of *Corynebacterium glutamicum* for l-lysine production. *Metabolic Engineering* 13 (2): 159–168. https://doi.org/10.1016/j.ymben.2011.01.003.

Cai, M., Zhao, Z., Li, X. et al. (2022). Development of a nonauxotrophic L-homoserine hyperproducer in *Escherichia coli* by systems metabolic engineering. *Metabolic Engineering* 73: 270–279. https://doi.org/10.1016/j.ymben.2022.08.003.

Cai, M., Liu, Z., Zhao, Z. et al. (2023). Microbial production of L-methionine and its precursors using systems metabolic engineering. *Biotechnology Advances* 69: 108260. https://doi.org/10.1016/j.biotechadv.2023.108260.

Carranza-Saavedra, D., Sánchez Henao, C.P., and Zapata Montoya, J.E. (2021). Kinetic analysis and modeling of l-valine production in fermentation batch from *E. coli* using glucose, lactose and whey as carbon sources. *Biotechnology Reports* 31: e00642. https://doi.org/10.1016/j.btre.2021.e00642.

Chai, M., Xu, X., Liu, Y. et al. (2023). Combinatorial protein engineering and metabolic engineering for efficient synthesis of l-histidine in *Corynebacterium glutamicum*. *ACS Synthetic Biology* 12 (4): 1275–1286. https://doi.org/10.1021/acssynbio.3c00026.

Chen, L. and Zeng, A.P. (2017). Rational design and metabolic analysis of *Escherichia coli* for effective production of L-tryptophan at high concentration. *Applied Microbiology and Biotechnology* 101 (2): 559–568. https://doi.org/10.1007/s00253-016-7772-5.

Chen, L., Chen, M., Ma, C. et al. (2018a). Discovery of feed-forward regulation in L-tryptophan biosynthesis and its use in metabolic engineering of *E. coli* for efficient tryptophan bioproduction. *Metabolic Engineering* 47: 434–444. https://doi.org/10.1016/j.ymben.2018.05.001.

Chen, Y., Liu, Y., Ding, D. et al. (2018b). Rational design and analysis of an *Escherichia coli* strain for high-efficiency tryptophan production. *Journal of Industrial Microbiology and Biotechnology* 45 (5): 357–367. https://doi.org/10.1007/s10295-018-2020-x.

Chen, M., Liang, H., Han, C. et al. (2022). Engineering of global transcription factor FruR to redirect the carbon flow in *Escherichia coli* for enhancing l-phenylalanine biosynthesis. *Microbial Cell Factories* 21 (1): 222. https://doi.org/10.1186/s12934-022-01954-7.

Chen, S.-L., Liu, T.-S., Zhang, W.-G. et al. (2023). An NADPH-auxotrophic *Corynebacterium glutamicum* recombinant strain and used it to construct L-leucine high-yielding strain. *International Microbiology* 26 (1): 11–24. https://doi.org/10.1007/s10123-022-00270-9.

Ding, X., Yang, W., Du, X. et al. (2023). High-level and -yield production of L-leucine in engineered *Escherichia coli* by multistep metabolic engineering. *Metabolic Engineering* 78: 128–136. https://doi.org/10.1016/j.ymben.2023.06.003.

Divya, K., Malothu, R., Dowlathabad, M.R. et al. (2022). Recent advances in engineering *Corynebacterium glutamicum* for production of Lysine. *Research Journal of Biotechnology* 17 (12): 190–206. https://doi.org/10.25303/1712rjbt1900206.

Dong, X., Zhao, Y., Hu, J. et al. (2016). Attenuating L-lysine production by deletion of ddh and lysE and their effect on L-threonine and L-isoleucine production in *Corynebacterium glutamicum*. *Enzyme and Microbial Technology* 93–94: 70–78. https://doi.org/10.1016/j.enzmictec.2016.07.013.

Du, L., Zhang, Z., Xu, Q. et al. (2019). Central metabolic pathway modification to improve L-tryptophan production in *Escherichia coli*. *Bioengineered* 10 (1): 59–70. https://doi.org/10.1080/21655979.2019.1592417.

Du, H., Zhao, Y., Wu, F. et al. (2020). Engineering *Halomonas bluephagenesis* for L-Threonine production. *Metabolic Engineering* 60: 119–127. https://doi.org/10.1016/j.ymben.2020.04.004.

Duan, M., Chen, S., Liu, X. et al. (2023). The application of *Corynebacterium glutamicum* in l-threonine biosynthesis. *Fermentation* 9 (9): 822. https://doi.org/10.3390/fermentation9090822.

Eggeling, L., Morbach, S., and Sahm, H. (1997). The fruits of molecular physiology: engineering the L-isoleucine biosynthesis pathway in *Corynebacterium glutamicum*. *Journal of Biotechnology* 56: 167–182. https://doi.org/10.1016/S0168-1656(97)00115-6.

Fang, Y., Wang, J., Ma, W. et al. (2020). Rebalancing microbial carbon distribution for L-threonine maximization using a thermal switch system. *Metabolic Engineering* 61: 33–46. https://doi.org/10.1016/j.ymben.2020.01.009.

Ferla, M.P. and Patrick, W.M. (2014). Bacterial methionine biosynthesis. *Microbiology (United Kingdom)* 160 (Part 8): 1571–1584. https://doi.org/10.1099/mic.0.077826-0.

Ferrer-Miralles, N., Corchero, J.L., Kumar, P. et al. (2011). Biological activities of histidine-rich peptides; merging biotechnology and nanomedicine. *Microbial Cell Factories* 10: 101. https://doi.org/10.1186/1475-2859-10-101.

Gao, H., Tuyishime, P., Zhang, X. et al. (2021). Engineering of microbial cells for L-valine production: challenges and opportunities. *Microbial Cell Factories* 20 (1): 172. https://doi.org/10.1186/s12934-021-01665-5.

Gruzdev, N., Hacham, Y., Haviv, H. et al. (2023). Conversion of methionine biosynthesis in *Escherichia coli* from trans- to direct-sulfurylation enhances extracellular methionine levels. *Microbial Cell Factories* 22 (1): 151. https://doi.org/10.1186/s12934-023-02150-x.

Gu, C., Mao, X., Chen, D. et al. (2019). Isoleucine plays an important role for maintaining immune function. *Current Protein & Peptide Science* 20 (7): 644–651. https://doi.org/10.2174/1389203720666190305163135.

Gu, P., Zhao, S., Li, C. et al. (2023). Construction of recombinant *Escherichia coli* with a high L-phenylalanine production yield from glucose. *Microbiology Research* 14 (3): 1185–1198. https://doi.org/10.3390/microbiolres14030079.

Guo, L., Ding, S., Liu, Y. et al. (2022). Enhancing tryptophan production by balancing precursors in *Escherichia coli. Biotechnology and Bioengineering* 119 (3): 983–993. https://doi.org/10.1002/bit.28019.

Guo, Y.F., Lu, Y., Song, L. et al. (2023). Identification of α-isopropylmalate synthase mutants capable of overproducing L-leucine in *Corynebacterium glutamicum. All Life* 16 (1): 2211237. https://doi.org/10.1080/26895293.2023.2211237.

Han, G., Xu, N., Sun, X. et al. (2020). Improvement of l-valine production by atmospheric and room temperature plasma mutagenesis and high-throughput screening in *Corynebacterium glutamicum. ACS Omega* 5 (10): 4751–4758. https://doi.org/10.1021/acsomega.9b02747.

Hanke, P., Parrello, B., Vasieva, O. et al. (2023). Engineering of increased L-Threonine production in bacteria by combinatorial cloning and machine learning. *Metabolic Engineering Communications* 17: e00225. https://doi.org/10.1016/j.mec.2023.e00225.

Hao, Y., Ma, Q., Liu, X. et al. (2020). High-yield production of L-valine in engineered *Escherichia coli* by a novel two-stage fermentation. *Metabolic Engineering* 62: 198–206. https://doi.org/10.1016/j.ymben.2020.09.007.

Hao, Y., Pan, X., Xing, R. et al. (2022). High-level production of L-valine in *Escherichia coli* using multi-modular engineering. *Bioresource Technology* 359: 127461. https://doi.org/10.1016/j.biortech.2022.127461.

Hao, Y., Pan, X., Li, G. et al. (2023). Construction of a plasmid-free l-leucine overproducing *Escherichia coli* strain through reprogramming of the metabolic flux. *Biotechnology for Biofuels and Bioproducts* 16 (1): https://doi.org/10.1186/s13068-023-02397-x.

Hasegawa, S., Uematsu, K., Natsuma, Y. et al. (2012). Improvement of the redox balance increases L-valine production by *Corynebacterium glutamicum* under oxygen deprivation conditions. *Applied and Environmental Microbiology* 78 (3): 865–875. https://doi.org/10.1128/AEM.07056-11.

He, X., Chen, K., Li, Y. et al. (2015). Enhanced l-lysine production from pretreated beet molasses by engineered *Escherichia coli* in fed-batch fermentation. *Bioprocess and Biosystems Engineering* 38 (8): 1615–1622. https://doi.org/10.1007/s00449-015-1403-x.

Hermann, T. (2003). Industrial production of amino acids by coryneform bacteria. *Journal of Biotechnology* 104 (1–3): 155–172. https://doi.org/10.1016/s0168-1656(03)00149-4.

Holeček, M. (2018). Branched-chain amino acids in health and disease: metabolism, alterations in blood plasma, and as supplements. *Nutrition and Metabolism* 15 (1): 33. https://doi.org/10.1186/s12986-018-0271-1.

Huang, J.F., Shen, Z.Y., Mao, Q.L. et al. (2018a). Systematic analysis of bottlenecks in a multibranched and multilevel regulated pathway: the molecular fundamentals of l-methionine biosynthesis in *Escherichia coli. ACS Synthetic Biology* 7 (11): 2577–2589. https://doi.org/10.1021/acssynbio.8b00249.

Huang, Q.G., Zeng, B.D., Liang, L. et al. (2018b). Genome shuffling and high-throughput screening of *Brevibacterium flavum* MDV1 for enhanced l-valine production. *World Journal of Microbiology and Biotechnology* 34 (8): 121. https://doi.org/10.1007/s11274-018-2502-z.

Humphries, P., Pretorius, E., and Naudé, H. (2008). Direct and indirect cellular effects of aspartame on the brain. *European Journal of Clinical Nutrition* 62 (4): 451–462. https://doi.org/10.1038/sj.ejcn.1602866.

Ivanov, K., Stoimenova, A., Obreshkova, D. et al. (2013). Biotechnology in the production of pharmaceutical industry ingredients: amino acids. *Biotechnology and Biotechnological Equipment* 27 (2): 3620–3626. https://doi.org/10.5504/bbeq.2012.0134.

Jojima, T., Igari, T., Noburyu, R. et al. (2021). Coexistence of the Entner–Doudoroff and Embden–Meyerhof–Parnas pathways enhances glucose consumption of ethanol-producing *Corynebacterium glutamicum*. *Biotechnology for Biofuels* 14 (1): 45. https://doi.org/10.1186/s13068-021-01876-3.

Kawai, R., Toya, Y., and Shimizu, H. (2022). Metabolic pathway design for growth-associated phenylalanine production using synthetically designed mutualism. *Bioprocess and Biosystems Engineering* 45 (9): 1539–1546. https://doi.org/10.1007/s00449-022-02762-4.

Kikuchi, A.M., Tanabe, A., and Iwahori, Y. (2021). A systematic review of the effect of L-tryptophan supplementation on mood and emotional functioning. *Journal of Dietary Supplements* 18 (3): 316–333. https://doi.org/10.1080/19390211.2020.1746725.

Kim, M., Oh, J.W., Jeong, D.W. et al. (2024). Biosynthesis of L-histidine from marine biomass-derived galactans in metabolically engineered *Corynebacterium glutamicum*. *Bioresource Technology* 391: 129963. https://doi.org/10.1016/j.biortech.2023.129963.

Kulis-Horn, R.K., Persicke, M., and Kalinowski, J. (2014). Histidine biosynthesis, its regulation and biotechnological application in *Corynebacterium glutamicum*. *Microbial Biotechnology* 7 (1): 5–25. https://doi.org/10.1111/1751-7915.12055.

Kumagai, H. (2014). Amino acid production. In: *The Prokaryotes: Applied Bacteriology and Biotechnology* (ed. E. Rosenberg, E.F. DeLong, S. Lory, et al.). Heidelberg: Springer-Verlag.

Leuchtenberger, W., Huthmacher, K., and Drauz, K. (2005). Biotechnological production of amino acids and derivatives: current status and prospects. *Applied Microbiology and Biotechnology* 69: 1–8. https://doi.org/10.1007/s00253-005-0155-y.

Li, B., Huang, L.G., Yang, Y.F. et al. (2023a). Metabolic engineering and pathway construction for O-acetyl-L-homoserine production in *Escherichia coli*. *3 Biotech* 13 (6): 173. https://doi.org/10.1007/s13205-023-03564-5.

Li, Z., Liu, Q., Sun, J. et al. (2023b). Multivariate modular metabolic engineering for enhanced l-methionine biosynthesis in *Escherichia coli*. *Biotechnology for Biofuels and Bioproducts* 16 (1): 101. https://doi.org/10.1186/s13068-023-02347-7.

Li, N., Lu, J., Wang, Z. et al. (2023c). Improving the regeneration rate of deep lethal mutant protoplasts by fusion to promote efficient L-lysine fermentation. *BMC Biotechnology* 23 (1): 22. https://doi.org/10.1186/s12896-023-00792-8.

Liu, J., Li, H., Xiong, H. et al. (2019). Two-stage carbon distribution and cofactor generation for improving l-threonine production of *Escherichia coli*. *Biotechnology and Bioengineering* 116 (1): 110–120. https://doi.org/10.1002/bit.26844.

Liu, L., Liu, X., Ma, Q. et al. (2021). Construction of recombinant *Escherichia coli* for production of l-phenylalanine-derived compounds. *World Journal of Microbiology and Biotechnology* 37 (5): 84. https://doi.org/10.1007/s11274-021-03050-1.

Liu, J., Xu, J.Z., Rao, Z.M. et al. (2022a). Industrial production of L-lysine in *Corynebacterium glutamicum*: progress and prospects. *Microbiological Research* 262: 127101. https://doi.org/10.1016/j.micres.2022.127101.

Liu, P., Xu, H., and Zhang, X. (2022b). Metabolic engineering of microorganisms for L-alanine production. *Journal of Industrial Microbiology and Biotechnology* 49 (2): kuab057. https://doi.org/10.1093/jimb/kuab057.

Liu, J., Ou, Y., Xu, J.Z. et al. (2023a). L-lysine production by systems metabolic engineering of an NADPH auto-regulated *Corynebacterium glutamicum*. *Bioresource Technology* 387: 129701. https://doi.org/10.1016/j.biortech.2023.129701.

Liu, S., Wang, B.B., Xu, J.Z. et al. (2023b). Engineering of shikimate pathway and terminal branch for efficient production of l-tryptophan in *Escherichia coli*. *International Journal of Molecular Sciences* 24 (14): 11866. https://doi.org/10.3390/ijms241411866.

Lo, E.K.K., Xu, J.H., Zhan, Q. et al. (2022). The emerging role of branched-chain amino acids in liver diseases. *Biomedicine* 10 (6): 1444. https://doi.org/10.3390/biomedicines10061444.

Luo, G., Zhao, N., Jiang, S. et al. (2021). Application of RecET-Cre/loxP system in *Corynebacterium glutamicum* ATCC14067 for l-leucine production. *Biotechnology Letters* 43 (1): 297–306. https://doi.org/10.1007/s10529-020-03000-1.

Ma, W., Wang, J., Li, Y. et al. (2016). Enhancing pentose phosphate pathway in *Corynebacterium glutamicum* to improve l-isoleucine production. *Biotechnology and Applied Biochemistry* 63 (6): 877–885. https://doi.org/10.1002/bab.1442.

Ma, Y., Cui, Y., Du, L. et al. (2018). Identification and application of a growth-regulated promoter for improving l-valine production in *Corynebacterium glutamicum*. *Microbial Cell Factories* 17 (1): 185. https://doi.org/10.1186/s12934-018-1031-7.

Meng, Y. and Wu, J. (2021). One-step and facile synthesis of poly(phenylalanine) as a robust drug carrier for enhanced cancer therapy. *ACS Applied Materials and Interfaces* 13 (42): 49658–49670. https://doi.org/10.1021/acsami.1c13013.

Moro, J., Tomé, D., Schmidely, P. et al. (2020). Histidine: a systematic review on metabolism and physiological effects in human and different animal species. *Nutrients* 12 (5): https://doi.org/10.3390/nu12051414.

Myles, J.G., Manoli, I., and Venditti, C.P. (2018). Effects of medical food leucine content in the management of methylmalonic and propionic acidemias. *Current Opinion in Clinical Nutrition and Metabolic Care* 21 (1): 42–48. https://doi.org/10.1097/MCO.0000000000000428.

Nakamori, S. (2017). Early history of the breeding of amino acid-producing strains. In: *Amino Acid Fermentation* (ed. A. Yokota and M. Ikeda). Tokyo: Springer.

Niu, H., Li, R., Liang, Q. et al. (2019). Metabolic engineering for improving l-tryptophan production in *Escherichia coli*. *Journal of Industrial Microbiology and Biotechnology* 46 (1): 55–65. https://doi.org/10.1007/s10295-018-2106-5.

Niu, K., Fu, Q., Mei, Z.L. et al. (2023). High-level production of l-methionine by dynamic deregulation of metabolism with engineered nonauxotroph *Escherichia coli*. *ACS Synthetic Biology* 12 (2): 492–501. https://doi.org/10.1021/acssynbio.2c00481.

Park, J.H., Oh, J.E., Lee, K.H. et al. (2012). Rational design of *Escherichia coli* for L-isoleucine production. *ACS Synthetic Biology* 1 (11): 532–540. https://doi.org/10.1021/sb300071a.

Pedroso, J.A.B., Zampieri, T.T., and Donato, J. (2015). Reviewing the effects of l-leucine supplementation in the regulation of food intake, energy balance, and glucose homeostasis. *Nutrients* 7 (5): 3914–3937. https://doi.org/10.3390/nu7053914.

Petit, C., Kim, Y., Lee, S.K. et al. (2018). Reduction of feedback inhibition in homoserine Kinase (ThrB) of *Corynebacterium glutamicum* enhances L-threonine biosynthesis. *ACS Omega* 3 (1): 1178–1186. https://doi.org/10.1021/acsomega.7b01597.

Pittard, J. and Yang, J. (2008). Biosynthesis of the aromatic amino acids. *EcoSal Plus* 3 (1): https://doi.org/10.1128/ecosalplus.3.6.1.8.

Sanchez, S., Rodríguez-Sanoja, R., Ramos, A. et al. (2018). Our microbes not only produce antibiotics, they also overproduce amino acids. *The Journal of Antibiotics* 71: 26–36. https://doi.org/10.1038/ja.2017.142.

Schwentner, A., Feith, A., Münch, E. et al. (2018). Metabolic engineering to guide evolution – Creating a novel mode for L-valine production with *Corynebacterium glutamicum*. *Metabolic Engineering* 47: 31–41. https://doi.org/10.1016/j.ymben.2018.02.015.

Schwentner, A., Feith, A., Münch, E. et al. (2019). Modular systems metabolic engineering enables balancing of relevant pathways for l-histidine production with *Corynebacterium glutamicum*. *Biotechnology for Biofuels* 12 (1): 65. https://doi.org/10.1186/s13068-019-1410-2.

Shang, X., Chai, X., Lu, X. et al. (2018). Native promoters of *Corynebacterium glutamicum* and its application in l-lysine production. *Biotechnology Letters* 40 (2): 383–391. https://doi.org/10.1007/s10529-017-2479-y.

Shen, Z.Y., Wang, Y.F., Wang, L.J. et al. (2023). Local metabolic response of *Escherichia coli* to the module genetic perturbations in L-methionine biosynthetic pathway. *Journal of Bioscience and Bioengineering* 135 (3): 217–223. https://doi.org/10.1016/j.jbiosc.2022.12.010.

Shi, F., Huan, X., Wang, X. et al. (2012). Overexpression of NAD kinases improves the l-isoleucine biosynthesis in *Corynebacterium glutamicum* ssp. lactofermentum. *Enzyme and Microbial Technology* 51 (2): 73–80. https://doi.org/10.1016/j.enzmictec.2012.04.003.

Shi, F., Li, K., Huan, X. et al. (2013). Expression of NAD(H) kinase and glucose-6-phosphate dehydrogenase improve NADPH supply and L-isoleucine biosynthesis in *Corynebacterium glutamicum* ssp. lactofermentum. *Applied Biochemistry and Biotechnology* 171 (2): 504–521. https://doi.org/10.1007/s12010-013-0389-6.

Shimomura, A., Matsui, I., Hamano, T. et al. (2014). Dietary L-lysine prevents arterial calcification in adenine-induced uremic rats. *Journal of the American Society of Nephrology: JASN* 25 (9): 1954–1965. https://doi.org/10.1681/ASN.2013090967.

Singh, S.P. and Upadhyay, S.K. (2023). *Microbial Bioreactors for Industrial Molecules*. Wiley. https://doi.org/10.1002/9781119874096.

Straits research (2023). Global amino acids market size, suppliers, Report to 2031. https://straitsresearch.com/report/amino-acids-market (accessed 28 December 2023).

Su, Y., Li, Q., Gu, S. et al. (2023). Development of biochar-based whole-cell biocatalysts for the production of l-tryptophan and l-phenylalanine. *ACS Sustainable Chemistry and Engineering* 11 (25): 9412–9423. https://doi.org/10.1021/acssuschemeng.3c01160.

Sutanto, C.N., Loh, W.W., and Kim, J.E. (2022). The impact of tryptophan supplementation on sleep quality: a systematic review, meta-analysis, and meta-regression. *Nutrition Reviews* 80 (2): 306–316. https://doi.org/10.1093/nutrit/nuab027.

Takpho, N., Watanabe, D., and Takagi, H. (2018). High-level production of valine by expression of the feedback inhibition-insensitive acetohydroxyacid synthase in *Saccharomyces cerevisiae*. *Metabolic Engineering* 46: 60–67. https://doi.org/10.1016/j.ymben.2018.02.011.

Tang, M., Pan, X., Yang, T. et al. (2023). Multidimensional engineering of *Escherichia coli* for efficient synthesis of L-tryptophan. *Bioresource Technology* 386: 129475. https://doi.org/10.1016/j.biortech.2023.129475.

Vogt, M., Haas, S., Klaffl, S. et al. (2014). Pushing product formation to its limit: metabolic engineering of *Corynebacterium glutamicum* for l-leucine overproduction. *Metabolic Engineering* 22: 40–52. https://doi.org/10.1016/j.ymben.2013.12.001.

Wang, X. (2019). Strategy for improving L-isoleucine production efficiency in *Corynebacterium glutamicum*. *Applied Microbiology and Biotechnology* 103 (5): 2101–2111. https://doi.org/10.1007/s00253-019-09632-2.

Wang, J., Wen, B., Wang, J. et al. (2013). Enhancing L-isoleucine production by thrABC overexpression combined with alaT deletion in *Corynebacterium glutamicum*. *Applied Biochemistry and Biotechnology* 171 (1): 20–30. https://doi.org/10.1007/s12010-013-0321-0.

Wang, X., Zhang, H., and Quinn, P.J. (2018). Production of l-valine from metabolically engineered *Corynebacterium glutamicum*. *Applied Microbiology and Biotechnology* 102 (10): 4319–4330. https://doi.org/10.1007/s00253-018-8952-2.

Wang, Y.Y., Xu, J.Z., and Zhang, W.G. (2019a). Metabolic engineering of l-leucine production in *Escherichia coli* and Corynebacterium glutamicum: a review. *Critical Reviews in Biotechnology* 39 (5): 633–647. https://doi.org/10.1080/07388551.2019.1577214.

Wang, Y.Y., Zhang, F., Xu, J.Z. et al. (2019b). Improvement of L-leucine production in *Corynebacterium glutamicum* by altering the redox flux. *International Journal of Molecular Sciences* 20 (8): 2020. https://doi.org/10.3390/ijms20082020.

Wang, Y.Y., Shi, K., Chen, P. et al. (2020a). Rational modification of the carbon metabolism of *Corynebacterium glutamicum* to enhance l-leucine production. *Journal of Industrial Microbiology and Biotechnology* 47 (6–7): 485–495. https://doi.org/10.1007/s10295-020-02282-8.

Wang, L., Yu, H., Xu, J. et al. (2020b). Deciphering the crucial roles of AraC-type transcriptional regulator Cgl2680 on NADPH metabolism and l-lysine production in *Corynebacterium glutamicum*. *World Journal of Microbiology and Biotechnology* 36 (6): 82. https://doi.org/10.1007/s11274-020-02861-y.

Wang, S., Fang, Y., Wang, Z. et al. (2021). Improving l-threonine production in *Escherichia coli* by elimination of transporters ProP and ProVWX. *Microbial Cell Factories* 20 (1): 58. https://doi.org/10.1186/s12934-021-01546-x.

Wang, Y., Xu, J., Jin, Z. et al. (2022). Improvement of acetyl-CoA supply and glucose utilization increases l-leucine production in *Corynebacterium glutamicum*. *Biotechnology Journal* 17 (8): e2100349. https://doi.org/10.1002/biot.202100349.

Wang, S., Chen, X., Jin, X. et al. (2023a). Creating polyploid *Escherichia coli* and its application in efficient L-threonine production. *Advanced Science* 10 (31): https://doi.org/10.1002/advs.202302417.

Wang, J., Wang, S., Zhao, S. et al. (2023b). Productivity enhancement in L-lysine fermentation using oxygen-enhanced bioreactor and oxygen vector. *Frontiers in Bioengineering and Biotechnology* 11: 1181963. https://doi.org/10.3389/fbioe.2023.1181963.

Wei, L., Xu, N., Wang, Y. et al. (2018). Promoter library-based module combination (PLMC) technology for optimization of threonine biosynthesis in *Corynebacterium glutamicum*. *Applied Microbiology and Biotechnology* 102 (9): 4117–4130. https://doi.org/10.1007/s00253-018-8911-y.

Weiner, M., Xu, N., Wang, Y. et al. (2017). Metabolic control analysis of L-phenylalanine production from glycerol with engineered *E. coli* using data from short-term steady-state perturbation experiments. *Biochemical Engineering Journal* 126: 86–100. https://doi.org/10.1016/j.bej.2017.06.016.

Wu, J., Liu, Y., Zhao, S. et al. (2019). Application of dynamic regulation to increase L-phenylalanine production in *Escherichia coli*. *Journal of Microbiology and Biotechnology* 29 (6): 923–932. https://doi.org/10.4014/jmb.1901.01058.

Wu, H., Tian, D., Fan, X. et al. (2020). Highly efficient production of l-histidine from glucose by metabolically engineered *Escherichia coli*. *ACS Synthetic Biology* 9 (7): 1813–1822. https://doi.org/10.1021/acssynbio.0c00163.

Xiao, J., Wang, D., Wang, L. et al. (2020). Increasing l-lysine production in *Corynebacterium glutamicum* by engineering amino acid transporters. *Amino Acids* 52 (10): 1363–1374. https://doi.org/10.1007/s00726-020-02893-6.

Xiong, B., Zhu, Y., Tian, D. et al. (2021). Flux redistribution of central carbon metabolism for efficient production of l-tryptophan in *Escherichia coli*. *Biotechnology and Bioengineering* 118 (3): 1393–1404. https://doi.org/10.1002/bit.27665.

Xu, Y., Liu, Y., Li, F. et al. (2017). Identification of a new gene yecC involved in threonine export in Escherichia coli. *FEMS Microbiology Letters* 364 (17): https://doi.org/10.1093/femsle/fnx174.

Xu, J.Z., Ruan, H.Z., Yu, H.B. et al. (2020). Metabolic engineering of carbohydrate metabolism systems in *Corynebacterium glutamicum* for improving the efficiency of l-lysine production from mixed sugar. *Microbial Cell Factories* 19 (1): 39. https://doi.org/10.1186/s12934-020-1294-7.

Yang, J., Fang, Y., Wang, J. et al. (2019). Deletion of regulator-encoding genes fadR, fabR and iclR to increase L-threonine production in *Escherichia coli. Applied Microbiology and Biotechnology* 103 (11): 4549, 46–4564. https://doi.org/10.1007/s00253-019-09818-8.

Yang, Q., Cai, D., Chen, W. et al. (2022). Combined metabolic analyses for the biosynthesis pathway of l-threonine in *Escherichia coli. Frontiers in Bioengineering and Biotechnology* 10: 1010931. https://doi.org/10.3389/fbioe.2022.1010931.

Yin, L., Zhao, J., Chen, C. et al. (2014). Enhancing the carbon flux and NADPH supply to increase L-isoleucine production in *Corynebacterium glutamicum. Biotechnology and Bioprocess Engineering* 19 (1): 132–142. https://doi.org/10.1007/s12257-013-0416-z.

Zhang, C., Zhang, J., Kang, Z. et al. (2015). Rational engineering of multiple module pathways for the production of l-phenylalanine in *Corynebacterium glutamicum. Journal of Industrial Microbiology and Biotechnology* 42 (5): 787–797. https://doi.org/10.1007/s10295-015-1593-x.

Zhang, H., Li, Y., Wang, C. et al. (2018). Understanding the high l-valine production in *Corynebacterium glutamicum* VWB-1 using transcriptomics and proteomics. *Scientific Reports* 8 (1): https://doi.org/10.1038/s41598-018-21926-5.

Zhang, Y., Chen, Z., Sun, P. et al. (2021a). Effect of low-level ultrasound treatment on the production of L-leucine by *Corynebacterium glutamicum* in fed-batch culture. *Bioengineered* 12 (1): 1078–1090. https://doi.org/10.1080/21655979.2021.1906028.

Zhang, Y., Liu, Y., Zhang, S. et al. (2021b). Metabolic engineering of *Corynebacterium glutamicum* WM001 to improve l-isoleucine production. *Biotechnology and Applied Biochemistry* 68 (3): 568–584. https://doi.org/10.1002/bab.1963.

Zhang, G., Ren, X., Liang, X. et al. (2021c). Improving the microbial production of amino acids: from conventional approaches to recent trends. *Biotechnology and Bioprocess Engineering* 26: 708–727. https://doi.org/10.1007/s12257-020-0390-1.

Zhang, Y., Xiong, H., Chen, Z. et al. (2021d). Effect of fed-batch and chemostat cultivation processes of *C. glutamicum* CP for L-leucine production. *Bioengineered* 12 (1): 426–439. https://doi.org/10.1080/21655979.2021.1874693.

Zhao, J., Hu, X., Li, Y. et al. (2015). Overexpression of ribosome elongation factor G and recycling factor increases l-isoleucine production in *Corynebacterium glutamicum. Applied Microbiology and Biotechnology* 99 (11): 4795–4805. https://doi.org/10.1007/s00253-015-6458-8.

Zhao, H., Fang, Y., Wang, X. et al. (2018). Increasing l-threonine production in *Escherichia coli* by engineering the glyoxylate shunt and the l-threonine biosynthesis pathway. *Applied Microbiology and Biotechnology* 102 (13): 5505–5518. https://doi.org/10.1007/s00253-018-9024-3.

Zhao, C., Fang, H., Wang, J. et al. (2020a). Application of fermentation process control to increase l-tryptophan production in *Escherichia coli. Biotechnology Progress* 36 (2): https://doi.org/10.1002/btpr.2944.

Zhao, L., Lu, Y., Yang, J. et al. (2020b). Expression regulation of multiple key genes to improve l-Threonine in *Escherichia coli. Microbial Cell Factories* 19 (1): https://doi.org/10.1186/s12934-020-01312-5.

3

Microbial Production of Dietary Short-Chain Fatty Acids

Alexandra Wallimann

CK-CARE, Christine Kühne-Center for Allergy Research and Education, 7265, Davos Wolfgang, Graubünden, Switzerland

3.1 Background

Microbes produce and secrete many metabolites, which possess beneficial properties and contribute to human health. Especially, metabolites produced by gut microbes have been studied extensively, as they are key mediators between the intestinal microbiota and the host. These metabolites, including secondary bile acids, succinate, indoles, polyamines (such as spermine, spermidine, and putrescine), and SCFAs are known to impact host physiology and metabolism (Lavelle and Sokol 2020; Krautkramer et al. 2021). They not only serve as an energy source for the host cells, in particular intestinal epithelial cells, but also affect cellular signaling and metabolism and show potent immunomodulatory capacities (Rooks and Garrett 2016). In addition, metabolites generated by microbes serve as nutrients for other microbes, thereby shaping the composition of the microbiota (Tang et al. 2019). In recent years, the term "postbiotic" has been introduced to summarize all the microbial-derived functional molecules that have a proposed interaction with the host, with potential health benefits especially due to their immunomodulatory capacities (Wegh et al. 2019).

SCFAs are among the most investigated microbial-derived metabolites. Although SCFAs are an important food constituent, of mainly oil or milk, in the form of triglycerides, the bacterial production in the gut using nondigestible fibers as substrate is the most important source of SCFA production (Schönfeld and Wojtczak 2016). Formate, acetate, propionate, butyrate, and pentanoate (valerate) build the class of SCFAs and consist of one to five carbon atoms respectively (Morrison and Preston 2016). The concentration of the different SCFAs varies across different colon sites and depends on the host diet, but generally lies between 10 and 100 mM, whereas in the blood circulation, levels are clearly lower, ranging from 1 to 10 μM (Yip et al. 2021).

SCFAs are produced upon anaerobic fermentation of nondigestible dietary fibers, including whole grains products and vegetables, which are digested by intestinal microbes (den Besten et al. 2013). Acetate can diffuse freely through cells, whereas propionate and butyrate are transported with carrier proteins located within the cells (Hosmer et al. 2023). *Via* systemic blood circulation, they can reach every organ of the organism and impact the physiology and metabolism thereof. SCFAs are generally known to have anti-inflammatory capacities and are considered as therapeutical in several chronic inflammatory diseases.

Microbial Nutraceuticals: Products and Processes, First Edition. Edited by Sudhir Pratap Singh and Santosh Kumar Upadhyay.

In the following chapter, the production, the mechanism of action, and the potential therapeutic use of SCFAs are discussed.

3.2 SCFA Generation and Its Producing Microbes

A variety of microbes is known to produce SCFAs from nondigestible carbohydrate polymers. Microbial SCFA production in the intestine is influenced by several factors, such as the pH or iron levels (Dostal et al. 2012, 2015; den Besten et al. 2013). In addition, the varying oxygen levels along the gastrointestinal tract have an indirect impact on SCFA fermentation, as this influences the microbial composition in the gut, and SCFAs are produced by bacteria preferring anaerobic conditions (Zheng et al. 2015). The introduction of novel sequencing technologies, especially the advancing of metagenomic shotgun sequencing, enabled the identification and characterization of SCFA-producing bacterial species (Morrison and Preston 2016).

Generally, bacteria from the *Bacteroidetes* phylum are known to be acetate and propionate producers, whereas bacteria from the *Firmicutes* phylum are rather known as butyrate producers (Puccetti et al. 2020).

3.2.1 Acetate

Many intestinal bacteria, including different species from the *Bacteroides*, *Bifidobacterium,* and *Prevotella* genus, are producing acetate (Koh et al. 2016). Therefore, acetate possesses the highest concentrations in the intestinal lumen (Louis et al. 2010). Acetogenic fibers, such as inulin, are digested by these bacteria, and two main pathways are used to produce acetate: pyruvate-derived acetogenesis and the reductive acetyl-CoA pathway, also termed Wood–Ljungdahl pathway, using CO_2 fixation (Ragsdale and Pierce 2008). These pathways often occur both in the same microorganism and, therefore, show functional redundancy.

3.2.2 Propionate

Gut microbes from the *Veillonella* and *Salmonella* genus, as well as *Coprococcus catus,* are producing propionate (Koh et al. 2016). Depending on the microbial species, three different metabolic pathways are used for propionate production. Propionate production *via* the succinate pathway occurs mainly in *Bacteroidetes* and *Negativicutes*, which convert succinate to methylmalonyl-CoA in the presence of Vitamin B12 (Fusco et al. 2023; Koh et al. 2016; Deleu et al. 2021). It was shown that this pathway, using hexose and pentose sugars, was the most dominant source of propionate production (Reichardt et al. 2014). The propanediol pathway is another metabolic pathway to produce propionate and is mainly used by *Lachnospiraceae*, namely *Ruminococcus obeum* and *Roseburia inulinivorans* (Reichardt et al. 2014). They mainly use fucose or rhamnose, which are deoxy carbohydrates. Finally, the acrylate pathway, where lactate is used to generate propionate, is another pathway used by microbes for propionate production (Koh et al. 2016). *C. catus* from the *Lachnospiraceae* family is known to use lactate to generate propionate *via* the aforementioned pathway (Reichardt et al. 2014).

3.2.3 Butyrate

Bacteria belonging to the phylum *Firmicutes* and the order *Clostridiales* are known to be dominant butyrate producers. The following species are major butyrate producers: *Faecalibacterium prausnitzii, Eubacterium rectale, Eubacterium hallii, Ruminococcus bromii, Coprococcus comes,* and *Coprococcus eutactus* (Fu et al. 2019; Louis et al. 2010). Two main key enzymes are used by these bacteria to produce butyrate, namely butyrate kinase and the butyryl-CoA:acetate-CoA transferase (Deleu et al. 2021). *F. prausnitzii, E. rectale,* and *E. hallii* use mainly the butyryl-CoA:acetate-CoA transferase. They synthesize butyrate from two molecules of acetyl-CoA, producing acetoacetyl-CoA, which is further converted to butyryl-CoA *via* beta-hydroxybutyric-CoA and crotonyl-CoA (Koh et al. 2016). In contrast, *C. comes* and *C. eutactus* are known to use butyrate kinase to produce butyrate (Morrison and Preston 2016; Deleu et al. 2021).

3.2.4 Valerate

Besides the three main SCFAs, acetate, propionate, and butyrate, valerate and formate also belong to the class of SCFAs, although they are rather poorly studied so far.

Valerate is known to be produced by bacteria from the *Prevotella* species including *Prevotella copri* and *Prevotella stercorea* (Hayashi et al. 2007; Isenring et al. 2022). Valerate generation needs propionate, which is subsequently coupled with ethanol (Oliphant and Allen-Vercoe 2019).

3.2.5 Formate

Formate is only detected at low concentrations within the human colon. Among others, *R. bromii* and *E. hallii* are dominant formate producers (Gomez et al. 2019) (Table 3.1).

Table 3.1 Microbes and pathways driving SCFA production.

SCFA	Producing microbes	Mechanisms and pathways	References
Formate	*Ruminococcus bromii* *Eubacterium hallii*	Pyruvate conversion to acetyl-CoA	Gomez et al. (2019)
Acetate	*Bacteroides* *Bifidobacterium* *Prevotella*	• Pyruvate-derived acetogenesis • Reductive acetyl-CoA pathway/Wood–Ljungdahl pathway	Louis et al. (2010); Ragsdale and Pierce (2008); Koh et al. (2016)
Propionate	*Veillonella* *Salmonella* *Coprococcus catus*	• Succinate pathway • Propanediol pathway • Acrylate pathway	Reichardt et al. (2014); Koh et al. (2016); Fusco et al. (2023)
Butyrate	*Faecalibacterium prausnitzii* *Eubacterium rectale* *Eubacterium hallii* *Ruminococcus bromii* *Coprococcus comes* *Coprococcus eutactus*	• Butyrate kinase • Butyryl-CoA:acetate-CoA transferase	Koh et al. (2016); Morrison and Preston (2016); Fu et al. (2019); Louis et al. (2010); Deleu et al. (2021)
Valerate	*Prevotella copri* *Prevotella stercorea*	Propionate coupling with ethanol	Hayashi et al. (2007); Oliphant and Allen-Vercoe (2019); Isenring et al. (2022)

3.3 Mechanism of Actions

SCFAs are transported by different transfer systems at the gut endothelial barrier to subsequently reach the bloodstream and affect distal organs. Carrier-mediated transporters, including sodium-coupled monocarboxylate transporter 1 (SMCT1) and monocarboxylate-transporter 1 (MCT1), which are expressed on intestinal epithelial cells, favor the transport of SCFAs from the intestinal lumen into the blood circulation, where they initially pass the portal vein (Sun et al. 2017). Furthermore, SCFAs can freely diffuse through intestinal epithelial cells to reach blood circulation. *Via* the blood circulation system, they can affect the metabolism and function of many distant organs including adipose tissue, lung, bone, and liver (Sun et al. 2017; Zaiss et al. 2019; Zhou and Fan 2019; Wallimann et al. 2021).

SCFAs impact the peripheral tissues and their cells mainly by using three different mechanisms: first, receptor-mediated, through GPCR, or free fatty acid receptors (FFAR), namely GPCR41 (FFAR3), GPCR43 (FFAR2) and GPCR109a signaling, second, *via* HDAC inhibition, and third, by directly affecting the metabolism of the cells.

GPCR41 and GPCR43 are activated by all SCFAs, whereas GPCR109a is mainly activated by butyrate (Sun et al. 2017). GPCRs, which are exclusively expressed on cell membranes, are found not only on many different immune cell types including monocytes/macrophages and lymphocytes, but also on intestinal epithelial cells, adipocytes, enteroendocrine cells, and pancreatic islets (Sun et al. 2017; Husted et al. 2017). GPCR41 and GPCR43 show around 50% similarity and identity, which explains why they are activated by all the different SCFAs; in contrast, GPCR41 shows a restricted specificity for carbon chain length, therefore preferring valerate (containing five carbon atoms) as a ligand (Brown et al. 2003). The binding of SCFAs to GPCRs triggers an intracellular release of Ca^{2+}, promoting activation of different downstream signaling pathways, including MAPK/ERK, p38, JNK, or Akt/PI3K kinase cascades, which regulate cell differentiation, proliferation, migration, and survival, as well as cytokine and chemokine production (Melhem et al. 2019).

SCFAs, mainly propionate and butyrate, were shown to act as HDAC inhibitors, and thus, regulate gene expression by modifying the epigenetic landscape, specifically acetylation of histones (Stilling et al. 2016). By removing acetyl groups from histone tails, HDACs generally promote a repressive chromatin structure, consequently reducing gene transcription of specific genes. Therefore, HDAC inhibitors, such as SCFAs, support histone acetylation and are particularly known to impact cell mitosis, which could explain their ability to reduce the proliferation of cancerous cells (Donohoe et al. 2012).

In addition, SCFAs supply anabolic pathways including gluconeogenesis and lipogenesis with carbon-containing precursors and are direct regulators of cellular metabolism (Schönfeld and Wojtczak 2016).

3.4 Impact on Host Health

As butyrate is an important fuel for intestinal epithelial cells by functioning as an energy substrate for oxidative generation of ATP, it directly supports the differentiation, growth, and survival of these cells (Venegas et al. 2019). Therefore, SCFAs, in particular butyrate, support epithelial barrier integrity in the gut, which is crucial to prevent translocation of microbial components, which would otherwise trigger a systemic chronic inflammation (Martel et al. 2022).

Several studies have investigated the different immunomodulatory activities of SCFAs, which are mostly described as being anti-inflammatory. In the intestine, SCFAs were shown to induce differentiation of T regulatory cells and production of the anti-inflammatory cytokine IL-10 (Luu and Visekruna 2019; Arpaia et al. 2013). Importantly, they also affect immune homeostasis at peripheral sites such as the musculoskeletal, respiratory, or nervous system (Ratajczak et al. 2019). Besides their capacity to promote differentiation of T regulatory cells, they also impact cells from the innate immune system, including monocytes/macrophages and neutrophils (Yao et al. 2022). In particular, SCFAs are known to inhibit pro-inflammatory tumor necrosis factor α (TNFα) production from monocytes and macrophages (Vinolo et al. 2011a; Usami et al. 2008).

The anti-inflammatory capacities of SCFAs are a key aspect of their biological importance. In particular, the capacity of butyrate to induce anti-inflammatory IL-10 and transforming growth factor β (TGF-β) was shown in several *in vitro* and *in vivo* studies. In contrast, butyrate was shown to inhibit the production of pro-inflammatory cytokines including IFNγ, TNFα, IL-1β, IL-6, and IL-8 by suppressing activation of the transcription factor NFκB (Liu et al. 2018). The induction of T regulatory cell differentiation by butyrate has been shown in the colon and spleen of mice (Furusawa et al. 2013; Hui et al. 2019). Furthermore, butyrate induced apoptosis in a variety of innate immune cells as well as lymphocytes, therefore potentially preventing excessive immune cell activation. (Aoyama et al. 2010; Kurita-Ochiai et al. 2001). In certain cases, SCFAs were shown to have pro-inflammatory effects by promoting chemotaxis in leukocytes, particularly in neutrophils (Vinolo et al. 2011b), or increasing the antimicrobial activity of macrophages by HDAC3 inhibition (Schulthess et al. 2019). Therefore, depending on the concentration and frequency of SCFA supplementation, SCFAs may differentially affect immune cell populations and inflammatory responses.

3.5 Potential of SCFAs as Therapeutics

Dietary interventions, in particular the application of prebiotics, probiotics, and postbiotics, show the potential to promote a healthy and balanced gut microbiota, which could prevent or even ameliorate the course of inflammatory disorders. Due to their anti-inflammatory capacities, SCFAs have already been considered as novel therapeutics in several chronic inflammatory diseases, for instance, inflammatory bowel disease and metabolic syndrome (Facchin et al. 2020; Cleophas et al. 2019).

Prebiotics, which are nondigestible fiber compounds, can promote the growth of beneficial microbes in the gut by serving as substrates for them. In contrast, probiotics are defined as live microorganisms with potential beneficial effects for the host if administered in an appropriate dosage. Finally, postbiotics, such as SCFAs, are considered as generated products from microorganisms, with a positive impact on host health. Postbiotics show many advantageous features. First, they act as natural modulators of the crosstalk between bacteria but do not repress other bacteria. Second, they can biologically act not only locally in the intestine but also systemically. Third, they are synthesized, absorbed, and transmitted in a suitable way and show low toxicity potential (Puccetti et al. 2020).

Intestinal SCFA levels and their production can be increased in different ways: first, indirectly *via* the application of prebiotics in terms of a fiber-rich diet, whose digestion triggers SCFA production in the gut, second, by application of bacteria, mainly probiotics, which are known to be potent SCFA producers, and third, through direct application of SCFAs. One study showed that daily intake of a high-fiber bar consisting of oat flakes, inulin, and other fibers for 28 days was enough to increase systemic SCFA levels and reduce proinflammatory cytokines in patients suffering from rheumatoid arthritis (Dürholz et al. 2020). Another study showed that daily sodium butyrate supplementation for four weeks in obese males was able to alter the overall inflammatory status in circulating monocytes (Cleophas et al. 2019). Similarly, oral supplementation with microencapsulated butyrate positively changed the gut microbiota in inflammatory bowel disease patients by supporting the growth of bacteria that are able to produce SCFAs and, therefore, potentially promote an anti-inflammatory environment in the intestine (Facchin et al. 2020). The broad benefits of SCFAs, in particular, their impact on host health, their capacity to serve as carbon sources for other microorganisms, and their usage as substrates for the chemical industry, arouse interest to produce SCFAs in an upscale format. Biotechnological SCFA production is based on the anaerobic degradation of organic substrates. Organic materials used for SCFA biosynthesis comprise food waste, agricultural residues, activated waste sludge, or lignocellulosic materials. The process mainly involves two stages: The first stage includes hydrolysis and acidogenesis, which actually leads to the generation of volatile SCFAs, whereas the second stage consists of acetogenesis and methanogenesis (Moestedt et al. 2019). Obviously, to accumulate the yield of SCFAs, the metabolization of SCFAs in the second stage must be prevented (Tomás-Pejó et al. 2023). The production rate and the composition of the generated SCFAs are dependent on several factors, which are currently still being optimized. First, the composition of the organic substrate and its continuous supply needs to be ensured. Second, the microbial cultures should consist of mixed microbial cultures, which are known to be strong SCFA producers, such as *Clostridium*, *Lactobacillus,* or *Bifidobacterium*. Third, fermentation conditions having a pH between 6.0 and 7.0, an organic loading rate of around 10 kgVS/m^{3}d, a hydraulic retention time shorter than 10 days, and thermophilic temperatures are keys to support efficient SCFA production (Moestedt et al. 2019; Strazzera et al. 2018). To finally purify and extract SCFAs, different methods such as ion exchange or liquid–liquid extraction are generally used (Strazzera et al. 2018).

3.6 Conclusions and Perspectives

To summarize, SCFAs are produced by a variety of different microbes in the gut upon digestion of dietary fibers. SCFAs not only serve as fuel for intestinal epithelial cells but can also reach the systemic circulation and affect peripheral tissues and cells. Due to their anti-inflammatory capacities, they are promising new therapeutic tools in preventing and treating chronic inflammatory diseases. However, further research in the field is required, to find optimal dosages and ways for continuous supply of SCFAs in the human body. Additionally, substantial progress was made in the biotechnological production of SCFAs, although further improvements in the fermentation process are required, to guarantee high enough production rates of SCFAs.

References

Aoyama, M., Kotani, J., and Usami, M. (2010). Butyrate and propionate induced activated or non-activated neutrophil apoptosis via Hdac inhibitor activity but without activating Gpr-41/Gpr-43 pathways. *Nutrition* 26 (6): 653–661. https://doi.org/10.1016/j.nut.2009.07.006.

Arpaia, N., Campbell, C., Fan, X. et al. (2013). Metabolites produced by commensal bacteria promote peripheral regulatory T-cell generation. *Nature* 504 (7480): 451–455. https://doi.org/10.1038/nature12726.

Brown, A.J., Goldsworthy, S.M., Barnes, A.A. et al. (2003). The orphan G protein-coupled receptors GPR41 and GPR43 are activated by propionate and other short chain carboxylic acids. *Journal of Biological Chemistry* 278 (13): 11312–11319. https://doi.org/10.1074/jbc.M211609200.

Cleophas, M.C.P., Ratter, J.M., Bekkering, S. et al. (2019). Effects of oral butyrate supplementation on inflammatory potential of circulating peripheral blood mononuclear cells in healthy and obese males. *Scientific Reports* 9 (1): 775. https://doi.org/10.1038/s41598-018-37246-7.

Deleu, S., Machiels, K., Raes, J. et al. (2021). Short chain fatty acids and its producing organisms: an overlooked therapy for IBD? *eBioMedicine* 66: 103293. https://doi.org/10.1016/j.ebiom.2021.103293.

den Besten, G., van Karen, E., Albert K, G. et al. (2013). The role of short-chain fatty acids in the interplay between diet, gut microbiota, and host energy metabolism. *Journal of Lipid Research* 54 (9): 2325–2340. https://doi.org/10.1194/jlr.R036012.

Donohoe, D.R., Collins, L.B., Wali, A. et al. (2012). The warburg effect dictates the mechanism of butyrate-mediated histone acetylation and cell proliferation. *Molecular Cell* 48 (4): 612–626. https://doi.org/10.1016/j.molcel.2012.08.033.

Dostal, A., Chassard, C., Hilty, F.M. et al. (2012). Iron depletion and repletion with ferrous sulfate or electrolytic iron modifies the composition and metabolic activity of the gut microbiota in rats3. *The Journal of Nutrition* 142 (2): 271–277. https://doi.org/10.3945/jn.111.148643.

Dostal, A., Lacroix, C., Bircher, L. et al. (2015). Iron modulates butyrate production by a child gut microbiota *in vitro*. *MBio* 6 (6): e01453. https://doi.org/10.1128/mBio.01453-15.

Dürholz, K., Hofmann, J., Iljazovic, A. et al. (2020). Dietary short-term fiber interventions in arthritis patients increase systemic SCFA levels and regulate inflammation. *Nutrients* 12 (10): 3207. https://doi.org/10.3390/nu12103207.

Facchin, S., Vitulo, N., Calgaro, M. et al. (2020). Microbiota changes induced by microencapsulated sodium butyrate in patients with inflammatory bowel disease. *Neurogastroenterology and Motility* 32 (10): e13914. https://doi.org/10.1111/nmo.13914.

Fu, X., Liu, Z., Zhu, C. et al. (2019). Nondigestible carbohydrates, butyrate, and butyrate-producing bacteria. *Critical Reviews in Food Science and Nutrition* 59 (sup1): S130–S152. https://doi.org/10.1080/10408398.2018.1542587.

Furusawa, Y., Obata, Y., Fukuda, S. et al. (2013). commensal microbe-derived butyrate induces the differentiation of colonic regulatory T cells. *Nature* 504 (7480): 446–450. https://doi.org/10.1038/nature12721.

Fusco, W., Lorenzo, M.B., Cintoni, M. et al. (2023). short-chain fatty-acid-producing bacteria: key components of the human gut microbiota. *Nutrients* 15 (9): 2211. https://doi.org/10.3390/nu15092211.

Gomez, J.A., Laverde, I.M., Duncan, S.H. et al. (2019). Formate cross-feeding and cooperative metabolic interactions revealed by transcriptomics in co-cultures of acetogenic and amylolytic human colonic bacteria. *Environmental Microbiology* 21 (1): 259–271. https://doi.org/10.1111/1462-2920.14454.

Hayashi, H., Shibata, K., Sakamoto, M. et al. (2007). *Prevotella copri* Sp. Nov. and *Prevotella stercorea* Sp. Nov., isolated from human faeces. *International Journal of Systematic and Evolutionary Microbiology* 57 (5): 941–946. https://doi.org/10.1099/ijs.0.64778-0.

Hosmer, J., McEwan, A.G., and Kappler, U. (2023). Bacterial acetate metabolism and its influence on human epithelia. *Emerging Topics in Life Sciences* 8 (1): 1–13. https://doi.org/10.1042/ETLS20220092.

Hugenholtz, F., Mullaney, J.A., Kleerebezem, M. et al. (2013). Modulation of the microbial fermentation in the gut by fermentable carbohydrates. *Bioactive Carbohydrates and Dietary Fibre* 2 (2): 133–142. https://doi.org/10.1016/j.bcdf.2013.09.008.

Hui, W., Dapeng, Y., Cao, Z., and Zhao, X. (2019). Butyrate inhibit collagen-induced arthritis via Treg/IL-10/Th17 Axis. *International Immunopharmacology* 68: 226–233. https://doi.org/10.1016/j.intimp.2019.01.018.

Husted, A.S., Trauelsen, M., Rudenko, O. et al. (2017). GPCR-mediated signaling of metabolites. *Cell Metabolism* 25 (4): 777–796. https://doi.org/10.1016/j.cmet.2017.03.008.

Isenring, J., Stevens, M.J.A., Jans, C. et al. (2022). Identification of valerate as carrying capacity modulator by analyzing *Lactiplantibacillus plantarum* colonization of colonic microbiota in vitro. *Frontiers in Microbiology* 13: 910609. https://doi.org/10.3389/fmicb.2022.910609.

Koh, A., De Vadder, F., Kovatcheva-Datchary, P., and Bäckhed, F. (2016). From dietary fiber to host physiology: short-chain fatty acids as key bacterial metabolites. *Cell* 165 (6): 1332–1345. https://doi.org/10.1016/j.cell.2016.05.041.

Krautkramer, K.A., Fan, J., and Bäckhed, F. (2021). Gut microbial metabolites as multi-kingdom intermediates. *Nature Reviews Microbiology* 19 (2): 77–94. https://doi.org/10.1038/s41579-020-0438-4.

Kurita-Ochiai, T., Ochiai, K., and Fukushima, K. (2001). Butyric acid-induced T-cell apoptosis is mediated by caspase-8 and -9 activation in a Fas-independent manner. *Clinical and Diagnostic Laboratory Immunology* 8 (2): 325–332. https://doi.org/10.1128/CDLI.8.2.325-332.2001.

Lavelle, A. and Sokol, H. (2020). Gut microbiota-derived metabolites as key actors in inflammatory bowel disease. *Nature Reviews Gastroenterology & Hepatology* 17 (4): 223–237. https://doi.org/10.1038/s41575-019-0258-z.

Liu, H., Wang, J., He, T. et al. (2018). Butyrate: a double-edged sword for health? *Advances in Nutrition* 9 (1): 21–29. https://doi.org/10.1093/advances/nmx009.

Louis, P., Young, P., Holtrop, G., and Flint, H.J. (2010). diversity of human colonic butyrate-producing bacteria revealed by analysis of the butyryl-CoA:acetate CoA-transferase gene. *Environmental Microbiology* 12 (2): 304–314. https://doi.org/10.1111/j.1462-2920.2009.02066.x.

Luu, M. and Visekruna, A. (2019). Short-chain fatty acids: bacterial messengers modulating the immunometabolism of T cells. *European Journal of Immunology* 49 (6): 842–848. https://doi.org/10.1002/eji.201848009.

Martel, J., Chang, S.-H., Ko, Y.-F. et al. (2022). Gut barrier disruption and chronic disease. *Trends in Endocrinology and Metabolism* 33 (4): 247–265. https://doi.org/10.1016/j.tem.2022.01.002.

Melhem, H., Berna Kaya, C., Ayata, K. et al. (2019). Metabolite-sensing g protein-coupled receptors connect the diet-microbiota-metabolites axis to inflammatory bowel disease. *Cells* 8 (5): 450. https://doi.org/10.3390/cells8050450.

Moestedt, J., Westerholm, M., Isaksson, S., and Schnürer, A. (2019). Inoculum source determines acetate and lactate production during anaerobic digestion of sewage sludge and food waste. *Bioengineering* 7 (1): 3. https://doi.org/10.3390/bioengineering7010003.

Morrison, D.J. and Preston, T. (2016). Formation of short chain fatty acids by the gut microbiota and their impact on human metabolism. *Gut Microbes* 7 (3): 189–200. https://doi.org/10.1080/19490976.2015.1134082.

Oliphant, K. and Allen-Vercoe, E. (2019). Macronutrient metabolism by the human gut microbiome: major fermentation by-products and their impact on host health. *Microbiome* 7 (1): 91. https://doi.org/10.1186/s40168-019-0704-8.

Puccetti, M., Xiroudaki, S., Ricci, M., and Giovagnoli, S. (2020). Postbiotic-enabled targeting of the host-microbiota-pathogen interface: hints of antibiotic decline? *Pharmaceutics* 12 (7): 624. https://doi.org/10.3390/pharmaceutics12070624.

Ragsdale, S.W. and Pierce, E. (2008). Acetogenesis and the wood–Ljungdahl pathway of CO_2 fixation. *Biochimica et Biophysica Acta (BBA) - Proteins and Proteomics* 1784 (12): 1873–1898. https://doi.org/10.1016/j.bbapap.2008.08.012.

Ratajczak, W., Ryl, A., Mizerski, A. et al. (2019). Immunomodulatory potential of gut microbiome-derived short-chain fatty acids (SCFAs). *Acta Biochimica Polonica* 66 (1): 1–12. https://doi.org/10.18388/abp.2018_2648.

Reichardt, N., Duncan, S.H., Young, P. et al. (2014). Phylogenetic distribution of three pathways for propionate production within the human gut microbiota. *The ISME Journal* 8 (6): 1323–1335. https://doi.org/10.1038/ismej.2014.14.

Rooks, M.G. and Garrett, W.S. (2016). Gut microbiota, metabolites and host immunity. *Nature Reviews Immunology* 16 (6): 341–352. https://doi.org/10.1038/nri.2016.42.

Schönfeld, P. and Wojtczak, L. (2016). Short- and medium-chain fatty acids in energy metabolism: the cellular perspective. *Journal of Lipid Research* 57 (6): 943–954. https://doi.org/10.1194/jlr.R067629.

Schulthess, J., Pandey, S., Capitani, M. et al. (2019). The short chain fatty acid butyrate imprints an antimicrobial program in macrophages. *Immunity* 50 (2): 432–445.e7. https://doi.org/10.1016/j.immuni.2018.12.018.

Stilling, R.M., van de Wouw, M., Clarke, G. et al. (2016). The neuropharmacology of butyrate: the bread and butter of the microbiota-gut-brain axis? *Neurochemistry International* 99: 110–132. https://doi.org/10.1016/j.neuint.2016.06.011.

Strazzera, G., Battista, F., Garcia, N.H. et al. (2018). Volatile fatty acids production from food wastes for biorefinery platforms: a review. *Journal of Environmental Management* 226: 278–288. https://doi.org/10.1016/j.jenvman.2018.08.039.

Sun, M., Wei, W., Liu, Z., and Cong, Y. (2017). Microbiota metabolite short chain fatty acids, GPCR, and inflammatory bowel diseases. *Journal of Gastroenterology* 52 (1): 1–8. https://doi.org/10.1007/s00535-016-1242-9.

Tang, Z.-Z., Chen, G., Hong, Q. et al. (2019). Multi-omic analysis of the microbiome and metabolome in healthy subjects reveals microbiome-dependent relationships between diet and metabolites. *Frontiers in Genetics* 10: 454. https://doi.org/10.3389/fgene.2019.00454.

Tomás-Pejó, E., González-Fernández, C., Greses, S. et al. (2023). Production of short-chain fatty acids (SCFAs) as chemicals or substrates for microbes to obtain biochemicals. *Biotechnology for Biofuels and Bioproducts* 16 (1): 96. https://doi.org/10.1186/s13068-023-02349-5.

Usami, M., Kishimoto, K., Ohata, A. et al. (2008). Butyrate and trichostatin a attenuate nuclear factor KB activation and tumor necrosis factor α secretion and increase prostaglandin E2 secretion in human peripheral blood mononuclear cells. *Nutrition Research* 28 (5): 321–328. https://doi.org/10.1016/j.nutres.2008.02.012.

Venegas, D.P., De la Fuente, M.K., Landskron, G. et al. (2019). Short chain fatty acids (SCFAs)-mediated gut epithelial and immune regulation and its relevance for inflammatory bowel diseases. *Frontiers in Immunology* 10: 277. https://doi.org/10.3389/fimmu.2019.00277.

Vinolo, M.A.R., John Ferguson, G., Kulkarni, S. et al. (2011a). SCFAs induce mouse neutrophil chemotaxis through the GPR43 receptor. *PLoS One* 6 (6): e21205. https://doi.org/10.1371/journal.pone.0021205.

Vinolo, M.A.R., Rodrigues, H.G., Nachbar, R.T., and Curi, R. (2011b). Regulation of inflammation by short chain fatty acids. *Nutrients* 3 (10): 858–876. https://doi.org/10.3390/nu3100858.

Wallimann, A., Magrath, W., Thompson, K. et al. (2021). Gut microbial-derived short-chain fatty acids and bone: a potential role in fracture healing. *European Cells and Materials* 41: 454–470. https://doi.org/10.22203/eCM.v041a29.

Wegh, G., Knol, R., and Belzer. (2019). Postbiotics and their potential applications in early life nutrition and beyond. *International Journal of Molecular Sciences* 20 (19): 4673. https://doi.org/10.3390/ijms20194673.

Yao, Y., Cai, X., Fei, W. et al. (2022). The role of short-chain fatty acids in immunity, inflammation and metabolism. *Critical Reviews in Food Science and Nutrition* 62 (1): 1–12. https://doi.org/10.1080/10408398.2020.1854675.

Yip, W., Hughes, M.R., Li, Y. et al. (2021). Butyrate shapes immune cell fate and function in allergic asthma. *Frontiers in Immunology* 12: 628453. https://doi.org/10.3389/fimmu.2021.628453.

Zaiss, M.M., Jones, R.M., Schett, G., and Pacifici, R. (2019). The gut-bone axis: how bacterial metabolites bridge the distance. *Journal of Clinical Investigation* 129 (8): 3018–3028. https://doi.org/10.1172/JCI128521.

Zheng, L., Kelly, C.J., and Colgan, S.P. (2015). Physiologic hypoxia and oxygen homeostasis in the healthy intestine. A review in the theme: cellular responses to hypoxia. *American Journal of Physiology-Cell Physiology* 309 (6): C350–C360. https://doi.org/10.1152/ajpcell.00191.2015.

Zhou, D. and Fan, J.-G. (2019). Microbial metabolites in non-alcoholic fatty liver disease. *World Journal of Gastroenterology* 25 (17): 2019–2028. https://doi.org/10.3748/wjg.v25.i17.2019.

4

Microbial Sources for Bioactive Peptides Conferring Health Benefits

Amin Abbasi[1], *Hadi Pourjafar*[2], *Hamideh Fathi Zavoshti*[3], *Sara Bazzaz*[1], *Parmis Mirzaei*[4], *Elham Sheykhsaran*[5], *and Hedayat Hosseini*[1]

[1] *Department of Food Science and Technology, National Nutrition and Food Technology Research Institute, Faculty of Nutrition Science and Food Technology, Shahid Beheshti University of Medical Sciences, 19395-4741, Tehran, Iran*
[2] *Dietary Supplements and Probiotic Research Center, Alborz University of Medical Sciences, 19395-4963, Karaj, Iran*
[3] *Department of Food Hygiene and Aquatics, Faculty of Veterinary Medicine, Tabriz University, 19294-2253, Tabriz, Iran*
[4] *Department of Clinical Nutrition & Dietetics, Faculty of Nutrition Science and Food Technology, Shahid Beheshti University of Medical Sciences, 19395-4741, Tehran, Iran*
[5] *Department of Medical Bacteriology and Virology, Tabriz University of Medical Sciences, 19294-4753, Tabriz, Iran*

4.1 Introduction

Food proteins have physiological functions that support health in addition to acting as nutrients. Peptide repeats encoded inside the parent protein, which become functional when they are cut into separate pieces, carry out the majority of the physiological functions of proteins (Rizzello et al. 2016). Various food manufacturing processes such as ripening, fermentation, cooking, and proteolysis by the enzymatic way during exposure to the digestive system conditions can release bioactive peptides from proteins. Because of their small dimensions and specialized nature, bioactive peptides are widely recognized for their capacity to hinder interactions between proteins. Nature continues to be the primary basis of bioactive peptides, with a variety of proteins found in animals, plants, fungi, microbes, and their byproducts. Over time, both classical and bioinformatic methods have identified bioactive peptides found in food. Using food-grade proteolytic enzymes, proteins are hydrolyzed in the classical method to discharge multiple peptides (Abdel-Hamid et al. 2017; Babini et al. 2017). As an alternative, bacteria could ferment proteins (Garcia-Tejedor et al. 2015; Aguilar-Toalá et al. 2017). Peptides are released into the hydrolysate by the bacteria's proteolytic enzymes, which hydrolyze the proteins.

The biological function of the hydrolysates is then examined under *in vitro* conditions. The hydrolysates are then verified by *in vivo* testing to see if they exhibit good bioactivity. The biologically active hydrolysate substrate can be further converted into a functional meal. For nonpharmacological therapy, the bioactive peptides could potentially be isolated and refined into nutraceutical supplements. Developing a cause-and-effect link between the ingestion of bioactive peptides and the desired health effects in humans has proven to be a substantial obstacle in the development of bioactive peptides for therapeutic purposes (Daliri et al. 2017) (Table 4.1). This chapter covers the production and processing of bioactive peptides by enzymatic hydrolysis and microbial

Microbial Nutraceuticals: Products and Processes, First Edition. Edited by Sudhir Pratap Singh and Santosh Kumar Upadhyay.

Table 4.1 Some bioactive peptides available in the market.

Source	Product type	Brand name	Amino acid sequence	Generation method	Health effects	References
Soy	Drink	Cholesterol block	CSPHP	Enzymatic hydrolysis	Cholesterol reduction	Chalamaiah et al. (2019)
Bonito fish	Tablet	Vasotensin®	LKPNM	Enzymatic hydrolysis by thermolysin and transformed to "LKP" in the gut milieu	Antihypertensive effect	Fujita and Yoshikawa (1999)
Milk	Ingredient	Capolac®	CPP	Isolation process	Calcium absorption	Chalamaiah et al. (2019)
Milk	Drink and capsules	Lacitum®	YLGYLEQLLR	Enzymatic hydrolysis with trypsin	Stress relief	Nagai et al. (2006)
Bonito	Capsules	Bonito peptide	LKPNM	Enzymatic hydrolysis with thermolysin	Benefits for controlling angiotensin-converting enzyme	(Fujita and Yoshikawa (1999)
Whey	Powder product	Lacprodan®	N/A	Enzymatic hydrolysis with alcalse	Adjusts blood sugar level	Chalamaiah et al. (2019)
Milk	Drink	Evolus®	VPP y IPP	Fermentation with *L. helveticus* LBK-16H	Antihypertensive effect	Authority EFS (2008)
Sour milk	Drink	Calpis	VPP y IPP	Fermentation with *S. cerevisiae* and *L. helveticus* CP790	Antihypertensive effect	Siltari et al. (2012)

fermentation pathways, the role of proteolytic systems of lactic acid bacteria in the liberation of bioactive peptides from the milk matrix, conventional purification and identification methods, promising health-promoting effects, and possible bioactive peptide applications. We also discuss the difficulties in developing bioactive peptides, how processing techniques affect peptides' bioactivity, the importance of cyclic peptides, and computer-based algorithms in peptide research (Figure 4.1).

4.2 Overview of Bioactive Peptides

In addition to their nutritional value, bioactive peptides are defined as "food-derived constituents (genuine or generated) that demonstrate a physiological effect in the host" (Vermeirssen et al. 2004). Mellander initially reported on bioactive peptides in 1950, stating that phosphorylated peptides derived from casein increased the amount of vitamin D-independent calcification in infants with rachitic conditions (Mellander 1950). Subsequent to that, fundamental investigation into the synthesis of bioactive peptides from different food protein molecules, with dairy proteins (milk) now being the key resource, has introduced a novel field of inquiry. Proteolysis is necessary for the discharge and stimulation of bioactive peptides, which are latent (or encrypted) within primary or

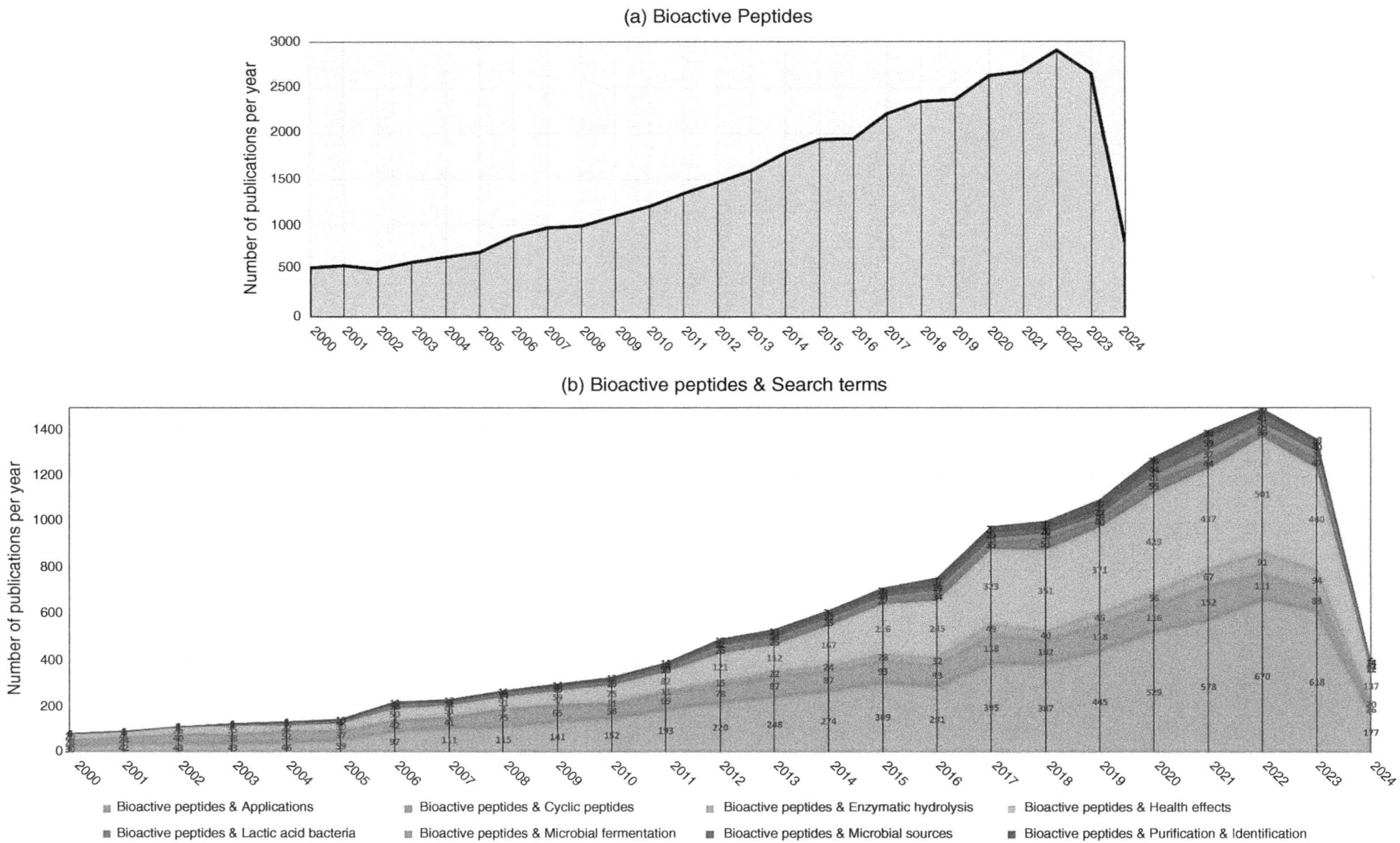

Figure 4.1 Illustrates the increase in the number of papers reporting research in the field of bioactive peptides. (a) Comparison between the number of publications per year based on the the search of bioactive peptides, (b) the search of the association between bioactive peptides and search terms ("Enzymatic hydrolysis"; "Microbial fermentation"; "Microbial sources"; "Lactic acid bacteria"; "Health effects"; "Purification"; "Identification"; "Cyclic peptides"; and "Application") in scientific databases.

parent proteins and can cause a physiological response in the body's systems (Gobbetti et al. 2004). By retaining calcium and other minerals, certain peptides also function as biocarriers, increasing bioavailability (Silva and Malcata 2005).

4.3 Production and Processing of Bioactive Peptides

Based on existing literature, the most popular techniques for generating bioactive peptides have been hydrolysis mediated by enzymes and fermentation of food proteins (Lee and Hur 2017). However, it has been discovered that water extracts of certain plant parts and mushrooms occasionally contain bioactive peptides (Geng et al. 2016).

4.3.1 Enzymatic Hydrolysis

This procedure involves subjecting the protein material to pH- and temperature-controlled enzyme hydrolysis (He et al. 2016; Huang et al. 2016; e Silva et al. 2017). Enzymatic hydrolysis is a more favored manner for generating bioactive peptides compared to microbial fermentation because of its short reaction time, predictability, and ease of scaling. The protein can be hydrolyzed by more than one proteolytic enzyme, either crude or purified, to produce a hydrolysate that contains brief peptide sequences. The ideal pH and temperature of the enzymes would determine whether to add them simultaneously or sequentially (Cai et al. 2017; Sangsawad et al. 2017). While specific bioactive peptides in food are not known to be produced by specific proteolytic enzymes, subtilisin hydrolysis typically results in biologically active peptides with low molecular weight. Huang et al. (2017), discovered that *Achatina fulica* snail foot muscle protein hydrolyzed with subtilisin contained more peptides with low molecular weight than specimens hydrolyzed with trypsin and papain (Huang et al. 2017). Zhang et al. (2012) additionally demonstrated that rice bran proteins hydrolyzed by subtilisin produced the greatest quantity of peptides with low molecular weight and displayed greater biological function than specimens hydrolyzed by papain, pepsin, and cysteine endopeptidase (Zhang et al. 2012). However, in order to achieve a good level of hydrolysis, it is crucial to take the enzyme-to-substrate ratio into account. Depending on the type of enzyme being used, different peptide sequences may have different biological activities (Mojica and De Mejía 2016). Proteases that generate peptides with a small molecular weight would be beneficial for the industrial synthesis of antihypertensive and antioxidant peptides. This is because peptides with a low molecular weight were discovered to possess superior biological properties (Wattanasiritham et al. 2016). Foods may contain a variety of additional nonprotein bioactive substances; therefore, it should maintain a separation between these compounds and the dietary proteins in order to avoid any possible interaction. For example, phenolic constituents have been shown to have antimicrobial (Maddox et al. 2010), antihypertensive (Singh et al. 2016), antioxidant (Alu'datt et al. 2016), and antidiabetic (Maddox et al. 2010) properties. As a result, they may affect the biological properties being measured when present in hydrolysates.

In order to prepare food proteins for enzymatic hydrolysis, phenolic substances can be extracted from the proteins utilizing techniques like acetone extraction (Do et al. 2014), ethanol extraction (Wattanasiritham et al. 2016), ultrasound-assisted extraction (Kumari et al. 2017), pressurized water extraction (Uzel 2017), and supercritical carbon dioxide extraction (Da Porto and Natolino 2017). The pH of the medium fluctuates as a result of protons released during proteolysis, which may have an impact on how well the hydrolytic process works. Although alkali or acid solutions can be added to change the pH (Do et al. 2014; e Silva et al. 2017), doing so typically

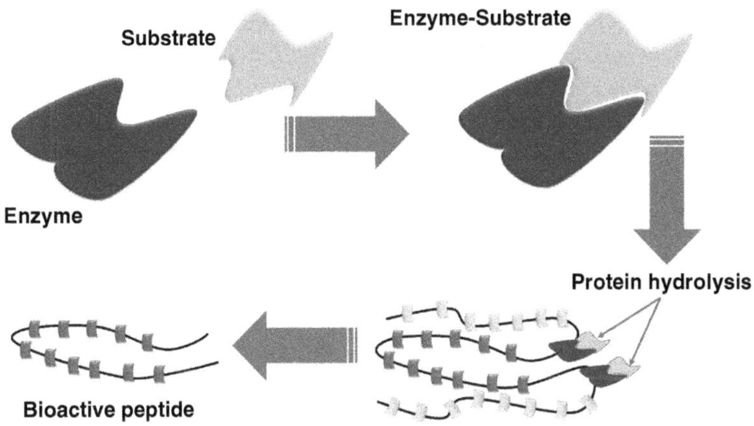

Figure 4.2 Schematic representation of the release of bioactive peptides mediated by enzymatic hydrolysis.

causes the hydrolysates to contain high concentrations of salt. For this reason, it is recommended to carry out the proteolysis in a buffer (Rahimi et al. 2016). The amount of hydrolysis and the kind of peptides produced in the hydrolysates are determined by the kind of enzyme utilized and the time and temperature applied during hydrolysis. Ferri et al. (2017) found that rice-derived protein hydrolyzed with bacillolysin exhibited more potent anti-tyrosinase and anti-inflammatory properties compared to specimens hydrolyzed by subtilisin; specimens hydrolyzed with leucyl aminopeptidase, cysteine endopeptidase, and papain demonstrated the minimum amount of bioactivity (Ferri et al. 2017). The significant anti-angiotensin converting enzyme (ACE) function was observed in bacillolysin-hydrolyzed samples, which were tailed *via* subtilisin and cysteine endopeptidase. Nevertheless, other research discovered that rice proteins hydrolyzed by cysteine endopeptidase or chymotrypsin had potent antioxidant properties (Dei Piu et al. 2014). The mixture is centrifuged following enzymatic hydrolysis in order to distinguish the precipitates from the supernatant that covers peptides with low molecular weight (Zhang et al. 2016) (Figure 4.2).

4.3.2 Microbial Fermentation

To facilitate the enzymatic hydrolysis of proteins during their growth, bacteria or yeast must be cultivated on protein substrates. Peptides are liberated from source proteins by the secretion of proteolytic enzymes from yeast or bacteria during their growth. The desired strain of bacteria is typically cultivated to its exponential stage in a solution in an environment that promotes bacterial growth (Hosseini et al. 2023). For inoculation onto a sterilized protein base, the cells are collected, purified, and floated in purified water, which usually includes glucose (Aguilar-Toalá et al. 2017; Rizzello et al. 2017). Depending on the protein type, fermentation duration, strain employed, and the degree of hydrolysis will vary. Scientists found that whey fermented with *Lactobacillus brevis* exhibited a greater capacity to inhibit ACE than whey fermented with *L. acidophilus, L. casei, L. reuteri, L. lactis, L. bifermentan, L. helveticus, L. paracasei,* and *L. plantarum* (Ahn et al. 2009). This indicates that because different microorganisms have distinct proteolytic systems, the efficacy of protein hydrolysates can differ among different cultures. (Daliri et al. 2017). El-Fattah et al. (2016) reported similar outcomes when they noticed that, following milk fermentation, various dairy starters showed varying degrees of ACE antioxidant capacity and proteolysis inhibition (El-Fattah et al. 2016). Different proteolytic capacities among bacteria belonging to the same species can lead to varying bioactivities. As an example, Sanjukta et al. (2015) noted that using *Bacillus subtilis* to ferment soybean protein compared to samples fermented with *B. subtilis* produced a greater level of free amino acids and hydrolysis (Sanjukta et al. 2015). Additionally, Chen et al. (2014) revealed that the degree of ACE inhibitory abilities, free amino nitrogen, titrable acidity, and hydrolysis of skim milk fermented with various distinct *L. helveticus* strains exhibited variations (Chen et al. 2014). In addition to bacteria starters, filamentous fungi (Lima et al. 2015) and yeast (Rai et al. 2016) have also been utilized in the generation of bioactive peptides. To speed up the proteolytic function, proteins may be co-cultured with various bacteria or even with yeast and bacteria (Chaves-López et al. 2014). The blend undergoes centrifugation to get the liquid portion that remains after the fermentation process. Proteolytic enzymes may then be applied to further hydrolyze the supernatant in order to generate shorter peptide sections (Babini et al. 2017). Otherwise, the peptides with low molecular weight present in the supernatant portion can be extracted employing solvent method or other manners, purified, and then mass spectrometry can be used to identify their amino acid sequences.

4.4 The Role of LAB Proteolytic Systems in the Liberation of Bioactive Peptides

Lactic acid bacteria (LAB) strains utilized in industry are proteolytic in nature and can be employed to generate bioactive peptides through microbial fermentation. The two enzymes that break down milk proteins are intracellular peptidases and cell-envelope proteinases. When milk proteins are used by proteolytic LAB, the process usually starts with the protein being broken down into oligopeptides by cell-envelope proteinases. These oligopeptides are then further broken down into amino acids and shorter peptides by the combined effect of intracellular peptidases to demonstrate proteolytic activities (Christensen et al. 1999). LABs offer a significant delivery system for di-, tri-, and oligopeptides and amino acids. This system is responsible for the generation of fermented milk such as Evolus® and Calpis®, which are Finnish and Japanese fermented milk goods known for their antihypertensive qualities. These properties are linked to the presence of tripeptides Ile-Pro-Pro (IPP) and Val-Pro-Pro (VPP), which are derived from β-casein, to have higher residual levels of peptides with biological activities such as ACE-inhibitory function. Additionally, some dairy goods like Festivo as a naturally ripened and low-fat cheese developed with *Bifidobacterium* spp. and *L. acidophilus* possess antihypertensive peptides (Ryhänen et al. 2001). Despite the comprehensive explanation of the proteolytic processes in LAB provided earlier, there is a scarcity of data on the synthesis of bioactive peptides throughout the fermentation of milk and the individual peptidases that are accountable for releasing these peptides throughout dairy processing. As an instance, Algaron et al. (2004) fermented milk containing a mutant *L. lactis* strain deficient in aminopeptidase N, PepX, or tripeptidase in order to assess the mutants' capacity to generate peptides with antihypertensive or immunomodulatory properties and establish a correlation between the mutation and bioactivity (Algaron et al. 2004). They discovered that, occasionally, the altered proteolytic structure of *L. lactis* produced a significantly different mixture of peptides, suggesting that the bioactive inherent of the synthesis peptides is influenced by the precise peptidase action of lactic acid bacteria. A gene encoding an endopeptidase was recently identified from *L. helveticus* and inserted into *E. coli via* cloning (Yamamoto et al. 2004).

The specific strain, E. coli CM4EP, underwent a transformation process with the endopeptidase gene *pepO*. This gene enabled the expression of endopeptidase operation, which has the capacity to discharge antihypertensive peptide VPP and IPP from oligopeptide parental patterns. The results indicate that the *pepO* gene output influences the synthesis of antihypertensive peptides that are produced from the microbial fermentation of dairy protein by *L. helveticus* (Yamamoto et al. 2004).

The presence of the gene *pepO* has been documented in *L. helveticus* CNRZ32, signifying that the endopeptidases responsible for generating VPP and IPP may exhibit specific variations (Chen and Steele 1998). Furthermore, an *L. helveticus*-derived endopeptidase was purified and characterized by Ueno and Yamamoto, and it was confirmed that propeptides serve as the substrate for this peptidase, which allows it to produce the tripeptides IPP and VPP (Ueno et al. 2004). The relationship between the proteolytic mechanism in microorganisms utilized for dairy fermenting and the biological characteristics of milk is essential in the development of tailored milk fermentation processes. This understanding will be advantageous in developing milk products that contain specific health-promoting compounds, which can be targeted toward niche sectors in the food and potentially pharmaceutical industries (Hayes et al. 2007).

4.5 Purification and Identification

Following the hydrolysis of the protein, which can be attained through the process of microbial fermentation or enzymatic hydrolysis, the bioactive peptides are liberated. Nevertheless, it is imperative to achieve effective purification of this element in order to ascertain its arrangement (Sridhar et al. 2021). Furthermore, the purification procedure is necessary to separate peptides with distinct functions from the targeted bioactive peptide, as certain peptides may exhibit antagonistic reactions. The process of purifying the peptides allows for the generation of higher levels of these compounds, resulting in high-efficiency goods that may be marketed as promising additives. The procedure for purification involves a sequence of stages that rely on the chemical and physical features of the peptides, including their specific noncovalent or covalent interactions, solubility, polarity, charge, and molecular size (Arumugam et al. 2018).

Membrane methods, including microfiltration, reverse osmosis, nanofiltration, and ultrafiltration, are employed for purifying purposes (Suwal et al. 2015). The approach usually used for bioactive peptides is ultrafiltration. This technology relies on the movement and storage of peptides dependent on their size and the dimensions of the membrane pores, which is facilitated by a pressure difference generated *via* a pump (Bazinet and Firdaous 2009). This enables the isolation of peptides based on their molecular weight threshold, using a rapid, cost-effective, and eco-friendly approach. It maintains the physiological characteristics of the peptide and can be easily expanded in size (Dullius et al. 2018). Several drawbacks of this practice include inadequate consistency, potential removal of peptides with molecular weights below an appropriate threshold, possible formation of reactions between membrane and hydrophobic peptides resulting in fouling of the membrane and blockage, challenges in acquiring pure peptides, and the necessity of an adequate specimen quantity (Acquah et al. 2019). In order to address these issues, a suggested line is to integrate them with additional methodologies such as chromatographic, multi-phase recycle membrane containers and electrodialysis procedures (Lafarga et al. 2020).

Capillary electrophoresis is a method used to identify and separate bioactive peptides (Kartsova et al. 2021). The method relies on the measurement of mass-to-charge ratio (m/z) and utilizes bonded silica capillary tubes (Sridhar et al. 2021). The technique is distinguished by its notable efficiency, minimal quantity of specimens, uncomplicated preparation of the samples, selectivity, and exceptional resolution (Kartsova et al. 2021). Chromatographic techniques can be complemented by this approach, although it suffers from issues such as limited reproducibility, detection limits, and sensitivity (Polikarpova et al. 2020).

Chromatographic approaches have the capability of separating and purifying peptides with more precision, resulting in the production of high-quality goods (Dullius et al. 2018). The peptides react with both the mobile and stationary phases. Commonly employed techniques include size exclusion, liquid–liquid partition, liquid–solid adsorption, and ion exchange (Muhire et al. 2024). Reversed-phase high-performance liquid chromatography (RP-HPLC) is a technique that can separate peptides by their hydrophobicity. Chromatography based on the ion exchange enables the segregation of peptides according to their electrical charge. Chromatography based on size exclusion enables the segregation of peptides according to their molecular weight. Thus, bioactive peptides with the appropriate molecular weight, hydrophobicity, and charge can likewise be concentrated utilizing these chromatographic procedures. The integration of chromatographic methods with fluorescence, mass detectors, diode arrays, and UV enables the detection and characterization of bioactive peptides (Sridhar et al. 2021). The drawbacks of these approaches include their time-consuming nature and high cost, which in turn increases the expenses associated with marketing these peptides. Following the purification process, it is crucial to determine its sequence. Various approaches can be utilized, with one prominent method being liquid chromatography-tandem mass spectrometry (LC-MS/MS). This technique is known for its precision and sensitivity in identifying bioactive peptides (Lafarga et al. 2020). Tandem mass spectrometry allows for the comprehensive analysis of peptides with unknown characteristics. Another alternative is surface-enhanced laser desorption/ionization time-of-flight mass spectrometry (SELDITOF-MS), or matrix-assisted laser desorption/ionization time-of-flight mass spectrometry (MALDI-TOF-MS/MS) (Zhou et al. 2020). These approaches are characterized by their speed, efficiency, and great sensitivity (Elam et al. 2021). MALDI-TOF-MS/MS utilizes UV laser pulses to produce ions, which are then analyzed depending on their ratio of mass to charge.

4.6 Promising Health-Promoting Effects

4.6.1 Hypocholesterolemic and Hypolipidemic Effects

Several bioactive peptides obtained from fish, kefir, milk, and soybean proteins have been discovered to exhibit hypocholesterolemic and hypolipidemic effects (Tung et al. 2018; Shojaee-Aliabadi et al. 2023). These peptides have the potential to lower the concentration of lipoproteins and lipids in the bloodstream by inhibiting the activity of lipase, thereby preventing the buildup of excessive cholesterol in blood vessels. Research has shown that inhibiting lipase in the stomach can lead to a diminution in fasting low-density lipoprotein (LDL) and postprandial triacylglycerol concentration, promote weight loss, avoid diabetes, and enhance metabolism (Bhandari et al. 2020; Abbasi et al. 2023). Möller et al. (2008) demonstrated that lipase suppression can be obtained through directly interacting with the colipase or lipase, or by hindering the enzyme from being absorbed through interaction with the interface of oil and water. They additionally demonstrated that the certain amino acid assembly of peptides can affect plasma cholesterol concentration (Möller et al. 2008). Proteins with low proportions of lysine–arginine and methionine–glycine, which are found in fish and soy proteins, have hypocholesterolemic action (Erdmann et al. 2008). Additionally, kefir peptides have been demonstrated to have hypolipidemic and anti-obesity effects in mice with weight gain caused by diets (Tung et al. 2018). According to Shi et al. (2019), yogurt peptides have been found to diminish body mass index and fat while also enhancing tolerance for glucose. Additionally, it was noted that the combination of cyanidin-3-O-β-glucoside and yogurt peptides had synergistic interactions, resulting in a drop in blood pressure and fat mass in the animal study (Shi et al. 2019). In addition, peptides that have a high concentration of hydrophobic amino acids possess the aptitude to attach to bile acids, resulting in an increase in the elimination of steroid hormones through feces (Erdmann et al. 2008). A study demonstrated that the fermentation of soymilk utilizing kefir grains has the potential to hinder and improve conditions such as liver–kidney toxicities, type II diabetes, obesity, and hyperlipidemia in rats that were fed a high-fat and high-fructose diet (Tiss et al. 2020).

4.6.2 Antithrombotic Effect

According to research, antithrombotic peptides range in length from three to twenty amino acid residues and resemble the fibrinogen gamma chain in certain ways. They work by competing with thrombocyte receptors to prevent platelets from forming or aggregating (Zhang 2016). Currently, the majority of antithrombotic peptides derived from food originate through hydrolysis by enzymes or digestive breakdown of proteins found in brewer's spent grain, amaranth grain, legume, blue mussel, and milk casein (Cian et al. 2018; Qiao et al. 2018). Compared to the process of breaking down proteins using enzymes, there have been limited investigations into fermenting dietary proteins to produce antithrombotic peptides. This presents an intriguing opportunity for investigators to conduct research. Rojas-Ronquillo et al. (2012) discovered the initial antithrombotic peptide (YQWPVLGPVRGPFPIIV, specifically the β-casein 193–209 segment) derived from bovine casein that was fermented *via L. casei*. The peptide exhibited a thrombin inhibitory effectiveness ratio of 4.6% per peptide concentration (μg/mL) (Rojas-Ronquillo et al. 2012). Following that, the investigators cited in the literature primarily employed either a single strain or a mixture of various strains of lactic acid bacteria, including *Lactococcus lactis*, *Lactobacillus johnsonii*, *L. casei*, and *Streptococcus thermophilus*, to ferment milk and produce antithrombotic peptides (Domínguez-González et al. 2014; Rendon-Rosales et al. 2019). Abd El-Fattah et al. (2018) formulated yogurt *via* skim milk fermenting enriched with whey protein concentrate utilizing only yogurt starter or a combination of yogurt starter and a highly proteolytic co-culture. Both yogurts were discovered to possess a maximum level of antithrombotic function, fluctuating from 42% to 74%. The significant antithrombotic action in the yogurt matrix is attributed to the physical resemblance between milk lactoferrin and the α-chain of fibrinogen, which has been proven to hinder platelet aggregation. In addition, caseinomacropeptide is structurally similar to the γ-chain of human fibrinogen, which contributes to the yogurt's antithrombotic activity. Furthermore, there is a notable molecular resemblance between the processes of milk coagulation and blood coagulation. Both processes include comparable interaction mechanisms, with κ-casein interacting with chymosin in milk coagulation and fibrinogen interacting with thrombin in blood coagulation (Abd El-Fattah et al. 2018).

4.6.3 Antihypertensive Activity

Hypertension is a persistent state characterized by a continual elevation of diastolic blood pressure (in excess of 90 mmHg) and systolic blood pressure (in excess of 140 mmHg). The adjustment of blood pressure is regulated by the kinin nitric oxide system (KNOS) and the renin-angiotensin-aldosterone signaling way. Renin cleaves angiotensinogen (DRVYIHPFHLVI) to generate angiotensin I (DRVYIHPFHL), which is an unresponsive precursor of ACE, a zinc metallopeptidase. Afterward, angiotensin I decapeptide (DRVYIHPFHL) is enzymatically broken *via* ACE to generate a highly potent vasoconstrictor called angiotensin II octapeptide (DRVYIHPF) (Li et al. 2004). At the same time, in KNOS, ACE could potentially break down a powerful vasodilator called bradykinin (RPPGFSPFR) (Li et al. 2004). The inhibition of ACE can result in a decrease in blood pressure and general markers for hypertension. Angiotensin-converting enzyme inhibition is the majority-investigated mechanism for developing antihypertensive medications. Some well-known drugs in this category include perindopril, alacepril, lisinopril, fosinopril, ramipril, zofenopril, enalapril, and captopril. However, the use of synthetic medications can have side effects ranging from mild symptoms like coughing and irritation to more serious issues like diminished kidney function and fetal defects. Thus, it is imperative to search for nonsynthetic ACE-I peptides that have milder or no adverse reactions after utilization, a greater duration of tissue retention, and a lower elimination time.

Investigations on structure–activity relationship (SAR) have demonstrated that the peptide sequence, width, and composition of amino acids play a crucial role in determining the actions of angiotensin-converting enzyme inhibitors. Peptides of smaller lengths, such as dipeptides and tripeptides, can efficiently traverse the epithelial cells of the intestine by utilizing peptide transporters. This mechanism helps to protect against digestion in the gastrointestinal tract and maintains the beneficial effects of ACE-I (Shen and Matsui 2019). ACE-I peptides commonly consist of 3 to 20 amino acid repeats and have a low molecular weight. The three amino acids located at the end of the protein chain, known as the C-terminus, have a significant impact on ACE inhibition. Specifically, there is a greater frequency of phenylalanine, proline, tryptophan, and tyrosine as hydrophobic aromatic amino acids at the C-terminus, while isoleucine, valine, alanine, and leucine as branched hydrophobic amino acids are more common at the N-terminus (Li et al. 2014). The quantitative SAR assessment indicates that dipeptides with large and hydrophobic sequences are favored. However, for tripeptides, the most common circumstances are (i) hydrophobic amino acids at the N-terminus, (ii) amino acids with a positive charge in the middle region, and (iii) aromatic amino acids at the C-terminus (Wu et al. 2006). Alternative viewpoints propose that negatively charged groups in peptides may aid in the removal of zinc ions from ACE, which is a vital aspect of the enzyme's optimal functioning (Andrews et al. 1985). It is logical to assume that there is a direct contact between the inhibitor and zinc throughout ACE suppression (Bicknell et al. 1987). The notable instances of ACE-I peptides generated from milk and milk products are IPP and VPP, which have recently been thoroughly reviewed (Lei et al. 2019). Upon investigation, it can be seen that both peptides satisfy the structural and functional criteria of having isoleucine and valine as branching hydrophobic amino acids at the N-terminus, as well as proline as an aromatic hydrophobic amino acid at the C-terminus. The combination of *R. philippinarum* clam and *B. natto* underwent fermentation, follow-on the creation of a new peptide (VISDEDGVTH) that is resistant to the digestive process and has ACE-I function with an IC_{50} value of 8.16 µM (Chen et al. 2018).

The thornback ray skin gelatin was subjected to enzymatic hydrolysis using two bacterial proteases, resulting in the production of two peptides (APGAP and GIPGAP). The IC_{50} values of these peptides were determined to be 170 and 27.9 µM, respectively (Lassoued et al. 2015). Both peptides have PGAP as a shared motif, with the existence of proline at the C-terminus and alanine at the N-terminus potentially enhancing ACE suppression. VEIS as an ACE-I peptide was obtained from the hydrosylate of silkworm pupa protein utilizing alkaline proteases. The ACE inhibition rate of the peptide was determined to be 54.27% (Li et al. 2014). Significantly, the ACE suppression rate increased to 94.7% during *in vitro* digestive metabolism, suggesting that the silkworm hydrosylate exhibits prodrug-like characteristics. An *in vivo* impact from ACE-I was demonstrated by employing Sprague-Dawley rats (Chen et al. 2018). The inclusion of valine as a branched amino acid at the beginning of the peptide and histidine as an aromatic amino acid could potentially enhance the ACE-I activity. The process of boiling abalone results in the generation of wastewater that is rich in proteins and has a substantial amount of salt. However, this wastewater is seldom reused. The ACE-I peptide (ATPGDEG) was extracted from the desalted wastewater and exhibited an IC_{50} of 0.39 mg/mL. This peptide demonstrated comparable antihypertensive action to captopril as the positive control (Qian et al. 2018). Evidently, the presence of alanine as the N-terminal amino acid in the peptide satisfies the specified criterion. Nevertheless, the carboxyl terminus harbors two amino acids with negative charges, which contradicts the current idea. Hence, it is necessary to perform additional investigations on the ACE-I activity of the peptide.

4.6.4 Mineral-Binding Activity

Peptides containing particular sequences have the ability to create complexes by attaching to minerals like phosphorus, calcium or other components present in a liquid media at the pH level seen in the intestines. Peptides possessing strong anionic characteristics have the ability to create soluble complexes with minerals. As a result, they become resistant to additional breakdown by intestinal proteolytic enzymes and enhance the bioaccessibility of the minerals. Wang et al. (2018) discovered a new peptide called FVDVT that has the ability to bind to calcium. This peptide was derived from the hydrosylate of wheat germ protein. The synthesis of the chelate involves the participation of the amino nitrogen and carboxyl oxygen atoms, which act as chelating sites. Additionally, the amino acids threonine and asparagine contribute to this process (Wang et al. 2018). Hou et al. (2018) discovered VLGYQIR as a calcium-binding peptide in the protein hydrosylate of Antarctic krill. The complex's chelating sites were found to interact with the amino or carbonyl groups of arginine, isoleucine, and glutamine as described (Hou et al. 2018). A recent investigation performed by Nwachukwu et al. (2019) has established that peptides produced from flaxseed exhibit a calmodulin-dependent suppressive impact on phosphodiesterase. This action has the potential to reduce chronic diseases, including cancer (Nwachukwu et al. 2019).

4.6.5 Opiate-Like Activity

Meperidine, codeine, and morphine are opioids frequently employed in therapeutic settings to treat both acute and chronic pain (Canizares et al. 2019). These peptides exhibit a strong attraction to an opiate receptor, specifically either the "traditional" δ-, μ-, and κ-type receptors or the "non-traditional" orphanin/nociception FQ peptide receptor (McDonald and Lambert 2013). Opioid peptides are compounds that bind to opioid receptors and can exhibit agonistic, partly agonistic, or antagonistic effects depending on their mode of action. The effectiveness of opioid peptides is determined by their reactivity to these receptors. Stimulation of μ-opioid peptide receptors is responsible for the majority of powerful pain-relieving effects, while δ-opioid peptide receptors are more ineffective in alleviating pain (Janecka et al. 2004). In addition, it has been discovered that distinct opioid receptors are accountable for various physiological consequences. The μ-receptors are responsible for regulating emotions and inhibiting the movement of the intestines. The δ-receptors play a role in influencing behavioral emotions. The κ-receptors are associated with relaxation and the consumption of food (Pihlanto-Leppälä 2000).

Opioid peptides can be classified into two primary classes, namely exogenous and endogenous according to their source. Endogenous opioid peptides, such as dynorphins, enkephalins, and endorphins, are spontaneously created in the body from bigger proteins called prodynorphin, proenkephalin, and proopiomelanocortin, respectively. This production occurs by enzymatic hydrolysis mediated *via* peptidases (Pihlanto-Leppälä 2000). Opioid peptides typically consist of between four and eight amino acids and exhibit a common structural pattern characterized by a shared N-terminal sequence, YGGF. These peptides are commonly referred to as conventional opioid peptides (Kaur et al. 2020). Researchers have attempted to enhance the effectiveness of these naturally occurring peptides by modifying them with techniques such as replacing, adding, deleting, forming cycles, or combining them with other compounds. These modifications aim to increase their potency for usage as pain-relieving medications in a clinical setting (Liu and Wang 2012; Garg et al. 2016). Nevertheless, opioids have a direct correlation with adverse effects including vomiting, drowsiness, physical dependency, respiratory depression, dizziness, constipation, nausea, and addiction. On the other hand, opioid peptides generated from food are comparatively less potent, which reduces the likelihood of adverse effects. Additionally, their synthesis cost is considerably lower.

External opioid peptides are derived from precursor proteins found in animals or plants by means of hydrolysis process. The food-origin peptides possessing opioid activities were initially identified *via* Zioudrou et al. in 1979 and were designated as exorphin, signifying that the peptides were sourced from an external source and had action similar to morphine (Pihlanto-Leppälä 2000). Atypical opioid peptides, also recognized as exogenous opioid peptides, exhibit different residues in their N-terminal location, with the tyrosine repeat remaining consistent at the terminus (Janecka et al. 2004). The N-terminal structure of these peptides has been discovered to be either Tyr-X1-X2-Phe or Tyr-X-Phe. The basic pattern of these peptides is significant because the presence of a residue of tyrosine at the amino terminus and additional aromatic amino acids at the third or fourth site enables the peptides to properly engage with the opioid receptors' binding site (Garg et al. 2016). Hartmann and Meisel (2007) emphasized the significance of the proline at the second situation in determining the three-dimensional arrangement of the phenylalanine and tyrosine lateral chains (Hartmann and Meisel 2007). Research has shown that opioid peptides generated by proteins from animals specifically attach to μ-opioid peptide receptors, while those developed from plant proteins specifically attach to δ-opioid peptide receptors (Yoshikawa et al. 2003). However, soymorphins, due to their receptor selectivity, deviate from this pattern and attach to μ-opioid peptide receptors (Liu and Udenigwe 2019). Opioid peptides have been detected in various dairy goods. They have also been isolated from fish, shellfish, meat, soybean, amaranth grain, and wheat (Ebner et al. 2015; Fan et al. 2019; Tovar-Pérez et al. 2019). Investigations have established that these opioid peptides result from food can have similar effects to the body's natural opioids, acting as agonists. Conversely, other peptides can have inhibitory properties comparable to naloxone, a powerful drug that blocks the effects of opioids on receptors (Hartmann and Meisel 2007). Dallas et al. (2016) utilized kefir grains for fermenting milk and successfully isolated five opioid peptides (YPFPGPIPN, YPFPGPIPNSLPQ, YPSYGLN, YPVEPF, and YPFPGPI). These peptides exhibit an ordinary architectural pattern found in opioid peptides, and their N-terminus motif consists of YPF or YP. This structural motif enables them to attach to the opioid-associated receptors and provides analgesic and sedative effects similar to those of opiates (Dallas et al. 2016). Several investigations have identified peptides with a similar pattern, but they have not investigated their possible opiate-similar effects. For example, Taha et al. (2017) gathered YPFPGPIPK from fermented buffalo milk (Taha et al. 2017), and Amorim et al. (2019) generated YPFPGPIPN from kefir (Amorim et al. 2019). This suggests that there is still a lot of potential for further research in this area.

4.7 The Impact of Processing Procedures on the Bioactivity of Peptides

The biological activity of bioactive peptides can be significantly influenced by food manufacturing approaches. Physical techniques such as irradiation, ultrasound, and heat have the potential to alter the functions and structure of proteins. These processing methods may also lead to Maillard reactions, resulting in the generation of allergenic constituents (Davis et al. 2001). Moreover, processing can make peptides more susceptible to immune system responses, gastrointestinal digestion, and absorption. Hence, it is crucial to establish the optimal circumstances for processing peptides/proteins in order to promote their bioactive functions. However, it is important to note that while some manufacturing procedures may decrease the performance of certain peptides, they may elevate the activity of others. In this regard, the antibacterial role of lysozyme (Takahashi et al. 2016) and α-lactalbumin (Agyei et al. 2016) augmented after being denaturized by a thermal process, whereas other peptides may miss their function under this condition. The boiling effects on the performance of biopeptides may vary depending on the treatment conditions of the parent protein and enzyme formulation. Studies have demonstrated that when raw casein is hydrolyzed with chymotrypsin, it produces antidiabetic peptides that have higher bioactivity compared to those obtained from boiled casein (Jan et al. 2016). Therefore, it is possible that hydrolysates of raw milk might be more efficient in managing diabetic conditions than pasteurized milk. Furthermore, the antioxidant capacity of cowpea-derived peptides decreases significantly during thermal processing. However, this exposure enhances the peptides' ability to hinder the cholesterol's solubility in micellar solutions when compared to raw specimens (Marques et al. 2015).

In other studies, the DPP-IV and α-glucosidase suppressor functions, as well as the ACE suppressor capacity of navy bean hydrolysates, did not show significant changes following cooking. However, the ability to inhibit α-amylase was significantly reduced. The ACE inhibitory peptides in beans, such as YAAGS and YAGGS, retained their biological properties despite being pre-cooked, suggesting that pre-cooked legumes can still offer advantageous health advantages (Mojica et al. 2015). The ACE inhibitory abilities of peptides AAPLAP, KAAAAP, IAGRP, KPVAAP, and KAAAATP from ham were also maintained after thermal processing at 117 °C for 6 minutes (Escudero et al. 2014). Likewise, collagen-origin ACE suppressor peptides maintained their performance after 2 hours of heating at 100 °C and within a pH range of 2–6 although their bioactivity decreased meaningfully in alkaline circumstances (Fu et al. 2015). In a separate investigation, the application of pulsed electric field (PEF) was utilized to enhance the efficacy of the antioxidant peptides KCHQP (derived from pine nuts) (Lin et al. 2017) and SHCMN (derived from soybean protein) (Lin et al. 2016). Although PEF did not alter the fundamental compositions of the peptides, it did noticeably decrease their zwitter potentials. Consequently, it is crucial to regulate the ideal circumstances under which bioactive peptides can be treated in order to preserve their bioactivity (Daliri et al. 2017).

4.8 Possible Bioactive Peptide Applications

It is widely recognized that numerous bioactive peptides inherently exist in conventional fermented meals that have been taken for a significant period of time prior to the establishment of the word "bioactive." These peptides can be produced through the fermentation of an extensive range of dietary proteins. Enzymatic processes in the stomach can lead to the generation of many peptides when food-containing protein substrates are ingested. The demand for functional foods and nutritional supplements is growing quickly, mostly because of their beneficial effects on human health and ability to avoid specific diseases. Bioactive peptides are incorporated or enhanced in these goods through alterations to the conventional manufacturing procedure, such as adjusting production settings or utilizing different starter cultures (Hartmann and Meisel 2007). Bioactive peptides provide several significant benefits in the medicinal sector. These include their minimal gathering in bodily tissues, strong bioactivity, specialized targeting to biological systems, a broad range of therapeutic effects, and anticipated low risk of causing allergies or toxicity. According to Álvarez García and Marcet (2018), pre- and probiotic beverages, energy drinks, and infant formula are currently popular products in the functional food business (Álvarez García and Marcet 2018).

Several readily accessible functional foods and components on the market contain bioactive peptides that are associated with various health benefits. In this regard, Davisco Foods International, United States, produces a whey protein hydrosylate that contains glycomacropeptide. This item is said to have anticarcinogenic, antibacterial, and antithrombotic properties. Furthermore, the bioactive peptides derived from fish proteins have been integrated into numerous items, which are subsequently marketed as food specifically intended for health purposes under the designation of FOSHU (Food for Specified Health Uses) in Japan. ValtyronR, a product from Japan, comprises sardine peptides that possess antihypertensive properties. It has been utilized in 33 various items, such as powdered soup, jelly, soft beverages, and dietary supplements. Calpis® and Evolus®, two renowned products from Japan and Finland, respectively, are fermented milk-based goods that include the tripeptides IPP and VPP. These products are reputed to possess antihypertensive properties. The protein hydrolysate of dry milk offered in the Netherlands, developed by DMV International, is said to include glutamine-rich peptides that have an immunomodulatory effect.

The understanding of the impact of bioactive peptides on the well-being of people has resulted in the creation of novel dietary supplements and medications (Abbasi et al. 2022). Additionally, hydrosylates have been utilized in the food industry owing to their numerous beneficial properties throughout the manufacturing process or in the final products. A study conducted by Xiao et al. (2018) discovered that fermented red bean flour exhibits several advantageous properties, including enhanced foaming stability, improved emulsion stability, higher fat absorption capacity, higher water holding capacity, and lower bulk density. These characteristics make it a valuable ingredient in the manufacturing of ice cream mixes, whipped toppings, weaning foods, infant foods, and bakery products (Xiao et al. 2018). Rao et al. (2012) performed an investigation that proposed the potential application of egg white hydrosylate as a food plasticizer (Rao et al. 2012). In addition, Yu et al. (2014) found that egg white protein has the potential to be employed as a plasticizer in cakes, while lysozyme can serve as an organic preservative for foods. In addition, egg white protein has been utilized as a clarifying agent for the purpose of eliminating sediment in alcoholic beverages (Yu et al. 2014). The study conducted by de Queiroz et al. (2017) demonstrated that the hydrosylate derived from goat viscera is a viable option for serving as an emulsion stabilizer (de Queiroz et al. 2017). The hydrosylates of grass carp fish skin have the potential to serve as functional food additives, specifically as emulsifiers and binder elements (Wasswa et al. 2007). In their study, Chalamaiah et al. (2010) discovered that the application of hydrosylates from meriga fish egg protein could enhance the emulsification capacity, protein solubility, and gelling properties of a food item (Chalamaiah et al. 2010). Peptides that are produced throughout the manufacturing of a product do not need to be included in the formulation or declared as an ingredient (Abbasi et al. 2022). This simplifies their marketing and makes it easier for consumers to accept them (Álvarez García and Marcet 2018).

4.9 One Advancement Over Linear Peptides with Cyclic Peptides

Several investigations have indicated that linear peptides have failed to meet objectives in certain applications. One significant disadvantage of linear peptides is their limited stability, which might negatively impact their use in therapy. Over the past few decades, researchers have begun exploring the construction of cyclic peptides as a means of addressing the limitations of linear peptides. Studies have shown that cyclic peptides are more resistant to being broken down by proteases (Hussack et al. 2011) and have a stronger binding affinity for receptors than their linear counterparts (Fukuzumi et al. 2012). The cyclization of peptides is thought to decrease their geographical

Table 4.2 Some cyclic peptides along with their respective activity and origin.

Peptide origin	Peptide name	Health effects	References
Amycolatopsis orientalis	Vancomycin	A broad range antibacterial substitute that is utilized in various bacterial infections	de Hoog et al. (2004)
Conus magus	Ziconotide	Significant pain killer; acts as an analgesic agent	Schmidtko et al. (2010)
Chromobacterium violaceum	Romidepsin	Anticancer activity (various types of malignant cell lines) mediated by apoptotic responses	Bertino and Otterson (2011); Campàs-Moya (2009); Abbasi et al. (2023)
Chemically synthesized from ciclosporin	Alisporivir	Hinders cyclophilin A, and it is supposed that it may have a possible effect on hepatitis C and Alzheimer's disease	Dubinin et al. (2021); Gallay and Lin (2013)
Tolypocladium inflatum	Cyclosporin A	As a calcineurin suppressor, it can diminish the activity of lymphocytes	Schreiber and Crabtree (1992)
Aplidium albicans	Plitidepsin	Antiviral, antitumor, and immunomodulation activities	Alonso-Álvarez et al. (2017)
Bacillus brevis	Tyrocidine	Antimicrobial function mediated by the disruption of the cell membrane	Qin et al. (2003)
Bacillus brevis	Gramicidin S	Antibiotic activity toward gram-positive and gram-negative and even some pathogenic fungi	Gause and Brazhnikova (1944); Kondejewski et al. (1996)

vibrations and enhance their surface area for engaging with the target, hence promoting their affinity to the target component and improving their metabolism (Cini et al. 2012). While there are various natural supplies of cyclic peptides, the solid-phase synthesis of these peptides has become a widely used manner in laboratories worldwide (Tapeinou et al. 2015). Approximately thirty years ago, the charge of producing these peptides was extremely high, making it nearly impossible to produce in large quantities. However, extensive work in peptides chemical manufacturing, the advancement of innovative purification techniques to produce peptides, and a substantial drop in the cost of amino acids as compared to previous years have resulted in an increased focus on cyclic peptides (Tapeinou et al. 2015) (Table 4.2).

4.10 Computer-based Methods for Peptide Research Utilization

Computational techniques, such as proteomic and peptidomic investigations, are highly beneficial for studying peptides. By employing computer-based methodologies, it is feasible to forecast the synthesis of peptides derived from particular dietary proteins. This manner enables the choice of appropriate parent proteins, high-efficiency proteolytic enzymes, and adaptable hydrolysis way, in addition to the investigation of the secondary organization and chemical and physical characteristics of the resulting peptides. The conventional approach to identifying and processing bioactive peptides is to perform digestion and chromatographic separation of the hydrolysis goods in a controlled laboratory setting. Following the bioactivity assessment, the peptide sequence will typically be determined. Typically, this method involves verifying the biological function of the chemically produced peptide sequence. The primary issue with the conventional approach is its low yield and the restriction on the amount of peptide specimens that may be analyzed simultaneously. Proteomics-based techniques utilize efficient protein digestion, and computer-based algorithms are used to forecast peptide function and offer physiological and chemometric data regarding the target peptide sequence. The essential stages in this procedure are as follows: Initially, the protein databases are scrutinized to choose the preferred proteins that possess well-established amino acid sequences. The proteins are thereafter broken down computationally utilizing the suitable proteolytic enzymes specific to the chosen protein. Peptides generated *via* computer simulations are subsequently analyzed for their structural characteristics and their biological functions, such as cytotoxicity and allergic reactions (Jenkins and Orsburn 2020). A comprehensive and practical compilation of many databases, including information on the molecular structure and physical characteristics of peptides, has been recently established (Usmani et al. 2018).

4.11 Challenges in Bioactive Peptide Development

Considering the encryption of bioactive peptides within foods, it becomes crucial to take into account the level of protein hydrolysis involved in their production. However, it should be noted that peptides discharged through proteolysis mediated by the fermentation and enzymatic ways are still vulnerable to additional hydrolysis as long as the fermentation process or enzymatic function is ongoing. Consequently, there is a possibility of diminished or entirely lost bioactivity as a result of the ongoing deterioration of the peptides. Developing a dynamic model for the process of protein hydrolysis under such circumstances presents a demanding task (Agyei et al. 2016). Additionally, the reproducibility of peptides derived from microbial fermentation utilizing wild microbes is limited. This is due to the unpredictable nature of live cells and their metabolic processes, including enzymes and enzyme levels, which cannot be controlled. Therefore, there is no assurance of a consistent amount of particular bioactive peptides generated following the process of fermentation.

To address this issue, the utilization of enhanced strains, genetically modified strains, or purified enzymes could be beneficial. However, it is crucial to emphasize that employing enzymes for protein hydrolysis is a more costly method compared to microbial fermentation. Moreover, bioactive peptides found in hydrolysates of food often exhibit enhanced performance due to cooperative influences with other elements present in the hydrolysate. Consequently, some peptides may display diminished bioactivity as nutraceuticals when tested individually. The stability of the peptides produced is a significant obstacle to the generation of bioactive peptides. Peptides derived from food are typically broken down easily in the digestive system, which hinders their ability to display any desired function in the body during *in vivo* setting. However, to overcome this issue, the pure isolated peptides can be stabilized by incorporating a structure-stimulating probe tail (Kaspar and Reichert 2013) and by modifying the peptide arrangement. Implementing these stabilizing procedures could enhance the bioaccessibility of the peptides once they are consumed.

4.12 Conclusions and Future Perspectives

Bioactive peptides found in food play a crucial role in maintaining a healthy diet and can be effective in preventing and treating diseases. The awareness among consumers about the impact of functional foods on overall health has prompted extensive research and the generation of bioactive peptides in various food goods. Among these, the milk-derived tripeptides VPP and IPP have been extensively studied and evidenced to yield beneficial outcomes in human trials as antihypertensive agents. This calls for further investigation into other bioactive peptides derived from food through human studies. Additionally, numerous hydrolysates with food origin have exhibited versatile and biologically active properties, yet the specific components responsible for these effects remain unidentified. It is vital to identify the peptides present in such hydrolysates in order to understand the mechanisms *via* which they exhibit their beneficial health-promoting influences. The exploration of microbial fermentation-mediated bioactive peptides will continue to be a hopeful and cost-effective method for producing biologically active peptides in food. This is because microbial proteolytic systems, which are generally considered safe, produce various peptides with varied capacities throughout the fermentation process. Soon after, the expansion and utilization of genetically modified microorganisms will be crucial, as they will discharge substantial volumes of enzymes with proteolytic activity to break down food proteins. In addition, bioactive peptides obtained directly from food sources will soon be easily accessible on the marketplace and commercialized as nutritional supplements. These peptides can be classified as medications since they will undergo thorough characterization, and their characteristics and precise action mechanisms will be determined.

References

Abbasi, A., Aghebati-Maleki, L., and Homayouni-Rad, A. (2022). The promising biological role of postbiotics derived from probiotic *Lactobacillus* species in reproductive health. *Critical Reviews in Food Science and Nutrition* 62 (32): 8829–8841.

Abbasi, A., Rad, A.H., Maleki, L.A. et al. (2023). Antigenotoxicity and cytotoxic potentials of cell-free supernatants derived from *Saccharomyces cerevisiae* var. *boulardii* on HT-29 human colon cancer cell lines. *Probiotics and Antimicrobial Proteins* 15 (6): 1583–1595.

Abbasi, A., Saadat, T.R., and Saadat, Y.R. (2022). Microbial exopolysaccharides–β-glucans–as promising postbiotic candidates in vaccine adjuvants. *International Journal of Biological Macromolecules* 223: 346–361.

Abbasi, A., Sanej, K.D., Moradi, S. et al. (2023). Bioactive compounds and biological activities of *Allium sativum* L. In: *Bioactive Compounds in the Storage Organs of Plants* (ed. H.N. Murthy, K.Y. Paek, and S.-Y. Park), 1–40. Springer.

Abd El-Fattah, A., Sakr, S., El-Dieb, S., and Elkashef, H. (2018). Developing functional yogurt rich in bioactive peptides and gamma-aminobutyric acid related to cardiovascular health. *LWT* 98: 390–397.

Abdel-Hamid, M., Otte, J., De Gobba, C. et al. (2017). Angiotensin I-converting enzyme inhibitory activity and antioxidant capacity of bioactive peptides derived from enzymatic hydrolysis of buffalo milk proteins. *International Dairy Journal* 66: 91–98.

Acquah, C., Chan, Y.W., Pan, S. et al. (2019). Structure-informed separation of bioactive peptides. *Journal of Food Biochemistry* 43 (1): e12765.

Aguilar-Toalá, J., Santiago-López, L., Peres, C. et al. (2017). Assessment of multifunctional activity of bioactive peptides derived from fermented milk by specific *Lactobacillus plantarum* strains. *Journal of Dairy Science* 100 (1): 65–75.

Agyei, D., Ongkudon, C.M., Wei, C.Y. et al. (2016). Bioprocess challenges to the isolation and purification of bioactive peptides. *Food and Bioproducts Processing* 98: 244–256.

Ahn, J., Park, S., Atwal, A. et al. (2009). Angiotensin I-converting enzyme (ACE) inhibitory peptides from whey fermented by lactobacillus species. *Journal of Food Biochemistry* 33 (4): 587–602.

Algaron, F., Miranda, G., Le Bars, D., and Monnet, V. (2004). Milk fermentation by *Lactococcus lactis* with modified proteolytic systems to accumulate potentially bio-active peptides. *Le Lait* 84 (1–2): 115–123.

Alonso-Álvarez, S., Pardal, E., Sánchez-Nieto, D. et al. (2017). Plitidepsin: design, development, and potential place in therapy. *Drug Design, Development and Therapy* 11: 253–264.

Alu'datt, M.H., Rababah, T., Johargy, A. et al. (2016). Extraction, optimisation and characterisation of phenolics from *Thymus vulgaris* L.: phenolic content and profiles in relation to antioxidant, antidiabetic and antihypertensive properties. *International Journal of Food Science and Technology* 51 (3): 720–730.

Álvarez García, C. and Marcet Manrique, I. (2018). Meat proteins as a potential source of bioactive ingredients for food and pharmaceutical use. In: *Novel Proteins for Food, Pharmaceuticals and Agriculture: Sources, Applications and Advances* (ed. M. Hayes), 29–49. Wiley-Blackwell.

Amorim, F.G., Coitinho, L.B., Dias, A.T. et al. (2019). Identification of new bioactive peptides from Kefir milk through proteopeptidomics: bioprospection of antihypertensive molecules. *Food Chemistry* 282: 109–119.

Andrews, P., Carson, J., Caselli, A. et al. (1985). Conformational analysis and active site modeling of angiotensin-converting enzyme inhibitors. *Journal of Medicinal Chemistry* 28 (3): 393–399.

Arumugam, V., Venkatesan, M., Ramachandran, S., and Sundaresan, U. (2018). Bioactive peptides from marine ascidians and future drug development–a review. *International Journal of Peptide Research and Therapeutics* 24: 13–18.

Authority EFS (2008). Evolus® and reduce arterial stiffness-Scientific substantiation of a health claim related to Lactobacillus helveticus fermented Evolus® low-fat milk products and reduction of arterial stiffness pursuant to Article 14 of the Regulation (EC) No 1924/2006-Scientific Opinion of the Panel on Dietetic Products, Nutrition and Allergies. *EFSA Journal* 6 (10): 824.

Babini, E., Tagliazucchi, D., Martini, S. et al. (2017). LC-ESI-QTOF-MS identification of novel antioxidant peptides obtained by enzymatic and microbial hydrolysis of vegetable proteins. *Food Chemistry* 228: 186–196.

Bazinet, L. and Firdaous, L. (2009). Membrane processes and devices for separation of bioactive peptides. *Recent Patents on Biotechnology* 3 (1): 61–72.

Bertino, E.M. and Otterson, G.A. (2011). Romidepsin: a novel histone deacetylase inhibitor for cancer. *Expert Opinion on Investigational Drugs* 20 (8): 1151–1158.

Bhandari, D., Rafiq, S., Gat, Y. et al. (2020). A review on bioactive peptides: physiological functions, bioavailability and safety. *International Journal of Peptide Research and Therapeutics* 26 (1): 139–150.

Bicknell, R., Holmquist, B., Lee, F.S. et al. (1987). Electronic spectroscopy of cobalt angiotensin converting enzyme and its inhibitor complexes. *Biochemistry* 26 (23): 7291–7297.

Cai, S., Ling, C., Lu, J. et al. (2017). EGAR, a food protein-derived tetrapeptide, reduces seizure activity in pentylenetetrazole-induced epilepsy models through α-amino-3-hydroxy-5-methyl-4-isoxazole propionate receptors. *Neurotherapeutics* 14: 212–226.

Campàs-Moya, C. (2009). Romidepsin for the treatment of cutaneous T-cell lymphoma. *Drugs of Today (Barcelona, Spain: 1998)* 45 (11): 787–795.

Canizares, M., Power, J.D., Rampersaud, Y.R., and Badley, E.M. (2019). Patterns of opioid use (codeine, morphine or meperidine) in the Canadian population over time: analysis of the Longitudinal National Population Health Survey 1994–2011. *BMJ Open* 9 (7): e029613.

Chalamaiah, M., Rao, G.N., Rao, D.G., and Jyothirmayi, T. (2010). Protein hydrolysates from meriga (*Cirrhinus mrigala*) egg and evaluation of their functional properties. *Food Chemistry* 120 (3): 652–657.

Chalamaiah, M., Ulug, S.K., Hong, H., and Wu, J. (2019). Regulatory requirements of bioactive peptides (protein hydrolysates) from food proteins. *Journal of Functional Foods* 58: 123–129.

Chaves-López, C., Serio, A., Paparella, A. et al. (2014). Impact of microbial cultures on proteolysis and release of bioactive peptides in fermented milk. *Food Microbiology* 42: 117–121.

Chen, Y., Gao, X., Wei, Y. et al. (2018). Isolation, purification and the anti-hypertensive effect of a novel angiotensin I-converting enzyme (ACE) inhibitory peptide from *Ruditapes philippinarum* fermented with *Bacillus natto*. *Food & Function* 9 (10): 5230–5237.

Chen, Y., Liu, W., Xue, J. et al. (2014). Angiotensin-converting enzyme inhibitory activity of *Lactobacillus helveticus* strains from traditional fermented dairy foods and antihypertensive effect of fermented milk of strain H9. *Journal of Dairy Science* 97 (11): 6680–6692.

Chen, Y.-S. and Steele, J.L. (1998). Genetic characterization and physiological role of endopeptidase O from *Lactobacillus helveticus* CNRZ32. *Applied and Environmental Microbiology* 64 (9): 3411–3415.

Christensen, J.E., Dudley, E.G., Pederson, J.A., and Steele, J.L. (1999). Peptidases and amino acid catabolism in lactic acid bacteria. *Antonie Van Leeuwenhoek* 76: 217–246.

Cian, R.E., Garzón, A.G., Martínez-Augustin, O. et al. (2018). Antithrombotic activity of brewers' spent grain peptides and their effects on blood coagulation pathways. *Plant Foods for Human Nutrition* 73: 241–246.

Cini, E., Bifulco, G., Menchi, G. et al. (2012). Synthesis of Enantiopure 7-Substituted Azepane-2-carboxylic acids as templates for conformationally constrained peptidomimetics. *European Journal of Organic Chemistry* 2012 (11): 2133–2141.

Da Porto, C. and Natolino, A. (2017). Supercritical fluid extraction of polyphenols from grape seed (Vitis vinifera): study on process variables and kinetics. *The Journal of Supercritical Fluids* 130: 239–245.

Daliri, E.B.-M., Lee, B.H., and Oh, D.H. (2017). Current perspectives on antihypertensive probiotics. *Probiotics and Antimicrobial Proteins* 9: 91–101.

Daliri, E.B.-M., Oh, D.H., and Lee, B.H. (2017). Bioactive peptides. *Foods* 6 (5): 32.

Dallas, D.C., Citerne, F., Tian, T. et al. (2016). Peptidomic analysis reveals proteolytic activity of kefir microorganisms on bovine milk proteins. *Food Chemistry* 197: 273–284.

Davis, P.J., Smales, C.M., and James, D.C. (2001). How can thermal processing modify the antigenicity of proteins? *Allergy* 56: 56–60.

Dei Piu, L., Tassoni, A., Serrazanetti, D.I. et al. (2014). Exploitation of starch industry liquid by-product to produce bioactive peptides from rice hydrolyzed proteins. *Food Chemistry* 155: 199–206.

Do, Q.D., Angkawijaya, A.E., Tran-Nguyen, P.L. et al. (2014). Effect of extraction solvent on total phenol content, total flavonoid content, and antioxidant activity of *Limnophila aromatica*. *Journal of Food and Drug Analysis* 22 (3): 296–302.

Domínguez-González, K.N., Cruz-Guerrero, A., González-Márquez, H. et al. (2014). Antihypertensive and antithrombotic activities of a commercial fermented milk product made with *Lactobacillus casei* Shirota and *Streptococcus thermophilus*. *International Journal of Dairy Technology* 67 (3): 358–364.

Dubinin, M.V., Starinets, V.S., Talanov, E.Y. et al. (2021). Alisporivir improves mitochondrial function in skeletal muscle of *mdx* mice but suppresses mitochondrial dynamics and biogenesis. *International Journal of Molecular Sciences* 22 (18): 9780.

Dullius, A., Goettert, M.I., and de Souza, C.F.V. (2018). Whey protein hydrolysates as a source of bioactive peptides for functional foods–Biotechnological facilitation of industrial scale-up. *Journal of Functional Foods* 42: 58–74.

e Silva, F.G.D., Hernández-Ledesma, B., Amigo, L. et al. (2017). Identification of peptides released from flaxseed (*Linum usitatissimum*) protein by Alcalase® hydrolysis: antioxidant activity. *LWT-Food Science and Technology* 76: 140–146.

Ebner, J., Arslan, A.A., Fedorova, M. et al. (2015). Peptide profiling of bovine kefir reveals 236 unique peptides released from caseins during its production by starter culture or kefir grains. *Journal of Proteomics* 117: 41–57.

Elam, E., Feng, J., Lv, Y.-M. et al. (2021). Recent advances on bioactive food derived anti-diabetic hydrolysates and peptides from natural resources. *Journal of Functional Foods* 86: 104674.
El-Fattah, A.A., Sakr, S., El-Dieb, S., and Elkashef, H. (2016). Angiotensin-converting enzyme inhibition and antioxidant activity of commercial dairy starter cultures. *Food Science and Biotechnology* 25: 1745–1751.
Erdmann, K., Cheung, B.W., and Schröder, H. (2008). The possible roles of food-derived bioactive peptides in reducing the risk of cardiovascular disease. *The Journal of Nutritional Biochemistry* 19 (10): 643–654.
Escudero, E., Mora, L., and Toldrá, F. (2014). Stability of ACE inhibitory ham peptides against heat treatment and in vitro digestion. *Food Chemistry* 161: 305–311.
Fan, M., Guo, T., Li, W. et al. (2019). Isolation and identification of novel casein-derived bioactive peptides and potential functions in fermented casein with *Lactobacillus helveticus*. *Food Science and Human Wellness* 8 (2): 156–176.
Ferri, M., Graen-Heedfeld, J., Bretz, K. et al. (2017). Peptide fractions obtained from rice by-products by means of an environment-friendly process show in vitro health-related bioactivities. *PLoS One* 12 (1): e0170954.
Fu, Y., Young, J.F., Dalsgaard, T.K., and Therkildsen, M. (2015). Separation of angiotensin I-converting enzyme inhibitory peptides from bovine connective tissue and their stability towards temperature, pH and digestive enzymes. *International Journal of Food Science and Technology* 50 (5): 1234–1243.
Fujita, H. and Yoshikawa, M. (1999). LKPNM: a prodrug-type ACE-inhibitory peptide derived from fish protein. *Immunopharmacology* 44 (1–2): 123–127.
Fukuzumi, T., Ju, L., and Bode, J.W. (2012). Chemoselective cyclization of unprotected linear peptides by α-ketoacid–hydroxylamine amide-ligation. *Organic & Biomolecular Chemistry* 10 (30): 5837–5844.
Gallay, P.A. and Lin, K. (2013). Profile of alisporivir and its potential in the treatment of hepatitis C. *Drug Design, Development and Therapy* 7: 105–115.
Garcia-Tejedor, A., Sanchez-Rivera, L., Recio, I. et al. (2015). Dairy *Debaryomyces hansenii* strains produce the antihypertensive casein-derived peptides LHLPLP and HLPLP. *LWT- Food Science and Technology* 61 (2): 550–556.
Garg, S., Nurgali, K., and Kumar, M.V. (2016). Food proteins as source of opioid peptides-a review. *Current Medicinal Chemistry* 23 (9): 893–910.
Gause, G.F. and Brazhnikova, M.G. (1944). Gramicidin *S* and its use in the treatment of infected wounds. *Nature* 154 (3918): 703.
Geng, X., Tian, G., Zhang, W. et al. (2016). A Tricholoma matsutake peptide with angiotensin converting enzyme inhibitory and antioxidative activities and antihypertensive effects in spontaneously hypertensive rats. *Scientific Reports* 6 (1): 24130.
Gobbetti, M., Minervini, F., and Rizzello, C.G. (2004). Angiotensin I-converting-enzyme-inhibitory and antimicrobial bioactive peptides. *International Journal of Dairy Technology* 57 (2–3): 173–188.
Hartmann, R. and Meisel, H. (2007). Food-derived peptides with biological activity: from research to food applications. *Current Opinion in Biotechnology* 18 (2): 163–169.
Hayes, M., Ross, R.P., Fitzgerald, G.F., and Stanton, C. (2007). Putting microbes to work: dairy fermentation, cell factories and bioactive peptides. Part I: overview. *Biotechnology Journal: Healthcare Nutrition Technology* 2 (4): 426–434.
He, R., Girgih, A.T., Rozoy, E. et al. (2016). Selective separation and concentration of antihypertensive peptides from rapeseed protein hydrolysate by electrodialysis with ultrafiltration membranes. *Food Chemistry* 197: 1008–1014.
de Hoog, M., Mouton, J.W., and van den Anker, J.N. (2004). Vancomycin: pharmacokinetics and administration regimens in neonates. *Clinical Pharmacokinetics* 43: 417–440.

Hosseini, H., Abbasi, A., Sabahi, S. et al. (2023). Assessing the potential biological activities of postbiotics derived from *Saccharomyces cerevisiae*: an in vitro study. *Probiotics and Antimicrobial Proteins* 1–17.

Hou, H., Wang, S., Zhu, X. et al. (2018). A novel calcium-binding peptide from Antarctic krill protein hydrolysates and identification of binding sites of calcium-peptide complex. *Food Chemistry* 243: 389–395.

Huang, J., Liu, Q., Xue, B. et al. (2016). Angiotensin-I-converting enzyme inhibitory activities and In vivo antihypertensive effects of sardine protein hydrolysate. *Journal of Food Science* 81 (11): H2831–H2840.

Huang, Y.-L., Ma, M.-F., Chow, C.-J., and Tsai, Y.-H. (2017). Angiotensin I-converting enzyme inhibitory and hypocholesterolemic activities: effects of protein hydrolysates prepared from *Achatina fulica* snail foot muscle. *International Journal of Food Properties* 20 (12): 3102–3111.

Hussack, G., Hirama, T., Ding, W. et al. (2011). Engineered single-domain antibodies with high protease resistance and thermal stability. *PLoS One* 6 (11): e28218.

Jan, F., Kumar, S., and Jha, R. (2016). Effect of boiling on the antidiabetic property of enzyme treated sheep milk casein. *Veterinary World* 9 (10): 1152.

Janecka, A., Fichna, J., and Janecki, T. (2004). Opioid receptors and their ligands. *Current Topics in Medicinal Chemistry* 4 (1): 1–17.

Jenkins, C. and Orsburn, B. (2020). In silico approach to accelerate the development of mass spectrometry-based proteomics methods for detection of viral proteins: application to COVID-19. *BioRxiv* https://doi.org/10.1101/2020.03.08.980383 Preprint.

Kartsova, L., Makeeva, D., Kravchenko, A. et al. (2021). Capillary electrophoresis as a powerful tool for the analyses of bacterial samples. *TrAC Trends in Analytical Chemistry* 134: 116110.

Kaspar, A.A. and Reichert, J.M. (2013). Future directions for peptide therapeutics development. *Drug Discovery Today* 18 (17–18): 807–817.

Kaur, J., Kumar, V., Sharma, K. et al. (2020). Opioid peptides: an overview of functional significance. *International Journal of Peptide Research and Therapeutics* 26: 33–41.

Kondejewski, L.H., Farmer, S.W., Wishart, D.S. et al. (1996). Modulation of structure and antibacterial and hemolytic activity by ring size in cyclic gramicidin *S* analogs. *Journal of Biological Chemistry* 271 (41): 25261–25268.

Kumari, B., Tiwari, B.K., Hossain, M.B. et al. (2017). Ultrasound-assisted extraction of polyphenols from potato peels: profiling and kinetic modelling. *International Journal of Food Science and Technology* 52 (6): 1432–1439.

Lafarga, T., Acién-Fernández, F.G., and Garcia-Vaquero, M. (2020). Bioactive peptides and carbohydrates from seaweed for food applications: natural occurrence, isolation, purification, and identification. *Algal Research* 48: 101909.

Lassoued, I., Mora, L., Barkia, A. et al. (2015). Bioactive peptides identified in thornback ray skin's gelatin hydrolysates by proteases from *Bacillus subtilis* and *Bacillus amyloliquefaciens*. *Journal of Proteomics* 128: 8–17.

Lee, S.Y. and Hur, S.J. (2017). Antihypertensive peptides from animal products, marine organisms, and plants. *Food Chemistry* 228: 506–517.

Lei, J., Sun, L., Huang, S. et al. (2019). The antimicrobial peptides and their potential clinical applications. *American Journal of Translational Research* 11 (7): 3919.

Li, G.-H., Le, G.-W., Shi, Y.-H., and Shrestha, S. (2004). Angiotensin I–converting enzyme inhibitory peptides derived from food proteins and their physiological and pharmacological effects. *Nutrition Research* 24 (7): 469–486.

Li, X., Li, Y., Huang, X. et al. (2014). Identification and characterization of a novel angiotensin I-converting enzyme inhibitory peptide (ACEIP) from silkworm pupa. *Food Science and Biotechnology* 23: 1017–1023.

Lima, C.A., Campos, J.F., Filho, J.L.L. et al. (2015). Antimicrobial and radical scavenging properties of bovine collagen hydrolysates produced by *Penicillium aurantiogriseum* URM 4622 collagenase. *Journal of Food Science and Technology* 52: 4459–4466.

Lin, S., Liang, R., Li, X. et al. (2016). Effect of pulsed electric field (PEF) on structures and antioxidant activity of soybean source peptides-SHCMN. *Food Chemistry* 213: 588–594.

Lin, S., Liang, R., Xue, P. et al. (2017). Antioxidant activity improvement of identified pine nut peptides by pulsed electric field (PEF) and the mechanism exploration. *Lwt* 75: 366–372.

Liu, Z. and Udenigwe, C.C. (2019). Role of food-derived opioid peptides in the central nervous and gastrointestinal systems. *Journal of Food Biochemistry* 43 (1): e12629.

Liu, W.X. and Wang, R. (2012). Endomorphins: potential roles and therapeutic indications in the development of opioid peptide analgesic drugs. *Medicinal Research Reviews* 32 (3): 536–580.

Maddox, C.E., Laur, L.M., and Tian, L. (2010). Antibacterial activity of phenolic compounds against the phytopathogen *Xylella fastidiosa*. *Current Microbiology* 60: 53–58.

Marques, M.R., Freitas, R.A.M.S., Carlos, A.C.C. et al. (2015). Peptides from cowpea present antioxidant activity, inhibit cholesterol synthesis and its solubilisation into micelles. *Food Chemistry* 168: 288–293.

McDonald, J. and Lambert, D.G. (2013). Opioid mechanisms and opioid drugs. *Anaesthesia & Intensive Care Medicine* 14 (11): 505–509.

Mellander, O. (1950). The physiological importance of the casein phosphopeptide calcium salts. II. Peroral calcium dosage of infants. Some aspects of the pathogenesis of rickets. *Acta Societatis Botanicorum Poloniae* 55: 247–257.

Mojica, L., Chen, K., and de Mejía, E.G. (2015). Impact of commercial precooking of common bean (*Phaseolus vulgaris*) on the generation of peptides, after pepsin–pancreatin hydrolysis, capable to inhibit dipeptidyl peptidase-IV. *Journal of Food Science* 80 (1): H188–H198.

Mojica, L. and De Mejía, E.G. (2016). Optimization of enzymatic production of anti-diabetic peptides from black bean (*Phaseolus vulgaris* L.) proteins, their characterization and biological potential. *Food & Function* 7 (2): 713–727.

Möller, N.P., Scholz-Ahrens, K.E., Roos, N., and Schrezenmeir, J. (2008). Bioactive peptides and proteins from foods: indication for health effects. *European Journal of Nutrition* 47 (4): 171–182.

Muhire, J., Sun, X., Zhang, F.X. et al. (2024). Recent trends in multidimensional countercurrent chromatography. *Journal of Separation Science* 47 (3): 2300768.

Nagai, T., Suzuki, N., and Nagashima, T. (2006). Antioxidative activities and angiotensin I-converting enzyme inhibitory activities of enzymatic hydrolysates from commercial kamaboko type samples. *Food Science and Technology International* 12 (4): 335–346.

Nwachukwu, I.D., Alashi, A.M., Zahradka, P.C., and Aluko, R.E. (2019). Transport, bioavailability, safety, and calmodulin-dependent-phosphodiesterase-inhibitory properties of flaxseed-derived bioactive peptides. *Journal of Agricultural and Food Chemistry* 67 (5): 1429–1436.

Pihlanto-Leppälä, A. (2000). Bioactive peptides derived from bovine whey proteins: opioid and ace-inhibitory peptides. *Trends in Food Science & Technology* 11 (9–10): 347–356.

Polikarpova, D., Makeeva, D., Kolotilina, N. et al. (2020). Nanosized cation exchanger for the electrophoretic separation and preconcentration of catecholamines and amino acids. *Electrophoresis* 41 (12): 1031–1038.

Qian, Z.J., Ryu, B., Kang, K.H. et al. (2018). Cellular properties of the fermented microalgae *Pavlova lutheri* and its isolated active peptide in osteoblastic differentiation of MG-63 cells. *Molecular Medicine Reports* 17 (1): 2044–2050.

Qiao, M., Tu, M., Wang, Z. et al. (2018). Identification and antithrombotic activity of peptides from blue mussel (*Mytilus edulis*) protein. *International Journal of Molecular Sciences* 19 (1): 138.

Qin, C., Bu, X., Wu, X., and Guo, Z. (2003). A chemical approach to generate molecular diversity based on the scaffold of cyclic decapeptide antibiotic tyrocidine A. *Journal of Combinatorial Chemistry* 5 (4): 353–355.

de Queiroz, A.L.M., Bezerra, T.K.A., de Freitas, P.S. et al. (2017). Functional protein hydrolysate from goat by-products: optimization and characterization studies. *Food Bioscience* 20: 19–27.

Rahimi, M., Ghaffari, S.M., Salami, M. et al. (2016). ACE-inhibitory and radical scavenging activities of bioactive peptides obtained from camel milk casein hydrolysis with proteinase K. *Dairy Science & Technology* 96: 489–499.

Rai, A.K., Kumari, R., Sanjukta, S., and Sahoo, D. (2016). Production of bioactive protein hydrolysate using the yeasts isolated from soft chhurpi. *Bioresource Technology* 219: 239–245.

Rao, Q., Rocca-Smith, J.R., Schoenfuss, T.C., and Labuza, T.P. (2012). Accelerated shelf-life testing of quality loss for a commercial hydrolysed hen egg white powder. *Food Chemistry* 135 (2): 464–472.

Rendon-Rosales, M.Á., Torres-Llanez, M.J., González-Córdova, A.F. et al. (2019). In vitro antithrombotic and hypocholesterolemic activities of milk fermented with specific strains of *Lactococcus lactis*. *Nutrients* 11 (9): 2150.

Rizzello, C.G., Lorusso, A., Russo, V. et al. (2017). Improving the antioxidant properties of quinoa flour through fermentation with selected autochthonous lactic acid bacteria. *International Journal of Food Microbiology* 241: 252–261.

Rizzello, C.G., Tagliazucchi, D., Babini, E. et al. (2016). Bioactive peptides from vegetable food matrices: research trends and novel biotechnologies for synthesis and recovery. *Journal of Functional Foods* 27: 549–569.

Rojas-Ronquillo, R., Cruz-Guerrero, A., Flores-Nájera, A. et al. (2012). Antithrombotic and angiotensin-converting enzyme inhibitory properties of peptides released from bovine casein by *Lactobacillus casei* Shirota. *International Dairy Journal* 26 (2): 147–154.

Ryhänen, E.-L., Pihlanto-Leppälä, A., and Pahkala, E. (2001). A new type of ripened, low-fat cheese with bioactive properties. *International Dairy Journal* 11 (4–7): 441–447.

Sangsawad, P., Roytrakul, S., and Yongsawatdigul, J. (2017). Angiotensin converting enzyme (ACE) inhibitory peptides derived from the simulated in vitro gastrointestinal digestion of cooked chicken breast. *Journal of Functional Foods* 29: 77–83.

Sanjukta, S., Rai, A.K., Muhammed, A. et al. (2015). Enhancement of antioxidant properties of two soybean varieties of Sikkim Himalayan region by proteolytic *Bacillus subtilis* fermentation. *Journal of Functional Foods* 14: 650–658.

Schmidtko, A., Lötsch, J., Freynhagen, R., and Geisslinger, G. (2010). Ziconotide for treatment of severe chronic pain. *The Lancet* 375 (9725): 1569–1577.

Schreiber, S.L. and Crabtree, G.R. (1992). The mechanism of action of cyclosporin A and FK506. *Immunology Today* 13 (4): 136–142.

Shen, W. and Matsui, T. (2019). Intestinal absorption of small peptides: a review. *International Journal of Food Science and Technology* 54 (6): 1942–1948.

Shi, M., Mathai, M.L., Xu, G. et al. (2019). The effects of supplementation with blueberry, cyanidin-3-O-β-glucoside, yoghurt and its peptides on obesity and related comorbidities in a diet-induced obese mouse model. *Journal of Functional Foods* 56: 92–101.

Shojaee-Aliabadi, S., Abbasi, A., Moradi, S. et al. (2023). Bioactive compounds and biological activities of Madonna lily (*Lilium candidum* L.). In: *Bioactive Compounds in the Storage Organs of Plants* (ed. H.N. Murthy, K.Y. Paek, and S.-Y. Park), 1–13. Springer.

Siltari, A., Kivimäki, A., Ehlers, P. et al. (2012). Effects of milk casein derived tripeptides on endothelial enzymes in vitro; a study with synthetic tripeptides. *Arzneimittel-Forschung* 62 (10): 477–481.

Silva, S.V. and Malcata, F.X. (2005). Caseins as source of bioactive peptides. *International Dairy Journal* 15 (1): 1–15.

Singh, B., Sharma, P., Kumar, A. et al. (2016). Antioxidant and in vivo genoprotective effects of phenolic compounds identified from an endophytic *Cladosporium velox* and their relationship with its host plant *Tinospora cordifolia*. *Journal of Ethnopharmacology* 194: 450–456.

Sridhar, K., Inbaraj, B.S., and Chen, B.-H. (2021). Recent developments on production, purification and biological activity of marine peptides. *Food Research International* 147: 110468.

Suwal, S., Doyen, A., and Bazinet, L. (2015). Characterization of protein, peptide and amino acid fouling on ion-exchange and filtration membranes: review of current and recently developed methods. *Journal of Membrane Science* 496: 267–283.

Taha, S., El Abd, M., De Gobba, C. et al. (2017). Antioxidant and antibacterial activities of bioactive peptides in buffalo's yoghurt fermented with different starter cultures. *Food Science and Biotechnology* 26: 1325–1332.

Takahashi, H., Tsuchiya, T., Takahashi, M. et al. (2016). Viability of murine norovirus in salads and dressings and its inactivation using heat-denatured lysozyme. *International Journal of Food Microbiology* 233: 29–33.

Tapeinou, A., Matsoukas, M.T., Simal, C., and Tselios, T. (2015). Review cyclic peptides on a merry-go-round; towards drug design. *Peptide Science* 104 (5): 453–461.

Tiss, M., Souiy, Z., Ben Abdeljelil, N. et al. (2020). Fermented soy milk prepared using kefir grains prevents and ameliorates obesity, type 2 diabetes, hyperlipidemia and Liver-Kidney toxicities in HFFD-rats. *Journal of Functional Foods* 67: 103869.

Tovar-Pérez, E.G., Lugo-Radillo, A., and Aguilera-Aguirre, S. (2019). Amaranth grain as a potential source of biologically active peptides: a review of their identification, production, bioactivity, and characterization. *Food Reviews International* 35 (3): 221–245.

Tung, Y.T., Chen, H.L., Wu, H.S. et al. (2018). Kefir peptides prevent hyperlipidemia and obesity in high-fat-diet-induced obese rats via lipid metabolism modulation. *Molecular Nutrition & Food Research* 62 (3): 1700505.

Ueno, K., Mizuno, S., and Yamamoto, N. (2004). Purification and characterization of an endopeptidase that has an important role in the carboxyl terminal processing of antihypertensive peptides in *Lactobacillus helveticus* CM4. *Letters in Applied Microbiology* 39 (4): 313–318.

Usmani, S.S., Kumar, R., Bhalla, S. et al. (2018). In silico tools and databases for designing peptide-based vaccine and drugs. *Advances in Protein Chemistry and Structural Biology* 112: 221–263.

Uzel, R.A. (2017). A practical method for isolation of phenolic compounds from black carrot utilizing pressurized water extraction with in-site particle generation in hot air assistance. *The Journal of Supercritical Fluids* 120: 320–327.

Vermeirssen, V., Van Camp, J., and Verstraete, W. (2004). Bioavailability of angiotensin I converting enzyme inhibitory peptides. *British Journal of Nutrition* 92 (3): 357–366.

Wang, L., Ding, Y., Zhang, X. et al. (2018). Isolation of a novel calcium-binding peptide from wheat germ protein hydrolysates and the prediction for its mechanism of combination. *Food Chemistry* 239: 416–426.

Wasswa, J., Tang, J., Gu, X.-h., and Yuan, X.-q. (2007). Influence of the extent of enzymatic hydrolysis on the functional properties of protein hydrolysate from grass carp (*Ctenopharyngodon idella*) skin. *Food Chemistry* 104 (4): 1698–1704.

Wattanasiritham, L., Theerakulkait, C., Wickramasekara, S. et al. (2016). Isolation and identification of antioxidant peptides from enzymatically hydrolyzed rice bran protein. *Food Chemistry* 192: 156–162.

Wu, J., Aluko, R.E., and Nakai, S. (2006). Structural requirements of angiotensin I-converting enzyme inhibitory peptides: quantitative structure– activity relationship study of di-and tripeptides. *Journal of Agricultural and Food Chemistry* 54 (3): 732–738.

Xiao, Y., Sun, M., Zhang, Q. et al. (2018). Effects of *Cordyceps militaris* (L.) Fr. fermentation on the nutritional, physicochemical, functional properties and angiotensin I converting enzyme inhibitory activity of red bean (*Phaseolus angularis* [Willd.] WF Wight.) flour. *Journal of Food Science and Technology* 55: 1244–1255.

Yamamoto, N., Shinoda, T., and Mizuno, S. (2004). Cloning and expression of an endopeptidase gene from *Lactobacillus helveticus* CM4 involved in processing antihypertensive peptides. *Milchwissenschaft* 59 (11): 593–597.

Yoshikawa, M., Takahashi, M., and Yang, S. (2003). Delta opioid peptides derived from plant proteins. *Current Pharmaceutical Design* 9 (16): 1325–1330.

Yu, Z., Yin, Y., Zhao, W. et al. (2014). Application and bioactive properties of proteins and peptides derived from hen eggs: opportunities and challenges. *Journal of the Science of Food and Agriculture* 94 (14): 2839–2845.

Zhang, S.B. (2016). In vitro antithrombotic activities of peanut protein hydrolysates. *Food Chemistry* 202: 1–8.

Zhang, H., Yokoyama, W.H., and Zhang, H. (2012). Concentration-dependent displacement of cholesterol in micelles by hydrophobic rice bran protein hydrolysates. *Journal of the Science of Food and Agriculture* 92 (7): 1395–1401.

Zhang, N., Zhang, C., Chen, Y., and Zheng, B. (2016). Purification and characterization of antioxidant peptides of *Pseudosciaena crocea* protein hydrolysates. *Molecules* 22 (1): 57.

Zhou, J., Chen, M., Wu, S. et al. (2020). A review on mushroom-derived bioactive peptides: preparation and biological activities. *Food Research International* 134: 109230.

5

Prebiotic Oligosaccharide Production in Microbial Cells

Avijeet S. Jaswal, Saroj Mishra, and R. Elangovan

Department of Biochemical Engineering and Biotechnology, Indian Institute of Technology Delhi, Hauz-Khas, New-Delhi, 110016, India

5.1 Oligosaccharides as Prebiotics

The use of oligosaccharides as prebiotics can be linked to the early studies indicating the health benefits conferred to infants fed on human milk *vis-à-vis* cow's milk as the gut microbiota in the former case demonstrated the richness of *Bifidobacterium* and a lower proportion of clostridia and enteric bacteria (Harmsen et al. 2000). Human milk contains nearly 200 nondigestible oligosaccharides (German et al. 2008). Long before this was demonstrated, an idea existed about the correlation between carbohydrates and their impact on gut microbes. The early studies in the 1980s and 1990s (Tanaka et al. 1983; Hidaka et al. 1991) reported the proliferation of bifidogenic bacteria in the fecal flora when inulin, oligofructose (derived from inulin), fructooligosaccharides (FOSs), galactooligosaccharides (GOSs), xylooligosaccharides (XOSs), or trans-galactosylated oligosaccharide were fed to humans, and their impact on human health was noted. Thus, the nondigestible nature of the oligosaccharides in the human gut and their impact on the gut microbes was more directly established. These along with the fiber polysaccharides are classified under prebiotics. It has recently been reported that there is both a reduction (by 95%) and a change in the intestinal flora with age (Arnold et al. 2021; Bosco and Noti 2021). Administration with the GOS can increase *Bifidobacterium* by 10-fold per gram of feces (in 5 weeks) in the age group of 65–80 years (Vulevic et al. 2015).

Prebiotics are "non-digestible functional ingredients which are selectively fermented and allow specific changes, both in the composition and/or activity of the gastrointestinal microflora that confers benefits upon host well-being and health" (Roberfroid 2011; Gibson et al. 2010). The currently well-known prebiotics are carbohydrates. While these are assumed to be nondigestible, that may not be entirely true. For these, it is important to reach the colon intact. "The key attribute of prebiotics, and the one that separates them from other dietary fibers, is that they are selectively metabolized by the more health-positive members of the colonic microbiota, thus influencing the ecological balance" (Rastall 2010). Enough evidence is now available that demonstrates that several diseases of the intestine; several diseases of circulatory, immunological, and nervous systems; metabolic disorders; and cancers of the colon are caused by an imbalance in the microbiota of the colon as demonstrated by the alleviation of these maladies by administration of either probiotics or prebiotics or a mixture of these, called as synbiotics (Tajadadi-Ebrahimi et al. 2014).

Microbial Nutraceuticals: Products and Processes, First Edition. Edited by Sudhir Pratap Singh and Santosh Kumar Upadhyay.

The prebiotic oligosaccharides are of great commercial interest with a prediction of about $10.55 billion by 2025 in the global market (Palaniappan et al. 2021). This is largely because the consumer is more focused on preventive healthcare and is aware of the gut health benefits. The lifestyle-related problems, such as diabetes, hypertension, and obesity have also increased the demand for oligosaccharides as these confer several benefits in managing these diseases (Nie et al. 2020). By market segmentation, the GOSs, the FOSs, and mannooligosaccharides (MNOSs) occupy the largest market share, while others such as the XOSs, chitooligosaccharides (COSs), and human milk oligosaccharides (HMOs) are classified as emerging prebiotics. Reviews on oligosaccharides in the past five years have largely focused on milk oligosaccharides and their impact on the neonatal microbiome, brain as well as immune benefits (Nolan et al. 2020; Pujari and Banerjee 2021; Chong et al. 2022; Fan et al. 2023; Lemoine et al. 2023; Miniello et al. 2023; Okburan and Kizilier 2023), and dairy-derived oligosaccharides (Quinn et al. 2020). The role of oligosaccharides as prebiotics on human metabolism, disease, and therapeutics has also been reviewed in several studies (Basavaiah and Gurudutt 2021; Husmann et al. 2022; Kennedy et al. 2023). Apart from this, the structural diversity of these molecules and specific roles and applications such as in plants, the health sector, or industrial applications have also been described (Ibrahim 2018).

Among the various categories, the GOSs (Mei et al. 2022; Wang et al. 2024) and the FOSs (Rahim et al. 2021; De Cosmi et al. 2022) have been the best studied in terms of their effect as prebiotics. The microbial enzymes involved in FOS synthesis have also been recently reviewed (Belmonte-Izquierdo et al. 2023). Some recent studies have also reported the prebiotic effect of nondigestible raffinose family oligosaccharides (RFOs) (Kanwal et al. 2023), the XOSs (Huang et al. 2022; Bouiche et al. 2023), and the potential of glucooligosaccharides as prebiotics (Zeng et al. 2023). The *in vitro* anti-pathogenic effects of some nondigestible oligosaccharides have also been published (Asadpoor et al. 2020). Mention has also been made of complex oligosaccharide utilization pathways in *Lactobacillus* sp. (Zúñiga et al. 2021), which should allow selection and diversification of other carbohydrates that show prebiotics potential.

Concerning their microbial production, the GOSs and the FOSs have been extensively researched for obtaining these in high purity levels (Srivastava and Mishra 2019; Maráz et al. 2022; de la Rosa et al. 2022). However, very little is covered on the metabolism and uptake of these by the gut microbiota and the organization of the genes encoding various enzyme systems. In this review, we first describe the structural diversity of the known prebiotics and their mechanism of action. This is followed by a description of the microbial whole cells and the enzymes involved in the synthesis of the low molecular weight GOSs and FOSs. Since the processivity of the concerned enzymes should be low to obtain well-defined low molecular weight oligosaccharides, a description is also given of the genetic engineering approaches that have been taken to produce these. Finally, a description of the new emerging prebiotics is given.

5.2 Structural Diversity of Prebiotic Oligosaccharides and Mechanism of Action

The realm of prebiotics encompasses a diverse array of oligosaccharides, each with unique structures and mechanisms of action. From the GOSs, derived from lactose, to the FOSs, synthesized from sucrose, these compounds selectively foster beneficial gut bacteria while suppressing pathogens. The COSs, derived from chitosan, MNOSs from mannans, and XOSs from xylose exhibit various biological activities beyond prebiotic effects. The RFOs, and pectooligosaccharides (POSs) further expand this diversity, offering a spectrum of health benefits from anti-inflammatory to

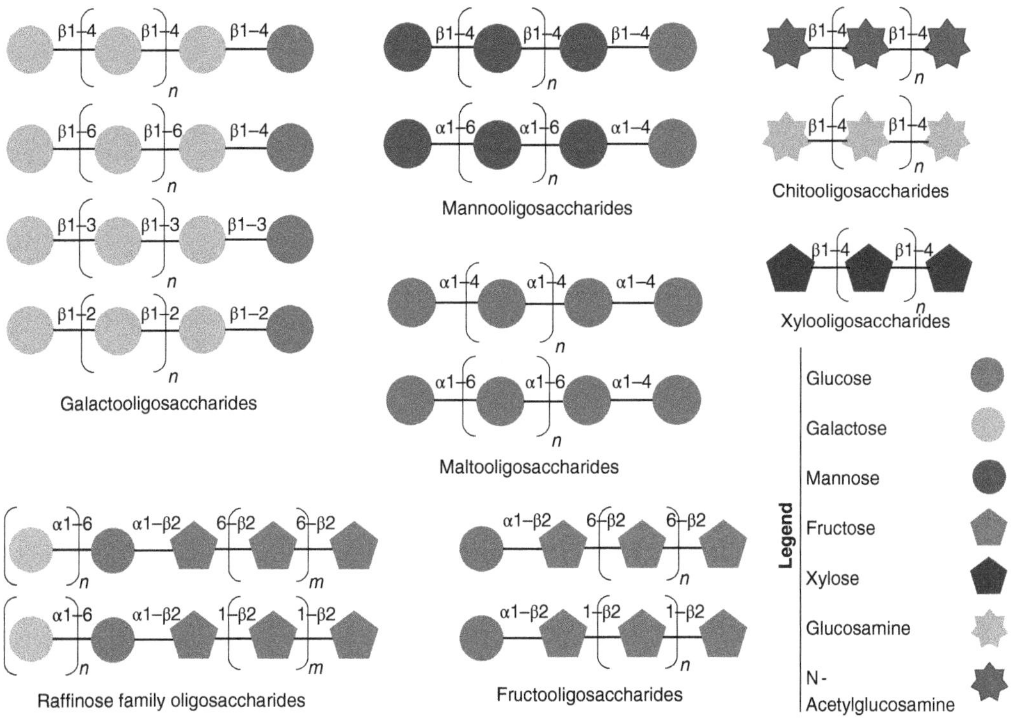

Figure 5.1 Structure and type of glycosidic linkages in different prebiotic oligosaccharides.

anticancer properties. The structural diversity exhibited by the existing and emerging prebiotics is illustrated in Figure 5.1.

5.2.1 Structures of Various Existing and Emerging Prebiotics

5.2.1.1 Galactooligosaccharides

The GOSs are a mixture of tri- to pentasaccharides, which has galactose in β-(1-6), β-(1-3), and/or β-(1-4) linkages (Figure 5.1) with the galactose of the terminal lactose molecule (Gibson et al. 2004; Vulevic et al. 2013). Those linked with a terminal lactulose (an isomer of lactose) also display prebiotic activities. Like other prebiotics, these have shown useful effects in the large intestine by selectively enhancing the growth of beneficial bacteria while reducing the number of pathogens and putrefactive bacteria. Although not as effective as the HMOs, these display many prebiotic properties like the human oligosaccharides (Salli et al. 2019). In addition to their role as prebiotics, these oligosaccharides serve various biological functions in humans, including anti-allergic, antidiabetic, and anti-obesity properties, and they contribute to the prevention of nonalcoholic fatty liver disease. (Elango et al. 2022).

5.2.1.2 Fructooligosaccharides

The FOSs are made up of mixtures of oligosaccharides containing a sucrose molecule in which the fructose unit is linked to additional fructose units. Different types of FOS series can result (Figure 5.1), which depend upon the linkage present between the monosaccharide residues. For instance, an additional fructose linked to sucrose by β-(2-1) linkage results in 1-kestose, two fructose added to sucrose results in 1,1-kestotetraose, while the addition of three fructose units results in 1,1,1-kestopentaose. On the other hand, an additional fructose unit linked by β-(2-6) linkage results in 6-kestose, two fructose added to sucrose in the same way results in 6,6-kestotetraose, while the addition of three fructose units results in 6,6,6-kestopentaose. Additional fructose units can also be linked to C6 of the glucose that results in the neo-FOS series. However, the neo-FOS series have not been explored since these are not produced by microbial fructofuranosidases (FFases) or may represent small amounts synthesized in the cells (for review, see Flores-Maltos et al. 2016; Singh et al. 2017).

5.2.1.3 Chitooligosaccharides

The COSs are derived from chitosan, which can be formed from the alkaline treatment of chitin. Chitin itself is a water-insoluble, cationic amino polysaccharide of β-1,4-linked *N*-acetylglucosamine (GlcNAc) and is the most abundant biomass after cellulose (Heggset et al. 2010; Mahata et al. 2014). Chitosan, thus, is a natural, nontoxic biopolymer and a linear polysaccharide consisting of β-1,4-GlcNAc and β-1,4-glucosamine (Figure 5.1) and is a completely or partially deacetylated form of chitin (Liaqat and Eltem 2018). Since chitosan contains unstable glycosidic bonds, it can be cleaved by hydrolyzing agents to produce oligomers that have variable degree of polymerization (DP). Those that have a DP <20 and an average MW <3.9 kDa are called COSs, chitosan oligomers, or chitooligomers (Lodhi et al. 2014; Mahata et al. 2014). These are water-soluble and possess a wide range of biological activities. These also find multiple applications in the areas of cosmetics, food, agriculture, and medicine (Liaqat and Eltem 2018).

5.2.1.4 Malto- and Isomaltooligosaccharides

Both malto- and isomaltooligosaccharides (MOSs, IMOs) contain different linkages and are derived from starch. MOSs consist of glucose units (typically 2–10) connected by α-(1-4) glycosidic bonds (Figure 5.1), whereas IMOs feature 2–10 glucose units linked together by α-(1-6) glycosidic bonds. (Shinde and Vamkudoth 2022). Since these are derived from starch, they are regarded as generally recognized as safe (GRAS) molecules (Ibrahim 2018). IMOs' definition has expanded to encompass α-(1-2) kojo-oligosaccharides or α-(1-3) nigero-oligosaccharides. These oligosaccharides are characterized not only by their DP and linkage type but also by their unique monosaccharide compositions and disaccharide sugar donors. Their emergence as vital contributors to health promotion extends across human consumers as well as poultry and ruminant animals, marking a significant stride in nutritional science (Shinde and Vamkudoth 2022).

5.2.1.5 Mannooligosaccharides

The MNOSs, derived through the partial hydrolysis of mannans, are characterized as nondigestible oligosaccharides (Faustino et al. 2021). Mannans consist of extensive chains of mannose residues and are sourced from a diverse array of origins, including vegetables, microorganisms, and seeds. Within this realm, MNOSs are divided into two primary cohorts: α- and β-MNOS. α-MNOS arise from the cleavage of α-(1-6) bonds within mannans sourced from yeast cell walls, while β-MNOS stem from plants abundant in mannans, cleaved at β-(1-4)-glycosidic bonds (Figure 5.1). Beyond their established role as prebiotics, MNOSs unveil a spectrum of specific biological impacts. They fortify the protective barriers of the intestinal mucosa, elevate immune vigor, and stand as stalwarts in the battle against mutagens and oxidative stress (Teng and Kim 2018). Furthermore, MNOSs wield a formidable shield against intestinal invaders, curbing the colonization prowess of notorious pathogens like *Salmonella*, *Escherichia coli*, and *Campylobacter* (Koc et al. 2010).

5.2.1.6 Raffinose Family Oligosaccharides

RFOs are in very high abundance in plants, only next to sucrose. These are abundant in the seeds of many plants, particularly crops such as chickpeas, lentils, and soybeans (Elango et al. 2022).

Sugars such as raffinose, stachyose, and verbascose belong to this family, and their synthesis occurs via galactinol-dependent and galactinol-independent pathways. Their constituent monomers are glucose, galactose, and fructose linked in different types of linkages (Figure 5.1). While these have well-defined roles in plants in terms of fighting abiotic stress, their importance is being realized as prebiotics in the human gut. Several probiotic strains including lactobacilli, bifidobacteria, *Lactococcus lactis*, *Streptococcus salivarius* ssp. *Thermophilus*, *Pediococcus* ssp., and *Enterococcus faecium* were evaluated for growth on the RFOs, and 75% of the strains utilized these sugars with some delay for the trisaccharide raffinose and the tetrasaccharide stachyose. However, limited fermentation was seen of the pentasaccharide verbascose (Zartl et al. 2018). These have been recommended in human diets to prevent cancer (Van den Ende 2013).

5.2.1.7 Xylooligosaccharides

The XOSs are produced from inexpensive and abundantly available renewable resources such as crops and their residues/by-products. The hemicellulose part of the plant is largely composed of the pentose sugar xylose. While the XOSs are naturally present in vegetables, fruits, honey, and milk, their quantity is less, making them unattractive for large-scale production. Structurally, the XOSs are linear chains of 2–10 D-xylose units bonded together by β-(1-4) glycosidic linkages (Figure 5.1). Accordingly, these are called xylobiose, xylotriose, xylotetraose, and so on (Carvalho et al. 2013). Apart from demonstrating prebiotic properties, these display sweetness, improve the immune system, and reduce the risk of cancer (Palaniappan et al. 2021).

5.2.2 General Mechanisms of Action of Prebiotics

While many publications exist on the structural diversity, applications, and manufacturing methods of prebiotics, there is a poor understanding of the way these are taken up and their mode of action in the gut space to allow the selective proliferation and activity of some microbes. Several possible explanations have been put forward and these are described below.

First, certain members of the colonic microbiome possess enzymes on their surface that can break (hydrolyze) the prebiotics. Second, these members possess specific transport systems allowing the uptake of the prebiotic and limited metabolism of the same or the hydrolyzed monosaccharides. Thus, a selective advantage is offered to a group of bacteria to proliferate. When it comes to GOSs, the intricacies of their chemical makeup, including linkage type, degree of branching, and DP, exhibit structural and functional variability influenced by reaction conditions and the specific microbial galactosidase employed during preparation (Kittibunchakul et al. 2020). Consequently, the gut microbiota must possess a high degree of adaptability to effectively metabolize this diverse array of molecules. This interplay between prebiotics and gut bacteria is depicted in Figure 5.2, illustrating the mechanism of transport exhibited by probiotic bacteria in the gut. This is best exemplified by the *Lactobacillus* genus. Research indicates that various types of GOSs exert differential effects on the growth of distinct *Lactobacillus* species. Two distinguishable GOS utilization profiles were found within this bacterial genus (Fuhren et al. 2020). A majority of strains (14 out of 21) demonstrated robust utilization of GOS substrates, particularly those with higher DP. Conversely, the remaining strains displayed varying efficiencies in metabolizing lactose and GOS disaccharides bonded via β-1,2 and β-1,3 linkages. Notably, studies involving *Lactobacillus rhamnosus* DR20 elucidated its ability to efficiently utilize both mono- and disaccharides (Gopal et al. 2001), contrasting with the preference of another probiotic bacterium, *Bifidobacterium lactis* DR10, for tri- and tetrasaccharides. In both instances, the utilization of these sugars was closely associated with the presence of β-galactosidase activity. Apart from β-galactosidases,

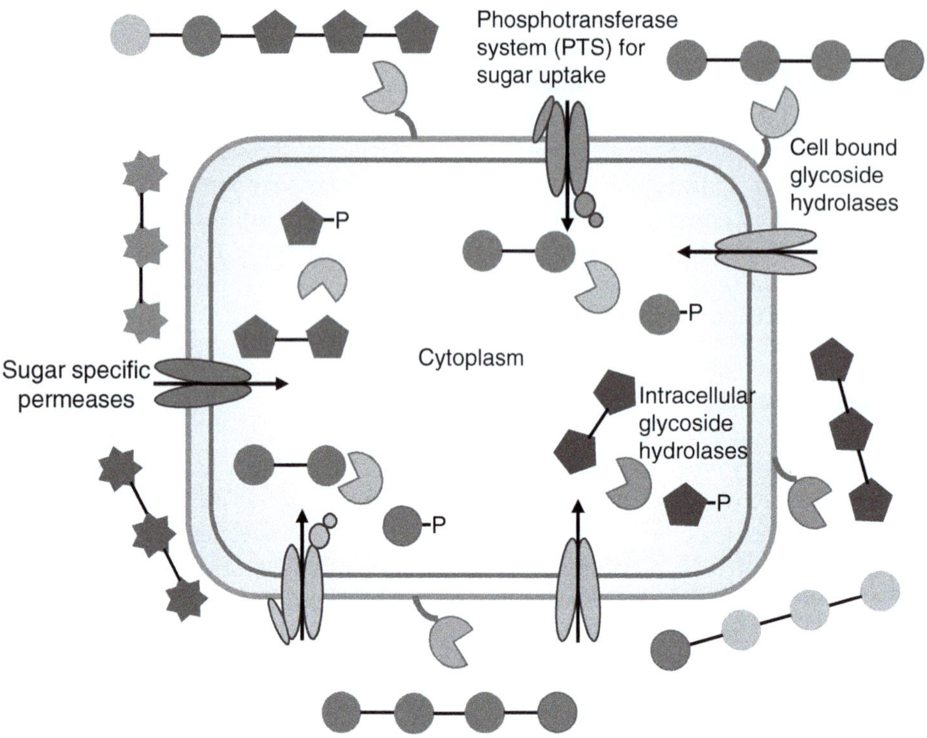

Figure 5.2 Mechanism of transports exhibited by the probiotic bacteria in the gut.

6-phospho-β-galactosidases are also involved in the breakdown of the imported GOSs, as seen with *L. rhamnosus* AMC143 (Arnold et al. 2018).

In lactobacilli, GOSs traverse the membrane through a specialized galactoside-pentose-hexuronide (GPH)-type LacS permease, followed by hydrolysis catalyzed by two β-galactosidases (LacA of the GH42 family and LacLM of the GH2 family), ultimately yielding glucose and galactose (Andersen et al. 2011). While lac operon structures vary across lactic acid bacteria (LAB), *lacS* and *lacA* genes are ubiquitous except in *L. delbrueckii* subsp. *Bulgaricus* (Yong and Klaenhammer 2015). Notably, phenotypic variations in GOS utilization are linked to specific genes, exemplified in *Lactobacillus plantarum*. Strains exhibit two distinct GOS utilization profiles, one favoring high DP GOS and the other demonstrating variable lactose breakdown and disaccharide preferences. The former harbors a lac operon comprising *lacA*, *lacS*, and a divergently oriented regulatory gene, *lacR*, while the latter lacks *lacR* (Fuhren et al. 2020). Deletion studies of *lacA* and/or *lacS* suggest a potential polar effect of *lacA* mutation on downstream *lacS*, impacting GOS utilization. The role of LacS in importing high DP GOS is evident, as observed in *Lactobacillus acidophilus* NCFM lacking a LacS homolog (Andersen et al. 2011). Concurrent deletion of *lacA* and *lacS* leads to diminished growth on GOSs. Additionally, other loci are likely to contribute to the transport process. Transcriptome and physiological analyses of *L. plantarum* and *L. plantarum* ΔccpA, metabolizing glucose or GOS, revealed differential gene expression, particularly implicating the CcpA-encoding gene (Chen et al. 2019). Notably, both *lac* and the *gal* gene clusters resemble those in *L. acidophilus* NCFM involved in GOS metabolism. Genome analysis of *L. plantarum* Y44 unveiled its capacity to metabolize 24 carbohydrate sources, alongside genes associated with exopolysaccharide and surface protein biosynthesis (Gao et al. 2020). Similar genetic signatures underpinning GOS utilization were identified in *Lactobacillus reuteri* ATCC PTA-6475 (Rattanaprasert et al. 2019).

For the FOSs, the low molecular weight FOSs are highly effective in evoking a response, so their generation, uptake, and metabolism need to be understood. In the case of FOSs, cell-associated β-FFases have been reported from several strains of *Bifidobacterium*, namely, *Bifidobacterium adolescentis* ATCC 15703, *Bifidobacterium infantis* (Imamura et al. 1994; Perrin et al. 2001), *Bifidobacterium longum* ATCC 15707, *Bifidobacterium thermophilum* ATCC 25525 (McKellar and Modler 1989). The maximum hydrolytic activity was seen with the FOSs that had a low (3–5) DP. While inulin induced nearly the same FFase (inulinase) activity in all three bacteria, *B. adolescentis* ATCC 15703 and *B. longum* ATCC 15707 also demonstrated constitutive activity (McKellar and Modler 1989). The inulinase production was regulated by catabolite repression. Induction by fructose has also been reported (Perrin et al. 2001) for the FFase of *B. infantis* ATCC 15697, which was accompanied by the production of acetate and lactate (Rastall 2010). Fructosyltransferases (FTases) typically possess signal peptide sequences at their *N*-terminus, facilitating secretion beyond the cell membrane into the periplasmic space, onto the cell surface, or into the extracellular milieu. In LAB, FTases often exhibit cell wall-anchoring motifs, located in the C-terminal region, indicating their presence on the cell surface (Zhang et al. 2022). However, their involvement in sucrose metabolism in LAB remains uncertain. In *L. reuteri* TMW 1.106, FTase expression was not induced by sucrose but correlated with cell aggregation and biofilm formation (Walter et al. 2008). By even lacking cell wall-anchoring motifs, FTases may associate with the cell wall *via* electrostatic interactions, as observed in *Lactobacillus crispatus* (Zhang et al. 2022). In *Streptococcus*, FTases bind to the cell surface, facilitated by C-terminal wall-associated domains rich in Pro-Gly-Ser-Thr residues, as well as hydrophobic domains (Rathsam and Jacques 1998). FTases from *Leuconostoc* feature Gly-Trp repeat sequences in the C terminus responsible for binding to the cell surface (Del Moral et al. 2008).

In *L. acidophilus*, the transportation and breakdown of FOSs are orchestrated by an adenosine triphosphate (ATP)-dependent binding cassette (ABC)-type transporter and an intracellular β-fructofuranosidase (β-FFase) enzyme, respectively (Barrangou et al. 2003). These enzymes are encoded within a single operon known as the multiple sugar metabolism (*msm*) operon, and a configuration reminiscent of *msm* operons is identified in *Streptococcus* species. Such operons typically comprise a LacI family transcriptional regulator (msmR), an ABC transporter (msmEFGK), a β-FFase (bfrA), and a sucrose phosphorylase. Expression of the operon is subject to catabolite repression and is specifically induced by FOSs and sucrose. Similar operons have been detected in various other *Lactobacillus* strains. Moreover, initial investigations suggest that *L. plantarum* utilizes a distinct mechanism for FOS internalization, involving the sucrose phosphoenolpyruvate-dependent phosphotransferase system (PTS), followed by intracellular hydrolysis by an endo-type cytoplasmic β-FFase (Saulnier et al. 2007). The sucrose PTS EIIBCA and β-FFase genes are clustered in a single locus, albeit divergently oriented. This cluster also encompasses genes encoding fructokinase, a sucrose operon repressor, and an α-glucosidase. Whole-genome sequencing of *L. plantarum* Y44 unveiled a repertoire of complete PTS complex genes, intact Embden-Meyerhof-Parnas pathway enzyme genes, as well as numerous carbohydrate-active enzyme (CAZy) genes, totaling 19 complete PTS complex genes (Gao et al. 2020). Interestingly, *L. plantarum* Y44 selectively metabolizes GFn-type FOS, particularly short-chain variants, for growth, as opposed to FFn-type FOS. Similarly, investigations on *L. plantarum* NRRL 1195 suggest a preference for tri- and tetrasaccharides, correlating with rapid sugar uptake. DNA microarray analyses revealed upregulation of FFase, sucrose PTS, fructokinase, α-glucosidase, and a sucrose operon repressor in response to low molecular weight FOSs, indicating a closely correlated uptake and metabolism process. Furthermore, *Bifidobacterium bifidum* exhibits robust growth on XOSs, suggestive of the presence of a specialized transport system for these substrates (Palframan et al. 2003).

Thus, it can be concluded that there exist many common features between GOS and FOS metabolism. Several differentially expressed genes have been identified that are involved in sugar metabolism, metabolite conversion, and changes in fatty acid biosynthesis (Chen et al. 2019). Once the GOSs or the FOSs are internalized, these allow a proliferation of specific bacteria that (i) show increased adhesion to the intestinal mucosa, (ii) enhance the epithelial barriers, (iii) inhibit pathogen adhesion by the production of biofilms, (iv) have metabolic pathways to produce short-chain fatty acids (SCFAs), and (v) modulate the immune system. Apart from supporting the growth of *Lactobacillus* and *Bifidobacterium*, the prebiotics promote the growth of other bacteria such as *Akkermansia*, *Eubacterium*, *Propionibacterium*, *Faecalibacterium*, *Leuconostoc*, and *Roseburia* (Rezende et al. 2021). Both facultative and obligate anaerobic bacteria metabolize sugars into lactic acid, butyric acid, and propionic acid. Formate and acetate are also produced through fermentation of GOSs (Morrison and Preston 2016; Wenzel et al. 2020). These acids exert multiple effects on the human body. For instance, the luminal pH is reduced, and this is reported to increase the bioavailability of calcium and iron. The prebiotics may also increase the paracellular calcium transport directly through the opening of the tight junctions as shown for melibiose and difructose anhydride in rat small and large intestinal epithelia (Mineo et al. 2004). Two major signaling pathways, namely, G-protein-coupled receptors and histone deacetylases, associated with the SCFAs are well recognized (He et al. 2020). Since SCFAs can diffuse through blood circulation through enterocytes, prebiotics not only affect the gastrointestinal tract but can also affect the distant site organs as shown for butyrate, which affects intestinal epithelial development (Hamer et al. 2008). In the airways and macrophages, propionate affects T helper 2 cells (Stinson et al. 2017). Apart from this, propionate has been reported to attenuate atherosclerosis by immune-dependent regulation of cholesterol metabolism in the intestines. It reduces total blood and low-density lipoprotein (LDL) cholesterol levels by reducing intestinal cholesterol absorption (Haghikia et al. 2022). Butyrate, from the SCFA, suppresses tumor growth and cancer cell metastasis. This is reported to operate *via* the regulation of the cell cycle, autophagy, cancer-related signaling pathways, and cancer cell metabolism (Son and Cho 2023). Thus, combination therapy using SCFAs and anticancer drugs has exhibited superior effects.

Apart from the production of SCFAs, peptidoglycans are also produced and these circulate throughout the body in the absence of infection. This is a part of the innate immunity system and prepares the host for any infections. Some of these circulating peptidoglycan fragments have been traced to the gut microbiota (Wolf and Underhill 2018). These can stimulate the innate immune system, enhancing the killing of *Streptococcus pneumoniae* and *Staphylococcus aureus* mediated by bone marrow-derived neutrophils. For this, the signaling mechanism is utilized *via* the pattern recognition receptor Nod1 (Clarke et al. 2010). As expected, the structure of the prebiotics determines both the microbial composition of the gut and the fermentation products (Fehlbaum et al. 2018).

5.3 Enzymes Involved in the Production of GOSs and FOSs

A common strategy for synthesizing GOSs and FOSs involves enzyme-catalyzed transfer reactions. In this process, readily available substrates such as lactose, maltose, or sucrose are employed, and specific glycosyltransferase or glycosidase enzymes are utilized to generate novel oligosaccharides.

This enzymatic approach enables the controlled production of GOSs and FOSs with tailored structures and properties, offering versatility for applications in various fields including food, pharmaceuticals, and biotechnology. Based on the substrate specificity, there are two types of enzymes as per IUBMB enzyme classification:

1) Glycoside hydrolase (GH): These can act on a wide variety of glycosyl donors but have competing specificity for water or sugar molecules as an acceptor, resulting in either hydrolysis or transglycosylation reaction.
2) Glycosyl transferase (GT): These are further classified into two distinct types based on the type of glycosyl donor:
   a) Non-Leloir enzymes: require nonactivated sugar donors like maltose, sucrose, lactose, and starch.
   b) Leloir enzymes: require activated donors like nucleotide-linked monosaccharides.

The enzymes that use nonactivated donors are of interest for the economical production of GOSs and FOSs. The general synthetic reaction mechanism of GH and GT follows a double displacement mechanism (Figure 5.3) involving glycosylation followed by de-glycosylation. During the glycosylation step, a conserved glutamate residue in the active site of the enzyme acts as a nucleophile and attacks the glycosidic bond which results in the formation of an activated enzyme-galactosyl (in the case of GOS synthesis) or enzyme-fructosyl (in the case of FOS synthesis) intermediate complex (Figure 5.3), and glucose is released as the leaving group. During the de-glycosylation step, either lactose, sucrose, or an oligosaccharide comes as an acceptor molecule which acts on the enzyme-hexose intermediate complex to synthesize the glycosidic bond between the acceptor and the donor molecules. Higher chain GOS synthesis thus ensues.

In the case of fructan synthesis, the reaction can be nonprocessive or processive depending upon the release of the newly formed fructan molecule from the active site of the enzyme. If the enzyme has a higher affinity for 1-kestose, the resulting fructan is high molecular weight, being an example of a processive enzyme. In processive enzymes, different linkages of high molecular weight fructan can result. Figure 5.4 shows different types of enzymes associated with the synthesis of different types of fructan molecules.

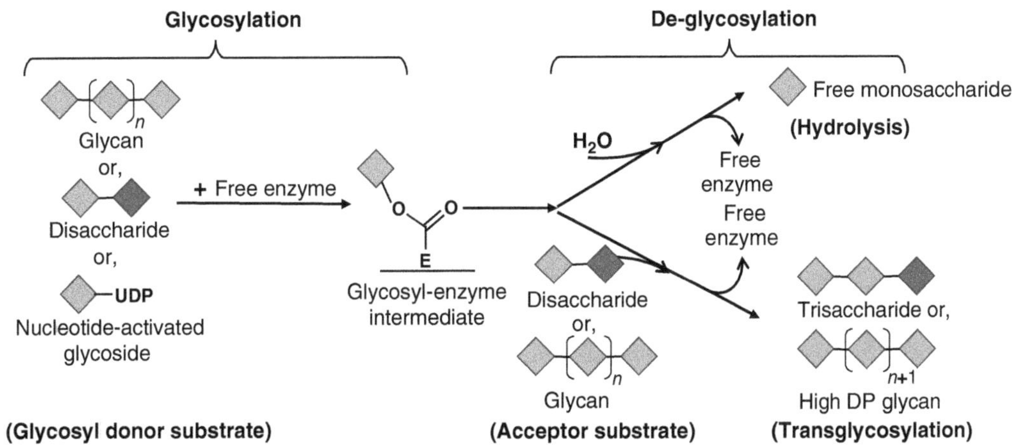

Figure 5.3 General scheme of enzyme-catalyzed transglycosylation reaction using non-Leloir/Leloir-type transferases and glycoside hydrolases.

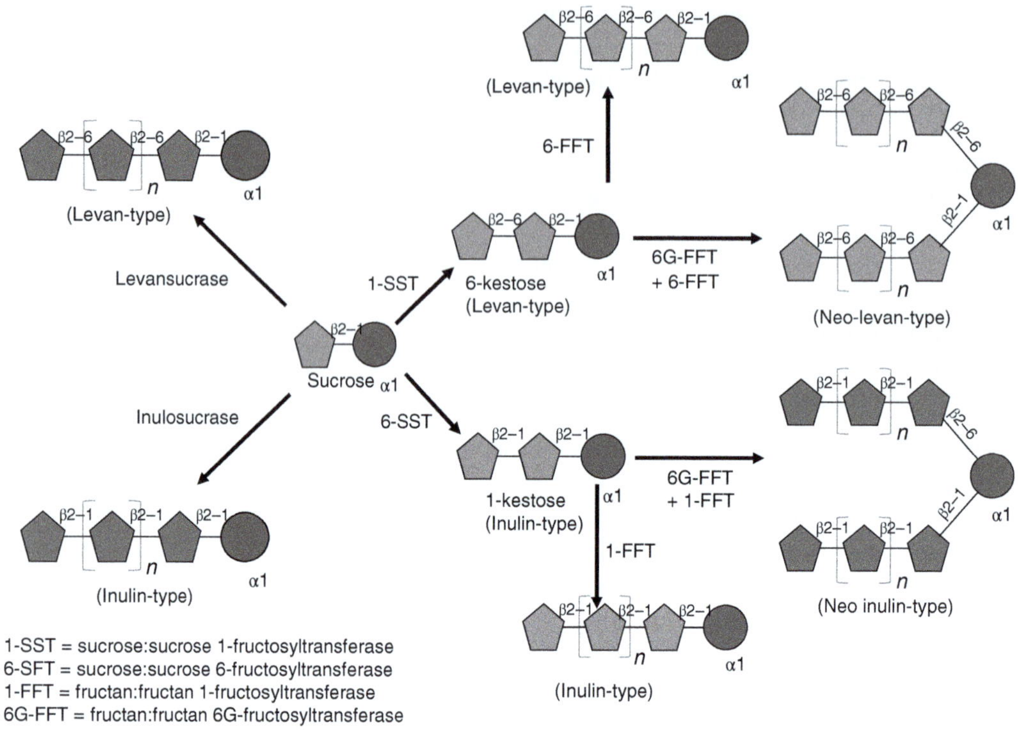

Figure 5.4 Type of fructans and associated enzymes for their synthesis.

5.4 Microbial Systems for the Synthesis of GOSs and FOSs

Both the whole cells and the purified enzymes have been used to produce GOSs and FOSs. The key advantage of using the whole cells is to exploit their ability to use cheap and abundantly available raw material and to catalyze either single- or multi-step reactions to transform the given substrate into the desired product. This strategy is more economical as it does not require purified enzymes (which adds to the cost of production), but the flip side is that it is dependent on several downstream processing steps such as cell removal and purification of the desired product from other cell leachates. If the substrate/product itself is metabolizable by the cells, the yields are compromised. The cells can be either the growing cells or the resting cells. Each approach offers distinct advantages and considerations, contributing to the versatility of microbial-based bioprocesses (Lin and Tao 2017). When biosynthesis of a molecule is not linked to cellular growth, the utilization of resting cells is favored over actively growing cells. Resting cells offer the advantage of utilizing cell-bound enzymes for biotransformation. This choice is motivated by the removal of unconsumed substrates, nutrients, and unwanted metabolites present in the growth medium, facilitating superior recovery rates of the desired product and streamlining downstream processing (de Carvalho 2017). Moreover, resting cells can achieve higher product yields, as the available carbon and energy resources are predominantly channeled toward product synthesis rather than biomass accumulation. However, it is essential to acclimatize resting cells if the added substrate is necessary to induce the required enzymes for the biosynthesis of the desired product (Lin and Tao 2017).

5.4.1 Production of GOSs Using Bacterial and Fungal Systems

The GOSs are produced through the transgalactosylation reaction catalyzed by β-galactosidase. Initially, lactose is hydrolyzed into glucose and galactose, which is then transferred to lactose (or other sugars as an acceptor) using any of the free hydroxyl groups depending on the specificity of the enzyme. Being hydrolytic, these enzymes bring about the synthesis reaction only under low water concentrations/activity. The extended reaction products serve as acceptors, initiating the formation of intricate linear and branched GOSs with DP typically ranging from two to six. The resultant product spectrum is influenced by both enzyme specificity and thermodynamic factors, ultimately determining the composition and structure of the synthesized GOSs. The β-galactosidase belongs to the GH42 family of the CAZy enzymes. The enzyme from *E. coli*, one of the most extensively studied enzymes, is a tetramer that contains four identical chains each containing 1023 amino acids. Each monomer within the structure is structured into five distinct domains, with the third domain, spanning amino acids 334–627, identified as the triose phosphate isomerase (TIM) domain (Juers et al. 2012). Domain 1 adopts a jelly-roll type β-barrel configuration, while Domains 2 and 4 exhibit fibronectin type III-like barrel structures. Domain 5 presents as a novel β-sandwich motif. The catalytic site is housed within the TIM domain, characterized by an α8β8 barrel structure, which classifies it under the α/β protein category. The active site forms a deep hole toward the C-terminus of the protein. Several other amino acids from other parts of the polypeptide chain also form elements of the active site. The enzyme undergoes two proposed transition states, occurring after each step of the reaction. If water is involved in the reaction, it leads to the formation of galactose. Alternatively, if water is absent and D-glucose acts as the acceptor, it results in the formation of polymerized products.

While the GOSs can be prepared from natural biological sources, commercial production from them is not feasible. Instead, these are prepared using lactose, a by-product of the dairy industry, using mostly bacterial β-galactosidases (as well as β-glucosidases and β-glycosidases) (Gosling et al. 2010), whole fungal cells, or fungal β-galactosidase. Very few reports appear on the production of GOSs using whole bacterial systems. Table 5.1 lists the GOS production by various bacterial and fungal enzyme systems. Commercially synthesized GOSs have 55% oligosaccharides, and the rest contains lactose, glucose, and galactose (Playne and Crittenden 2009). Among the many β-galactosidases, the ones most studied and industrially relevant are either from bacterial systems such as *Bacillus circulans*, *Lactobacillus*, *Pantoea anthrophila*, and *Streptococcus thermophilus* or from the fungi *Aspergillus oryzae* and *Kluyveromyces lactis*. The non-lacZ β-galactosidases from *B. circulans* are particularly important as these show strong transgalactosylation activity (Choi et al. 2020). High-purity GOSs have been obtained using *L. plantarum* CICC 22186 derived β-galactosidase for synthesis followed by fermentation of the residual lactose by *K. lactis* 1772. The total yield of GOSs was 10%, but the product obtained was very pure (Zhang et al. 2021). The recombinant crude β-galactosidase from *P. anthrophila*, expressed in *E. coli* was used for GOS synthesis from either lactose or milk whey. A maximum GOS (with β, (1–6) and β, (1-3) galactosyl linkages) yield of 40% (w/w), as measured by high-performance anion-exchange chromatography with pulsed amperometric detection (HPAEC-PAD), was obtained with 400 g/L lactose corresponding to 86% conversion. On comparing GOS synthesis between milk whey and pure lactose, a yield of 38% (60% of lactose conversion) was obtained, with the same product profile, at the same concentration of 300 g/L (Yañez-Ñeco et al. 2021). The reactions with both disaccharides produced, besides hydrolysis into their glucose and galactose monosaccharides, the synthesis of GOSs by transgalactosylation.

A recombinant *L. lactis*, containing a synthetic codon-optimized β-galactosidase gene from *Sulfolobus solfataricus* was used. The killed and permeabilized cells, producing a hyper-thermostable

Table 5.1 GOS synthesis using bacterial and fungal systems.

Organism	Fermentation conditions temp (°C), pH	Initial lactose (g/L)	Yield (%)	References
Bacteria				
Bifidobacterium bifidum	(39, 6.8)	500	20	Tzortzis et al. (2005)
Bifidobacterium bifidum	(40, 6.8)	500	40	Goulas et al. (2007)
Bifidobacterium longum	(40, 7)	257	63	Schwab et al. (2011)
Lactobacillus plantarum	(50, 7)	400	34	Gobinath and Prapulla (2014)
Lactococcus lactis	(85, 6)	400	20	Yu and O'Sullivan (2014)
Fungi				
Kluyveromyces lactis	(40, 7)	400	25	Chockchaisawasdee et al. (2005)
Talaromyces thermophilus	(40, 6.5)	200	50	Nakkharat et al. (2006)
Kluyveromyces lactis	(40, 7)	400	44	Rodriguez-Colinas et al. (2011)
Kluyveromyces lactis	(40, 7)	400	30	Lisboa et al. (2012)
Pichia kluyveri	(30, 7)	400	14	Fai et al. (2014)
Pseudozyma tsukubaensis	(30, 7)	400	16	Fai et al. (2014)
Kluyveromyces marxianus	(40, 6.5)	200	36	Srivastava et al. (2015)

β-galactosidase, were used for the synthesis of GOSs. A maximum of 197 g/L of GOSs (tri- and tetrasaccharides) were produced from 40% initial lactose, after 55 hours of incubation (Yu and O'Sullivan 2014). Further, treatment with UV, as well as an immobilization method, was applied to produce GOSs from lactose. This allowed for the removal of the DNA contaminant as well as reutilization of the whole cells. Production of 150 g/L of GOSs (tri-oligosaccharides and tetra-oligosaccharides) was achieved with minor contaminations by di-oligosaccharides (Yu and O'Sullivan 2018). More recently, GOS synthesis has been reported from lactose and lactulose by dairy propionibacteria, specifically, *Propionibacterium acidipropionici* LET 120, when the culture supernatant of this bacterium was used along with lactozym (Sabater et al. 2019), a yield of 26.8% of GOSs and 26.1% of lactulose oligosaccharides was obtained after 24 hours. The GOS yield was in the range reported with *L. reuteri* and *Bifidobacterium* species.

Several well-developed fungi, including yeast systems, have been developed to produce GOSs. The genera most investigated belong to *Aspergillus*, *Kluyveromyces*, and *Saccharomyces* but the reported yields have been low. Among many yeast strains that were screened, *Kluyveromyces marxianus* NCIM 3551 exhibited the highest β-galactosidase activity and was thus selected to produce GOSs. Cell permeabilization allowed efficient bioconversion by β-galactosidase. A maximum GOS yield of 36% (w/w) of total sugars was achieved with the predominant products being tri- and tetra-GOS (Srivastava et al. 2015). A productivity of 24 g/L/h was obtained under the optimized conditions at a lactose concentration of 20% (w/v). This data was translated to the development of a small-scale process in which *K. marxianus* NCIM 3551 was first used for the synthesis of GOSs followed by selective fermentation of the sugar reaction mixture containing lactose and glucose as impurities. In the total reaction that contained 38% of GOSs (w/w of total carbohydrates), fermentation of these sugars with *Saccharomyces cerevisiae* followed by *K. lactis* var. *lactis* resulted in GOS enrichment by 7% and 38% respectively (Srivastava and Mishra 2019). An efficient cell-based biotransformation system, containing whole cells of *Sporobolomyces singularis* MTCC 5491 (providing the β-galactosidase) and a yeast isolate NUTIDY007 (for removal of glucose), was used to produce GOSs (tri- to pentasaccharides) of high purity (≥85%). Continuous production of GOSs was also investigated using whole cells in a reactor with a microfiltration membrane system (Saravanan et al. 2017).

5.4.2 Production of FOSs Using Bacterial and Fungal Systems

Unlike the GOSs, the FOSs can be synthesized using whole cells in a single step. The FOS production has been reported using cells of *Microbacterium paraoxydans* (Ojha et al. 2016). With sucrose-induced cells, a maximum FOS yield of 155 g/L (corresponding to a product yield of 0.38 g/g initial substrate) was obtained with 40% w/v initial sucrose. The tri-saccharide, 1-kestose, and the tetrasaccharide, 1,1-kestotetraose were identified as the major products. Optimization of several parameters (enzyme equivalent cells, substrate concentration) led to an overall yield of 0.44 g/g initial substrate. Importantly, most other bacterial strains contain either levansucrase (LS) or inulosucrase (IS), which determine the linkages in the adjacent fructose molecules that are either β(2-6) or β(1-2), respectively (Jadaun et al. 2019). The synthesis by bacterial enzymes occurs by sequential transfructosylation of fructose units to a sucrose molecule resulting in high molecular weight fructans of the order of 10^6 to 10^8 Da (Porras-Domínguez et al. 2014; Santos-Moriano et al. 2015). Since the LSs display hydrolytic activity, these can release free fructose, which can be used by the same enzymes to produce levan oligomers (levanbiose and levantriose) (De Siqueira and Oner 2023). Very little work has been done on bacterial ISs as the yield of FOSs with these enzymes is low (Ko et al. 2022).

The FOSs can also be synthesized from sucrose by glycosyl transfer reactions using the FFase as well as the FTases (Tonozuka et al. 2020). Both the hydrolase-type and the transferase-type (non-Leloir-type) enzymes employ a similar anomer-retaining catalytic mechanism with a catalytic triad located at the base of the active site (Reddy and Maley 1996; Pons et al. 2004). At low sucrose concentrations, the FFases exhibit hydrolytic activity while the transfructosylating activity is displayed at high sucrose concentrations (Ganaie et al. 2014). Typically, the bacterial FTases (IS and LS) are grouped under the GH68 family while the FFases and the FTases from the fructan-producing plant and fungal species belong to the GH32 family. Both share a common topological structure of the five-bladed β-propeller fold and are grouped in the glycoside hydrolase (GH)-J clan. The catalytic triad contains a cluster of three amino acids (Asp-Asp-Glu) which participate in the formation of the fructosyl-enzyme complex. Several other conserved residues exist that participate in either the substrate binding or the receptor binding (van Hijum et al. 2006). Enzymes responsible for FOS synthesis are produced by both gram-positive and gram-negative bacteria, with particular focus on the genera *Lactobacillus*, *Leuconostoc*, and *Streptococcus* within gram-positive bacteria. The FTases from the LAB typically comprises four domains: the signal peptide, the *N*-terminal region, the catalytic domain, and the C-terminal region. Interaction between the receptor substrate 1-kestose and InuJ, a specific InuJ substrate (IS), predominantly occurs at subsite 21, which contains conserved residues unique to the GH68 family (Ozimek et al. 2006; Pijning et al. 2011). Additionally, numerous conserved residues located distally from the active sites of *Lactobacillus* significantly influence enzymatic activity. Very recently, a novel FTase (called InuCA) that has a much shorter C-terminus compared to the other FTases from the LAB, has been identified in *L. crispatus*. Although not present in the gut but in the vaginal microbiota of humans, it has the highest isoelectric point that allows binding to the cell surface. This FTase displayed catalytic activity like other FTases and synthesized FOSs using sucrose (Zhang et al. 2022).

There has been considerable interest in recent years in establishing a structure–activity relationship in the FTases, which is expected to form a basis for producing well-defined FOSs. For instance, the trisaccharide 1-kestose, the smallest FOS, comprises a major portion of commercially available FOS formulations and has been shown to exert a superior prebiotic effect (Suzuki et al. 2006; Tochio et al. 2018; Endo et al. 2020; Chen and Chen 2023) compared to the high molecular weight FOSs by allowing a rapid proliferation of butyrate-producing *Bifidobacterium*, *Lactobacillus* spp., and *Faecalibacterium prausnitzii* over other microbes like *Bacteroides* spp. and *Campylobacter* spp. Along with the prebiotic effect, 1-kestose can also be used as a noncaloric and noncariogenic sweetener, which makes it one of the most valuable short-chain FOSs to be used as functional food ingredient. Hence, engineering of FTases to selectively produce short-chain FOSs is desirable. Some advances have been made in this direction. It has been shown that not only the active site (e.g. subsite -1) but also the amino acids far away from the active site play an important role in determining the DP of the synthesized FOSs. Several residues have been identified at the primary, secondary, and tertiary levels that play an important role in determining the specificity and also the product profile of the enzymes (Charoenwongpaiboon et al. 2019; Ortiz-Soto et al. 2019). In *L. crispatus*, deletion of the C-terminal segment of InuCA led to the synthesis of FOSs with longer fructose chains compared to the wild type, suggesting an association between the C-terminal sequences of InuCA and the enzyme's binding pocket (Zhang et al. 2022). Conversely, in the case of *L. reuteri* IS, the C-terminal amino acid residue (D689) had no impact on the product profile, likely serving as a supporting amino acid for receptor oligosaccharide binding (Charoenwongpaiboon et al. 2019). SacB, a levansucrase from *Bacillus subtilis*, naturally lacks a C-terminal region and synthesizes high molecular weight levan (Porras-Dominguez et al. 2015). However, a mutant with an extended C-terminal part synthesized larger fructan chains than the wild type, indicating the involvement of the C-terminal portion in receptor binding (Chambert et al. 1992). Moreover, the surface motifs surrounding the substrate binding pockets of LSs have been shown to influence fructan chain length (Strube et al. 2011). Structural comparisons among various LSs revealed highly variable sizes and orientations of surface loops, indicating variability in the central substrate binding cavities, which ultimately determine FOS chain lengths or molecular weight of levan (Ortiz-Soto et al. 2019).

As mentioned above, the fungal FTases are classified in GH32 of the glycosidase hydrolase families and are reported from several species of *Aspergillus*, *Aureobasidium*, *Kluyveromyces*, and *Penicillium* (Chuankhayan et al. 2010). These are evolutionarily closer to the plant fructan-hydrolyzing enzymes (FFases such as invertase) than to the bacterial FTases (Lammens et al. 2009). Several fungal FFases have also been studied extensively, such as those from *Aspergillus niger* ATCC 20611.9), *Aspergillus japonicus*, *Aspergillus kawachi* (Nagaya et al. 2017) as these exhibit strong transfructosylation activity leading to the production of mainly 1-kestose and nystose. Structurally, the FFases contain an additional β-sandwich domain apart from the four domains present in the members of the GH68 family as revealed in *A. kawachi*. Comparison of the GH32 *A. kawachi* FFase with several members of the GH68 FFases indicated that their structures are highly dissimilar except for the β-propeller domain considering that these proteins are categorized into the same GH-J clan (Nagaya et al. 2017).

5.4.2.1 FOSs Production in Bacterial Systems

In the realm of bacterial systems, considerable research efforts have focused on optimizing the production of the high molecular weight FOSs, generally termed fructans. Key genera of bacteria involved in this intensive exploration include *Acetobacter*, *Bacillus*, *Erwinia*, *Gluconacetobacter*, *Halomonas*, *Microbacterium*, *Pseudomonas*, *Streptococcus*, and *Zymomonas*. Optimal culture conditions, essential for both microbial growth and fructan production, including parameters such as pH, temperature, aeration, and medium composition, need to be carefully defined and fine-tuned for each microorganism to achieve enhanced yields. The concentrations of sources of carbon, nitrogen, and other essential nutrients are critical factors influencing both microbial growth and the synthesis of microbial levan. Table 5.2 lists the culture conditions and yield of fructan from different bacteria. As seen in Table 5.2, *B. subtilis* (natto) Takahashi, demonstrated efficiencies ranging from 47.87% to 91.81%, with immobilized whole-cell synthesis proving more efficient. *Pseudomonas fluorescens* showcased a bioefficiency of 54.68% in whole-cell synthesis (Jathore et al. 2012). Erkorkmaz et al. (2022) demonstrated a bioefficiency of 76.60%, under specific conditions, for *Halomonas smyrnensis* AAD6.

5.4.2.2 FOSs Production in Fungal Systems

Many fungal systems have been investigated using fungal-based fermentation systems belonging predominantly to the genera of *Aspergillus*, *Aureobasidium*, *Fusarium*, and *Penicillium* (Table 5.3). The production yield of FOSs, in % g FOS/g sucrose, by different fungal strains varies and appears to be dependent on the method of production and experimental conditions. *Aureobasidium pullulans*, a commonly used microorganism for FOS production, exhibited a yield of 38% when grown at 30 °C for 30 hours with an initial sucrose concentration of 400 g/L (Hayashi et al. 1992). *Aspergillus niger*, on the other hand, demonstrated a wider range, with reported values between 24.0% and 70.0% (Nguyen et al. 1999). For *Au. pullulans*, the yield of FOSs varied from 44.0% (Shin et al. 2004) to 62.0% (Yoshikawa et al. 2008). *Aspergillus japonicus* exhibited varying yields, ranging from 55.8% (Wang and Zhou 2006) to 61.0% (Chiang et al. 1997). The yield of FOSs using *A. oryzae* whole cells varied between 53.0% and 57.4% (Sangeetha et al. 2005). The pH of the medium has been reported to play a crucial role in the growth of the fungal systems and controls the production of the underlying enzymes (for instance, the relevant FTase) that bring about the conversion of sucrose into FOSs. Studies indicate that an initial pH of 5.5 was optimum for FTase production by *A. oryzae* CFR202 (Sangeetha et al. 2005), *A. japonicus* JN19 (Wang and Zhou 2006), and *Penicillium purpurogenum* (Dhake and Patil 2007). *Aspergillus aculeatus* exhibited a yield of 61% when grown at 60 °C for 36 hours with an initial sucrose concentration of 630 g/L (Ghazi et al. 2005). *Au. pullulans* showed a yield of 61% when grown at 25 °C for 12 hours with a lower initial sucrose concentration of 200 g/L (Liang et al. 2021). The yield of FOSs from *A. japonicus* was 61% when grown at 37 °C for 24 hours with an initial sucrose concentration of 400 g/L (Chien et al. 2001), and 64% when grown at 30 °C for 21 hours with an initial sucrose concentration of 200 g/L (Mussatto et al. 2009). In *Au. pullulans*, both immobilization and increased temperature (50 °C) increased the yield of FOSs (Jung et al. 2011). Different strains of *A. niger* showed a yield of 51–55% under different fermentation conditions (Lin and Lee 2008; Zeng et al. 2016; Wang et al. 2021). *Penicillium citrinum* exhibited a yield of 56% when grown at 30 °C for 40 hours with an initial sucrose concentration of 700 g/L (Nobre et al. 2016).

The fermentative production of FOSs is limited due to the accumulation of inhibitory by-products like glucose, which diminish the FOS yield. Continuous removal of glucose, either through nano-filtration or co-fermentation with non-saccharolytic yeast, offers an alternative approach that streamlines downstream processes by obviating the need for enzymatic extraction and purification methods. Nishizawa et al. (2001) demonstrated the efficacy of nanofiltration membranes in eliminating glucose, a key inhibitor of FTase, during FOS production from sucrose. This approach resulted in a remarkable 90% FOS yield, surpassing yields achieved *via* conventional batch production systems (55–60%). Intriguingly, Nobre et al. (2016) outlined an integrated system

Table 5.2 FOS production in various bacterial cells.

Organism	Bio-efficiency (% of the theoretical maximum)	Process	pH, temp (°C), incubation time	Fructan obtained (g/L)	Initial sucrose (g/L)	References
Bacillus polymyxa (NRRL B-18475)	51.06	Whole-cell synthesis	7, 30, 10 d	36	150	Han and Clarke (1990)
Bacillus subtilis (natto) Takahashi	47.87	Whole-cell synthesis	7, 37, 21 h	45	200	Shih et al. (2005)
Bacillus subtilis (natto) Takahashi	91.81	Immobilized whole-cell synthesis	7, 37, 72 h	86.3	200	Shih et al. (2010)
Pseudomonas fluorescens	54.68	Whole-cell synthesis	7, 28, 6 d	15.42	60	Jathore et al. (2012)
Bacillus subtilis (natto) Takahashi	51.06	Whole-cell synthesis	7, 37, 24 h	60	250	Wu et al. (2013)
Gluconacetobacter diazotrophicus	37.23	Whole-cell synthesis	6, 30, 48 h	14	80	Molinari and Boiardi, (2013)
Zymomonas mobilis B-14023	79.43	Immobilized whole-cell synthesis	7, 30, 18 h	112	300	Lorenzetti et al. (2015)
Zymomonas mobilis B-14023	28.37	Immobilized whole-cell synthesis	6, 28, 42 h	40	300	Silbir et al. (2014)
Acetobacter xylinum NCIM2526	44.52	Whole-cell synthesis	6.8, 30, 60 h	12.7	60.7	Srikanth et al. (2015)
Bacillus lentus V8	43.40	Whole-cell synthesis	6.5, 30, 48 h	51	250	Abou-Taleb et al. (2015)
Bacillus licheniformis	52.66	Whole-cell synthesis	7.4, 37, 72 h	99	400	Kekez et al. (2015)
Bacillus atrophaeus	6.21	Whole-cell synthesis	6, 30, 30 h	3.5	120	González-Garcinuño et al. (2017)
Acinetobacter nectaris	5.32	Whole-cell synthesis	6, 30, 40 h	3	120	González-Garcinuño et al. (2017)
Halomonas smyrnensis AAD6	76.60	Whole-cell synthesis	7, 37, 175 h	18	50	Erkorkmaz et al. (2022)

Table 5.3 FOS production in fungal systems.

Organism	Initial sucrose (g/L)	Fermentation conditions	Yield (%)	References
Aureobasidium pullulans	400	30 °C (30 h)	38	Hayashi et al. (1992)
Aureobasidium pullulans	770	50 °C (48 h)	55	Yun and Song (1996)
Aspergillus japonicus	400	37 °C (24 h)	61	Chien et al. (2001)
Aspergillus oryzae	600	55 °C (16 h)	53	Sangeetha et al. (2005)
Aspergillus aculeatus	630	60 °C (36 h)	61	Ghazi et al. (2005)
Aspergillus niger	300	30 °C (18 h)	55	Lin and Lee (2008)
Aspergillus japonicus	200	30 °C (21 h)	64	Mussatto et al. (2009)
Aureobasidium pullulans	770	50 °C (24 h)	23	Jung et al. (2011)
Aspergillus japonicus	300	30 °C (24 h)	62	Sheu et al. (2013)
Aspergillus niger	600	30 °C (40 h)	51	Zeng et al. (2016)
Penicillium citrinum	700	30 °C (40 h)	56	Nobre et al. (2016)
Aureobasidium pullulans	500	30 °C (3 h)	61	Khatun et al. (2020)
Aspergillus niger	500	50 °C (30 h)	51	Wang et al. (2021)
Aureobasidium pullulans	200	25 °C (12 h)	61	Liang et al. (2021)

for FOS production and enrichment through co-culture fermentation involving a FOS-producing strain (*Aspergillus ibericus*) and a yeast strain (*S. cerevisiae* YIL162 W) capable of removing small saccharides. This approach yielded FOS mixtures with a purity of 97% (w/w). Another strategy involves expressing glucose oxidase in *A. niger* mycelia, co-localizing it with FFase. This modification resulted in an increased purity of FOSs, from 52% to 60%. Furthermore, co-expression of peroxidase in the fungal system facilitated the removal of potential inhibitors of glucose oxidase, leading to enhanced FOS yield and purity (Wan et al. 2024).

5.5 Novel Prebiotic Oligosaccharides

Apart from the emerging prebiotics (COSs, HMOs, XOSs) mentioned in Section 5.1, several others have been categorized as novel prebiotics, and these are listed below.

5.5.1 Pectic Oligosaccharides

The oligosaccharides that originate from the polysaccharide pectin are called POSs. They are based on the extension of galacturonic acid (homogalacturonan) or rhamnose (rhamnogalacturonan I). Methyl esterification occurs at the carboxylic groups, and the structure can be acetylated at C2 or C3. Various types of sugars (e.g. arabinose, galactose, and xylose) or ferulic acid are linked to the side chains, and depending on the source of the POSs, their structures are highly variable. The POSs, generated from apple pectin (with different degrees of esterification) were evaluated for their immuno- and bacterio-modulatory properties (Singh et al. 2019) and a significant reduction in the inflammatory response (triggered by the lipopolysaccharides of *E. coli*) was observed.

A promotion in the growth of several beneficial gut bacteria was also seen. Detailed analysis indicated the DP of the POSs to play an important role in the biological activity. These have also been to exhibit several antibacterial, antidiabetic, anticholesterolemic, anti-obesity, antitumor, and antioxidant activities (Gullón et al. 2021).

5.5.2 Resistant Starch

Resistant starch (RS) occurs as one of the components of the dietary fiber, nonstarch carbohydrates (cellulose, hemicellulose, pectin, inulin, and fructans), nondigestible oligosaccharides (FOSs, GOSs), and lignin, being the other components. RS is sourced from whole grains, seeds, legumes, and other starch-containing foods. As per the digestibility of starch, the RS (one of the three classifieds) is not hydrolyzed even after 2 hours of residence in the small intestine (Dupuis and Liu 2019). The other two (rapidly digestible and slowly digestible) are hydrolyzed to dextrins by α-amylase enzyme within 20 minutes and 20–120 minutes, respectively, after consumption. Like other prebiotics, RS is resistant to digestion in the upper gut and promotes the proliferation of several groups of *Firmicutes*. Butyrate is produced as a partial degradation product of RS and can also be metabolized by *B. adolescentis* and *Ruminococcus bromii* and to a lesser extent by *Eubacterium rectale* and *Bacteroides thetaiotaomicron* (Włodarczyk and Śliżewska 2021). The presence of *R. bromii* was found to be essential for the degradation of RS in mixed culture fermentations. Based on several studies conducted, it can be concluded that the commercial RS types, (type II, type III, and type IV) and the RS types II and III (isolated from food) have the prebiotic capacity and can be used as a prebiotic food (Tekin and Dincer 2023).

5.5.3 Polydextrose

Randomly polymerized glucose with some sorbitol end-groups is called polydextrose (PD), and this is marketed as a dietary fiber. It has an average DP of ~12 and the DP can exceed 30. Developed by Pfizer originally, it is a low-calorie bulking agent used in conjunction with intense sweeteners in human food and beverage products around the world. Due to its complex structure, it is only partially digested in the human gut (Fava et al. 2007). This is followed by slow fermentation in the distal part of the colon that gives rise to moderate amounts of fermentation products such as SCFAs and gas (Röytiö and Ouwehand 2014). These further modify the microbiota composition, impact appetite and satiety, increase mineral absorption, and improve gastric function. It has also been shown that the circulating fatty acid concentrations were associated with higher fat oxidation which led to decreased fat storage and thus impact weight. There is high evidence of PD being involved in postprandial benefits, energy intake, and bowel function (do Carmo et al. 2016).

5.5.4 Polyphenols and Flavanols

The polyphenols present in the diet also exhibit significant biological properties that depend on their bioavailability which, in turn, is dependent on their DP. The bioavailability is primarily determined by the gut microbiota, which play a significant role in their production after the consumption of food containing high molecular weight polyphenols (Cardona et al. 2013). There are over 8000 known polyphenols that have been reported from plants, vegetables, and fruits. Many of these reach the colon intact and can be utilized by the gut microbiota (Cardona et al. 2013). Polyphenols derived from cranberry-rich extracts have been shown to stimulate the growth of *Akkermansia muciniphila*, a naturally occurring prebiotic in the human gastrointestinal tract

(Anhê et al. 2015). Antimicrobial action has also been demonstrated by several polyphenolics (Singh et al. 2019). Evidence exists to demonstrate the effect of dietary polyphenols on the modulation of the composition and the activity of the colonic microbiota, although this two-way relationship is still poorly understood.

The flavanols are noncarbohydrate, polyphenolic secondary metabolites that make a subclass under a larger group of plant compounds, the flavonoids. These have been recommended as prebiotics as these stimulate, both *in vitro* and *in vivo*, the growth of LAB (Tzounis et al. 2011). Out of the total polyphenolic compounds present in cocoa beans, 60% is made of epicatechin (with up to 35% of polyphenol content), catechin, and procyanidin B2 (epicatechin-(4β-8)-epicatechin) (all being monomeric), and oligomeric flavanols (Rusconi and Conti 2010). While epicatechin and catechin are absorbed in the small intestine, recirculated to body tissues, and transformed in the liver for generating bioactive substances, a large fraction of the polymeric procyanidins (not absorbed in the upper gastrointestinal) tract can reach the colon and are used by the local microbiota (Jaramillo Flores 2019). These produce several low–low-molecular-weight metabolites that have immunomodulatory effects.

5.5.5 Lactulose

Lactulose is a synthetic disaccharide containing fructose and galactose. It is obtained from lactose *via* an isomerization process in which glucose is converted into fructose, with the sugars galactose and fructose linked together by a β-(1-4) glycosidic linkage. It is found in small amounts in normal milk. Lactulose is primarily metabolized by the LAB, which prefers lactulose as a food source and converts it into SCFAs (Gibson et al. 2017; Karakan et al. 2021). The major product of the fermentation of lactulose is acetate. Currently, it is used at high levels as a detoxifying agent for the treatment of hepatic encephalopathy (Annex 1999). This action is based on its ability to alter the gut microbiota to decrease ammonia production and absorption (Clausen and Mortensen 1997). There is enough preclinical and clinical evidence to support that low-dose lactulose stimulates the growth of health-promoting gut bacteria while inhibiting pathogenic bacterial growth (Karakan et al. 2021). It has shown an overall increase in the production of beneficial metabolites and improvements in mineral absorption with good tolerability. It meets all the markers of a prebiotic.

5.5.6 Human Milk Oligosaccharides

The involvement of nondigestible HMOs in the generation of healthy gut microbiota has been recognized for several decades and can be easily said to have resulted in the emergence of the field of prebiotics. In infants and adults, HMOs have been demonstrated to specifically modulate the intestinal microbiota by way of allowing the proliferation of bifidobacteria and inhibiting the growth of pathogenic bacteria. Utilization of the HMOs has been shown to result in the production of SCFAs, which are considered key molecules for gut barrier and immune function. These are currently under active research and several of these are being produced on a large scale by microbial fermentation (Bych et al. 2019).

5.5.7 Synbiotics

Apart from the prebiotics, a mixture of prebiotics and probiotics has been demonstrated (Śliżewska et al. 2019) to be effective in improving human and animal health. Such mixtures are called synbiotics and in this, the probiotics bacteria utilize the prebiotics and use them for growth. According to the International Scientific Association for Probiotics and Prebiotics (ISAPP), synbiotics are of two types, namely, synergistic, and complementary. As the name suggests, synergistic synbiotics contain a substrate that is utilized by co-administered probiotics while a complementary synbiotic contains a probiotic and a prebiotic that together confer health benefits but are not co-dependent (Swanson et al. 2020).

5.5.8 Mushrooms

In general, mushrooms contain bioactive polysaccharides, such as chitin, mannans, galactans, xylans, glucans, krestin, lentinan, and hemicellulose. These are considered rich sources of fiber, amino acids, antioxidants, and medicinal foods. Several mushroom species like *Agaricus bisporus*, *Ganoderma lucidum*, *Hericium erinaceus*, *Lentinula edodes*, and *Pleurotus ostreatus* have shown prebiotic effects (Kostić et al. 2023), and thus can be repurposed. The mushroom wastes are also considered rich sources of prebiotics.

The polyphenols and other carbohydrates (RS, PD, XOSs, POSs, and the HMOs) that modulate the intestinal microbiome are still looked at as prebiotic candidates and have not met the ISAPP consensus definition (Gibson et al. 2017). The fermentable fibers (such as β-glucans, guar gums, pectins, RSII), vitamins, minerals, and fatty acids themselves are being viewed as overlapping and adjacent to the current boundaries of prebiotics (Cunningham et al. 2021).

5.6 Future Perspectives

With the growing knowledge of how dietary nondigestible molecules (carbohydrates and noncarbohydrate sources) modulate the human gut, there is considerable interest in the identification of novel sources of both microbes and the prebiotics that influence human health. The isolation of these microbes is not likely to be restricted to the gut but to other parts of the human body (such as skin, lungs, and urogenital tracts), fermented foods, and the environment. The prebiotics will also be prepared from novel sources as a focus on sustainability, cost, and scale emerges. These include the 1.3 billion tons of food waste generated annually in the food chain, which represents a rich and sustainable source of natural bioactive ingredients. Side streams emanating from the processing of fruit, vegetables, and grain are also enriched for potential prebiotics. Major potential prebiotics from these waste streams include POSs from orange peel and arabinoxylans from brewing and distillery wastes. It has also been proposed that chemical and/or structural modification of existing prebiotics by way of acidic/enzymatic/oxidation treatments and sonication/applying pressure can be applied to modify functionality. A combination of existing prebiotics in optimized ratios can be tried to create a specific effect (Lam and Cheung 2019). The commercialization of these new and existing prebiotics would mean the standardization of products that require homogeneity and consistency. This would include sourcing these, the extraction protocols, the modifications applied, safety evaluation, and various mixing strategies (Bych et al. 2019). The major disease targets for the administration of prebiotic therapy are asthma, cognition, hypercholesterolemia, liver disease, metabolic disease, mood disorders, obesity, oral health, and subfertility (for review, see Cunningham et al. 2021). Targeted delivery of the prebiotics (or a combination of prebiotics) to the distal part of the colon for the treatment of cancer and selective proliferation of carbohydrate-degrading genera is also expected, which will reduce proteolysis and the generation of harmful metabolites.

References

Abou-Taleb, K.A., Abdel-Monem, M.O., Yassin, M.H., and Draz, A.A. (2015). Production, purification and characterization of levan polymer from *Bacillus lentus* V8 strain. *British Microbiology Research Journal* 5: 22–32. https://doi.org/10.9734/BMRJ/2015/12448.

Andersen, J.M., Barrangou, R., Hachem, M.A. et al. (2011). Transcriptional and functional analysis of galactooligosaccharide uptake by lacS in *Lactobacillus acidophilus*. *Proceedings of the Natural Academic Science USA* 108: 17785–17790. https://doi.org/10.1073/pnas.111415210.

Anhê, F.F., Roy, D., Pilon, G. et al. (2015). A polyphenol-rich cranberry extract protects from diet-induced obesity, insulin resistance and intestinal inflammation in association with increased *Akkermansia* spp. population in the gut microbiota of mice. *Gut* 64: 872–883. https://doi.org/10.1136/gutjnl-2014-307142.

Annex I (1999). Summary of product characteristics. Committee for Proprietary Medicinal Products. The European Public Assessment Report (EPAR). Stocrin. The European Agency for the Evaluation of Medicinal Products. London.

Arnold, J.W., Roach, J., Fabela, S. et al. (2021). The pleiotropic effects of prebiotic galacto-oligosaccharides on the aging gut. *Microbiome* 9: 31. https://doi.org/10.1186/s40168-020-00980-0.

Arnold, J.W., Simpson, J.B., Roach, J. et al. (2018). Prebiotics for lactose intolerance: variability in galacto-oligosaccharide utilization by intestinal *Lactobacillus rhamnosus*. *Nutrients* 10: 1517. https://doi.org/10.3390/nu10101517.

Asadpoor, M., Peeters, C., Henricks, P.A. et al. (2020). Anti-pathogenic functions of non-digestible oligosaccharides in vitro. *Nutrients* 12: 1789. https://doi.org/10.3390/nu12061789.

Barrangou, R., Altermann, E., Hutkins, R. et al. (2003). Functional and comparative genomic analyses of an operon involved in fructooligosaccharide utilization by *Lactobacillus acidophilus*. *Proceedings of the National Academy of Sciences* 100: 8957–8962. https://doi.org/10.1073/pnas.1332765100.

Basavaiah, R. and Gurudutt, P.S. (2021). Prebiotic carbohydrates for therapeutics. *Endocrine, Metabolic & Immune Disorders Drug Targets* 21: 230–245. https://doi.org/10.2174/1871530320666200929140522.

Belmonte-Izquierdo, Y., Salomé-Abarca, L.F., González-Hernández, J.C., and López, M.G. (2023). Fructooligosaccharides (FOS) production by microorganisms with fructosyltransferase activity. *Fermentation* 9: 968. https://doi.org/10.3390/fermentation9110968.

Bosco, N. and Noti, M. (2021). The aging gut microbiome and its impact on host immunity. *Genes and Immunity* 22: 289–303. https://doi.org/10.1038/s41435-021-00126-8.

Bouiche, C., Houfani, A.A., Azzouz, Z. et al. (2023). Exploring the potential of xylooligosaccharides as prebiotics: insights from CAZymes families and their emerging health benefits. *Biologia* 78: 1789–1806. https://doi.org/10.1007/s11756-023-01440-41-18.

Bych, K., Mikš, M.H., Johanson, T. et al. (2019). Production of HMOs using microbial hosts - from cell engineering to large scale production. *Current Opinion in Biotechnology* 56: 130–137. https://doi.org/10.1016/j.copbio.2018.11.003.

Cardona, F., Andrés-Lacueva, C., Tulipani, S. et al. (2013). Benefits of polyphenols on gut microbiota and implications in human health. *The Journal of Nutritional Biochemistry* 24: 1415–1422. https://doi.org/10.1016/j.jnutbio.2013.05.001.

Carvalho, A.F.A., Neto, P.d.O., da Silva, D.F., and Pastore, G.M. (2013). Xylo-oligosaccharides from lignocellulosic materials: chemical structure, health benefits and production by chemical and enzymatic hydrolysis. *Food Research International* 51: 75–85. https://doi.org/10.1016/j.foodres.2012.11.021.

Chambert, R., Rain-Guion, M.C., and Petit-Glatron, M.F. (1992). Readthrough of the *Bacillus subtilis* stop codon produces an extended enzyme displaying a higher polymerase activity. *Biochimica et Biophysica Acta* 1132: 145–153. https://doi.org/10.1016/0167-4781(92)90005-k.

Charoenwongpaiboon, T., Sitthiyotha, T., Na Ayutthaya, P.P. et al. (2019). Modulation of fructooligosaccharide chain length and insight into the product binding motif of *Lactobacillus reuteri* 121 inulosucrase. *Carbohydrate Polymers* 209: 111–121. https://doi.org/10.1016/j.carbpol.2018.12.078.

Chen, P. and Chen, F. (2023). 1-Kestose, the smallest fructooligosaccharide component, protection for mild to moderate ulcerative colitis patients. *Alimentary Pharmacology & Therapeutics* 57: 1349–1350. https://doi.org/10.1111/apt.17420.

Chen, C., Wang, L., Lu, Y. et al. (2019). Comparative transcriptional analysis of *Lactobacillus plantarum* and its ccpA-knockout mutant under galactooligosaccharides and glucose conditions. *Frontiers in Microbiology* 10: 1–12. https://doi.org/10.3389/fmicb.2019.01584.

Chiang, C.J., Lee, W.C., Sheu, D.C., and Duan, K.J. (1997). Immobilization of β-fructofuranosidases from *Aspergillus* on methacrylamide-based polymeric beads for production of fructooligosaccharides. *Biotechnology Progress* 13: 577–582. https://doi.org/10.1021/bp970067z.

Chien, C.S., Lee, W.C., and Lin, T.J. (2001). Immobilization of *Aspergillus japonicus* by entrapping cells in gluten for production of fructooligosaccharides. *Enzyme and Microbial Technology* 29: 252–257. https://doi.org/10.1016/S0141-0229(01)00384-2.

Chockchaisawasdee, S., Athanasopoulos, V.I., Niranjan, K., and Rastall, R.A. (2005). Synthesis of galacto-oligosaccharide from lactose using β-galactosidase from *Kluyveromyces lactis*: studies on batch and continuous UF membrane-fitted bioreactors. *Biotechnology and Bioengineering* 89: 434–443. https://doi.org/10.1002/bit.20357.

Choi, J.Y., Hong, H., Seo, H. et al. (2020). High galacto-oligosaccharide production and a structural model for transgalactosylation of β-galactosidase II from *Bacillus circulans*. *Journal of Agricultural and Food Chemistry* 68: 13806–13814. https://doi.org/10.1021/acs.jafc.0c5871.

Chong, H.Y., Tan, L.T.H., Law, J.W.F. et al. (2022). Exploring the potential of human milk and formula milk on infants' gut and health. *Nutrients* 14: 3554. https://doi.org/10.3390/nu14173554.

Chuankhayan, P., Hsieh, C.Y., Huang, Y.C. et al. (2010). Crystal structures of *Aspergillus japonicus* fructosyltransferase complex with donor/acceptor substrates reveal complete subsites in the active site for catalysis. *The Journal of Biological Chemistry* 285: 23251–23264. https://doi.org/10.1074/jbc.M110.113027.

Clarke, T.B., Davis, K.M., Lysenko, E.S. et al. (2010). Recognition of peptidoglycan from the microbiota by Nod1 enhances systemic innate immunity. *Nature Medicine* 16: 228–231. https://doi.org/10.1038/nm.2087.

Clausen, M.R. and Mortensen, P.B. (1997). Lactulose, disaccharides and colonic flora. Clinical consequences. *Drugs* 53: 930–942. https://doi.org/10.2165/00003495-199753060-00003.

Cunningham, M., Azcarate-Peril, M.A., Barnard, A. et al. (2021). Shaping the future of probiotics and prebiotics. *Trends in Microbiology* 29: 667–685. https://doi.org/10.1016/j.tim.2021.01.003.

de Carvalho, C.C. (2017). Whole cell biocatalysts: essential workers from nature to the industry. *Microbial Biotechnology* 10: 250–263. https://doi.org/10.1111/1751-7915.12363.

De Cosmi, V., Mazzocchi, A., Agostoni, C., and Visioli, F. (2022). Fructooligosaccharides: from breast milk components to potential supplements. A systematic review. *Advances in Nutrition* 13: 318–327. https://doi.org/10.1093/advances/nmab102.

de la Rosa, O., Flores-Gallegos, A.C., Muñíz-Márquez, D. et al. (2022). Successive fermentation of aguamiel and molasses by *Aspergillus oryzae* and *Saccharomyces cerevisiae* to obtain high-purity fructooligosaccharides. *Food* 11: 1786. https://doi.org/10.3390/foods11121786.

De Siqueira, E.C. and Öner, E.T. (2023). Co-production of levan with other high-value bioproducts: a review. *International Journal of Biological Macromolecules* 235: 123800. https://doi.org/10.1016/j.ijbiomac.2023.123800.

Del Moral, S., Olvera, C., Rodriguez, M.E., and Munguia, A.L. (2008). Functional role of the additional domains in inulosucrase (IslA) from *Leuconostoc citreum* CW28. *BMC Biochemistry* 9: 6–15. https://doi.org/10.1186/1471-2091-9-6.

Dhake, A.B. and Patil, M.B. (2007). Effect of substrate feeding on production of fructosyltransferase by *Penicillium purpurogenum*. *Brazilian Journal of Microbiology* 38: 194–199. https://doi.org/10.1590/S1517-83822007000200002.

do Carmo, M.M., Walker, J.C., Novello, D. et al. (2016). Polydextrose: physiological function, and effects on health. *Nutrients* 8: 553. https://doi.org/10.3390/nu8090553.

Dupuis, J.H. and Liu, Q. (2019). Potato starch: a review of physicochemical, functional and nutritional properties. *American Journal of Potato Research* 96: 127–138. https://doi.org/10.1007/s12230-018-09696-2.

Elango, D., Rajendran, K., Van der Laan, L. et al. (2022). Raffinose family oligosaccharides: friend or foe for human and plant health? *Frontiers in Plant Science* 13: 829118. https://doi.org/10.3389/fpls.2022.829118.

Endo, A., Hirano, K., Ose, R. et al. (2020). Impact of kestose supplementation on the healthy adult microbiota in vitro fecal batch cultures. *Anaerobe* 61: 102076. https://doi.org/10.1016/j.anaerobe.2019.102076.

Erkorkmaz, B.A., Kırtel, O., Abaramak, G. et al. (2022). UV and chemically induced *Halomonas smyrnensis* mutants for enhanced levan productivity. *Journal of Biotechnology* 356: 19–29. https://doi.org/10.1016/j.jbiotec.2022.07.005.

Fai, A.E.C., da Silva, J.B., de Andrade, C.J. et al. (2014). Production of prebiotic galactooligosaccharides from lactose by *Pseudozyma tsukubaensis* and *Pichia kluyveri*. *Biocatalysis and Agricultural Biotechnology* 3: 343–350. https://doi.org/10.1016/j.bcab.2014.04.005.

Fan, Y., McMath, A.L., and Donovan, S.M. (2023). Review on the impact of milk oligosaccharides on the brain and neurocognitive development in early life. *Nutrients* 15: 3743. https://doi.org/10.3390/nu15173743.

Faustino, M., Durão, J., Pereira, C.F. et al. (2021). Mannans and mannan oligosaccharides (MOS) from *Saccharomyces cerevisiae* – A sustainable source of functional ingredients. *Carbohydrate Polymers* 272: 118467. https://doi.org/10.1016/j.carbpol.2021.118467.

Fava, F., Mäkivuokko, H., Siljander-Rasi, H. et al. (2007). Effect of polydextrose on intestinal microbes and immune functions in pigs. *British Journal of Nutrition* 98: 123–133. https://doi.org/10.1017/S0007114507691818.

Fehlbaum, S., Prudence, K., Kieboom, J. et al. (2018). In vitro fermentation of selected prebiotics and their effects on the composition and activity of the adult gut microbiota. *International Journal of Molecular Sciences* 19: 3097. https://doi.org/10.3390/ijms19103097.

Flores-Maltos, D.A., Mussatto, S.I., Contreras-Esquivel, J.C. et al. (2016). Biotechnological production and application of fructooligosaccharides. *Critical Reviews in Biotechnology* 36: 259–267. https://doi.org/10.3109/07388551.2014.953443.

Fuhren, J., Schwalbe, M., Peralta-Marzal, L. et al. (2020). Phenotypic and genetic characterization of differential galacto-oligosaccharide utilization in *Lactobacillus plantarum*. *Scientific Reports* 10: 21657. https://doi.org/10.1038/s41598-020-78721-4.

Ganaie, M.A., Rawat, H.K., Wani, O.A. et al. (2014). Immobilization of fructosyltransferase by chitosan and alginate for efficient production of fructooligosaccharides. *Process Biochemistry* 49: 840–844. https://doi.org/10.1016/j.procbio.2014.01.026.

Gao, Y., Liu, Y., Sun, M. et al. (2020). Physiological function analysis of *Lactobacillus plantarum* Y44 based on genotypic and phenotypic characteristics. *Journal of Dairy Science* 103: 5916–5930. https://doi.org/10.3168/jds.2019-18047.

German, J.B., Freeman, S.L., Lebrilla, C.B., and Mills, D.A. (2008). Human milk oligosaccharides: evolution, structures and bioselectivity as substrates for intestinal bacteria. *Nestlé Nutrition Workshop Series. Paediatric Programme* 62: 205–218. discussion 218-22. https://doi.org/10.1159/000146322.

Ghazi, I., De Segura, A.G., Fernández-Arrojo, L. et al. (2005). Immobilisation of fructosyltransferase from *Aspergillus aculeatus* on epoxy-activated Sepabeads EC for the synthesis of fructo-oligosaccharides. *Journal of Molecular Catalysis B: Enzymatic* 35: 19–27. https://doi.org/10.1016/j.molcatb.2005.04.013.

Gibson, G., Hutkins, R., Sanders, M. et al. (2017). Expert consensus document: the International Scientific Association for Probiotics and Prebiotics (ISAPP) consensus statement on the definition and scope of prebiotics. *Nature Reviews. Gastroenterology & Hepatology* 14: 491–502. https://doi.org/10.1038/nrgastro.2017.75.

Gibson, G.R., Rabiu, B., Rycroft, C.E., and Rastall, R.A. (2004). Trans-galactooligosaccharides as prebiotics. In: *Handbook of Functional Dairy Products*, 91–108. Boca Raton: CRC Press https://doi.org/10.5555/20053034541.

Gibson, G.R., Scott, K.P., Rastall, R.A. et al. (2010). Dietary prebiotics: current status and new definition. *Food Science Technology Bulletin Functional Foods* 7: 1–19. https://doi.org/10.1616/1476-2137.15880.

Gobinath, D. and Prapulla, S.G. (2014). Permeabilized probiotic *Lactobacillus plantarum* as a source of β-galactosidase for the synthesis of prebiotic galactooligosaccharides. *Biotechnology Letters* 36: 153–157. https://doi.org/10.1007/s10529-013-1345-9.

González-Garcinuño, Á., Tabernero, A., Sánchez-Álvarez, J.M. et al. (2017). Effect of bacteria type and sucrose concentration on levan yield and its molecular weight. *Microbial Cell Factories* 16: 1–11. https://doi.org/10.1186/s12934-017-0703-z.

Gopal, P.K., Sullivan, P.A., and Smart, J.B. (2001). Utilisation of galactooligosaccharides as selective substrates for growth by lactic acid bacteria including *Bifidobacterium lactis* DR10 and *Lactobacillus rhamnosus* DR20. *International Dairy Journal* 11: 19–25. https://doi.org/10.1016/S0958-6946(01)00026-7.

Gosling, A., Stevens, G.W., Barber, A.R. et al. (2010). Recent advances refining galactooligosaccharide production from lactose. *Food Chemistry* 121: 307–318. https://doi.org/10.1016/j.foodchem.2009.12.063.

Goulas, A., Tzortzis, G., and Gibson, G.R. (2007). Development of a process for the production and purification of α-and β-galactooligosaccharides from *Bifidobacterium bifidum* NCIMB 41171. *International Dairy Journal* 17 (6): 648–656. https://doi.org/10.1016/j.idairyj.2006.08.010.

Gullón, P., del Río, P.G., Gullón, B. et al. (2021). Pectooligosaccharides as emerging functional ingredients: sources, extraction technologies, and biological activities. In: *Sustainable Production Technology in Food* (ed. J.M. Lorenzo, P.E.S. Munekata, and F.J. Barba), 71–92. Academic Press https://doi.org/10.1016/B978-0-12-821233-2.00004-6.

Haghikia, A., Zimmermann, F., Schumann, P. et al. (2022). Propionate attenuates atherosclerosis by immune-dependent regulation of intestinal cholesterol metabolism. *European Heart Journal* 43: 518–533. https://doi.org/10.1093/eurheartj/ehab644.

Hamer, H.M., Jonkers, D., Venema, K. et al. (2008). Review article: the role of butyrate on colonic function. *Alimentary Pharmacology & Therapeutics* 27: 104–119. https://doi.org/10.1111/j.1365-2036.2007.03562.x.

Han, Y.W. and Clarke, M.A. (1990). Production and characterization of microbial levan. *Journal of Agricultural and Food Chemistry* 38: 393–396. https://doi.org/10.1021/jf00092a011.

Harmsen, H.J.M., Gibson, G.R., Elfferich, P. et al. (2000). Comparison of viable cell counts and fluorescence in situ hybridization using specific rRNA-based probes for the quantification of human fecal bacteria. *FEMS Microbiology Letters* 183: 125–129. https://doi.org/10.1111/j.1574-6968.2000.tb08945.x.

Hayashi, S., Hayashi, T., Kinoshita, J. et al. (1992). Immobilization of β-fructofuranosidase from *Aureobasidium* sp. ATCC 20524 on porous silica. *Journal of Industrial Microbiology & Biotechnology* 9: 247–250. https://doi.org/10.1007/BF01569631.

He, J., Zhang, P., Shen, L. et al. (2020). Short-chain fatty acids and their association with signalling pathways in inflammation, glucose and lipid metabolism. *International Journal of Molecular Sciences* 21: 6356. https://doi.org/10.3390/ijms21176356.

Heggset, E.B., Dybvik, A.I., Hoell, I.A. et al. (2010). Degradation of chitosans with a family 46 chitosanase from *Streptomyces coelicolor* A3 (2). *Biomacromolecules* 11: 2487–2497. https://doi.org/10.1021/bm1006745.

Hidaka, H., Tashiro, Y., and Eida, T. (1991). Proliferation of bifidobacteria by oligosaccharides and their useful effect on human health. *Bifidobacteria and Microflora* 10: 65–79. https://doi.org/10.12938/bifidus1982.10.1_65.

Huang, C., Yu, Y., Li, Z. et al. (2022). The preparation technology and application of xylo-oligosaccharide as prebiotics in different fields: a review. *Frontiers in Nutrition* 9: 996811. https://doi.org/10.3389/fnut.2022.996811.

Husmann, F.M., Zimmermann, M.B., and Herter-Aeberli, I. (2022). The effect of prebiotics on human iron absorption: a review. *Advances in Nutrition* 13: 2296–2304. https://doi.org/10.1093/advances/nmac079.

Ibrahim, O.O. (2018). Functional oligo-saccharides: chemicals structure, manufacturing, health benefits, applications and regulations. *Journal Of Food Chemistry And Nanotechnology* 4: 65–76. https://doi.org/10.17756/jfcn.2018-060.

Imamura, L., Hisamitsu, K., and Kobashi, K. (1994). Purification and characterization of β-fructofuranosidase from *Bifidobacterium infantis*. *Biological & Pharmaceutical Bulletin* 17: 596–602. https://doi.org/10.1248/bpb.17.596.

Jadaun, J.S., Narnoliya, L.K., Agarwal, N., and Singh, S.P. (2019). Catalytic biosynthesis of levan and short-chain fructooligosaccharides from sucrose-containing feedstocks by employing the levansucrase from Leuconostoc mesenteroides MTCC10508. *International Journal of Biological Macromolecules* 127: 486–495. https://doi.org/10.1016/j.ijbiomac.2019.01.070.

Jaramillo Flores, M.E. (2019). Cocoa flavanols: natural agents with attenuating effects on metabolic syndrome risk factors. *Nutrients* 11: 751. https://doi.org/10.3390/nu11040751.

Jathore, N.R., Bule, M.V., Tilay, A.V., and Annapure, U.S. (2012). Microbial levan from *Pseudomonas fluorescens*: characterization and medium optimization for enhanced production. *Food Science and Biotechnology* 21: 1045–1053. https://doi.org/10.1007/s10068-012-0136-8.

Juers, D.H., Matthews, B.W., and Huber, R.E. (2012). LacZ β-galactosidase: structure and function of an enzyme of historical and molecular biological importance. *Protein Science* 21: 1792–1807. https://doi.org/10.1002/pro.2165.

Jung, K.H., Bang, S.H., Oh, T.K., and Park, H.J. (2011). Industrial production of fructooligosaccharides by immobilized cells of *Aureobasidium pullulans* in a packed bed reactor. *Biotechnology Letters* 33: 1621–1624. https://doi.org/10.1007/s10529-011-0606-8.

Kanwal, F., Ren, D., Kanwal, W. et al. (2023). The potential role of nondigestible Raffinose family oligosaccharides as prebiotics. *Glycobiology* 33: 274–288. https://doi.org/10.1093/glycob/cwad015.

Karakan, T., Tuohy, K.M., and Janssen-van Solingen, G. (2021). Low-dose lactulose as a prebiotic for improved gut health and enhanced mineral absorption. *Frontiers in Nutrition* 8: 672925. https://doi.org/10.3389/fnut.2021.672925.

Kekez, B.D., Gojgic-Cvijovic, G.D., Jakovljevic, D.M. et al. (2015). High levan production by *Bacillus licheniformis* NS032 using ammonium chloride as the sole nitrogen source. *Applied Biochemistry and Biotechnology* 175: 3068–3083. https://doi.org/10.1007/s12010-015-1475-8.

Kennedy, J.M., De Silva, A., Walton, G.E., and Gibson, G.R. (2023). A review on the use of prebiotics in ulcerative colitis. *Trends in Microbiology* S0966–842x. https://doi.org/10.1016/j.tim.2023.11.007.

Khatun, M.S., Harrison, M.D., Speight, R.E. et al. (2020). Efficient production of fructo-oligosaccharides from sucrose and molasses by a novel *Aureobasidium pullulan* strain. *Biochemical Engineering Journal* 163: 107747. https://doi.org/10.1016/j.bej.2020.107747.

Kittibunchakul, S., van Leeuwen, S.S., Dijkhuizen, L. et al. (2020). Structural comparison of different galacto-oligosaccharide mixtures formed by β-galactosidases from lactic acid bacteria and bifidobacteria. *Journal of Agricultural and Food Chemistry* 68: 4437–4446. https://doi.org/10.1021/acs.jafc.9b08156.

Ko, H., Sung, B.H., Kim, M.J. et al. (2022). Fructan biosynthesis by yeast cell factories. *Journal of Microbiology and Biotechnology* 32: 1373. https://doi.org/10.4014/2Fjmb.2207.07062.

Koc, F., Samli, H., Okur, A. et al. (2010). Effects of *Saccharomyces cerevisiae* and/or mannanoligosaccharide on performance, blood parameters and intestinal microbiota of broiler chicks. *Bulgarian Journal of Agricultural Science* 16: 643–650.

Kostić, M., Ivanov, M., Fernandes, Â. et al. (2023). A comparative study of *Lactarius* mushrooms: chemical characterization, antibacterial, antibiofilm, antioxidant and cytotoxic activity. *Journal of Fungi* 9: 70. https://doi.org/10.3390/jof9010070.

Lam, K.L. and Cheung, P.C. (2019). Carbohydrate-based prebiotics in targeted modulation of gut microbiome. *Journal of Agricultural and Food Chemistry* 67: 12335–12340. https://doi.org/10.1021/acs.jafc.9b04811.

Lammens, W., Le Roy, K., Schroeven, L. et al. (2009). Structural insights into glycoside hydrolase family 32 and 68 enzymes: functional implications. *Journal of Experimental Botany* 60: 727–740. https://doi.org/10.1093/jxb/ern333.

Lemoine, A., Tounian, P., Adel-Patient, K., and Thomas, M. (2023). Pre-, pro-, syn-, and postbiotics in infant formulas: what are the immune benefits for infants? *Nutrients* 15: 1231. https://doi.org/10.3390/nu15051231.

Liang, X., Li, C., Cao, W. et al. (2021). Fermentative production of fructooligosaccharides using *Aureobasidium pullulans*: effect of dissolved oxygen concentration and fermentation mode. *Molecules* 26: 3867. https://doi.org/10.3390/molecules26133867.

Liaqat, F. and Eltem, R. (2018). Chitooligosaccharides and their biological activities: a comprehensive review. *Carbohydrate Polymers* 184: 243–259. https://doi.org/10.1016/j.carbpol.2017.12.067.

Lin, T.J. and Lee, Y.C. (2008). High-content fructooligosaccharides production using two immobilized microorganisms in an internal-loop airlift bioreactor. *Journal of the Chinese Institute of Chemical Engineers* 39: 211–217. https://doi.org/10.1016/j.jcice.2008.01.006.

Lin, B. and Tao, Y. (2017). Whole-cell biocatalysts by design. *Microbial Cell Factories* 16: 106. https://doi.org/10.1186/s12934-017-0724-7.

Lisboa, C.R., de Simoni Martinez, L., Trindade, R.A. et al. (2012). Response surface methodology applied to the enzymatic synthesis of galacto-oligosaccharides from cheese whey. *Food Science and Biotechnology* 21: 1519–1524. https://doi.org/10.1007/s10068-012-0202-2.

Lodhi, G., Kim, Y.S., Hwang, J.W. et al. (2014). Chitooligosaccharide and its derivatives: preparation and biological applications. *BioMed Research International* 2014: 654913. https://doi.org/10.1155/2014/654913.

Lorenzetti, M.F., Moro, M.R., and García-Cruz, C.H. (2015). Alginate/PVA beads for levan production by *Zymomonas mobilis*. *Journal of Food Process Engineering* 38: 31–36. https://doi.org/10.1111/jfpe.12123.
Mahata, M., Shinya, S., Masaki, E. et al. (2014). Production of chitooligosaccharides from *Rhizopus oligosporus* NRRL2710 cells by chitosanase digestion. *Carbohydrate Research* 383: 27–33. https://doi.org/10.1016/j.carres.2013.06.002.
Maráz, A., Kovács, Z., Benjamins, E., and Pázmándi, M. (2022). Recent developments in microbial production of high-purity galacto-oligosaccharides. *World Journal of Microbiology and Biotechnology* 38: 95. https://doi.org/10.1007/s11274-022-03279-4.
McKellar, R.C. and Modler, H.W. (1989). Metabolism of fructo-oligosaccharides by *Bifidobacterium* spp. *Applied Microbiology and Biotechnology* 31: 537–541. https://doi.org/10.1007/BF00270790.
Mei, Z., Yuan, J., and Li, D. (2022). Biological activity of galacto-oligosaccharides: a review. *Frontiers in Microbiology* 13: 993052. https://doi.org/10.3389/fmicb.2022.993052.
Mineo, H., Amano, M., Chiji, H. et al. (2004). Indigestible disaccharides open tight junctions and enhance net calcium, magnesium, and zinc absorption in isolated rat small and large intestinal epithelium. *Digestive Diseases and Sciences* 49: 122–132. https://doi.org/10.1023/B:DDAS.0000011613.11671.c7.
Miniello, V.L., Miniello, A., Ficele, L. et al. (2023). Gut immunobiosis and biomodulators. *Nutrients* 15: 2114. https://doi.org/10.3390/nu15092114.
Molinari, M.L. and Boiardi, J.L. (2013). Levans production by *Gluconacetobacter diazotrophicus*. *Electronic Journal of Biotechnology* 16: 7–17. https://doi.org/10.2225/vol16-issue3-fulltext-9.
Morrison, D.J. and Preston, T. (2016). Formation of short chain fatty acids by the gut microbiota and their impact on human metabolism. *Gut Microbes* 7: 189–200. https://doi.org/10.1080/19490976.2015.1134082.
Mussatto, S.I., Aguilar, C.N., Rodrigues, L.R., and Teixeira, J.A. (2009). Fructooligosaccharides and β-fructofuranosidase production by *Aspergillus japonicus* immobilized on lignocellulosic materials. *Journal of Molecular Catalysis B: Enzymatic* 59: 76–81. https://doi.org/10.1016/j.molcatb.2009.01.005.
Nagaya, M., Kimura, M., Gozu, Y. et al. (2017). Crystal structure of a β-fructofuranosidase with high transfructosylation activity from *Aspergillus kawachii*. *Bioscience, Biotechnology, and Biochemistry* 81: 1786–1795. https://doi.org/10.1080/09168451.2017.1353405.
Nakkharat, P., Kulbe, K.D., Yamabhai, M., and Haltrich, D. (2006). Formation of galacto-oligosaccharides during lactose hydrolysis by a novel β-galactosidase from the moderately thermophilic fungus *Talaromyces thermophilus*. *Biotechnology Journal* 1: 633–638. https://doi.org/10.1002/biot.200600013.
Nguyen, Q.D., Mattes, F., Hoschke, Á. et al. (1999). Production, purification and identification of fructooligosaccharides produced by β-fructofuranosidase from *Aspergillus niger* IMI 303386. *Biotechnology Letters* 21: 183–186. https://doi.org/10.1023/A:1005429525865.
Nie, Q., Chen, H., Hu, J. et al. (2020). Effects of nondigestible oligosaccharides on obesity. *Annual Review of Food Science and Technology* 11: 205–233. https://doi.org/10.1146/annurev-food-032519-051743.
Nishizawa, K., Nakajima, M., and Nabetani, H. (2001). Kinetic study on transfructosylation by β-fructofuranosidase from *Aspergillus niger* ATCC 20611 and availability of a membrane reactor for fructooligosaccharide production. *Food Science and Technology Research* 7: 39–44. https://doi.org/10.3136/fstr.7.39.
Nobre, C., Castro, C.C., Hantson, A.L. et al. (2016). Strategies for the production of high-content fructo-oligosaccharides through the removal of small saccharides by co-culture or successive fermentation with yeast. *Carbohydrate Polymers* 136: 274–281. https://doi.org/10.1016/j.carbpol.2015.08.088.

Nolan, L.S., Rimer, J.M., and Good, M. (2020). The role of human milk oligosaccharides and probiotics on the neonatal microbiome and risk of necrotizing enterocolitis: a narrative review. *Nutrients* 12: 3052. https://doi.org/10.3390/nu12103052.

Ojha, S., Rana, N., and Mishra, S. (2016). Fructo-oligosaccharide synthesis by whole cells of Microbacterium paraoxydans. *Tetrahedron: Asymmetry* 27: 1245–1252. https://doi.org/10.1016/j.tetasy.2016.10.002.

Okburan, G. and Kızıler, S. (2023). Human milk oligosaccharides as prebiotics. *Pediatrics and Neonatology* 64: 231–238. https://doi.org/10.1016/j.pedneo.2022.09.017.

Ortiz-Soto, M.E., Porras-Dominguez, J.R., Seibel, J., and Lopez-Munguia, A. (2019). A close look at the structural features and reaction conditions that modulate the synthesis of low and high molecular weight fructans by levansucrases. *Carbohydrate Polymers* 219: 130–142. https://doi.org/10.1016/j.carbpol.2019.05.014.

Ozimek, L.K., Kralj, S., Van der Maarel, M.J., and Dijkhuizen, L. (2006). The levansucrase and inulosucrase enzymes of *Lactobacillus reuteri* 121 catalyse processive and non-processive transglycosylation reactions. *Microbiologica* 152: 1187–1196. https://doi.org/10.1099/mic.0.28484-0.

Palaniappan, A., Antony, U., and Emmambux, M.N. (2021). Current status of xylooligosaccharides: production, characterization, health benefits and food application. *Trends in Food Science and Technology* 111: 506–519. https://doi.org/10.1016/j.tifs.2021.02.047.

Palframan, R.J., Gibson, G.R., and Rastall, R.A. (2003). Carbohydrate preferences of *Bifidobacterium* species isolated from the human gut. *Current Issues in Intestinal Microbiology* 4: 71–75.

Perrin, S., Warchol, M., Grill, J.P., and Schneider, F. (2001). Fermentations of fructo-oligosaccharides and their components by *Bifidobacterium infantis* ATCC 15697 on batch culture in semi-synthetic medium. *Journal of Applied Microbiology* 90: 859–865. https://doi.org/10.1046/j.1365-2672.2001.01317.x.

Pijning, T., Anwar, M.A., Böger, M. et al. (2011). Crystal structure of inulosucrase from *Lactobacillus*: insights into the substrate specificity and product specificity of GH68 fructansucrases. *Journal of Molecular Biology* 412: 80–93. https://doi.org/10.1016/j.jmb.2011.07.031.

Playne, M.J. and Crittenden, R.G. (2009). Galacto-oligosaccharides and other products derived from lactose. In: *Advanced Dairy Chemistry: Volume 3: Lactose, Water, Salts and Minor Constituents*, 121–201. New York, NY: Springer New York https://doi.org/10.1007/978-0-387-84865-5_5.

Pons, T., Naumoff, D.G., Martínez-Fleites, C., and Hernández, L. (2004). Three acidic residues are at the active site of a β-propeller architecture in glycoside hydrolase families 32, 43, 62, and 68. *Proteins: Structure, Function, and Bioinformatics* 54: 424–432. https://doi.org/10.1002/prot.10604.

Porras-Dominguez, J.R., Avila-Fernandez, A., Miranda-Molina, A. et al. (2015). *Bacillus subtilis* 168 levansucrase (SacB) activity affects average levan molecular weight. *Carbohydrate Polymers* 132: 338–344. https://doi.org/10.1016/j.carbpol.2015.06.056.

Porras-Domínguez, J.R., Ávila-Fernández, Á., Rodríguez-Alegría, M.E. et al. (2014). Levan-type FOS production using a *Bacillus licheniformis* endolevanase. *Process Biochemistry* 49: 783–790. https://doi.org/10.1016/j.procbio.2014.02.005.

Pujari, R. and Banerjee, G. (2021). Impact of prebiotics on immune response: from the bench to the clinic. *Immunology and Cell Biology* 99: 255–273. https://doi.org/10.1111/imcb.12409.

Quinn, E.M., Joshi, L., and Hickey, R.M. (2020). Symposium review: dairy-derived oligosaccharides—their influence on host–microbe interactions in the gastrointestinal tract of infants. *Journal of Dairy Science* 103: 3816–3827. https://doi.org/10.3168/jds.2019-17645.

Rahim, M.A., Saeed, F., Khalid, W. et al. (2021). Functional and nutraceutical properties of fructo-oligosaccharides derivatives: a review. *International Journal of Food Properties* 24: 1588–1602. https://doi.org/10.1080/10942912.2021.1986520.

Rastall, R. A. (2010). Functional oligosaccharides: application and manufacture. *Annual Review of Food Science and Technology* 1: 305–339. https://doi.org/10.1146/annurev.food.080708.100746.

Rathsam, C. and Jacques, N.A. (1998). Role of C-terminal domains in surface attachment of the fructosyltransferase of *Streptococcus salivarius* ATCC 25975. *Journal of Bacteriology* 180: 6400–6403. https://doi.org/10.1128/JB.180.23.6400-6403.1998.

Rattanaprasert, M., van Pijkeren, J.P., Ramer-Tait, A.E. et al. (2019). Genes involved in galactooligosaccharide metabolism in Lactobacillus reuteri and their ecological role in the gastrointestinal tract. *Applied and Environmental Microbiology* 85: e01788–e01719. https://doi.org/10.1128/AEM.01788-19.

Reddy, A. and Maley, F. (1996). Studies on identifying the catalytic role of Glu-204 in the active site of yeast invertase. *The Journal of Biological Chemistry* 271: 13953–13958. https://doi.org/10.1074/jbc.271.24.13953.

Rezende, E.S.V., Lima, G.C., and Naves, M.M.V. (2021). Dietary fibers as beneficial microbiota modulators: a proposed classification by prebiotic categories. *Nutrition* 89: 111217. https://doi.org/10.1016/j.nut.2021.111217.

Roberfroid, M. (2011). Defining functional foods and associated claims. *Journal of Functional Foods* 3–24. https://doi.org/10.1533/9780857092557.1.3.

Rodriguez-Colinas, B., de Abreu, M.A., Fernandez-Arrojo, L. et al. (2011). Production of galacto-oligosaccharides by the β-galactosidase from *Kluyveromyces lactis*: comparative analysis of permeabilized cells versus soluble enzyme. *Journal of Agricultural and Food Chemistry* 59: 10477–10484. https://doi.org/10.1021/jf2022012.

Röytiö, H. and Ouwehand, A.C. (2014). The fermentation of polydextrose in the large intestine and its beneficial effects. *Beneficial Microbes* 5: 305–313. https://doi.org/10.3920/BM2013.0065.

Rusconi, M. and Conti, A. (2010). *Theobroma cacao* L., the Food of the Gods: a scientific approach beyond myths and claims. *Pharmacological Research* 61: 5–13. https://doi.org/10.1016/j.phrs.2009.08.008.

Sabater, C., Fara, A., Palacios, J. et al. (2019). Synthesis of prebiotic galactooligosaccharides from lactose and lactulose by dairy propionibacteria. *Food Microbiology* 77: 93–105. https://doi.org/10.1016/j.fm.2018.08.014.

Salli, K., Anglenius, H., Hirvonen, J. et al. (2019). The effect of 2′-fucosyllactose on simulated infant gut microbiome and metabolites: a pilot study in comparison to GOS and lactose. *Scientific Reports* 9: 13232. https://doi.org/10.1038/s41598-019-49497-z.

Sangeetha, P. T., Ramesh, M. N., & Prapulla, S. G. (2005). Fructooligosaccharide production using fructosyl transferase obtained from recycling culture of *Aspergillus oryzae* CFR 202. *Process Biochemistry* 40: 1085–1088. https://doi.org/10.1016/j.procbio.2004.03.009.

Santos-Moriano, P., Fernandez-Arrojo, L., Poveda, A. et al. (2015). Levan versus fructooligosaccharide synthesis using the levansucrase from *Zymomonas mobilis*: effect of reaction conditions. *Journal of Molecular Catalysis B: Enzymatic* 119: 18–25. https://doi.org/10.1016/j.molcatb.2015.05.011.

Saravanan, R., Shubethar, S., Narayanan, S. et al. (2017). A novel process for the production of high-purity galactooligosaccharides (GOS) using consortium of microbes. *Preparative Biochemistry & Biotechnology* 47: 245–253. https://doi.org/10.1080/10826068.2016.1207082.

Saulnier, D.M.A., Molenaar, D., de Vos, W.M. et al. (2007). Identification of fructooligosaccharide metabolism in *Lactobacillus plantarum* WCSF1 through microarrays. *Applied and Environmental Microbiology* 73: 1753–1765. https://doi.org/10.1128/aem.01151-06.

Schwab, C., Lee, V., Sørensen, K.I., and Gänzle, M.G. (2011). Production of galactooligosaccharides and heterooligosaccharides with disrupted cell extracts and whole cells of lactic acid bacteria and *bifidobacteria*. *International Dairy Journal* 21: 748–754. https://doi.org/10.1016/j.idairyj.2011.04.010.

Sheu, D.C., Chang, J.Y., Wang, C.Y. et al. (2013). Continuous production of high-purity fructooligosaccharides and ethanol by immobilized *Aspergillus japonicus* and *Pichia heimii*. *Bioprocess and Biosystems Engineering* 36: 1745–1751. https://doi.org/10.1007/s00449-013-0949-8.

Shih, L., Chen, L.D., and Wu, J.Y. (2010). Levan production using *Bacillus subtilis* natto cells immobilized on alginate. *Carbohydrate Polymers* 82 (1): 111–117. https://doi.org/10.1016/j.carbpol.2010.04.030.

Shih, I.L., Yu, Y.T., Shieh, C.J., and Hsieh, C.Y. (2005). Selective production and characterization of levan by *Bacillus subtilis* (Natto) Takahashi. *Journal of Agricultural and Food Chemistry* 53: 8211–8215. https://doi.org/10.1021/jf058084o.

Shin, H.T., Baig, S.Y., Lee, S.W. et al. (2004). Production of fructo-oligosaccharides from molasses by *Aureobasidium pullulans* cells. *Bioresource Technology* 93: 59–62. https://doi.org/10.1016/j.biortech.2003.10.008.

Shinde, V.K. and Vamkudoth, K.R. (2022). Maltooligosaccharide forming amylases and their applications in food and pharma industry. *Journal of Food Science and Technology* 59: 3733–3744. https://doi.org/10.1007/s13197-021-05262-7.

Silbir, S., Dagbagli, S., Yegin, S. et al. (2014). Levan production by *Zymomonas mobilis* in batch and continuous fermentation systems. *Carbohydrate Polymers* 99: 454–461. https://doi.org/10.1016/j.carbpol.2013.08.031.

Singh, A.K., Cabral, C., Ramesh Kumar, R. et al. (2019). Beneficial effects of dietary polyphenols on gut microbiota and strategies to improve delivery efficiency. *Nutrients* 11: 2216. https://doi.org/10.3390/nu11092216.

Singh, S.P., Jadaun, J.S., Narnoliya, L.K. et al. (2017). Prebiotic oligosaccharides: special focus on fructooligosaccharides, its biosynthesis and bioactivity. *Applied Biochemistry and Biotechnology* 183 (2): 613–635. https://doi.org/10.1007/s12010-017-2605-2.

Śliżewska, K., Markowiak, P., Żbikowski, A., and Szeleszczuk, P. (2019). Effects of synbiotics on the gut microbiota, blood and rearing parameters of chickens. *FEMS Microbiology Letters* 366: fnz116. https://doi.org/10.1093/femsle/fnz116.

Son, M.Y. and Cho, H.S. (2023). Anticancer effects of gut microbiota-derived short-chain fatty acids in cancers. *Journal of Microbiology and Biotechnology* 33: 849–856. https://doi.org/10.4014/jmb.2301.01031.

Srikanth, R., Siddartha, G., Reddy, C.H.S. et al. (2015). Antioxidant and anti-inflammatory levan produced from *Acetobacter xylinum* NCIM2526 and its statistical optimization. *Carbohydrate Polymers* 123: 8–16. https://doi.org/10.1016/j.carbpol.2014.12.079.

Srivastava, A. and Mishra, S. (2019). Enrichment and evaluation of galacto-oligosaccharides produced by whole cell treatment of sugar reaction mixture. *Molecular Biology Reports* 46: 1181–1188. https://doi.org/10.1007/s11033-019-04585-1.

Srivastava, A., Mishra, S., and Chand, S. (2015). Transgalactosylation of lactose for synthesis of galacto-oligosaccharides using *Kluyveromyces marxianus* NCIM 3551. *New Biotechnology* 32: 412–418. https://doi.org/10.1016/j.nbt.2015.04.004.

Stinson, L.F., Payne, M.S., and Keelan, J.A. (2017). Planting the seed: origins, composition, and postnatal health significance of the fetal gastrointestinal microbiota. *Critical Reviews in Microbiology* 43: 352–369. https://doi.org/10.1080/1040841X.2016.1211088.

Strube, C.P., Homann, A., Gamer, M. et al. (2011). Polysaccharide synthesis of the levansucrase SacB from *Bacillus megaterium* is controlled by distinct surface motifs. *The Journal of Biological Chemistry* 286: 17593–17600. https://doi.org/10.1074/jbc.M110.203166.

Suzuki, N., Aiba, Y., Takeda, H. et al. (2006). Superiority of 1-kestose, the smallest fructooligosaccharide, to a synthetic mixture of fructooligosaccharides in the selective stimulating activity on bifidobacteria. *Bioscience & Microflora* 25: 109–116. https://doi.org/10.12938/bifidus.25.109.

Swanson, K.S., Gibson, G.R., Hutkins, R. et al. (2020). The International Scientific Association for Probiotics and Prebiotics (ISAPP) consensus statement on the definition and scope of synbiotics. *Nature Reviews. Gastroenterology & Hepatology* 17: 687–701. https://doi.org/10.1038/s41575-020-0344-2.

Tajadadi-Ebrahimi, M., Bahmani, F., Shakeri, H. et al. (2014). Effects of daily consumption of synbiotic bread on insulin metabolism and serum high-sensitivity C-reactive protein among diabetic patients: a double-blind, randomized, controlled clinical trial. *Annals of Nutrition & Metabolism* 65: 34–41. https://doi.org/10.1159/000365153.

Tanaka, R., Takayama, H., Morotomi, M. et al. (1983). Effects of administration of TOS and *Bifidobacterium breve* 4006 on the human fecal flora. *Bifidobacteria Microflora* 2: 17–24. https://doi.org/10.12938/bifidus1982.2.1_17.

Tekin, T. and Dincer, E. (2023). Effect of resistant starch types as a prebiotic. *Applied Microbiology and Biotechnology* 107: 491–515. https://doi.org/10.1007/s00253-022-12325-y.

Teng, P.-Y. and Kim, W.K. (2018). Review: roles of prebiotics in intestinal ecosystem of broilers. *Front Veternary Sci* 5: 245. https://doi.org/10.3389/fvets.2018.002.

Tochio, T., Kadota, Y., Tanaka, T., and Koga, Y. (2018). 1-Kestose, the smallest fructooligosaccharide component, which efficiently stimulates *Faecalibacterium prausnitzii* as well as Bifidobacteria in humans. *Food* 7: 140. https://doi.org/10.3390/foods7090140.

Tonozuka, T., Kitamura, J., Nagaya, M. et al. (2020). Crystal structure of a glycoside hydrolase family 68 β-fructosyltransferase from *Beijerinckia indica* subsp. indica in complex with fructose. *Bioscience, Biotechnology, and Biochemistry* 84: 2508–2520. https://doi.org/10.1080/09168451.2020.1804317.

Tzortzis, G., Goulas, A.K., and Gibson, G.R. (2005). Synthesis of prebiotic galactooligosaccharides using whole cells of a novel strain, *Bifidobacterium bifidum* NCIMB 41171. *Applied Microbiology and Biotechnology* 68: 412–416. https://doi.org/10.1007/s00253-005-1919-0.

Tzounis, X., Rodriguez-Mateos, A., Vulevic, J. et al. (2011). Prebiotic evaluation of cocoa-derived flavanols in healthy humans by using a randomized, controlled, double-blind, crossover intervention study. *The American Journal of Clinical Nutrition* 93: 62–72. https://doi.org/10.3945/ajcn.110.000075.

Van den Ende, W. (2013). Multifunctional fructans and raffinose family oligosaccharides. *Frontiers in Plant Science* 4: 247. https://doi.org/10.3389/fpls.2013.00247.

van Hijum, S.A., Kralj, S., Ozimek, L.K. et al. (2006). Structure-function relationships of glucansucrase and fructansucrase enzymes from lactic acid bacteria. *Microbiology and Molecular Biology Reviews* 70: 157–176. https://doi.org/10.1128/mmbr.70.1.157-176.2006.

Vulevic, J., Juric, A., Tzortzis, G., and Gibson, G.R. (2013). A mixture of trans-galactooligosaccharides reduces markers of metabolic syndrome and modulates the fecal microbiota and immune function of overweight adults. *The Journal of Nutrition* 143: 324–331. https://doi.org/10.3945/jn.112.166132.

Vulevic, J., Juric, A., Walton, G.E. et al. (2015). Influence of galacto-oligosaccharide mixture (B-GOS) on gut microbiota, immune parameters and metabonomics in elderly persons. *The British Journal of Nutrition* 114: 586–595. https://doi.org/10.1017/S0007114515001889.

Walter, J., Schwab, C., Loach, D.M. et al. (2008). Glucosyltransferase A (GtfA) and inulosucrase (Inu) of *Lactobacillus reuteri* TMW1.106 contribute to cell aggregation, in vitro biofilm formation, and colonization of the mouse gastrointestinal tract. *Microbiol (Reading)* 154: 72–80. https://doi.org/10.1099/mic.0.2007/010637-0.

Wan, X., Wang, L., Chang, J. et al. (2024). Effective synthesis of high-content fructooligosaccharides in engineered *Aspergillus niger*. *Microbial Cell Factories* 23: 76. https://doi.org/10.1186/s12934-024-02353-w.

Wang, K., Duan, F., Sun, T. et al. (2024). Galactooligosaccharides: synthesis, metabolism, bioactivities and food applications. *Critical Reviews in Food Science and Nutrition* 64 (18): 6060–6176. https://doi.org/10.1080/10408398.2022.2164244.

Wang, J., Zhang, J., Wang, L. et al. (2021). Continuous production of fructooligosaccharides by recycling of the thermal-stable β-fructofuranosidase produced by *Aspergillus niger*. *Biotechnology Letters* 43: 1175–1182. https://doi.org/10.1007/s10529-021-03099-w.

Wang, L.M. and Zhou, H.M. (2006). Isolation and identification of a novel *Aspergillus japonicus* JN19 producing β-fructofuranosidase and characterization of the enzyme. *Journal of Food Biochemistry* 30: 641–658. https://doi.org/10.1111/j.1745-4514.2006.00085.x.

Wenzel, T.J., Gates, E.J., Ranger, A.L., and Klegeris, A. (2020). Short-chain fatty acids (SCFAs) alone or in combination regulate select immune functions of microglia-like cells. *Molecular and Cellular Neurosciences* 105: 103493. https://doi.org/10.1016/j.mcn.2020.103493.

Włodarczyk, M. and Śliżewska, K. (2021). Efficiency of resistant starch and dextrins as prebiotics: a review of the existing evidence and clinical trials. *Nutrients* 13: 3808. https://doi.org/10.3390/nu13113808.

Wolf, A.and Underhill, D. (2018). Peptidoglycan recognition by the innate immune system. *Nature Reviews. Immunology* 18: 243–254. https://doi.org/10.1038/nri.2017.136.

Wu, F.C., Chou, S.Z., and Shih, L. (2013). Factors affecting the production and molecular weight of levan of *Bacillus subtilis* natto in batch and fed-batch culture in fermenter. *Journal of the Taiwan Institute of Chemical Engineers* 44: 846–853. https://doi.org/10.1016/j.jtice.2013.03.009.

Yañez-Ñeco, C.V., Cervantes, F.V., Amaya-Delgado, L. et al. (2021). Synthesis of β(1 → 3) and β(1 → 6) galactooligosaccharides from lactose and whey using a recombinant β-galactosidase from *Pantoea anthophila*. *Electronic Journal of Biotechnology* 49: 14–21. https://doi.org/10.1016/j.ejbt.2020.10.004.

Yong, J.G. and Klaenhammer, T.R. (2015). Genetic mechanisms of prebiotic oligosaccharide metabolism in probiotic microbes. *Annual Review of Food Science and Technology* 6: 137–156. https://doi.org/10.1146/annurev-food-022814-015706.

Yoshikawa, J., Amachi, S., Shinoyama, H., and Fujii, T. (2008). Production of fructooligosaccharides by crude enzyme preparations of β-fructofuranosidase from *Aureobasidium pullulans*. *Biotechnology Letters* 30: 535–539. https://doi.org/10.1007/s10529-007-9568-2.

Yu, L. and O'Sullivan, D.J. (2014). Production of galactooligosaccharides using a hyperthermophilic β-galactosidase in permeabilized whole cells of *Lactococcus lactis*. *Journal of Dairy Science* 97: 694–703. https://doi.org/10.3168/jds.2013-7492.

Yu, L. and O'Sullivan, D.J. (2018). Immobilization of whole cells of *Lactococcus lactis* containing high levels of a hyperthermostable β-galactosidase enzyme in chitosan beads for efficient galacto-oligosaccharide production. *Journal of Dairy Science* 101: 2974–2983. https://doi.org/10.3168/jds.2017-13770.

Yun, J.W. and Song, S.K. (1996). Continuous production of fructooligosaccharides using fructosyltransferase immobilized on ion exchange resin. *Biotechnology and Bioprocess Engineering* 1: 18–21. https://doi.org/10.1007/BF02949138.

Zartl, B., Silberbauer, K., Loeppert, R. et al. (2018). Fermentation of non-digestible raffinose family oligosaccharides and galactomannans by probiotics. *Food & Function* 9: 1638–1646. https://doi.org/10.1039/c7fo01887h.

Zeng, M., van Pijkeren, J.P., and Pan, X. (2023). Gluco-oligosaccharides as potential prebiotics: synthesis, purification, structural characterization, and evaluation of prebiotic effect. *Comprehensive Reviews in Food Science* 22: 2611–2651. https://doi.org/10.1111/1541-4337.13156.

Zeng, X.A., Zhou, K., Liu, D.M. et al. (2016). Preparation of fructooligosaccharides using *Aspergillus niger* 6640 whole-cell as catalyst for bio-transformation. *LWT-Food Science and Technology* 65: 1072–1079. https://doi.org/10.1016/j.lwt.2015.09.031.

Zhang, J., Li, L., Gu, S. et al. (2022). Characterization of a novel fructosyltransferase from *Lactobacillus crispatus*, InuCA, that attaches to the cell surface by electrostatic interaction. *Applied and Environmental Microbiology* 88: e02399–e02321. https://doi.org/10.1128/aem.02399-21.

Zhang, X., Yao, C., Wang, T. et al. (2021). Production of high-purity galacto-oligosaccharides (GOS) by *Lactobacillus*-derived β-galactosidase. *European Food Research and Technology* 247: 1501–1510. https://doi.org/10.1007/s00217-021-03727-9.

Zúñiga, M., Yebra, M.J., and Monedero, V. (2021). Complex oligosaccharide utilization pathways in *Lactobacillus*. *Current Issues in Molecular Biology* 40: 49–80. https://doi.org/10.21775/cimb.040.049.

6

Bio-production of Rare Sugars, Applications, Safety, and Health Benefits

Sweety Sharma[1], Satya Narayan Patel[2], Suresh D. Pillai[2], Jyoti Yadav[3], and Sudhir Pratap Singh[3]

[1] *Indian Institute of Science Education and Research Mohali, SAS Nagar, Punjab, 140306, India*
[2] *Department of Food Science and Technology, Texas A&M University, College Station, Texas, 77843, United States*
[3] *Department of Industrial Biotechnology, Gujarat Biotechnology University, GIFT City, Gandhinagar, Gujarat, 382355, India*

6.1 Introduction

In recent years, rare sugars have attracted significant attention due to their unique biochemical and functional properties. These molecules occur in very limited quantities in nature, exhibiting some excellent properties, making them highly desirable in the food, pharmaceutical, and cosmetics industries (Xu et al. 2022). Due to increasing demands for commercialization, these molecules must be produced at a large scale. With the advancements in protein engineering, gene editing, and fermentation technologies, microbial production has emerged as an efficient and sustainable method for scaling up the production of these sugars (Mijailovic et al. 2021). Some of the essential examples of rare sugars include D-allulose, D-allose, D-tagatose, D-talose, turanose, and trehalose. Each of these functional molecules exhibits unique properties such as antidiabetic, antioxidant, and prebiotic, making them attractive to a large population of health-conscious people and several industries (Li et al. 2024).

Increasing reports of these molecules in food and health benefits extending from low-calorie sweeteners to therapeutic agents make them highly desirable in the near future (Ahmed et al. 2022; Smith et al. 2022). Along with improvements in microbial production, the assignment of generally recognized as safe (GRAS) status to some of these molecules like D-allulose, D-tagatose, and trehalose provides proof for the safety of their consumption (Hossain et al. 2015; Roy et al. 2018). The present review briefly summarizes the microbial production, applications, and safety along with the key benefits of rare sugars of high importance, including D-allulose, D-allose, D-tagatose, D-talose, turanose, and trehalose (Figure 6.1). This study becomes significant for assessing these sugars, which hold immense potential as a substitute for table sugar and potent bioactive molecules.

6.2 D-Allulose

D-allulose, also known as D-psicose, falls into the category of simple sugars. It is a rare ketohexose, an epimer of D-fructose with molecular formula $C_6H_{12}O_6$ and molecular weight 180.16. The only difference between D-allulose from D-fructose is the paired structure between C2 and C3 positions.

Microbial Nutraceuticals: Products and Processes, First Edition. Edited by Sudhir Pratap Singh and Santosh Kumar Upadhyay.

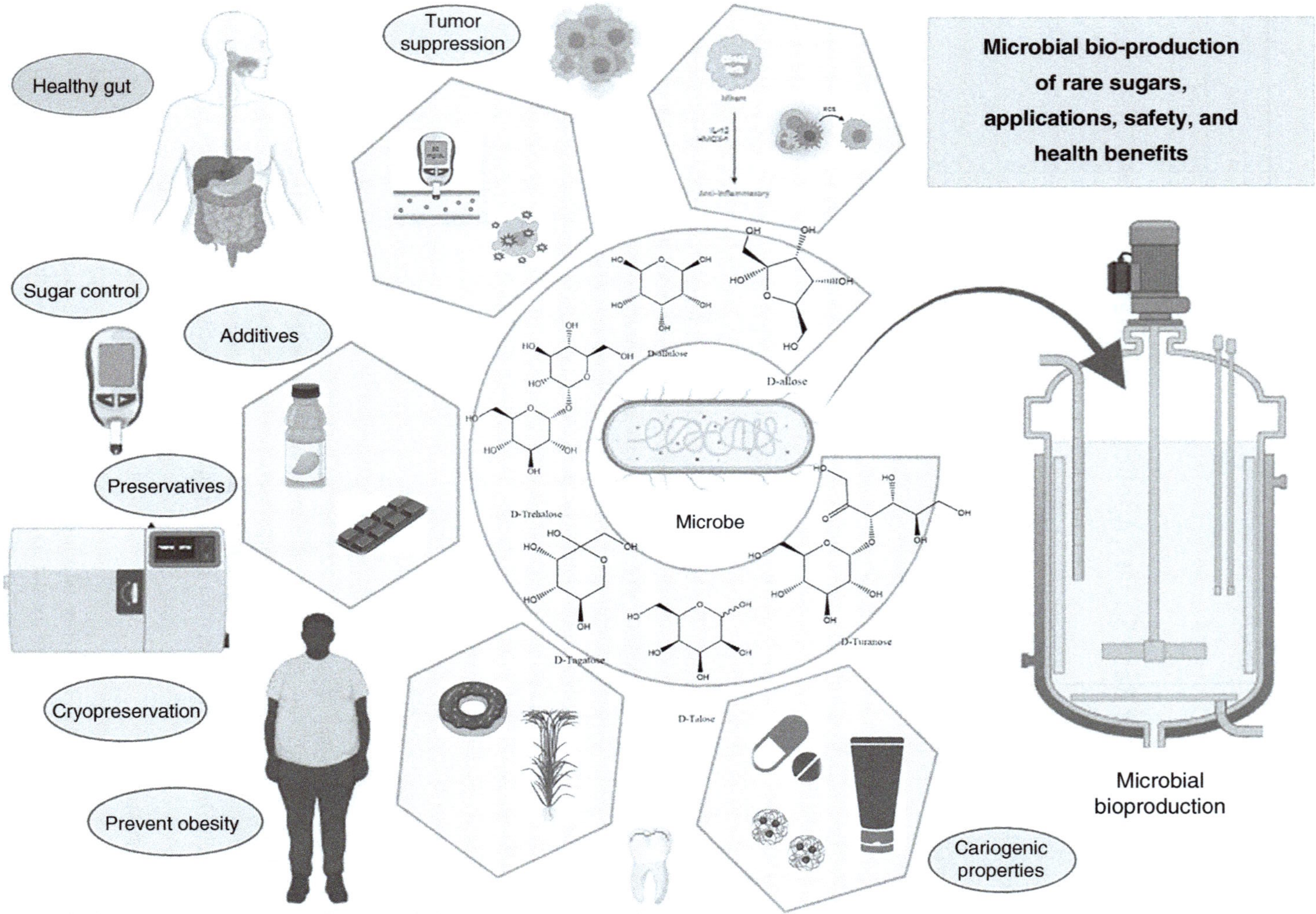

Figure 6.1 Bio-production of rare sugars, applications, safety, and health benefits.

It is the most studied molecule among the various classes of characterized rare sugars. D-allulose is a low-calorie monosaccharide available in very small amounts in natural products such as *Itea* plant, wheat, processed beet, and cane molasses (Jiang et al. 2020; Patra et al. 2017). Based on its physical and biochemical properties, it retains 70% of the sweetness of sucrose, but unlike common sugars, it possesses nearly zero calories and low glycemic values (Au-Yeung et al. 2023; Patel et al. 2016; Patel et al. 2023). D-allulose appears as a white crystalline powder, exhibits low hygroscopicity and high water solubility, and is odorless. The US Food and Drug Administration (FDA) has approved the GRAS status to D-allulose (Zhang et al. 2017). It exhibits almost similar texture and taste compared to other dietary sugars, but its properties that stand out are its nearly zero calorie value and low glycemic values, which are the biggest concern for people globally living in a modern lifestyle (Granström et al. 2004). This makes D-allulose a commodity of valuable importance for its application in food, pharmaceutical, and healthcare science.

D-allulose can be synthesized through both chemical and biological routes. The chemical synthesis mainly starts with 1,2:4,5-di-O-isopropylidene-β-D-fructopyranose, molybdate ion, and pyridine as a chemo-catalysts. However, the production of numerous toxic by-products in the intermediate steps makes the final product purification difficult and expensive (Chen et al. 2021a; Patel et al. 2021; Bilik and Tihlarik 1973; Doner 1979). Most importantly, its chemical synthesis produces hazardous by-products, and therefore, this process is not environmentally friendly. Comparatively, the biological production of D-allulose is more efficient, with mild enzymatic reaction conditions and no by-product generation, which makes it a more sustainable process. Mainly four types of ketose 3-epimerase are known for the biosynthesis of D-allulose, including D-tagatose 3-epimerase, L-ribulose 3-epimerase, D-fructose 3-epimerase, and D-allulose 3-epimerase, which have been characterized from various microbial sources. D-allulose 3-epimerase (DAEase) displays relatively higher specificity towards D-allulose, which is the reason researchers are using DAEase for the biosynthesis of D-allulose (Chen et al. 2021a; Patel et al. 2018; Patel et al. 2021). The first DAEase was reported from the *Agrobacterium tumefaciens*. Promising progress has been noted in the biosynthesis of D-allulose from D-fructose in the last 15 years. Till now, more than 30 DAEases have been identified and characterized from various microbial sources. Some of them are from *Pseudomonas cichorii, Christensenella minuta, Clostridium cellulolyticum, Ruminococcus* sp., *Clostridium scindens, Clostridium bolteae, Caballeronia fortuita, A. tumefaciens, Clostridium* sp., *Desmospora* sp., *Treponema primitia, Staphylococcus aureus, Thermoclostridium caenicola, Pirellula* sp., *Rhodopirellula baltica, Dorea* sp., *Halanaerobium congolense, Flavonifractor plautii*, DaeM (from metagenome), *Bacillus* sp. (DaeB), *Rhodobacter sphaeroides, Christensenellaceae bacterium, Blautia produca*, and *Ruminiclostridium papyrosolvens* (Table 6.1). The temperature and pH activities of ketose epimerases are reported in the range of 40–88 °C and 6–9 pH ranges (Table 6.1). D-allulose 3-epimerase from the DaeM (from metagenome) showed higher thermal stability at 70 °C (3240 minutes) compared to the other reported D-allulose 3-epimerases (Patel et al. 2020). Another enzyme, DaeB, exhibits excessive stability at 50 °C (Patel et al. 2021). Dual enzyme catalysis, involving xylose isomerase and D-allulose 3-epimerase, can be adopted for the production of D-allulose from D-glucose (Patel et al. 2025).

6.2.1 Physiological Functions and Health Benefits

Many studies in the past have shown various physiological functions of D-allulose, namely anti-obesity, antidiabetic, anti-hyperlipidemic, anti-inflammatory, and antioxidative effects. D-allulose is recommended as a natural ingredient with healthcare attributes (Shintani et al. 2017; Franchi et al. 2021). D-allulose offers potential application as a prebiotic additive to develop new synbiotic preparation and monitor diet-induced obesity and lipid metabolism (ÖZGÜR et al. 2022).

Table 6.1 Biochemical properties of ketose 3-epimerases reported from microbial sources for D-allulose biosynthesis.

Microbial sources	Optimum pH	Optimum temperature (°C)	Half-life (minute)	Bioconversion (%)	References
Pseudomonas cichorii	7.5	60	NR	20:80 (30 °C)	Ishida et al. (1997)
Christensenella minuta	6.0	50	40 (50 °C)	30:70	Wang et al. (2020)
Clostridium cellulolyticum	8.0	55	408 (60 °C)	32:68	Mu et al. (2011)
Ruminococcus sp.	7.5–8.0	60	~ 96 (60 °C)	28:72	Zhu et al. (2012)
Clostridium scindens	7.5	60	50 (60 °C) 108 (50 °C)	28:72	Zhang et al. (2013)
Clostridium bolteae	7.0	55	156 (55 °C)	32:68 (60 °C)	Jia et al. (2014)
Caballeronia fortuita	7.5	65	63 (60 °C) 307 (55°C)	28.1:71.9	Li et al. (2019)
Agrobacterium tumefaciens	8.0	50	8.9 (55°C) 64 (50 °C)	32:68 (30 °C)	Kim et al. (2006)
Clostridium sp.	8.0	65	15 (60 °C)	28:72 (65 °C)	Mu et al. (2013)
Desmospora sp.	7.5	60	120 (50 °C)	30:70	Zhang et al. (2013)
Treponema primitia	8.0	70	30 (50 °C)	28:72	Zhang et al. (2016)
Staphylococcus aureus	7.5	70	120 (70°C)	38.9:61.1	Zhu et al. (2019)
Thermoclostridium caenicola	7.5	65	816 (50 °C)	28:72	Chen et al. 2021
Rhodopirellula baltica	8.0	60	52 (60 °C)	28.6:61.4	Mao et al. (2020)
Pirellula sp.	7.5	60	360 (60 °C)	30:70	Li et al. (2021)
Dorea sp.	6.0	70	30 (60 °C)	30:70 (70 °C)	Zhang et al. (2015)
Halanaerobium congolense	8.0	70	66 (70 °C)	35:65	Zhu et al. (2021)
Flavonifractor plautii	7.0	65	40 (65 °C)	31:69	Park et al. (2016a)
DaeM (from metagenome)	7.0	80	9900 (60 °C) 3240 (70 °C) 49 (80 °C)	31:69	Patel et al. (2020)
Bacillus sp. (DaeB)	8.0	55	36,000 (50 °C) 1320 (55 °C)	28.5:71.5	Patel et al. (2021)
Rhodobacter sphaeroides	9.0	40	60 (50 °C)	23:77 (40 °C)	Zhang et al. (2009)
Christensenellaceae bacterium	7.5	55	150 (55 °C)	31:69	Guan et al. (2024)
Blautia produca	8	55	180 (55 °C)	30:70	Tang et al. (2022)
Ruminiclostridium papyrosolvens	7.5	60	NR	29.9:70.1	Yang et al. (2022)

6.2.1.1 Anti-obesity and Antidiabetic Effects

Obesity-related type-2 diabetes is the main risk factor associated with mortality globally. The zero-calorie sucrose substitute, D-allulose, has shown suppressive levels of postprandial plasma glucose with anti-obesity and antidiabetic effects during animal and clinical trials in Asian and American population (Matsuo et al. 2002; Matsuo and Izumori 2009; Hayashi et al. 2010; Braunstein et al. 2018; Franchi et al. 2021).

6.2.1.2 Anti-hyperlipidemic Effects

The biological activities of D-allulose have been focused around reducing the postprandial glucose release in blood. However, it is well known that D-allulose also regulates the enzymes involved in lipid metabolism, thereby reducing hyperaccumulation of abdominal fat tissue and altering cholesterol metabolism (Han et al. 2016; Shintani et al. 2017; Do et al. 2019).

6.2.1.3 Anti-inflammatory and Antioxidative Effects

D-allulose acts on suppressing the pro-inflammatory cytokines' (such as TNF-α, monocytes, and interleukin-6) release in serum, thereby shows anti-inflammatory properties (Moller and Berger 2003; Kim et al. 2017). It also has antioxidant effects protecting 6-hydroxydopamine-induced apoptosis or testicular injury, by scavenging reactive oxygen species (ROS) (Takata et al. 2005; Suna et al. 2007). The monosaccharide can also extend lifespan by increasing superoxide dismutase (SOD) and catalase (CAT) activity (Shintani et al. 2017).

It is noteworthy that the above-mentioned physiological functions of D-allulose make it a potential pharmaceutical agent for the treatment of obesity-related diseases. Meanwhile, it has been called "relatively harmless" according to the toxicity rating chart, so its consumption is considered safe (Mu et al. 2011). It also holds promising market potential in the food and beverage industry worldwide. As it is a zero-calorie sugar with reducing properties, it helps to improve the textural properties of processed products through Maillard reaction (Sun et al. 2008). The molecule also shows potential in fermentation and dairy industry as it is known for regulating the lactic acid bacteria production, therefore regulating acid production of over-fermented milk and can be used for the development of probiotic products from dairy substrates (Kimoto-Nira et al. 2017).

6.3 D-Allose

D-Allose ($C_6H_{12}O_6$) is a rare sugar that is present in limited quantities in nature (Chen et al. 2018a). It is an aldohexose with the opposite configuration of the hydroxyl group at the C3 position of D-glucose and an aldose-ketose isomer of D-allulose. It is a white crystalline powder, is highly soluble in water, and can cyclize to form either a five-membered furanose ring or a six-membered pyranose ring (more common). It consists of 80% sweetness of sucrose with almost zero calories and is non-toxic (Xu et al. 2017). D-Allose has diverse physiological functions, including potent anticancer and antitumor effects, which have attracted keen interest in its potential clinical applications. It has shown efficacy against multiple cancer cell lines and is also reported to enhance the effects of traditional cancer treatments like radiotherapy and chemotherapy (Samukawa et al. 2024). Other excellent physiological functions shown by the D-allose molecule include antioxidant, anti-inflammatory, anti-hypertensive, and immunosuppressive properties, and it has proven to be active against ischemia-reperfusion injury of the liver (Morimoto et al. 2006; Tang et al. 2024).

D-allose has been identified in some plant systems like *Protea rubropilosa* beard (Perold et al. 1973), *Veronica filiformis* (Chari et al. 1981), *Mentzelia* (Jensen et al. 1981), and potato leaves

(Weckwerth et al. 2004), and some other plant sources like *Halodule pinifolia* with a yield of 3.67% (Kannan et al. 2012). However, the limited availability of D-allose in nature makes its extraction non-economical. On the other hand, many difficulties are associated with its chemical synthesis, such as complicated steps, low transformation rate, low selectivity, by-product formation, and chemical pollution (Choi et al. 2021). Henceforth, biological methods, including enzymatic conversion pathways, are gaining more attention for large-scale production. These pathways offer advantages over chemical methods, including higher specificity, improved activity under mild conditions, and reduced by-product formation (Chen et al. 2021b; Lim and Oh 2011).

D-Allose can be synthesized from D-allulose by one-step conversion using various aldose-ketose isomerases, including L-rhamnose isomerase (EC 5.3.1.14), galactose-6-phosphate isomerase (EC 5.3.1.26), glucose-6-phosphate isomerase, and ribose-5-phosphate isomerase (EC 5.3.1.6) (Chen et al. 2018b, 2021b) (Table 6.2). Different microbial sources have been harnessed for the production of D-allose. Some of these include *Pseudomonas stutzeri*, *Bacillus pallidus* Y25, *Thermoanaerobacterium saccharolyticum* NTOU1, *Caldicellulosiruptor saccharolyticus* ATCC 43494, *Bacillus subtilis* WB600, *Thermobacillus composti* KWC4, *Caldicellulosiruptor obsidiansis* OB47, *Clostridium stercorarium* ATCC 35414, *Shinella zoogloeoides* NN6, *Clostridium thermocellum*, *Thermotoga lettingae* TMO, *Lactococcus lactis*, *Chloroflexus islandicus*, and *Pyrococcus furiosus*. Based on physiochemical properties, L-rhamnose isomerase encoded by a gene from hot-spring metagenome outperforms other characterized genes as it shows the highest D-allulose to D-allose conversion, without any by-product. Also, it works well at acidic to neutral pH and high temperature with one of the highest temperature tolerance (Table 6.2).

However, due to the high cost of substrate, that is, D-allulose, which is another rare sugar, there is a growing interest in developing more efficient production techniques using one-pot reaction methods to produce D-allose from D-fructose by employing specific enzymes (Tang et al. 2024). In this technique, first, D-fructose is converted to D-allulose by D-allulose-3-epimerase, and then, a second enzyme, L-rhamnose isomerase, converts the synthesized D-allulose into D-allose. Another study reports a one-pot synthesis of D-allose by employing D-allulose-3-epimerase and ribose-5-phosphate isomerase (Lee et al. 2018). Moreover, there is a study where three enzymes, namely D-glucose isomerase (DGI), D-allulose 3-epimerase (DAE), and ribose-5-phosphate isomerase (RPI), have been co-expressed in *Escherichia coli* to convert D-glucose into D-fructose, D-fructose into D-allulose, and ultimately D-allulose into D-allose, respectively (Zheng et al. 2022).

The nontoxic nature and low-calorie profile, along with its diverse biological activities, make D-allose an important next-generation sweetener and therapeutic agent. As research progresses, the continuous production and application of D-allose could significantly broaden its industrial and clinical uses.

6.3.1 Physiological Functions and Health Benefits

6.3.1.1 Anticancer and Antitumor Properties

D-Allose has gained significant attention for its anticancer properties. It has been shown to inhibit the growth and metastasis of various cancer cells, including those in the ovary, pancreas, liver, prostate, head and neck, and more (Sui et al. 2005; Malm et al. 2015; Yamaguchi et al. 2008; Naha et al. 2008; Hoshikawa et al. 2018). It has also been proven to be effective in inhibiting bladder and brain cancer (Tohi et al. 2022; Suzuki et al. 2023). D-Allose mediates its anticancer effects primarily through the upregulation of thioredoxin-interacting protein (TXNIP) and the stabilization of $p27^{kip1}$ protein, leading to cell cycle arrest and reduced glucose uptake (Noguchi et al. 2016). Furthermore, it has demonstrated synergistic effects when combined with chemotherapy and radiotherapy, enhancing the efficacy of these treatments (Samukawa et al. 2024; Kanaji et al. 2018).

Table 6.2 L-Rhamnose isomerases for D-allose production characterized from different microbial sources.

Enzymes	Source	Optimum operational conditions	k_{cat}/K_m (mM^{-1} s^{-1})	Half-life (h)	Equilibrium ratio (D-allulose :D-allose :D-altrose)	References
L-Rhamnose isomerase	*Pseudomonas stutzeri*	pH 9.0, 42 °C, 1mM Mn^{2+}	NR	NR	67:25:08	Menavuvu et al. (2006)
		pH 9.0, 42 °C, 10 mM Mn^{2+}	NR	NR	68.5:30:1.3	Morimoto et al. (2006)
	Bacillus pallidus Y25	pH 7.0, 50 °C, 1 mM Mn^{2+}	0.825	1 (65 °C)	65:35:00	Poonperm et al. (2007)
	Thermoanaerobacterium saccharolyticum NTOU1	pH 7.0, 65 °C, 1 mM Co^{2+}	0.28	>2 (75 °C)	66:34:00	Lin et al. (2011)
	Caldicellulosiruptor saccharolyticus ATCC 43494	pH 7.0, 90 °C, 1 mM Co^{2+}	4.77	~1 (90 °C)	67:33:00	Lin et al. (2011)
	Bacillus subtilis WB600	pH 9.0, 60 °C, 1 mM Mn^{2+}	0.12	6 (65 °C)	62.5:37.5:00	Bai et al. (2015)
	Thermobacillus composti KWC4	pH 7.5, 65 °C, 1 mM Mn^{2+}	0.035	3.65 (65 °C)	77:23:00	Xu et al. (2017)
	Caldicellulosiruptor obsidiansis OB47	pH 8.0, 60 °C, 1 mM Co^{2+}	0.44	3.3 (85 °C)	68:32:00	Chen et al., (2018a)
	Clostridium stercorarium ATCC 35414	pH 7.0, 70 °C, 1 mM Mn^{2+}	2.11	22.8 (65 °C)	67:33:00	Seo et al. (2018)
	Shinella zoogloeoides NN6	pH 9.0, 60 °C, 1 mM Mn^{2+}	201	NR	69:21:00	Morimoto et al. (2006)
	Chloroflexus islandicus	pH 7.0, 75 °C, 1 mM Mn^{2+}	2.981	288 (65 °C)	60:40:00	Sharma et al. (2024)
Ribose-5-phosphate isomerase	*Clostridium thermocellum*	pH 7.5, 50 °C	NR	96 (50 °C), 4.7 (65 °C)	67:33:00	Park et al. (2007)
	Thermotoga lettingae TMO	pH 8.0, 75 °C	0.0018	3.3 (75 °C)	72:28:00	Feng et al. (2013)
Galactose-6-phosphate isomerase	*Lactococcus lactis*	pH 7.0, 30 °C	0.00077	NR	62:25:13	Park et al. (2007)
Glucose-6-phosphaate isomerase	*Pyrococcus furiosus*	pH 9.0, 95 °C, 1.5 mM Co^{2+}	1.58	7.9 (95 °C)	66:32:02	Yoon et al. (2009)

6.3.1.2 Antioxidant Properties

D-Allose exhibits strong antioxidant properties by scavenging reactive oxygen species (ROS) and inhibiting their production (Ishihara et al. 2011). This antioxidant action is beneficial in reducing the negative effects of ischemia/reperfusion (I/R) injury and improving outcomes in conditions like focal cerebral ischemia (Gao et al. 2011). The competitive inhibition of ROS formation by D-allose against D-glucose in mitochondria underscores its role in reducing oxidative stress (Noguchi et al. 2016).

6.3.1.3 Anti-inflammatory Effects

D-Allose has significant anti-inflammatory properties, as evidenced by its ability to attenuate I/R injury in various tissues (Muntean et al. 2016). Studies have shown that D-allose can decrease brain damage in forebrain ischemia by inhibiting inflammatory responses and oxidative stress. It has also been effective in ameliorating non-alcoholic steatohepatitis disease by inhibiting inflammation and upregulating peroxisome proliferator-activated receptor γ (PPARγ) (Huang et al. 2016).

6.3.1.4 Cryoprotective, Immunosuppressive, and Other Characteristics

D-Allose possesses cryoprotective properties similar to trehalose, protecting cells during freezing (Sui et al. 2007). Its immunosuppressive effects are particularly useful in organ transplantation, a procedure which helps to reduce tissue injury and prevents immunological responses without any detectable side effects (Hossain et al. 2000; Shintani et al. 2020). It can enhance the effectiveness of metronidazole against trichomonads, reduce metronidazole resistance, and prevent osteoporosis by inhibiting osteoclast differentiation (Harada et al. 2012; Yamaguchi et al. 2008).

6.3.1.5 Sweetener and Food Additive

D-Allose, a monosaccharide with approximately 80% of the sweetness of sucrose, is a low-calorie, non-nutritive sweetener suitable for food preparations (Yun-Zhan et al. 2020). D-Allose can be used as a substitute for sucrose to help reduce weight. It enhances taste, color, and flavor through Maillard reactions, working as a reducing sugar and glycosylating proteins effectively (Sun et al. 2006). Various experiments on rats have confirmed its nontoxic nature, demonstrating no sub-chronic or acute toxicity (Iga et al. 2010).

6.3.1.6 Benefits of D-Allose in Plants

D-Allose has the potential to be used as a novel herbicide. The dual activity as both a growth inhibitor and an inducer of disease resistance makes it a valuable candidate for sustainable agricultural practices (Kano et al. 2010). D-Allose induces the upregulation of defense-related genes in plants, enhancing their resistance to various pathogens. For instance, it has been shown to enhance the resistance of rice to bacterial blight caused by *Xanthomonas oryzae pv. oryzae*. It is also reported to affect the expression of gibberellin-responsive genes and modulate the expression of abscisic acid signaling genes through an HXK-dependent pathway (Fukumoto et al. 2011).

6.4 Trehalose

Trehalose ($C_{12}H_{22}O_{11}$) is an uncommon disaccharide composed of two glucose molecules linked by α-(1$\rightarrow$1)-glycosidic bond in a head-to-head position, making it a non-reducing sugar. It is a white, odorless powder that is nontoxic, and the sweetness is around 45% relative to sucrose, even after exhibiting the same caloric content as sucrose (Cai et al. 2018). The properties like lower sweetness and bioactive nature make it an important ingredient in numerous food products. It naturally occurs in sources like mushrooms, seaweed, honey, and yeast but has not been reported from mammalian cells (Agarwal and Singh 2021). In yeast, it is synthesized in two steps: (i) formation of trehalose-6-phosphate from UDP-glucose and glucose-6-phosphate catalyzed by trehalose-phosphate synthase followed by (ii) removal of phosphate by trehalose-6-phosphate phosphatase to produce trehalose. The synthesis in other reported organisms occurs in a similar manner (Paul 2007).

The disaccharide has recently gained much scientific attention due to its potential as a stress-responsive compound and food stabilizer (Chen et al. 2022). It has also been reported as a therapeutic agent because it holds the capacity to manage several metabolic diseases, and as a prebiotic molecule, it can improve the gut microbiome (Chen and Gibney 2023). There are several reports of trehalose for its ability to stabilize proteins and lipids, making it helpful in combating various types of stresses such as salinity, desiccation, and temperature fluctuation (Iordachescu and Imai 2011). Several theories are put forth to explain these properties, such as glass formation (vitrification) and water replacement, which contribute to maintaining the structure and function of biomolecules. This explains its importance in stress tolerance and makes it a potential target for genetic engineering to enhance stress resistance in plants (Hu et al. 2023).

Trehalose has been granted "GRAS" status by FDA in the year 2000 (GRN No. 00045). It is being used in many food items, from baked goods to processed vegetables and even seafood. Several additional benefits like cryopreservation, the ability to reduce oxidation, improving texture in baked foods, and extending shelf life by binding moisture are increasing its demand day by day (Hu et al. 2023). Medical research is also exploring trehalose's role in diseases like Parkinson's and Huntington's due to its activity in autophagy, and some groups are searching for its possible role in controlling diabetes (Pupyshev et al. 2022).

Trehalose synthase is a well-known enzyme for enzymatic trehalose production. This enzyme catalyzes one step, reversible conversion of maltose into trehalose with high yield. There are various microbial sources for trehalose synthase gene for the production of trehalose directly from maltose. Some of these include *Pimelobacter* sp. R48, *Mycobacterium smegmatis*, *Thermobifida fusca* (DSM 43792), *P. stutzeri*, *Picrophilus torridus* (DSM 9790), *Thermus thermophilus* (ATCC 33923), *Deinococcus radiodurans* (ATCC 13939), *Enterobacter hormaechei*, *Paenarthrobacter aurescens*, *Meiothermus ruber*, *Corynebacterium glutamicum* (ATCC 13032), *Deinococcus geothermalis* (DSMZ 11300), *Thermomonospora curvata* (DSM 43183), *Pseudomonas* sp. P8005, *Rhodococcus opacus* (ATCC 41021), Metagenomic sequence from saline-alkali soil, *Thermus thermophilus* HB27, *Thermobaculum terrenum*, *Thermus antranikianii*, gene from hot spring metagenome, and *Pseudarthrobacter* sp. (Table 6.3). Among these, the trehalose synthase gene from *Thermus thermophilus* (ATCC 33923) seems to be the best as it produces high trehalose and zero by-products under efficient operational conditions.

6.4.1 Physiological Functions and Health Benefits

6.4.1.1 Cryopreservation

Trehalose is widely used as a cryoprotective agent (CPA) to protect cells and tissues like human platelets, sperm, adipose tissue, yeast cells, and extracellular vesicles (Gläfke et al. 2012; Ahmad & Aksoy 2012); Pu et al. 2005; Diniz-Mendes et al. 1999; Bosch et al. 2016) This molecule can be used along with other CPAs like DMSO, glycerol, L-proline for the preservation of fetal skin and human skin fibroblasts, adipose-derived stem cells (ADSCs), and red blood cells (RBCs), respectively (Yang et al. 2017). There are several non-permeable CPAs which are also used for cryopreservation purposes, like glycopeptides, liposomes, and hydroxylethyl starch in combination with trehalose,

Table 6.3 Trehalose synthases for trehalose production characterized from different microbial sources.

Source microorganism	Optimum operational conditions	Half-life (h)	Equilibrium ratio (maltose:trehalose: glucose)	Time to attain equilibrium (h)	Trehalose yield (%)	By-product (D-glucose; %)	References
Pimelobacter sp. R48	pH 7.5, 20 °C	50% (1 h, 30 °C)	17.2:76.7:6.1	~20	76.7	6.1	Nishimoto et al. (1996)
Mycobacterium smegmatis	pH 7.0, NA	NA	~45:45:10	6	42–45	8–10	Pan et al. (2004)
Thermobifida fusca (DSM 43792)	pH 6.5, 25 °C	NR	Between 35:55:10 and 20:65:15	24	55–65	10–15	Wei et al. (2004)
Pseudomonas stutzeri	pH 8.5, 35 °C	40% (1 h, 55 °C)	28:72:00	19	72	0	Lee et al. (2005b)
Picrophilus torridus (DSM 9790)	pH 6.0, 45 °C	90% (20 min, 65 °C)	31.2:61:7.8	72	61	7.8	Chen et al. (2006)
Thermus thermophilus(ATCC 33923)	pH 6.5, 65 °C	88% (30 min, 65 °C)	20:80:00	48 (30 °C)	80	NR	Wang et al. (2007)
Deinococcus radiodurans (ATCC 13939)	pH 6.5, 15 °C	90% (30 min, 40 °C)	08:92:00	48 (5 °C)	92	NR	Wang et al. (2007)
Enterobacter hormaechei	pH 6.0, 37 °C	NR	52:48:00	NR	48	NR	Yue et al. (2009)
Arthrobacter aurescens	pH 6.5, 35 °C	~100% (20 min, 20–30 °C)	27.3:59.5:13.2	8 (30 °C)	59.5	13.2	Xiuli et al. (2009)
Meiothermus ruber	pH 6.5, 50 °C	90% (5 h, 60 °C)	34.6:47:18.4	24	47	18.4	Zhu et al. (2010)
Corynebacterium glutamicum (ATCC 13032)	pH 7.0, 35 °C	80% (30 min, 40 °C)	19.3:59:21.7	4	59	21.7	Kim et al. (2010)
Deinococcus geothermalis (DSMZ 11300)	pH 7.6, 40 °C	57% (8 h, 55 °C)	31:60.4:8.6	24	60.4	8.6	Filipkowski (2012)
Thermomonospora curvata (DSM 43183)	pH 6.5, 35 °C	NR	22:70:8	NR	70	8	Liang et al. (2013)
Pseudomonas sp. P8005	pH 7.2, 37 °C	>80% (30 min, 10-40°C)	22:70:8:00	12	70	8	Gao et al. (2013)
Rhodococcus opacus (ATCC 41021)	pH 7.0, 25 °C	~100% (1 h, 15-45°C)	21:67:12:00	NR	67	12	Yan et al. (2013)
Metagenomic sequence from saline-alkali soil	pH 9.0, 45 °C	NR	22:78:00	NR	78	NR	Jiang et al. (2013)
Thermus thermophilus HB27	pH 9.0, 50 °C	NR	26:74:00	10	74	NR	Li et al. (2015)
Thermobaculum terrenum	pH 7.5, 45 °C	80% (30 min, 70 °C)	30:70:00	10	70	NR	Wang et al. (2017)
Thermus antranikianii	pH 7.0, 60 °C	NR	37.4:62.6:00	2	62.6	NR	Lin et al. (2020)
Gene from hot spring metagenome	pH 7.0, 45 °C	90% (48 h, 50 °C)	32:63:05	3	63	5	Agarwal and Singh (2021)
Pseudarthrobacter sp.	pH 7.0, 20 °C	>90% (6 h, 40 °C)	23.8:71.7:4.5	19.84	75.16	5	Trakarnpaiboon et al. (2023)

which is majorly used for the preservation of RBC (Hu et al. 2023). The nontoxic characteristics of trehalose, no requirement of removal after cryopreservation, and its ability to prevent ice crystal formation make it the most suitable CPA over the traditional CPAs (Pu et al. 2005).

6.4.1.2 Blood Sugar and Insulin Response

Trehalose exhibits a low glycemic index (GI) value; thus, it keeps blood sugar levels stable through slower glucose release via a significant reduction in insulin secretion and active incretin secretion compared to other common sugars like glucose (Yoshizane et al. 2017). Lower insulin secretion prevents a rapid increase in blood glucose levels, which is crucial for individuals with type-2 diabetes or those who are pre-diabetic, preventing insulin resistance and maintaining overall metabolic health. Trehalose and glucose are metabolized differently, where trehalose is believed to impede glucose transport via the solute carrier 2A (SLC2A) transporter (Chen and Gibney 2023). Additionally, upon investigation, it has been found that not all trehalose molecules are broken down into glucose and taken up, which could be one of the possible reasons for the independent physiological activities of trehalose apart from glucose (Iturriaga et al. 2009).

6.4.1.3 Regulation of Glucose Homeostasis and Lipid Metabolism

As discussed above, trehalose modulates insulin sensitivity by regulating the glucose signaling pathway, which ultimately reduces the fat cell buildup (Arai et al. 2013). The maintenance of lipid levels by this uncommon sugar molecule is mainly attributed to the enhanced function of pancreatic islets, leading to improved and healthy pancreatic beta cells. This is helpful in controlling insulin secretion and prevention of untimely beta cell apoptosis, which is vital for long-term diabetes management (Yaribeygi et al. 2019). Overall, trehalose can modulate lipid metabolism by regulating postprandial insulin secretion into the blood, which possibly help to reduce fat accumulation and promote healthier lipid profiles (Yaribeygi et al. 2019).

6.4.1.4 Antioxidant and Anti-inflammatory Effects

Trehalose has shown to attenuate oxidative stress by scavenging reactive oxygen species (ROS), reducing oxidative stress within the cells. This helps in the prevention of cellular damage caused by free radicals generated during metabolic processes and in response to various environmental stresses (Zhang et al. 2023). This molecule also shows anti-inflammatory responses by enhancing inflammatory mediators. This helps in improving chronic inflammatory responses, which are the root cause of many metabolic diseases like diabetes (Yu et al. 2023).

6.4.1.5 Gut Microbiome Modulation

Trehalose acts as a potential prebiotic molecule as it stimulates the growth of enormous gut bacteria which are beneficial for improving gut health. It particularly triggers the growth of bacteriocin-producing lactic acid bacteria (LABs) and reduces the harmful ones improving the overall digestive function and leads to better immune function (Chen and Gibney 2023).

6.4.1.6 Dental Health and Weight Management

Trehalose is one of the most promising sugars in dental health due to its non-cariogenic properties; that is, it does not promote tooth decay. In fact, it inhibits the adherence of cariogenic bacteria like *Streptococcus mutans* to tooth surfaces, thus preventing risks of dental caries (Neta et al. 2000). As mentioned earlier, it has a low glycemic index (GI) and lower digestion rate, which can help in a slower rise of blood glucose levels. This can aid in weight management by enhancing satiety and reducing overall food consumption (Hamada et al. 2022).

6.4.1.7 Stress Regulator in Plants

Apart from animals, this disaccharide also plays a critical role in plants' adaptation and tolerance against various abiotic stresses (Kosar et al. 2019). It acts as a bioactive molecule mediating diverse molecular, biochemical, and physiological processes in plants against stress stimulation (Hassan et al. 2023). These processes help the plant adjust and maintain tolerance against unfavorable conditions. In most cases, it acts as a signaling molecule, stimulating different processes and stress-responsive pathways (Morgutti et al. 2019). Nowadays, various research is being carried out to elucidate the mechanisms by which trehalose-mediated stress tolerance takes place in plants.

6.5 D-Tagatose

D-Tagatose is a ketohexose sugar, a C-4 epimer of D-fructose. It is also known as a structural isomer of D-galactose. It is found naturally in fruits such as apples and oranges (Vastenavond et al. 2011) and gum exudates from trees such as cacao tree and *Rocella* spp (Livesey and Brown 1996; Bär et al. 1999). Its presence has also been noticed in dairy products (Vastenavond et al. 2011). The presence of tagatose naturally with other existing sugars is very low, about <3 mg/g, and hence is considered rare in nature (Vastenavond et al. 2011). The calculated caloric value of D-tagatose is 1.5 kcal/g, and it is 8% less sweet than sucrose, sharing almost similar physical properties to sucrose, such as color, texture, and mouth feel (Lu et al. 2008). Notably, rare sugar D-tagatose has been documented for several physiological functions such as anti-obesity, prebiotic, non-carcinogenic, weight loss, and low glycemic effect besides anti-biofilm formation. The US FDA have tagged D-tagatose GRAS status with no hazard to human health (Rai et al. 2023).

The commercial-scale production of rare sugar with a cost-efficient process is required to meet the global market needs. The first commercial-scale D-tagatose production was started by Arla Food Industry Private Limited using a chemical route (Rai et al. 2023). This sugar can be produced using galactose via chemical or biological synthesis methods. The chemical synthesis method involved the direct isomerization of D-galactose into D-tagatose using metal hydroxides (Shehata et al. 2023). However, the usage of chemicals in the isomerization step for D-tagatose synthesis has always been a concern for human health safety (Roy et al. 2018) and high cost due to subsequent purification steps, which are the major disadvantage of the method (Shehata et al. 2023). Therefore, researchers have focused on enzyme-mediated method for D-tagatose production by employing bacterial L-arabinose isomerase (L-AI) to convert the D-galactose to D-tagatose (Kim 2004). In the past 20 years, several L-AIs have been discovered and characterized from various microorganisms for the biosynthesis of D-tagatose. Several microbial species were identified to produce L-AI including *Anoxybacillus flavithermus, Lactobacillus sakei* 23 K, *Lactobacillus fermentum*, *Klebsiella pneumoniae*, *L. lactis*, *Shewanella* sp., *B. subtilis*, *Enterococcus faecium*, *Alicyclobacillus hesperidum* URH17-3-68, *Clostridium hylemonae, Arthrobacter* sp. 22c, *Bacillus thermoglucosidasius* KCTC 1828, *Lactobacillus plantarum*, *Geobacillus stearothermophilus, Pediococcus pentosaceus* PC-5, *Shigella flexneri,* and *Pseudoalteromonas haloplanktis* ATCC 14,393 (Shehata et al. 2023). Table 6.4 shows the sources and properties of L-arabinose isomerase from microbial sources. Most of the characterized L-AIs enzyme catalytic activity depends on the cofactors such as Co^{2+} and Mn^{2+}.

D-Tagatose, with all the potential properties highlighted above, is widely used in the industry as a low-calorie sweetener and bulking agent (Zehner and Biospherics Inc 1988). A lot of physiological properties have been considered for D-tagatose, with its branches extended from humans to plants.

Table 6.4 Sources and properties of L-arbinose isomerase.

Microbial sources	pH optima	Temperature optima (°C)	Metal ions	k_{cat}/K_m (mM^{-1} min^{-1})	References
Bacillus halodurans	7.5–8	50	Mn^{2+}	0.4	Lee et al. (2005a)
Lactobacillus sakei 23 K	5–7	30–40	Mg^{2+}, Mn^{2+}	10.3	Rhimi et al. (2010)
Shewanella sp. ANA-3	5.5–6.5	15–35	Mn^{2+}	NA	Rhimi et al. (2011)
Lactobacillus fermentum	6.5	65	Mn^{2+}	9	Xu et al. (2011)
Arthrobacter sp.	5–9	47–52	NR	NA	Wanarska and Kur (2012)
Bacillus thermoglucosidasius	7	40	Mn^{2+}	2.8	Seo (2013)
Enterococcus faecium DBFIQ E36	7	50	NA	NA	Torres et al. (2014)
Pediococcus pentosaceus PC-5	6	50	Co^{2+}, Mn^{2+}	2.9	Men et al. (2014)
Bacillus coagulans NL01	7.5	60	Co^{2+}, Mn^{2+}	1	Mei et al. (2016)
Thermoanaerobacter mathranii	8	65	Mn^{2+}	NA	Jørgensen et al. (2004)
Geobacillus thermodenitrificans	7.5	60	Co^{2+}, Mn^{2+}	0.5	Kim and Oh (2005)
Bacillus stearothermophilus IAM1101	7.5	65	Mn^{2+}	NA	Cheng et al. (2010a)
Acidothermus cellulolytics ATCC 43068	7.5	75	Co^{2+}, Mn^{2+}	9.3	Cheng et al. (2010b)
Alicyclobacillus hesperidum URH17-3-68	7	70	Co^{2+}	1.2	Fan et al. (2014)
Thermoanaerobacterium saccharolyticum NTOU1	7–7.5	70	Co^{2+}, Mn^{2+}	2.41	Hung et al. (2014)
Bacillus coagulans NL01 AI (BCAI)	7.5	60	Mn^{2+}	8.7	Mei et al. (2016)
Bacillus amyloliquefaciens CAAI	7.5	45	NA	2.34	Shehata et al. (2023)
Klebsiella pneumoniae	8	40	Mn^{2+}	NA	Shin et al. (2022)

6.5.1 Physiological Functions and Health Benefits

6.5.1.1 Oral Health

D-Tagatose possesses cryoprotective and antioxidative properties that show positive effects on dental health by inhibiting dental plaque and biofilm formation by microbes like *Streptococcus* and *Actinomycetes* and reduces associated systemic allergies like asthma and rhinitis (Vitale et al. 2023).

6.5.1.2 Prebiotic and Systemic Health

D-Tagatose being a low-calorie sugar has low digestibility that reduces its glycemic and insulinemic response. As it escapes digestion by the pancreatic enzymes, it favors the growth of intestinal microbiota involved in its fermentation in the lower intestine (Lu et al. 2008). Therefore, it is widely used as an alternative to common sugars to help control diabetes and obesity (Carocho et al. 2017).

6.5.1.3 Antiaging

The low-energy sugar is capable of efficiently managing the energy intake despite of compromising the original sweetness. Therefore, it helps to reduce protein glycosylation with sugar molecules in tissues of muscle and brain that ultimately slows down aging (Bunn and Higgins 1981).

6.5.1.4 D-Tagatose Restricts Plant Pathogen

D-Tagatose has shown to be active against a wide range of fungal pathogens of plants. Solution of D-tagatose has been used to treat seeds for the purpose of sterilization and prevent any fungal contamination (Corneo et al. 2021; Ohara et al. 2015).

6.6 D-Talose

D-Talose is one of the rare aldohexose sugars; it rarely exists in natural resources. It is a C-4 epimer of mannose and C-2 epimer of galactose. Its chemical formula is $C_6H_{12}O_6$, with a molar mass of 180.16 g/mol. This sugar is highly soluble in water and slightly soluble in ethanol, methanol, and other organic solvents. D-Talose and its glycoconjugate derivatives are well known for antitumor and antimicrobial activities (Singh et al. 2021; Singh et al. 2022). D-Talose has a potential application in the pharmaceutical industry. For example, it can serve as a precursor molecule for various antibiotics, of which Caminoside A is the best example, an antimicrobial compound extracted from the sponge *Caminus sphaeroconia*, holding a 6-deoxyribose group. Moreover, its O2 and O3-methylated forms in sub-millimolar have been demonstrated as inhibitors of galactose-binding galectin-4 protein and galectin-8 proteins, which are involved in cancer and inflammation (Van Overtveldt et al. 2018). Therefore, the development of an efficient process is needed to produce this expensive sugar in bulk. Due to the aforementioned beneficial applications, including as a raw material to create value-added products and as a food additive, it has immense applications in the pharmaceutical and food industries.

Currently, D-talose chemical synthesis is inefficient and complicated and requires a high amount of dibutyltin oxide or five consecutive chemical steps and a higher number of solvents. Therefore, scientists are exploring the biological route for efficient bio-production of D-talose. However, only a few studies are available for enzymatic biosynthesis of D-talose (Van Overtveldt et al. 2018). The *Rhodothermus marinus* cellobiose 2-epimerase has been evaluated to produce D-talose using the D-galactose. Using the cellobiose 2-epimerase, the D-talose can be produced with a yield of 8.5% and purity of 86%. The L-ribose isomerase (L-RI) also catalyzes the D-tagatose to D-talose conversion. A limited number of L-RIs have been reported from *Actinotalea fermentans* ATCC 43,279 (Tseng et al. 2017), *Acinetobacter* sp., (Mizanur et al. 2001), *Cellulomonas parahominis* MB4264 (Morimoto et al. 2013), *Mycetocola miduiensis* (Mahmood et al. 2020), and *Geodermatophilus obscurus* (Hung et al. 2015). Researchers have also demonstrated the immobilization of L-RI onto carbon-based nanomaterials such as multiwalled carbon nanotubes and graphene oxide via physical absorption and covalent conjugation to preserve the unique properties of the enzyme. The bioconversion of D-tagatose to D-talose by immobilized L-RI was 12–14% (Singh et al. 2021). In another study, L-RI was immobilized on zeolite-based metal framework. After immobilization, the stability and reusability of the enzyme increased compared to its free form of the enzyme. In this report, the conversion was reported to be 15.2% (Singh et al. 2022).

6.7 Turanose

The carbohydrate turanose is a disaccharide consisting of two sugar molecules, namely D-glucose, linked to D-fructose through α-1-3-glycosidic bond. The chemical formula for D-turanose is $C_{12}H_{22}O_{11}$. The natural source for D-turanose is honey, where it is found in a concentration of 4.7% (Kaškonienė et al. 2010; Belay et al. 2017; Kamboj et al. 2020). Turanose contains 50% sweetness of sucrose and substitutes regular table sugar with low-calorie value (Pikis et al. 2002; Park et al. 2019). In the year 1889, Alekhine discovered turanose along with glucose during the hydrolysis of trisaccharide melezitose with weak acids (Moelwyn-Hughes 1929). However, the first crystalline structure of D-turanose was discovered in 1906 (Moelwyn-Hughes 1929). The synthesis of D-turanose can be achieved through both chemical and biological routes. However, chemical synthesis of D-turanose is a temperature-sensitive and labor-intensive procedure, thus making the route difficult for large-scale synthesis (Lichtenthaler and Rönninger 1990). Therefore, scientists have adopted *in vitro* enzymatically synthesis of D-turanose. Amylosucrase (ASase) is the most widely used enzyme belonging to the hydrolases family. The ASase enzyme uses sucrose as a substrate to synthesize linear glucan polymers; however, if the concentration of sucrose is increased in the reaction medium, ASase turns on the synthesis of turanose (Wang et al. 2012). So, the large-scale production of turanose deploying enzymes has been used at the industrial level (Table 6.5).

6.7.1 Physiological Functions

Turanose, being an unfermentable, low-calorie, non-cariogenic sugar, has a variety of physiological functions. It possesses several advantageous qualities, such as low glycemic value, high stability during food processing, and prebiotic properties, and it can be utilized during the fermentation of valuable compounds. To name a few, potential applications of turanose are in the fermentation industries, pharmaceutical industries, food and beverage industries, and functional foods. Some of the various applications are explained below in brief.

6.7.1.1 Blood Sugar Control and Weight Management

Like other sugars, turanose also serves as a source of energy, but unlike common sugar, it is metabolized slowly. This property leads to the gradual release of glucose into the blood, which prevents sudden spikes in glucose levels, helping in better glycemic control. Thus, it provides a slower glycemic response and sustained energy release, making it suitable for people with type-2 diabetes. Also, it helps in better regulation of hunger and gives satiety, reducing craving and overeating, thus helping in weight management (Chung et al. 2017b; Ponnurangam and Balaji 2024).

6.7.1.2 Anti-inflammatory Properties

Several studies have demonstrated the potential application of turanose as an anti-inflammatory agent. In one such study, its effects have been tested for inflammatory bowel disease (IBD), one of the common diseases in many developing countries. It regulates several signaling pathways in the colon tissues helping reduce inflammation (Kim et al. 2019). Another study suggested that turanose exhibits both *in vitro* and *in vivo* inflammatory effects including the clinical trials (Chung et al. 2017a).

6.7.1.3 Prebiotic Effects

Similar to other prebiotics, turanose might promote the growth of beneficial gut bacteria, including *Bifidobacterium* and Lactobacilli, promoting improved gut health and boosting immunity (Ruiz-Aceituno et al. 2018).

Table 6.5 Biochemical properties of amylosucrase characterize from different microbial sources.

Organism	Optimum pH	Optimum temperature (°C)	Thermal stability (min)	Turanose yield	References
Bifidobacterium thermophilum	7	50	23,104 h (45 °C), 577 h (50 °C), 70.7 h (55 °C)	43.3% (1 M sucrose + 1 M fructose) at 48 h, 35 °C	Choi et al. (2019)
Deinococcus radiodurans	8	50	900 (30 °C)	15%, 100 mM sucrose, 24 h	Pizzut-Serin et al. (2005)
Neisseria subflava	8	45	23,104 (40 °C), 546 (45°C)	29% (1 M sucrose) in 24 h	Park et al. (2018)
Alteromonas macleodii	8	45	708 (40 °C), 30 (45 °C)	NR	Ha et al. (2009)
Arthrobacter chlorophenicolicus	8	NR	104 (40 °C), 3 (45 °C)	23%, 1 M sucrose	Seo et al. (2012)
Cellulomonas carbonis	7	40	16,620 (40 °C), 44 (45 °C)	NR	Wang et al. (2017)
Methylobacillus flagellatus	8.5	45	20 (50 °C), 4 (55 °C)	15%, 48 h	Jeong et al. (2014)
Neisseria polysaccharea	8	37	1260 (30 °C), 25 (45 °C)	73.7% (2 M sucrose + 0.75 M fructose) in 80 h, 35 °C	Park et al. (2016b)
Synechococcus sp.	6.5–7	30	NR	NR	Perez-Cenci and Solerno (2014)
Deinococcus geothermalis	8	45	1686 (50 °C), 408 (55 °C)	22%	Guérin et al. (2012)
Truepera radiovictrix			2400 (55 °C), 384 (60 °C), 73 (65 °C)	NR	Zhu et al. (2018)
Deinococcus radiopugnans			25,260 (40 °C), 198 (45 °C)	NR	Kim et al. (2014)
As_{met}.	9	60	60 (60 °C), 390 (55 °C)	47% (1.5 M sucrose + 0.5 M fructose)	Agarwal et al. (2019)
Deinococcus cellulosilyticus	6–8	40	239 (45 °C)	15.69% (0.5 M sucrose)	Lee et al. (2024)

6.7.1.4 Gut and Dental Health

As already discussed, due to the prebiotic effects exhibited by turanose, it promotes good bacteria, leading to better digestion and possibly protecting against gastrointestinal disorders (Kim et al. 2019; Moynihan 1998).

6.7.1.5 Pathogen Detection

Based on the fermentation-based properties of turanose, it is used for the identification of methicillin-resistant as well as susceptible strains of *S. aureus* (Archer and Niemeyer 1994). In another case, turanose metabolism is used to differentiate between vancomycin-resistant (VR) and vancomycin-susceptible (VS) *Enterococci* isolates (Raeisi et al. 2017). Besides, along with other methods, this sugar is also utilized for the identification of *Staphylococcus saprophyticus*, which causes the most common urinary tract infections (UTIs) in females (Morgan 1983).

6.7.1.6 Honey Authentication

One of the unique applications of turanose is in the detection of honey adulteration by evaluating sugar ratios like maltose/turanose, maltose/isomaltose, sucrose/turanose, and fructose/glucose (Nozal et al. 2005; Hoxha et al. 2019). Though the sugar content in honey varies depending on its botanical source and season, making it essential to account for these factors when assessing quality, Horvath and Molnár-Perl (1997) provided a method. This group provided specific ranges for these sugar ratios (e.g. sucrose/turanose = 0.05–1.9), which can help differentiate pure honey from adulterated samples containing fructose syrup, cane syrup, or corn syrup (Horvath and Molnár-Perl 1997). Turanose plays a crucial role in identifying honey purity and ensuring safety standards.

6.7.1.7 Food Processing and Osmoprotection

Like other reports of rare sugars, turanose also helps enhance the texture, stability, and oil absorption properties of processed foods like instant noodles (Oh et al. 2019). It also serves the role of an osmoprotectant preventing organisms from dehydration or osmotic stress (Han et al. 2021).

6.8 Conclusion

The microbial bio-production of rare sugars like allulose, allose, tagatose, talose, turanose, and trehalose is significant in health-related and industrial applications. The day-by-day improvement in the bio-catalytic and fermentation technologies offers their economic and sustainable production. All the molecules exhibit excellent physiological functions due to their unique properties, from low calorie to metabolic control, obesity control, and diabetes management. Although few of these molecules have been recognized as safe for consumption, most of them still need a lot of research to ensure their safety for large-scale and long-term usage. The ongoing advancements in biotechnological production and further research into the roles of these sugars in various aspects present them as promising future alternatives for food and beverages. Also, beneficial physiological functions make them healthier options that fulfill consumer demands and address various health concerns. Further studies about microbial production technologies, safety issues, and other applications of these sugar molecules will be beneficial in commercial and therapeutic applications.

References

Agarwal, N. and Singh, S.P. (2021). A novel trehalose synthase for the production of trehalose and trehalulose. *Microbiology Spectrum* 9 (3): e01333–e01321.

Agarwal, N., Narnoliya, L.K., and Singh, S.P. (2019). Characterization of a novel amylosucrase gene from the metagenome of a thermal aquatic habitat, and its use in turanose production from sucrose biomass. *Enzyme and Microbial Technology* 131: 109372.

Ahmad, E. and Aksoy, M. (2012). Trehalose as a cryoprotective agent for the sperm cells: a mini review. *Animal Health, Production and Hygiene* 1 (2): 123–129.

Ahmed, A., Khan, T.A., Dan Ramdath, D. et al. (2022). Rare sugars and their health effects in humans: a systematic review and narrative synthesis of the evidence from human trials. *Nutrition Reviews* 80 (2): 255–270.

Arai, C., MIYAkE, M., Matsumoto, Y. et al. (2013). Trehalose prevents adipocyte hypertrophy and mitigates insulin resistance in mice with established obesity. *Journal of Nutritional Science and Vitaminology* 59 (5): 393–401.

Archer, G.L. and Niemeyer, D.M. (1994). Origin and evolution of DNA associated with resistance to methicillin in *staphylococci. Trends in Microbiology* 2 (10): 343–347.

Au-Yeung, F., Jenkins, A.L., Prancevic, S. et al. (2023). Comparison of postprandial glycemic and insulinemic response of allulose when consumed alone or when added to sucrose: a randomized controlled trial. *Journal of Functional Foods* 105: 105569.

Bai, W., Shen, J., Zhu, Y. et al. (2015). Characteristics and kinetic properties of L-rhamnose isomerase from *Bacillus subtilis* by isothermal titration calorimetry for the production of D-allose. *Food Science and Technology Research* 21 (1): 13–22.

Bär, A., Lina, B.A., De Groot, D.M. et al. (1999). Effect of d-tagatose on liver weight and glycogen content of rats. *Regulatory Toxicology and Pharmacology* 29 (2): S11–S28.

Belay, A., Haki, G.D., Birringer, M. et al. (2017). Sugar profile and physicochemical properties of Ethiopian monofloral honey. *International Journal of Food Properties* 20 (11): 2855–2866.

Bilik, V. and Tihlarik, K. (1973). Reaction of saccharides catalyzed by molybdate ions IX. Epimerization of ketohexoses. *Chem Zvesti* 28: 106–109.

Bosch, S., de Beaurepaire, L., Allard, M. et al. (2016). Trehalose prevents aggregation of exosomes and cryodamage. *Scientific Reports* 6 (1): 36162.

Braunstein, C.R., Noronha, J.C., Glenn, A.J. et al. (2018). A double-blind, randomized controlled, acute feeding equivalence trial of small, catalytic doses of fructose and allulose on postprandial blood glucose metabolism in healthy participants: the Fructose and Allulose Catalytic Effects (FACE) Trial. *Nutrients* 10 (6): 750.

Bunn, H.F. and Higgins, P.J. (1981). Reaction of monosaccharides with proteins: possible evolutionary significance. *Science* 213 (4504): 222–224.

Cai, X., Seitl, I., Mu, W. et al. (2018). Biotechnical production of trehalose through the trehalose synthase pathway: current status and future prospects. *Applied Microbiology and Biotechnology* 102: 2965–2976.

Carocho, M., Morales, P., and Ferreira, I.C. (2017). Sweeteners as food additives in the XXI century: a review of what is known, and what is to come. *Food and Chemical Toxicology* 107: 302–317.

Chari, V.M., Grayer-Barkmeijer, R.J., Harborne, J.B., and Österdahl, B.G. (1981). An acylated allose-containing 8-hydroxyflavone glycoside from *Veronica filiformis. Phytochemistry* 20 (8): 1977–1979.

Chen, A. and Gibney, P.A. (2023). Dietary trehalose as a bioactive nutrient. *Nutrients* 15 (6): 1393.

Chen, Y.S., Lee, G.C., and Shaw, J.F. (2006). Gene cloning, expression, and biochemical characterization of a recombinant trehalose synthase from *Picrophilus torridus* in *Escherichia coli. Journal of Agricultural and Food Chemistry 54* (19): 7098–7104.

Chen, Z., Chen, J., Zhang, W. et al. (2018a). Improving thermostability and catalytic behavior of L-rhamnose isomerase from *Caldicellulosiruptor obsidiansis* OB47 toward D-allulose by site-directed mutagenesis. *Journal of Agricultural and Food Chemistry* 66 (45): 12017–12024.

Chen, Z., Xu, W., Zhang, W. et al. (2018b). Characterization of a thermostable recombinant l-rhamnose isomerase from *Caldicellulosiruptor obsidiansis* OB47 and its application for the production of l-fructose and l-rhamnulose. *Journal of the Science of Food and Agriculture* 98 (6): 2184–2193.

Chen, J., Chen, D., Ke, M. et al. (2021a). Characterization of a recombinant D-allulose 3-epimerase from *Thermoclostridium caenicola* with potential application in D-allulose production. *Molecular Biotechnology* 63: 534–543.

Chen, Z., Zhang, W., and Mu, W. (2021b). Various enzymes for the biotechnological production of D-allose. In: *Novel Enzymes for Functional Carbohydrates Production: From Scientific Research to Application in Health Food Industry*, 85–104. Springer.

Chen, A., Tapia, H., Goddard, J.M., and Gibney, P.A. (2022). Trehalose and its applications in the food industry. *Comprehensive Reviews in Food Science and Food Safety* 21 (6): 5004–5037.

Cheng, L., Mu, W., and Jiang, B. (2010a). Thermostable L-arabinose isomerase from *Bacillus stearothermophilus* IAM 11001 for D-tagatose production: gene cloning, purification and characterisation. *Journal of the Science of Food and Agriculture* 90 (8): 1327–1333.

Cheng, L., Mu, W., Zhang, T., and Jiang, B. (2010b). An L-arabinose isomerase from *Acidothermus cellulolytics* ATCC 43068: cloning, expression, purification, and characterization. *Applied Microbiology and Biotechnology* 86: 1089–1097.

Choi, S.W., Lee, J.A., and Yoo, S.H. (2019). Sucrose-based biosynthetic process for chain-length-defined α-glucan and functional sweetener by *Bifidobacterium amylosucrase*. *Carbohydrate Polymers* 205: 581–588.

Choi, M.N., Shin, K.C., Kim, D.W. et al. (2021). Production of D-allose from D-allulose using commercial immobilized glucose isomerase. *Frontiers in Bioengineering and Biotechnology* 9: 681253.

Chung, J.Y., Kim, Y.S., Kim, Y., and Yoo, S.H. (2017a). Regulation of inflammation by sucrose isomer, turanose, in raw 264.7 cells. *Journal of Cancer Prevention* 22 (3): 195.

Chung, J.Y., Lee, J., Lee, D. et al. (2017b). Acute and 13-week subchronic toxicological evaluations of turanose in mice. *Nutrition Research and Practice* 11 (6): 452–460.

Corneo, P.E., Jermini, M., Nadalini, S. et al. (2021). Foliar and root applications of the rare sugar tagatose control powdery mildew in soilless grown cucumbers. *Crop Protection* 149: 105753.

Diniz-Mendes, L., Bernardes, E., De Araujo, P.S. et al. (1999). Preservation of frozen yeast cells by trehalose. *Biotechnology and Bioengineering* 65 (5): 572–578.

Do, G.Y., Kwon, E.Y., Kim, Y.J. et al. (2019). Supplementation of non-dairy creamer-enriched high-fat diet with D-allulose ameliorated blood glucose and body fat accumulation in C57BL/6J mice. *Applied Sciences* 9 (13): 2750.

Doner, L.W. (1979). Isomerization of D-fructose by base: liquid-chromatographic evaluation and the isolation of D-psicose. *Carbohydrate Research* 70 (2): 209–216.

Fan, C., Liu, K., Zhang, T. et al. (2014). Biochemical characterization of a thermostable l-arabinose isomerase from a thermoacidophilic bacterium, *Alicyclobacillus hesperidum* URH17-3-68. *Journal of Molecular Catalysis B: Enzymatic* 102: 120–126.

Feng, Z., Mu, W., and Jiang, B. (2013). Characterization of ribose-5-phosphate isomerase converting D-psicose to D-allose from *Thermotoga lettingae* TMO. *Biotechnology Letters* 35: 719–724.

Filipkowski, P., Panek, A., Agnieszka, F. et al. (2012). Expression of *Deinococcus geothermalis* trehalose synthase gene in *Escherichia coli* and its enzymatic properties. *African Journal of Biotechnology* 11 (67): 13131–13139.

Franchi, F., Yaranov, D.M., Rollini, F. et al. (2021). Effects of D-allulose on glucose tolerance and insulin response to a standard oral sucrose load: results of a prospective, randomized, crossover study. *BMJ Open Diabetes Research & Care* 9 (1): e001939.

Fukumoto, T., Kano, A., Ohtani, K. et al. (2011). Rare sugar d-allose suppresses gibberellin signaling through hexokinase-dependent pathway in Oryza sativa L. *Planta* 234: 1083–1095.

Gao, D., Kawai, N., and Tamiya, T. (2011). The anti-inflammatory effects of D-allose contribute to attenuation of cerebral ischemia–reperfusion injury. *Medical Hypotheses* 76 (6): 911–913.

Gao, Y., Xi, Y., Lu, X.L. et al. (2013). Cloning, expression and functional characterization of a novel trehalose synthase from marine *Pseudomonas* sp. P8005. *World Journal of Microbiology and Biotechnology* 29: 2195–2206.

Gläfke, C., Akhoondi, M., Oldenhof, H. et al. (2012). Cryopreservation of platelets using trehalose: the role of membrane phase behavior during freezing. *Biotechnology Progress* 28 (5): 1347–1354.

Granström, T.B., Takata, G., Tokuda, M., and Izumori, K. (2004). Izumoring: a novel and complete strategy for bioproduction of rare sugars. *Journal of Bioscience and Bioengineering* 97 (2): 89–94.

Guan, L., Zhu, L., Wang, K. et al. (2024). Biochemical characterization, structure-guided mutagenesis, and application of a recombinant D-allulose 3-epimerase from *Christensenellaceae bacterium* for the biocatalytic production of D-allulose. *Frontiers in Bioengineering and Biotechnology* 12: 1365814.

Guérin, F., Barbe, S., Pizzut-Serin, S. et al. (2012). Structural investigation of the thermostability and product specificity of amylosucrase from the bacterium *Deinococcus geothermalis*. *Journal of Biological Chemistry* 287 (9): 6642–6654.

Ha, S.J., Seo, D.H., Jung, J.H. et al. (2009). Molecular cloning and functional expression of a new amylosucrase from *Alteromonas macleodii*. *Bioscience, Biotechnology, and Biochemistry* 73 (7): 1505–1512.

Hamada, N., Wadazumi, T., Hirata, Y. et al. (2022). Effects of trehalose solutions at different concentrations on high-intensity intermittent exercise performance. *Nutrients* 14 (9): 1776.

Han, Y., Han, H.J., Kim, A.H. et al. (2016). d-Allulose supplementation normalized the body weight and fat-pad mass in diet-induced obese mice via the regulation of lipid metabolism under isocaloric fed condition. *Molecular Nutrition & Food Research* 60 (7): 1695–1706.

Han, D.J., Lee, B.H., and Yoo, S.H. (2021). Physicochemical properties of turanose and its potential applications as a sucrose substitute. *Food Science and Biotechnology* 30: 433–441.

Harada, M., Kondo, E., Hayashi, H. et al. (2012). D-Allose and D-psicose reinforce the action of metronidazole on trichomonad. *Parasitology Research* 110: 1565–1567.

Hassan, M.U., Nawaz, M., Shah, A.N. et al. (2023). Trehalose: a key player in plant growth regulation and tolerance to abiotic stresses. *Journal of Plant Growth Regulation* 42 (8): 4935–4957.

Hayashi, N., Iida, T., Yamada, T. et al. (2010). Study on the postprandial blood glucose suppression effect of D-psicose in borderline diabetes and the safety of long-term ingestion by normal human subjects. *Bioscience, Biotechnology, and Biochemistry* 74 (3): 510–519.

Horvath, K. and Molnár-Perl, I. (1997). Simultaneous quantitation of mono-, di-and trisaccharides by GC-MS of their TMS ether oxime derivatives: II in honey. *Chromatographia* 45: 328–335.

Hoshikawa, H., Kamitori, K., Indo, K. et al. (2018). Combined treatment with D-allose, docetaxel and radiation inhibits the tumor growth in an in vivo model of head and neck cancer. *Oncology Letters* 15 (3): 3422–3428.

Hossain, M.A., Wakabayashi, H., Goda, F. et al. (2000). Effect of the immunosuppressants FK506 and D-allose on allogenic orthotopic liver transplantation in rats. *Transplantation Proceedings* 32 (7): 2021–2023.

Hossain, A., Yamaguchi, F., Matsuo, T. et al. (2015). Rare sugar D-allulose: potential role and therapeutic monitoring in maintaining obesity and type 2 diabetes mellitus. *Pharmacology & Therapeutics* 155: 49–59.

Hoxha, F., Kongoli, R., and Malollari, I. (2019). Using ratio of the main sugars and some oligosaccharides content to indicate market's honey authenticity. *European Journal of Engineering and Technology* 7 (3): 1.

Hu, Y., Liu, X., Liu, F. et al. (2023). Trehalose in biomedical cryopreservation–properties, mechanisms, delivery methods, applications, benefits, and problems. *ACS Biomaterials Science & Engineering* 9 (3): 1190–1204.

Huang, T., Gao, D., Hei, Y. et al. (2016). D-allose protects the blood brain barrier through PPARγ-mediated anti-inflammatory pathway in the mice model of ischemia reperfusion injury. *Brain Research* 1642: 478–486.

Hung, X.G., Tseng, W.C., Liu, S.M. et al. (2014). Characterization of a thermophilic L-arabinose isomerase from *Thermoanaerobacterium saccharolyticum* NTOU1. *Biochemical Engineering Journal* 83: 121–128.

Hung, X.G., Yu, M.Y., Chen, Y.C., and Fang, T.Y. (2015). Characterization of a recombinant L-ribose isomerase from *Geodermatophilus obscurus* DSM 43160 and application of this enzyme to the production of L-ribose from L-arabinose. *Journal of Marine Science and Technology* 23 (4): 20.

Iga, Y., Nakamichi, K., Shirai, Y., and Matsuo, T. (2010). Acute and sub-chronic toxicity of D-allose in rats. *Bioscience, Biotechnology, and Biochemistry* 74 (7): 1476–1478.

Iordachescu, M. and Imai, R. (2011). Trehalose and abiotic stress in biological systems. *Abiotic Stress in Plants-Mechanisms and Adaptations* 22: 215–234.

Ishida, Y., Kamiya, T., Itoh, H. et al. (1997). Cloning and characterization of the D-tagatose 3-epimerase gene from *Pseudomonas cichorii* ST-24. *Journal of Fermentation and Bioengineering* 83 (6): 529–534.

Ishihara, Y., Katayama, K., Sakabe, M. et al. (2011). Antioxidant properties of rare sugar D-allose: effects on mitochondrial reactive oxygen species production in neuro2A cells. *Journal of Bioscience and Bioengineering* 112 (6): 638–642.

Iturriaga, G., Suárez, R., and Nova-Franco, B. (2009). Trehalose metabolism: from osmoprotection to signaling. *International Journal of Molecular Sciences* 10 (9): 3793–3810.

Jensen, S.R., Mikkelsen, C.B., and Nielsen, B.J. (1981). Iridoid mono-and di-glycosides in *Mentzelia*. *Phytochemistry 20* (1): 71–83.

Jeong, J.W., Seo, D.H., Jung, J.H. et al. (2014). Biosynthesis of glucosyl glycerol, a compatible solute, using intermolecular transglycosylation activity of amylosucrase from *Methylobacillus flagellatus* KT. *Applied Biochemistry and Biotechnology* 173: 904–917.

Jia, M., Mu, W., Chu, F. et al. (2014). A D-psicose 3-epimerase with neutral pH optimum from *Clostridium bolteae* for D-psicose production: cloning, expression, purification, and characterization. *Applied Microbiology and Biotechnology* 98: 717–725.

Jiang, L., Lin, M., Zhang, Y. et al. (2013). Identification and characterization of a novel trehalose synthase gene derived from saline-alkali soil metagenomes. *PLoS One* 8 (10): e77437.

Jiang, S., Xiao, W., Zhu, X. et al. (2020). Review on D-allulose: in vivo metabolism, catalytic mechanism, engineering strain construction, bio-production technology. *Frontiers in Bioengineering and Biotechnology* 8: 26.

Jørgensen, F., Hansen, O.C., and Stougaard, P. (2004). Enzymatic conversion of D-galactose to D-tagatose: heterologous expression and characterisation of a thermostable L-arabinose isomerase from *Thermoanaerobacter mathranii*. *Applied Microbiology and Biotechnology* 64: 816–822.

Kamboj, R., Nayik, G.A., Bera, M.B., and Nanda, V. (2020). Sugar profile and rheological behaviour of four different Indian honey varieties. *Journal of Food Science and Technology* 57: 2985–2993.

Kanaji, N., Kamitori, K., Hossain, A. et al. (2018). Additive antitumour effect of D-allose in combination with cisplatin in non-small cell lung cancer cells. *Oncology Reports* 39 (3): 1292–1298.

Kannan, R.R.R., Arumugam, R., and Anantharaman, P. (2012). Chemical composition and antibacterial activity of Indian seagrasses against urinary tract pathogens. *Food Chemistry* 135 (4): 2470–2473.

Kano, A., Gomi, K., Yamasaki-Kokudo, Y. et al. (2010). A rare sugar, D-allose, confers resistance to rice bacterial blight with upregulation of defense-related genes in Oryza sativa. *Phytopathology* 100 (1): 85–90.

Kaškonienė, V., Venskutonis, P.R., and Čeksterytė, V. (2010). Carbohydrate composition and electrical conductivity of different origin honeys from Lithuania. *LWT- Food Science and Technology* 43 (5): 801–807.

Kim, P. (2004). Current studies on biological tagatose production using L-arabinose isomerase: a review and future perspective. *Applied Microbiology and Biotechnology* 65: 243–249.

Kim, H.J. and Oh, D.K. (2005). Purification and characterization of an L-arabinose isomerase from an isolated strain of *Geobacillus thermodenitrificans* producing D-tagatose. *Journal of Biotechnology* 120 (2): 162–173.

Kim, H.J., Hyun, E.K., Kim, Y.S. et al. (2006). Characterization of an *Agrobacterium tumefaciens* D-psicose 3-epimerase that converts D-fructose to D-psicose. *Applied and Environmental Microbiology* 72 (2): 981–985.c.

Kim, T.K., Jang, J.H., Cho, H.Y. et al. (2010). Gene cloning and characterization of a trehalose synthase from *Corynebacterium glutamicum* ATCC13032. *Food Science and Biotechnology* 19: 565–569.

Kim, M.D., Seo, D.H., Jung, J.H. et al. (2014). Molecular cloning and expression of amylosucrase from highly radiation-resistant *Deinococcus radiopugnans. Food Science and Biotechnology 23*: 2007–2012.

Kim, S.E., Kim, S.J., Kim, H.J., and Sung, M.K. (2017). D-Psicose, a sugar substitute, suppresses body fat deposition by altering networks of inflammatory response and lipid metabolism in C57BL/6J-ob/ob mice. *Journal of Functional Foods* 28: 265–274.

Kim, E., Bae, J., Lee, J. et al. (2019). Purification and characterization of turanose, a sucrose isomer and its anti-inflammatory effects in dextran sulfate sodium (DSS)-induced colitis model. *Journal of Functional Foods* 63: 103570.

Kimoto-Nira, H., Moriya, N., Hayakawa, S. et al. (2017). Effects of rare sugar D-allulose on acid production and probiotic activities of dairy lactic acid bacteria. *Journal of Dairy Science 100* (7): 5936–5944.

Kosar, F., Akram, N.A., Sadiq, M. et al. (2019). Trehalose: a key organic osmolyte effectively involved in plant abiotic stress tolerance. *Journal of Plant Growth Regulation* 38: 606–618.

Lee, D.W., Choe, E.A., Kim, S.B. et al. (2005a). Distinct metal dependence for catalytic and structural functions in the L-arabinose isomerases from the mesophilic *Bacillus halodurans* and the thermophilic *Geobacillus stearothermophilus. Archives of Biochemistry and Biophysics* 434 (2): 333–343.

Lee, J.H., Lee, K.H., Kim, C.G. et al. (2005b). Cloning and expression of a trehalose synthase from *Pseudomonas stutzeri* CJ38 in *Escherichia coli* for the production of trehalose. *Applied Microbiology and Biotechnology* 68: 213–219.

Lee, T.E., Shin, K.C., and Oh, D.K. (2018). Biotransformation of fructose to allose by a one-pot reaction using *Flavonifractor plautii* D-allulose 3-epimerase and *Clostridium thermocellum* ribose 5-phosphate isomerase. *Journal of Microbial Biotechnology* 28 (3): 418–424.

Lee, C.Y., So, Y.S., Lim, M.C. et al. (2024). Characterization of a unique pH-dependent amylosucrase from *Deinococcus cellulosilyticus. International Journal of Biological Macromolecules* 269: 131834.

Li, Y., Sun, X., Feng, Y., and Yuan, Q. (2015). Cloning, expression and activity optimization of trehalose synthase from *Thermus thermophilus* HB27. *Chemical Engineering Science* 135: 323–329.

Li, S., Chen, Z., Zhang, W. et al. (2019). Characterization of a D-tagatose 3-epimerase from *Caballeronia fortuita* and its application in rare sugar production. *International Journal of Biological Macromolecules* 138: 536–545.

Li, C., Li, L., Feng, Z. et al. (2021). Two-step biosynthesis of D-allulose via a multienzyme cascade for the bioconversion of fruit juices. *Food Chemistry* 357: 129746.

Li, J., Dai, Q., Zhu, Y. et al. (2024). Low-calorie bulk sweeteners: recent advances in physical benefits, applications, and bioproduction. *Critical Reviews in Food Science and Nutrition* 64 (19): 6581–6595.

Liang, J., Huang, R., Huang, Y. et al. (2013). Cloning, expression, properties, and functional amino acid residues of new trehalose synthase from *Thermomonospora curvata* DSM 43183. *Journal of Molecular Catalysis B: Enzymatic* 90: 26–32.

Lichtenthaler, F.W. and Rönninger, S. (1990). α-D-Glucopyranosyl-D-fructoses: distribution of furanoid and pyranoid tautomers in water, dimethyl sulphoxide, and pyridine. Studies on ketoses. Part 4. *Journal of the Chemical Society, Perkin Transactions* 2 (8): 1489–1497.

Lim, Y.R. and Oh, D.K. (2011). Microbial metabolism and biotechnological production of D-allose. *Applied Microbiology and Biotechnology* 91: 229–235.

Lin, C.J., Tseng, W.C., and Fang, T.Y. (2011). Characterization of a thermophilic L-rhamnose isomerase from *Caldicellulosiruptor saccharolyticus* ATCC 43494. *Journal of Agricultural and Food Chemistry* 59 (16): 8702–8708.

Lin, Y.F., Su, P.C., and Chen, P.T. (2020). Production and characterization of a recombinant thermophilic trehalose synthase from *Thermus antranikianii*. *Journal of Bioscience and Bioengineering* 129 (4): 418–422.

Livesey, G. and Brown, J.C. (1996). D-Tagatose is a bulk sweetener with zero energy determined in rats. *The Journal of Nutrition* 126 (6): 1601–1609.

Lu, Y., Levin, G.V., and Donner, T.W. (2008). Tagatose, a new antidiabetic and obesity control drug. *Diabetes, Obesity and Metabolism* 10 (2): 109–134.

Mahmood, S., Iqbal, M.W., Riaz, T. et al. (2020). Characterization of a recombinant L-ribose isomerase from *Mycetocola miduiensis* and its application for the production of L-ribulose. *Enzyme and Microbial Technology* 135: 109510.

Malm, S.W., Hanke, N.T., Gill, A. et al. (2015). The anti-tumor efficacy of 2-deoxyglucose and D-allose are enhanced with p38 inhibition in pancreatic and ovarian cell lines. *Journal of Experimental & Clinical Cancer Research* 34: 1–12.

Mao, S., Cheng, X., Zhu, Z. et al. (2020). Engineering a thermostable version of D-allulose 3-epimerase from *Rhodopirellula baltica* via site-directed mutagenesis based on B-factors analysis. *Enzyme and Microbial Technology* 132: 109441.

Matsuo, T. and Izumori, K. (2009). Retraction: D-psicose inhibits intestinal α-glucosidase and suppresses the glycemic response after ingestion of carbohydrates in rats. *Journal of Clinical Biochemistry and Nutrition* 45 (2): 202–206.

Matsuo, T., Suzuki, H., Hashiguchi, M., and Izumori, K. (2002). D-Psicose is a rare sugar that provides no energy to growing rats. *Journal of Nutritional Science and Vitaminology* 48 (1): 77–80.

Mei, W., Wang, L., Zang, Y. et al. (2016). Characterization of an L-arabinose isomerase from *Bacillus coagulans* NL01 and its application for D-tagatose production. *BMC Biotechnology* 16: 1–11.

Men, Y., Zhu, Y., Zhang, L. et al. (2014). Enzymatic conversion of D-galactose to D-tagatose: cloning, overexpression and characterization of L-arabinose isomerase from *Pediococcus pentosaceus* PC-5. *Microbiological Research* 169 (2–3): 171–178.

Menavuvu, B.T., Poonperm, W., Leang, K. et al. (2006). Efficient biosynthesis of D-allose from D-psicose by cross-linked recombinant L-rhamnose isomerase: separation of product by ethanol crystallization. *Journal of Bioscience and Bioengineering* 101 (4): 340–345.

Mijailovic, N., Nesler, A., Perazzolli, M. et al. (2021). Rare sugars: recent advances and their potential role in sustainable crop protection. *Molecules* 26 (6): 1720.

Mizanur, R.M., Takata, G., and Izumori, K. (2001). Cloning and characterization of a novel gene encoding L-ribose isomerase from *Acinetobacter* sp. strain DL-28 in *Escherichia coli*. *Biochimica et Biophysica Acta (BBA)-Gene Structure and Expression* 1521 (1–3): 141–145.

Moelwyn-Hughes, E.A. (1929). The kinetics of the hydrolysis of certain glucosides, part III; β-methylglucoside, cellobiose, melibiose, and turanose. *Transactions of the Faraday Society* 25: 503–520.

Moller, D.E. and Berger, J.P. (2003). Role of PPARs in the regulation of obesity-related insulin sensitivity and inflammation. *International Journal of Obesity* 27 (3): S17–S21.

Morgan, J.W. (1983). Abbreviated scheme for presumptive identification of *Staphylococcus saprophyticus* from urine cultures. *Journal of Clinical Microbiology* 18 (5): 1272–1274.

Morgutti, S., Negrini, N., Pucciariello, C., and Sacchi, G.A. (2019). Role of trehalose and regulation of its levels as a signal molecule to abiotic stresses in plants. In: *Plant Signaling Molecules*, 235–255. Woodhead Publishing.

Morimoto, K., Park, C.S., Ozaki, M. et al. (2006). Large scale production of D-allose from D-psicose using continuous bioreactor and separation system. *Enzyme and Microbial Technology* 38 (6): 855–859.

Morimoto, K., Terami, Y., Maeda, Y.I. et al. (2013). Cloning and characterization of the L-ribose isomerase gene from *Cellulomonas parahominis* MB426. *Journal of Bioscience and Bioengineering* 115 (4): 377–381.

Moynihan, P.J. (1998). Update on the nomenclature of carbohydrates and their dental effects. *Journal of Dentistry* 26 (3): 209–218.

Mu, W., Chu, F., Xing, Q. et al. (2011). Cloning, expression, and characterization of a D-psicose 3-epimerase from *Clostridium cellulolyticum* H10. *Journal of Agricultural and Food Chemistry* 59 (14): 7785–7792.

Mu, W., Zhang, W., Fang, D. et al. (2013). Characterization of ad-psicose-producing enzyme, d-psicose 3-epimerase, from *Clostridium* sp. *Biotechnology Letters* 35: 1481–1486.

Muntean, D.M., Sturza, A., Dănilă, M.D. et al. (2016). The role of mitochondrial reactive oxygen species in cardiovascular injury and protective strategies. *Oxidative Medicine and Cellular Longevity* 2016 (1): 8254942.

Naha, N., Lee, H.Y., Jo, M.J. et al. (2008). Rare sugar D-allose induces programmed cell death in hormone refractory prostate cancer cells. *Apoptosis 13*: 1121–1134.

Neta, T., Takada, K., and Hirasawa, M. (2000). Low-cariogenicity of trehalose as a substrate. *Journal of Dentistry* 28 (8): 571–576.

Nishimoto, T., Nakano, M., Nakada, T. et al. (1996). Purification and properties of a novel enzyme, trehalose synthase, from *Pimelobacter* sp. R48. *Bioscience, Biotechnology, and Biochemistry 60* (4): 640–644.

Noguchi, C., Kamitori, K., Hossain, A. et al. (2016). D-allose inhibits cancer cell growth by reducing GLUT1 expression. *The Tohoku Journal of Experimental Medicine* 238 (2): 131–141.

Nozal, M.J., Bernal, J.L., Toribio, L. et al. (2005). The use of carbohydrate profiles and chemometrics in the characterization of natural honeys of identical geographical origin. *Journal of Agricultural and Food Chemistry* 53 (8): 3095–3100.

Oh, I., Park, Y., and Lee, S. (2019). Effect of turanose on the rheology and oil uptake of instant fried noodles. *International Journal of Food Science & Technology* 55 (3): 1336–1342.

Ohara, T., Ishida, Y., Kudou, R., et al. (2015). Plant disease control agent comprising D-tagatose as active ingredient, and plant disease control method. US Patent 9,125,409, filed 18 August 2019 and issued 8 September 2015.

ÖZGÜR, M., ÖZGÜR, E.B., and DİNÇOĞLU, A.H. (2022). Prebiotic effect of D-allulose (D-psicose): traditional review. *Turkiye Klinikleri Journal of Health Sciences/Türkiye Klinikleri Sağlık Bilimleri Dergisi 7* (2): 573–577.

Pan, Y.T., Koroth Edavana, V., Jourdian, W.J. et al. (2004). Trehalose synthase of *Mycobacterium smegmatis*: purification, cloning, expression, and properties of the enzyme. *European Journal of Biochemistry* 271 (21): 4259–4269.

Park, H.Y., Park, C.S., Kim, H.J., and Oh, D.K. (2007). Substrate specificity of a galactose 6-phosphate isomerase from *Lactococcus lactis* that produces D-allose from D-psicose. *Journal of Biotechnology* 132 (1): 88–95.

Park, C.S., Kim, T., Hong, S.H. et al. (2016a). D-allulose production from D-fructose by permeabilized recombinant cells of *Corynebacterium glutamicum* cells expressing D-allulose 3-epimerase *Flavonifractor plautii*. *PLoS One* 11 (7): e0160044.

Park, M.-O., Lee, B.-H., Lim, E. et al. (2016b). Enzymatic process for high-yield turanose production and its potential property as an adipogenesis regulator. *Journal of Agricultural Food Chemistry* 64 (23): 4758–4764.

Park, M.O., Chandrasekaran, M., and Yoo, S.H. (2019). Production and characterization of low-calorie turanose and digestion-resistant starch by an amylosucrase from *Neisseria subflava*. *Food Chemistry* 300: 125225.

Patel, S.N., Sharma, M., Lata, K. et al. (2016). Improved operational stability of d-psicose 3-epimerase by a novel protein engineering strategy, and d-psicose production from fruit and vegetable residues. *Bioresource Technology* 1 (216): 121–127.

Patel, S.N., Singh, V., Sharma, M. et al. (2018). Development of a thermo-stable and recyclable magnetic nanobiocatalyst for bioprocessing of fruit processing residues and D-allulose synthesis. *Bioresource Technology* 247: 633–639.

Patel, S.N., Kaushal, G., and Singh, S.P. (2020). A novel d-Allulose 3-epimerase gene from the metagenome of a thermal aquatic habitat and d-Allulose production by *Bacillus subtilis* whole-cell catalysis. *Applied and Environmental Microbiology* 86 (5).

Patel, S.N., Kaushal, G., and Singh, S.P. (2021). d-Allulose 3-epimerase of *Bacillus* sp. origin manifests profuse heat-stability and noteworthy potential of d-fructose epimerization. *Microbial Cell Factories* 20: 1–16.

Patel, S.N., Sharma, S., Singh, A.K., and Singh, S.P. (2023). Potential microbial bioresources for functional Sugar molecules. *Microbial Bioreactors for Industrial Molecules*. 18: 211–236.

Patel, S.N., Sharma, S., Gossai, N. et al. (2025). A novel xylose isomerase suitable for D-fructose production and synergistic catalysis with D-allulose 3-epimerase for the biosynthesis of D-allulose. Process Biochemistry 151: 52–64.

Patra, F., Patel, A., and Shah, N. (2017). Microbial production of low-calorie sugars. In: *Microbial Production of Food Ingredients and Additives*, 259–290. Academic Press.

Paul, M. (2007). Trehalose 6-phosphate. *Current Opinion in Plant Biology* 10 (3): 303–309.

Perez-Cenci, M. and Salerno, G.L. (2014). Functional characterization of *Synechococcus* amylosucrase and fructokinase encoding genes discovers two novel actors on the stage of cyanobacterial sucrose metabolism. *Plant Science* 224: 95–102.

Perold, G.W., Beylis, P., and Howard, A.S. (1973). Metabolites of proteaceae. Part VIII. The occurrence of (+)-D-allose in nature: rubropilosin and pilorubrosin from *Protea rubropilosa* beard. *Journal of the Chemical Society, Perkin Transactions* 1: 643–649.

Pikis, A., Immel, S., Robrish, S.A., and Thompson, J. (2002). Metabolism of sucrose and its five isomers by *Fusobacterium mortiferum*. *Microbiology* 148 (3): 843–852.

Pizzut-Serin, S., Potocki-Véronèse, G., van der Veen, B.A. et al. (2005). Characterisation of a novel amylosucrase from *Deinococcus radiodurans*. *FEBS Letters* 579 (6): 1405–1410.

Ponnurangam, M. and Balaji, S. (2024). Tune in to the terrific applications of turanose. *European Food Research and Technology* 250 (2): 375–387.

Poonperm, W., Takata, G., Okada, H. et al. (2007). Cloning, sequencing, overexpression and characterization of L-rhamnose isomerase from *Bacillus pallidus* Y25 for rare sugar production. *Applied Microbiology and Biotechnology* 76: 1297–1307.

Pu, L.L., Cui, X., Fink, B.F. et al. (2005). Cryopreservation of adipose tissues: the role of trehalose. *Aesthetic Surgery Journal* 25 (2): 126–131.

Pupyshev, A.B., Klyushnik, T.P., Akopyan, A.A. et al. (2022). Disaccharide trehalose in experimental therapies for neurodegenerative disorders: molecular targets and translational potential. *Pharmacological Research* 183: 106373.

Raeisi, J., Saifi, M., Pourshafie, M.R. et al. (2017). Rapid identification of vancomycin resistant *Enterococcus faecalis* clinical isolates using a sugar fermentation method. *Journal of Clinical and Diagnostic Research: JCDR* 11 (3): DC14.

Rai, S.K., Singh, A., Kauldhar, B.S., and Yadav, S.K. (2023). Robust nano-enzyme conjugates for the sustainable synthesis of a rare sugar D-tagatose. *International Journal of Biological Macromolecules* 231: 123406.

Rhimi, M., Ilhammami, R., Bajic, G. et al. (2010). The acid tolerant L-arabinose isomerase from the food grade *Lactobacillus sakei* 23K is an attractive D-tagatose producer. *Bioresource Technology* 101 (23): 9171–9177.

Rhimi, M., Bajic, G., Ilhammami, R. et al. (2011). The acid-tolerant L-arabinose isomerase from the mesophilic *Shewanella* sp. ANA-3 is highly active at low temperatures. *Microbial Cell Factories* 10: 1–11.

Roy, S., Chikkerur, J., Roy, S.C. et al. (2018). Tagatose as a potential nutraceutical: production, properties, biological roles, and applications. *Journal of Food Science* 83 (11): 2699–2709.

Ruiz-Aceituno, L., Hernandez-Hernandez, O., Kolida, S. et al. (2018). Sweetness and sensory properties of commercial and novel oligosaccharides of prebiotic potential. *Lwt* 97: 476–482.

Samukawa, Y., Ouchi, Y., Miyashita, T. et al. (2024). Systemically aminstered D-allose inhibits the tumor energy pathway and exerts synergistic effects with radiation. *Anticancer Research* 44 (5): 1895–1903.

Seo, M.J. (2013). Characterization of an L-arabinose isomerase from *Bacillus thermoglucosidasius* for D-tagatose production. *Bioscience, Biotechnology, and Biochemistry* 77 (2): 385–388.

Seo, D.H., Jung, J.H., Choi, H.C. et al. (2012). Functional expression of amylosucrase, a glucan-synthesizing enzyme, from *Arthrobacter chlorophenolicus* A6. *Journal of Microbiology and Biotechnology* 22 (9): 1253–1257.

Seo, M.J., Choi, J.H., Kang, S.H. et al. (2018). Characterization of L-rhamnose isomerase from *Clostridium stercorarium* and its application to the production of D-allose from D-allulose (D-psicose). *Biotechnology Letters* 40: 325–334.

Sharma, S., Patel, S.N., and Singh, S.P. (2024). A novel thermotolerant l-rhamnose isomerase variant for biocatalytic conversion of d-allulose to d-allose. *Applied Microbiology and Biotechnology* 108 (1): 279.

Shehata, H.M., Abd El-Ghany, M.N., Hamdi, S.A. et al. (2023). Characterization of a metallic-ions-independent L-arabinose isomerase from endophytic *Bacillus amyloliquefaciens* for production of D-tagatose as a functional sweetener. *Fermentation* 9 (8): 749.

Shin, K.C., Seo, M.J., Kim, S.J. et al. (2022). Characterization of l-arabinose isomerase from *Klebsiella pneumoniae* and its application in the production of d-tagatose from d-galactose. *Applied Sciences* 12 (9): 4696.

Shintani, T., Yamada, T., Hayashi, N. et al. (2017). Rare sugar syrup containing d-allulose but not high-fructose corn syrup maintains glucose tolerance and insulin sensitivity partly via hepatic glucokinase translocation in Wistar rats. *Journal of Agricultural and Food Chemistry* 65 (13): 2888–2894.

Shintani, H., Shintani, T., and Sato, M. (2020). D-Allose, a trace component in human serum, and its pharmaceutical applicability. *International Journal of Applied Biology and Pharmaceutical Technology* 11 (3): 200–213.

Singh, A., Rai, S.K., Manisha, M., and Yadav, S.K. (2021). Immobilized L-ribose isomerase for the sustained synthesis of a rare sugar D-talose. *Molecular Catalysis* 511: 111723.

Singh, A., Rai, S.K., and Yadav, S.K. (2022). Metal-based micro-composite of L-arabinose isomerase and L-ribose isomerase for the sustainable synthesis of L-ribose and D-talose. *Colloids and Surfaces B: Biointerfaces* 217: 112637.

Smith, A., Avery, A., Ford, R. et al. (2022). Rare sugars: metabolic impacts and mechanisms of action: a scoping review. *British Journal of Nutrition* 128 (3): 389–406.

Sui, L., Dong, Y., Watanabe, Y. et al. (2005). Growth inhibitory effect of D-allose on human ovarian carcinoma cells in vitro. *Anticancer Research* 25 (4): 2639–2644.

Sui, L., Nomura, R., Dong, Y. et al. (2007). Cryoprotective effects of D-allose on mammalian cells. *Cryobiology* 55 (2): 87–92.

Sun, Y., Hayakawa, S., Puangmanee, S., and Izumori, K. (2006). Chemical properties and antioxidative activity of glycated α-lactalbumin with a rare sugar, D-allose, by Maillard reaction. *Food Chemistry* 95 (3): 509–517.

Sun, Y., Hayakawa, S., Ogawa, M. et al. (2008). Influence of a rare sugar, D-psicose, on the physicochemical and functional properties of an aerated food system containing egg albumen. *Journal of Agricultural and Food Chemistry* 56 (12): 4789–4796.

Suna, S., Yamaguchi, F., Kimura, S. et al. (2007). Preventive effect of D-psicose, one of rare ketohexoses, on di-(2-ethylhexyl) phthalate (DEHP)-induced testicular injury in rat. *Toxicology Letters* 173 (2): 107–117.

Suzuki, K., Ogawa, D., Kanda, T. et al. (2023). Antiproliferative effects of D-allose associated with reduced cell division frequency in glioblastoma. *Scientific Reports* 13 (1): 19515.

Takata, M.K., Yamaguchi, F., Nakanose, K. et al. (2005). Neuroprotective effect of D-psicose on 6-hydroxydopamine-induced apoptosis in rat pheochromocytoma (PC12) cells. *Journal of Bioscience and Bioengineering* 100 (5): 511–516.

Tang, X., An, Y., Iqbal, M.W. et al. (2022). The characterization of a novel D-allulose 3-epimerase from *Blautia produca* and its application in D-allulose production. *Food* 11 (20): 3225.

Tang, X., Ravikumar, Y., Zhang, G. et al. (2024). D-allose, a typical rare sugar: properties, applications, and biosynthetic advances and challenges. *Critical Reviews in Food Science and Nutrition* 1–28.

Tohi, Y., Taoka, R., Zhang, X. et al. (2022). Antitumor effects of orally administered rare sugar D-allose in bladder cancer. *International Journal of Molecular Sciences* 23 (12): 6771.

Torres, P.R., Manzo, R.M., Rubiolo, A.C. et al. (2014). Purification of an l-arabinose isomerase from *Enterococcus faecium* DBFIQ E36 employing a biospecific affinity strategy. *Journal of Molecular Catalysis B: Enzymatic* 102: 99–105.

Trakarnpaiboon, S., Bunterngsook, B., Lekakarn, H. et al. (2023). Characterization of cold-active trehalose synthase from *Pseudarthrobacter* sp. for trehalose bioproduction. *Bioresources and Bioprocessing* 10 (1): 65.

Tseng, W.C., Wu, T.J., Chang, Y.J. et al. (2017). Overexpression and characterization of a recombinant L-ribose isomerase from *Actinotalea fermentans* ATCC 43279. *Journal of Biotechnology* 259: 168–174.

Van Overtveldt, S., Gevaert, O., Cherlet, M. et al. (2018). Converting galactose into the rare sugar talose with cellobiose 2-epimerase as biocatalyst. *Molecules* 23 (10): 2519.

Vastenavond, C., Bertelsen, H., Hansen, S. et al. (2011). Alternative sweeteners. *Agro Food Industry Hi Tech* 15 (4): 39–41.

Vitale, M.C., Pascadopoli, M., Gallo, S. et al. (2023). Oral findings in pediatric patients with allergic rhinitis and asthma: a cohort study of an Italian setting. *Journal of Clinical Pediatric Dentistry* 47 (6): 51–58.

Wanarska, M. and Kur, J. (2012). A method for the production of D-tagatose using a recombinant Pichia pastoris strain secreting β-D-galactosidase from *Arthrobacter chlorophenolicus* and a recombinant L-arabinose isomerase from *Arthrobacter* sp. 22c. *Microbial Cell Factories* 11: 1–15.

Wang, J.H., Tsai, M.Y., Chen, J.J. et al. (2007). Role of the C-terminal domain of *Thermus thermophilus* trehalose synthase in the thermophilicity, thermostability, and efficient production of trehalose. *Journal of Agricultural and Food Chemistry* 55 (9): 3435–3443.

Wang, R., Bae, J.S., Kim, J.H. et al. (2012). Development of an efficient bioprocess for turanose production by sucrose isomerisation reaction of amylosucrase. *Food Chemistry* 132 (2): 773–779.

Wang, J., Ren, X., Wang, R. et al. (2017). Structural characteristics and function of a new kind of thermostable trehalose synthase from *Thermobaculum terrenum*. *Journal of Agricultural and Food Chemistry* 65 (35): 7726–7735.

Wang, Y., Ravikumar, Y., Zhang, G. et al. (2020). Biocatalytic synthesis of D-allulose using novel D-tagatose 3-epimerase from *Christensenella minuta*. *Frontiers in Chemistry* 8: 622325.

Weckwerth, W., Loureiro, M.E., Wenzel, K., and Fiehn, O. (2004). Differential metabolic networks unravel the effects of silent plant phenotypes. *Proceedings of the National Academy of Sciences* 101 (20): 7809–7814.

Wei, Y.T., Zhu, Q.X., Luo, Z.F. et al. (2004). Cloning, expression and identification of a new trehalose synthase gene from *Thermobifida fusca* genome. *Acta Biochimica et Biophysica Sinica* 36 (7): 477–484.

Xiuli, W., Hongbiao, D., Ming, Y., and Yu, Q. (2009). Gene cloning, expression, and characterization of a novel trehalose synthase from Arthrobacter aurescens. *Applied Microbiology and Biotechnology* 83:x477–482.

Xu, Z., Qing, Y., Li, S. et al. (2011). A novel L-arabinose isomerase from *Lactobacillus fermentum* CGMCC2921 for D-tagatose production: gene cloning, purification and characterization. *Journal of Molecular Catalysis B: Enzymatic* 70 (1–2): 1–7.

Xu, W., Zhang, W., Tian, Y. et al. (2017). Characterization of a novel thermostable L-rhamnose isomerase from *Thermobacillus composti* KWC4 and its application for production of D-allose. *Process Biochemistry* 53: 153–161.

Xu, Y., Wu, Y., Liu, Y. et al. (2022). Sustainable bioproduction of natural sugar substitutes: strategies and challenges. *Trends in Food Science & Technology* 129: 512–527.

Yamaguchi, F., Takata, M., Kamitori, K. et al. (2008). Rare sugar D-allose induces specific up-regulation of TXNIP and subsequent G1 cell cycle arrest in hepatocellular carcinoma cells by stabilization of p27kip1. *International Journal of Oncology* 32 (2): 377–385.

Yan, J., Qiao, Y., Hu, J., and Ding, H. (2013). Cloning, expression and characterization of a trehalose synthase gene from *Rhodococcus opacus*. *The Protein Journal* 32: 223–229.

Yang, J., Pan, C., Zhang, J. et al. (2017). Exploring the potential of biocompatible osmoprotectants as highly efficient cryoprotectants. *ACS Applied Materials & Interfaces* 9 (49): 42516–42524.

Yang, J., Fan, D., Zhao, F. et al. (2022). Characterization of D-allulose-3-epimerase from *Ruminiclostridium papyrosolvens* and immobilization within metal-organic frameworks. *Frontiers in Bioengineering and Biotechnology* 10: 869536.

Yaribeygi, H., Yaribeygi, A., Sathyapalan, T., and Sahebkar, A. (2019). Molecular mechanisms of trehalose in modulating glucose homeostasis in diabetes. *Diabetes and Metabolic Syndrome: Clinical Research and Reviews* 13 (3): 2214–2218.

Yoon, R.Y., Yeom, S.J., Park, C.S., and Oh, D.K. (2009). Substrate specificity of a glucose-6-phosphate isomerase from *Pyrococcus furiosus* for monosaccharides. *Applied Microbiology and Biotechnology* 83: 295–303.

Yoshizane, C., Mizote, A., Yamada, M. et al. (2017). Glycemic, insulinemic and incretin responses after oral trehalose ingestion in healthy subjects. *Nutrition Journal* 16: 1–6.

Yu, S., Park, H., and Kim, W. (2023). Trehalose inhibits inflammatory responses through mitochondrial reprogramming in raw 264.7 macrophages. *Antioxidants 12* (6): 1166.

Yue, M., Wu, X.L., Gong, W.N., and Ding, H.B. (2009). Molecular cloning and expression of a novel trehalose synthase gene from *Enterobacter hormaechei*. *Microbial Cell Factories* 8: 1–7.

Yun-Zhan, N.I., Chun-Jun, Q.I., Wen-Bin, S.U. et al. (2020). Anomeric configuration-dependence of the Lattrell-Dax epimerization from D-glucose to synthetically useful D-allose derivatives. *Chinese Journal of Natural Medicines* 18 (10): 723–728.

Zehner, L.R. and Biospherics Inc. (1988). *D-tagatose as a low-calorie carbohydrate sweetener and bulking agent.* U.S. Patent 4,786,722, files 29 August 1986, and issued 22 November 1988.

Zhang, L., Mu, W., Jiang, B., and Zhang, T. (2009). Characterization of D-tagatose-3-epimerase from *Rhodobacter sphaeroides* that converts D-fructose into D-psicose. *Biotechnology Letters* 31: 857–862.

Zhang, W., Fang, D., Xing, Q. et al. (2013). Characterization of a novel metal-dependent D-psicose 3-epimerase from *Clostridium scindens* 35704. *PLoS One* 8 (4): e62987.

Zhang, W., Li, H., Zhang, T. et al. (2015). Characterization of a D-psicose 3-epimerase from *Dorea* sp. CAG317 with an acidic pH optimum and a high specific activity. *Journal of Molecular Catalysis B: Enzymatic* 120: 68–74.

Zhang, W., Zhang, T., Jiang, B., and Mu, W. (2016). Biochemical characterization of a d-psicose 3-epimerase from *Treponema primitia* ZAS-1 and its application on enzymatic production of d-psicose. *Journal of the Science of Food and Agriculture* 96 (1): 49–56.

Zhang, W., Zhang, T., Jiang, B., and Mu, W. (2017). Enzymatic approaches to rare sugar production. *Biotechnology Advances* 35 (2): 267–274.

Zhang, S., Qiu, X., Zhang, Y. et al. (2023). Metabolomic analysis of trehalose alleviating oxidative stress in myoblasts. *International Journal of Molecular Sciences* 24 (17): 13346.

Zheng, L.J., Guo, Q., Zhang, Y.X. et al. (2022). Engineering of *Escherichia coli* for D-allose fermentative synthesis from D-glucose through izumoring cascade epimerization. *Frontiers in Bioengineering and Biotechnology* 10: 1050808.

Zhu, Y., Wei, D., Zhang, J. et al. (2010). Overexpression and characterization of a thermostable trehalose synthase from *Meiothermus ruber*. *Extremophiles* 14: 1–8.

Zhu, Y., Men, Y., Bai, W. et al. (2012). Overexpression of d-psicose 3-epimerase from *Ruminococcus* sp. in *Escherichia coli* and its potential application in d-psicose production. *Biotechnology Letters* 34: 1901–1906.

Zhu, X., Tian, Y., Xu, W. et al. (2018). Biochemical characterization of a highly thermostable amylosucrase from *Truepera radiovictrix* DSM 17093. *International Journal of Biological Macromolecules* 116: 744–752.

Zhu, Z., Gao, D., Li, C. et al. (2019). Redesign of a novel D-allulose 3-epimerase from *Staphylococcus aureus* for thermostability and efficient biocatalytic production of D-allulose. *Microbial Cell Factories* 18: 1–10.

Zhu, Z., Li, L., Zhang, W. et al. (2021). Improving the enzyme property of D-allulose 3-epimerase from a thermophilic organism of *Halanaerobium congolense* through rational design. *Enzyme and Microbial Technology* 149: 109850.

7

Microbial Engineering for the Production of High-value Polyphenolics

Deepak Sharma[1], *Shweta Kamboj*[1], *Maninder Jeet Kaur*[1], *Ranju Kumari Rathour*[1], *and Nitish Sharma*[2]

[1] *Department of Biotechnology, Chandigarh College of Technology, Chandigarh Group of Colleges, Landran, Mohali, Punjab, 140301, India*
[2] *Department of Genetics, Waksman institute of Microbiology, Rutgers University (A State University of New Jersey), New Brunswick, NJ, 08901, USA*

7.1 Introduction

Plants are known to be the natural resource of a diverse range of secondary metabolites, which are recognized as the potent antioxidants and anti-inflammatory agents and having myriad significant health benefits (Upadhyay and Singh, 2021, 2023, 2024). Polyphenols are such naturally occurring bioactive compound having aromatic ring and contain several hydroxyl groups (Chojnacka et al. 2021; Zhang et al. 2022). A wide range of flavonoids, phenolics, stilbenes, and lignans are most commonly present in polyphenols. These bioactive compounds play a significant in plant defense mechanism against pathogens (Watson 2018). Besides being beneficial for plants, these compounds have also gained a considerable attention as these have been associated with providing therapeutic potential and prevent various cardiovascular diseases, cancers, and neurodegenerative disorders. These compounds are also reported to boost the digestive system and type-2 diabetes (Sharma et al. 2019, 2024; Chiriac et al. 2021). Berries, tea, dark chocolates, and red wine are the best-known or some of the most common sources of polyphenols. However, despite being greatly important, the concentrations of polyphenols are very low in plants. As plants are the primary and most common source of polyphenols, the effectiveness and yield of polyphenols also vary with the seasonal, environmental, and geographical conditions (Kumar et al. 2019). These bioactive compounds are extracted from plants using traditional methods. The conventional methods have certain limitations with respect to their sustainability, feasibility, low yield, scalability, efficiency, and cost-effectiveness. Another major limitation of using plant as primary source is additional land demand for cultivation of polyphenol-rich plants, long growth, and maturation cycle (Olszewska et al. 2020; Albuquerque et al. 2021; Zagoskina et al. 2023). Thus, there is a need to find alternative source with maximum yield in short span of time and cost-effective methods of extraction.

Emerging interest in polyphenols and their beneficial health impact have forced scientific community to find microbial alternative as potent producers of polyphenols. Various microbial sources specifically gut microbiota *Lactobacillus* sp., *S. cerevisiae*, *Candida glabrata*, etc., have been reported to produce polyphenols *via* fermentation (Piekarska-Radzik and Klewicka 2021;

Microbial Nutraceuticals: Products and Processes, First Edition. Edited by Sudhir Pratap Singh and Santosh Kumar Upadhyay.

Sharma et al. 2022). There are certain fermented products (beer, wine, kefir, yoghurt) documented for the polyphenols with enhanced efficacy and bioavailability enumerated through microbial metabolism. It has also been reported that the production of polyphenols *via* fermentation not only increases the bioactivity but also leads to the production of novel and unique bioactive compounds with enhanced properties and health benefits (Yang et al. 2023). The gut microbiota plays a significant role in converting dietary polyphenols into high-value metabolites (Catalkaya et al. 2020). Polyphenols and gut microbiota interact to modulate inflammatory disorders, oxidative stress, and digestion of lipids, which in turn affects gut health, energy metabolism, and general well-being (Gade and Kumar 2023; Sarkar et al. 2022).

Native microorganisms though play a significant role in production of value-added products, in bioremediation, and in biotechnology at large scale but also have certain limitations with respect to their effectiveness, specificity, and efficiency, especially, when used in the crude ways (Singh and Upadhyay 2021, 2023). Therefore, manipulation of their genetic makeup to tailor a strain with much yield under varied conditions can be of greater importance. Engineering of microbes with specific traits offer higher yield and efficiency compared to the native producers (Rathour et al. 2022). Advances in synthetic biology, genetic engineering, recombinant DNA technology, and bio-processing have enabled the scientific community to alter the metabolic machinery for higher production of value-added products (Milke et al. 2018). Certain microbial hosts like *E. coli* and probiotic bacteria (*S. cerevisiae, Corynebacterium glutamicum*, and *Lactococcus lactis*) are most preferred microbes engineered in such a way that they can produce polyphenols with higher yield (Dudnik et al. 2018). Besides yield, the other major benefit of microbial polyphenols is related to health, as during microbial fermentation, the polyphenols are transferred from bounded to free form that helps in improving human health by increasing the bioavailability of these high-value biomolecules. Furthermore, the production of polyphenols using gut microbiota also boosts the host health (Sun et al. 2021; Sarkar et al. 2022).

Despite being useful in the production of polyphenols, there are certain challenges associated with microbial production processes, for example, complex metabolic pathways, toxic intermediates, overall cost of process, and down streaming. Thus, in this chapter, authors have provided a comprehensive overview of production of microbial polyphenols. Authors have also highlighted different genetic and metabolic approaches used to build microbial cell factories and point out recent developments in the field and future prospects of this least explored arena of study. Furthermore, the potential of microbial engineering in transforming the bounded high-value polyphenolics for more scalable and sustainable biotechnological solutions and health benefits has also been emphasized.

7.2 Properties and Classification of Polyphenols

The presence of one or more hydroxyl groups and at least a phenyl ring are the main characteristic features of polyphenols. The structural composition may vary from a single aromatic/benzene ring to the multiple complexes (Stromsnes et al. 2021). A phenol having saccharide group attached through C—C bond, a puerarin, and 10-atom benzopyran isoflavone are the major components of fussed structure of polyphenols (Zagoskina et al. 2023). The major function as well as properties of these stewed ring secondary metabolites is to protect plants from pathogens and also from the harsh ultraviolet radiations of the sun. Besides this, these complex compounds also help in maintaining color, odor, oxidative stability, flavor, and astringency of plant-based food stuffs (Pan et al. 2023). There are many reports which confirm the anticancerous and antidiabetic properties of polyphenols along with their applicability in the treatment of cardiovascular as well as neurodegenerative diseases (Ebrahimi and Lante 2021; Subhan et al. 2024). Shikimic acid and phenylalanine are main

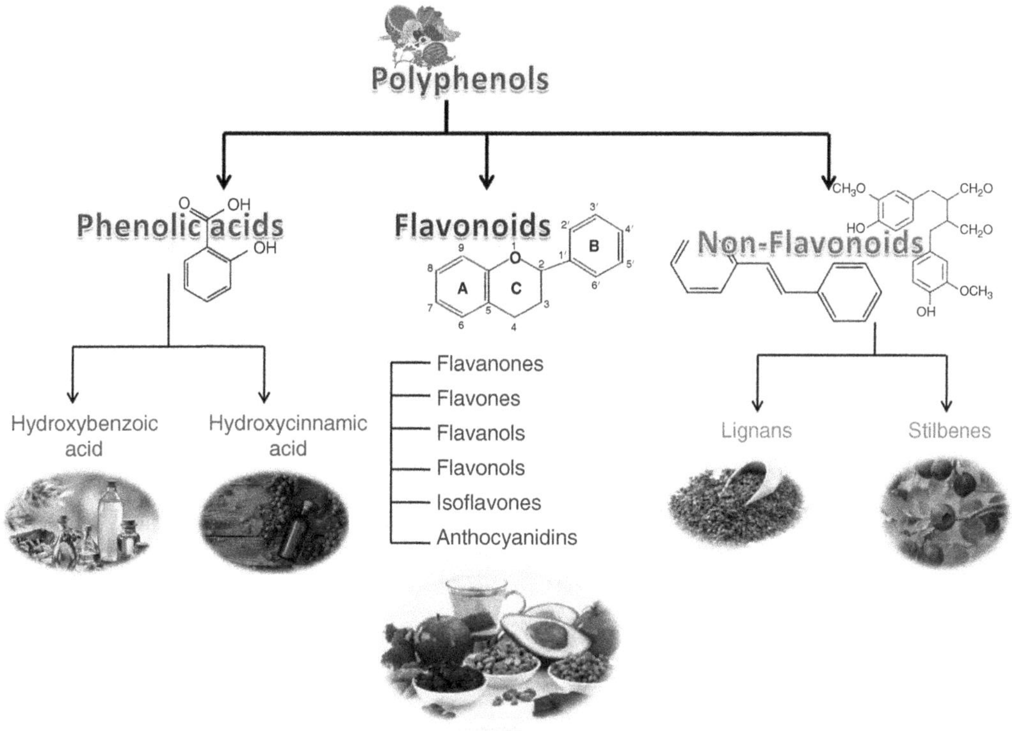

Figure 7.1 Source, structure, and functional attributes of polyphenols and their classification. volff/Adobe Stock Photos, NikoG/Adobe Stock, robynmac/Adobe Stock Photos, Subbotina Anna/Adobe Stock Photos, Viktor Koldunov/Adobe Stock Photos.

precursor and intermediate of phenolic compounds in both plants and microorganism (Pan et al. 2023; Bertelli et al. 2021).

There are number of polyphenolics extracted and identified from plants. On the basis of number of benzene rings, functional groups, and structural elements, polyphenols are broadly categorized as phenolic acids, flavonoids, and non-flavonoids (including stilbenes and lignans) (Figure 7.1) (Stromsnes et al. 2021).

7.2.1 Phenolic Acid

Phenolic acids are one of the major subgroups of diverse phytochemicals called as polyphenols. They mainly contribute about 30% of total polyphenolic content of plants. The chemical structure of phenolic acids has multiple alcohol groups attached to the aromatic ring. More specifically, an aromatic ring with hydroxyl group (single or multiple) and a carboxylic acid are known as phenolic acids (Vidal-Casanella et al. 2021). These phyto-constituents are mainly reported to have health benefits as well as antioxidant properties and largely present in fruits, grains, and vegetables. Phenolic acids are gaining much research interest due to their certain chemical, biological, and nutritional benefits. Phenolic acids are useful in prevention of cardiovascular and cancerous diseases and also possess anti-inflammatory properties (Rashmi and Negi 2020). Recently, Caruso et al. (2022) evaluated the neuroprotective role of phenolic acids as treatment for Alzheimer's and cognitive disorders.

Broadly, phenolic acids are classified as hydroxybenzoic and hydroxycinnamic acid. Hydroxybenzoic acid contains a single hydroxyl entity and also known as monohydroxybenzoic acid, which is a benzoic acid derivative. This phenolic acid is most commonly reported in plants of *Vitex* genus (*V. agnus, V. negundo*), *Hypericum perforatum*, and *Spongiochloris* sp.; microbial sources of hydrobenzoic acids are *Ganodermalucidum* and *Cryptanaerobacterphenolicus* (Kalinowska et al. 2021). Gallic acid and protocatechuic acid (PCA) are two common hydroxybenzoic acids largely found in gallnuts, hazel nuts, pears, tea leaves, various vegetables, and fruits (Gong and Zha 2023). Besides this, hydroxycinnamic acid is derivative of cinnamic acid having C_3 to C_6 carbon structure. Caffeic and ferulic acids are two common examples of this type of phenolic acids found in coffee, cereals, and fruits (Sova and Saso 2020).

7.2.2 Flavonoids

Like phenolic acids, flavonoids are also a subclass of polyphenols present in majority of plants including fruits, vegetables, tea tree, and wines. The 60% of total polyphenols are flavonoids. Flavonoids have 15 carbons and two aromatic rings joined together by C≡C bonds, a phenylchromen-4-1 scaffold (may contain hydroxyl or methoxy group) at C2 or C3, and a ketone entity at C4. The flavonoids are divided into six different groups, namely flavanones, flavones, flavanols, flavonols, isoflavones, and anthocyanidins (Safe et al. 2021). Phenylalanine and malonyl-Co A are two major precursors of flavonoids. Until 2020, approximately 8000 different flavonoids have been reported in plants involved in different functions (Alseekh et al. 2021). Quercetin, apigenin, kaempferol, hesperidin, catechins, cyaniding, naringenin, etc., are some of the common examples of flavonoids. These kinds of polyphenols also possess free radical scavenging activity, act as anti-inflammatory agents, and helps in the reduction of oxidative stress (Liu et al. 2019; Tyagi et al. 2024).

7.2.3 Non-flavonoids

These categories of polyphenols mainly include lignans and stilbenes. Lignans have a dibenzylbutane backbone structure that basically has two C3 and C6 carbon phenylpropane bonded by C—C bond. This structure is also known as bicycle-octane structure as it having two octanes (González-Sarrías et al. 2020). The presence of hydroxyl may vary (one to multiple) and directly contribute to antioxidant property. The biological activities of lignan molecule depend upon its structural stereoisomeric forms and distance between central C—C bonds (Hazafa et al. 2022). Lignans can act as important ingredient of balanced diet and used supplement to facilitate gut microbiota. Flaxseeds, rye, oats, chickpeas, lentils, barriers, and broccoli are the primary source of lignans in the diet. Lignans are reported to boost digestive system, maintain hormonal balance, improve cardiovascular disorder, and haveanticancerous activity (Dwita et al. 2022). Stilbenes are also an important non-flavonoid polyphenol present in plants having health benefits similar to those oflignans and other polyphenols. Methylene bridge formed in its structure helps in binding two phenyl rings of stilbenes (Navarro-Orcajada et al. 2023). Resveratrol is most common and well-known stilbene presents in the berries, grapes, and wine, specifically the red wine having cardiovascular, antioxidant, and anti-inflammatory activity (Pecyna et al. 2020).

7.3 Sources of Polyphenols

Polyphenols are widespread and diverse phenolic molecules with molecular weights nearly equals to 30 KDa (Rudrapal et al. 2024). The size of these molecules significantly varies from species to species in plants. Initially, plants were considered as the most common source of polyphenols, and they have indeed played a critical role in diet and nutrition (Zagoskina et al. 2023). However, the food-feed crises, limited land availability, and excess market demand have forced the scientific community to look for the more promising microbial alternatives as compared to the plant-based products. Microorganisms are increasingly being used as hosts for the heterologous synthesis of natural products due to their rapid growth, ease of cultivation, and suitability for industrial processes (Pham et al. 2019). Most of the polyphenols remain in the colon, where gut microbiota could transform them into the metabolites that enter systemic circulation and undergo further metabolism.

7.3.1 Plant as a Source for Polyphenols

Plant-based foods are rich in polyphenols, having wide variety of complex structures and contribute to majority of dietary metabolites. Polyphenols, as discussed earlier, are secondary metabolites of plants, and act as important phytochemicals due to their variety of applications. However, the concentrations of different polyphenols may vary depending on environmental factors, harvesting time, and plant species (Lacko-Bartošová et al. 2023). For instance, flavonoids have been reported in very low concentration in red wines, while onions are rich in flavonoid content and contain nearly 1.2 g of flavanols per kg of fresh sample, followed by leeks, broccoli, and berries (Panche et al. 2016). Fruits (apple, barriers, grapes, oranges, and citrus fruits), herbs (cloves and turmeric), vegetables (artichokes, onion, and spinach), medicinal plants, nuts, and different beverages including tea, coffee, and wine are rich sources of different types of polyphenols discovered so far. As polyphenols have been reported to promote plant growth and to improve the defense system, these molecules have been also reported in every part of plant body including budding tissue, flowering bud, fresh leaves, bark, and even stems (Marranzano et al. 2018; Bisht et al. 2024). Most recently, 1302 mg/L 3-Feroylquinic acid (polyphenol) has been reported by Mueed and co-workers (2023) in 03 (*Menthalongifolia*, *Cassia absus*, and *Camellia sinensis*) out of 08 selected medicinal plants. Further evaluation of these plants resulted in the presence of gallic acid, vanillic acid, ferulic acid, catechins, and other phenolic compounds.

Among different fruits, grapes, citrus fruits, and apples are the richest source of polyphenols. Grapes are the common fruits cultivated worldwide and used as the material for wine manufacturing while being rich in other important biomolecules such as tannins (malvidin-3-O-monoglucoside), hydroxytyrosol, and phenolic acids, which majorly contribute to the antioxidant potential of wines (Zhou et al. 2023a). Certain vegetables are also rich in polyphenols with onions at the top. A number of polyphenols including quercetin, myricetin, and kaempferol have been extracted from onions using deep eutectic solvent extraction method (Shang et al. 2024). Millets, cereals, and oil-containing plants have been reported to contain polyphenols as discussed in Table 7.1. Although plants are the richest source of polyphenols, there are certain limitations associated with them for being as sources, and that is primarily the yield (which may vary with environment conditions), degradation of polyphenols (due to excess exposure to light), heat and oxygen, and post-harvesting handling errors. Thus, the microbial production of polyphenols itself becomes the greater interest and an easy utility to maintain the standardized product yield.

7.3.2 Microbes as Polyphenol Source

Unlike plant sources, microbial counterparts utilized for the production of polyphenols have very limited availability till now. However, the ongoing advancements in biotechnological interventions have increased the fermentative production of polyphenols using microbes to overcome the

Table 7.1 Polyphenols reported in different plant sources.

Sources	Examples	Polyphenols	References
Fruits	Grapes	Catechin and resveratrol	Balanov et al. (2021)
	Apples	Epicatechin, chlorogenic acid, and phloridzin	Preti and Tarola (2021); Asma et al. (2023)
	Citrus fruits	Hesperidin, narirutin, naringin, and eriocitrin.	Liu et al. (2022); Borghi and Pavanelli (2023)
	Berries	Anthocyanins andellagitannins	Luís et al. (2018)
	Pomegranate	Ellagic acid andpunicalagin	Buenrostro-Figueroa et al. (2023)
	Cherries	Anthocyanins andhydroxycinnamic acids	Antognoni et al. (2020)
	Bananas	Catecholamines anddopamine	Athanasiadis et al. (2023)
Vegetables	Onion	Quercetin and its derivates	Moradi et al. (2022)
	Potato	Chlorogenic acid, caffeic acid, and ferulic acid	Ren and Zhou (2021)
	Tomato	Naringenin, quercetin, and kaempferol	López-Téllez and del PilarCañizares-Macías (2024)
	Spinach	Patuletin, spinacetin, and jaceidin	Díaz-Batalla et al. (2006)
	Broccoli	Kaempferol and isorhamnetin	Duan et al., (2021)
Cereals and millets	Wheat	Ferulic acid, coumaric acid, and caffeic acid	Değirmencioğlu and Gökmen (2018)
	Rice	PCA, catechins, and flavonoids	Tiyagi et al., (2022)
	Maize	Ferulic acid, coumaric acid, and anthocyanins	Taira and Tokuji (2018); Luna-Vital et al., (2020)
	Barley	Catechins, procyanidins, and ferulic acid	Phan et al. (2019)
	Oats	Avenanthramides, ferulic acid, and caffeic acid	Peterson (2017); Holtekjølen et al. (2018)
	Millet	Quercetin, kaempferol, and ferulic acid	Mahendran et al. (2021); Cai et al. (2022)
Nuts	Almonds	Catechin andcaffeic acid	Monagas et al. (2017)
	Walnuts	Ellagic acid	MoosaviDolatabadi et al. (2018)
	Pistachios	Anthocyanins and flavonoids	Hosseinian et al. (2019)
	Cashews	Anacardic acids, flavonoids, and phenolic acids	Naculet al. (2021)
Medicinal Plants	Turmeric	Curcumin, demethoxycurcumin, and bisdemethoxycurcumin.	Somparn et al. (2010)
	Ginger	Gingerol, shogaol, and paradol	Sharma et al. (2023), Al-Suhaimi and Al-Riziza (2018)
	Garlic	Allicin, alliin, and diallyl sulfide	Wang et al., (2017b)
	Aloe vera	Aloesin, aloin, catechins, and flavonoids	Ernst (2021)

Table 7.1 (Continued)

Sources	Examples	Polyphenols	References
Oil-bearing plants	Olive	Hydroxytyrosol, tyrosol, oleuropein, and ligstroside	Bulotta and Celano (2014)
	Flaxseed	Secoisolariciresinoldiglucoside	Hano and Renault (2017)
	Sesame	Sesamin, sesamol, and sesamolin	Jin et al. (2020)
	Pumpkin seed	Catechins, PCA, and caffeic acid.	Kačániová et al. (2019)
	Chia seed	Quercetin, kaempferol, and chlorogenic acid	Martínez-Cruz and Paredes-López (2014)

limitations associated with plant-based polyphenol production. Generally, *de novo* biotransformation methods are used by the microbes to synthesize polyphenolic compounds from simple substrates (Pham et al. 2019; Sharma et al. 2022). In the recent past, different bacterial as well as fungal species like *Lactobacillus* sp., *Bifidobacterium* sp., *E. coli, Clostridium* sp., *Bacteroides* sp., *Enterococcus* sp., *Aspergillus* sp., *Candida* sp., *Streptomyces* sp., *Bacillus* sp., *C. utilis*, *Aspergillus* sp., *Penicillium* sp., *Bacteroides* sp., and *Rhizopus* sp. have been reported to produce the polyphenols (Sharma et al. 2022; Piekarska-Radzik and Klewicka 2021; Makarewicz et al. 2021). In a previous study, Hu et al. (2022) has reported that the concentration of phenolic content increased up to threefold when rose residue substrate was fermented using *Lactiplantibacillus plantarum* B7 and *B. subtilis natto*. A considerable increase in polyphenol concentrations was reported during co-fermentation of guava leaves with *Monascusanka* and *Bacillus* sp. (Wang et al. 2017a). Among well-known probiotic bacteria, *Lactobacillus* sp. and *Bifidobacterium* sp. have been reported to convert the polyphenols through enzymatic transformations, whereas the *Clostridium* sp. breakdowns the flavonoids in their gut environment (Zhou et al. 2023b). Gut microbiota is crucial for host physiology and metabolism. A well-balanced composition of gut microbiota is vital for maintaining an individual's health and overall well-being (Krishnamurthy et al. 2023). The diverse groups of gut microorganisms possess enzymatic capabilities that enable them to degrade and modify dietary polyphenols into smaller metabolites with enhanced bioavailability and bioactivity (Makarewicz et al. 2021). Besides the production of polyphenols, gut microbes can also produce certain enzymes which modify their structures through tinkering different bonds or conjugates like glycosides, glucuronides, sulfates, amides, esters, lactones, etc., and also involve the hydrolysis of their rings, as well as demethylation, decarboxylation, and reduction of dietary polyphenols to facilitate their digestion (Plamada and Vodnar 2021).

Fermentation of the polyphenols can increase the growth of *Bifidobacteria* and selectively alters the Firmicutes-to-Bacteroidetes ratio compared to the controls. Additionally, polyphenols can also enhance the production of short-chain fatty acids by endogenous factors such as microbiota and digestive enzymes. Together with the characteristics of food matrix, the aforementioned factors can considerably affect the bio-accessibility, uptake, and further metabolism of phenolic compounds (Lippolis et al. 2023). Similarly, *Bifidobacterium* sp. has also been reported to digest the lignans by converting them into secoisolariciresinol (SECO) and also metabolize the isoflavones such as daidzin and genistin found in soy extracts (Peirotén et al. 2019). Furthermore, in another study, Kwaw et al. (2018) also reported an increase in polyphenolic content when mulberry juice was subjected to the fermentation using strains of *Lactobacillius* sp. Microorganisms are largely involved in the metabolism and modification of polyphenols, but there is currently very little or

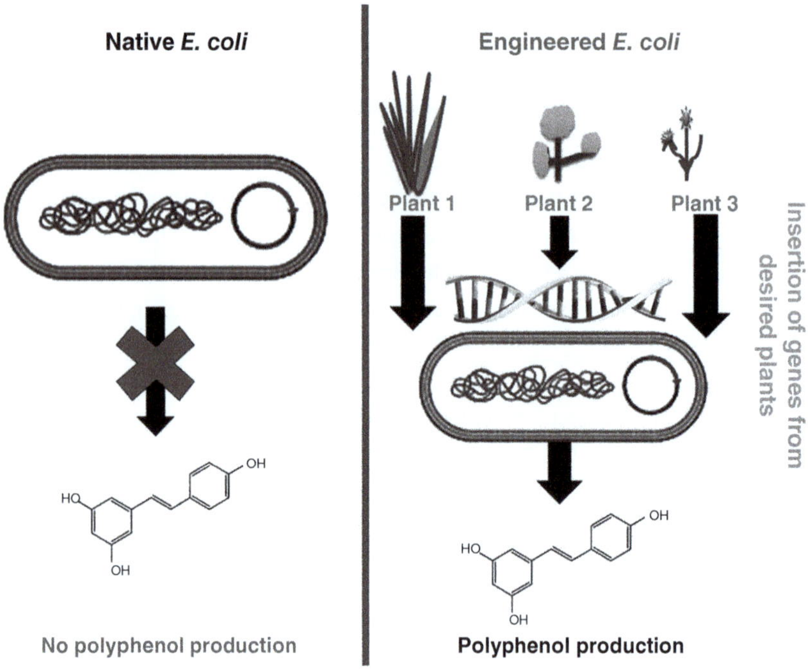

Figure 7.2 A comparison between native and engineered *E. coli* for the polyphenol production.

limited literature available on direct participation of microorganisms in polyphenol production. However, the need of the polyphenols as food, pharmaceuticals, and nutraceuticals is increasing tremendously, and industries frequently sought the availability of polyphenols from cheaper natural resources, especially when it comes to the particular compositions and qualities. To address this gap, the development of genetically modified organisms (GMOs) certainly holds promises. Because, the GMOs can be engineered with precise metabolic pathways to produce the desired polyphenols more efficiently and sustainably (Figure 7.2).

7.4 Metabolic Engineering of Bacteria for Polyphenol Production

Microbial polyphenol production represents an effective option for manufacturing natural compounds in substantial quantities as well as in an environment-friendly manner. However, bacteria do not have native machinery to build plant-derived polyphenols. Therefore, to accomplish this involves the integration of heterologous plant processes *via* genetic and metabolic engineering approaches. Genetic engineering of bacteria for polyphenol biosynthesis includes altering bacterial genetic material to enable it for polyphenol synthesis. It begins with the selection of target polyphenols, followed by the identification of biosynthetic pathways, and the use of a suitable genetic engineering technology to achieve desired polyphenol synthesis (Figure 7.3) (Chu et al. 2024). Furthermore, developments in synthetic biology and metabolic engineering approaches have offered some new options to synthesize the plant polyphenol compounds in microbes (Li et al. 2024a). Polyphenolic compounds and their direct precursors have been effectively manufactured in common biomolecule production hosts such as *E. coli, S. cerevisiae, Corynebacterium glutamicum,* and *Lactococcus lactis* (Dudnik et al. 2018, Sun et al. 2022).

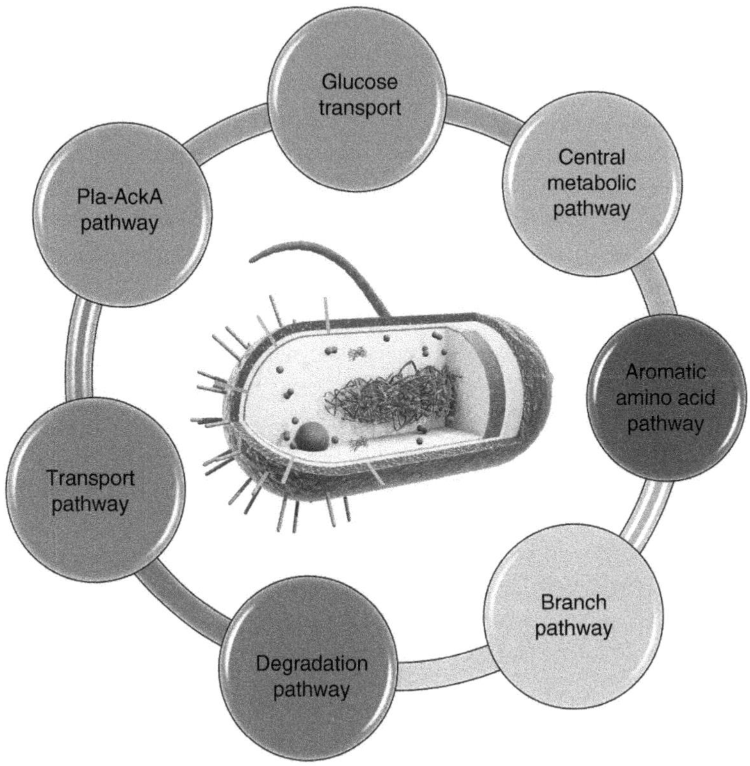

Figure 7.3 Different metabolic engineering pathways used for the production of polyphenols.

7.4.1 Genetic Engineering Approach for Polyphenol Production in Bacteria

In the recent past, *E. coli* has been used to produce a variety of flavonoids. In 2003, Hwang et al. has published the first study on the synthesis of plant-derived flavonoids in *E. coli* by engineering its strain with three enzymes from diverse sources (phenylalanine ammonia lyase (*PAL*) from *Rhodotorularubra*, coumarate: coenzyme A ligase (*4CL*) from *Streptomyces coelicolor*, and chalcone synthase (*CHS*) from *Glycyrrhizaechinate*) (Hwang et al. 2003). The supplementation of engineered *E. coli* with tyrosine or phenylalanine resulted in the naringeninchalcone and pinocembrin chalcone production, respectively. This technique included different genes cloned in vectors having T7 promoter and ribosome-binding sequence (RBS) that regulate all three genes (*PAL*, *4CL*, and *CHS*). The heterologous expression of genes leads to the production of maximum 0.45 mg/L of naringenin. However, in 2005, Miyahisa and coworkers increased gene expression to 60 mg/L by integrating *PAL*, *CHS*, and *4CL* with chalcone isomerase (CHI) in single vector system (Miyahisa et al. 2005). In another work to achieve the high production of isoquercetin, a single gene deletion in a strain of *E. coli* MEC367/Δ*pgi* expressing UGT73B3 from *A. thaliana*, the production titer was improved in 1 L controlled bioreactor by providing a higher oxygen transfer rate. Finally, the production titer of isoquercetin reached 3.9 g/L in 56 h with 30 g/L glucose as the sole carbon source and 5 g/L quercetin as the substrate (Xia and Eiteman 2017). Recently, an *E. coli* strain was modified to generate a high concentration of caffeic acid from glucose (Sakae et al. 2023). To improve l-tyrosine synthesis, RgTAL tyrosine ammonia-lyase and SeC3H p-coumaric acid 3-hydroxylase were combined with a feedback-resistant chorismate mutase/prephenate dehydrogenase. The resultant strain, CA3, generated 1.58 g/L of caffeic acid from glucose without requiring any tyrosine supplementation in the medium. To further minimize p-coumaric acid accumulation, 4-hydroxyphenylacetate 3-hydroxylase from *Pseudomonas aeruginosa* (*PaHpaBC*) was used. The final designed strain, CA8, produced 6.17 g/L of caffeic acid directly from the glucose. Similarly, Wang et al. (2023a) attempted to increase the caffeic acid production in *E. coli* by inhibiting the competing l-tyrosine synthesis routes. The manufacturing route of the cofactor FAD (Flavin Adenine Dinucleotide) was enhanced, and the expression of previously described polyphenol transporters to promote the caffeic acid production was raised through this approach. Transcriptomics investigation of *E. coli* exposed to the caffeic acid and ferulic acid stress identified that genes for 19 transporters were overexpressed. Furthermore, overexpression of *ycjP*, a sugar ABC transporter permease, boosted the caffeic acid concentration to 775.7 mg/L, and this titer was increased up to 7922.0 mg/L in a 5-L fermenter. Another work reported the *de novo* biosynthesis route for chlorogenic acid in *E. coli* using modular engineering (Wang et al. 2023b). Researches strengthened the shikimate pathway and studied the effect of pyruvate metabolism on chlorogenic acid synthesis. To support the important enzymes such as quinate/shikimate 5-dehydrogenase (*YdiB*) and 4-hydroxyphenylacetate 3-monooxygenase (*HpaBC*), a cofactor regeneration system has been devised. The engineered *E. coli* produced a titer of 2789.2 mg/L in 5-L fermenter. More recently, co-expression of the MVA (Mevalonate) route for DMAPP (Dimethylallyl pyrophosphate) production with a prenyltransferase in *E. coli* facilitated *in vivo*prenylation of resveratrol (Wang et al. 2024). Employing metabolic flux engineering to accumulate acetyl-CoA and malonyl-CoA in *E. coli*, a considerable titer of 68.4 mg/L was obtained in flask cultures. In summary, genetic and metabolic engineering has advanced the use of microorganisms as factories or resources for polyphenol biosynthesis.

7.4.2 Genetic Engineering of Fungi for Polyphenol Production

Bio-based products derived from fungus have become extensively investigated due to their increasing relevance in the industrial sector, healthcare, and food applications. Edible fungus and different fungal strains have been regarded as the promising biological resources for high-value compounds. Metabolites include food additives, pigments, colors, polyphenols, industrial chemicals, antibiotics, and other useful substances (Tiwari and Dufossé 2023). Several yeasts, including *Pichia pastoris*, *S. cerevisiae*, and *Yarrowia lipolytica*, besides several other fungal strains, have been used in the commercial production of high-value metabolites (Baggio et al. 2019). Shang et al. (2023) has altered the *Y. lipolytica* to make polydatin. The resveratrol synthesis route was first developed in *Y. lipolytica*, and then, by boosting the flow of the shikimate pathway, shifting carbon metabolism, and increasing the copies of essential genes, *Y. lipolytica* produced a resveratrol output of 487.77 mg/L. Furthermore, by inhibiting polydatin breakdown in microbial metabolism, it was made possible to accumulate in higher concentrations. Following optimization, *Y. lipolytica* produced a maximum polydatin yield of 6.88 g/L. Similarly, Kulasekaran et al. (2024) reported the assembly and construction of a resveratrol manufacturing pathway in *S. cerevisiae* for *de novo* resveratrol synthesis with seaweed extract as the fermentation medium. The researchers inserted genes from *Flavobacterium johnsoniae* (FjTAL), *Arabidopsis thaliana* (4CL1), and *Vitisvinifera* (VvSTS) into low, high, and integrative vectors. These vectors were then transformed into *S. cerevisiae* W303-1a. The modified strain generated 13.6 ± 0.54 mg/L of resveratrol in the seaweed extract medium. In the same pipeline, Meng et al. (2023) aimed to increase the resveratrol synthesis in *S. cerevisiae* by developing a biosynthetic route that combined the phenylalanine and tyrosine mechanisms. They used a bifunctional phenylalanine/tyrosine ammonia lyase from *Rhodotorulatoruloides*, which resulted in a 462% increase in resveratrol synthesis. The yield of 1155.0 mg/L was achieved by integrating multi-copy biosynthetic pathway genes, increasing metabolic flow to aromatic amino acids and malonyl-CoA, and eliminating by-product route genes. In addition to this, the strain was optimized for resveratrol synthesis on minimal medium without the requirement of any exogenous amino acid, resulting in the highest resveratrol titer of 4.1 g/L in *S. cerevisiae*. Recently, Li et al. (2024b) has used a system engineering method to produce substantial amounts of (2S)-naringenin in *S. cerevisiae*. Initially, they optimized the synthesis routes for shikimic acid and aromatic amino acids. Carbon flow was diverted into the (2S)-naringenin biosynthesis route by increasing the use of aromatic amino acids and adding a heterologous phosphoketolase pathway. To decrease byproduct buildup, they substituted the endogenous enoyl-CoA reductase with a plant-homologous enzyme. Then, they overexpressed the natural malonyl-CoA synthesis pathway genes and inserted the acetaldehyde dehydrogenase (acylating) pathway to increase (2S)-naringenin production. In shaking flasks, regulating the subcellular organelle carbon trafficking mechanism led in a 2S-naringenin output of 986.2 mg/L. However, fermentation in a 5-L bioreactor resulted in the 3420.6 mg/L (2S)-naringenin titer. Similarly, Mejía-Manzano et al. (2024) utilized a combined approach of systems biology, enzyme selection criteria, and molecular engineering strategy for naringenin production in *S. cerevisiae* utilizing glucose. The batch fermentation resulted in 40.67 ± 3.47 μg/L/h naringenin production with a titer of 6.10 ± 0.52 mg/L (22.41 ± 1.91 μM) and a yield of 3.26 ± 1.36 mg/g ($Y_{P/S}$). Zhang et al. (2024a) appropriated a similar technique, reconstructing the *de novo* synthesis process of p-coumaric acid in *S. cerevisiae* and fine-tuning the metabolic pathway *via* promoter and terminator engineering to get high levels of (2S)-naringenin. After this, they successfully synthesized (2S)-eriodictyol (ERD) by incorporating the *ThF3'H* gene from *Tricyrtishirta*. Through increasing the copy number of the *ThF3'H* gene, ERD synthesis was increased further, reaching 132.08 mg/L. In the similar pursuit of yield enhancement, Zhang et al. (2024b) described the *de novo* production of genistein in modified *S. cerevisiae*. They used various combinations of isoflavone synthase, cytochrome P450 reductase, and 2-hydroxyisoflavone dehydratase to synthesize genistein, which was then integrated into the pathway later. The catalytic activity of isoflavone synthase was increased by adding heme and raising the intracellular $NADPH/NADP^+$ ratio. Through modifying the fatty acid synthesis route, Zhang et al. (2024b) were also able to redistribute malonyl-CoA flow and balance metabolic fluxes, yielding 23.33 mg/L genistein. Furthermore, isoflavoneglycosyl transferases were introduced into *S. cerevisiae*, and the optimized strain generated 15.80 mg/L of genistin and 10.03 mg/L of genistein-8-C-glucoside.

7.5 Model Organisms for Polyphenol Production

The organisms used to study a complex biological operation and could also be genetically modified to synthesize the desired product are known as model organisms. They play a significant role in synthetic biology and bioprocess engineering. Several organisms are considered models for polyphenol production due to their genetic tractability, ease of cultivation, and the ability to manipulate their metabolic pathways. These organisms are frequently used in research studies to know about the polyphenol biosynthesis, enhance production, and explore their potential applications. Among various organism, *S. cerevisiae*, *E. coli*, and *Corynebacterium glutamicum* are key model organisms reported for polyphenol production. Table 7.2 summarized different model organism used for production of polyphenols. Detailed description of some of the widely utilized microorganisms has been described below.

Table 7.2 Different model organism used for the production of polyphenols.

Microorganism	Polyphenol produced	Production method/application	References
S. cerevisiae	Resveratrol	Genetic engineering of yeast to express stilbene synthase gene for resveratrol production	Kim et al. (2012)
	Epicatechin andcatechin	Fermentation of tea leaves (*Camellia sinensis*) extract	Rivas et al. (2020)
Pichia pastoris	Quercetin	Transformation of *P. pastoris* with genes for flavonoid biosynthesis	Hong et al. (2019)
Kluyveromyces lactis	Rutin	Bioconversion of quercetin to rutin using engineered *K. lactis* strains	Babaei et al. (2017)
Debaryomyces hansenii	Anthocyanins	Fermentation of fruit juices	Pérez-Magariño et al. (2018)
Yarrowia lipolytica	Caffeic acid andferulicacid	Genetic modification for phenolic acid production	Ledesma-Amaro et al., (2015)
E. Coli	Resveratrol	Expression of stilbene synthase gene	Liu et al. (2023)
	Quercetin	Expression of flavonoid biosynthesis pathway genes	Park et al. (2023)
	Naringenin	Biosynthesis pathway optimization	Zhang et al. (2024a)
	Catechins	Metabolic engineering for precursor biosynthesis	Wang et al. (2023a)
Corynebacterium glutamicum	Resveratrol	Overexpression of resveratrol biosynthetic genes	Kallscheuer et al. (2019)
	Quercetin	Optimization of metabolic pathways for quercetin	Wolf et al. (2021)
	Luteolin	Introduction of luteolin biosynthetic pathway	Chu et al. (2024)
	Apigenin	Genetic manipulation to enhance apigenin production	Amoah et al. (2024)

7.5.1 Yeast

The production of polyphenols by yeast, especially *S. cerevisiae*, represents a promising avenue in the realm of biotechnological research and industrial applications. Yeast as model organism offers numerous advantages for polyphenol production, including its well-established genetics, ease in genetic manipulation, and scalability in fermentation processes. Researchers modify yeast strains to optimize metabolic pathways and enhance the production of specific polyphenols through various biotechnological approaches (Peng et al. 2020).

Fermentation studies have been crucial in harnessing yeast's capacity to synthesize polyphenols under controlled conditions using bioreactors (Zhang et al. 2018). These investigations optimize parameters like nutrient availability, pH, and amount of oxygen to achieve the highest polyphenol yields while maintaining product purity and efficacy. Studies comprising the fermentation approach have produced polyphenols with natural sources and confirm the bioactivity and possible therapeutic benefits of yeast-produced polyphenols, particularly their antioxidant properties (Choi et al. 2019). High-throughput screening methods recognize strains with more advanced polyphenol production traits, including stress tolerance and metabolic efficiency during fermentation processes (Zhang et al. 2021). Such developments are important for creating sustainable production methods that minimize environmental impact and reduce production costs (Huang et al. 2020).

7.5.2 *Escherichia coli*

E. coli has become a flexible and resilient system for producing polyphenols, utilizing its established genetic tools, fast growth, and scalability in biotechnology. Polyphenols, various secondary metabolites found in plants, are valued for their antioxidant, anti-inflammatory, and potential therapeutic effects. *E. coli's* adaptability as a microbial host for polyphenol production arises from its capacity to be engineered to express external pathways essential for polyphenol synthesis.

7.5.3 *Corynebacterium Glutamicum*

A Gram-positive bacterium well-known for its industrial applications in the production of amino acids has recently emerged as a promising model organism for synthesizing polyphenols. Polyphenols are bioactive compounds which are highly valuable for their antioxidant properties and potential health benefits. Hence, researchers have successfully altered *C. glutamicum*'s metabolic pathways to facilitate polyphenols production through strategic modifications. This strategy includes introducing or enhancing the expression of enzymes crucial in polyphenol biosynthesis, thereby redirecting the bacterium's natural metabolic flux toward the production of these valuable compounds (Kallscheuer et al. 2020). *C. glutamicum*is advantageous as a model organism for polyphenol production for number of reasons., Its extensively studied genetics and physiology enable manipulation of metabolic pathways to optimize polyphenol yields. Secondly, the bacterium's robust growth traits and its capability to thrive in industrial fermentation environments make it a well-suited host for scaling up production processes (Wolf et al. 2021). The ability to tailor polyphenol production in *C. glutamicum* through genetic engineering opens doors to create novel derivatives with enhanced bioactivity and stability (Chu et al. 2024).

7.6 Examples of Some Important Polyphenols Produced in *E. coli*

Resveratrol: Resveratrol, a polyphenol presents in grapes and other plants, has been effectively synthesized in *E. coli* by integrating genes encoding stilbene synthase. These genes play a vital role in catalyzing resveratrol biosynthesis from basic carbon sources like glucose. Through genetic engineering, the expression of these genes has been optimized to achieve significant resveratrol yields. This advancement shows potential for applications in pharmaceuticals and dietary supplements, owing to resveratrol's essential health benefits (Liu et al. 2023).

Quercetin: Quercetinis another extensively researched polyphenol which has been produced through *E. coli* by incorporating genes that encode enzymes from the flavonoid biosynthetic pathway. The integrated pathway facilitates the transformation of precursor molecules into quercetin, recognized for its antioxidant properties and potential anticancer effects. To optimizemetabolic flux and genetic regulation, scientists have enhanced quercetin production, demonstrating *E. coli's* ability to produce complex polyphenols using sustainable carbon sources such as glucose (Park et al. 2023).

Naringenin: Naringeninis a flavanone valued with applications in pharmaceuticals and cosmetics, which has also been successfully synthesized in *E. coli*. This involves different engineering metabolic pathways to redirect carbon flux towardnaringenin biosynthesis, along with optimizing precursor availability and enzyme activities. These efforts have led to the efficient production of naringenin, highlighting *E. coli's* potential to produce high-value polyphenols of significant industrial importance (Zhang et al. 2024a). Apart from these examples, there is a vast weaponry of such important molecules hidden in nature which needs to be explored to check and utilize their health benefits.

7.7 Conclusion and Future Directions

The application aspect of polyphenols is very diverse as it covers pharmaceuticals (as potent anticancerous and anti-inflammatory agent), cosmetic industry (due to their antioxidant potential), and nutrition (as important dietary fiber that promotes growth of gut microbiota). Plants as traditional source of polyphenols have certain limitation, for example, seasonal variability, expensive extraction methods, and environment destruction. Thus, microbial production of polyphenols using inexpensive waste residues becomes imperative and offers a sustainable alternative to the traditional methods. Engineering of metabolic/biosynthetic pathways of microbes using advanced gene manipulation techniques is of greater importance. These engineered microbes not only surpass various limitations associated with natural sources but also helps in the enhancement of yield, stability, and efficiency of the product. However, challenges associated with strain improvement, modification of pathways, optimization of conditions, integrated market strategies, and effective bioprocess development require further research and development activities and advances in green sustainable biotechnology.

References

Albuquerque, B.R., Heleno, S.A., Oliveira, M.B.P. et al. (2021). Phenolic compounds: current industrial applications, limitations and future challenges. *Food & Function* 12 (1): 14–29.

Alseekh, S., Perez de Souza, L., Benina, M., and Fernie, A.R. (2021). The style and substance of plant flavonoid decoration; towards defining both structure and function. *Phytochemistry* 174: 112347.

Al-Suhaimi, E.A. and Al-Riziza, N.A. (2018). The pharmacological properties of gingerol and its derivatives: potential anticancer activities. *Journal of Medicinal Plant Research* 12 (17): 222–230.

Amoah, O.J., Thapa, S.B., Ma, S.Y. et al. (2024). Biosynthesis of apigenin glucosides in engineered Corynebacterium glutamicum. *Journal of Microbiology and Biotechnology* 34 (5): 1154.

Antognoni, F., Potente, G., Mandrioli, R. et al. (2020). Fruit quality characterization of new sweet cherry cultivars as a good source of bioactive phenolic compounds with antioxidant and neuroprotective potential. *Antioxidants* 9 (8): 677.

Asma, U., Morozova, K., Ferrentino, G., and Scampicchio, M. (2023). Apples and apple by-products: antioxidant properties and food applications. *Antioxidants* 12 (7): 1456.

Athanasiadis, V., Chatzimitakos, T., Mantiniotou, M. et al. (2023). Investigation of the polyphenol recovery of overripe banana peel extract utilizing cloud point extraction. *Engineering* 4 (4): 3026–3038.

Babaei, M., Barzegar, A., Asadollahi, M.A., and Arjmand, S. (2017). Rutin production by biotransformation of quercetin using engineered *Kluyveromyces lactis*. *Applied Biochemistry and Biotechnology* 183 (1): 337–349.

Baggio, L.M., Panagio, L.A., Gasparin, F.G.M. et al. (2019). Production of fibrinogenolytic and fibrinolytic enzymes by a strain of *Penicillium* sp. isolated from contaminated soil with industrial eluent. *ActaScientiarum. Health Sciences* 41: 40606.

Balanov, P.E., Smotraeva, I.V., Abdullaeva, M.S. et al. (2021). Study on resveratrol content in grapes and wine products. In: *E3S Web of Conferences*, vol. 247, 01063. EDP Sciences.

Bertelli, A., Biagi, M., Corsini, M. et al. (2021). Polyphenols: from theory to practice. *Foods* 10 (11): 2595.

Bisht, D., Bhatt, V., Arya, R.K.K. et al. (2024). Plants Based Nutraceuticals: An Overview. Herbal Nutraceuticals; Wiley; Portico. https://doi.org/10.1002/9781394241576.ch1.

Borghi, S.M. and Pavanelli, W.R. (2023). Antioxidant compounds and health benefits of citrus fruits. *Antioxidants* 12 (8): 1526.

Buenrostro-Figueroa, J.J., Nevárez-Moorillón, G.V., Chávez-González, M.L. et al. (2023). Improved extraction of high value-added polyphenols from pomegranate peel by solid-state fermentation. *Fermentation* 9 (6): 530.

Bulotta, S. and Celano, M. (2014). Hydroxytyrosol, tyrosol and derivatives and their potential effects on human health. *Molecules* 19 (10): 16973–16998.

Cai, Y., Li, D., Liao, X. et al. (2022). Quercetin and kaempferol in millet grains and their antioxidant activity. *Food Chemistry* 369: 130527.

Caruso, G., Godos, J., Privitera, A. et al. (2022). Phenolic acids and prevention of cognitive decline: polyphenols with a neuroprotective role in cognitive disorders and Alzheimer's disease. *Nutrients* 14 (4): 819.

Catalkaya, G., Venema, K., Lucini, L. et al. (2020). Interaction of dietary polyphenols and gut microbiota: microbial metabolism of polyphenols, influence on the gut microbiota, and implications on host health. *Food Frontiers* 1 (2): 109–133.

Chiriac, E.R., Chiţescu, C.L., Geană, E.I. et al. (2021). Advanced analytical approaches for the analysis of polyphenols in plants matrices—a review. *Separations* 8 (5): 65.

Choi, J., Son, J., and Lee, S.Y. (2019). Metabolic engineering of microorganisms for the production of natural compounds. *Trends in Biotechnology* 37 (6): 548–562.

Chojnacka, K., Skrzypczak, D., Izydorczyk, G. et al. (2021). Antiviral properties of polyphenols from plants. *Foods* 10 (10): 2277.

Chu, L.L., Tran, C.T., Pham, D.T. et al. (2024). Metabolic engineering of *Corynebacteriumglutamicum* for the production of flavonoids and stilbenoids. *Molecules* 29 (10): 2252.

Değirmencioğlu, N. and Gökmen, V. (2018). Comparative study of phenolic content and antioxidant activity of various types of whole wheat and wheat bran fractions. *Journal of Agricultural and Food Chemistry* 66 (26): 6630–6638.

Díaz-Batalla, L., Widholm, J.M., Fahey, G.C. et al. (2006). Flavonoid content and antioxidant capacity of spinach genotypes determined by high-performance liquid chromatography. *Journal of the Science of Food and Agriculture* 86 (3): 468–476.

Duan, Y., Santiago, F.E.M., Dos Reis, A.R. et al. (2021). Genotypic variation of flavonols and antioxidant capacity in broccoli. *Food Chemistry* 338: 127997.

Dudnik, A., Gaspar, P., Neves, A.R., and Forster, J. (2018). Engineering of microbial cell factories for the production of plant polyphenols with health-beneficial properties. *Current Pharmaceutical Design* 24 (19): 2208–2225.

Dwita, L.P., Iwo, M.I., Mauludin, R., and Elfahmi. (2022). Neuroprotective potential of lignan-rich fraction of *Piper cubeba* L. by improving antioxidant capacity in the rat's brain. *Brazilian Journal of Biology* 82: e266573.

Ebrahimi, P. and Lante, A. (2021). Polyphenols: a comprehensive review of their nutritional properties. *The Open Biotechnology Journal* 15 (15): 164–172.

Ernst, E. (2021). Aloe vera: a review of toxicity and adverse clinical effects. *Journal of Clinical Pharmacy and Therapeutics* 46 (1): 24–39.

Gade, A. and Kumar, M.S. (2023). Gut microbial metabolites of dietary polyphenols and their potential role in human health and diseases. *Journal of Physiology and Biochemistry* 79 (4): 695–718.

Gong, D. and Zha, Z. (2023). Hydroxybenzoicacids, Handbook of food bioactive ingredients: properties and applications. In: *Handbook of Food Bioactive Ingredients* (ed. S.M. Jafari, A. Rashidinejad, and J. Simal-Gandara), 29–58. Cham: Springer International Publishing.

González-Sarrías, A., Tomás-Barberán, F.A., and García-Villalba, R. (2020). Structural diversity of polyphenols and distribution in foods. In: *Dietary Polyphenols: Their Metabolism and Health Effects* (ed. F.A. Tomás-Barberán, A. González-Sarrías, and R. García-Villalba), 1–29. Wiley Online.

Hano, C. and Renault, J.H. (2017). Identification and quantification of secoisolariciresinoldiglucoside and its hydrolysates in flaxseed (*Linumusitatissimum* L.). *Industrial Crops and Products* 108: 558–568.

Hazafa, A., Iqbal, M.O., Javaid, U. et al. (2022). Inhibitory effect of polyphenols (phenolic acids, lignans, and stilbenes) on cancer by regulating signal transduction pathways: a review. *Clinical and Translational Oncology* 24: 1–14.

Holtekjølen, A.K., Uhlen, A.K., Bråthen, E., and Sahlstrøm, S. (2018). Phenolic acids in oats – a review. *Journal of Cereal Science* 79: 301–309.

Hong, S.Y., Kim, K.H., Lee, H.J. et al. (2019). Enhanced production of quercetin in *Pichia pastoris* by expression of flavonoid biosynthetic pathway genes. *Journal of Agricultural and Food Chemistry* 67 (2): 689–698.

Hosseinian, F.S., Li, W., Beta, T., and Jianping, W. (2019). Anthocyanin, flavonoid, and antioxidant capacity changes in purple pistachio (*Pistaciavera* L.) as affected by storage time. *Food Chemistry* 285: 72–81.

Hu, Y., Wang, X.Y., Qin, C.Q. et al. (2022). Fermentation of rose residue by *Lactiplantibacillusplantarum* B7 and *Bacillus subtilisnatto* promotes polyphenol content and beneficial bioactivity. *Journal of Bioscience and Bioengineering* 134 (6): 501–507.

Huang, W., Ma, T., and Zhong, J. (2020). Metabolic engineering of *Saccharomyces cerevisiae* for production of polyphenols. *Frontiers in Bioengineering and Biotechnology* 8 (531).

Hwang, E.I., Kaneko, M., Ohnishi, Y., and Horinouchi, S. (2003). Production of plant specific flavanones by *Escherichia coli* containing an artificial gene cluster. *Applied and Environmental Microbiology* 69: 2699–2706.

Jin, J., Liu, Z., and Jiang, X. (2020). Sesame lignans and their potential health benefits: an umbrella review of systematic reviews and meta-analyses. *Food Chemistry* 309: 125699.

Kačániová, M., Kunová, S., Horská, E., and Štefániková, J. (2019). Phenolic composition, antioxidant and enzyme inhibitory activities of pumpkin seed extracts. *Journal of Food Science and Technology* 56 (9): 4137–4144.

Kalinowska, M., Gołębiewska, E., Świderski, G. et al. (2021). Plant-derived and dietary hydroxybenzoic acids—a comprehensive study of structural, anti-/pro-oxidant, lipophilic, antimicrobial, and cytotoxic activity in MDA-MB-231 and MCF-7 cell lines. *Nutrients* 13 (9): 3107.

Kallscheuer, N., Menezes, R., Foito, A. et al. (2019). Identification and microbial production of the raspberry phenol salidroside that is active against Huntington's disease. *Plant physiology* 179 (3): 969–985.

Kim, Y.H., Kim, Y.S., Roh, G.S. et al. (2012). Resveratrol blocks diabetes-induced early vascular lesions and vascular endothelial growth factor induction in mouse retinas. *Acta Ophthalmologica* 90 (1): e31–e37.

Krishnamurthy, H.K., Pereira, M., Bosco, J. et al. (2023). Gut commensals and their metabolites in health and disease. *Frontiers in Microbiology* 14: 1244293. https://doi.org/10.3389/fmicb.2023.1244293.

Kulasekaran, N.T., Thilakam, M.L., Gopal, D. et al. (2024). Denovo production of resveratrol by engineered *Saccharomyces cerevisiae* W303-1a using pretreated *Gracilariacorticata* extracts. *Biotechnology Letters* 46 (1): 19–28.

Kumar, V., Sharma, A., Kohli, S.K. et al. (2019). Differential distribution of polyphenols in plants using multivariate techniques. *Biotechnology Research and Innovation* 3 (1): 1–21.

Kwaw, E., Ma, Y., Tchabo, W. et al. (2018). Effect of *Lactobacillus* strains on phenolic profile, color attributes and antioxidant activities of lactic-acid-fermented mulberry juice. *Food Chemistry* 250: 148–154.

Lacko-Bartošová, M., Kaur, A., Lacko-Bartošová, L. et al. (2023). Concentration of phenolic compounds and phenolic acids of various spelt cultivars in response to growing years. *Agriculture* 13 (10): 2024.

Ledesma-Amaro, R., Nicaud, J.M., and Rossignol, T. (2015). Production of antioxidant compounds by *Yarrowia lipolytica* strains cultivated on glycerol. *Microbial Cell Factories* 14 (1): 1–12.

Li, J., Lu, X., Zou, X., and Ye, B.C. (2024a). Recent advances in microbial metabolic engineering for production of natural phenolic acids. *Journal of Agricultural and Food Chemistry* 72 (9): 4538–4551.

Li, H., Ma, W., Wang, W. et al. (2024b). Synergetic engineering of multiple pathways for De Novo (2S)-naringenin biosynthesis in *Saccharomyces cerevisiae*. *ACS Sustainable Chemistry & Engineering* 12 (1): 59–71.

Lippolis, T., Cofano, M., Caponio, G.R. et al. (2023). Bioaccessibility and bioavailability of diet polyphenols and their modulation of gut microbiota. *International Journal of Molecular Sciences* 24 (4): 3813.

Liu, Y., Weng, W., Gao, R., and Liu, Y. (2019). New insights for cellular and molecular mechanisms of aging and aging-related diseases: herbal medicine as potential therapeutic approach. *Oxidative Medicine and Cellular Longevity* 1: 4598167.

Liu, S., Lou, Y., Li, Y. et al. (2022). Review of phytochemical and nutritional characteristics and food applications of *Citrus* L. fruits. *Frontiers in Nutrition* 9: 968604.

Liu, Y., Tan, T., Wu, H. et al. (2023). Engineering *Escherichia coli* for de novo resveratrol biosynthesis from simple carbon sources. *Metabolic Engineering* 101: 123–134.

López-Téllez, J.M. and del PilarCañizares-Macías, M. (2024). Evaluation of tomato (*Lycopersiconesculentum* P. Mill.) by-product extracts obtained by different extraction methods as exploitation strategy of high-value polyphenols. *Food and Bioprocess Technology* 17 (10): 3217–3235.

Luís, Â., Duarte, A.P., Pereira, L. et al. (2018). Interactions between the major bioactive polyphenols of berries: effects on antioxidant properties. *European Food Research and Technology* 244 (2): 175–185.

Luna-Vital, D., Luzardo-Ocampo, I., Cuellar-Nuñez, M.L. et al. (2020). Maize extract rich in ferulic acid and anthocyanins prevents high-fat-induced obesity in mice by modulating SIRT1, AMPK and IL-6 associated metabolic and inflammatory pathways. *Journal of Nutritional Biochemistry* 79: 108343.

Mahendran, G., Thangaraj, P., Sathya, M., and Sivakumar, T. (2021). Quantitative analysis of ferulic acid, kaempferol and quercetin in millet (*Panicummiliaceum* L.) grains. *Food Chemistry* 342: 128055.

Makarewicz, M., Drożdż, I., Tarko, T., and Duda-Chodak, A. (2021). The interactions between polyphenols and microorganisms, especially gut microbiota. *Antioxidants* 10 (2): 188.

Marranzano, M., Rosa, R.L., Malaguarnera, M. et al. (2018). Polyphenols: plant sources and food industry applications. *Current Pharmaceutical Design* 24 (35): 4125–4130.

Martínez-Cruz, O. and Paredes-López, O. (2014). Polyphenol composition and antioxidant activity of chia seeds grown across different geographical regions. *Plant Foods for Human Nutrition* 69 (2): 196–202.

Mejía-Manzano, L.A., Ortiz-Alcaráz, C.I., Parra Daza, L.E. et al. (2024). *Saccharomyces cerevisiae*biofactory to produce naringenin using a systems biology approach and a bicistronic vector expression strategy in flavonoid production. *Microbiology Spectrum* 12 (1): e03374–e03323.
Meng, L., Diao, M., Wang, Q. et al. (2023). Efficient biosynthesis of resveratrol via combining phenylalanine and tyrosine pathways in *Saccharomyces cerevisiae*. *Microbial Cell Factories* 22 (1): 46.
Milke, L., Aschenbrenner, J., Marienhagen, J., and Kallscheuer, N. (2018). Production of plant-derived polyphenols in microorganisms: current state and perspectives. *Applied Microbiology and Biotechnology* 102: 1575–1585.
Miyahisa, I., Kaneko, M., Funa, N. et al. (2005). Efficient production of (2S)–flavanones by *Escherichia coli* contain in gin artificial biosynthetic gene cluster. *Applied Microbiology and Biotechnology* 68: 498–504.
Monagas, M., Garrido, I., Lebrón-Aguilar, R. et al. (2017). Identification and quantification of polyphenols in almond skins by HPLC-DAD-ESI-MS. *Journal of Food Composition and Analysis* 61: 54–60.
Moosavi Dolatabadi, K., Dehghan, G., and Hosseini, S.M.H. (2018). Characterization of polyphenols and antioxidant capacity of Persian walnut (*Juglansregia* L.) during kernel development. *Food Chemistry* 252: 331–339.
Moradi, D., Ramezan, Y., Eskandari, S. et al. (2022). The enhanced yield of polyphenols and antioxidant activity from onion peel (*Allium cepa* L.) and its incorporation into low-density polyethylene films. *Food Measure* 16: 1325–1339.
Mueed, A., Shibli, S., Al-Quwaie, D.A. et al. (2023). Extraction, characterization of polyphenols from certain medicinal plants and evaluation of their antioxidant, antitumor, antidiabetic, antimicrobial properties, and potential use in human nutrition. *Frontiers in Nutrition* 10: 1125106.
Nacul, A., Sánchez-Sánchez, J., Gomes, T. et al. (2021). Composition of cashew kernels from different genotypes cultivated in Mozambique and their potential for industrial applications. *Food Chemistry* 347: 127606.
Navarro-Orcajada, S., Conesa, I., Vidal-Sánchez, F.J. et al. (2023). Stilbenes: characterization, bioactivity, encapsulation and structural modifications. A review of their current limitations and promising approaches. *Critical Reviews in Food Science and Nutrition* 63 (25): 7269–7287.
Olszewska, M.A., Gędas, A., and Simões, M. (2020). Antimicrobial polyphenol-rich extracts: applications and limitations in the food industry. *Food Research International* 134: 109214.
Pan, Y., Zhang, Y., Hou, M. et al. (2023). Properties of polyphenols and polyphenol-containing wastewaters and their treatment by Fenton/Fenton-like reactions. *Separation and Purification Technology* 317: 123905.
Panche, A.N., Diwan, A.D., and Chandra, S.R. (2016). Flavonoids: an overview. *Journal of Nutritional Science* 5: e47.
Park, S.H., Jang, J.W., Yoon, J.H. et al. (2023). Production of quercetin from glucose by metabolically engineered *Escherichia coli*. *Journal of Biotechnology* 345: 89–98.
Pecyna, P., Wargula, J., Murias, M., and Kucinska, M. (2020). More than resveratrol: new insights into stilbene-based compounds. *Biomolecules* 10 (8): 1111.
Peirotén, Á., Álvarez, I., and Landete, J.M. (2019). Production of flavonoid and lignanaglycones from flaxseed and soy extracts by *Bifidobacterium* strains. *International Journal of Food Science & Technology* 55 (5): 2122–2131.
Peng, B., Su, C., and Huang, S. (2020). Optimization of fermentation conditions for polyphenol production by *Saccharomyces cerevisiae*. *Journal of Agricultural and Food Chemistry* 68 (30): 7998–8006.
Pérez-Magariño, S., Ortega-Heras, M., Garcinuño-Martínez, R. et al. (2018). Anthocyanin production by *Debaryomyces hansenii* strains in different fruit juices. *Food Chemistry* 240: 1158–1165.

Peterson, D.M. (2017). Avenanthramides in oats: sources, structures, bioactivities and responses to processing and storage. *Phytochemistry Reviews* 16 (5): 527–534.

Pham, J.V., Yilma, M.A., Feliz, A. et al. (2019). A review of the microbial production of bioactive natural products and biologics. *Frontiers in Microbiology* 10: 1404.

Phan, M.A.T., Bucknall, M.P., Arcot, J., and O'Hara, I.M. (2019). Analysis of phenolic compounds in barley cultivars and their contributions to antioxidant activities. *Journal of Agricultural and Food Chemistry* https://doi.org/10.1021/acs.jafc.9b04617.

Piekarska-Radzik, L. and Klewicka, E. (2021). Mutual influence of polyphenols and *Lactobacillus* spp. bacteria in food: a review. *European Food Research and Technology* 247 (1): 9–24.

Plamada, D. and Vodnar, D.C. (2021). Polyphenols-Gut microbiota interrelationship: atransition to a new generation of prebiotics. *Nutrients* 14 (1): 137.

Preti, R. and Tarola, A.M. (2021). Study of polyphenols, antioxidant capacity, and minerals for the valorisation of ancient apple cultivars from Northeast Italy. *European Food Research and Technology* 247 (2): 273–283.

Rashmi, H.B. and Negi, P.S. (2020). Phenolic acids from vegetables: a review on processing stability and health benefits. *Food Research International* 136: 109298.

Rathour, R.K., Sharma, D., Sharma, N. et al. (2022). Engineered microorganisms for bioremediation. In: *Current Developments in Biotechnology and Bioengineering* (ed. S. Joshi, A. Pandey, R. Sirohi, and S.H. Park), 335–361. Elsevier.

Ren, F. and Zhou, S. (2021). Phenolic components and health beneficial properties of onions. *Agriculture* 11 (9): 872.

Rivas, B., Torre, P., Monteagudo-Mera, A. et al. (2020). Production of catechins and epicatechins through bioconversion of green tea leaves by *Saccharomyces cerevisiae*. *Foods* 9 (3): 348.

Rudrapal, M., Rakshit, G., Singh, R.P. et al. (2024). Dietary polyphenols: review on chemistry/sources, bioavailability/metabolism, antioxidant effects, and their role in disease management. *Antioxidants* 13 (4): 429.

Safe, S., Jayaraman, A., Chapkin, R.S. et al. (2021). Flavonoids: structure–function and mechanisms of action and opportunities for drug development. *Toxicological Research* 37: 147–162.

Sakae, K., Nonaka, D., Kishida, M. et al. (2023). Caffeic acid production from glucose using metabolically engineered *Escherichia coli*. *Enzyme and Microbial Technology* 164: 110193.

Sarkar, P., Abedin, M.M., Singh, S.P. et al. (2022). Microbial production and transformation of polyphenols. In: *Current Developments in Biotechnology and Bioengineering* (ed. S. Joshi, A. Pandey, R. Sirohi, and S.H. Park), 189–208. Elsevier.

Shang, Y., Zhang, P., Wei, W. et al. (2023). Metabolic engineering for the high-yield production of polydatin in *Yarrowialipolytica*. *Bioresource Technology* 381: 129129.

Shang, X.-c., Zhang, Y.-q., Zheng, Y.-f., and Li, Y. (2024). Temperature-responsive deep eutectic solvents as eco-friendly and recyclable media for microwave extraction of flavonoid compounds from waste onion (*Allium cepa* L.) skins. *Biomass Conversion and Biorefinery* 14 (3): 3729–3738.

Sharma, A., Shahzad, B., Rehman, A. et al. (2019). Response of phenylpropanoid pathway and the role of polyphenols in plants under abiotic stress. *Molecules* 24 (13): 2452.

Sharma, R., Diwan, B., Singh, B.P., and Kulshrestha, S. (2022). Probiotic fermentation of polyphenols: potential sources of novel functional foods. *Food Production, Processing and Nutrition* 4 (1): 21.

Sharma, S., Shukla, M.K., Sharma, K.C. et al. (2023). Revisiting the therapeutic potential of gingerols against different pharmacological activities. Naunyn-Schmiedeberg's Archives of Pharmacology 396 (4): 633 –647. https://doi.org/10.1007/s00210-022-02372-7.

Sharma, A., Madhu, Kondalkar, S.A. et al. (2024). Herbal Nutraceutical as Alternative Medicine. Herbal Nutraceuticals; Wiley; Portico. https://doi.org/10.1002/9781394241576.ch2.

Singh, S.P. and Upadhyay, S.K. (2021). Bioprospecting of Microorganism-Based Industrial Molecules. Wiley. https://doi.org/10.1002/9781119717317.

Singh, S.P. and Upadhyay, S.K. (2023). Microbial Bioreactors for Industrial Molecules. Wiley. https://doi.org/10.1002/9781119874096.

Somparn, P., Phisalaphong, C., Nakornchai, S. et al. (2010). Curcumin, demethoxycurcumin, bisdemethoxycurcumin, tetrahydrocurcumin and turmerones differentially regulate anti-inflammatory and anti-proliferative responses through a ROS-independent, PKC-dependent pathway. *Molecules* 15 (1): 457–468.

Sova, M. and Saso, L. (2020). Natural sources, pharmacokinetics, biological activities and health benefits of hydroxycinnamic acids and their metabolites. *Nutrients* 12 (8): 2190.

Stromsnes, K., Lagzdina, R., Olaso-Gonzalez, G. et al. (2021). Pharmacological properties of polyphenols: bioavailability, mechanisms of action, and biological effects in in vitro studies, animal models, and humans. *Biomedicines* 9 (8): 1074.

Subhan, S., Zeeshan, M., Rahman, A.U., and Yaseen, M. (2024). Fundamentals of polyphenols: nomenclature, classification and properties. In: *Science and Engineering of Polyphenols: Fundamentals and Industrial Scale Applications* (ed. C. Verma), 1–36.

Sun, X., Li, X., Shen, X. et al. (2021). Recent advances in microbial production of phenolic compounds. *Chinese Journal of Chemical Engineering* 30: 54–61.

Sun, J., Sun, W., Zhang, G. et al. (2022). High efficient production of plant flavonoids by microbial cell factories: challenges and opportunities. *Metabolic Engineering* 70: 143–154.

Taira, J. and Tokuji, Y. (2018). Phenolic compounds in maize grains and their health-promoting effects. *Journal of Nutritional Science and Vitaminology* 64 (4): 371–376.

Tiwari, P. and Dufossé, L. (2013). Focus and insights into the synthetic biology-mediated chassis of economically important fungi for the production of high-value metabolites. *Microorganisms* 11 (5): 1141.

Tyagi, A., Lim, M.J., Kim, N.H. et al. (2022). Quantification of amino acids, phenolic compounds profiling from nine rice varieties and their antioxidant potential. *Antioxidants* 11 (5): 839.

Tyagi, S., Shumayla, Singh, S. et al. (2024). TaCCS1-B expression modulates copper, enzymatic antioxidants and polyphenols contents and provides abiotic stress tolerance in transgenic Arabidopsis. Physiologia Plantarum 176 (6): e14645. https://doi.org/10.1111/ppl.14645.

Upadhyay, S.K. and Singh, S.P. (2021). Bioprospecting of Plant Biodiversity for Industrial Molecules. Wiley. https://doi.org/10.1002/9781119718017.

Upadhyay, S.K. and Singh, S.P. (2023). Plants as Bioreactors for Industrial Molecules. Wiley. https://doi.org/10.1002/9781119875116.

Upadhyay, S.K. and Singh, S.P. (2024). Herbal Nutraceuticals: Products and Processes. Wiley. https://doi.org/10.1002/9781394241576.

Vidal-Casanella, O., Núñez, O., Granados, M. et al. (2021). Analytical methods for exploring nutraceuticals based on phenolic acids and polyphenols. *Applied Sciences* 11 (18): 8276.

Wang, L., Bei, Q., Wu, Y. et al. (2017a). Characterization of soluble and insoluble-bound polyphenols from *Psidiumguajava* L. leaves co-fermented with *Monascusanka* and *Bacillus* sp. and their bio-activities. *Journal of Functional Foods* 32: 149–159.

Wang, H., Cao, G., and Prior, R.L. (2017b). An update on bioactive compounds in garlic (*Allium sativum* L.): from bulbs to green tissues. *Critical Reviews in Food Science and Nutrition* 57 (8): 1667–1680.

Wang, L., Li, N., Yu, S., and Zhou, J. (2023a). Enhancing caffeic acid production in *Escherichia coli* by engineering the biosynthesis pathway and transporter. *Bioresource Technology* 368: 128320.

Wang, L., Wang, H., Chen, J. et al. (2023b). Efficient production of chlorogenic acid in *Escherichia coli* via modular pathway and cofactor engineering. *Journal of Agricultural and Food Chemistry* 71 (41): 15204–15212.

Wang, H., Zhou, T., Liu, H. et al. (2024). Heterologous biosynthesis of prenylated resveratrol through multiplex metabolic engineering in *Escherichia coli*. *Green Chemistry* 26 (8): 4792–4802.

Watson, R.R. (ed.) (2018). *Polyphenols in Plants: Isolation, Purification and Extract Preparation*. Academic Press.

Wolf, S., Becker, J., Tsuge, Y. et al. (2021). Advances in metabolic engineering of *Corynebacterium glutamicum* to produce high-value active ingredients for food, feed, human health, and well-being. *Essays in Biochemistry* 65 (2): 197–212.

Xia, T. and Eiteman, M.A. (2017). Quercetin glucoside production by engineered *Escherichia coli*. *Applied Biochemistry and Biotechnology* 182: 1358–1370.

Yang, F., Chen, C., Ni, D. et al. (2023). Effects of fermentation on bioactivity and the composition of polyphenols contained in polyphenol-rich foods: areview. *Foods* 12 (17): 3315.

Zagoskina, N.V., Zubova, M.Y., Nechaeva, T.L. et al. (2023). Polyphenols in plants: structure, biosynthesis, abiotic stress regulation, and practical applications (review). *International Journal of Molecular Sciences* 24 (18): 13874.

Zhang, L., Zhu, L., and Zhang, H. (2018). Fermentation strategies for the production of plant polyphenols: challenges and prospects. *Critical Reviews in Biotechnology* 38 (5): 694–709.

Zhang, Y., Hu, M., and Gao, K. (2021). Screening and improvement of yeast strains for enhanced polyphenol production. *Food Chemistry* 346: 128979.

Zhang, Z., Li, X., Sang, S. et al. (2022). Polyphenols as plant-based nutraceuticals: health effects, encapsulation, nano-delivery, and application. *Foods* 11 (15): 2189.

Zhang, S., Liu, J., Xiao, Z. et al. (2024a). Systems metabolic engineering of *Saccharomyces cerevisiae* for the high-level production of (2 S)-eriodictyol. *Journal of Fungi* 10 (2): 119.

Zhang, X., Li, H., Sun, J. et al. (2024b). Metabolic engineering of *Escherichia coli* for high-yield production of naringenin. *Applied Microbiology and Biotechnology* 108 (3): 1201–1214.

Zhou, M., Fakayode, O.A., and Li, H. (2023a). Green extraction of polyphenols via deep eutectic solvents and assisted technologies from agri-food by-products. *Molecules* 28 (19): 6852.

Zhou, Y., Wei, Y., Jiang, L. et al. (2023b). Anaerobic phloroglucinol degradation by *Clostridium scatologenes*. *MBio* 14 (4): e0109923.

8

Microbial Approaches for Lactose Transformation into High-value Rare Sugars

Ashutosh Kumar Singh[1], *Amit Kumar Rai*[2], *and Sudhir Pratap Singh*[3]

[1]*Department of Biotechnology, Panjab University, Chandigarh, 160014, India*
[2]*Food and Nutritional Biotechnology Division, National Agri-Food and Biomanufacturing Institute, BRIC-NABI, Sector-81 (Knowledge city) SAS Nagar, Mohali, Panjab, 140306, India*
[3]*Department of Industrial Biotechnology, Gujarat Biotechnology University, GIFT city, Gandhinagar, Gujarat, 382355, India*

8.1 Introduction

Lactose is referred to as a milk sugar of galactose and glucose monomer present in milk and milk products like butter, cheese, and yogurt (Gambelli 2017). Its abundance is 2–7% in milk. The human milk contains the highest level of lactose, i.e., about 7% (w/v). Bovine milk contains an average of 4.8% (w/v) lactose, while milk of monotremes, such as echidna and platypus, contains less than 0.1% (w/v) lactose. Interestingly, Pacific pinnipeds like California sea lioness have nil lactose content in milk (Morrissey 1985).

Milk is consumed across the globe. India is one of the leading consumers of milk. As per the report of OECD-FAO, the increment of 1.6% milk production is speculated annually from the years 2022 to 2029. Milk consumption was maximum in India (81 million tons), followed by the European Union (33.4 million tons), the USA (21.2 million tons), China (12 million tons), and Brazil (11 million tons) in the year 2021. The increasing global population, rapid urbanization, nutritional, and health awareness demand high productivity of dairy products. At the same time, an increase in milk productivity has increased dairy waste, too. To protect the environmental interest, proper management of dairy waste is required. Whey disposal increases biological and chemical oxygen demand for water bodies. Lactose (45–52 g/L) in whey is the main component responsible for high BOD and COD of cheese waste disposal to water bodies (Güven et al. 2008). Due to sweetening properties and low solubility, direct utilization of whey is difficult (Seki and Saito 2012). Whey can be valorized by converting lactose into its rare sugar derivatives. Lactose can be transformed into high-value rare sugars *via* microbial approaches such as microbial fermentation or enzymatic reaction. The rare sugars synthesized from lactose have proved useful in dealing with health issues like obesity, diabetes, hyperammonia, and chronic constipation (Bass and Dennis 1981; Lu et al. 2008; Mu et al. 2013).

Rare sugars are monosaccharides or disaccharides found in low abundance in certain natural grains, fruits, and vegetables but exert critical physiological and metabolic effects on human health (Mijailovic et al. 2021; Ahmed et al. 2022). The consumption of rare sugar has been reported to provide several health benefits like reduction of body fat, reduction in body weight, decrease in

Microbial Nutraceuticals: Products and Processes, First Edition. Edited by Sudhir Pratap Singh and Santosh Kumar Upadhyay.

LDL cholesterol and total cholesterol level, and improved glycemic control (Ensor et al. 2015; Patel et al. 2020; Agarwal and Singh 2021; Smith et al. 2022). Rare sugars have also been found to stimulate the reactive oxygen species, modulate gut microbiome, improve mineral absorption and inflammation, etc. (Hou et al. 2022). The growing diabetes and obesity globally have emphasized the essentiality of low glycemic and healthy sugar consumption. Despite rare sugar's fascinating health benefits and other applications in agriculture (crop protection) and the cosmetic industry, its high cost of production hampers the industrialization of rare sugar production (Zhang et al. 2023). The current global rare sugar market is expected to grow to $1.83 billion in the year 2023 with a compound annual growth rate (CAGR) of 5.4%, and it is projected to reach $2.1 billion by 2027 with a CAGR of 4.3%.

Lactose from whey is a commodity of low value that should be efficiently utilized for the biosynthesis of high-value rare sugars like lactulose, tagatose, lactosucrose, and epilactose (Vera et al. 2022). Lactose-derived rare sugar has several applications in the pharmaceutical, cosmetic, and agricultural sectors. Microbial approaches are economical, environment-friendly, and sustainable for rare sugar production in comparison with chemical synthesis (Panesar and Kumari 2011). This chapter discusses the microbial approaches for lactose transformation into high-value rare sugar.

8.2 Lactose-derived Rare Sugar Production Through Microbial Approach

8.2.1 Lactosucrose

Lactosucrose is a lactose-derived white solid synthetic trisaccharide consisting of glucose, galactose, and fructose monomer (Xu et al. 2021). It is a prebiotic molecule that has a higher laxative threshold than other lactose-derived prebiotics, including galacto-oligosaccharides and lactulose (Silvério et al. 2015). In Japan, it is used in the preparation of functional food products and is listed as food for specific health uses (FOSHU) (Silvério et al. 2015). Different enzymes like β-fructofuranosidase, levansucrase, and β-galactosidase can catalyze both hydrolysis and transfer reactions that can be employed for lactosucrose production. Lactose and sucrose serve as substrates for the catalytic synthesis of lactosucrose.

Levansucrase belongs to the glycoside hydrolase family (GH 68) (Xu et al. 2021). The mechanism of catalysis involves hydrolysis of the glycosidic bond of fructose donor (generally sucrose) to form a covalent enzyme fructosyl intermediate and release of glucose molecule (Paulis and Barry 1969). Further, the fructose moiety is transferred to the acceptor molecule (generally lactose) through enzyme, and subsequently, the disaccharide is extended with one fructosyl unit. Lactosucrose production can be executed *via* cell-free levansucrase or whole-cell catalysis. Cell-free method of lactosucrose production could be a promising approach for higher productivity, as the whole-cell process may release many unwanted by-products in addition to the desired product. *Aerobacter levanicum* levensucrase was first characterized for the production of lactosucrose (Avigad et al. 1956). Later on, levensucrase was characterized from many other sources, e.g., *Bacillus subtilis* natto and *Paenibacillus polymyxa*, for lactosucrose production (Choi et al. 2004; Dos Santos et al. 2013). However, heterologous expression is an efficient method to produce a target enzyme. Heterologous expression was achieved by using levansucrase enzyme from *B. subtilis* NCIMB 11871, *Zymomonas mobilis, Bacillus methylotrophicus, Brenneria goodwinii, Pseudomonas aurantiaca, Leuconostoc mesenteroides,* etc.(Han et al. 2007; Han et al. 2009; Li et al. 2015; Wu et al. 2015; Xu et al. 2018). Levansucrase from *L. mesenteroides* is a promising biocatalyst that can achieve high productivity of lactosucrose (224 g/L/h) (Li et al. 2015). Whole-cell catalysis of lactosucrose has also been attempted for lactosucrose production. In the case of whole-cell catalysis, the enzyme tends to be more stable and integrated than the cell-free system. One advantage of whole-cell catalysis is the continuous production of the desired product. The first report on the microbial synthesis of lactosucrose was from *P. polymyxa* IFO 3020 strain (Choi et al. 2004). *B. subtilis* KCCM 32835 has been reported to yield high productivity of lactosucrose (18.3 g/L/h) using whole-cell catalysis (Park et al. 2005). Table 8.1 provides a summary of lactosucrose production by levansucrase enzyme through native, recombinant, and whole-cell catalysis.

β-fructofuranosidase is known as invertase or saccharase, that catalyzes the hydrolysis of sucrose into glucose and fructose in an equimolar ratio. The β-fructofuranosidase belongs to enzyme families GH32 and GH68. It is widely present in bacteria, fungi, and plants (Banguela and Hernández 2006). Both GH32 and GH68 family fructofuranosidases catalyze the hydrolysis of the non-reducing terminal of fructofuranoside residue. At high concentrations of sucrose, this enzyme exhibits fructosyltransferase activity, catalyzing cleavage of β-(1-2) linkage of sucrose followed by the transfer of fructose onto another sucrose molecule (Gutiérrez-Alonso et al. 2009). This results in the formation of kestose (glucose-fructose 2), 1-nystose (glucose-fructose 3), and 1-fructosyl nystose (glucose-fructose 4). β-Fructofuranosidase-mediated catalysis results in the synthesis of lactosucrose from sucrose and lactose. The high-level productivity of lactosucrose has been demonstrated in the case of β-fructofuranosidases enzyme from the soil isolate, *Arthrobacter* sp. K-1 (Fujita et al. 1990). β-fructofuranosidases do hydrolysis of lactosucrose, causing decrement in overall productivity. Table 8.2 summarizes lactosucrose production by β-fructofuranosidases enzyme.

β-galactosidase is a glycosyl hydrolase (GH35) lactase that catalyzes hydrolysis of lactose into glucose and galactose (Thakur et al. 2022). β-Galactosidase is widely used in the milk industry for lactose hydrolysis and for the synthesis of bioactive molecules like lactulose and galactooligosaccharides. In addition to hydrolysis, it can also catalyze transgalactosylation of galactosyl moiety to galactose acceptor like sucrose. In transgalactosylation, β-galactosidase hydrolyzes lactose into glucose and galactose and forms a β-(1-4) linkage between galactose and glucose moiety of sucrose to produce lactosucrose (Yan et al. 2021). β-Galactosidase from *Bacillus circulans* and *Lactobacillus plantarum* FMNP01 has been characterized for lactosucrose production (Liao et al. 2023).

8.2.2 Tagatose

Tagatose is an FDA-approved and GRAS status molecule that can be used as a natural sugar substitute (Roh et al. 1999). It is a naturally occurring rare ketohexose isomer of galactose and epimer of fructose. Tagatose has 92% relative sweetness of sucrose (Saraiva et al. 2020). Its first natural occurrence was noticed in the gum exudates of *Sterculia setigera* (a tropical deciduous tree species) lichens in low quantity. Later, it was also detected in trace amount in sterilized milk, yogurt, cheese, hot cocoa, some fruits, white beans, etc. (Roy et al. 2018). Tagatose intake helps to control diabetes and manage weight loss, as this rare sugar has a lower glycemic index and lower caloric value than table sugar (Guerrero-Wyss et al. 2018). It is a rare sugar with many health-promoting attributes like antihyperglycemic, anticarcinogenic and antioxidant effects. However, excessive consumption of tagatose-containing products (30 grams per day) causes symptoms like belching, blotting, laxation, and flatulence (Lu et al. 2008). Meanwhile, the above-discussed gastrointestinal symptoms are mild, transient, and not experienced by all consumers. There is also a chance of tolerance development if consumed frequently. Canada, Korea, Australia, New Zealand, the United States,

Table 8.1 Lactosucrose production by levansucrase enzyme.

Bacterial origin	Enzyme form	Temperature (°C)	pH	Maximum lactosucrose production (g/L)	Productivity (g/L/h)	References
Aerobacter levanicum	Native	30	5.4	NR	NR	Avigad (1957)
Bacillus subtilis natto	Native	35	6.2	53	26.5	Takahama et al. (1991)
B. subtilis NCIMB 11871	Native	NR	6.0	131	21.9	Seibel et al. (2006)
Bacillus methylotrophicus	Recombinant	37	6.5	143	7.15	Wu et al. (2015)
Leuconostoc mesenteroides	Recombinant	50	6.5	224	224	Li et al. (2015)
Brenneria goodwinii	Recombinant	35	6.0	100	50	Xu et al. (2018)
Pseudomonas aurantiaca	Recombinant	45	4.0	285	142	Han et al. (2007)
Paenibacillus polymyxa	Recombinant	45	6.0	140	210	Choi et al. (2004)
Zymomonas mobilis	Recombinant	23	7.0	103	25.8	Han et al. (2009)
B. subtilis KCCM 32835	Whole-cell catalysis	55	6.0	183	18.3	Park et al. (2005)
Bacillus amyloliquefaciens IFO 15535	Whole-cell catalysis	55	6.0	4.8	4.8	
Geobacillus stearothermophilus ATCC 12980	Whole-cell catalysis	55	6.0	10.4	10.4	
P. polymyxa	Whole-cell catalysis	55	6.0	170	28	
Klebsiella pneumoniae ATCC 25306	Whole-cell catalysis	55	6.0	16	1.1	
Pseudomonas syringae IFO 14086	Whole-cell catalysis	40	6.0	3.9	3.9	
Rahnella aquatilis KTC 2858	Whole-cell catalysis	40	6.0	5.6	5.6	
R. aquatilis ATCC 55046	Whole-cell catalysis	50	6.0	79	5.3	Lee et al. (2007)
Sterigmatomyces elviae ATCC 18894	Whole-cell catalysis	50	6.0	91	6.1	
S. elviae ATCC 18894 (mutant)	Whole-cell catalysis	50	6.0	184	12.3	
S. elviae ATCC 18894 (immobilized cell on calcium alginate beads)	Whole-cell catalysis	50	6.0	192	9.6	

Table 8.2 Lactosucrose production by β-fructofuranosidase enzyme.

Bacterial origin	Enzyme form	Temperature (°C)	pH	Maximum lactosucrose production (g/L)	Productivity (g/L/h)	References
Bacillus methanolicus LB-1	Native	37	7.0	110	3.92	Liu et al. (2023)
Arthrobacter sp. K-1	Native	55	NR	135	63.4	Pilgrim et al. (2001)
Bacillus sp. 417	Native	45	5–6	54	6.75	Ikegaki and Park (1997)
Arthrobacter sp. 10137	Immobilized bi-enzymes	40		160	NR	Long et al. (2019)
Arthrobacter sp. 10138	Recombinant	50	6	109	681	Chen et al. (2020)

and the European Union have approved tagatose for human consumption. For tagatose biosynthesis, lactose needs to be hydrolyzed into glucose and galactose, and then, later galactose is used to produce tagatose. Galactitol 2-dehydrogenase and L-arabinose isomerase are used to produce tagatose (Jayamuthunagai et al. 2017). Galactitol 2-dehydrogenase produces tagatose from galactitol (a reduced product of galactose). Galactitol 2-dehydrogenase oxidizes galactitol into tagatose in the presence of NAD^+. Yet, the use of galactitol for the synthesis of tagatose is limited because of its high price and expensive cofactor (NAD^+) requirement (Su et al. 2021).

L-Arabinose isomerase (AI) is an intracellular catalyst mostly used for the isomerization of galactose into tagatose, but it has greater catalytic efficiency for L-arabinose than for D-galactose. In microbes, AI is encoded by araA (L-arabinose component) gene for the conversion of L-arabinose to L-ribulose (Jayamuthunagai et al. 2017). Currently, more than 20 L-AIases have been isolated and characterized from various environment-adapted microbes like psychrophiles, mesophiles, thermophiles, hyperthermophiles, acido-thermophiles, alkalophiles, and acidophiles. The microbes isolated and characterized for the production of L-AIs are *Lactobacillus fermentum* from Chinese pickles, *Pediococcus pentosaceus* PC-5 from fermented pickle, *L. plantarum* CY.6 from kimchi, *Lactobacillus brevis* from fermented cabbage, *Enterococcus faecium* from raw cow milk, *Bacillus stearothermophilus* from hot spring soil of Tunisia, *Alicyclobacillus* sp. TP7 from Indonesia hot spring, *Thermotoga neapolitana* from submarine thermal vent, *Geobacillus thermodenitrificans* from Korean compost sample, *Arthrobacter* sp. 22c from Antarctic soil sample, etc. In most cases, AIs have been characterized from thermophilic and hyperthermophilic sources. The reaction rate of high-temperature operating systems is better than that of their low-temperature counterparts because of the higher solubility of the substrate and low viscosity at high temperatures. However, browning reaction and by-product formation at elevated temperatures restricted their application. Many AIs variants have been characterized from thermophilic sources (*Thermotoga Neapolitan, Anoxybacillus flavithermus, Thermotoga. Neapolitana, Bacillus stearothermophilus, Thermotoga maritime*, etc.), but they have high optimum temperatures (75–95°C), which generates by-product and exerts browning effect on sugar (Shen and Wu 2004). Therefore, screening of efficient AIs from mesophilic microbial sources is required. AIs from mesophilic sources have lower optimum temperatures and low requirements for metallic cofactors. Recently, researchers have explored AIs from mesophilic *Bacillus* strains, but their catalytic efficiency is less for galactose, and they show a high affinity for arabinose. AI from *Bacillus halodurans* exhibits catalytic efficiency of 0.4 mM/min

Table 8.3 Tagatose production by recombinant L-arabinose isomerase.

Bacterial origin	Temperature (°C)	pH	Maximum tagatose production (g/L)	Productivity (g/L/h)	References
Thermotoga neapolitana	85	7.0	NR	NR	Kim et al. (2002)
Escherichia coli	30	8.0	17.7	0.10	Yoon et al. (2003)
Thermoanaerobacter mathranii	65	8.0	75	NR	Jørgensen et al. (2004)
T. maritime	90	7.5	NR	NR	Lee et al. (2004)
Geobacillus thermodenitrificans	70	8.5	NR	NR	Kim and Oh (2005)
Bacillus stearothermophilus	7.5–8.0	80	NR	NR	Rhimi and Bejar (2006)
Lactobacillus plantarum	7.5	60	25	1.2	Chouayekh et al. (2007)
Bacillus licheniformis	7.5	50	NR	NR	Prabhu et al. (2008)
Lactobacillus sakei	5.0–7.0	30–40	NR	NR	Rhimi et al. (2010)
Bacillus subtilis	7.5	32	NR	NR	Kim et al. (2010)
Lactobacillus fermentum	6.5	65	NR	NR	Xu et al. (2011)
Shewanella sp.	5.5–6.5	15–35	NR	NR	Rhimi et al. (2011)
Anoxybacillus flavithermus	10.5	95	NR	NR	Li et al. (2011)
Pediococcus pentosaceus	6.0	50	NR	NR	Men et al. (2014)
Alicyclobacillus hesperidium	7.0	70	NR	NR	Fan et al. (2014)
Clostridium hylemonae	7.0–7.5	50	NR	NR	Nguyen et al. (2018)
Lactobacillus brevis	7.0	65	NR	NR	Du et al. (2019)
Enterococcus faecium	5.5	50	NR	NR	Manzo et al. (2019)
Arthrobacter psychrolactophilus B7	7.0–7.5	60	NR	NR	Nirwantono et al. (2024)
Bifidobacterium adolescentis	6.5	55	NR	NR	Zhang et al. (2020)

toward D-galactose (Lee et al. 2005). *Shewanella sp.* AI exhibits relatively high catalytic efficiency for galactose transformation (22 mM/min) (Jayaraman et al. 2021). Its catalytic efficiency was further enhanced by the substitution of bulky phenylalanine residues with smaller hydrophilic residues. Table 8.3 provides the details of AI enzyme used for tagatose production.

Phosphoglucose isomerase from *Pseudomonas aeruginosa* PAO catalyzes isomerization of galactose into tagatose, but its high km value makes it unsuitable choice (Patel et al. 2016). Fructose-6-phosphate-4-epimerase catalyzes c4 epimerization of fructose-6-phosphate to tagatose-6-phosphate, and later phosphate can be removed by the application of phosphoglycolate phosphatase (Dai et al. 2020).

However, isomerases have the fundamental limitation of thermodynamic equilibria in which the concentration of product and temperature do not change with time, a limiting factor for a low conversion rate of tagatose from galactose (Liu et al. 2019). Unlike isomerase, redox reaction catalyzed by oxidoreductase involves two steps: reduction and oxidation in the presence of cofactors. Oxidoreductase reverse reaction is controlled by thermodynamic coupling of cofactors, which prevents the thermodynamic reaction equilibria issue caused by isomerase and epimerase enzymes. Galactitol 2-dehydrogenase belongs to an oxidoreductase class of enzymes, but the high cost of the substrate, galactitol, makes the whole process expensive for its industrial application (Jagtap et al. 2014). Recently, oxidoreductase enzyme has been harnessed to produce tagatose in yeast and *Escherichia coli* systems. In yeast, the whole-cell conversion process was developed to overcome the thermodynamic limitation of isomerases. Lactose is hydrolyzed by β-galactosidase, and further by the application of xylose reductase and galactitol-2-hydrogenase, it was converted into galactitol and tagatose, respectively (Zhang et al. 2023). This is the smart approach to tagatose production where one-half of the substrate is used for cell growth and maintenance, and the most efficient oxidoreductase utilizes the rest for tagatose production (Liu et al. 2019).

8.2.3 Lactulose

Lactulose is a disaccharide of galactose and fructose that can be isomerized from lactose (Montgomery and Hudson 1930; Olano and Corzo 2009). It is the first commercially available derivative of lactose. It is 60% sweet in comparison with sucrose and 150% sweeter than lactose (Mukherjee and John 2022). It is a lactose-derived rare sugar with prebiotic properties that can promote the growth of health-promoting bacteria like *Lactobacillus* and *Bifidobacteria*. Further, it inhibits the growth of pathogenic microbes like *Shigella and Salmonella,* in the gastrointestinal tract. Lactulose is used in the pharmaceutical industry for the treatment of chronic constipation, hepatic encephalopathy, hyperammonemia, tumor prevention, and inhibition of secondary bile acid formation (Medical 2002).

Currently, the microbial approach to lactulose production from lactose is carried out either by controlled transgalactosylation by glycosyl hydrolases (β-galactosidase and β-glycosidase) or by direct isomerization of lactose through cellobiose 2-epimerase enzymes (Letsididi et al. 2018; de Albuquerque et al. 2021). β-Glycosidase can catalyze the synthesis of lactulose in the presence of nucleophile acceptor (fructose). Galactosyl moiety is transferred to the added nucleophile acceptor (i.e., fructose) by β-glycosidase for the generation of lactulose. However, β-glycosidase-catalyzed transgalactosylation has certain limitations that hinder the industrial translation of this enzymatic process, e.g., high fructose addition in the reaction media by-product generation at the high reaction temperature (75°C) and low yield of lactulose (Fattahi et al. 2010; Aburto et al. 2020). At the same time, β-galactosidase operates under milder conditions (30–50°C) that support industrial usage of β-galactosidase than β-glycosidase. β-galactosidase hydrolyzes lactose and generates a mixture of glucose and β-galactosidase–galactose intermediate. Afterward, the intermediate has two fates: (i) In the presence of water nucleophile, the intermediate hydrolyzes and galactose is left with a trace amount of galactooligosaccharides (GOS); (ii) in the presence of fructose, controlled transgalactosylation leads to the formation of lactulose. Besides fructose, other nucleophile acceptors, e.g., lactose, GOS, glucose, galactose, and lactulose, may lead to synthesis of other transgalactosylated products (Kim et al. 2006; Tang et al. 2011). Lactulose production *via* glycosyl hydrolase enzyme has the economic disadvantage of additional fructose requirement, low yield (~19%), and purification burden of unwanted products.

The direct isomerization of lactose into lactulose through *N*-acyl-D-glucosamine 2-epimerases (AGE) superfamily belonging to cellobiose 2-epimerase (CE) is a desirable approach. Cellobiose 2-epimerase was first reported from anaerobic ruminal bacterium *Ruminococcus albus* (Ito et al. 2008). Since then, over twenty CEs have been characterized from different bacterial sources, e.g., *Ruminococcus* albus (Fujiwara et al. 2013), *Rhodothermus marinus* (Ojima et al. 2011), *Spirochaeta thermophila* (Park et al. 2013), *Cellulosilyticum lentocellum* (Krewinkel et al. 2015), *Dysgonomonas gadei* (Krewinkel et al. 2015), *Dictyoglomus turgidum* (Kim et al. 2012), *Thermoanaerobacterium saccharolyticum* (Hung et al. 2011), *Clostridium disporicum* (Wang et al. 2024), *Saccharophagus degradans* (Krewinkel et al. 2015), *Caldicellulosiruptor obsidiansis* (Chen et al. 2017), *Teredinibacter turnerae* (Krewinkel et al. 2015), *Dyadobacter fermentans* (Krewinkel et al. 2015), *Flavobacterium johnsoniae* (Krewinkel et al. 2015), *Eubacterium cellulosolvens* (Taguchi et al. 2008), *Caldicellulosiruptor saccharolyticus* (Kim and Oh 2012), *Herpetosiphon aurantiacus* (Krewinkel et al. 2015), *Pedobacter heparinus* (Krewinkel et al. 2015), *Spirosoma linguale* (Wang et al. 2016), *Roseburia faecis* (Jameson et al. 2021), *and Bacteroides fragilis* (Senoura et al. 2009). CE can either do isomerization of glucose moiety into fructose or it can do epimerization of glucose moiety into mannose to produce lactulose and epilactose, respectively. Most characterized CEs have only epimerization activity, whereas CEs from *Clostridium disporicum, C. saccharolyticus, Caldicellulosiruptor bescii, C. obsidiansis, Spirochaeta thermophile,* and *D. turgidum* exhibit both epimerization and isomerization activities (Jameson et al. 2021). Among characterized CEs, *C. saccharolyticus* CE exhibits relatively high productivity for lactulose conversion. In addition, the application of (1 : 1) borate-lactose molar ratio enhances the lactulose yield at the neutral conditions from 58% to 88% (Kim et al. 2013). However, the application of borate to increase productivity is a matter of safety concern in food supplement products as it can cause nausea, vomiting, abdominal pain, etc. Apart from thermostable CEs for the production of lactulose, there is a research gap in identifying highly catalytic efficient cold-active CEs for applications in the dairy industry. Therefore, exploring cold-active CEs for efficient functionality at low temperatures remains an open area of research.

8.2.4 Epilactose

Epilactose is a valuable lactose derivative synthesized *via* epimerization of lactose (Seki and Saito 2012). It is a rare sugar, non-digestible disaccharide with prebiotic properties and resistant to rat intestinal enzyme (Nishimukai et al. 2008). Epilactose has therapeutic application in the treatment of post-gastrectomy osteopenia and anemia through the promotion of calcium and ferrum absorption, respectively (Suzuki et al. 2010). Epilactose is very expensive because its chemical synthesis and purification are very tough. At present, epilactose can be biotechnologically produced through cellobiose 2-epimerase enzyme. Over 20 CEs have been characterized for lactose transformation into lactulose and epilactose, as discussed above. Mostly, CEs from mesophilic sources

Table 8.4 Lactulose and epilactose production by recombinant cellobiose 2-epimerase enzyme.

Enzyme source	Optimum pH range	Optimum temperature range (°C)	k_{cat} (s^{-1})	k_m (mM)	k_{cat}/K_m (mM/s)	References
Caldicellulosiruptor obsidiansis OB47	7.5	70	233.4	301.7	0.773	Chen et al. (2017)
Caldicellulosiruptor sp. Rt8. B8	8.0–9.0	70	197.6	35.8	5.5	Liangfei et al. (2022)
Ruminococcus albus	7.5	30	52.1	33	1.6	Ito et al. (2007)
Caldicellulosiruptor saccharolyticus	7.5	75	NR	NR	NR	Kim and Oh (2012)
Rhodothermus marinus	6.3	80	111	28.8	3.85	Ojima et al. (2011)
Cellulosilyticum lentocellum	8.0	40	NR	129	NR	Krewinkel et al. (2015)
Pedobacter heparinus	6.3	35	5.43	24.5	0.222	Krewinkel et al. (2015)
Eubacterium cellulosolvens	7.5	35	32.5	72	0.451	Taguchi et al. (2008)
Herpetosiphon aurantiacus	7.3	45	14	51.9	0.27	Krewinkel et al. (2015)
Dictyoglomus turgidum	7.0	70	89.4	79.6	1.12	Kim et al. (2012)
Saccharophagus degradans	7.7	35	7.82	29.2	0.268	Krewinkel et al. (2015)
Thermoanaerobacterium saccharolyticum	7.0	60	30.9	124.7	0.248	Chen et al. (2015)
Bacteroides fragilis	7.5	45	79.5	6.56	12.1	Senoura et al. (2009)
Spirochaeta thermophile	7.0	60	NR	NR	NR	Park et al. (2013)
Teredinibacter turnerae	8.8	35	175	238	0.735	Krewinkel et al. (2015)
Dysgonomonas gadei	7.5	40	NR	108	NR	Chen et al. (2015)
Spirosoma linguale	7.7	45	92.1	206	0.447	Krewinkel et al. (2015)
Dyadobacter fermentans	7.7	50	44.9	95.7	0.469	Krewinkel et al. (2015)
Flavobacterium johnsoniae	8.4	35	17	34.9	0.501	Krewinkel et al. (2015)
Caldicellulosiruptor bescii	7.5	70	NR	NR	NR	Jameson et al. (2021)
Roseburia faecis	8	50	NR	NR	NR	Jameson et al. (2021)

exhibit optimum temperature in the range of 30–50°C and CEs from thermophilic sources show optimum temperature above 50°C. CE from *Bacteroides fragilis* is a catalytically efficient enzyme for epilactose production with a catalytic efficiency of 12.1 mmol/s. CE from *C. saccharolyticus* is an efficient enzyme for lactulose and epilactose transformation, with a yield of ~58% and ~15%, respectively. However, to be industrially suitable, CEs need to show high catalytic activity at low temperatures and in mild acidic environments with high stability. Recently, our group has characterized a novel CE gene from a thermal aquatic habitat that encodes an enzyme that can transform lactose to epilactose in a wide range of temperatures (4 to 70°C) and pH (6 to 10) (Singh et al. 2025). Generally, whey pH is slightly acidic, and the reported CEs have an optimum pH range above 6.0. Therefore, cold-active and acid-tolerant enzyme mining is a major research gap in CE screening. *Roseburia intestinalis* CE has shown the epimerization activity of lactose at 8°C, with a yield of ~21% (Chen et al. 2020). Table 8.4 summarizes the lactulose and epilactose production *via* cellobiose 2-epimerase enzyme. Epilactose is non-GRAS (generally recognized as safe) status molecule. Hence, enough data related to epilactose metabolism, physiological function, toxicity to animals and humans, etc., need to be evaluated.

8.3 Conclusion

Lactose valorization into high-value rare sugar derivatives offers an industrially significant solution to utilize dairy industry byproduct, e.g., whey. The biological processes of rare sugar production are considered green, sustainable, eco-friendly, and economical approach for lactose transformation. Tagatose, epilactose, lactulose, and lactosucrose have a great future in the food, nutraceuticals, pharmaceuticals, and agriculture sectors. For upscale production of rare sugar molecules, industrial processes should be engineered by employing efficient and stable biocatalysts and/or potential microbial strains.

Acknowledgements

The authors acknowledge the Department of Biotechnology (DBT), Government of India, for financial support via the DBT research project, BT/PR31829/NDB/39/649/2019.

References

Aburto, C., Guerrero, C., Vera, C., and Illanes, A. (2020). Improvement in the yield and selectivity of lactulose synthesis with *Bacillus circulans* β-galactosidase. *LWT* 118: 108746.

Agarwal, N. and Singh, S.P. (2021). A novel trehalose synthase for the production of trehalose and trehalulose. *Microbiology Spectrum* 9: e01333–e01321.

Ahmed, A., Khan, T.A., Dan Ramdath, D. et al. (2022). Rare sugars and their health effects in humans: a systematic review and narrative synthesis of the evidence from human trials. *Nutrition Reviews* 80: 255–270.

de Albuquerque, T.L., de Sousa, M., e Silva, N.C.G. et al. (2021). β-Galactosidase from *Kluyveromyces lactis*: characterization, production, immobilization and applications-a review. *International Journal of Biological Macromolecules* 191: 881–898.

Avigad, G. (1957). Enzymatic synthesis and characterization of a new trisaccharide, α-lactosyl-β-fructofuranoside. *Journal of Biological Chemistry* 229: 121–129.

Avigad, G., Feingold, D.S., and Hestrin, S. (1956). The mechanism of polysaccharide production from sucrose. 3. Donor-acceptor specificity of levansucrase from *Aerobacter levanicum*. *Biochemical Journal* 64: 340–351.

Banguela, A. and Hernández, L. (2006). Fructans: from natural sources to transgenic plants. *Biotecnologia Aplicada* 23: 202–210.

Bass, P. and Dennis, S. (1981). The laxative effects of lactulose in normal and constipated subjects. *Journal of Clinical Gastroenterology* 3: 23–28.

Chen, C., Deng, J., Lv, X. et al. (2020). Biocatalytic synthesis of lactosucrose using a recombinant thermostable β-fructofuranosidase from *Arthrobacter* sp. *Bioengineered* 10138 (11): 416–427.

Chen, Q., Levin, R., Zhang, W. et al. (2017). Characterisation of a novel cellobiose 2-epimerase from thermophilic *Caldicellulosiruptor obsidiansis* for lactulose production. *Journal of the Science of Food and Agriculture* 97: 3095–3105.

Chen, Q., Xiao, Y., Zhang, W. et al. (2020). Computer-aided search for a cold-active cellobiose 2-epimerase. *Journal of Dairy Science* 103: 7730–7741.

Chen, Q., Zhang, W., Zhang, T. et al. (2015). Characterization of an epilactose-producing cellobiose 2-epimerase from *Thermoanaerobacterium saccharolyticum*. *Journal of Molecular Catalysis B: Enzymatic* 116: 39–44.

Choi, H.J., Kim, C.S., Kim, P. et al. (2004). Lactosucrose bioconversion from lactose and sucrose by whole cells of *Paenibacillus polymyxa* harboring levansucrase activity. *Biotechnology Progress* 20: 1876–1879.

Chouayekh, H., Bejar, W., Rhimi, M. et al. (2007). Characterization of an l-arabinose isomerase from the *Lactobacillus plantarum* NC8 strain showing pronounced stability at acidic pH. *FEMS Microbiology Letters* 277: 260–267.

Dai, Y., Zhang, J., Zhang, T. et al. (2020). Characteristics of a fructose 6-phosphate 4-epimerase from *Caldilinea aerophila* DSM 14535 and its application for biosynthesis of tagatose. *Enzyme and Microbial Technology* 139: 109594.

Dos Santos, L.F., De Melo, F., Paiva, W. et al. (2013). Characterization and optimization of levan production by *Bacillus subtilis* NATTO. *Romanian Biotechnological Letters* 18: 8413–8422.

Du, M., Zhao, D., Cheng, S. et al. (2019). Towards efficient enzymatic conversion of d-galactose to d-tagatose: purification and characterization of l-arabinose isomerase from *Lactobacillus brevis*. *Bioprocess and Biosystems Engineering* 42: 107–116.

Ensor, M., Banfield, A.B., Smith, R.R. et al. (2015). Safety and efficacy of d-tagatose in glycemic control in subjects with type 2 diabetes. *Journal of Endocrinology, Diabetes and Obesity* 3.

Fan, C., Liu, K., Zhang, T. et al. (2014). Biochemical characterization of a thermostable l-arabinose isomerase from a thermoacidophilic bacterium, *Alicyclobacillus hesperidum* URH17-3-68. *Journal of Molecular Catalysis B: Enzymatic* 102: 120–126.

Fattahi, H., Ashtiani, F.Z., Bonakdarpour, B. et al. (2010). Enzymatic synthesis of lactulose by commercial ß-galactosidase from *Klyveromyces lactis*. *Afinidad* 67.

Fujita, K., Hara, K., Hashimoto, H., and Kitahata, S. (1990). Studies on. BETA.-fructofuranosidase from *Arthrobacter* sp. K-1. Part I. Purification and some properties of. BETA.-fructofuranosidase I from *Arthrobacter* sp. K-1. *Agricultural and Biological Chemistry* 54: 913–919.

Fujiwara, T., Saburi, W., Inoue, S. et al. (2013). Crystal structure of *Ruminococcus albus* cellobiose 2-epimerase: structural insights into epimerization of unmodified sugar. *FEBS Letters* 587: 840–846.

Gambelli, L. (2017). Milk and its sugar-lactose: a picture of evaluation methodologies. *Beverages* 3 (4): 35.

Guerrero-Wyss, M., Durán Agüero, S., and Angarita Dávila, L. (2018). d-tagatose is a promising sweetener to control glycaemia: a new functional food. *BioMed Research International* 2018: 8718053.

Gutiérrez-Alonso, P., Fernández-Arrojo, L., Plou, F.J., and Fernández-Lobato, M. (2009). Biochemical characterization of a β-fructofuranosidase from *Rhodotorula dairenensis* with transfructosylating activity. *FEMS Yeast Research* 9: 768–773.

Güven, G., Perendeci, A., and Tanyolaç, A. (2008). Electrochemical treatment of deproteinated whey wastewater and optimization of treatment conditions with response surface methodology. *Journal of Hazardous Materials* 157: 69–78.

Han, W.-C., Byun, S.-H., Kim, M.-H. et al. (2009). Production of lactosucrose from sucrose and lactose by a levansucrase from *Zymomonas mobilis*. *Journal of Microbiology and Biotechnology* 19: 1153–1160.

Han, W.-C., Byun, S.-H., Lee, J.-C. et al. (2007). Synthesis of lactosucrose formed by levansucrase from *Pseudomonas aurantiaca*. *Journal of Biotechnology* 131: S113.

Hou, K., Wu, Z.-X., Chen, X.-Y. et al. (2022). Microbiota in health and diseases. *Signal Transduction and Targeted Therapy* 7: 135.

Hung, K.-S., Liu, S.-M., Tzou, W.-S. et al. (2011). Characterization of a novel GH10 thermostable, halophilic xylanase from the marine bacterium *Thermoanaerobacterium saccharolyticum* NTOU1. *Process Biochemistry* 46: 1257–1263.

Ikegaki, M. and Park, Y.K. (1997). Produção de lactosacarose por β-frutofuranosidase de *bacillus* sp n°417 a partir de lactose e sacarose. *Food Science and Technology* 17: 188–191.

Ito, S., Hamada, S., Yamaguchi, K. et al. (2007). Cloning and sequencing of the cellobiose 2-epimerase gene from an obligatory anaerobe, *Ruminococcus albus*. *Biochemical and Biophysical Research Communications* 360: 640–645.

Ito, S., Taguchi, H., Hamada, S. et al. (2008). Enzymatic properties of cellobiose 2-epimerase from *Ruminococcus albus* and the synthesis of rare oligosaccharides by the enzyme. *Applied Microbiology and Biotechnology* 79: 433–441.

Jagtap, S.S., Singh, R., Kang, Y.C. et al. (2014). Cloning and characterization of a galactitol 2-dehydrogenase from *Rhizobium legumenosarum* and its application in d-tagatose production. *Enzyme and Microbial Technology* 58: 44–51.

Jameson, J.K., Mathiesen, G., Pope, P.B. et al. (2021). Biochemical characterization of two cellobiose 2-epimerases and application for efficient production of lactulose and epilactose. *Current Research in Biotechnology* 3: 57–64.

Jayamuthunagai, J., Gautam, P., Srisowmeya, G., and Chakravarthy, M. (2017). Biocatalytic production of d-tagatose: a potential rare sugar with versatile applications. *Critical Reviews in Food Science and Nutrition* 57: 3430–3437.

Jayaraman, A.B., Kandasamy, T., Venkataraman, D., and Meenakshisundaram, S. (2021). Rational design of *Shewanella* sp. l-arabinose isomerase for d-galactose isomerase activity under mesophilic conditions. *Enzyme and Microbial Technology* 147: 109796.

Jørgensen, F., Hansen, O., and Stougaard, P. (2004). Enzymatic conversion of d-galactose to d-tagatose: heterologous expression and characterisation of a thermostable l-arabinose isomerase from *Thermoanaerobacter mathranii*. *Applied Microbiology and Biotechnology* 64: 816–822.

Kim, J.-E., Kim, Y.-S., Kang, L.-W., and Oh, D.-K. (2012). Characterization of a recombinant cellobiose 2-epimerase from *Dictyoglomus turgidum* that epimerizes and isomerizes β-1, 4-and α-1, 4-gluco-oligosaccharides. *Biotechnology Letters* 34: 2061–2068.

Kim, Y.-S., Kim, J.-E., and Oh, D.-K. (2013). Borate enhances the production of lactulose from lactose by cellobiose 2-epimerase from *Caldicellulosiruptor saccharolyticus*. *Bioresource Technology* 128: 809–812.

Kim, B.-C., Lee, Y.-H., Lee, H.-S. et al. (2002). Cloning, expression and characterization of l-arabinose isomerase from *Thermotoga neapolitana*: bioconversion of d-galactose to d-tagatose using the enzyme. *FEMS Microbiology Letters* 212: 121–126.

Kim, H.-J. and Oh, D.-K. (2005). Purification and characterization of an l-arabinose isomerase from an isolated strain of *Geobacillus thermodenitrificans* producing d-tagatose. *Journal of Biotechnology* 120: 162–173.

Kim, Y.-S. and Oh, D.-K. (2012). Lactulose production from lactose as a single substrate by a thermostable cellobiose 2-epimerase from *Caldicellulosiruptor saccharolyticus*. *Bioresource Technology* 104: 668–672.

Kim, Y.-S., Park, C.-S., and Oh, D.-K. (2006). Lactulose production from lactose and fructose by a thermostable β-galactosidase from *Sulfolobus solfataricus*. *Enzyme and Microbial Technology* 39: 903–908.

Kim, J.-H., Prabhu, P., Jeya, M. et al. (2010). Characterization of an l-arabinose isomerase from *Bacillus subtilis*. *Applied Microbiology and Biotechnology* 85: 1839–1847.

Krewinkel, M., Kaiser, J., Merz, M. et al. (2015). Novel cellobiose 2-epimerases for the production of epilactose from milk ultrafiltrate containing lactose. *Journal of Dairy Science* 98: 3665–3678.

Lee, D.-W., Choe, E.-A., Kim, S.-B. et al. (2005). Distinct metal dependence for catalytic and structural functions in the l-arabinose isomerases from the mesophilic *Bacillus halodurans* and the thermophilic *Geobacillus stearothermophilus*. *Archives of Biochemistry and Biophysics* 434: 333–343.

Lee, D.-W., Jang, H.-J., Choe, E.-A. et al. (2004). Characterization of a thermostable l-arabinose (d-galactose) isomerase from the hyperthermophilic eubacterium *Thermotoga maritima*. *Applied and Environmental Microbiology* 70: 1397–1404.

Lee, J.-H., Lim, J.-S., Park, C.-H. et al. (2007). Continuous production of lactosucrose by immobilized *Sterigmatomyces elviae* mutant. *Journal of Microbiology and Biotechnology* 17: 1533–1537.

Letsididi, R., Hassanin, H.A., Koko, M.Y. et al. (2018). Lactulose production by a thermostable glycoside hydrolase from the hyperthermophilic archaeon *Caldivirga maquilingensis* IC-167. *Journal of the Science of Food and Agriculture* 98: 928–937.

Li, W., Yu, S., Zhang, T. et al. (2015). Efficient biosynthesis of lactosucrose from sucrose and lactose by the purified recombinant levansucrase from *Leuconostoc mesenteroides* B-512 FMC. *Journal of Agricultural and Food Chemistry* 63: 9755–9763.

Li, Y., Zhu, Y., Liu, A., and Sun, Y. (2011). Identification and characterization of a novel l-arabinose isomerase from *Anoxybacillus flavithermus* useful in d-tagatose production. *Extremophiles* 15: 441–450.

Liangfei, L., Yafeng, Z., Kai, X., and Zheng, X. (2022). Identification of a thermostable cellobiose 2-epimerase from *Caldicellulosiruptor* sp. Rt8. B8 and production of epilactose using *Bacillus subtilis*. *Journal of the Science of Food and Agriculture* 102: 85–94.

Liao, X., Li, Y., Li, Y. et al. (2023). Optimization of the production conditions of tri-GOS and lactosucrose from lactose and sucrose with recombinant β-galactosidase. *Preparative Biochemistry & Biotechnology* 53: 401–411.

Liu, J., Ma, Y., Zhang, M. et al. (2023). Biosynthesis of lactosucrose by a new source of β-fructofuranosidase from *Bacillus methanolicus* LB-1. *Journal of Bioscience and Bioengineering* 135: 118–126.

Liu, J.-J., Zhang, G.-C., Kwak, S. et al. (2019). Overcoming the thermodynamic equilibrium of an isomerization reaction through oxidoreductive reactions for biotransformation. *Nature Communications* 10: 1356.

Long, J., Pan, T., Xie, Z. et al. (2019). Effective production of lactosucrose using β-fructofuranosidase and glucose oxidase co-immobilized by sol–gel encapsulation. *Food Science & Nutrition* 7: 3302–3316.

Lu, Y., Levin, G., and Donner, T.W. (2008). Tagatose, a new antidiabetic and obesity control drug. *Diabetes, Obesity and Metabolism* 10: 109–134.

Manzo, R.M., Antunes, A.S.L.M., de Sousa, M.J. et al. (2019). Biochemical characterization of heat-tolerant recombinant l-arabinose isomerase from *Enterococcus faecium* DBFIQ E36 strain with feasible applications in d-tagatose production. *Molecular Biotechnology* 61: 385–399.

Medical, S.C. (2002). nutritional and technological properties of lactulose. An update. *European Journal of Nutrition* 41: i17–i25.

Men, Y., Zhu, Y., Zhang, L. et al. (2014). Enzymatic conversion of d-galactose to d-tagatose: cloning, overexpression and characterization of l-arabinose isomerase from *Pediococcus pentosaceus* PC-5. *Microbiological Research* 169: 171–178.

Mijailovic, N., Nesler, A., Perazzolli, M. et al. (2021). Rare sugars: recent advances and their potential role in sustainable crop protection. *Molecules* 26: 1720.

Montgomery, E.M. and Hudson, C. (1930). Relations between rotatory power and structure in the sugar group. XXVII. Synthesis of a new disaccharide ketose (lactulose) from lactose1. *Journal of the American Chemical Society* 52: 2101–2106.

Morrissey, P. (1985). Lactose: chemical and physicochemical properties. In: *Developments in Dairy Chemistry—3: Lactose and Minor Constituents* (ed. P.F. Fox), 1–34. Springer.

Mu, W., Chen, Q., Wang, X. et al. (2013). Current studies on physiological functions and biological production of lactosucrose. *Applied Microbiology and Biotechnology* 97: 7073–7080.

Mukherjee, S. and John, S. (2022). Lactulose. In: *StatPearls [Internet]* (ed. S. John). StatPearls Publishing.

Nguyen, T.-K., Hong, M.-G., Chang, P.-S. et al. (2018). Biochemical properties of l-arabinose isomerase from *Clostridium hylemonae* to produce d-tagatose as a functional sweetener. *PLoS One* 13: e0196099.

Nirwantono, R., Laksmi, F.A., Nuryana, I. et al. (2024). Exploring an l-arabinose isomerase from cryophile bacteria *Arthrobacter psychrolactophilus* B7 for d-tagatose production. *International Journal of Biological Macromolecules* 254: 127781.

Nishimukai, M., Watanabe, J., Taguchi, H. et al. (2008). Effects of epilactose on calcium absorption and serum lipid metabolism in rats. *Journal of Agricultural and Food Chemistry* 56: 10340–10345.

Ojima, T., Saburi, W., Sato, H. et al. (2011). Biochemical characterization of a thermophilic cellobiose 2-epimerase from a thermohalophilic bacterium, *Rhodothermus marinus* JCM9785. *Bioscience, Biotechnology, and Biochemistry* 75: 2162–2168.

Olano, A. and Corzo, N. (2009). Lactulose as a food ingredient. *Journal of the Science of Food and Agriculture* 89: 1987–1990.

Panesar, P.S. and Kumari, S. (2011). Lactulose: production, purification and potential applications. *Biotechnology Advances* 29: 940–948.

Park, N.-H., Choi, H.-J., and Oh, D.-K. (2005). Lactosucrose production by various microorganisms harboring levansucrase activity. *Biotechnology Letters* 27: 495–497.

Park, C.-S., Kim, J.-E., Lee, S.-H. et al. (2013). Characterization of a recombinant mannobiose 2-epimerase from *Spirochaeta thermophila* that is suggested to be a cellobiose 2-epimerase. *Biotechnology Letters* 35: 1873–1880.

Patel, S.N., Kaushal, G., and Singh, S.P. (2020). A novel d-allulose 3-epimerase gene from the metagenome of a thermal aquatic habitat and d-allulose production by *Bacillus subtilis* whole-cell atalysis. *Applied and Environmental Microbiology* 86.

Patel, M.J., Patel, A.T., Akhani, R. et al. (2016). Bioproduction of d-tagatose from d-galactose using phosphoglucose isomerase from *Pseudomonas aeruginosa* PAO1. *Applied Biochemistry and Biotechnology* 179: 715–727.

Paulis, C. and Barry, S. (1969). A preliminary study of the incorporation of glucose-C14 in the enzymatic synthesis of sucrose and levan by the levansucrase system of *Streptococcus salivarius*. *Journal of the Baltimore College of Dental Surgery* 24: 27–36.

Pilgrim, A., Kawase, M., Ohashi, M. et al. (2001). Reaction kinetics and modeling of the enzyme-catalyzed production of lactosucrose using β-fructofuranosidase from *Arthrobacter* sp. K-1. *Bioscience, Biotechnology, and Biochemistry* 65: 758–765.

Prabhu, P., Kumar Tiwari, M., Jeya, M. et al. (2008). Cloning and characterization of a novel l-arabinose isomerase from *Bacillus licheniformis*. *Applied Microbiology and Biotechnology* 81: 283–290.

Rhimi, M., Bajic, G., Ilhammami, R. et al. (2011). The acid-tolerant L-arabinose isomerase from the mesophilic *Shewanella* sp. ANA-3 is highly active at low temperatures. *Microbial Cell Factories* 10: 1–11.

Rhimi, M. and Bejar, S. (2006). Cloning, purification and biochemical characterization of metallic-ions independent and thermoactive l-arabinose isomerase from the *Bacillus stearothermophilus* US100 strain. *Biochimica et Biophysica Acta (BBA) - General Subjects* 1760: 191–199.

Rhimi, M., Ilhammami, R., Bajic, G. et al. (2010). The acid tolerant l-arabinose isomerase from the food grade *Lactobacillus sakei* 23K is an attractive d-tagatose producer. *Bioresource Technology* 101: 9171–9177.

Roh, H.-J., Kim, S.-Y., Kim, S.-S. et al. (1999). Physicochemical properties of a low calorie sweetener, tagatose. *Korean Journal of Food Science and Technology* 31: 24–29.

Roy, S., Chikkerur, J., Roy, S.C. et al. (2018). Tagatose as a potential nutraceutical: production, properties, biological roles, and applications. *Journal of Food Science* 83: 2699–2709.

Saraiva, A., Carrascosa, C., Raheem, D. et al. (2020). Natural sweeteners: the relevance of food naturalness for consumers, food security aspects, sustainability and health impacts. *International Journal of Environmental Research and Public Health* 17: 6285.

Seibel, J., Moraru, R., Götze, S. et al. (2006). Synthesis of sucrose analogues and the mechanism of action of *Bacillus subtilis* fructosyltransferase (levansucrase). *Carbohydrate Research* 341: 2335–2349.

Seki, N. and Saito, H.J. (2012). Lactose as a source for lactulose and other functional lactose derivatives. *International Dairy Journal* 22: 110–115.

Senoura, T., Taguchi, H., Ito, S. et al. (2009). Identification of the cellobiose 2-epimerase gene in the genome of *Bacteroides fragilis* NCTC. *Bioscience, Biotechnology, and Biochemistry* 9343 (73): 400–406.

Shen, S.-C. and Wu, J. (2004). Maillard browning in ethanolic solution. *Journal of Food Science* 69: FCT273–FCT279.

Silvério, S.C., Macedo, E.A., Teixeira, J.A., and Rodrigues, L.R. (2015). Perspectives on the biotechnological production and potential applications of lactosucrose: a review. *Journal of Functional Foods* 19: 74–90.

Singh, A.K., Rai, A.K., Sathaye, S.B., and Singh, S.P. (2025). A novel cellobiose 2-epimerase from thermal aquatic metagenome for the production of epilactose. *Journal of Agricultural and Food Chemistry*. https://doi.org/10.1021/acs.jafc.4c09753.

Smith, A., Avery, A., Ford, R. et al. (2022). Rare sugars: metabolic impacts and mechanisms of action: a scoping review. *British Journal of Nutrition* 128: 389–406.

Su, W., Li, F., Li, X. et al. (2021). Using galactitol dehydrogenase coupled with water-forming NADH oxidase for efficient enzymatic synthesis of l-tagatose. *New Biotechnology* 62: 18–25.

Suzuki, T., Nishimukai, M., Shinoki, A. et al. (2010). Ingestion of epilactose, a non-digestible disaccharide, improves postgastrectomy osteopenia and anemia in rats through the promotion of intestinal calcium and iron absorption. *Journal of Agricultural and Food Chemistry* 58: 10787–10792.

Taguchi, H., Senoura, T., Hamada, S. et al. (2008). Cloning and sequencing of the gene for cellobiose 2-epimerase from a ruminal strain of *Eubacterium cellulosolvens*. *FEMS Microbiology Letters* 287: 34–40.

Takahama, A., Kuze, J., Okano, S. et al. (1991). Production of lactosucrose by *Bacillus natto* levansucrase and some properties of the enzyme. *Nippon Shokuhin Kogyo Gakkaishi* 38: 789–796.

Tang, L., Li, Z.-A., Dong, X.-X. et al. (2011). Lactulose biosynthesis by β-galactosidase from a newly isolated *Arthrobacter* sp. *Journal of Industrial Microbiology and Biotechnology* 38: 471–476.

Thakur, M., Kumar Rai, A., and Singh, S.P. (2022). An acid-tolerant and cold-active β-galactosidase potentially suitable to process milk and whey samples. *Applied Microbiology and Biotechnology* 106 (9): 3599–3610.

Vera, C., Guerrero, C., and Illanes, A. (2022). Trends in lactose-derived bioactives: synthesis and purification. *Systems Microbiology and Biomanufacturing* 2: 393–412.

Wang, D., Kim, D.H., Seo, N. et al. (2016). A novel glycoside hydrolase family 5 beta-1,3-1,6-endoglucanase from *Saccharophagus degradans* 2-40T and its transglycosylase activity. *Applied and Environmental Microbiology* 82: 4340–4349.

Wang, B., Lei, S., Li, Q., and Luo, Y. (2024). Production of lactulose from lactose using a novel cellobiose 2-epimerase from *Clostridium disporicum. Enzyme and Microbial Technology* 179: 110466.

Wu, C., Zhang, T., Mu, W. et al. (2015). Biosynthesis of lactosylfructoside by an intracellular levansucrase from *Bacillus methylotrophicus* SK 21.002. *Carbohydrate Research* 401: 122–126.

Xu, W., Liu, Q., Yu, S. et al. (2018). Synthesis of lactosucrose using a recombinant levansucrase from *Brenneria goodwinii. Applied Biochemistry and Biotechnology* 186: 292–305.

Xu, Z., Qing, Y., Li, S. et al. (2011). A novel l-arabinose isomerase from *Lactobacillus fermentum* CGMCC2921 for d-tagatose production: gene cloning, purification and characterization. *Journal of Molecular Catalysis B: Enzymatic* 70: 1–7.

Xu, W., Zhang, W., and Wu, H. (2021). Enzymatic production of lactosucrose by levansucrase, β-fructofuranosidase, and β-galactosidase. In: *Novel Enzymes for Functional Carbohydrates Production: From Scientific Research to Application in Health Food Industry* (ed. W. Mu, W. Zhang, and Q. Chen), 125–146.

Yan, Y., Guan, W., Li, X. et al. (2021). β-galactosidase GALA from *Bacillus circulans* with high transgalactosylation activity. *Bioengineered* 12: 8908–8919.

Yoon, S.-H., Kim, P., and Oh, D.-K. (2003). Properties of l-arabinose isomerase from *Escherichia coli* as biocatalyst for tagatose production. *World Journal of Microbiology and Biotechnology* 19: 47–51.

Zhang, G., An, Y., Parvez, A. et al. (2020). Exploring a highly d-galactose specific l-arabinose isomerase from *Bifidobacterium adolescentis* for d-tagatose production. *Frontiers in Bioengineering and Biotechnology* 8: 377.

Zhang, G., An, Y., Zabed, H.M. et al. (2023). Rewiring *Bacillus subtilis* and bioprocess optimization for oxidoreductive reaction-mediated biosynthesis of d-tagatose. *Bioresource Technology* 389: 129843.

Zhang, W., Chen, D., Chen, J. et al. (2023). d-allulose, a versatile rare sugar: recent biotechnological advances and challenges. *Critical Reviews in Food Science and Nutrition* 63: 5661–5679.

9

Engineering Microbial Pathways for the Production of 2′-Fucosyllactose

Vijaya Bharathi Srinivasan, Balvinder Singh, and Govindan Rajamohan

Council of Scientific Industrial Research-Institute of Microbial Technology, Chandigarh, 160036, India

9.1 Introduction

9.1.1 Human Milk Oligosaccharides (HMOs)

The carbohydrates are energy-charged macromolecule abundantly found in the ecosystem and are essentially required for all life forms. The major constituents of the human milk were lactose (70 g/L), lipids (40 g/L), and human milk oligosaccharides (HMOs) (5–15 g/L) (Bode 2009; Zivkovic et al. 2011; Zhu et al. 2022). The HMO, a non-lactose carbohydrate, is found abundantly in the human breast milk, in addition to the primary sugar component lactose. Currently, more than 200 distinct HMO structures were identified and it is expected to be identified more with emerging analytical methods (German et al. 2008; Kobata 2010; Wu et al. 2010; Wu et al. 2011; Zhu et al. 2022). The *N*-acetylglucosamine, D-glucose, D-galactose, L-fucose, and *N*-acetylneuraminic acid are monosaccharides required for HMOs. All the HMOs have lactose motif at their reducing end and can be extended by glycosidic linkages. Moreover, L-fucose and *N*-acetyl neuraminic acid contribute to the formation of complex oligosaccharide structures found in milk (Han et al. 2012).

The HMOs are nondigestible and constitute monosaccharides to oligosaccharides, which contributes to structural carbohydrate diversity. Majority of the HMOs are fucosylated that include 2′-fucosyllactose (2′-FL), 3′-fucosyllactose (3′-FL), lacto-*N*-fucopentaose, and lacto-*N*-difucohexaose, which are demonstrated to have significant health benefits for infants and newborns (Figure 9.1; Wicinski et al. 2020; Zhu et al. 2022; Palur et al. 2023).

More than 30% of total HMOs consist of 2′-FL, a trisaccharide made up of lactose and fucose linked by α-(1,2)-linkage (Figure 9.2), with an immense physiological function (Chaturvedi et al. 2001; Morrow et al. 2004; Wicinski et al. 2020; Zhu et al. 2022; Palur et al. 2023). Therefore, the World Health Organization and the United Nations Children's Fund consider human milk to be standard infant nutrition for the first semester (Garwolinska et al. 2018).

The HMOs are attractive target due to its innate biological functions in preventing and treating diseases in children and adults (Bode et al. 2014; Wicinski et al. 2020). However, due to limited studies in deciphering the source and mechanisms of individual HMOs, there is an urgent need to

Microbial Nutraceuticals: Products and Processes, First Edition. Edited by Sudhir Pratap Singh and Santosh Kumar Upadhyay.

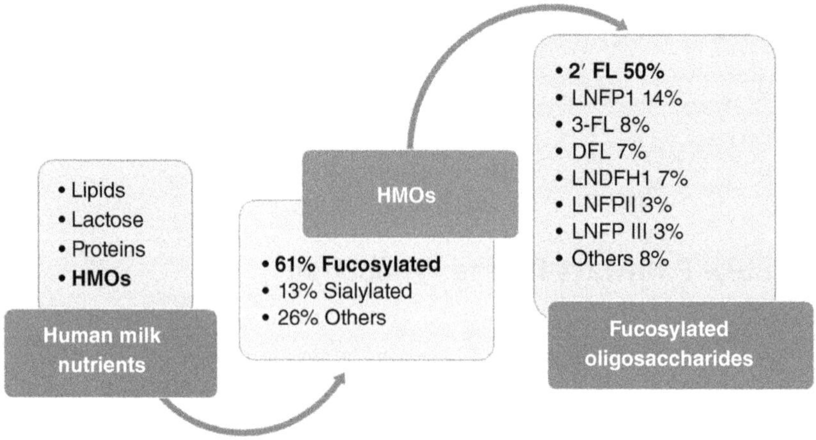

Figure 9.1 Human milk composition. The general human milk nutrients comprised of lipids, lactose, proteins, and HMO. Majority of the HMOs are fucosylated and sialylated. The fucosylated oligosaccharides consist of 2′FL, 3′-FL, difucosyllactose (DFL), lacto-*N*-fucopentaose I (LNFP-I), lacto-*N*-fucopentaose II (LNFP-II), lacto-*N*-fucopentaose III (LNFP-III), and lacto-*N*-difucohexaose I (LNDFH-I).

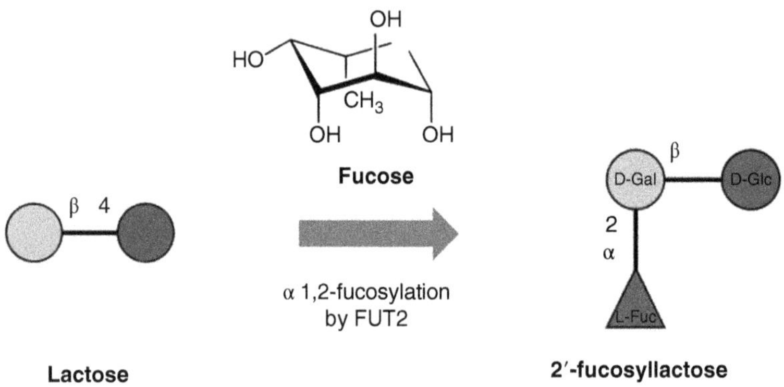

Figure 9.2 Biosynthesis of 2′-FL from lactose. The trisaccharide 2′-FL composed of disaccharide D-galactose (D-Gal), D-glucose (D-Glc), and monosaccharide L-fucose (L-Fuc) through the α-(1→2) bond linkage.

enhance the production of active bioactive compounds through either enzymatic synthesis or modulating or engineering the pathways responsible for the production of complex carbohydrate molecules.

9.1.2 Biological Properties and Functions of 2′-FL

A significant oligosaccharide found in human breast milk, 2′-FL, is thought to be a novel food and promising addition to infant formula because of its several health benefits in the host immune and nerve systems, growth enhancement of beneficial gut commensals and inhibition of the potential priority pathogens (Lucas et al. 1990; Eiwegger et al. 2004; Newburg et al., 2005; Sprenger et al. 2022; Zhu et al. 2022). Additionally, it gained prime importance after the approvals from the Food and Drug Administration (FDA) and European Food Safety Authority (EFSA) for the use as nutritional supplement in infant formulas, nutraceuticals, and food additives (Han et al. 2012; Marriage et al. 2015; Goehring et al. 2016; Wicinski et al. 2020; Sprenger et al. 2022). With the global approvals to be produced in large scale for the commercial applications especially in infant formula milks, it gained great attention with celebrity status. The accessibility of 2′-FL in the worldwide market space was restricted due to its high costs.

Several studies had demonstrated the production of 2′-FL achieved by extracting the human milk and the use of chemical and enzymatic synthesis procedures. However, these methods were not viable for the large-scale production due to multiple protection and deprotection steps with low yield. Therefore, the use of genetic and metabolic engineering strategies in microorganisms will be feasible and cost-effective for the large-scale production of 2′-FL (Wicinski et al. 2020; Sprenger et al. 2022; Palur et al. 2023).

9.2 Human Milk Microbiome

Research studies uncovered that the human milk microbiome have gained attention and recognition as an important mammalian reproductive microbiome that primarily influences the survival and development of offspring (Togo et al. 2019). Ojo-Okunola et al. (2018) described the core bacteriome in human milk, which includes nine genera including *Staphylococcus, Streptococcus, Serratia, Pseudomonas, Corynebacterium, Ralstonia, Propionibacterium, Sphingomonas,* and *Bradyrhizobium*. However, the abundance of these genera varies among humans and decreased diversity during breastfeeding course compared with colostrum and transition milk (Diez-Sampedro et al. 2019). Interestingly, several studies highlighted the presence of fungal species such as *Candida, Alternaria, Rhodotorula, Saccharomyces cerevisiae, Aspergillus glaucus,* and *Penicillium rubens* in the human milk (Jiménez et al. 2015; Moossavi et al. 2020; Moubareck 2021). Mounting literature evidence indicates that viruses such as bacteriophages in human milk transmit from mother to neonates to colonize the infant gastrointestinal tract (Pannaraj et al. 2018; Liang et al. 2020; Mohandas and Pannaraj 2020; Bushman and Liang 2021). While studies have identified the partial composition and prevalence of milk microbiome, however, the role of these microbiota in shaping the evolution of human microbiome remains to be enigmatic. Wang et al. (2024) demonstrated the enriched isolation of 865 strains representing 4 phyla, 21 genera, and 54 species from human milk and able to cultivate over 94% bacterial species with culturomics strategies. It is anticipated that with modern technologies, establishment of complete microbiome analysis and culturable methods for dominant microbiota of human milk will be possible in near future.

The regulatory role of 2′-FL has been demonstrated in the establishment of early gut microbiota in infants and also in the growth promotion of probiotic bacteria namely *Bifidobacterium longum, B. bifidum,* and *B. infantis*. These bacteria have the ability to utilize HMOs through extra- and intracellular pathways. The *B. bifidum* produce secretory or extracellular α-1,2-fucosidase to hydrolyze 2′-FL, while *B. infantis and B. longum* express ABC transporter on the cell surface for the translocation of oligosaccharide 2′-FL into the cell and further hydrolyzed by the intracellularly expressed α-1,2-fucosidase (Garrido et al. 2012; Matsuki et al. 2016; Zabel et al. 2019).

With the available information on human milk microbiota, efficient microbial and metagenomic screening will aid to identify efficient enzymes, pathways, and microbial natural hosts that may be amenable for the overproduction of 2′FL at commercial scale.

9.2.1 Chemical Synthesis of 2′-FL

Human milk is a primary source for 2′FL and can be isolated using different purification steps. However, the two concerning factors, low yield and limited supply and availability for the production, will be always challenging. The 2′-FL was chemically produced by benzylation derivatives, and the synthesis requires side chain protection and deprotection stringent procedures (Zhu et al. 2022). Although, 2′-FL could be synthesized but with strict requirement of protection and deprotection steps, it will not be viable route for the scale up due to challenging steps led to high cost and with few toxic reagents for glycosylation reactions. In 2019, Agoston et al. reported that the FDA approved production of 2′-FL through established chemical synthesis method by Glycom A/S. The process involves multiple steps, with L-fucose as starting substrate to synthesize the fucosyl donor, further from lactose to synthesize lactose acceptor, and to finally synthesize pure fucosyllactose using one-step column chromatography. The chemical synthesis method is preferably used for the preclinical and clinical research and development activities.

9.2.2 Enzymatic Synthesis of 2′-FL

The enzymatic synthesis of 2′-FL, a process achieved mostly through fucosyltransferases and α-L-fucosidases (glycoside hydrolase family 29) enzymes. The advantages include stereo-selectivity and regio-selectivity. The synthesis necessitates costly donor GDP-fucose, low catalytic activity of α-1,2-fucosyltransferase, and very low-yield products following the transglycosylation reaction (Zhou et al., 2021). Albermann et al. (2001) demonstrated that with recombinant Gmd and WcaG, donor GDP-fucose was obtained with 78% yield and subsequently to 2′-FL by a α-1,2-fucosyltransferase (FucT2) with 65% yield on milligram scale. The exoglycoside hydrolase, α-L-fucosidase, hydrolyze fucose attached to fucosyl oligosaccharides. These enzymes are found in many microbial species like *Bacteroides, Paenibacillus, Tannerella., Xanthomonas, Thermotoga, Bifidobacterium,* and *Lactobacillus;* animals; and plants (Escamilla-Lozano, et al. 2019; Guzmán-Rodríguez et al. 2019). Broadly, mechanisms of fucosidases are classified into inversion and retention types, while the later has potential transglycosylation activity for the synthesis of FL. Due to low transglycosylation efficiency, the α-L-fucosidase from *Thermotoga maritima* and *Fusarium graminearum* and metagenome clonal library were resulted in 32%, 14%, and 6% low yield of FL, respectively, from different substrates (Lezyk et al. 2016; Zeuner et al. 2018; Guzmán-Rodríguez et al. 2019). Therefore, there is a demand to identify and generate the enzyme with evolved high transglycosidic activity and good stereoselectivity and regioselectivity for the large-scale production of FL.

Lezyk et al. (2016) reported that by increasing the fucosylation reaction effectiveness, they screened and found novel α-L-fucosidases from metagenome which produced 3.6% FL. *Thermotoga maritima* yielded 6.4% α-L-fucosidases. Sugiyama et al. (2016) identified mutants (N423H and N423D/D766N) exhibited highest activity in the presence of 10 mM β-fucosyl fluoride and 10 mM lactose at pH 5.0 and pH 5.5 and produce 8.5 mM 2′FL compared to D766G mutant which produces less than 1 mM 2′-FL. Therefore, this approach is not cost-effective and favorable for the production of 2′-FL at commercial scale due to low yield with high cost.

With an advent of modern molecular engineering techniques, enzymes without transglycosidic activity may be mutated to have the ability of an efficient transglycosidases activity using glycosynthase approach (Mackenzie et al. 1998). One such mutated inverse fucosidase is fucosynthase, a promising strategy for the synthesis of oligosaccharides at large scale. Generation of mutant enzyme with transglycosidic activity can catalyze the synthesis of 2′-FL from β-L-Fucosyl fluoride triacetate (FucF) and lactose with a yield of 6% (Wada et al. 2008), and with improved combinatorial mutations, it is resulted in 88% yield (Sugiyama et al. 2016). Using directed evolution strategies, the generated mutant enzyme exhibits 7–60% increased transglycosidic activity (Osanjo et al. 2007).

9.2.3 Biological Production of 2′-FL Through Genetic Engineering Strategies

The HMOs exhibit an excellent health benefit properties for infant, child, and adult populations. Due to limited access of HMOs, its potential functional benefits linked with high market value and demand promoted the researchers and industries to explore and develop genetic engineering strategies to overproduce 2′-FL at commercial scale. Thus, the biosynthesis of 2′-FL through microbial host will be cost-effective and genetically amenable for a large-scale production than chemical and enzymatic synthesis (Han et al., 2012; Li et al. 2021; Lu et al. 2021; Li et al. 2022; Liu et al. 2022; Zhu et al. 2022; Palur et al. 2023). Therefore, manipulating the biosynthetic pathways for the overproduction of protein of commercial interest in diverse microbial systems has become an important research focus for future.

9.2.4 Engineering Gram-Negative Bacterial Host [*Escherichia coli*] for 2′-FL Production

With an advent of recombinant technologies, genetic engineering strategies such as metabolic engineering and pathway engineering have become most powerful and promising techniques for the production of valuable chemicals from any sources with higher production efficiency and reduced cost. Additionally, bacteria and yeast with generally recognized as safe (GRAS) status approved by global agencies could be employed in the production of biochemicals. The 2′-FL synthesis may be achieved through synthetic and system biology tools for the production of HMOs. The recombinant *E. coli* BL21 strain producing 2′-FL are marketed as novel food ingredient (EFSA Panel on Nutrition, 2023). Therefore, using *E. coli* as a cell factory will be an advantageous and approved host for commercial production of 2′-FL.

Fucosyltransferases, a lactose acceptor for a glycosylation process, are necessary for the synthesis of 2′-FL. Studies have demonstrated that the lactose catabolism can be reduced by inhibiting the galactosidase (LacZ), while the lactose permeabilization can be increased by overexpressing lactose permease (LacY). In general, the synthesis of GDP-fucose, a donor of fucosyltransferases, could be achieved by two metabolic pathways: *de novo* and salvage pathways. *E. coli* is the most common host for the generation of recombinant 2′-FL. The *de novo* pathway in *E. coli* produces endogenous GDP-fucose (a critical precursor of colonic acid for bacterial cell wall synthesis) and involves several reactions catalyzed by –phosphomannomutase (ManB), mannose-1-phosphomannomutase (ManC), mannose-1-phosphate guanylyltransferase (ManC), GDP-d-mannose-4,6-dehydratase (Gmd) and WcaG codes for GDP-fucose synthase (Lu et al. 2021). Wan et al. (2020) reported that hyperexpression of genes *gmd, wcaG, manB,* and *manC* that contributes to changes in metabolic flux and combinatorial pathways increases the biosynthesis of GDP-fucose. Additionally, recombinant *E. coli* JM109 strain produced twice the yield of 2′-FL, 380 versus 200 mg/g of CDW; after the integration of *de novo* and salvage pathway, genes were created together with *futC* (Baumgärtner et al. 2013). To increase the availability of GDP-fucose for the 2′-FL production, guanosine nucleotides and GDP-mannose metabolism pathways were modified with an addition of cofactor NADPH recycle pathway (Zhu et al. 2022).

E. coli was engineered utilizing the glycerol and lactose for the production of 2′-FL (Lee et al. 2012). Chin et al. (2015) modified the endogenous lac operon to reduce the intrinsic beta-galactosidase activity, resulted in 6.4 g/L titer of 2′-FL. Additionally, Huang et al. (2017) engineered *E. coli* strain to overexpress the mannose operon genes involved in de novo pathway and removed *wcaJ, lacZ,* and *lon* genes, resulted in 9.12 g/L of 2′-FL.

Another attractive alternate strategy is salvage pathway; it requires only overexpression of one gene *fkp*, instead of multiple genes as in *de novo* pathway. The GDP-fucose is synthesized from the supplemented fucose through catalysis of fucose kinase or a bifunctional enzyme fucose-1-phosphate guanylyltransferase (Fkp) and subsequently transferred to lactose *via* heterologously overexpressed α-1,2-fucosyltransferase to produce 2′-FL. Overall, using two plasmids expressing *de novo* and salvage pathways enhances the 2′-FL production in engineered *E. coli* host. Further, deletion of rhamnose isomerase and arabinose isomerase in recombinant *E. coli* enhances the production of 2′-FL from fucose (Jung et al. 2019). In general, *E. coli* is considered as best choice for the production of 2′-FL, due to the presence of *de novo* pathway harboring GDP-fucose synthesis gene cluster. Further, LacZ that inactivated host *E. coli* JM107 was used for the expression of α-1,2-fucosyltransferase. The overexpression of a positive regulator for the colonic acid operator gene *rcsA* and removal of *wcaJ* encodes for enzyme catalyzing conversion from GDP-fucose to colanic acid led to the gram-scale production of 2′FL (Drouillard et al. 2006; Lu et al. 2021). To improve the production, the *Bacteroides fragilis* fucokinase/GDP-fucose pyrophosphorylase (Fkp) and *Helicobacter pylori fucT2*, α-1,2-fucosyltransferase, were introduced into salvage pathway to produce 23.1 g/L of 2′-FL in fed-batch fermentation (Chin et al. 2016). Additionally, overexpression of *fkp* from *B. fragilis, fucT2* gene, and deleting *lacZ, fucI, fucK,* and *rhaA* yielded 47.0 g/L of 2′-FL (Jung et al. 2019). In addition to *E. coli* host, many GRAS-classified organisms were used for 2′-FL production.

The synthesis of 2′-FL was achieved through simulation of microbial fermentation by the conversion of synthesized GDP to fucose mediated through α-1,2-fucosyltransferase (FucT).The key factor in the synthesis of 2′-FL is exogenously expressed α-1,2-fucosyltransferase (Faijes et al. 2019). The overproduction of intracellular GDP and α-1,2-fucosyltransferase is an important bottleneck step for the efficient production of 2′FL in microbial systems. Many groups have screened microbial sources for the isolation of stable, hyperactive enzyme, while *futC* are being employed for industrial production of 2′-FL (Huang et al. 2017). Another important factor is to increase the concentration of GDP-fucose by overexpressing or mutating the genes in the *de novo* and salvage pathways (Dumon et al. 2001; Baumgärtner et al. 2013; Chin et al. 2016). The microbial biosynthesis of 2′-FL will take several days and may be amenable for a large-scale production compared to chemical and/or enzymatic methods (Han et al., 2012). The metabolic engineering strategies are most promising and employed to increase the synthesis and production of 2′-FL.

Recombinant *E. coli* strains were constructed using metabolic engineering strategies to produce 2′-FL. These strategies involve optimizing co-expression systems for key enzymes namely –phosphomannomutase (ManB), mannose-1-phosphate guanylytransferase (ManC), GDP-d-mannose 4,6-dehydratase (Gmd), GDP-l-fucose synthase (WcaG), and α-1,2-fucosyltransferase (FucT; Figure 9.3).

Due to availability of genetic tools and simple growth conditions for *Escherichia coli*, it is a preferred Gram-negative host for all the genetic engineering studies. *Helicobacter pylori,* α-1,2-fucosyltransferase (FucT2), was cloned and expressed in *E. coli,* resulted in 3.4-fold increased production of 2′-FL in batch fermentation, compared to 1.23 g/L in wild type (Chin et al. 2015; Li et al. 2021; Lu et al. 2021). Another immersive strategy was metabolic pathway integration and cofactor engineering; two GDP-L-fucose pathways were incorporated into the *E. coli* genome for the enhanced production of 2′-FL in fed-batch fermentation (Baumgärtner et al. 2013). The NADPH and GTP cofactors are produced from pentose phosphate pathway and the tricarboxylic acid cycle, by expressing *E. coli wcaJ, zwf,* and *gsk* for the enhanced production of GDP-L-fucose (Wan et al. 2020; Li et al. 2022). Therefore, GDP-L-fucose production could be achieved by enhancing the intracellular redox state of host bacteria with an increased target yield.

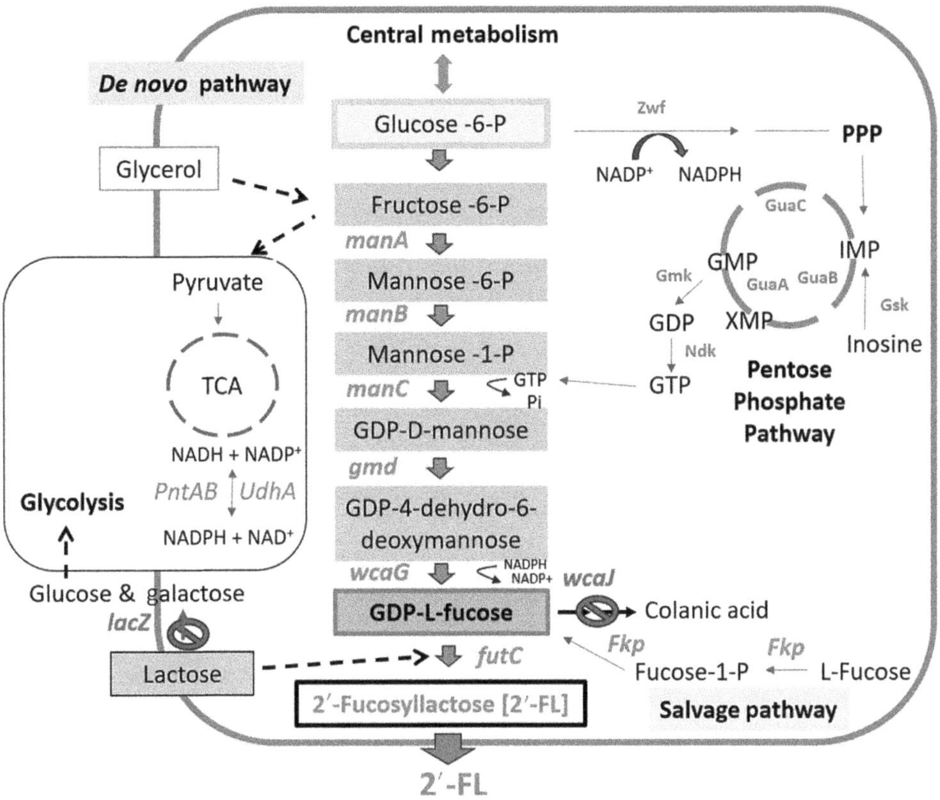

Figure 9.3 Model illustrates the *de novo* and salvage pathways for the synthesis of 2′-FL in recombinant *E. coli*. The over expressed genes are *zwf, pntAB, udhA, gsk, guaA, guaB, guaC, gmk, ndk*, genes associated with de novo and salvage pathways were *manA, manB, manC, gmd, wcaG, futC, fkp* and the deleted genes are *lacZ* and *wcaJ* represented as stop symbol.

9.2.5 Engineering Gram-Positive Bacterial Host for 2′-FL Production

The GRAS strain *Bacillus subtilis* has been used for the production of 2′-FL. Deng et al. (2019a) developed engineered *B. subtilis* strain for efficient production of 2′-FL, by incorporating the genes related to GDP-fucose production, overproduction of sugar transporter glcP, LAC12 lactose permease, and deletion of pathway and engineered GTP cofactor regeneration systems' related genes *guaC, yfkN,* and *yesZ*. The engineered GRAS *B. subtilis* able to produce 5.01 g/L of 2′-FL in fed-batch fermentation and with aptamer regulatory switch able to enhance the production of 2′-FL (Deng et al. 2019a; Deng et al. 2019b). Overall, *Bacillus* strains were used as expression platform host for the production of 2′-FL (Lu et al. 2021; Zhang et al. 2022).

9.2.6 Engineering Yeast for 2′-FL Production

Saccharomyces cerevisiae was engineered to produce 2′-FL by overexpressing genes (*fkp*) involved in salvage pathway (Lu et al. 2021). The *lac12* and *futT2* genes from *Kluyveromyces lactis* were expressed in food-grade yeast *S. cerevisiae* for the synthesize of 2′-FL; however, the fermentation result was not satisfactory (Yu et al. 2018). Hollands et al. (2019) demonstrated the engineering the de novo pathway to produce GDP-mannose to GDP-fucose in *S. cerevisiae* (Figure 9.4) and *Yarrowia lipolytica* led to overproduction of 2′-FL. Furthermore, overexpressing *K. lactis* LAC12 gene for the

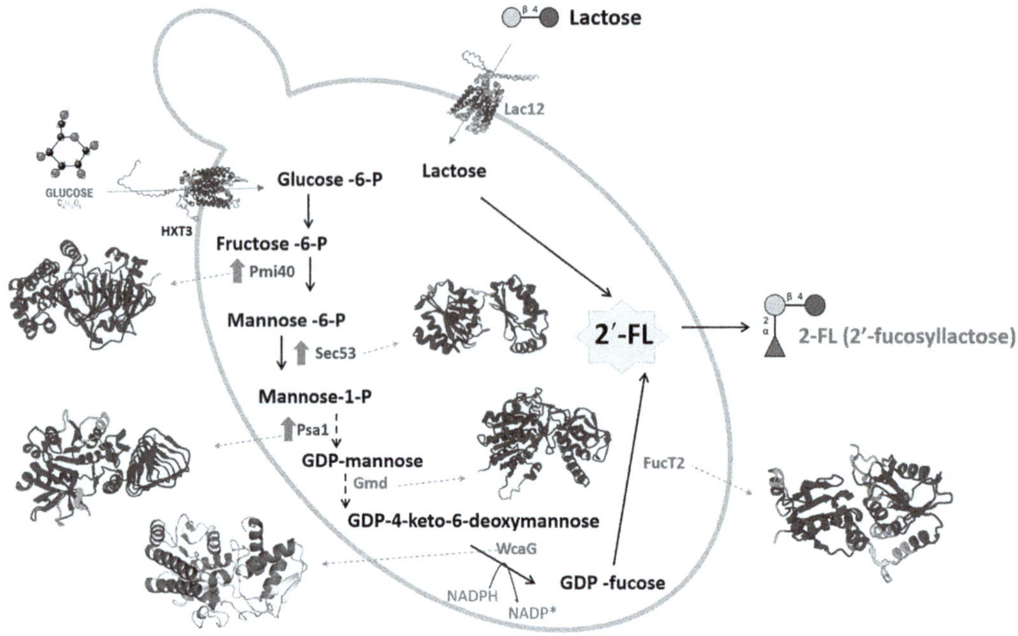

Figure 9.4 Schematic representation of 2′-FL biosynthesis in pathway engineered *S. cerevisiae*. The overexpressed genes (mannose-6-phosphate isomerase–*pmi40*; phosphomannomutase–*sec53*; GDP-mannose pyrophosphorylase–*psa1*); are indicated as upward arrow for the increased supply of GDP-mannose. Subsequently, by expressing genes from *Escherichia coli* K12 [GDP-mannose-4, 6-dehydrogenase-*gmd* and GDP-L-fucose synthase/GDP-4-keto-6-deoxymannose-3, 5-epimerase-4-reductase–*wcaG*], *Helicobacter pylori* [*fucT2*], and *K. lactis* [lactose importer *lac12*] for the overproduction of 2′-FL in recombinant *S. cerevisiae*. The alpha-fold predicted protein structures and crystal structure of WcaG (PDB 1GFS) were represented in the figure.

import of lactose and α-1,2-fucosyltransfereases for the biosynthesis of 2′-FL. Several transporters were identified from yeast, and the best-described example is fungal *Neurospora crassa* cellobiose transporter (cellodextrin transporter CDT2) for the 2′-FL efflux.

9.2.7 Global Regulatory Approval, Commercialization, Market Value, and Application of 2′-FL

The general schematic representation of modules involved in the development of recombinant technology products from petri plate to market are shown in Figure 9.5. The engineered strains will be subcultured in petri plates for the production of recombinant proteins from shake flask to large-scale cultivation. Upon standardized conditions for the upscaling and downstream, the products will be subjected to regulatory compliance and approval for the marketing of recombinant proteins.

The 2′-FL is the most important component of HMO having immense health potential and significant commercial value. This unique complex carbohydrate HMOs play vital role in gut health and immunity. The global market is estimated to reach US$ 100.8 million in 2024 with the increased sales at a CAGR of 6.5%. The market is expanding rapidly and estimated to reach value of US$ 161.6 million by 2034. In European Union (EU), the HMOs were approved as novel foods, while in the United States (USA), the first HMOs, 2′-FL, were certified to be considered as "GRAS" substances.

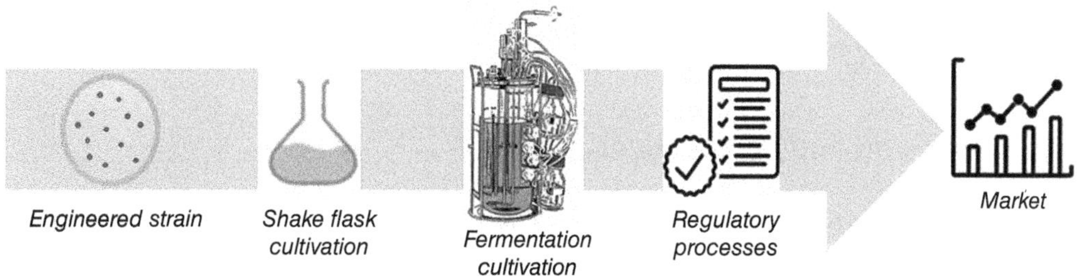

Figure 9.5 Schematic illustration of essential modules for the production of recombinant products to market.

In 2015, Glycom A/S produced 2′-FL through chemical synthesis method; subsequently, increased market demand was observed in both EU and USA. The Royal DSM, a global science-based company, holds regulatory clearance for the highest number of HMOs, including GlyCare 2′-FL, has been approved in Canada and China for use in infant formula and nutritional supplements for toddler. Currently, two HMOs from DSM-Firmenich company were authorized for use in Chinese market as formulated milk powder and newborn formula for special medical purposes. These include a common HMO in human milk, 2′-FL, and a complex sugar lacto-*N*-neotetraose (LNnT) essentially required for the early development of the infant gut flora.

HMOs are an important functional component of human breast milk and have pivotal role in supporting the infant health, gut health, immune function, and cognitive development. With growing evidence of its immense human health benefits and along with increased demand in global market propelled the researcher to develop genetic strategies for an efficient industrial production of 2′-FL. In summary, 2′-FL has made significant strides in commercialization, benefiting both infant formula and scientific research.

9.3 Challenges or Future Outlook

The microbial production of industrially important enzymes and proteins at large scale are always challenging task. The major limitations are choosing appropriate microbial host for overproduction, availability of genetic tools for manipulation, generation of stable expression platforms or modules, and development of facile downstream processing at large scale. Several reports demonstrate the production of 2′-FL, an important component of HMO, in microbial systems with low to high yield. However, developing an efficient expression platform or modular systems for stable overproduction of industrially important enzymes in diverse microbial systems with facile upstream and downstream processing is need of hour. Emerging modern biology approaches and strategies will be useful in generating microbial strains with enhanced protein production by systematically altering the genes, pathways, and export systems. The 2′-FL biosynthetic pathways may be altered by replacing the rate limiting enzymes with rationally designed or generated enzymes. The catalytic efficiency of these genes may be achieved through directed evolution strategies, grafting the loop regions from other proteins, active site engineering are other attractive methods. The sugar- or oxygen- or light-inducible microbial expression platforms, generation of modular expression cassettes with integron technologies for multicopy, and multisite integration in microbial systems could be developed. Engineering the pathways in the GRAS group of microbes is attractive alternative for downstream processing, yield, and regulatory procedures. The screening and selection of effective enzymes from pristine metagenomes for the construction of novel and efficient expression modular platform for different chassis development for the expression and higher production of 2′-FL in large scale fermentation batches.

9.4 Conclusion and Perspectives

The HMOs are rich carbohydrates predominantly found in human milk in contrast to farmed animals, which have very small amounts of oligosaccharides. These HMOs have potential benefit to infant health and development. The HMOs composed of variety of oligosaccharide structures, and most of them were fucosylated. One of the most important fucosylated HMOs is 2′-FL, a promising oligosaccharide with multiple biological functions including pre- and probiotic activities, which is considered as novel foods and being used far beyond as infant food ingredient and dietary supplements.

More than decades of research evidence unparallelly demonstrated the diverse physiological functions of soluble milk glycan 2′-FL in infant and human health. In 2015, Glycom A/S received a first regulatory approval in United States, and by 2016, it was available in the market as infant formula. Thereafter, 2′-FL has become an attractive research focus for scientific community. Importantly, in recent years, the 2′-FL was approved as GRAS by FDA and as a novel food by the EFSA (EFSA Panel on Nutrition, 2023).

Although 2-FL has remarkable biological functions with regulatory approvals to use commercially in different countries as food ingredients in infant formula, dietary supplements, and novel foods, the production of soluble milk glycan 2′-FL at large production scale is still challenging concern, due to lower yield in different microbial biosynthesis methodologies. The fucosylated glycan 2′-FL has remarkable biological functions which closely mimics that human milk has gained significant attention and commercial demand.

The production of 2′-FL was achieved by major three methods: chemical, chemo-enzymatic, and microbial cell factories. The chemical method is not feasible at commercial scale due to low stereoselectivity and use of toxic solvents. Similarly, the chemo-enzymatic approaches require rate-limiting enzymes and expensive substrates. With an advent of modern genetic engineering techniques and strategies, the production of 2′FL in diverse microbial modular expression systems at commercial scale looks possible in coming years.

The microbial synthesis of 2′-FL was achieved with low yield essentially through supplementation of fucose in salvage pathway and use of inexpensive substrates in de novo pathways. However, achieving the high efficiency and high production with low cost is an important parameter in the successful commercial production of soluble glycan 2′-FL.

The production of 2′-FL was achieved in genetically amenable microbial hosts such as *E. coli* by modulating α-1,2-fucosyltransferase obtained 112.56 g/L; *Bacillus subtilis* with pathway engineering strategies resulted in 88.3 g/L, and the recombinant *Saccharomyces cerevisiae* and *Yarrowia lipolytica* have reported to produce 32.05 g/L and 24 g/L of 2′-FL, respectively.

The use of food-grade microbes such as *B. subtilis, S. cerevisiae,* and *Y. lipolytica* will be advantageous for upstream and downstream processing and qualifies the food safety regulatory guidelines compared to toxin-producing bacteria. Importantly, these bacteria lack endogenous lactose permeases for the lactose utilization; hence, addressing this bottleneck through directed evolution or protein engineering strategies for better lactose utilization will enhance production and yield of 2′-FL. Overall, to improve the production of 2′-FL, various cellular and metabolic engineering strategies have been implemented, namely overexpressing essential rate-limiting enzymes, precursors, and cofactor engineering. Addressing these critical factors through emerging technologies will be of commercial value with high yield. Increasing the catalytic efficiency of the rate-limiting enzyme α-1,2-fucosyltransferase (FutC) and α-1,6-fucosyltransferase (FucT) through DNA shuffling, error-prone PCR, and/or protein engineering approaches will be good starting point for the efficient production of 2′-FL. The overexpression of enzymes involved in GDP-D-mannose synthesis pathway and NADPH regeneration pathways enhanced the accumulation of 2′-FL. The GDP-D-mannose is an essential building block for the GDP-L-fucose and synthesis of 2′-FL. These pathways are catalyzed by mannose-6-phosphate isomerase (Pmi40), phosphomannomutase (Sec53), and mannose-1-phosphate guanylyltransferase (Psa1) enzymes. The NADPH is an essential cofactor for the synthesis of 2′-FL; optimizing the synthesis of precursor and cofactors will be an efficient production platform for 2′-FL. Another promising strategy is cofactor engineering, which might be useful in generating feedback resistant genes to overcome the feedback inhibition of metabolites for the improved yield. Microbial chassis development has great potential to enhance the production of commercially important proteins. Research progress in this area of research continues to boost the production of 2′-FL and HMOs at laboratory scale. However, in future, concerted efforts should be focused on screening using novel approaches, generating newer enzyme variants, increasing transport of oligosaccharides, and establishing microbial designer platforms for the improved industrial scale production.

Acknowledgement

We thank the Director, CSIR-Institute of Microbial Technology for the excellent facilities and support. We would like to thank all the researchers who have contributed to this field of study. However, due to limitations, we are unable to include all research contributions in this review.

References

Agoston, K., Hederos, M.J., Bajza, I., and Dekany, G. (2019). Kilogram scale chemical synthesis of 2′-fucosyllactose. *Carbohydrate Research* 476: 71–77.

Albermann, C., Piepersberg, W., and Wehmeier, U.F. (2001). Synthesis of the milk oligosaccharide 2′-fucosyllactose using recombinant bacterial enzymes. *Carbohydrate Research* 334: 97–103.

Baumgärtner, F., Seitz, L., Sprenger, G.A., and Albermann, C. (2013). Construction of *Escherichia coli* strains with chromosomally integrated expression cassettes for the synthesis of 2′-fucosyllactose. *Microbial Cell Factories* 12: 40.

Bode, L. (2009). Human milk oligosaccharides: prebiotics and beyond. *Nutrition Reviews* 67 (Supplement 2): S183–S191.

Bode, L., McGuire, M., Rodriguez, J.M. et al. (2014). It's alive: microbes and cells in human milk and their potential benefits to mother and infant. *Advances in Nutrition* 5: 571–573.

Bushman, F. and Liang, G. (2021). Assembly of the virome in newborn human infants. *Current Opinion in Virology* 48: 17–22.

Chaturvedi, P., Warren, C.D., Altaye, M. et al. (2001). Fucosylated human milk oligosaccharides vary between individuals and over the course of lactation. *Glycobiology* 11: 365–372.

Chin, Y.W., Kim, J.Y., Lee, W.H., and Seo, J.H. (2015). Enhanced production of 2′- fucosyllactose in engineered Escherichia coli BL21star(DE3) by modulation of lactose metabolism and fuco syltransferase. *Journal of Biotechnology* 210: 107–115.

Chin, Y.W., Seo, N., Kim, J.H., and Seo, J.H. (2016). Metabolic engineering of *Escherichia coli* to produce 2′-fucosyllactose via salvage pathway of guanosine 5′-diphosphate (GDP)-l-fucose. *Biotechnology and Bioengineering* 113: 2443–2452.

EFSA Panel on Nutrition, Novel Foods and Food Allergens (NDA), Turck, D., Bohn, T. et al. (2023). Safety of 2'-fucosyllactose (2'-FL) produced by a derivative strain (*Escherichia coli* SGR5) of *E. coli* W (ATCC 9637) as a Novel Food pursuant to Regulation (EU) 2015/2283. *EFSA Journal* 21 (11): e08333.

Deng, J., Chen, C., Gu, Y. et al. (2019a). Creating an *in vivo* bifunctional gene expression circuit through an aptamer-based regulatory mechanism for dynamic metabolic engineering in *Bacillus subtilis*. *Metabolic Engineering* 55: 179–190.

Deng, J., Gu, L., Chen, T. et al. (2019b). Engineering the substrate transport and cofactor regeneration systems for enhancing 2′-fucosyllactose synthesis in *Bacillus subtilis*. *ACS Synthetic Biology* 8 (10): 2418–2427.

Diez-Sampedro, A., Flowers, M., Olenick, M. et al. (2019). Women's choice regarding breastfeeding and its effect on well-being. *Nursing for Women's Health* 23: 383–389.

Drouillard, S., Driguez, H., and Samain, E. (2006). Large-scale synthesis of H-antigen oligosaccharides by expressing *Helicobacter pylori* alpha1, 2-fucosyltransferase in metabolically engineered *Escherichia coli* cells. *Angewandte Chemie (International Ed. in English)* 45 (11): 1778–1780.

Dumon, C., Priem, B., Martin, S.L. et al. (2001). In vivo fucosylation of lacto-*N*-neotetraose and lacto-*N*-neohexaose by heterologous expression of *Helicobacter pylori* alpha-1,3 fucosyltransferase in engineered *Escherichia coli*. *Glycoconjugate Journal* 18: 465–474.

Eiwegger, T., Stahl, B., Schmitt, J. et al. (2004). Human milk--derived oligosaccharides and plant-derived oligosaccharides stimulate cytokine production of cord blood T-cells in vitro. *Pediatric Research* 56: 536–540.

Escamilla-Lozano, Y., Guzmán-Rodríguez, F., Alatorre-Santamaría, S. et al. (2019). Synthesis of fucosyl-oligosaccharides using α-l-fucosidase from *Lactobacillus rhamnosus* GG. *Molecules* 24: 2402.

Faijes, M., Castejón-Vilatersana, M., Val-Cid, C., and Planas, A. (2019). Enzymatic and cell factory approaches to the production of human milk oligosaccharides. *Biotechnology Advances* 37: 667–697.

Garrido, D., Barile, D., and Mills, D.A. (2012). A molecular basis for bifidobacterial enrichment in the infant gastrointestinal tract. *Advances in Nutrition* 3: 415S–421S.

Garwolinska, D., Namiesnik, J., Kot-Wasik, A., and Hewelt-Belka, W. (2018). Chemistry of human breast milk—a comprehensive review of the composition and role of milk metabolites in child development. *Journal of Agricultural and Food Chemistry* 66: 11881–11896.

German, J.B., Freeman, S.L., Lebrilla, C.B., and Mills, D.A. (2008). Human milk oligosaccharides: evolution, structures and bioselectivity as substrates for intestinal bacteria. *Nestlé Nutrition Workshop Series. Paediatric Programme* 62:205-18; discussion: 218–222.

Goehring, K.C., Marriage, B.J., Oliver, J.S. et al. (2016). 512 Similar to those who are breastfed, infants fed a formula containing 2′-fucosyllactose have lower inflammatory cytokines in a randomized controlled trial. *The Journal of Nutrition* 146: 2559–2566.

Guzmán-Rodríguez, F., Alatorre-Santamaría, S., Gómez-Ruiz, L. et al. (2019). Employment of fucosidases for the synthesis of fucosylated oligosaccharides with biological potential. *Biotechnology and Applied Biochemistry* 66: 172–191.

Han, N.S., Kim, T.J., Park, Y.C. et al. (2012). Biotechnological production of human milk oligosaccharides. *Biotechnology Advances* 30: 1268–1278.

Hollands, K., Baron, C.M., Gibson, K.J. et al. (2019). Engineering two species of yeast as cell factories for 2′-fucosyllactose. *Metabolic Engineering* 52: 232–242.

Huang, D., Yang, K., Liu, J. et al. (2017). Metabolic engineering of *Escherichia coli* for the production of 2′-fucosyllactose and 3-fucosyllactose through modular pathway enhancement. *Metabolic Engineering* 41: 23–38.

Jiménez, E., de Andrés, J., Manrique, M. et al. (2015). Metagenomic analysis of milk of healthy and mastitis-suffering women. *Journal of Human Lactation* 31 (3): 406–415.

Jung, S.M., Chin, Y.W., Lee, Y.G., and Seo, J.H. (2019). Enhanced production of 2′-fucosyllactose from fucose by elimination of rhamnose isomerase and arabinose isomerase in engineered *Escherichia coli. BiotechnolBioeng.* 116: 2412–2417.

Kobata, A. (2010). Structures and application of oligosaccharides in human milk. *Proceedings of the Japan Academy. Series B, Physical and Biological Sciences* 86: 731–747.

Lee, W.H., Pathanibul, P., Quarterman, J. et al. (2012). Whole cell biosynthesis of a functional oligosaccharide, 2′-fucosyllactose, using engineered *Escherichia coli. Microbial Cell Factories* 11: 48.

Lezyk, M., Jers, C., Kjaerulff, L. et al. (2016). Novel α-L-fucosidases from a soil metagenome for production of fucosylated human milk oligosaccharides. *PLoS One* 11: e0147438.

Li, M., Li, C., Hu, M., and Zhang, T. (2022). Metabolic engineering strategies of de novo pathway for enhancing 2′-fucosyllactose synthesis in *Escherichia coli. Microbial Biotechnology* 15 (5): 1561–1573.

Li, W., Zhu, Y., Wan, L. et al. (2021). Pathway optimization of 2′-fucosyllactose production in engineered *Escherichia coli. Journal of Agricultural and Food Chemistry* 69 (5): 1567–1577.

Liang, G., Zhao, C., Zhang, H. et al. (2020). The stepwise assembly of the neonatal virome is modulated by breastfeeding. *Nature* 581: 470–474.

Liu, Y., Zhu, Y., Wang, H. et al. (2022). Strategies for enhancing microbial production of 2′-fucosyllactose, the most abundant human milk oligosaccharide. *Journal of Agricultural and Food Chemistry* 70 (37): 11481–11499.

Lu, M., Mosleh, I., and Abbaspourrad, A. (2021). Engineered microbial routes for human milk oligosaccharides synthesis. *ACS Synthetic Biology* 10 (5): 923–938.

Lucas, A., Morley, R., Cole, T.J. et al. (1990). Early diet in preterm babies and developmental status at 18 months. *Lancet* 335: 1477–1481.

Mackenzie, L.F., Wang, Q., Antony, R. et al. (1998). Glycosynthases: mutant glycosidases for oligosaccharide synthesis. *Journal of the American Chemical Society* 120: 5583–5584.

Marriage, B.J., Buck, R.H., Goehring, K.C. et al. (2015). Infants fed a lower calorie formula with 2′FL show growth and 2′FL uptake like breast-fed infants. *Journal of Pediatric Gastroenterology and Nutrition* 61: 649–658.

Matsuki, T., Yahagi, K., Mori, H. et al. (2016). A key genetic factor for fucosyllactose utilization affects infant gut microbiota development. *Nature Communications* 7: 11939.

Mohandas, S. and Pannaraj, P.S. (2020). Beyond the bacterial microbiome: virome of human milk and effects on the developing infant. *Milk, Mucosal Immunity and the Microbiome: Impact on the Neonate* 94: 86–93.

Moossavi, S., Fehr, K., Derakhshani, H. et al. (2020). Human milk fungi: environmental determinants and inter-kingdom associations with milk bacteria in the CHILD Cohort Study. *BMC Microbiology* 20: 146.

Morrow, A.L., Ruiz-Palacios, G.M., Altaye, M. et al. (2004). Human milk oligosaccharides are associated with protection against diarrhea in breast-fed infants. *The Journal of Pediatrics* 145: 297–303.

Moubareck, C.A. (2021). Human milk microbiota and oligosaccharides: a glimpse into benefits, diversity, and correlations. *Nutrients* 13: 1123.

Newburg, D.S., Ruiz-Palacios, G.M., and Morrow, A.L. (2005). Human milk glycans protect infants against enteric pathogens. *Annual Review of Nutrition* 25: 37–58.

Ojo-Okunola, A., Nicol, M., and du Toit, E. (2018). Human breast milk bacteriome in health and disease. *Nutrients* 10: 1643.

Osanjo, G., Dion, M., Drone, J. et al. (2007). Directed evolution of the alpha-L-fucosidase from *Thermotoga maritima* into an alpha-L-transfucosidase. *Biochemistry* 46: 1022–1033.

Palur, D.S.K., Pressley, S.R., and Atsumi, S. (2023). Microbial production of human milk oligosaccharides. *Molecules* 28 (3): 1491.

Pannaraj, P.S., Ly, M., Cerini, C. et al. (2018). Shared and distinct features of human milk and infant stool viromes. *Frontiers in Microbiology* 9: 1162.

Sprenger, N., Tytgat, H.L.P., Binia, A. et al. (2022). Biology of human milk oligosaccharides: from basic science to clinical evidence. *Journal of Human Nutrition and Dietetics* 35 (2): 280–299.

Sugiyama, Y., Gotoh, A., Katoh, T. et al. (2016). Introduction of H-antigens into oligosaccharides and sugar chains of glycoproteins using highly efficient 1,2-α-l-fucosynthase. *Glycobiology* 26 (11): 1235–1247.

Togo, A., Dufour, J.C., Lagier, J.C. et al. (2019). Repertoire of human breast and milk microbiota: a systematic review. *Future Microbiology* 14: 623–641.

Wada, J., Honda, Y., Nagae, M. et al. (2008). 1,2-alpha-l-fucosynthase: a glycosynthase derived from an inverting alpha-glycosidase with an unusual reaction mechanism. *FEBS Letters* 582: 3739–3743.

Wan, L., Zhu, Y., Li, W. et al. (2020). Combinatorial modular pathway engineering for guanosine 5′-diphosphate-l-fucose production in recombinant *Escherichia coli*. *Journal of Agricultural and Food Chemistry* 68 (20): 5668–5675.

Wang, F., Yu, L., Ren, Y. et al. (2024). An optimized culturomics strategy for isolation of human milk microbiota. *Frontiers in Microbiology* 15: 1272062.

Wicinski, M., Sawicka, E., Gębalski, J. et al. (2020). Human milk oligosaccharides: health benefits, potential applications in infant formulas, and pharmacology. *Nutrients* 12: 266.

Wu, S., Grimm, R., German, J.B., and Lebrilla, C.B. (2011). Annotation and structural analysis of sialylated human milk oligosaccharides. *Journal of Proteome Research* 10: 856–868.

Wu, S., Tao, N., German, J.B. et al. (2010). Development of an annotated library of neutral human milk oligosaccharides. *Journal of Proteome Research* 9: 4138–4151.

Yu, S., Liu, J.J., Yun, E.J. et al. (2018). Production of a human milk oligosaccharide 2′-fucosyllactose by metabolically engineered *Saccharomyces cerevisiae*. *Microbial Cell Factories* 17 (1): 101.

Zabel, B., Yde, C.C., Roos, P. et al. (2019). Novel genes and metabolite trends in *Bifidobacterium longum* subsp. *infantis* Bi-26 metabolism of human milk oligosaccharide 2′-fucosyllactose. *Scientific Reports* 9 (1): 7983.

Zeuner, B., Muschiol, J., Holck, J. et al. (2018). Substrate specificity and transfucosylation activity of GH29 α-l-fucosidases for enzymatic production of human milk oligosaccharides. *New Biotechnology* 41: 34–45.

Zhang, Q., Liu, Z., Xia, H. et al. (2022). Engineered *Bacillus subtilis* for the de novo production of 2′-fucosyllactose. *Microbial Cell Factories* 21 (1): 110.

Zhou, W., Jiang, H., Wang, L. et al. (2021). Biotechnological production of 2′-fucosyllactose: a prevalent fucosylated human milk oligosaccharide. *ACS Synthetic Biology* 10 (3): 447–458.

Zhu, Y., Wan, L., Li, W. et al. (2022). Recent advances on 2′-fucosyllactose: physiological properties, applications, and production approaches. *Critical Reviews in Food Science and Nutrition* 62 (8): 2083–2092.

Zivkovic, A.M., German, J.B., Lebrilla, C.B., and Mills, D.A. (2011). Human milk glycobiome and its impact on the infant gastrointestinal microbiota. *Proceedings of the National Academy of Sciences of the United States of America* 108 (Supplement 1): 4653–4658.

10

Microbial Production of Human Milk Oligosaccharides (HMOs)

Prakram Singh Chauhan[1], *Tripti Dadheech*[2], *and Arunika Saxena*[3]

[1] *DBT-IOC Centre for Advanced Bioenergy Research, R&D Centre, Indian Oil Corporation Ltd., Sector-13, Faridabad, Haryana, 121007, India*
[2] *Department of Chemistry, Queen's University, Kingston, Ontario, K7L 3N6, Canada*
[3] *Biology Division, DC Model School Group of Institution, Sector – 9, Faridabad, Haryana, 121006, India*

10.1 Introduction

Human milk is a valuable nutritional resource for newborns due to its role in building the immune system and nourishing gut microorganisms. Micronutrients such as vitamins and minerals are found in human milk, along with lactose, lipids, proteins, and micronutrients such as lactose (Lu et al. 2021). Lactose is the primary supply of carbohydrates in human milk, with a mean content material of 70 g/L. Lipids are found in quantities ranging from 30 to 60 g/L. Human milk includes glycans, which might be the third most generic stable element after sugars and fats. The number one glycans are human milk oligosaccharides (HMOs). They offer various health advantages.

Antibiotics can decorate antibiotic characteristics, function as soluble bait receptors to reduce pathogen infections, improve the immune system, enhance intestinal permeability, and reduce necrotizing enterocolitis (Li et al. 2020). Researchers are interested in producing HMOs due to their role in newborn development and new characterization methodologies. Several ways for producing HMOs have been documented, including extraction, chemical synthesis, chemoenzymatic synthesis, and tailored microbial approaches (Pérez-Escalante et al. 2022). Human milk contains many HMOs; however, there are only a limited number of donors. Researchers have mostly concentrated on agricultural animals, including cows and goats. In theory, the US dairy industry's waste by-products may yield 220,000 kg of bovine milk oligosaccharides yearly. It is estimated that colostrum contains 0.7–1 g/L of bovine milk oligosaccharides, whereas mature milk contains considerably less. Bovine milk includes just 8 neutral and 10 sialylated oligosaccharides, indicating a low level of complexity (Bode et al. 2016). Keeping apart HMOs from donors preserves their native systems but an absence of human milk donors prevents massive-scale production. Farm animal oligosaccharides lack the structural variety of HMOs. The extraction technology cannot match market demand for HMOs. Chemical synthesis can be used for large-scale manufacture; however, the complex chemical method and poor yield of target HMOs make the approach arduous and inefficient. To solve these issues, chemoenzymatic synthesis was created. Chemoenzymatic route can provide structural variety to chemical production of HMOs, but often results in poor

Microbial Nutraceuticals: Products and Processes, First Edition. Edited by Sudhir Pratap Singh and Santosh Kumar Upadhyay.

milligram yields (Saumonneau et al. 2016). Enzymes help produce desired compounds; however, chemoenzymatic cascades do not easier the process. Chemoenzymatic synthesis techniques are not suitable for clinical applications or food formulations due to the use of hazardous chemicals and pricey substrates in the enzymatic reaction pathway. New chemical synthesis processes have led to kilogram-scale manufacturing, decreased purification stages, and increased yields (Bandara et al. 2020). However, the FDA has not accorded these procedures generally recognized as safe (GRAS) approval. The designed microbial pathway has been intensively researched as an alternate technique for producing and synthesizing HMOs on a large scale, overcoming the aforementioned limitations. Using an appropriate host strain and microbial fermentation, HMOs may be produced, including 2′-fucosyllactose (2′-FL), 3′-fucosyllactose (3′-FL), and lacto-*N*-neotetraose (LNnT) (Bych et al. 2019). Researchers choose host cells based on precursor absorption, HMO excretion, and heterologous glycosyltransferase (GT) gene overexpression. To increase HMO production, we can use molecular biology technology to introduce necessary metabolic genes like lactose permease and GT, while removing unwanted genes like the colanic acid pathway that regulate artificial promoters. The use of metabolic pathway assembly technology can also enhance designed microbial pathways by optimizing strength intake and substrate delivery. Cutting-edge changed microbial hosts consist of *Escherichia coli*, *Saccharomyces cerevisiae*, *Bacillus subtilis*, and *Corynebacterium glutamicum*. *E. coli* is usually employed in HMO synthesis attributable to its facile DNA creation, high yield, and brief production.

Furthermore, the thorough information of *E. coli's* genetics, genomes, and metabolic strategies makes it an excellent host (Cronan 2014). Currently, most research of employing altered *E. coli* to create HMOs are 2′-FL as a result of its simple structure and high yield in HMOs. Better yields have been produced with the aid of genetic engineering and metabolic pathway adjustments. Other HMOs, including three-FL, LNT, LNnT, lacto-*N*-fucopentaose I (LNFP I), and lacto-*N*-neofucopentaose I (LNnFP I), with complex systems, had been produced in changed *E. coli*, despite the fact that the number of reports is smaller than 2′-FL (Seydametova et al. 2019). However, *E. coli* is not the right host pressure because of its endotoxin content. Numerous strains, which include *S. cerevisiae*, *B. subtilis*, *Yarrowia lipolytica*, and *C. glutamicum*, had been diagnosed as hosts (Hollands et al. 2019). Although these strains are safe, there is limited evidence of their application in HMO generation. This chapter provides up-to-date information on recent production methods, different strains, including their metabolic routes and production techniques, challenges in upscaling of milk oligosaccharides and global demands, and their market trend.

10.2 Type and Structure of HMOs

HMOs are the most abundant biomolecules in milk after lactose and lipids, and their amount in milk varies between 5 and 20 g/L. Up to 1% HMO is absorbed from the intestine and enters the body. This diversity and richness is unique to humans and is not found in other species. The composition of HMOs generally varies depending on the level of breastfeeding, the mother's nutrition, genetic predisposition, and even the region where the breastfeeding mother lives and her health history (Ayechu-Muruzabal et al. 2018). Although HMOs consist of only five different monosaccharides, the complex structure found in human milk is unique (Thurl et al. 2017). Monosaccharides used as building blocks of HMOs include glucose (Glc), galactose (Gal), *N*-acetylglucosamine (GlcNAc), fucose (Fuc), and sialic acid (Neu5Ac). These monosaccharides are linked together by various methods such as glycosidic bonds. With only some exceptions, HMO constructions adhere to a precise constructing layout (discern 1). The lactose synthase protein complex catalyzes the

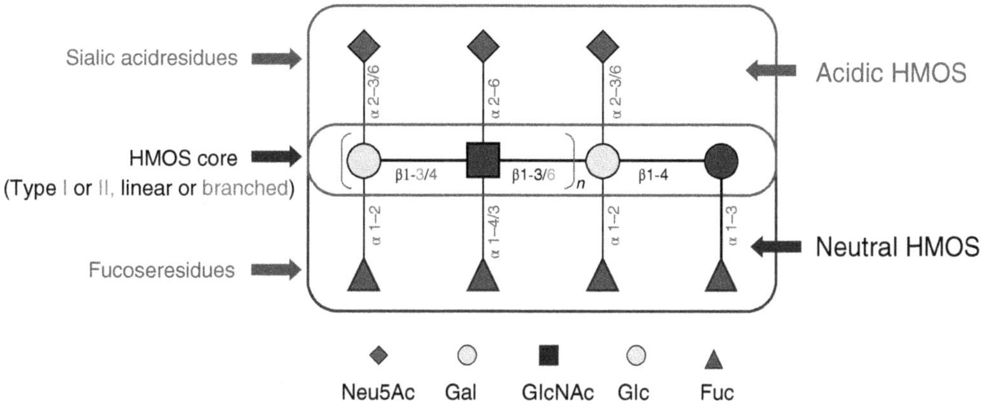

Figure 10.1 HMO's general building plan. Lactose and sort I (Gal (β1-3)GlcNAc-R) or kind II (Gal (β1-4) GlcNAc-R) HMOs' core structures may be linearly enlarged by including Gal-GlcNAc constructing blocks to terminal galactoses through β1-3 or β1-6 glycosidic connections, respectively. The latter case causes branching of the HMOs' structure. Fucoses (Fuc) and/or sialic acid (Neu5Ac) residues may be brought to the (elongated) HMOs' center structures utilizing a variety of criteria. Symbolic representation of monosaccharides using CFG requirements. *Source:* Adapted from Varki et al. (2015).

manufacturing of a β1-4 glycosidic bond between galactose and glucose, resulting in the lactose unit "Gal (β1-4) Glc" (Kuhn et al. 1980). Many trisaccharides can be produced by adding galactose or fucose to reduced or unreduced lactose residue using galactosyl- or fucosyltransferase enzymes. Ingredients include 3â²-galactosyllactose (Gal(β1-tri)Gal(β1-4)Glc), tetraâ²-galactosylllactose (Gal(β1-tetra)Gal(β1-4)Glc), 6â²-galactosyllactose (Gal(β1-6)Gal(β1-4)Glc), 2′-FL (Fuc(α1-2) Gal(β1-4)Glc), and trifucosyllactose (tri-FL) (Gal(β1-4)[Fuc α1-3] Glc). If sialic acid binds to the nonreducing side of lactose via sialyltransferase, then 3-2-sialyllactose (tri-2-SL; Neu5Ac (α2-3) Gal(β1-tetra)Glc) and hexa-2-sialyllactose (6-SL; Neu5Ac(α2-6)Gal(β1-4)Glc) is formed (Figure 10.1).

Lactose may be elongated in addition by adding Gal (β1-x)GlcNAc devices of kind I (Gal (β1-3) GlcNAc, lacto-*N*-biose) or type II (Gal (β1-4)GlcNAc, *N*-acetyllactosamine) to its unfastened three-OH group. Presently, 19 wonderful HMO center systems had been identified. These core systems may be linear or branching, with additional fucose or sialic acid residues. It is well-known that how an extensive variety of HMOs' systems formed within the human mammary gland. HMOs' synthesis is concept to take location within the Golgi equipment of the mammary gland epithelium. Along with other influences early in life, the mother's genetics also have a significant impact on HMOs' milk production. Table 10.1 shows the expression patterns of alleles of the Lewis (Le) and Secretor (Se) genes encoding notable fucosyltransferases (FUTs). These FUTs have been shown to catalyze the fucosylation of lactose and other important human milk proteins. The active Se gene encodes FUT2, which transports fucose via 1-2 glycosidic bonds. Important HMOs found in FUT2 include 2′-FL and LNFP I. LNFP I and other glycans with reduced terminus Fuc (α1-2) Gal (β1-3)GlcNAc are classified as Led or H class 1 antigens. H-type antigen links the Le/Se gadget to the blood AKI system (Oriol et al. 1986). In contrast, the active Le gene encodes FUT3, which promotes fucosylation of α1-3 or α1-4 glycosidic linkages. FUT3 models include LNFP II and LNFP III. LNFP III is an example of a Lewisx (Lex) structural motif, while LNFP II represents a Lewisa (Lea) epitope. The Lea epitope is important in the carbohydrate series. Gal (β1-3)[Fuc (α1-4)] GlcNAc-R. Lex antigens have a type II system with the following residues: Gal (β1-4) [Fuc (α1-3)] GlcNAc-R. Although all Se and Le genes are active, the HMOs' system can undergo fucosylation of α1-2, α1-3, or α1-4 glycosidic linkages.

Table 10.1 Glycosyltransferases with their sugar nucleotides.

Glycosyltransferase	Sugar nucleotide
Glucosyltransferase	UDP-Glc
Galactosyltransferase	UDP-Gal
N-acetylglucosaminyltransferase	UDP-GlcNAc
N-acetylgalactosaminyltransferase	UDP-GalNAc
Mannnosyltransferase	GDP-Man
Fucosyltransferase	GDP-Fuc
Sialyltransferase	CMP-NeuAc

Lacto-*N*-difucohexaose I (LNDFH I) is similar to the Lewis b epitope with a monosaccharide structure. Fuc (α1-2) = Gal (β1-tri) [Fuc (α1-4)]GlcNAc-R is a well-established metabolite with the combination of FUT2 and FUT3. Importantly, additional Le/Se-type neutral fucosyltransferases may contribute to the development of 1-trifucosylated HMOs together with tri-FL or LNFP V. Size exclusion chromatography (SEC) and cross-linking index (RI) can be used to determine the complex and relative concentrations of various HMOs present in milk. In general, the similarity of matrix-assisted laser desorption/ionization (MALDI)-mass spectrometry (MS) analysis of SEC HMOs will lead to more accurate information about HMOs' complexity and monosaccharide composition (Stahl et al. 1994). The acidic subcomponent expands the entire repertoire of HMOs. The total number of neutral and acidic HMO samples based on MALDI-MS analysis of milk carbohydrate SEC often exceeds 1000 systems (Boehm and Stahl 2007). The milk mass of four specific milk types was considered as the parental Le/Se prestige, and specific fucosylated HMO systems were identified in the corresponding milk (Thurl et al. 1997). Therefore, by detecting specific fucosylated HMOs and 2′-FL, DFL, LNFP I, LNFP II, LNDFH I, and LNDFH II and using the appropriate analytical procedure, male or female milk type can also be determined.

10.3 Different Methods for HMO Production

According to industry estimates, the international trade for HMOs will be rated at $199,000,000 by 2022. The popularity of nutraceuticals and declining human-milk-feeding rates make HMO an increasingly attractive food additive. Therefore, a low-cost, large-scale platform is needed to synthesize them. On an industrial scale, HMO can be produced through fermenting production strains, biochemical manufacturing, and other enzymatic processes. Though, current approaches for HMOs production mainly include microbial cell factories. For industrial applications, chemical and enzymatic syntheses have also been considered. Nevertheless, these techniques have solely been verified for gram-scale manufacturing in controlled laboratory environments. They are incapable of attaining extensive production on a kilogram scale or beyond.

10.3.1 Chemical Synthesis

Numerous fruitful approaches became evolved to produce HMOs via chemical reaction. In 1979, researchers established one of the foremost chemical production procedures of lacto-*N*-tetraose

(LNT) (Takamura et al. 1980). Over the course of time, chemical synthetic routes have been demonstrated to create complex HMO structures, resulting in over 15 diverse structures varying from oligosaccharides with 4 glycosyl residues to those with 8 glycosyl residues (Jennum et al. 2014; Chen 2015). The solid-phase techniques have been used to assemble the HMO backbone (Jennum et al. 2014). The manufacturing of convoluted hexasaccharide lacto-*N*-hexaose (LNH) has been facilitated by the advances in solid-supported oligosaccharide synthesis, which involves coupling sugar moieties to solidified support (Knerr and Schmidt 2000). Currently, gram-scale generation of LNT has been acquired using low-cost raw resources including D-galactose, D-glucosamine, and lactose, paving the door for a better realization and elucidation of the sugar's role in human milk (Craft and Townsend 2017).

Moreover, fucosylation has resulted in the production of fucosyllactose 2ʹ-FL (recent report of kilogram-scale synthesis by Agoston et al. 2019), LNT (gram-scale synthesis by Craft and Townsend 2017), LNFP I (laboratory-scale synthesis by Jennum et al. 2014). During the chemical synthesis process, it is necessary to modify the hydroxyl groups in sugars using protecting groups in order to achieve regiospecific glycosidic bond formation. Subsequently, these protecting groups must be removed after the synthesis. The process is complicated by the steps of protection, deprotection, and purification. Glycom A/S managed to chemically synthesize 2′-FL and LNnT and incorporated these HMOs into clinical and nonclinical investigations from 2005 to 2012 (Agoston et al. 2019). The synthesis of 2′-FL involves the use of donor and acceptor as 1-S-phenylfucose and lactose 6′-O-pivaloate ester, respectively. Notably, this method exists with its distinctive ability to interact with a natural acceptor beyond the need for crystallization and purification. As a result, the aforementioned approach authorizes for the production to be completed with least possible purification attempts (Figure 10.2) (Ujihara and Fukumoto 2023).

Albeit, the primary obstacle in chemical production of HMOs lies in the extensive level of functionalization (regio- and stereoselective hydroxylations) and the requirement for numerous protecting group manipulations. These manipulations involve iterative steps of introducing and removing protective group introduction and removal of protective groups, and their complexity increases with the length of the chain or branching patterns of the target oligosaccharide (Das and Mukhopadhyay 2016). This results in a time-consuming chemical synthesis and/or comparatively low product yields due to huge number of intermediate purifications steps, making it cost-inefficient overall. To date, only smaller HMOs have been provided in adequate quantities.

While advancements in the solid-supported oligosaccharide production has been carried off (Fair et al. 2015), industrial scale production of HMO still looks extremely difficult at this time due to the employment of harmful substances like transition metal enzymes, something that are unacceptable regarding usage in food (Petschacher and Nidetzky 2016). As a result, chemical production of intricate carbohydrates may nevermore always be an economical approach of synthesizing HMOs on a large scale. Reducing the reaction intermediates and product distillation stages is critical to developing a viable, scalable process (Agoston et al. 2019). Although chemically produced LNnT and 2ʹ-FL were first approved as innovative components in infant formula (Bode et al. 2016), microbial cell factories are now the sole technique used for industrial synthesis of these HMOs, since they are more economically viable (Bych et al. 2019). Nevertheless, chemical synthesis continues to play a crucial role in analytical applications and in providing the necessary synthesizing units for chemoenzymatic processes. In fact, in order to wholly realize the prospects of chemical combination for HMO generation, the above-mentioned approach is currently routinely paired with synthetic approaches using enzymes, notably the use of regio- and stereospecific GTs (Xiao et al. 2016). More lately, a chemoenzymatic pathway utilizing a glycosidase was employed for the generation of HMO progenitor configuration lacto-*N*-triose II (GlcNAc-(β1,3)-Lac; LNT2)

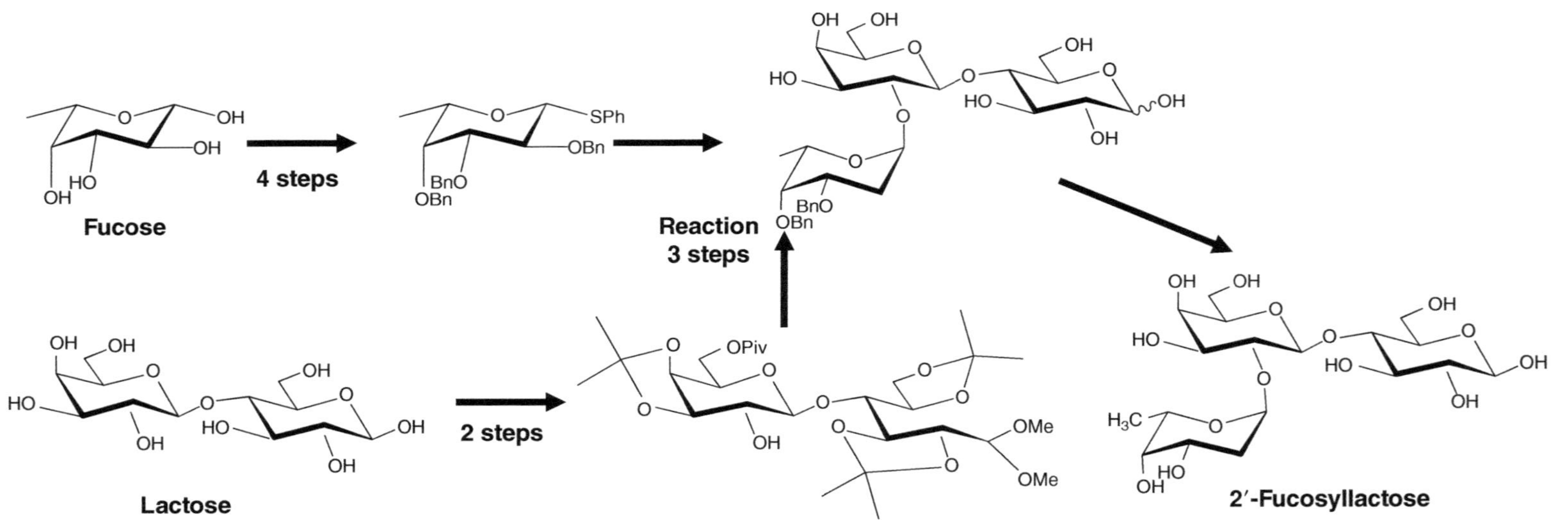

Figure 10.2 2′-FL generation by chemical synthesis. *Source:* Ujihara and Fukumoto (2023)/Glycoforum.

(Muschiol and Meyer 2019) and the same strategy was used for the synthesis of LNT (Schmölzer et al. 2019). Such chemoenzymatic techniques are very beneficial for constructing vast libraries of HMO structures that may be employed in bioactivity research (Xiao et al. 2016; Prudden et al. 2017).

10.3.2 Enzymatic Synthesis (Chemoenzymatic HMO Synthesis)

The chemoenzymatic generation of naturally occurring products has appeared as an option to prevailing chemical synthesis methods. Even though chemical synthesis has advanced, enzymatic techniques continue to be studied as valuable strategies for HMO generation due to their preciseness. This strategy unites the adaptability of chemical combination procedures with great stereo- and regioselectivity of enzyme-activated reactions, eradicating the requisite regarding de-protection and protection stages. In recent years, 31 HMOs have been synthesized using this approach (Xiao et al. 2016).

Enzymes from diverse categories including transglycosidases, glycosidases (glycosylhydrolases), GTs, and glycosynthases turned out to be utilized for chemoenzymatic synthesis of HMOs (Chen 2015). Several techniques have been developed that involve glycosidases or GTs. High-energy donor substrates that are activated to transfer sugar moieties on acceptor substrates by GTs induced the formation of disaccharides, oligosaccharides, and polysaccharides (Palcic 2011). A Leloir glycosyltransferase (GT) enzyme generates HMOs from sugar-nucleotide donor substrates in humans. Nucleotide-activated sugars are used as donors by Leloir-type GTs to synthesize HMO both *in vitro* and *in vivo*. Glycoside hydrolases (GHs) offer an substitutive method for catalyzing HMO production compared to sugar-nucleotide-dependent GTs. Unlike GTs, GHs have the ability to utilize broad range of inexpensive substrates and their enzymes are frequently characterized by robustness and ease of expression.

Most GTs exhibit a high degree of specificity in terms of both region and stereochemistry, allowing effective and precise synthesis of glycosides in a single reaction. These enzymes have been effectively used for HMO production in microbial cell factories. Thus, *in vitro* utilization of GTs is deemed further challenging. In fact, this is primarily due to their reliance on sugar nucleotides, high-energy donor substrates for GTs, which necessitates establishment regarding multienzymatic cascade processes to recycle nucleotides and maximize process economy. Still, achieving satisfactory expression levels of GTs remains difficult. On the basis of these discoveries, a method for multigram-scale production system of LNT synthesis was devised depending on the one-pot multienzyme strategy (OPME) (McArthur et al. 2019). OPME comprehends executing multi-step enzyme-catalyzed reactions within a single-reaction process, getting rid of essential requirement for extracting chemical intermediates (Zheng et al. 2022). The availability of various HMOs through chemical and chemoenzymatic synthesis has reached unprecedented levels. However, the huge price of purified enzymes and the requirement for high-priced cofactors like ATP make chemoenzymatic synthesis a less economically practicable approach for producing HMOs at commercial scale (Zeuner et al. 2019). The biosynthesis of HMOs through enzymatic reactions or fermentation is a complicated, time-consuming, and frequently low-yield process. Due to cost-constraints, biosynthesis using microbial cell factories is presently the sole technology employed in the industrial production of HMOs.

10.3.2.1 Glycosyltransferase

As opposed to strategy employing GHs, usage about GTs in the production approach using GTs, which participate in order to form glycan structures inside living organisms, offers a notable benefits to create saccharide polymer with three to ten sugar moieties having desirable configuration alongside great region- and stereospecific characteristics, while avoiding the generation of by-products or the need for protective groups. Nevertheless, there have been two barriers to the utilization of GTs. The first barrier was the limited availability of these enzymes compared to glycosidases. The second barrier involved the requirement for sugar nucleotides as high-energy substrates (Table 10.1).

There have previously been reports of many mammalian GTs with different specificities. Since these enzymes are sourced from higher eukaryotes, it was challenging to execute recombinant expression of GTs in *E. coli* until recently. However, a breakthrough came in the 2000s, which discovered oligosaccharide structures in glycocalyx found in bacteria and compared these structures to the structures found in mammalian glycans. This revelation suggests bacteria as viable origin of GTs that can be successfully expressed in *E. coli*.

Bacterial GTs are produced using *E. coli* mostly and have gained a lot of attention. Emergence of effective activity-screening protocols and quick advances in bacterial genomic sequence analysis has led to the discovery of potential GT genes in several pathogens. Consequently, numerous scientific studies have reported the identification and characterization of these genes. Numerous of these genes, including galactosyltransferase, *N*-acetylglucosaminyltransferase (Wakarchuk et al. 1996), and sialyltransferase (Gilbert et al. 1997) from a bacterium causing meningococcal disease, *Neisseria meningitidis;* sialyltransferase from a characteristic pathogen related to food-borne illness; *Campylobacter jejuni* (Gilbert et al. 2002); and fucosyltransferase from *Helicobacter pylori* (Wang et al. 1999), have been successfully expressed as functional soluble proteins in *E. coli*.

In bacteria, GTs are essential for the biogenesis of extracellular polysaccharides like lipopolysaccharide (LPS) and other cell surface saccharides. In some cases, the referred saccharides often bear a striking resemblance to polysaccharides found in mammals and are also frequently affiliated with cell attachment (as in the case of *H. pylori*) and/or antigenic escape of host (mimicry) (Appelmelk et al. 1997). Because of these bacteria's broad substrate selectivity, GTs have been shown to be a useful approach for producing HMOs (Xiao et al. 2016; Palcic 2011). However, these enzymes have the drawback of being found only in infective taxa like *Helicobacter, Neisseria,* or *Pasteurella,* with few exceptions. Such limitation clearly restricts their accessibility, leading to the utilization of recombinant enzymes created by expressing the relevant genes in genetic variants of *E. coli* such as B or K-12. Unlike GTs found in mammals, GTs sourced from bacteria are no longer glycosylated, making process of recombinant enzyme preparation easier. To date, several enzymes, primarily sourced from bacteria, including (β1,3)-*N*-acetyl-glucosaminyltransferases, (β1,3)-galactosyltransferases, (β1,4)-galactosyltransferases, (α2,3)-sialyltransferases, (α2,6)-sialyltransferases, (α1,2)-fucosyltransferases, (α1,3)-fucosyltransferases, and (α1,4)-fucosyltransferases, turned out to be essential catalysts regarding enzymatic biosynthesis of HMOs (Choi et al. 2016).

The secondary issue that arises is the requirement for sugar nucleotides considering that *E. coli* is known to express potential GTs. The synthesis of oligosaccharides via GTs is hampered by the limited ability of sugar nucleotides to withstand changes, which are usually overpriced. However, by using quite cheap 2-acetamino-2-deoxy-β-D-glucose (GlcNAc) and cytidine monophosphate (CMP), the South Korean company GeneChem has successfully synthesized CMP-sialic acid (CMP-NeuAc), a sugar nucleotide for creating 3′-sialyllactulose (Ujihara and Fukumoto 2023; U.S. FDA 2018) (Figure 10.3).

In view of this procedure, cytidine monophosphate generated via transferase reaction is effectively converted back to cytidine triphosphate within a reprocessing setup utilizing acetyl dihydrogen phosphate (acetyl-p). An alternative team created a hybrid enzyme by combining cytidine 5′-monophosphate-sialic acid synthase (CMP-CSS) with (α2,3)-sialyltransferase ((α2,3)-siaT) and, through the use of this protein, the research team demonstrated the ability to produce 3′-SL on a

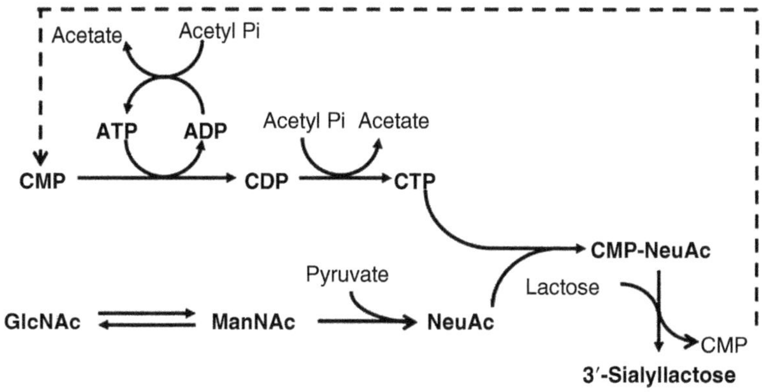

Figure 10.3 3′-SL generation by using glycosyltransferase's activity. ATP, adenosine triphosphate; ADP, adenosine diphosphate; CMP, cytidine monophosphate; CDP, cytidine diphosphate; CTP, cytidine triphosphate; GlcNAc, *N*-acetylglucosamine; ManNAc, *N*-acetylmannosamine; NeuAc, *N*-acetylneuraminic acid. *Source:* Ujihara and Fukumoto (2023)/Glycoforum.

large scale of 100 g (Gilbert et al. 1998). Thus, bulk production of oligosaccharides utilizing GTs has become feasible. However, adoption of these technologies for the commercial production of oligosaccharides were even now restricted due to the necessity of distinct processing conditions for enzyme construction and sugar-nucleotide recreation.

Numerous transferases are membrane-associated enzymes, posing challenges for the solubility and expression of recombinant enzymes (Danby and Withers 2016). Little solvability observed in recombinant enzymes and poor expression of these enzymes in microbial hosts frequently impedes cloning and expression of enzymes, namely fucosyl-transferases (FucTs) of bacterial or mammalian sources (Choi et al. 2016; Danby and Withers 2016). Enzyme structural characteristics that mostly contribute to this includes multiple repetition of hydrophobic amino acid as every seven residues and α-helical segments containing nonpolar residues with positive charge. This trait was particularly evident in FutA expression, a 1,3-FucTs found in *Helicobacter* strains. However, recent advancements have addressed this limitation by systematically deleting C-terminal end of FutA, leaving merely one heptad repeat, and optimizing codons, resulting in enhanced solubility (Choi et al. 2016).

Various instances of multienzyme cascade systems, such as GTs for HMO production, can be found. Sialyltransferases (SiaTs) with variable selectivity are utilized in the production of sialylated milk oligosaccharides including 3^J-sialyllactose (3^J-SL), 6^J-sialyllactose (6^J-SL), and disialyllacto-*N*-tetraose (DSLNT) (Yu et al. 2017). A diverse collection of 60 distinct HMO configurations was recently generated by expressing a range of human GTs in mammalian cell lines (Prudden et al. 2017). These instances highlight the effectiveness of GTs in synthesizing lucid HMO configurations with variable lengths and disparate fucosylation, sialylation, and branching patterns for use in analytical applications, notably bioactivity research. Nevertheless, these reactions cannot yet be expandable to yield levels in industrial sector (Bode et al. 2016).

10.3.2.2 Glycosidase

Glycosidase-based approaches take use of the reverse hydrolysis process of oligosaccharides. These procedures are regularly employed in small-scale synthesis in laboratories due to the ready availability of glycosidases in many instances. The underlying concept of this approach hinges on the premise that, when the reaction mixture is rich in sugars (the byproducts of hydrolysis), the reverse synthetic process will persist until reaching balance. Nonetheless, this approach is inherently incompetent, resulting in significant quantities of substrates lingering in the blend of reactants. A research team solved the above indicated problem via proficient transposition of galactose sugar to *N*-acetylglucosamine through altering donor sugar with 4-nitrophenol as well as subsequently breaking down it with water molecules by β-D-galactoside galactohydrolase enzyme in the existence of a substantial quantity of *N*-acetylglucosamine acceptor (Ajisaka 2017).

The primary issue with employing glycosylhydrolases in transglycosylation reaction exists in enzyme's intrinsic hydrolytic activity, which often affects the products as well as the substrates involved. Although there are multiple indigenous transglycosidases (Bissaro et al. 2015), which are GHs that never catalyze hydrolytic reactions, the exclusive human milk oligosaccharide-pertinent instance is the TcTS, a transsialidase found in *Trypanosoma cruzi* (Holck et al. 2014). As a result, the major attention has shifted toward GHs that can catalyze transglycosylation while competing with hydrolysis. The conditions of the reaction can be improved to promote transglycosylation, for as by raising substrate concentration or altering pH (Lundemo et al. 2017). However, enzyme engineering is the most effective technique for reducing hydrolysis activity and improving transglycosylation activity in glycosylhydrolases. Transglycosylation via enzymes has demonstrated tremendous promise in the production of HMOs, particularly for LNT or lactose fucosylation or sialylation (Holck et al. 2014) and for the generation of HMO's core structures (Jamek et al. 2018). This approach has the potential to significantly increase the existing restricted range of HMOs' repertoire on an industrial scale, but its success largely depends on the identification of innovative transglycosidase enzymes or the genetic modification of existing catalysts to boost transglycosylation efficacy.

10.3.3 Microbial Cell Factories (Whole-Cell Reaction Method)

Microorganisms can function as whole-cell catalysts for the synthesis of commercially significant substances (Zeuner et al. 2019). Microbial cell factories are ideal candidates for large-scale industrial production due to their ability to utilize inexpensive carbon feedstocks and their high specificity as biocatalysts (Cho et al. 2022). Moreover, the temperature at which the production strains operate is proportionately low as well as they do not need expensive enzyme purification and cofactors (Cho et al. 2022).

Advancements in metabolic engineering and synthetic biology approaches have made microbial production platforms a sustainable substitute for procedures based on chemicals in the synthesis of HMOs. A team of researchers has successfully created around 42 different HMO structures using microbial cell factories (Walsh et al. 2020). *Escherichia coli* is now the top choice for producing HMOs because of its rapid growth, genetic manipulability, and scalability of culture volumes (Bych et al. 2019). *E. coli* lacks its own GT enzymes for HMO synthesis. Yet, the existence of an effective biosynthetic route for sugar nucleotide generation and a progenitor for HMO generation (Woo et al. 2019), establishes this organism as a feasible candidate for HMO production. In the past, HMOs were obtained through whole cell reactions, enzyme-based synthesis, and milk extraction. However, in recent times, the direct fermentation method has gained widespread popularity as the preferred approach for synthesizing HMOs.

The whole-cell reaction approach has been established for the synthesis of oligosaccharides to enhance efficiency by streamlining the production of sugar nucleotides. This strategy employs bacterial cells as catalysts rather than purified enzymes, eliminating the need for enzyme purification. By treating microbial strains using carbon-based solvents or surface active agents, their plasma membrane become penetrable to chemicals, allowing for the completion of complex multistep reactions in a single vessel by mingling bacterial cells with the required functional characteristics. In contrast to the enzymatic reaction system, where ATP (adenosine triphosphate) must be

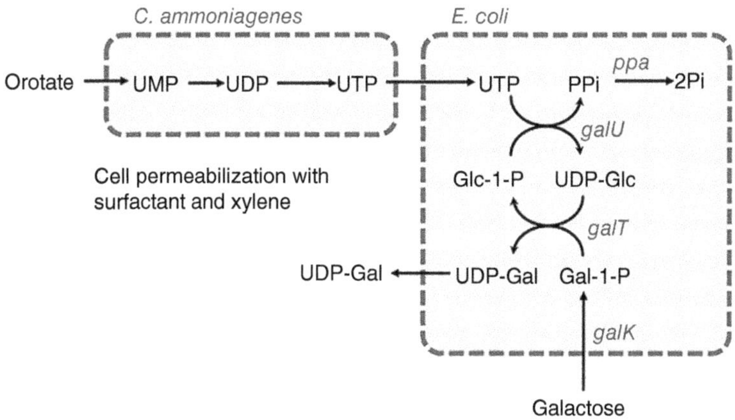

Figure 10.4 UDP-galactose synthesis via cell reaction process. UMP, uridine monophosphate; UDP, uridine diphosphate; UTP, uridine triphosphate; UDP-Gal, UDP-Galactose; UDP-Glc, UDP-Glucose; PPi, inorganic pyrophosphate; Pi, inorganic phosphate; ppa, pyrophosphatase; galT, galactose-1-phosphate uridyltransferase (GALT); galU, UDP-glucose pyrophosphorylase; galK, galactokinase. *Source:* Ujihara and Fukumoto (2023)/Glycoforum.

externally provided or recycled using acetyl phosphate to regenerate phosphorylated nucleotides, the whole-cell system stands out as adenosine triphosphate may be synthesized metabolically using low-cost glucose. For instance, one system for uridine diphosphate galactose synthesis which requires orotic acid and galactose, a low-priced origin of nucleic acids acting as reactants, has been created utilizing engineered *E. coli* bacteria producing biocatalysts associated with uridine diphosphate galactose biosynthesis as well as *Corynebacterium ammoniagenes* known for their high-level production capacity of nucleoside triphosphate. This resulted in the generation of uridine diphosphate galactose at the rate of 44 g/L in just 21 hours (Koizumi et al. 1998) (Figure 10.4). Likewise, by mixing engineered *E. coli* bacteria producing cytidine 5′-monophosphate-sialic acid synthase and cytidine triphosphate synthase (CTPS) with cells of *C. ammoniagenes*, 17 g/L of Cytidine 5′-monophospho-*N*-acetylneuraminic acid was produced in 27 hours, beginning from *N*-acetylneuraminic acids (Neu5Ac) and orotic acid (Endo et al. 2000).

Hence, by integrating *E. coli* cells that express GTs with those harboring sugar nucleotide production system, we could establish a platform for directly generating oligosaccharides by coupling systems of whole-cell multiple reactions. By integrating *E. coli* cells expressing the *Neisseria gonorrhoeae* (α2,3)-siaT gene with cells harboring the cytidine 5′-monophospho-*N*-acetylneuraminic acid production system, 3′-SL production reached 33 g/L in 11 hours (Endo et al. 2000) (Figure 10.5). Furthermore, the accumulation of globotriaose, the oligosaccharide configuration identified by Shiga toxins synthesized in pathogenic *E. coli* O-157 strain (Galα1-4Galβ 1-4Glc), was observed to be 188 g/L in 36 hours when *E. coli* cells expressed the *N. gonorrhoeae* (α1,4)-galactosyltransferase gene once they are coupled with cells possessing uridine diphosphate galactose production system (Koizumi et al. 1998).

Earlier, HMOs were separated from breast milk in order to analyze their structure and conduct research and development activities. Still, this method is not appropriate for large-scale industrial generation of HMOs. The complexity of the process, the number of reaction steps involved, restricted accessibility, and huge price of starting substances render chemical synthesis of HMOs excessively costly for use in nutrition and dairy sector. On the other hand, companies like Kyowa Hakko Kogyo Co. and Abbott Laboratories have utilized these biosynthetic capabilities of microbes

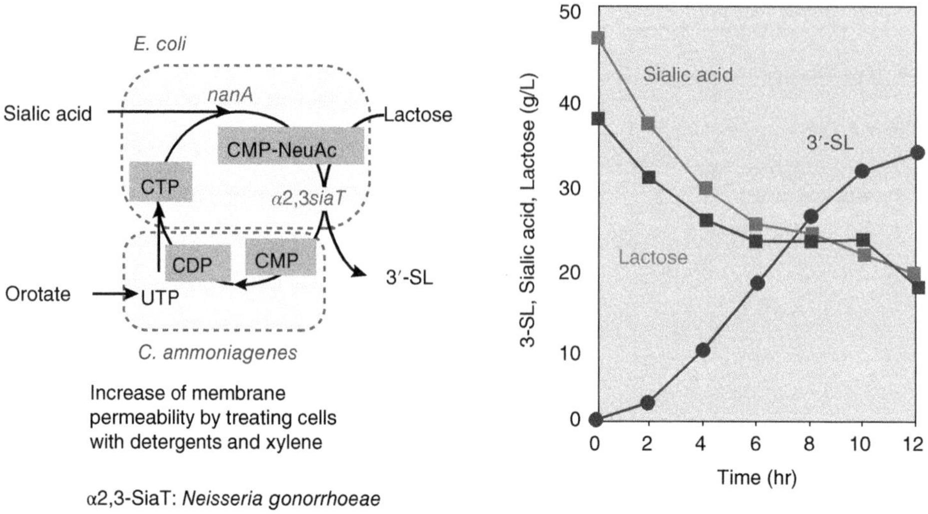

Figure 10.5 3′-SL generation via cell reaction process. NanA, *N*-acetylneuraminic acid; (α2,3)-siaT, (α2,3)-sialyltransferase; 3′-SL, 3′-sialyllactulose; CMP, cytidine monophosphate; CDP, cytidine diphosphate; CTP, cytidine triphosphate; CMP-NeuAc, cytidine 5′-monophospho-*N*-acetylneuraminic acid; UTP, uridine triphosphate *Source:* Ujihara and Fukumoto (2023)/Glycoforum.

by combining different cell homogenates, resulting in successful production of huge amount of LNnT and fucosylated oligosaccharides (adapted from Prieto 2005).

The initial two HMOs were launched in the United States and Europe in the year 2015 and 2016 amid fierce competitiveness from Danish Glycom A/S, German Jennewein Biotechnologie GmbH, and American Abbott Laboratories (Bych et al. 2019; Bode et al. 2016). After extensive research in metabolic pathway engineering, fucosyllactose and LNnT emerged as a principal HMOs synthesized at a larger scale via fermentation of genetically modified *E. coli*. Both substances have received approval for safety from the European Food Safety Authority (EFSA) and the US Food and Drug Administration (FDA) (Bych et al. 2019) and are presently accessible in newborn formula in over 30 countries. Fermentation titers for fucosyllactose have turned out to be 180 g/L (Bych et al. 2019). In a clinical experiment involving infants aged up to 6 months, the combination of 2J-FL and LNnT in infant formula was shown to be secured and well-endured, with parents reporting decreased illness after ingestion (Puccio et al. 2017). Beneficial impacts on adult microbiota have also been shown. Furthermore, attempts have been made to genetically modify *S. cerevisiae* (Liu et al. 2018) or *Lactococcus lactis* (Li et al. 2017)—organisms with the GRAS designation (generally regarded as safe)—to create HMOs.

Nevertheless, despite the fact that the significant advancements in metabolic engineering and fermentation strategy over the past 20 years have led to the increase in HMO synthesis, there are still limitations to the HMO synthesis. Fermentation titers for extensively intricate HMOs are frequently lower, and products are confined within the cells themselves (Faijes et al. 2019). For commercial-scale synthesis, extracellular products appreciably facilitate process of purification (Bych et al. 2019).

10.3.3.1 2′-Fucosyllactose

The synthesis of 2′-FL requires the action of an (α1,2)-fucosyltransferase enzyme, which adds guanosine diphosphate fucose to the galactose component of lactose. Guanosine diphosphate

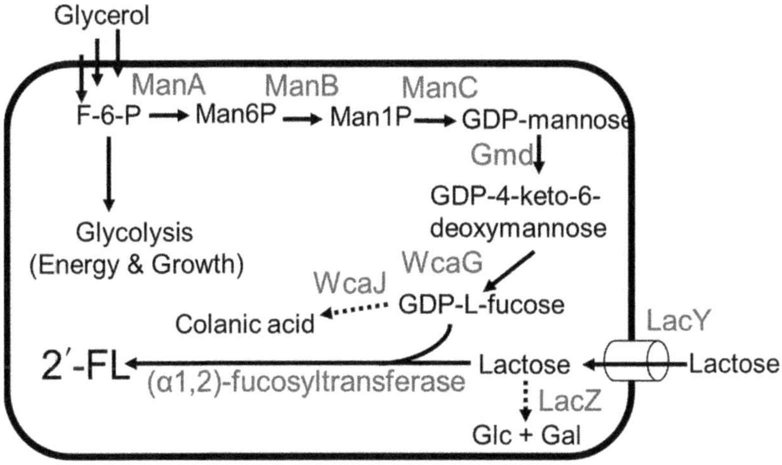

Figure 10.6 2′FL synthesis from lactose and glycerol. Arrows with dots designate competing pathways. ManA, mannose-6-phosphate isomerase; ManB, phosphomannomutase; ManC, α-D-mannose-1-phosphate guanyltransferase; GDP, Guanosine diphosphate; Gmd, GDP-D-mannose-4,6-dehyratase; WcaG, GDP-L-fucose synthetase; LacZ, β-galactosidase; WcaJ, UDP-glucose:undecaprenyl-phosphate glucose-1-phosphate transferase; LacY, lactose permease; LacZ, β-galactosidase. *Source:* Palur et al. (2023)/MDPI/CC BY 4.0.

fucose is produced via two different biochemical routes: salvage pathway and *de novo* pathway (Wan et al. 2020; Zhai et al. 2015). Out of these, the *de novo* pathway is widespread in eukaryotes and prokaryotes (Wan et al. 2020). The transition from D-fructose-6-p to guanosine diphosphate fucose has been performed through 5 enzyme-based reactions (Figure 10.6) (Wan et al. 2020; Palur et al. 2023).

Phosphomannose isomerase (PMI; ManA) transforms D-fructose-6-p to α-D-mannose-6-p. Subsequently, phosphomannomutase (ManB) converts α-D-mannose-6-p to mannose-1-p. Further actions involve the action of GDP-mannose pyrophosphorylase (also known as ManC), which transforms mannose-1-p into guanosine diphosphomannose (Figure 10.6). Following this, guanosine diphosphomannose undergoes further modification and is converted into guanosine diphosphate-4-keto-6-deoxy-D-mannose by the enzyme guanosine diphosphomannose 4,6-dehydratase. Eventually, guanosine diphosphate fucose is synthesized from guanosine diphosphate-4-keto-6-deoxy-D-mannose via action of GDP-4-keto-6-deoxy-D-mannose-3,5-epimerase-4-reductase (Wan et al. 2020).

Salvage pathway exists typically in multicellular organisms and has recently been identified in *Bacteroides fragilis*, a common bacterium in GI tract (Zhai et al. 2015). This biosynthetic route uses an enzyme with dual-function GDP-fucose pyrophosphorylase (Fkp) to convert externally obtained hexose deoxy sugar into guanosine diphosphate fucose through a dual-step enzyme-based process (Figure 10.7) (Zhai et al. 2015; Palur et al. 2023). Fkp first transforms L-fucose to 6-deoxy-L-galactose 1-phosphate and then transformed into guanosine diphosphate fucose. Several (α1,2)-fucosyltransferases have been discovered for the last step in the 2′-FL biosynthesis route (Seydametova et al. 2019).

Sucrose was utilized as the exclusive carbon source to establish a fucosyllactose synthesis system within *E. coli* (Figure 10.7; Parschat et al. 2020). The breakdown of sucrose resulted in lactose sugar generation and guanosine diphosphate fucose as a reaction intermediate in the production of sugar (Figure 10.7). *Pasteurella multocida*-derived enzyme GalTpm1141, a β-*N*-acetylglucosaminyl-glycopeptide (β1,4)-galactosyltransferase, exhibited remarkable functional activity regarding

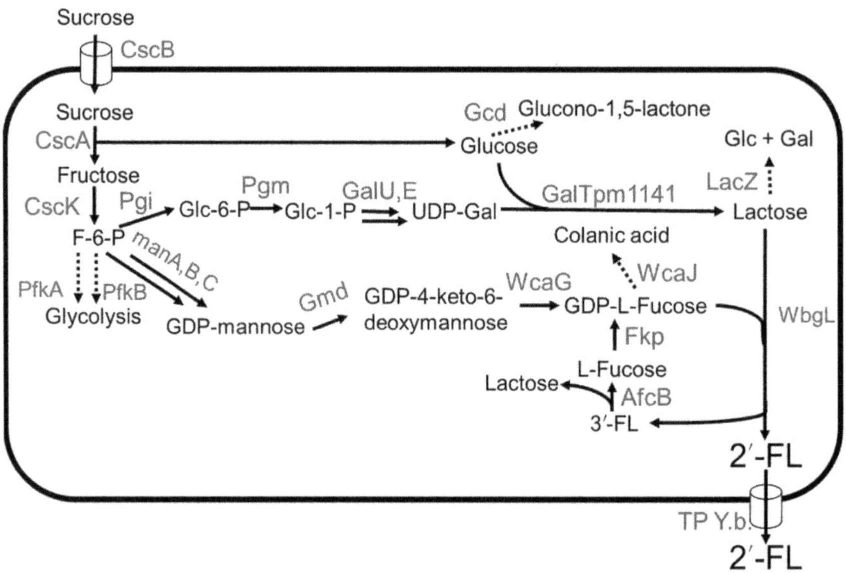

Figure 10.7 2′-FL synthesis via sucrose utilization as substrate. Arrows with dots designate competing pathways. ManA, mannose-6-phosphate isomerase; ManB, phosphomannomutase; ManC, α-D-mannose-1-phosphate guanyltransferase; Gmd, GDP-D-mannose-4,6-dehyratase; WcaG, GDP-L-fucose synthetase; Pgi, glucose-6-phosphate isomerase; Pgm, phosphoglucomutase; GalU, UTP-glucose-1-phosphate uridylyltransferase; GalE, UDP-glucose-4-epimerase; WcaJ, UDP-glucose:undecaprenyl-phosphate glucose-1-phosphate transferase; Gcd, quinoprotein glucose dehydrogenase; LacZ, β-galactosidase; Fkp, bifunctional fucokinase/fucose-1-phosphate guanylyltransferase; AfcB, (α1−3)-fucosidase; GalTpm1141, (β1−4)-galactosyltransferase; WbgL, (α1,2)-fucosyltransferase; TP Y.b., sugar efflux transporter; CscB, sucrose permease; CscA, sucrose hydrolase; CscK, fructokinase; PfkA, 6-phosphofructokinase 1; PfkB, 6-phosphofructokinase 2. *Source:* Palur et al. (2023)/MDPI/CC BY 4.0.

synthesis of sugar lactose from uridine diphosphate galactose and dextrose (Parschat et al. 2020). To facilitate the growth of genetically modified microbes using sucrose and augment the generation of guanosine diphosphate fucose and lactose as substrates in the metabolic route, the genes associated with *de novo* synthesis of guanosine diphosphate fucose and cscABKR from *E. coli* W were integrated into the chromosome.

LacZ, gnd, wcaJ, and pfkA deletions enhanced the generation of fucosyllactose. TP Y.b. gene sourced from *Yersinia bercovieri* ATCC43970, which encodes a sugar efflux transporter, was expressed to enable the modified genetic variant to move fucosyllactose out of the cell. By establishing modified genetic variant to express wbgL, which encodes for *E. coli* O126 2-fucosyltransferase (Engels and Elling 2014), a significant amount of fucosyllactose, reaching 3 g/L, was produced inside a fed-batch bioreactor using sucrose as only carbon source. Owing to promiscuous expression of WbgL, the generation of by-products 3′-FL and lactodifucotetraose (LDFT) occurred. For minimizing 3′-FL generation, the production strain was modified by inserting afcB, which encodes an α-L-fucoside fucohydrolase, sourced from *Bifidobacterium bifidum* as well as a gene from salvage purine synthesis pathway, fkp into it. Alpha-L-fucoside fucohydrolase catalyzes the generation of L-fucose and lactose by hydrolyzing 3′-FL (Figure 10.7). The salvage pathway allows Fkp to convert free 6-deoxy-L-galactose to guanosine diphosphate fucose through 6-deoxy-L-galactose 1-phosphate. Inside a fed-batch 3-L fermenter, utilizing sucrose as a single carbon source, the modified strain achieved approximately 60 g/L of 2′-FL as product (Parschat et al. 2020).

The biosynthetic pathway reported for fucosyllactose synthesis in *E. coli* has been constructed using *de novo* pathway for guanosine diphosphate fucose (Figure 10.6) (Liu et al. 2022). To enhance the generation of guanosine diphosphate fucose, genes associated with *de novo* nucleotide biosynthetic pathway were overexpressed with a high-copy-number plasmid. LacY has also been transcribed and translated to facilitate import of lactose. LacZ has been knocked out to slow down lactose degradation. Knocking out of wcaJ restricts the passage of metabolites through this route over time from guanosine diphosphate fucose to colonic acid (Figure 10.6), leading to increased guanosine diphosphate fucose production. Transgenesis of the wbgL gene, which encodes for 2-fucosyltransferase, into the chromosome allowed the engineered microbial genetic variant to create 2′-FL. Final resulted microbial cell factories producing wbgL accumulated 79 g/L of 2′-FL in a 3-L fermenter employing glycerol and lactose as carbon sources (Liu et al. 2022).

The FutC gene from *H. pylori*, which encodes for (α1,2)-fucosyltransferase (Seydametova et al. 2019), was employed to produce 2′-FL in *E. coli* which is caused by its strong selectivity for sugar lactose (Li et al. 2022). The gene expression levels responsible for lactose absorption and *de novo* assimilation of 6-deoxy-L-galactose have been altered in order to increase the production of lactose, guanosine diphosphate fucose, and reaction intermediates for 2′-FL synthesis. The metabolic route regarding 2′-FL synthesis made use of only one NADPH and GTP molecule. By overexpressing zwf, pntAB, and gsk genes, the synthesis of 2′-FL was considerably increased while reducing cellular burden through a balanced supply of GTP and NADPH. Ultimately, this engineered strain achieved a yield of 22 g/L of 2′-FL using lactose as substrate inside 3-L fermenter (Li et al. 2022).

Saccharomyces cerevisiae underwent genetic modification to synthesize 2′-FL from lactose and xylose (Lee et al. 2020). The usage of yeast as a host microorganism eliminates the chances associated with invasion of bacteria by phages in the course of industrial-scale synthesis, potential endotoxin contamination, and negative perception of consumer. The *E. coli* O126 wbgL and *Kluyveromyces lactis* LAC12 (a lactose transporter gene) were inserted inside *S. cerevisiae* genome. Additionally, *E. coli* gmd and wcaG genes were inserted inside *S. cerevisiae* genome to permit *de novo* generation of guanosine diphosphate fucose. This recombinant genetic variant of host organism accumulated 26 g/L of 2′-FL from lactose and xylose inside a fed-batch 1-L fermentor (Lee et al. 2020). Furthermore, *Bacillus cereus* FutBc, a purported (α1,2)-fucosyltransferase, was too employed in order to generate 2′-FL. This FutBc has been expressed leading toward 1.8 times increase in 2′-FL synthesis in comparison with *H. pylori's* FutC. Both LAC12 and FutBc genes were integrated into *S. cerevisiae*, resulting in the generation of 27 g/L of 2′-FL using lactose as well as sucrose within a 5-L fermentor utilizing fed batch culture techniques (Xu et al. 2021).

10.4 Strategies for Enhanced HMO Production

Presently, engineered strains of *C. glutamicum, S. cerevisiae, E. coli,* and *B. subtilis* can successfully generate HMOs through their metabolism. A large commercial potential exists for new food enhancers and preservatives like LNT, LNnT, 2′-FL, and 3′-FL. 2′-FL and LNnT have been certified for production at large commercial scale. Despite the fact that advancements in metabolic engineering and synthetic biology have facilitated HMO generation, undefined metabolic characteristics and choices regarding suitable microbial host genetic variants remain primary focal point of modern research. Enhancing cytidine 5′-monophospho-*N*-acetylneuraminic acid or guanosine diphosphate fucose accessibility via genetic engineering approaches is crucial for improving the activity of salvage and *de novo* synthesis route. Biosynthesis of donors is inexpensive yet less efficient in the *de novo* route. For this reason, few researchers incorporate salvage into route of *de novo* synthesis to upsurge the rate of metabolism.

Escherichia coli is often utilized as a strain for engineering metabolic pathways, despite its inherent toxicity that complicates downstream processes. Scientists are exploring other palatable host strains including *B. subtilis, C. glutamicum,* and *S. cerevisiae* as potential alternatives. However, the production models and yields are currently meager. Due to the genetic differences between these alternative strains and *E. coli*, a different approach is necessary. For instance, in order to employ *S. cerevisiae*, a particular gene which encodes for lactose permease must be inserted inside this genetic variant of *Saccharomyces* since it is unable to import and break down lactose. Similarly, *C. glutamicum* requires LacY, WcaG, and Gmd genes in active state for biosynthesis of HMOs. Furthermore, *E. coli* could enhance production route of guanosine diphosphate fucose through intervening with colanic acid biochemical pathway, which is not present in *S. cerevisiae*. As a result, more research to study the metabolic patterns of various microorganisms is required. By correlating these patterns with the well-studied *E. coli*, identifying identical biochemical pathways or functionally active genes in other possible strains could prove to be a successful strategy for improving HMO synthesis in these strains.

10.4.1 Designing Cell Factories for Commercial Synthesis

Appealing qualities for potential microbial hosts comprise effectual metabolic routes for generation of nucleotide sugar donors, effective absorption of lactose and HMO transport out of cells. Other valuable characters consist of access to genetic engineering techniques, proven protection, rapid progression, and appropriateness for scaling up. Additionally, traits that impact downstream operational costs involve minimal biomass emergence; minimal cell lysis rates (viscosity); minimal levels of contaminants like protein, DNA, and endotoxins (LPS); and an enhanced carbohydrate contamination profile. While *E. coli* possesses a few of these attributes, it falls short in others. A smaller number of initiatives to investigate other possible hosts such as *C. glutamicum* (Seo et al. 2017) and *S. cerevisiae* (Yu et al. 2018) have begun to appear. However, as of now, there has been no persistent engineering work undertaken using other possible host to effectively compete with present *E. coli* engineered subtypes.

10.4.2 Modification of Metabolic Pathway

The metabolic pathway engineering approach for the uncomplicated and topmost prevalent HMOs using *E. coli* as the microbial cell factory is demonstrated below.

10.4.2.1 Exploitation of Lactose Substrate for Producing HMOs

So far, the majority of fermentation-driven HMO synthesis relies upon lactose as externally derived glycosyl acceptor. Native *E. coli* strains can effectively take in and break down lactose. As a result, all scientific teams involved in HMO synthesis have either removed (Dumon et al. 2001; Chin et al. 2016) or modified the expression of lacZ gene encoding β-galactosidase hydrolyzing lactose (Chin et al. 2015).

10.4.2.2 Engineering of GDP-L-Fucose Pool Occurring Inside a Cell

Guanosine diphosphate fucose is synthesized in *E. coli* as a progenitor to colanic acid, which is a polysaccharide that exists outside of cells and is not worthy. There are three distinct strategies, or associations of strategies, that can be employed to boost the accessibility of GDP-fucose.

1) Enhanced expression rate of guanosine diphosphate fucose route through employing powerful transcription initiation sequences (Baumgärtner et al. 2013), plasmid-based systems (Lee et al. 2009), or via excessive expression of RscA gene, which encodes for an activator protein of colanic acid production pathway (Huang et al. 2017).
2) Removal of single or numerous genes located downstream of guanosine diphosphate fucose like wcaJ to halt the continuation of conversion process to colanic acid (Huang et al. 2017).
3) Aiming at potent restricting cofactors like NADPH and GTP (Lee et al. 2012).

10.4.2.3 Transferase Expression and Engineering

Sugar transferases are the primary catalysts of HMO production strains, transferring sugar component from sugar nucleotides to lactose or alternative HMOs. Regarding 2′-FL, this fucosyltransferase type utilized impacts both the frequency of 2′-FL formation and by-product profiles (Huang et al. 2017). Chin et al. reported that (α1,2)-fucosyltransferase, wcfB with an elevated level of 2′-FL formation, does not generate side product difucosyl-lactose (DFL) (Chin et al. 2015) as other (α1,2)-fucosyltransferases like FutC do.

Identifying a suitable (α1,3)-fucosyltransferase for 3′-FL synthesis has been challenging. This is owing to problems with soluble enzyme expression in *E. coli* and the enzyme's preference for substrates other than lactose (Zhang et al. 2010). Recent research (Choi et al. 2016) found that altering four amino acids (A128N/H129E/Y132I/S46F) boosted the *in vitro* activity of specific (α1,3)-fucosyltransferase by 15 times compared to the original version. Though, inconsistent results have been observed when engineering the (α1,3)-fucosyltransferase for *in vitro* versus *in vivo* applications (Choi et al. 2016). Thus, it is vital to confirm *in vitro* discoveries *in vivo* settings.

10.4.2.4 Exporting Product Outside Cell

Enhancing the transportation of HMOs from the cell would boost the speed of product formation, minimize the generation of unwanted by-products, and facilitate subsequent purification process. Despite limited knowledge on transport of HMOs by *E. coli* outside the cell, a group of sugar efflux transporters was identified to move 3′-FL out of the cell (Jennewein et al. 2010). Though, these exporters have the drawback of also exporting lactose (Liu et al. 1999). More advanced studies are required to decide whether upregulated expression of these transporters could increase HMO synthesis.

10.4.3 Process of Fermentation and Scaling-up

Large fermenter of volume over 200 m^3 provides issues, notably the significantly extended blending hours. These issues induce variations in oxygen, carbon source, and pH level, which has deleterious influence on physiology and stability of strains, as well as HMO titers (Takors 2012). The reduced production of biomass, usually confronted in industrial settings, could be allocated to synthesis and subsequent absorption of metabolites or the rapid changes in the environment that triggers starvation and additional stress responses (Lara et al. 2009).

To minimize harmful overflow and anaerobic metabolism, carbon source feed rates must be tailored based on the microorganism's physiology and the fermenter's oxygen transfer rate capacity. One well-established method for producing strains in large fermenters is through modulation of central metabolic pathway including mixed-acid fermentation and Warburg effect. This strategy results in lower metabolite production and increased biomass production, thus enabling better effectual energy usage for product generation (Lee and Kim 2015). Furthermore, the composition of trace metals in media may be tailored to facilitate the deterioration of deterrent metabolites like formic acid, which would under other conditions aggregated to toxic levels (Soini et al. 2008). As a result, it is recommended to have an excellent cut-down imitation of the manufacturing conditions in laboratory-scale setup (Takors 2012). Additional facts on scaling up of large-scale processes in *E. coli* could be tracked down in research analysis (Takors 2012; Lee and Kim 2015).

At the strain level, a fundamental difficulty in scaling up is the intrinsic chromosomal fickleness of high-producing strains, largely triggered by the metabolic stress of producing the desired product. Over numerous generations in an industrial fermentation process, significant evolutionary pressure as opposed to increased generation of products might disrupt metabolic pathways, resulting in the dominance of nonproducers (Rugbjerg et al. 2018a). This is why relying solely on plasmid-free, fully chromosomal strains, which is the industrial norm, is insufficient to guarantee complete genetic uniformity. Thus, most novel approaches to prolong the productivity of production strains has begun to come out, including this notion of synthetic product addiction (Rugbjerg et al. 2018b).

Another key emphasis zone of focus is the utilization of resources, including their processing and sterilization on an industrial scale. Lactose is an essential component in HMO synthesis. While there have been suggestions for *in vivo* lactose generation, the favored procedure appears to be direct addition. But, lactose possesses challenges due to its low solubility and tendency to isomerize into lactulose, making it difficult to handle and store concentrated solutions of lactose. Homogenous thoughts relate to the usage of the chief carbon and energy resources. Carbon sources like glycerol, sucrose, and glucose were disclosed to generate HMOs. Respective carbon resources have unique benefits and drawbacks in industrial environment. Sucrose is appealing alternative with respect to its affordability, accessibility, solvability, and thermodynamic stability, although this strategy needs modeling of biochemical routes in *E. coli* for utilization (Samain et al. 2015). *E. coli* can easily metabolize glucose and glycerol; however, glucose can produce undesirable side products in the course of heat sterilization, while glycerol is costly to a greater extent. In contrast, glycerol may be gripped as a concentrated liquid, nearly 100%, thus avoiding unnecessary dilution in the method and assures greater strength, which benefits downstream processes.

10.4.4 Quality of the Product and Downstream Processes

The high-grade HMO final products necessitate exceptional consideration for the design with regard to an effective downstream processing (DSP). These microbial production strains and engineered proteins for expression, DNA/RNA, LPSs, and additional biomacromolecules must be eliminated entirely. Functioning of DSP unit must be conscientiously chosen to meet these needs. HMO products must be of excellent purity because they are mostly used for infant formula supplements. Nevertheless, the existing production techniques result in contamination to some extent. Hence, a competent purification methodology is indispensable for the purification and conversion of these products into powder form.

10.5 Purification Methods

HMO products must be of excellent purity because they are mostly used for newborn formula supplements. However, the manufacturing processes discussed thus far produce contaminants in variable degrees. As a result, they must be purified and powdered using effective purification methods. The HMO produced all through the fermentation manner is dissolved within the subculture medium, which incorporates contaminants inclusive of lifestyle medium components and microorganism. Bacteria are usually removed by membrane separation and/or centrifugation, followed by removal (or deionization) by ion exchange resin or electrodialysis. During the process, charged salts and impurities are separated and removed. To put off pigments, the product is decolorized the use of activated carbon or synthetic adsorbents before being concentrated. In step with the sensible software documents filed by using companies, the manner of getting rid of contaminants every now and then includes the chromatographic separation technique. The effectiveness of chromatographic separation depends on the type of cargo. For example, for acidic HMOs including 3′-SL and 6′-SL that transfer negative charges, ion exchange resins can perform the separation, but for neutral HMOs including 2′-FL and 6′-FL, LNnT can best predict the removal of compounds and techniques such as simulated changing bed chromatography and gel filtration chromatography are needed for similar impurity separation, which is the first issue of product development (Jennewein and Helfrich 2015). Crystallization and spray drying are the chosen options for the final powder of purified concentrated drug. Crystallization is a good way to increase the purity of neutral HMOs and was previously in the spotlight when companies competed for patent packages, but many companies now use it for spray drying (Matwiejuk et al. 2015). Spray drying is a process in which a nonliquid material is sprayed into hot air, quickly evaporating moisture and forming a dry powder.

Robinson et al. (2018) proposed a solid-phase extraction (SPE) setup to improve the technology's ability to separate oligosaccharides (OS) from lactose. Wash the solid phase with a small amount of acetonitrile (ACN) and trifluoroacetic acid (TFA). Lactose removal and retention of various OSs were greatly improved when 4% ACN/0.1% TFA was used instead of more concentrated detergents. Acid and neutral operating systems behave differently. The radical SPE method complements the benefits of milk OS in applications that primarily do not require lactose. Dijkstra et al. (2022) devised a technique for purifying an impartial HMO through microbial fermentation. The discovery relates to a system for the purification of an impartial HMO from a fermentation broth, the system comprising the steps of (i) separating biomass from the fermentation broth to offer a crude solution and (ii) treating the crude answer with (a) a cation-exchange cloth, (b) an anion-trade material, and (c) a cation-alternate adsorbent resin; thereby acquiring a clear answer containing the neutral HMO. Furthermore, the invention describes a technique for fermentatively generating HMO in a fermentation broth and purifying the HMO from the broth. Wang et al. (2022) mounted an effective purification technique for 2′-FL with the use of membrane filtering and activated carbon adsorption. Based totally on the typical *E. coli* biosynthesis of 2′-FL fermentation broth, most people of the contaminants were eliminated and focused utilizing membrane filtering generation following easy flocculation. The goal of 2′-FL becomes eluted selectively utilizing activated carbon adsorption and ethanol-gradient elution approach. The 2′-FL made of 90% or extra purity can be easily manufactured. This observation investigated a new technique to the industrial synthesis of 2′-FL. The absence of powerful downstream era for organic fermentation, particularly purification technology, has become the primary limiting component in its adoption (Figure 10.8).

10.6 Global Demand and Recent Market Aspects of HMOs

The worldwide HMO market became worth USD 177.91 million in 2022 and is expected to growth at a common annual growth rate (CAGR) of 9.7% at some point of the forecast duration (2024–2031), from USD 195.17 million in 2023 to USD 409.32 million in 2031. The marketplace for HMO is being greater diagnosed as an essential factor of infant health and nutrition. HMOs (complex carbohydrates) are clearly found in breast milk and are thought to have some health benefits. These bioactive substances contribute to baby protection, fetal health, and cognitive development.

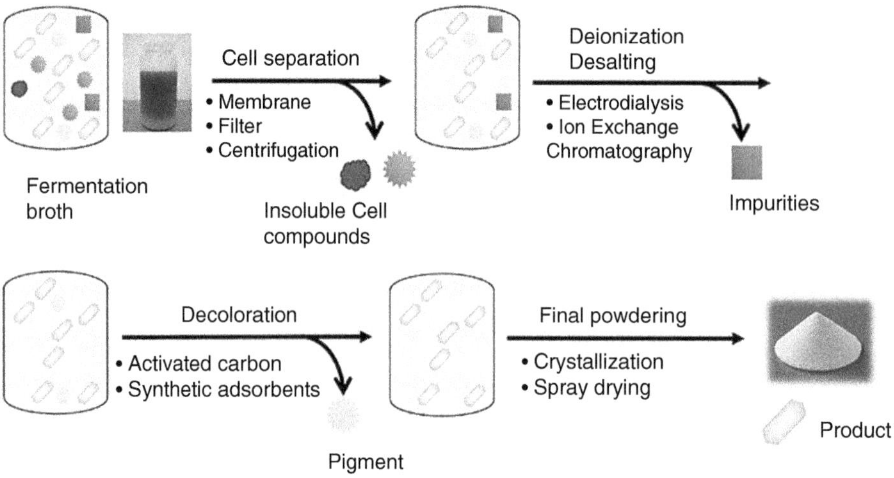

Figure 10.8 Overview of HMO purification technology. *Source:* Ujihara and Fukumoto (2023)/Glycoforum.

As our knowledge of HMOs continues to grow, there is a growing market beyond infant nutrition to include supplements, dietary supplements, and prebiotics. Increasing public interest in the health benefits of HMOs and the goal of producing infant formula similar to breast milk is a major factor in the growth of the industry. The ability of HMOs to improve the immune system and gut health is consistent with the growing need for herbal and holistic treatment plans. Other research focuses on identifying and designing specific HMOs to achieve specific health outcomes. However, questions include the problems and cost of creating HMOs, especially in reconstructing the type of HMOs found in milk. Management issues and the efficiency and effectiveness of boundary tools also affect business dynamics. As HMO research advances, the market is showing the potential not only to be the best in nutrition for children but also to extend health across the lifespan. The ability to change the context of food and health reflects the HMO market's position in shaping the future of health and nutrition (Ujihara and Fukumoto 2023; SKYQUEST 2024; Corona et al. 2021).

10.6.1 HMOs' Market Segmental Analysis

HMOs' marketplace is split into three segments: kind, software, and place. The market is divided into four segments primarily based on type: 2′-FL and 3′-FL, sialyllactose, LNT, and LNnT. The market is divided into four segments primarily based on utility: toddler formulation, useful foods and liquids, meals/scientific dietary supplements, and others (Table 10.2).

10.6.2 HMO Market Analysis by Product

In HMOs, "2′-FL and 3′-FL" work effectively in global market. These HMOs are clearly present in milk and have been extensively studied for their beneficial and anti-inflammatory properties. The importance of 2′-FL and 3′-FL stems from their presence in breast milk and their ability to support intestinal health and help improve the immune system of young children (Cheng et al. 2021). Simultaneously, the sialyllactose category is increasing rapidly and is regarded as the fastest developing section in the HMO marketplace. Sialyllactose is essential for brain improvement and cognitive feature, for this reason, it is far a topic of exceptional hobby to look at. As scientific information improves, call for sialyllactose-primarily-based merchandise in child formulation and functional food rises, leading to the segment's fast growth (Gan et al. 2023).

Table 10.2 Type of HMOs and their demand in world wide.

By type	By application	By concentration	By region
• 2′-FL • 3′-FL • 3′-SL • 6′-SL	• Infant formula • Functional food and beverages • Food supplements • Other applications include clinical nutrition and pharmaceuticals	• Neutral • Acidic	• North America • Europe • Asia pacific • South America • Row (Africa and Middle East)

10.6.3 HMOs' Market Regional Analyzes None

HMOs' market exhibited sizable geographical variances, with North America having the largest market proportion. North America's dominance is driven by its superior infant formula market, research capabilities, and growing consumer demand for HMO-fortified products. The region's emphasis on supporting good infant health and development has led to the use of HMOs in infant formula and related products. Europe is now the second largest market for HMOs after North America. A healthy baby and vitamin balance, as well as good environmental management that supports innovation in the food and beverage industry, have a positive impact on the EU economy. The usage of HMOs in infant formulation and different applications has been boosted via emergency care unit (ECU) clients' inclination for beneficial and health-enhancing products.

10.6.4 Factors Affecting the HMOs' Market

- A better understanding of the health benefits of HMOs is important. Call for HMO-fortified items has grown because of HMOs' involvement in supporting toddler immune device, cognitive development, and gastrointestinal fitness, in addition to immersive health advantage for adults.
- The use of HMOs in system products stems from patron demand for new child system that carefully replicates the fitness benefits of breast milk. HMOs have emerged as a critical element of newborn nutrition as parents attempt to offer the best nutrition feasible for their children.

10.6.5 Dairy Oligosaccharide Industry Restrictions

- HMO synthesis is a complex and difficult process. The technology and luxury required to multiply different instances of HMO in milk have an impact on the efficiency and effectiveness of HMO production.
- Clients have to apprehend the blessings of HMOs, specially for non-toddler usage. It might be tough, but boosting expertise and consciousness of HMO talents is important for the marketplace's boom.

10.6.6 Competition Landscape of the Global Human Milk Oligosaccharides' (HMOs') Market

The HMOs' market is a very dynamic market where existing companies and new entrants compete for market share due to the growing demand for HMO-fortified products. Major competitors include DuPont Vitamin & Biosciences, Nestlé SA, Glycosyn LLC, Jennewein Biotechnologie GmbH, and Elicityl due to their dedication to developing HMOs, extensive research, and building relationships in the infant formula and food industry.

10.6.7 Latest Trends in the HMO Market

- In January 2023, Danone introduced the launch of a brand new little one components providing HMOs. Aptamil Pronutra HA is designed to help toddler's immune device and digestive fitness.
- In February 2023, Nestlé introduced a brand new toddler system with HMOs. The component, Nido FortifiGrow+, is supposed to assist toddlers' highbrow development.
- In March 2023, a studies posted within the magazine "Nature" discovered that HMOs might also help defend neonates from respiratory illnesses. It was found that newborn babies fed with HMO-fortified formula had a lower incidence of respiratory diseases than babies fed with HMO-free formula.
- In April 2023, research in the journal Pediatrics found that HMOs may help improve bowel movements in children. Newborns receiving HMOs were found to have a more balanced and diverse gut microbiota than children not receiving HMOs (Bührer et al. 2022).

10.6.8 Highlights of Global HMOs' Market

- HMOs are being studied for feasible use beyond infant vitamins, as they are clearly found in breast milk. HMOs have been investigated for their use of nutritional supplements, diets, and even prescription medications to address the health concerns of individuals of all ages.
- As medical expertise of HMOs develops, more profiles are being evolved that carefully reflect unique moms' breast milk. Since each breast has a unique HMO component, this change allows babies to receive certain vitamins.

10.7 Applications of HMOs

It is believed that HMOs protect intestinal infections and promote the proliferation of bifidobacteria in the alimentary tract (Daddaoua et al. 2006). HMOs are thought to enact their participation in newborn development, besides their potential prebiotic effects, and might have anti-adhesion properties which limit adherence of infectious microorganisms to enterocytes (Lane et al. 2010). Furthermore, HMOs modulate immune processes within gut-associated lymphoid tissue (GALT) (Guarner 2009) and can transform leaky gut in premature newborn during the first month after birth in a dose-dependent manner (Taylor et al. 2009).

Initially recognized as bifidogenic component in breast milk, HMOs were largely acknowledged for their prebiotic or bifidogenic properties. Yet, recent research has revealed that HMOs have a broader role than just being a substrate for bacterial formation. According to some researchers (Ignatova-Ivanova et al. 2010), the occurrence of galactooligosaccharides in HMOs promotes increased growth and generation of biocides.

Researchers (Martín-Sosa et al. 2002) proposed HMOs to have a positive effect in reducing urinary system inflammations. On the other hand, Abrahams and Labbok (2011) stated that suckling newborns had a lower likelihood of forming otitis introduced via pneumococcus as well as the decreased possibility of human respiratory syncytial virus (hRSV) and human orthopneumovirus infection. This is attributed to the fact that human milk frequently coats the mucous membrane of uppermost portion of pharynx in newborns as well as seldom penetrates the nostrils, nasal cavity, mouth, pharynx, and larynx upper respiratory tract amid breastfeeding.

Most prevalent HMOs occurring in human milk are referred to as 2′-FL. The potential activity of 2′-FL has drawn an ample interest due to its plentifulness. 2′-FL works as a prebiotic, supports primeval formation of a healthy gut microbes dominated by *Bifidobacterium*, and protects neonates from certain diarrheal illnesses (Van den Abbeele et al. 2019). LDFT is the uttermost prevalent fucosylated HMO in human milk, with a concentration of around 0.43 g/L during the first year of breastfeeding. Immunosuppressive and immunostimulatory microbes have the ability to digest LDFT, which promotes newborn's well-balanced digestive tract microbiota. Sialylated HMOs are of relevance because they provide a dietary source of sialic acid, which helps prevent intestinal infections and fostering optimal brain function in humans (Wang and Brand-Miller 2003).

10.7.1 Functions of HMOs

HMOs serve a variety of functions, including (i) acting as prebiotics and stimulating favorable microorganism to colonize, (ii) exerting defense system against infectious agents and protecting newborns from infectious diseases, (iii) acting as ligands as well as interacting with living cells invaded by pathogens, (iv) acting as anti-inflammatories and immunosuppressants, and (v) acting as nutriments for infant brain maturation (Figure 10.9; Ray et al. 2019).

10.7.2 Involvement of HMOs as if Prebiotics

HMOs, with their complicated combination of oligosaccharides, allure both mutually beneficial dependent species which are adapted to mucus and HMOs adapted bifidobacteria to the newborn digestive tract. Various investigations occurring outside host organisms displayed bifidobacterial species to thrive on HMOs (Jantscher-Krenn and Bode 2012). This sort of commensal feeding significantly impacts immune system development and the composition of digestive tract flora in newborns (Yu et al. 2013). Bulk of HMOs are not absorbed by the intestine, and instead, they are fermented into short-chain fatty acids by some bacterial populations producing acidic conditions. This acidic condition in the intestine promotes progression of additional *bifidobacterial* strains with benefits. HMOs prevent the progression of potent infectious bacteria by promoting the flourishing of *Bifidobacterium* species (Bode 2012). Breast-fed newborns appear to have a greater quantity of beneficial bifidobacteria compared with milk powder-fed newborns (Collado et al. 2012).

The analysis of *Bifidobacterium longum* subsp. *infantis* genome suggests its compatibility with employing HMOs in the gut of neonates. The conserved gene clusters across numerous strains approve the mechanisms of genomic rearrangements for this particular type of phylotype linked with neonates. *B. longum* subsp. *infantis* proportionately absorbs type 1 and 2 HMOs, whereas most bifidobacterial species favor type 1 HMOs.

A coevolution with humans is suggested by the preponderance of HMOs with type 1 configuration as well as the preservation of the galacto-*N*-biose (GNB)/lacto-*N*-biose (LNB) pathway in bifidobacteria (Asakuma et al. 2011). Based on MS analysis, glycoprofiling of HMO usage patterns revealed that *Bacteroides vulgatus* and *B. infantis* have an exclusive inclination toward fucosylated oligosaccharides (Marcobal et al. 2010). *B. longum biovar infantis* ATCC15697 prefers small-mass HMOs with fucosidase and sialidase activities, which are mostly released in colostrum and during the first month of nursing (LoCascio et al. 2007). A riffled disciplined study of healthy term neonates revealed that formulas containing 2′FL and LNnT, two types of HMOs, can potentially swing the composition of gut microbiota to resemble that of suckling newborns (Steenhout et al. 2016). Neonates born to nonsecretor mothers experience a delay in the development of a gut microbiome rich in bifidobacteria (Lewis et al. 2015). The secretor status of infants and mothers influences the composition of the newborn's microbiota toward the age of 2–3 years (Smith-Brown et al. 2016).

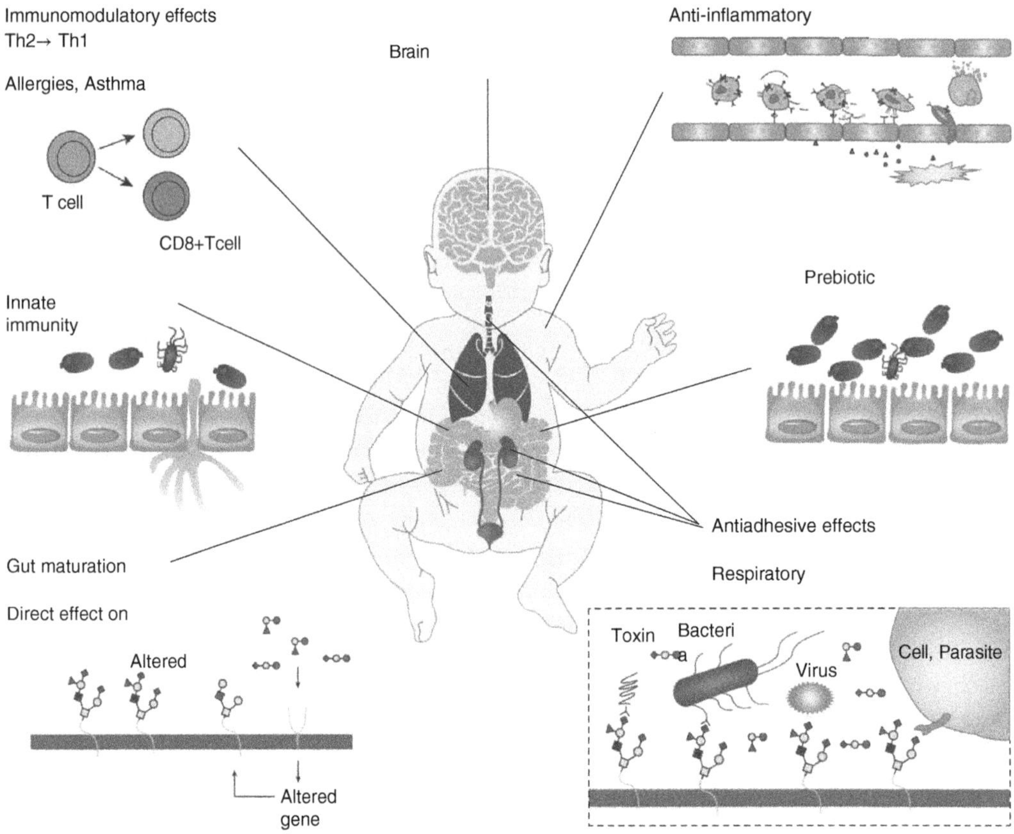

Figure 10.9 Prospective advantages for suckling infants. HMOs can act as prebiotics, immunomodulators, and signaling molecules to improve the gut immunity of infants. *Source:* Ray et al. (2019)/John Wiley & Sons/CC BY 4.0.

10.7.3 Antiadhesive and Antimicrobial Characteristics of HMOs

The innate immune mechanism relies on HMOs as its primary element to safeguard neonates from intestinal and other harmful infectious agents (Newburg et al. 2005). HMOs block the pathogen adhesions by acting as dissolvable glycan receptor entices. Additionally, HMOs function as antiadhesive and antimicrobial agent, directly minimizing microbial infections. Moreover, they indirectly control infectious agents by promoting the growth of nonpathogenic commensals, giving them competing benefits. HMOs can imitate viral receptors configurations and inhibit adhesion to target cells, thus averting infections (Morozov et al. 2018).

The sialylated fraction of HMOs effectively inhibits agglutination of red blood cells caused by uropathogenic *Escherichia coli* (UPEC) and enterotoxigenic *Escherichia coli* (ETEC) (Martín-Sosa et al. 2002). The HMO fraction of breast milk hinders the attachment of *Haemophilus influenzae* and *Streptococcus pneumoniae* to human buccal epithelial cells or pharynx (Andersson et al. 1986). Receptors on pig milk proteins, that is, sialyl Lewis X and B antigens inhibit *H. pylori* colonization (Gustafsson et al. 2006). HMOs at biological levels obstruct type-1 human immunodeficiency virus (HIV)'s envelope glycoprotein attachment to the dendritic cell-specific intercellular adhesion molecule 3-grabbing non-integrin (DC-SIGN) on human antigen-presenting cells (APC) by over 80% (Hong et al. 2009). DC-SIGN recognizes certain subsets of oligomannose- and complex-type *N*-glycans as well as fucosylated glycans in HMOs (van Liempt et al. 2006). Breast milk's Lewis X motif may adhere to DC-SIGN, blocking the uptake of HIV-1 and its subsequent transmission to CD4+ T cells (Naarding et al. 2005).

10.7.4 HMO's Impact on Intestinal Epithelial Cells

The interaction between HMOs and infectious microorganisms inhibits the microbes from adhering to the intestinal epithelial cells (Kunz et al. 2000). By modulating gene expression of intestinal epithelial cells, HMOs directly impact these cells, altering their cell surface glycans and triggering other responses. Research report has validated that in human intestinal epithelial cells cultured *in vitro*, HMOs caused differentiation and death and inhibited cell proliferation (Bode 2012).

HMOs, particularly disialyllacto-*N*-tetraose, provide defense against necrotizing enterocolitis (Jantscher-Krenn et al. 2012). HMOs have been shown to influence multiple phases of intestinal epithelial cells throughout digestive tract formation outside cells (Kuntz et al. 2008). HMOs reduced intestinal cell multiplication and transformed cell-division cycle flux via influencing relevant regulatory genes and MAP kinase signaling (Kuntz et al. 2009). Acidic HMOs help to reduce the occurrence of inflammation causing conditions like necrotizing enterocolitis in suckling neonates (Bode et al. 2004b).

10.7.5 HMO's Influence on Immune Cells

It has been shown that the antiphlogistic characteristics of HMOs reduce the development of platelet-neutrophil complexes, which decreases the expression of neutrophil β2 integrin (Bode et al. 2004b). Under changing circumstances, certain HMOs prevent leukocyte rolling as well as adherence to endothelial cells, which has antiphlogistic properties (Bode et al. 2004a). *In vitro* research has demonstrated that acidic HMOs can impact cytokine synthesis and T-cell activation derived from cord blood (Eiwegger et al. 2004). HMOs aid to build the immune system, perhaps heading toward a further steady Th1/Th2 response.

Eiwegger et al. (2010) suggested that acidic HMOs could potentially regulate postpartum allergen-specific immune responses in atopy-susceptible persons by suppressing Th2 type responses. Meanwhile, Goehring et al. (2016) found that in a riffled disciplined experiment, the healthy term newborns consuming formulae enriched with 2′-FL exhibited decreased rates of cytokine comparable to those in a suckling control group. Newborns born via caesarean delivery with an elevated genetic predisposition to allergies might possess decreased likelihood of developing IgE-associated eczema at 2 years if they are provided breast milk containing FUT2-dependent HMOs (Sprenger et al. 2017).

10.8 Conclusion and Future Outlook

Currently, engineered strains including *E. coli*, *S. cerevisiae*, *B. subtilis*, and *C. glutamicum* can produce HMOs. New food additives including 2′-FL, 3′-FL, LNnT, and LNT have significant market potential. Industrial-scale production of 2′-FL and LNnT is now allowed. Current research focuses on the unknown metabolic pattern and selection of suitable host strains, despite advancements in metabolic engineering and synthetic biology that enable HMO manufacturing. Gene modification approaches can improve the availability of GDP-fucose or CMP-Neu5Ac, enhancing the salvage or *de novo* pathway activity. While cheaper, *de novo* donor manufacturing is inefficient. Some researchers use salvage in the *de novo* synthesis route to boost metabolic rate. Although *E. coli* is commonly utilized for metabolic engineering, its inherent toxicity can hinder downstream processes. Researchers are exploring edible host strains such *S. cerevisiae*, *B. subtilis*, and *C. glutamicum* as alternatives, but production methods and yields remain limited. Alternative strains' genetic traits differ from *E. coli*, necessitating a unique technique. To employ *S. cerevisiae*, the lactose permease gene must be introduced into the strain, as it cannot absorb or metabolize lactose. *Corynebacterium glutamicum* requires Gmd, WcaG, and LacY genes for proper HMO synthesis. *E. coli* can enhance the GDP-L-fucose production pathway by inhibiting colanic acid metabolism. More research is needed to better understand the metabolic processes of various bacteria. Identifying functional genes or metabolic pathways in alternative strains, comparable to those explored in *E. coli*, can boost HMO production. Synthesis, isolation, and purification of complex HMOs remain challenging, limiting their output. Fucosylation catalysis is a significant obstacle for effective microbial production.

However, limitations such as poor yield, limited enzyme activity, and high costs for substrates like GDP-L-fucose and p-nitrophenyl-α-fucose (pNP-α-fucose). Using microbial systems to produce fucosyltransferase appears to be more successful. Designing metabolic pathways for 2′-FL generation, which is the most researched kind of HMO, can solve possible issues in producing complex structures. To manufacture complicated HMOs with high yield, 2′-FL can serve as a model product. Modifying engineered microbial pathways to improve GDP-fucose synthesis might boost production of 2′-FL significantly. Improving fucosylation efficiency is crucial for synthesizing complicated, structured HMOs with high yields. Developing fucosylase by protein engineering can enhance transfucosylation by reducing superfluous hydrolysis activity. Additionally, the catalytic effectiveness of fucosylation enzymes from diverse sources varies. As a result, the discovery of efficient transfucosylases and protein engineering technologies will be considered as complimentary techniques to boost utilization efficiency.

As we learn more about microbial metabolic processes and cell factories, we can apply more edible strains to constructed pathways. Combining molecular biology and genetic engineering technologies improves HMO production from modified strains. Meanwhile, advancements in protein engineering will enable the creation of complex HMOs. Engineered microbial methods provide significant potential for future HMO production.

References

Abrahams, S.W. and Labbok, M.H. (2011). Breastfeeding and otitis media: a review of recent evidence. *Current Allergy and Asthma Reports* 11: 508–512.

Agoston, K., Hederos, M.J., Bajza, I., and Dekany, G. (2019). Kilogram scale chemical synthesis of 2′-fucosyllactose. *Carbohydrate Research* 476: 71–77.

Ajisaka, K. (2017). Preparation of milk oligosaccharides for the investigation of their functions. *Milk Science* 66 (3): 205–217.

Andersson, B., Porras, O., Hanson, L.Å. et al. (1986). Inhibition of attachment of Streptococcus pneumoniae and Haemophilus influenzae by human milk and receptor oligosaccharides. *Journal of Infectious Diseases* 153 (2): 232–237.

Appelmelk, B.J., Negrini, R., Moran, A.P., and Kuipers, E.J. (1997). Molecular mimicry between *Helicobacter pylori* and the host. *Trends in Microbiology* 5 (2): 70–73.

Asakuma, S., Hatakeyama, E., Urashima, T. et al. (2011). Physiology of consumption of human milk oligosaccharides by infant gut-associated bifidobacteria. *Journal of Biological Chemistry* 286 (40): 34583–34592.

Ayechu-Muruzabal, V., van Stigt, A.H., Mank, M. et al. (2018). Diversity of human milk oligosaccharides and effects on early life immune development. *Frontiers in Pediatrics* 2018 (6): 239.

Bandara, M.D., Stine, K.J., and Demchenko, A.V. (2020). Chemical synthesis of human milk oligosaccharides: lacto-*N*-neohexaose (Galβ1→ 4GlcNAcβ1→) 2 3, 6Galβ1→ 4Glc. *Organic & Biomolecular Chemistry* 18 (9): 1747–1753.

Baumgärtner, F., Seitz, L., Sprenger, G.A., and Albermann, C. (2013). Construction of *Escherichia coli* strains with chromosomally integrated expression cassettes for the synthesis of 2′-fucosyllactose. *Microbial Cell Factories* 12: 1–13.

Bissaro, B., Monsan, P., Fauré, R., and O'Donohue, M.J. (2015). Glycosynthesis in a waterworld: new insight into the molecular basis of transglycosylation in retaining glycoside hydrolases. *Biochemical Journal* 467 (1): 17–35.

Bode, L. (2012). Human milk oligosaccharides: every baby needs a sugar mama. *Glycobiology* 22 (9): 1147–1162.

Bode, L., Contractor, N., Barile, D. et al. (2016). Overcoming the limited availability of human milk oligosaccharides: challenges and opportunities for research and application. *Nutrition Reviews* 74 (10): 635–644.

Bode, L., Kunz, C., Muhly-Reinholz, M. et al. (2004a). Inhibition of monocyte, lymphocyte, and neutrophil adhesion to endothelial cells by human milk oligosaccharides. *Thrombosis and Haemostasis* 92 (12): 1402–1410.

Bode, L., Rudloff, S., Kunz, C. et al. (2004b). Human milk oligosaccharides reduce platelet-neutrophil complex formation leading to a decrease in neutrophil β 2 integrin expression. *Journal of Leukocyte Biology* 76 (4): 820–826.

Boehm, G. and Stahl, B. (2007). Oligosaccharides from milk1. *The Journal of Nutrition* 137 (3): 847S–849S.

Bührer, C., Ensenauer, R., Jochum, F. et al. (2022). Infant formulas with synthetic oligosaccharides and respective marketing practices: position Statement of the German Society for Child and Adolescent Medicine eV (DGKJ), Commission for Nutrition. *Molecular and Cellular Pediatrics* 9 (1): 14.

Bych, K., Mikš, M.H., Johanson, T. et al. (2019). Production of HMOs using microbial hosts—from cell engineering to large scale production. *Current Opinion in Biotechnology* 56: 130–137.

Chen, X. (2015). Human milk oligosaccharides (HMOS): structure, function, and enzyme-catalyzed synthesis. *Advances in Carbohydrate Chemistry and Biochemistry* 72: 113–190.

Cheng, L., Akkerman, R., Kong, C. et al. (2021). More than sugar in the milk: human milk oligosaccharides as essential bioactive molecules in breast milk and current insight in beneficial effects. *Critical Reviews in Food Science and Nutrition* 61 (7): 1184–1200.

Chin, Y.W., Kim, J.Y., Lee, W.H., and Seo, J.H. (2015). Enhanced production of 2′-fucosyllactose in engineered *Escherichia coli* BL21star (DE3) by modulation of lactose metabolism and fucosyltransferase. *Journal of Biotechnology* 210: 107–115.

Chin, Y.W., Seo, N., Kim, J.H., and Seo, J.H. (2016). Metabolic engineering of *Escherichia coli* to produce 2′-fucosyllactose via salvage pathway of guanosine 5′-diphosphate (GDP)-l-fucose. *Biotechnology and Bioengineering* 113 (11): 2443–2452.

Cho, J.S., Kim, G.B., Eun, H. et al. (2022). Designing microbial cell factories for the production of chemicals. *JACS Au* 2 (8): 1781–1799.

Choi, Y.H., Kim, J.H., Park, B.S., and Kim, B.-G. (2016). Solubilization and iterative saturation mutagenesis of α1,3-fucosyltransferase from *Helicobacter pylori* to enhance its catalytic efficiency. *Biotechnology and Bioengineering* 113: 1666–1675.

Collado, M.C., Cernada, M., Baüerl, C. et al. (2012). Microbial ecology and host-microbiota interactions during early life stages. *Gut Microbes* 3 (4): 352–365.

Corona, L., Lussu, A., Bosco, A. et al. (2021). Human milk oligosaccharides: a comprehensive review towards metabolomics. *Children* 8 (9): 804.

Craft, K.M. and Townsend, S.D. (2017). Synthesis of lacto-N-tetraose. *Carbohydrate Research* 440: 43–50.

Cronan, J.E. (2014). *Escherichia coli* as an experimental organism. *eLS* https://doi.org/10.1038/npg.els.0002026.

Daddaoua, A., Puerta, V., Requena, P. et al. (2006). Goat milk oligosaccharides are anti-inflammatory in rats with hapten-induced colitis. *The Journal of Nutrition* 136 (3): 672–676.

Danby, P.M. and Withers, S.G. (2016). Advances in enzymatic glycoside synthesis. *ACS Chemical Biology* 11 (7): 1784–1794.

Das, R. and Mukhopadhyay, B. (2016). Chemical O-glycosylations: an overview. *ChemistryOpen* 5 (5): 401–433.

Dijkstra, Z. J., Jeurissen, F. J. H., Van Genuchten, M. H. M., & Huisman, J. (2022). US Patent 11,377,462. Washington, DC: U.S. Patent and Trademark Office.

Dumon, C., Priem, B., Martin, S.L. et al. (2001). In vivo fucosylation of lacto-N-neotetraose and lacto-N-neohexaose by heterologous expression of *Helicobacter pylori* α-1, 3 fucosyltransferase in engineered *Escherichia coli*. *Glycoconjugate Journal* 18: 465–474.

Eiwegger, T., Stahl, B., Haidl, P. et al. (2010). Prebiotic oligosaccharides: in vitro evidence for gastrointestinal epithelial transfer and immunomodulatory properties. *Pediatric Allergy and Immunology* 21 (8): 1179–1188.

Eiwegger, T., Stahl, B., Schmitt, J. et al. (2004). Human milk–derived oligosaccharides and plant-derived oligosaccharides stimulate cytokine production of cord blood T-cells in vitro. *Pediatric Research* 56 (4): 536–540.

Endo, T., Koizumi, S., Tabata, K., and Ozaki, A. (2000). Large-scale production of CMP-NeuAc and sialylated oligosaccharides through bacterial coupling. *Applied Microbiology and Biotechnology* 53: 257–261.

Engels, L. and Elling, L. (2014). WbgL: a novel bacterial α1, 2-fucosyltransferase for the synthesis of 2′-fucosyllactose. *Glycobiology* 24 (2): 170–178.

Faijes, M., Castejón-Vilatersana, M., Val-Cid, C., and Planas, A. (2019). Enzymatic and cell factory approaches to the production of human milk oligosaccharides. *Biotechnology Advances* 37 (5): 667–697.

Fair, R.J., Hahm, H.S., and Seeberger, P.H. (2015). Combination of automated solid-phase and enzymatic oligosaccharide synthesis provides access to α (2, 3)-sialylated glycans. *Chemical Communications* 51 (28): 6183–6185.

Gan, J., Cao, C., Stahl, B. et al. (2023). Advances and challenges for obtaining human milk oligosaccharides: extraction from natural sources and synthesis by intentional design. *Trends in Food Science & Technology* 141: 104203.

Gilbert, M., Bayer, R., Cunningham, A.M. et al. (1998). The synthesis of sialylated oligosaccharides using a CMP-Neu5Ac synthetase/sialyltransferase fusion. *Nature Biotechnology* 16 (8): 769–772.

Gilbert, M., Cunningham, A.M., Watson, D.C. et al. (1997). Characterization of a recombinant *Neisseria Meningitides* α-2, 3-sialyltransferase and its acceptor specificity. *European Journal of Biochemistry* 249 (1): 187–194.

Gilbert, M., Karwaski, M.F., Bernatchez, S. et al. (2002). The genetic bases for the variation in the lipo-oligosaccharide of the mucosal pathogen, *Campylobacter jejuni*: biosynthesis of sialylated ganglioside mimics in the core oligosaccharide. *Journal of Biological Chemistry* 277 (1): 327–337.

Goehring, K.C., Marriage, B.J., Oliver, J.S. et al. (2016). Similar to those who are breastfed, infants fed a formula containing 2′-fucosyllactose have lower inflammatory cytokines in a randomized controlled trial. *The Journal of Nutrition* 146 (12): 2559–2566.

Guarner, F. (2009). Prebiotics, probiotics and helminths: the 'natural' solution? *Digestive Diseases* 27 (3): 412–417.

Gustafsson, A., Hultberg, A., Sjöström, R. et al. (2006). Carbohydrate-dependent inhibition of *Helicobacter pylori* colonization using porcine milk. *Glycobiology* 16 (1): 1–10.

Holck, J., Larsen, D.M., Michalak, M. et al. (2014). Enzyme catalysed production of sialylated human milk oligosaccharides and galactooligosaccharides by *Trypanosoma cruzi* trans-sialidase. *New Biotechnology* 31 (2): 156–165.

Hollands, K., Baron, C.M., Gibson, K.J. et al. (2019). Engineering two species of yeast as cell factories for 2′-fucosyllactose. *Metabolic Engineering* 52: 232–242.

Hong, P., Ninonuevo, M.R., Lee, B. et al. (2009). Human milk oligosaccharides reduce HIV-1-gp120 binding to dendritic cell-specific ICAM3-grabbing non-integrin (DC-SIGN). *British Journal of Nutrition* 101 (4): 482–486.

Huang, D., Yang, K., Liu, J. et al. (2017). Metabolic engineering of *Escherichia coli* for the production of 2′-fucosyllactose and 3-fucosyllactose through modular pathway enhancement. *Metabolic Engineering* 41: 23–38.

Ignatova-Ivanova, T., Ivanov, R., Iliev, I., and Ivanova, I. (2010). Influence of galactooligosaccharides on the growth and antimicrobial activity of Lactobacillus fermentum I-5. *Biotechnology & Biotechnological Equipment* 24 (supl): 484–487.

Jamek, S.B., Muschiol, J., Holck, J. et al. (2018). Loop protein engineering for improved transglycosylation activity of a β-N-acetylhexosaminidase. *Chembiochem* 19 (17): 1858–1865.

Jantscher-Krenn, E. and Bode, L.J.M.P. (2012). Human milk oligosaccharides and their potential benefits for the breast-fed neonate. *Minerva Pediatrica* 64 (1): 83–99.

Jantscher-Krenn, E., Zherebtsov, M., Nissan, C. et al. (2012). The human milk oligosaccharide disialyllacto-N-tetraose prevents necrotising enterocolitis in neonatal rats. *Gut* 61 (10): 1417–1425.

Jennewein, S., & Helfrich, M. (2015). Process for purification of 2′-fucosyllactose using simulated moving bed chromatography. Patent EP2857410A1, filed 4 October 2013.

Jennewein, S., Hu¨fner, E., Parkot, J. (2010). Hmo synthesis. Patent (WO2010142305), filed 8 June 2009 and issued 16 December 2010.

Jennum, C.A., Fenger, T.H., Bruun, L.M., and Madsen, R. (2014). One-pot glycosylations in the synthesis of human milk oligosaccharides. *European Journal of Organic Chemistry* 2014 (15): 3232–3241.

Knerr, L. and Schmidt, R.R. (2000). Solid-phase synthesis of a branched hexasaccharide related to lacto-N-hexaose. *European Journal of Organic Chemistry* 2000 (15): 2803–2808.

Koizumi, S., Endo, T., Tabata, K., and Ozaki, A. (1998). Large-scale production of UDP-galactose and globotriose by coupling metabolically engineered bacteria. *Nature Biotechnology* 16 (9): 847–850.

Kuhn, N.J., Carrick, D.T., and Wilde, C.J. (1980). Lactose synthesis: the possibilities of regulation. *Journal of Dairy Science* 63 (2): 328–336.

Kuntz, S., Kunz, C., and Rudloff, S. (2009). Oligosaccharides from human milk induce growth arrest via G2/M by influencing growth-related cell cycle genes in intestinal epithelial cells. *British Journal of Nutrition* 101 (9): 1306–1315.

Kuntz, S., Rudloff, S., and Kunz, C. (2008). Oligosaccharides from human milk influence growth-related characteristics of intestinally transformed and non-transformed intestinal cells. *British Journal of Nutrition* 99 (3): 462–471.

Kunz, C., Rudloff, S., Baier, W. et al. (2000). Oligosaccharides in human milk: structural, functional, and metabolic aspects. *Annual Review of Nutrition* 20 (1): 699–722.

Lane, J.A., Mehra, R.K., Carrington, S.D., and Hickey, R.M. (2010). The food glycome: a source of protection against pathogen colonization in the gastrointestinal tract. *International Journal of Food Microbiology* 142 (1–2): 1–13.

Lara, A.R., Taymaz-Nikerel, H., Mashego, M.R. et al. (2009). Fast dynamic response of the fermentative metabolism of *Escherichia coli* to aerobic and anaerobic glucose pulses. *Biotechnology and Bioengineering* 104 (6): 1153–1161.

Lee, W.H., Han, N.S., Park, Y.C., and Seo, J.H. (2009). Modulation of guanosine 5′-diphosphate-D-mannose metabolism in recombinant *Escherichia coli* for production of guanosine 5′-diphosphate-L-fucose. *Bioresource Technology* 100 (24): 6143–6148.

Lee, S.Y. and Kim, H.U. (2015). Systems strategies for developing industrial microbial strains. *Nature Biotechnology* 33 (10): 1061–1072.

Lee, J.W., Kwak, S., Liu, J.J. et al. (2020). Enhanced 2′-Fucosyllactose production by engineered *Saccharomyces cerevisiae* using xylose as a co-substrate. *Metabolic Engineering* 62: 322–329.

Lee, W.H., Shin, S.Y., Kim, M.D. et al. (2012). Modulation of guanosine nucleotides biosynthetic pathways enhanced GDP-L-fucose production in recombinant *Escherichia coli*. *Applied Microbiology and Biotechnology* 93: 2327–2334.

Lewis, Z.T., Totten, S.M., Smilowitz, J.T. et al. (2015). Maternal fucosyltransferase 2 status affects the gut bifidobacterial communities of breastfed infants. *Microbiome* 3: 1–21.

Li, L., Kim, S.A., Heo, J.E. et al. (2017). One-pot synthesis of GDP-l-fucose by a four-enzyme cascade expressed in *Lactococcus lactis*. *Journal of Biotechnology* 264: 1–7.

Li, M., Li, C., Hu, M., and Zhang, T. (2022). Metabolic engineering strategies of de novo pathway for enhancing 2′-fucosyllactose synthesis in *Escherichia coli*. *Microbial Biotechnology* 15 (5): 1561–1573.

Li, A.L., Ni, W.W., Li, Y. et al. (2020). Effect of 2′-fucosyllactose supplementation on intestinal flora in mice with intestinal inflammatory diseases. *International Dairy Journal* 110: 104797.

Liu, J.J., Kwak, S., Pathanibul, P. et al. (2018). Biosynthesis of a functional human milk oligosaccharide, 2′-fucosyllactose, and l-fucose using engineered *Saccharomyces cerevisiae*. *ACS Synthetic Biology* 7 (11): 2529–2536.

Liu, J.Y., Miller, P.F., Willard, J., and Olson, E.R. (1999). Functional and biochemical characterization of *Escherichia coli* sugar efflux transporters. *Journal of Biological Chemistry* 274 (33): 22977–22984.

Liu, Y., Zhu, Y., Wan, L. et al. (2022). High-level de novo biosynthesis of 2′-fucosyllactose by metabolically engineered *Escherichia coli*. *Journal of Agricultural and Food Chemistry* 70 (29): 9017–9025.

LoCascio, R.G., Ninonuevo, M.R., Freeman, S.L. et al. (2007). Glycoprofiling of bifidobacterial consumption of human milk oligosaccharides demonstrates strain specific, preferential consumption of small chain glycans secreted in early human lactation. *Journal of Agricultural and Food Chemistry* 55 (22): 8914–8919.

Lu, M., Mosleh, I., and Abbaspourrad, A. (2021). Engineered microbial routes for human milk oligosaccharides synthesis. *ACS Synthetic Biology* 10 (5): 923–938.

Lundemo, P., Karlsson, E.N., and Adlercreutz, P. (2017). Eliminating hydrolytic activity without affecting the transglycosylation of a GH1 β-glucosidase. *Applied Microbiology and Biotechnology* 101: 1121–1131.

Marcobal, A., Barboza, M., Froehlich, J.W. et al. (2010). Consumption of human milk oligosaccharides by gut-related microbes. *Journal of Agricultural and Food Chemistry* 58 (9): 5334–5340.

Martín-Sosa, S., Martín, M.J., and Hueso, P. (2002). The sialylated fraction of milk oligosaccharides is partially responsible for binding to enterotoxigenic and uropathogenic *Escherichia coli* human strains. *The Journal of Nutrition* 132 (10): 3067–3072.

Matwiejuk, M., Chassagne, P., Osztrovszky, G., Horváth, F., & Dekany, G. 2015. Separation of 2'-O-fucosyllactose from fermentation broth. Patent WO2015/188834Al, filed 11 June 2015 and issued 17 December 2015.

McArthur, J.B., Yu, H., and Chen, X. (2019). A bacterial β1–3-galactosyltransferase enables multigram-scale synthesis of human milk lacto-N-tetraose (LNT) and its fucosides. *ACS Catalysis* 9 (12): 10721–10726.

Morozov, V., Hansman, G., Hanisch, F.G. et al. (2018). Human milk oligosaccharides as promising antivirals. *Molecular Nutrition & Food Research* 62 (6): 1700679.

Muschiol, J. and Meyer, A.S. (2019). A chemo-enzymatic approach for the synthesis of human milk oligosaccharide backbone structures. *Zeitschrift für Naturforschung. Section C* 74 (3–4): 85–89.

Naarding, M.A., Ludwig, I.S., Groot, F. et al. (2005). Lewis X component in human milk binds DC-SIGN and inhibits HIV-1 transfer to CD4+ T lymphocytes. *The Journal of Clinical Investigation* 115 (11): 3256–3264.

Newburg, D.S., Ruiz-Palacios, G.M., and Morrow, A.L. (2005). Human milk glycans protect infants against enteric pathogens. *Annual Review of Nutrition* 25: 37–58.

Oriol, R., Le Pendu, J., and Mollicone, R. (1986). Genetics of ABO, H, Lewis, X and related antigens. *Vox Sanguinis* 51 (3): 161–171.

Palcic, M.M. (2011). Glycosyltransferases as biocatalysts. *Current Opinion in Chemical Biology* 15 (2): 226–233.

Palur, D.S.K., Pressley, S.R., and Atsumi, S. (2023). Microbial production of human milk oligosaccharides. *Molecules* 28 (3): 1491.

Parschat, K., Schreiber, S., Wartenberg, D. et al. (2020). High-titer de novo biosynthesis of the predominant human milk oligosaccharide 2′-fucosyllactose from sucrose in *Escherichia coli*. *ACS Synthetic Biology* 9 (10): 2784–2796.

Pérez-Escalante, E., Alatorre-Santamaría, S., Castañeda-Ovando, A. et al. (2022). Human milk oligosaccharides as bioactive compounds in infant formula: recent advances and trends in synthetic methods. *Critical Reviews in Food Science and Nutrition* 62 (1): 181–214.

Petschacher, B. and Nidetzky, B. (2016). Biotechnological production of fucosylated human milk oligosaccharides: prokaryotic fucosyltransferases and their use in biocatalytic cascades or whole cell conversion systems. *Journal of Biotechnology* 235: 61–83.

Prieto, P.A. (2005). In vitro and clinical experiences with a human milk oligosaccharide, lacto-N-neotetraose, and fructooligosaccharides. *Foods & Food Ingredients Journal of Japan* 2010: 1018–1030.

Prudden, A.R., Liu, L., Capicciotti, C.J. et al. (2017). Synthesis of asymmetrical multiantennary human milk oligosaccharides. *Proceedings of the National Academy of Sciences* 114 (27): 6954–6959.

Puccio, G., Alliet, P., Cajozzo, C. et al. (2017). Effects of infant formula with human milk oligosaccharides on growth and morbidity: a randomized multicenter trial. *Journal of Pediatric Gastroenterology and Nutrition* 64 (4): 624–631.

Ray, C., Kerketta, J.A., Rao, S. et al. (2019). Human milk oligosaccharides: the journey ahead. *International Journal of Pediatrics* 2019: 2390240.

Robinson, R.C., Colet, E., Tian, T. et al. (2018). An improved method for the purification of milk oligosaccharides by graphitised carbon-solid phase extraction. *International Dairy Journal* 80: 62–68.

Rugbjerg, P., Myling-Petersen, N., Porse, A. et al. (2018a). Diverse genetic error modes constrain large-scale bio-based production. *Nature Communications* 9 (1): 787.

Rugbjerg, P., Sarup-Lytzen, K., Nagy, M., and Sommer, M.O.A. (2018b). Synthetic addiction extends the productive life time of engineered *Escherichia coli* populations. *Proceedings of the National Academy of Sciences* 115 (10): 2347–2352.

Samain, E., Peltier-Pain, P., Bych, K., Johanson, T., Champion, E., Dekany, G. (2015). Oligosaccharide production. Patent (US20170175154), filed 12 July 2021 and issued 17 May 2023.

Saumonneau, A., Champion, E., Peltier-Pain, P. et al. (2016). Design of an α-L-transfucosidase for the synthesis of fucosylated HMOs. *Glycobiology* 26 (3): 261–269.

Schmölzer, K., Weingarten, M., Baldenius, K., and Nidetzky, B. (2019). Lacto-N-tetraose synthesis by wild-type and glycosynthase variants of the β-N-hexosaminidase from *Bifidobacterium bifidum. Organic & Biomolecular Chemistry* 17 (23): 5661–5665.

Seo, J., Chin, Y., & Jo, H. (2017). Method for producing 2′-fucosyllactose by using *Corynebacterium glutamicum.* Patent (WO2017188684), filed 24 April 2017 and issued 2 November 2017.

Seydametova, E., Yu, J., Shin, J. et al. (2019). Search for bacterial α1, 2-fucosyltransferases for whole-cell biosynthesis of 2′-fucosyllactose in recombinant *Escherichia coli. Microbiological Research* 222: 35–42.

SKYQUEST (2024). Human milk oligosaccharides (HMO) market. https://www.skyquestt.com/report/human-milk-oligosaccharides-market#:~:text=Global%20Human%20Milk%20Oligosaccharides%20(HMO)%20Market%20size%20was%20valued%20at,period%20(2024%2D2031) (Accessed 2024).

Smith-Brown, P., Morrison, M., Krause, L., and Davies, P.S. (2016). Mothers secretor status affects development of childrens microbiota composition and function: a pilot study. *PLoS One* 11 (9): e0161211.

Soini, J., Ukkonen, K., and Neubauer, P. (2008). High cell density media for *Escherichia coli* are generally designed for aerobic cultivations–consequences for large-scale bioprocesses and shake flask cultures. *Microbial Cell Factories* 7: 1–11.

Sprenger, N., Odenwald, H., Kukkonen, A.K. et al. (2017). FUT2-dependent breast milk oligosaccharides and allergy at 2 and 5 years of age in infants with high hereditary allergy risk. *European Journal of Nutrition* 56: 1293–1301.

Stahl, B., Thurl, S., Zeng, J.R. et al. (1994). Oligosaccharides from human milk as revealed by matrix-assisted laser desorption/ionization mass spectrometry. *Analytical Biochemistry* 223 (2): 218–226.

Steenhout, P., Sperisen, P., Martin, F.P. et al. (2016). Term infant formula supplemented with human milk oligosaccharides (2′ Fucosyllactose and lacto-N-neotetraose) shifts stool microbiota and metabolic signatures closer to that of breastfed infants. *The FASEB Journal* 30: 275–277.

Takamura, T., Chiba, T., Ishihara, H., and Tejima, S. (1980). Chemical modification of lactose. XIII. Synthesis of lacto-N-tetraose. *Chemical and Pharmaceutical Bulletin* 28 (6): 1804–1809.

Takors, R. (2012). Scale-up of microbial processes:Impacts, tools and open questions. *Journal of Biotechnology* 160 (1–2): 3–9.

Taylor, S.N., Basile, L.A., Ebeling, M., and Wagner, C.L. (2009). Intestinal permeability in preterm infants by feeding type: mother's milk versus formula. *Breastfeeding Medicine* 4 (1): 11–15.

Thurl, S., Henker, J., Siegel, M. et al. (1997). Detection of four human milk groups with respect to Lewis blood group dependent oligosaccharides. *Glycoconjugate Journal* 14: 795–799.

Thurl, S., Munzert, M., Boehm, G. et al. (2017). Systematic review of the concentrations of oligosaccharides in human milk. *Nutrition Reviews* 75 (11): 920–933.

Ujihara, T. and Fukumoto, K. (2023). History and prospects of methods for industrially producing human milk oligosaccharides. *Glycoforum* 26: A18.

Van den Abbeele, P., Duysburgh, C., Vazquez, E. et al. (2019). 2′-Fucosyllactose alters the composition and activity of gut microbiota from formula-fed infants receiving complementary feeding in a validated intestinal model. *Journal of Functional Foods* 61: 103484.

van Liempt, E., Bank, C.M., Mehta, P. et al. (2006). Specificity of DC-SIGN for mannose-and fucose-containing glycans. *FEBS Letters* 580 (26): 6123–6131.

Varki, A., Cummings, R.D., Aebi, M. et al. (2015). Symbol nomenclature for graphical representations of glycans. *Glycobiology* 25 (12): 1323–1324.

Wakarchuk, W., Martin, A., Jennings, M.P. et al. (1996). Functional relationships of the genetic locus encoding the glycosyltransferase enzymes involved in expression of the lacto-N-neotetraose terminal lipopolysaccharide structure in *Neisseria meningitidis*. *Journal of Biological Chemistry* 271 (32): 19166–19173.

Walsh, C., Lane, J.A., van Sinderen, D., and Hickey, R.M. (2020). From lab bench to formulated ingredient: characterization, production, and commercialization of human milk oligosaccharides. *Journal of Functional Foods* 72: 104052.

Wan, L., Zhu, Y., Li, W. et al. (2020). Combinatorial modular pathway engineering for guanosine 5′-diphosphate-l-fucose production in recombinant *Escherichia coli*. *Journal of Agricultural and Food Chemistry* 68 (20): 5668–5675.

Wang, B. and Brand-Miller, J. (2003). The role and potential of sialic acid in human nutrition. *European Journal of Clinical Nutrition* 57 (11): 1351–1369.

Wang, Y., Hu, J., Sun, M. et al. (2022). Efficient purification of 2′-fucosyllactose by membrane filtration and activated carbon adsorption. *Fermentation* 8 (11): 655.

Wang, G., Rasko, D.A., Sherburne, R., and Taylor, D.E. (1999). Molecular genetic basis for the variable expression of Lewis Y antigen in *Helicobacter pylori*: analysis of the α (1, 2) fucosyltransferase gene. *Molecular Microbiology* 31 (4): 1265–1274.

Woo, J.E., Seong, H.J., Lee, S.Y., and Jang, Y.S. (2019). Metabolic engineering of *Escherichia coli* for the production of hyaluronic acid from glucose and galactose. *Frontiers in Bioengineering and Biotechnology* 7: 351.

Xiao, Z., Guo, Y., Liu, Y. et al. (2016). Chemoenzymatic synthesis of a library of human milk oligosaccharides. *The Journal of Organic Chemistry* 81 (14): 5851–5865.

Xu, M., Meng, X., Zhang, W. et al. (2021). Improved production of 2′-fucosyllactose in engineered *Saccharomyces cerevisiae* expressing a putative α-1, 2-fucosyltransferase from Bacillus cereus. *Microbial Cell Factories* 20: 1–13.

Yu, Z.T., Chen, C., and Newburg, D.S. (2013). Utilization of major fucosylated and sialylated human milk oligosaccharides by isolated human gut microbes. *Glycobiology* 23 (11): 1281–1292.

Yu, S., Liu, J.J., Yun, E.J. et al. (2018). Production of a human milk oligosaccharide 2′-fucosyllactose by metabolically engineered *Saccharomyces cerevisiae*. *Microbial Cell Factories* 17 (1): 1–10.

Yu, H., Yan, X., Autran, C.A. et al. (2017). Enzymatic and chemoenzymatic syntheses of disialyl glycans and their necrotizing enterocolitis preventing effects. *The Journal of Organic Chemistry* 82 (24): 13152–13160.

Zeuner, B., Teze, D., Muschiol, J., and Meyer, A.S. (2019). Synthesis of human milk oligosaccharides: protein engineering strategies for improved enzymatic transglycosylation. *Molecules* 24 (11): 2033.

Zhai, Y., Han, D., Pan, Y. et al. (2015). Enhancing GDP-fucose production in recombinant *Escherichia coli* by metabolic pathway engineering. *Enzyme and Microbial Technology* 69: 38–45.

Zhang, L., Lau, K., Cheng, J. et al. (2010). Helicobacter hepaticus Hh0072 gene encodes a novel α1-3-fucosyltransferase belonging to CAZy GT11 family. *Glycobiology* 20 (9): 1077–1088.

Zheng, J., Xu, H., Fang, J., and Zhang, X. (2022). Enzymatic and chemoenzymatic synthesis of human milk oligosaccharides and derivatives. *Carbohydrate Polymers* 291: 119564.

11

Beta (β)-glucan as Microbial Polymer with Nutraceutical Potential: Chemistry, Biosynthesis, Extraction, Identification, and Industrial Production of Bioactive Compound for Human Health

Pawan Prabhakar[1,3], Deepak Kumar Verma[2], and Mamoni Banerjee[1]

[1] *Bio-Research Laboratory, Rajendra Mishra School of Engineering Entrepreneurship, Indian Institute of Technology Kharagpur, Kharagpur, West Bengal, 721302, India*
[2] *Agricultural and Food Engineering Department, Indian Institute of Technology Kharagpur, Kharagpur, West Bengal, 721302, India*
[3] *School of Business, Woxsen University, Hyderabad, Telangana, 502345, India*

11.1 Introduction

Nutraceuticals are physiologically active components that combine nutrition and medication to support optimal health and provide benefits. These items have a crucial impact on human healthcare and its long-term sustainability, particularly for the advancement of future therapies (Jacobo-Velázquez et al. 2023). Food items are nutritious substances that are consumed, ingested, or otherwise ingested to support life, supply energy, and facilitate growth. The demand for foods with health-promoting features has significantly grown due to the rising prevalence of obesity, high blood pressure, coronary heart disease, and other chronic illnesses (Adefegha 2018; Makkar et al. 2020; Bisht et al. 2024). In a few decades, researchers across the world have reported a high correlation between a consumer's purchasing choice and the health benefits promoted in product advertisements. Therefore, food manufacturers throughout the globe are actively seeking ways to include components that have beneficial impacts on health in their goods. Nutraceuticals have gained acknowledgment for their nutritional advantages, as well as their medicinal properties and safety record (Prabhakar and Banerjee 2020; Jadhav et al. 2023; Upadhyay and Singh 2024). The terms "nutraceuticals," "food supplements," and "dietary supplements" emerged following the conceptualization by Dr De Felice in "Nutrition" and "Pharmaceutical' in 1989. However, the word has been excessively utilized in marketing efforts within the food sector and pharmacy for supplements. Additionally, some countries have established regulatory definitions for this term. While a unified definition is still lacking, in the majority of nations, the term often pertains to nutritional supplements (Bruno et al. 2022; Vishvakarma et al. 2023).

Nevertheless, several authors make distinctions between the categories of nutraceuticals, functional foods, and dietary supplements. Functional foods are meals that are prepared or cooked using scientific methodologies to maximize their nutritional advantages. These foods play a crucial part in maintaining good health. The regulatory authorities across the world do not provide a clear distinction between food supplements and nutraceuticals (Alvarez-Mon et al. 2021; Puri et al. 2022; Saeed and Zafar 2024). Nutraceuticals are seeing global growth in the realm of services such as healthcare

Microbial Nutraceuticals: Products and Processes, First Edition. Edited by Sudhir Pratap Singh and Santosh Kumar Upadhyay.

promotion and disease reduction. People worldwide have been utilizing nutraceuticals derived from different bacteria, fungi, and microalgae for centuries, either with awareness or without realizing it. Specific strains of bacteria have historically been ingested as probiotics, while other types of large fungi have been utilized as traditional remedies for many years (Alali et al. 2021; Bhagea et al. 2022).

Nutraceuticals often lack patent protection and are sold in the market, while pharmaceutical products owe under strict patents and copyrights to manufacture and sell in the market. Both pharmaceutical and nutraceutical substances can be used for the treatment or prevention of illnesses; however, only pharmaceutical compounds are officially approved by the government (Prabhakar and Mamoni 2021; Chandra et al. 2022; Nori et al. 2023; Sharma et al. 2024). These products may encompass vitamins, minerals, botanical extracts, functional foods, and specifically bioactive chemicals. Nutraceuticals have the objective of enhancing health, preventing chronic illnesses, slowing down the aging process, and promoting the structure and function of the body. These chemicals help restore normal digestion, improve food uptake, facilitate cellular detoxification, and inhibit harmful biological processes (Dhanjal et al. 2020; Alali et al. 2021). Furthermore, they promote the growth of good gut microbiota and aid in the evacuation of waste from the body. In addition to their structural and biochemical improvements, nutraceuticals also have notable anticancer, anti-aging, and antioxidant characteristics. They enhance the regulation of the immune system, strengthen the immunological response, enhance the process of engulfing and destroying foreign particles, reduce excessive reactions of the immune system, and suppress the immune system's attack on its own cells (Rowland et al. 2018; Wang et al. 2021).

The adaptability of nutraceuticals shows potential in the prevention and treatment of contemporary health issues such as diabetes, hypertension, and cancer, as well as disorders associated with oxidative stress, neurodegeneration, and cardiovascular disease. Therefore, nutraceuticals have emerged as a viable option for meeting the nutritional needs of the human body. Significantly, nutraceuticals exhibit antibacterial characteristics that successfully prolong the freshness of many food items (Garza-Juárez et al. 2023; Dama et al. 2024). Polysaccharides, also known as glycans, are the most abundant biopolymers found in nature. Bacteria and fungi are present in both plants and animals, playing many crucial functions in supporting the survival of living species (Fontana and Widmalm 2023). They function as reservoirs and repositories of energy and constitute the structural tissue of plants and certain animals (such as crustaceans and insects). Polysaccharides are water-loving and prone to hydrolysis when exposed to acidic water, breaking down into their individual monosaccharide components (Mahmoud et al. 2021; Lefter et al. 2024). The health-promoting activities of microbial polysaccharides have led to their recognition as a valuable source for nutraceuticals. Microbial polymeric compounds, such as polysaccharides, exopolysaccharides (EPS), polyhydroxyalkanoates (PHAs), and lipopeptides, are becoming increasingly popular as potential nutraceuticals because of their varied structures and extensive health advantages (Ahuja et al. 2023; Mouro et al. 2024). Among them, β-glucan is such polysaccharide which is recognized for its ability to modify the immune system, act as antioxidants, and promote the growth of beneficial bacteria in the gut. β-glucan is complex carbohydrates composed of individual D-glucose molecules that are connected by β-glycosidic linkages. It is present in several natural sources including yeast, mushrooms, bacteria, algae, barley, and oat. Based on their origin, β-glucan can be categorized as either originating from cereal or non-cereal sources. Oats and barley are cereal sources of β-glucan, whereas non-cereal sources can include mushrooms, algae, bacteria, and seaweed (Zhu et al. 2016; Jayachandran et al. 2018). They are part of a specific group of dietary fiber and serve as a vital part of the cell wall in certain harmful bacteria (*Aspergillus fumigatus, Pneumocystis carinii, Histoplasma capsulatum, Cryptococcus neoformans, Candida albicans*) and fungi belonging to the Basidiomycetes class and baker's yeast (*Saccharomyces cerevisiae*) (Kaur et al. 2019; Philippini et al. 2019). β-glucan derived from the cell wall of *S. cerevisiae* has demonstrated an in vitro binding capability to zearalenone. The complex was shown to have both hydrogen bonds and van der Waals interactions (Rawling et al. 2023). The tea polyphenols complexes with oat β-glucan demonstrated significantly greater levels of superoxide dismutase and glutathione peroxidase activity in the liver compared to the physical combination. Further research demonstrated the use of β-glucan microparticles with a high level of porosity, which was encapsulated with an antigen, for the purpose of delivering drugs (Veverka et al. 2014). β-glucan has several appealing uses, serving as both a dietary additive and a preventative treatment, and functions as a homopolymer for drug administration. The distinctive characteristics of β-glucan certainly stem from their intrinsic and very pronounced ability to form helical structures, a trait not found in other polysaccharides (Bai et al. 2019; Han et al. 2020). Its rheological qualities, such as gel formation, emulsification, and thickening, make it suitable for the development of unique nutraceutical food products, in addition to its health benefits (Chen et al. 2022; Lante et al. 2023). Therefore, β-glucan possesses the capacity to alter the functional characteristics of food items, including sensory attributes, rheology, texture, and viscosity. The biological activity of β-glucan varies significantly because of changes in sources, growing conditions, extraction, and purification procedures, notwithstanding their immense potential (Sofi et al. 2017; Kaur et al. 2019). β-glucan may be extracted from many sources such as yeast, microalgae, bacteria, and mycelium, which are non-cereal sources rich in microbial content. Although it has been demonstrated to have health-promoting effects, its potential as a nutraceutical element is not being fully used. Frequently, it is just added to food goods for its technological advantages, such as its ability to thicken liquids. Furthermore, the wide acceptance of polymeric nutraceuticals is impeded by limitations in their accessibility and difficulties in extracting them from these microorganisms. Therefore, this chapter offers a thorough review of β-glucan including its chemistry, biosynthesis, nutraceutical and medicinal values, and various biotechnological processing for high production β-glucan using microorganisms in controlled conditions together with the challenges that remain in the large-scale production and regulatory approval with future perspectives.

11.2 Classification, Chemistry, and Biosynthesis of β-glucan

β-glucan is a kind of non-starch homopolysaccharide made up of D-glucose molecules that exist naturally. β-glucan can be classified into two types, either cereal-derived or non-cereal-derived, based on their origin. Oat, rice, and barley are cereal sources of β-glucan, whereas mushrooms, bacteria, algae, and seaweed are non-cereal sources and play a crucial role in the composition of the cell wall and function as a source of energy storage in many organisms such as algae, plants (specifically oat and barley), bacteria (primarily probiotic bacteria), and fungi (including yeast and edible mushrooms) (Zhu et al. 2016; Jayachandran et al. 2018; Yi et al. 2021). β-glucans have varying characteristics, including solubility, branching, molecular mass, and shape, depending on their origin. These polysaccharides consist of long- or short-chain linear polymers made up of glucose subunits that are connected by β-(1,3) or β-(1,4) bonds, with additional β-(1,6) branches (Utama et al. 2020; Yi et al. 2021) as shown in Figure 11.1(a), (b). The biochemical activity of β-glucan is influenced by factors such as the positioning of glycosidic linkages, the extent and length of branching, conformation, and solubility (Han et al. 2020). The β-glucan structure contains different types of glycosidic bonds and forms the walls of fungi and yeast, consisting of β-D-glucopyranose molecules connected by β-1,3- and β-1,6-glycosidic bonds, which generates a branched structure. The β-glucan derived

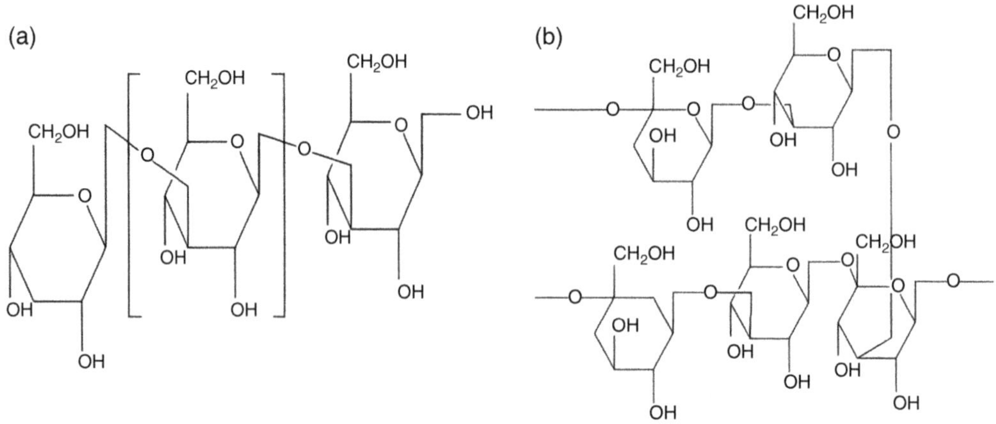

Figure 11.1 Branched structure of (a) bacterial β-glucan (linked with β-(1,3)-glycosidic bonds) and (b) yeast β-glucan (linked with β-(1,3)- and β-(1,6)-glycosidic bonds).

from yeast (*S. cerevisiae*) mostly has branching β-(1,3)-linkage bonds and often has higher molecular weight, while the bacterial β-glucan from *Agrobacterium sp.* is mostly composed of linear β-(1,3)-glycosidic linkages (Ciecierska et al. 2019). The other isomer is found in unaltered grain products and consists of unbranched chains made up of β-D-glucopyranose monomers linked by β-1,3- and β-1,4-glycosidic bonds (Schmidt 2022).

11.2.1 Biosynthesis of β-glucan in Bacteria

The production of β-glucan in bacteria is an intricate process that requires the involvement of several enzymes and regulatory systems. β-Glucan is a complex carbohydrate made up of glucose molecules connected by β-glycosidic linkages. They have significant functions in the cell walls of bacteria, contributing to the structure, integrity, and protection of the cells. The biosynthesis begins with the formation of the precursor uridine diphosphate (UDP)-glucose, which is synthesized from glucose-1-phosphate and uridine triphosphate (UTP) by the enzyme UDP-glucose pyrophosphorylase. The β-glucan synthase complex, which includes enzymes like glucosyltransferases, initiates and elongates the β-glucan chain. The glucose units are transferred from UDP glucose to the growing polysaccharide chain. The enzyme β-glucan synthase polymerizes the glucose monomers into long β-glucan chains. This enzyme complex is often membrane-associated, facilitating the translocation of the growing polysaccharide chain across the bacterial membrane (Bulmer et al. 2021; Ubiparip et al. 2021; Singh and Bhardwaj 2023). Some β-glucans are branched, and enzymes like branching glycosyltransferases can introduce side chains. Other enzymes may modify the β-glucan, affecting its properties and function. Genes encoding enzymes involved in β-glucan biosynthesis can be regulated at the transcriptional level by various factors responsive to environmental and cellular signals. Enzymes involved in β-glucan synthesis can be regulated by post-translational modifications, such as phosphorylation, which can affect their activity and stability (Singh et al. 2020; Ubiparip et al. 2021). In addition, β-glucan, functioning as a hydrocolloid, can modify the rheological and textural characteristics of many food items. The molecular structure of β-glucan in oats, including factors such as the DP3: DP4 ratio, number of cellulose oligomers, ratio of β-(1→3) or β-(1→6) connections, and molecular mass, sets it apart from another cereal β-glucan (Lante et al. 2023).

11.2.2 Biosynthesis of β-glucan in Fungi

The biosynthesis of β-glucan in fungi is a vital process for the formation of the fungal cell wall, which provides structural integrity and protection. β-Glucans in fungi are primarily composed of glucose monomers linked by β-1,3-glycosidic bonds, with some β-1,6 branches (Figure 11.1b). The yeast *S. cerevisiae* is a model organism for better understanding β-glucan biosynthesis. Its cell wall contains significant amounts of β-1,3-glucan with β-1,6 branches (Yoshimi et al. 2017; Ruiz-Herrera and Ortiz-Castellanos 2019). Like bacteria, the biosynthesis of fungal β-glucan starts with the formation of UDP glucose, synthesized from glucose-1-phosphate and UTP by UDP-glucose pyrophosphorylase. The β-glucan synthase complex, primarily composed of the catalytic subunit Fks1p and the regulatory subunit Rho1p, initiates and elongates the β-glucan chain. The enzyme complex transfers glucose units from UDP glucose to the growing β-glucan polymer. The Fks1p enzyme is responsible for polymerizing the glucose monomers into β-1,3-glucan chains. This process occurs at the plasma membrane, where the newly synthesized glucan is extruded into the extracellular space, forming part of the cell wall. β-1,3-glucan chains can be branched by β-1,6-glucan. The enzymes involved in branching include β-1,6-glucan synthases, and these branches contribute to the structural complexity and function of the fungal cell wall (Chhetri et al. 2020; Wagener et al. 2020; Liu et al. 2024). The regulation of β-glucan biosynthesis in fungi is critical for maintaining cell wall integrity and responding to environmental changes. Enzymes involved in β-glucan synthesis may undergo post-translational modifications, such as phosphorylation or glycosylation, affecting their activity and stability. The cell wall integrity (CWI) signaling pathway, involving proteins like Rho1p, Pkc1p, and MAP kinase cascade components, plays a significant role in regulating β-glucan synthesis in response to stress (Papaspyridi et al. 2018; Ibe and Munro 2021).

11.2.3 Biosynthesis of β-glucan in Microalgae

Microalgae are autotrophic organisms that inhabit marine, freshwater, and soil habitats. They synthesize organic compounds by photosynthesis. Their remarkable metabolic flexibility, ability to adapt to different culture conditions, and quick development make them a subject of increasing interest for investigations on their potential as a source of physiologically useful compounds (Siddik et al. 2024). Microalgae live in aquatic environments, including lakes, rivers, and oceans. They may also be farmed using many types of agricultural, home, and industrial effluents. The size of microalgae ranges significantly, spanning from a few mm to 100 mm. They possess a wide-ranging capacity to adjust to many habitats, including variations in temperature, pH levels, salinity, and light conditions in terms of intensity, wavelength, and availability (Dineshbabu et al. 2019; Nagarajan et al. 2021). Microalgae are crucial components of planetary ecosystems, contributing to more than 50% of the total yearly primary output worldwide. Microalgae provide several benefits compared to macroalgae and plant species due to their superior photosynthetic efficiency, rapid growth rate, increased levels of important chemicals, and reduced need for culture space or arable lands (Prasad et al. 2021; Ampofo and Abbey 2022). β-Glucans are often found in mushrooms, yeasts, and bacteria, but rarely occur in microalgae, have a broad range of uses, and possess significant therapeutic and commercial promise. While β-glucans such as paramylon from *Euglena gracilis*, an algae-like organism, have been extensively studied, there is less knowledge on the β-glucan content of microalgae. Only a few species of cultivated microalgae have been examined to determine the composition of their biomass, specifically in terms of carbohydrates like β-glucan (Reitan et al. 2021; Murphy et al. 2023). Polysaccharides in microalgae are produced in the chloroplast and cytosol by the Calvin cycle. The activated glucose units are subsequently moved to create β-glucan, which is then followed by the formation of condensed chrysolaminarin. During the Calvin cycle, carbon dioxide is assimilated by the enzyme ribulose-1,5-biphosphate carboxylase oxygenase (Rubisco). Rubisco undergoes carboxylation to produce 3-phosphoglycerate, which is then used in the synthesis of carbohydrates. Polysaccharide production and sulfation occur in the Golgi apparatus. Gibberellic acid (GA) is produced by the arrangement of stacked cisternae, also known as dictyosomes, and membrane tetrads. These structures are recognized as multienzymatic complexes (Costa et al. 2021; Yi et al. 2021). Polysaccharide production in microalgae can be affected by several variables, such as the kind of species and strains, as well as the circumstances in which they are cultivated, their physiological state, and the age of the culture. The process of synthesizing and storing carbohydrates in diatoms has been well understood (Babich et al. 2022; Moreira et al. 2022). Like many other species, glucan production includes the condensation of nucleoside diphosphate sugars. UDP glucose, the active precursor of chrysolaminarin, is synthesized from glucose-6 phosphate by the action of phosphoglucomutase (PGM) and UDP-sugar pyrophosphorylase (UPP). These reactions are expected to occur in the cytosol. UDP glucose is added to the linear b-1,3 polymer by an enzyme called b-1,3 glucan synthase (BGS). The chrysolaminarin synthase of *Phaeodactylum tricornutum* was shown to be present in the tonoplast. This indicates that the enzyme utilizes cytosolic uridine diphosphate glucose (UDP-Glc) for the synthesis of glucan and facilitates the movement of the resulting polysaccharide into the vacuole. In addition, the b-1,3-glucan undergoes branching by the action of tonoplast-located 1,6-b-transglycosylases (TGS), which add a glucose unit to the C-6 site of the polymer (Figure 11.2). This process results in the formation of chrysolaminarin (de Jesus Raposo et al. 2014; Yang et al. 2018; Russo et al. 2023). These factors can result in significant variations in the composition of the polysaccharides. Polysaccharides in microalgal cells exist primarily as storage and structural polymers, with their composition and structure varying depending on the monosaccharides involved. Storage polysaccharides (SPSs) serve as sources of energy for metabolic activities, whereas cell-wall-related polysaccharides (CWPSs) have functions such as providing mechanical support, facilitating biorecognition, and enabling ion-exchange capacity (Bernaerts et al. 2018; Ullah et al. 2019; Yi et al. 2021).

11.3 Extraction, Isolation, and Identification of β-glucan from Microbial Source

Extraction is one of the crucial steps for obtaining the phytocompounds or metabolites from plants and microbes using suitable solvent and extraction techniques (Chávez-González et al. 2020; Prabhakar et al. 2023a). The formation of β-glucan results in a significant rise in the thickness of the liquid medium due to the accumulation of biopolymers and the proliferation of microorganisms. An important factor that impacts the technical feasibility of β-glucan synthesis is the possible disruptions caused by the release of β-glucan during the extraction process (Thomas et al. 2022). Like other phytochemical extraction, the extraction of β-glucan is critically dependent on several elements, such as the choice of solvent, extraction method, duration and temperature of extraction, specific microbial species selected for extraction, and the β-glucan content present in that species. The selection of solvent and appropriate extraction method is crucial for achieving a high yield of compound (Maheshwari et al. 2017; Prabhakar et al. 2023b). β-Glucan extraction is done by conventional procedures such as acidic extraction, alkaline extraction, accelerated solvent extraction, and hot water extraction. Non-conventional or modern techniques are occasionally referred to as green technologies since they align with criteria of the Environmental Protection

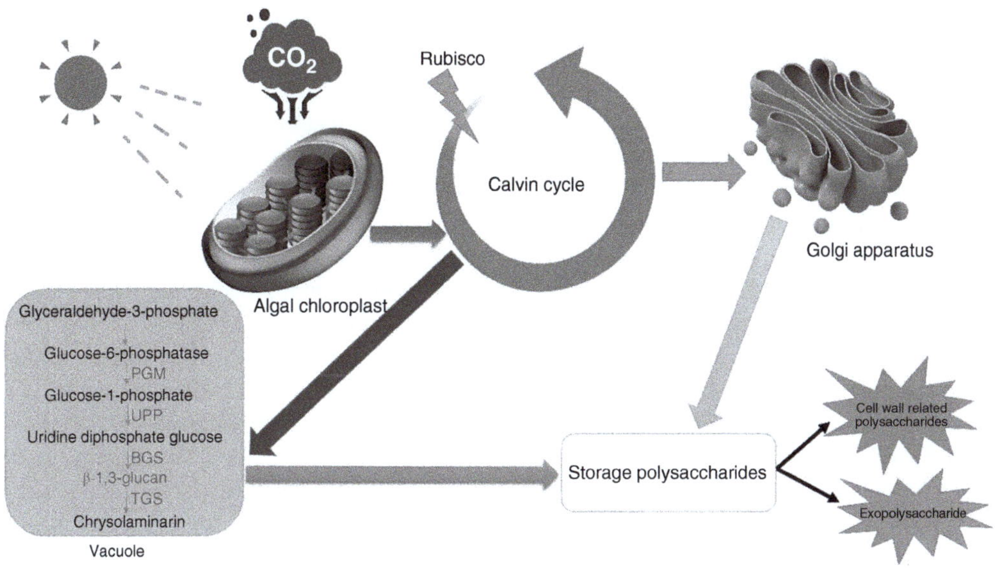

Figure 11.2 A schematic view of biosynthesis of polysaccharides in microalgae: Glucose units that have been activated in the cytosol are combined to form a polymer and then moved into the vacuole. In the vacuole, the β-1,3-glucan is altered by adding saccharide branches, resulting in the formation of chrysolaminarin, which is used as storage polysaccharides (cell-wall-related polysaccharides and exopolysaccharides). Rubisco: Ribulose-1,5-biphosphate carboxylase oxygenase; PGM: phosphoglucomutase; UPP: UDP-sugar pyrophosphorylase; BGS: 1,3-β-glucan synthase; TGS: 1,6-b-transglycosylase.

Agency, United States (Du et al. 2014; Climova et al. 2021; Prabhakar et al. 2024a). Several non-conventional extraction techniques are available, such as ultrasound-assisted extraction (UAE), microwave-assisted extraction (MAE), pressurized liquid extraction (PLE), pulsed electric field (PEF) extraction, enzyme-assisted extraction (EAE), pressurized low polarity water extraction, supercritical fluid extraction (SFE), and molecular distillation (Verma and Thakur 2021; Gupta et al. 2024; Prabhakar et al. 2024b). There are four methods for isolating β-glucan: alkaline extraction, acidic extraction, enzymatic extraction, and water extraction. Some other sophisticated techniques encompass ultrasound-assisted extraction, microwave-assisted extraction, and superheated water extraction (Singla et al. 2024). Prior to extraction, the pre-treatment process is necessary to prepare the sample that may help to preserve and even improve the β-glucan profile of the microbial sample. Major pre-treatment processes include drying, centrifugation, homogenization, deproteinization, starch removal by enzymes, and filtration to achieve higher β-glucan content (Zhu et al. 2016; Kumar et al. 2022). The extraction technique significantly impacts the structure and molecular weight of β-glucan. Furthermore, the isolation/purification of β-glucan is performed using various separation techniques like column chromatography equipped with specific column for β-glucan such as affinity chromatography on cellulose column, anion-exchange chromatography, cellulose column chromatography, coagulate and filtration, and isoelectric precipitation (Benito-Román et al. 2014; Shi 2016; Fazio et al. 2020). The identification and characterization of β-glucan require the use of sophisticated analytical techniques to ascertain the chemical structure and characteristics of chemicals generated from microorganisms. Methods like thin-layer chromatography provide a first means of identifying compounds such as β-glucan. Fourier transform infrared spectroscopy (FTIR) is a technique used to identify functional groups and verify the presence of β-glucan. Other important identification techniques, including high-performance liquid chromatography (HPLC), liquid chromatography-mass spectrometry (LC-MS), and nuclear magnetic resonance (NMR), are used to measure the β-glucan concentration, determine its structure, and evaluate its purity (Upadhyay et al. 2017; Chioru and Chirsanova 2023) (Figure 11.3). The variations in estimations of β-glucan molecular weight may arise from changes in the techniques employed for extraction and purification, as well as from processes of aggregation and depolymerization that occur during the extraction process (Singla et al. 2024). The optimal extraction technique is contingent upon the origins and compositions of β-glucan. The most prevalent method for extracting β-glucan was found to be hot water extraction, while other approaches are alkali extraction, ultrasound-assisted extraction, enzyme-assisted extraction, and microwave-assisted extraction. Alkaline extraction is a commonly employed procedure across numerous industries for extracting compounds from plant or animal sources (Maheshwari et al. 2017; Liu et al. 2023).

The procedure includes using an alkaline solution, such as sodium hydroxide (NaOH) or potassium hydroxide (KOH), to disintegrate the cell walls and dissolve the required constituents at an elevated temperature to dissolve β-glucan. This method is preferable for the extraction of β-glucan from cereals (Yu et al. 2021). The most prevalent and convenient technique for extracting fungal polysaccharides that are soluble in water is the process of extraction in hot or boiling water (Zhu et al. 2016). Sen et al. (2013) obtained β-glucan from the fruiting bodies of a hybrid fungus, pfls1h of *Pleurotus florida* and *Lentinus squarrosulus* (Mont.) Singer, by using distilled water as the extraction solvent. The unrefined polysaccharide was separated using Sepharose 6B gel permeation chromatography, with water as the eluent, at a flow rate of 0.5 mL/min, which obtained two fractions of pure polysaccharide (Sen et al. 2013). Bhanja et al. (2014) obtained and separated two glucans that are not soluble in water from the fruiting bodies of the Ramaria botrytis mushroom. The structure of the linear (1/3)-a-D-glucan and the nonlinear b-(1/3)-D-glucan, which is branched at O-6, was determined using chemical and NMR research. Lentinan was obtained and purified from the fruiting body of *Lentinus edodes* using a process including ethanol precipitation, acetic acid solubilization, and diethylaminoethyl cellulose column chromatography. In another experiment, Schulze et al. (2016a,b) used ultrasonic-assisted extraction to produce β-glucan from two microalgae, *Scenedesmus ovalternus* SAG 52.80 and *Porphyridium purpureum* SAG 1380-1d, which were determined to be the most efficient producers of β-glucan, with a yield exceeding 20% of their dry weight, when cultivated under standard circumstances.

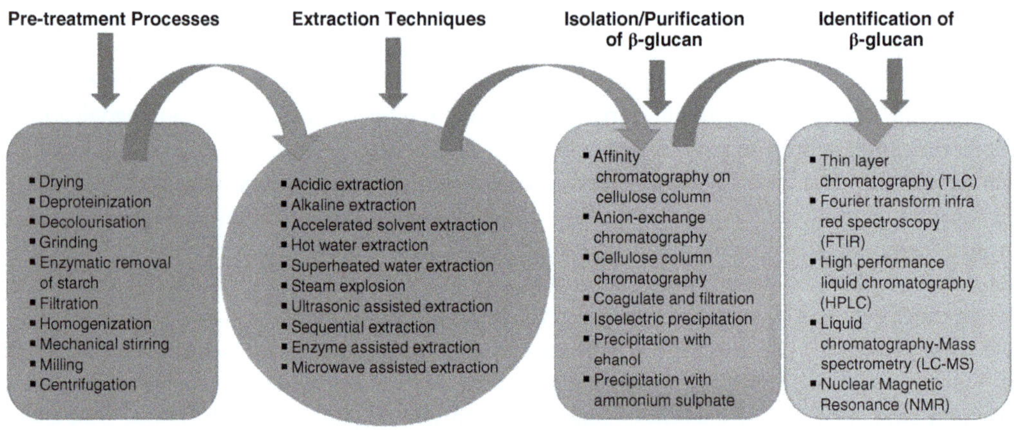

Figure 11.3 Sequential step in extraction, isolation, and identification of β-glucan from microbial sources.

11.4 Biotechnological Process for the Production of β-glucan from Microbes

The β-glucan consists of a core made up of interconnected glucose units, with a 1,3-β-linkage being important to their function. Nevertheless, there are discrepancies in the branching pattern along the main structure, as certain molecules exhibit branching at different locations, while others lack any branching altogether (Caseiro et al. 2022). β-Glucan can be classified as either cereal or non-cereal, exhibiting distinct branching topologies, lengths, and variances based on their origin. β-Glucans generated from cereal are commonly branched at the 1,4 position, but β-glucans from fungal and yeast sources tend to have branching at the 1,6 position (Choi and Kim 2023). Biotechnological processing provides enhanced control and repeatability in the manufacture of β-glucan obtained from microorganisms. β-glucans are obtained from several plentiful sources, such as microorganisms including bacteria, yeast, microalgae, and fungal mycelium (Zhu et al. 2016). Biotechnological processing allows for precise regulation and consistency of several environmental factors, such as growth conditions, pH levels, additives, foam production, and aeration. Depending on the specific system, all of these factors can be automated. The circumstances can be intentionally altered to exert control over production. The biotechnological production of β-glucan from microbes, including both bacteria and fungi, involves several critical steps: strain selection, fermentation, downstream processing, purification, and quality control. The microbial strains with high β-glucan production capabilities are selected for this purpose. The most common bacteria include *Agrobacterium* and *Lactobacillus*, while common fungi include *S. cerevisiae,* Aspergillus, and Candida species and are preferred for biotechnological production of β-glucan (Philippini et al. 2019; Murphy et al. 2023).

11.4.1 Bacteria

β-Glucans are produced via secondary metabolites in the bacteria *Bacillus natto and Xanthomonas campestris*, rendering an increase in population or biomass ineffectual for enhancing yield (Utama et al. 2020). Crognale et al. (2007) generated β-glucan from *Botryosphaeria rhodina* DABAC-P82 by investigating the presence and location of glucan-hydrolytic enzymes, the rheology of the growth medium, and the transport of oxygen. The researchers determined that the highest concentration of β-glucan in the bioreactor was linked to the levels of nitrogen and dissolved oxygen. Stack et al. (2010) reported that the formation of β-glucan from *Lactobacillus paracasei* NFBC 338 was examined by studying the activity of the membrane-associated glycosyltransferase enzyme, which is represented by the gtf gene. The findings suggested that the formation of a β-glucan exopolysaccharide by strains intended for use as probiotics might enhance their effectiveness in cultivation, processing, and ingestion. Therefore, the expression of the gtf gene might potentially be a simple method to enhance strains that may otherwise be susceptible in these applications. Sato et al. (2012) produced a water-soluble β-glucan with low molecular weight from *Aureobasidium pullulans* 1A1 using alkaline extraction and ethanol precipitation. The bacteria produce glucose via synthesis, which serves as a secondary metabolite in the generation of β-glucan molecules. The pH level also has a crucial impact on the growth of a large number of cells, with the most optimum production of β-glucan happening within a pH range of 5.5–7.0. Kalyanasundaram et al. (2012) conducted a study on the synthesis of curdlan, a type of β-glucan by *Agrobacterium* sp. ATCC 31750 in both shake flask cultures and bioreactors. The researchers employed a two-stage culture approach to enhance the production of curdlan in the mutant strain by optimizing the pH value.

The findings revealed that when the pH level increased, the cells exhibited a higher consumption of sucrose and a greater production of curdlan. Microorganisms have adaptation periods when they are exposed to new environmental conditions, which can cause a delay in β-glucan synthesis, particularly if the conditions are unfavorable. A comparative research on the adaptation stages of bacteria and yeast revealed that *X. campestris* and yeast both exhibited a lag phase lasting 72 hours; however, *B. natto* had a shorter lag phase of only 24 hours. The lag phase is influenced by several factors such as the initial cell count, the composition of the culture media, the growth conditions, the temperature, the duration of incubation, the pH level, and the substrate concentration (Aquinas et al. 2022; Murphy et al. 2023). The various biological techniques and reaction conditions for producing β-glucan from bacterial species are given in Table 11.1.

11.4.2 Fungi

In order to meet the commercial need for β-glucan, it may be synthesized by microbial fermentation using bacteria like *Lasiodiplodia theobromae, B. rhodina, and Ganoderma lucidum* (Mykhalevych et al. 2022). Kim et al. (2013) developed mutant strains of *Sparassis crispa* that produce high levels of β-glucan. Through additional culture optimization, the productivity of β-glucan in these mutant strains was further enhanced. In another work, Park et al. (2014) increased the amount of β-glucan in the development of cauliflower mushroom (*Sparassis latifolia*) by adding three different enzymes (chitinase, β-glucuronidase, and lysing enzyme complex) as elicitors to the sawdust-based medium. Various enzymatic treatments led to a significant upsurge in the concentration of β-glucan, with an increase of up to 31% compared to the control. The findings suggest that the use of biotic elicitors can increase the concentration of β-glucan in cauliflower mushrooms by stimulating the fungal cell wall. The most prevalent and practical technique for extracting fungal polysaccharides that are soluble in water is by extraction in hot or boiling water. Bhanja et al. (2014) obtained and separated two glucans that are not soluble in water from the fruiting bodies of the *Ramaria botrytis* mushroom. The structure of the linear (1,3)-β-D-glucan and the nonlinear β-(1,3)-D-glucan, which is branched at O-6, was determined using chemical and NMR research. Lentinan was obtained and purified from the fruiting body of Lentinus edodes using a process including ethanol precipitation, acetic acid solubilization, and diethylaminoethyl cellulose column chromatography. Yeast bacto glycerol (YBG) is widely regarded as the ideal growth medium for yeast cell growth. Glucose is a vital component necessary for growth and the formation of β-glucan, as yeast cells utilize glucose as a building block in their cell walls. Zhou et al. (2019) modified the system in *S. cerevisiae* to increase the accumulation of β-glucan by introducing a bacterial 1-6 β-glucan synthase and overexpressing phosphoglucomutase and RHO1. Due to this, there was a 43% rise in the β-glucan concentration. Studies have demonstrated that media can influence both the quantity and effectiveness of β-glucan. Aguilar-Uscanga and Francois (2003) examined the impact of growing circumstances on the composition of the cell wall and the concentration of β-glucan in yeast cells. The researchers took into account many aspects like the carbon supply, temperature, pH, and aeration. They also compared the culture of shaking flasks to that of controlled batch reactors. The results indicated that the carbon source had a significant effect on the synthesis of β-glucan. Notable differences were seen between shaking flask cultures and batch fermenters. The fluctuation in pH throughout the growth phase resulted in a significant reduction of up to 40% in β-glucan concentrations, emphasizing the crucial need of maintaining a stable pH throughout the fermentation process. Jaehrig et al. (2008) evaluated several media, such as wort, yeast peptone, and glucose, and found variations in the β-glucan concentration and bioactive qualities, including antioxidant activities. The biotechnological processes applied on different species of fungi for higher production of β-glucan with their experimental conditions are shown in Table 11.1.

Table 11.1 Experimental conditions (temperature, pH, time, rpm) to facilitate biotechnological and industrial production of β-glucan from microbes.

Organisms	Species	Growth medium/extraction process	Experimental conditions	β-glucan yield	References
Agrobacterium sp. CFR 24 Strain	Bacteria	Coconut water	6.0, 30°C, 150 rpm	11.50 g/L	Shivakumar and Vijayendra (2006)
Pleurotus ostreatus	Fungi	Yeast extract—5 g/L, glucose—40 g/L	30°C, pH 6.0, 120 rpm	17.12 mg/L	Gern et al. (2008)
P. ostreatus	Fungi	Glucose and wheat extracts infused corn steep liquor extract of 20 g/L and 40 g/L, respectively	30°C, pH 4.5, 120 rpm	Total polysaccharide content = 17.01 ± 4.15 mg/L	Gern et al. (2008)
Agrobacterium sp.	Bacteria	Sucrose—20 g/L, peptone—5 g/L, and yeast extract—5 g/L	30°C, pH 7.0, bioreactor capacity—7.5 L, aeration rate—1.0 vvm, agitation speed—700 rpm	66 g/L	Kalyanasundaram et al. (2012)
Ganoderma lucidum	Fungi	Extraction with NaOH (dil.) using Sephadex G-15	50°C, 12 h	18%	Kao et al. (2012)
Agaricus bisporus	Fungi	Extraction using ultrasonic waves, separation by adding ethanol, and separation by centrifugation	70°C, 62 min, 230 W	6.02%	Tian et al. (2012)
Rhizopus oryzae	Fungi	Potato dextrose broth and yeast extract (2 g/L)	25°C, 4 d, speed—125 rpm	6.99%	Sutter et al. (2017)
Aspergillus oryzae	Fungi	Potato dextrose broth and yeast extract (2 g/L)	25°C, 4 d, speed—125 rpm	18.41%	Sutter et al. (2017)
Saccharomyces cerevisiae	Yeast	Yeast extract—glucose broth containing 15 g/L glucose, 5.2 g/L K_2HPO_4, 3.18 g/L KH_2PO_4, 0.12 g/L $MgSO_4$, 0.5 g/L, and 0.54 g/L NH_4Cl	30°C, 48 h, 200 rpm	41.69%	Pengkumsri et al. (2016)
G. lucidum	Fungi	Potato dextrose broth	25° C, 7 d	13.26%	Park and kim (2018)
Euglena gracilis var. saccharophila	Microalgae	Euglena liquid medium (EM) (1 g sodium acetate, 1 g peptone, 2 g tryptone, 2 g yeast extract, and 0.01 g $CaCl_2{\cdot}2H_2O$), glucose paramylon media consists of 7.7 g glucose, 10 g yeast extract, 0.2 g $CaCO_3$, 0.5 g $MgSO_4$, 0.4 g $(NH_4)_2HPO_4$, 0.2 g KH_2PO_4, 1.81 g NH_4Cl, 2 mL of trace mineral stock A, and 1 mL of trace mineral stock B. Trace mineral stock A was prepared by dissolving 2.2 gm of $ZnSO_4{\cdot}7H_2O$, 2 gm of $MnSO_4{\cdot}4\ H_2O$, 0.5 gm $Na_2MoO_4{\cdot}2\ H_2O$, and 0.04 gm $CoCl_2{\cdot}6\ H_2O$ in 3 mL HCl (conc.). The resulting solution was then diluted with double distilled water to a final volume of 50 mL	12 h (dark), 12 h (light), 1500 lumens	8.1 g/L diurnal cycle, 4.9 g/L	Sun et al. (2018)

(Continued)

Table 11.1 (Continued)

Organisms	Species	Growth medium/extraction process	Experimental conditions	β-glucan yield	References
Euglena gracilis Z strain	Microalgae	Euglena liquid medium (EM) (1 g sodium acetate, 1 g peptone, 2 g tryptone, 2 g yeast extract, and 0.01 g $CaCl_2 \cdot 2H_2O$), glucose paramylon media consists of 7.7 g glucose, 10 g yeast extract, 0.2 g $CaCO_3$, 0.5 g $MgSO_4$, 0.4 g $(NH_4)_2HPO_4$, 0.2 g KH_2PO_4, 1.81 g NH_4Cl, 2 mL of trace mineral stock A, and 1 mL of trace mineral stock B. Trace mineral stock A was prepared by dissolving 2.2 gm of $ZnSO_4 \bullet 7H_2O$, 2 gm of $MnSO_4 \bullet 4\ H_2O$, 0.5 gm $Na_2MoO_4 \bullet 2\ H_2O$, and 0.04 gm $CoCl_2 \bullet 6\ H_2O$ in 3 mL HCl (conc.). The resulting solution was then diluted with double distilled water to a final volume of 50 mL	12 h (dark), 12 h (light), 1500 lumens	8.1 g/L diurnal cycle, 4.9 g/L	Sun et al. (2018)
Levilactobacillus brevis	Bacteria	10 g of peptone, 5 g of yeast extract, 5 g of meat extract, 4 g of K_2HPO_4, 2.6 g of $KH_2PO_4 \cdot 3\ H_2O$, 3 g of NH_4Cl, 1 g of Tween 80, 0.5 g of cysteine-HCl, 10 g of maltose, 5 g of glucose, 5 g of fructose, 0.2 g of $MgSO_4 \cdot 7\ H_2O$, and 0.038 g of $MnSO_4 \cdot H_2O$, wheat and rye doughs	35°C, 24 h	51.8 g/kg	Bockwoldt et al. (2021)
Pediococcus claussenii	Bacteria	10 g of peptone, 5 g of yeast extract, 5 g of meat extract, 4 g of K_2HPO_4, 2.6 g of $KH_2PO_4 \bullet 3\ H_2O$, 3 g of NH_4Cl, 1 g of Tween 80, 0.5 g of cysteine-HCl, 10 g of maltose, 5 g of glucose, 5 g of fructose, 0.2 g of $MgSO_4 \bullet 7\ H_2O$, and 0.038 g of $MnSO_4 \bullet H2O$, wheat and rye doughs	25°C, 24 h	47.5 g/kg	Bockwoldt et al. (2021)
S. cerevisiae	Yeast	Potato dextrose agar	30°C, 120 h, 200 rpm	3.95 mg	Utama et al. (2021)
Bacillus natto	Bacteria	Nutrient agar, yeast extract glucose (15 g/L glucose, 5.2 g/L K_2HPO_4, 3.18 g/L KH2PO4, 0.12 g/L $MgSO_4$, 0.5 g/L yeast extract, and 0.54 g/L NH_4Cl)	30°C, 120 h, 200 rpm	1.345 mg	Utama et al. (2021)
Xanthomonas campestris	Fungi	Nutrient agar, yeast extract glucose (15 g/L glucose, 5.2 g/L K_2HPO_4, 3.18 g/L KH_2PO_4, 0.12 g/L $MgSO_4$, 0.5 g/L yeast extract, and 0.54 g/L NH_4Cl)	30°C, 120 h, 200 rpm	0.785 mg	Utama et al. (2021)
Rhizopus oligosporus	Fungi	Potato dextrose agar	35°C, 35 h	0.25% w/w	Rizal et al. (2021)
Phellinus linteus	Fungi	Yeast malt extract glucose	pH 4–5, 40–50 rpm	14.16% w/w	Shin et al. (2021)

11.4.3 Microalgae

Microalgae are primarily organisms that can produce their own food through photosynthesis. However, some microalgae may also be grown under settings where they rely on external sources of carbon, either in the absence of light (heterotrophic conditions) or in the presence of both light and external carbon sources (Khan et al. 2018a). Several research reports have identified that the microalgae are regarded as a rich source of microbial β-glucan. *E. gracilis* is a widely studied microalgal species known for its production of β-glucan. *E. gracilis* is a single-celled organism that can do photosynthesis. It belongs to the group of microalgae and can store a polysaccharide called paramylon, which is a type of 1-3 β-glucan, as a reserve for energy (Gissibl et al. 2019; Bedard et al. 2024). Paramylon is a high-molecular-weight polymer composed of linear, unbranched 1-3 β-glucan. The molecular weight is predicted to be between the range of 100–500 kilodaltons (kDa). *E. gracilis* can store significant quantities of paramylon, which can make up to 90% of its overall cell mass. The production of paramylon was compared in two strains of *Euglena, E. gracilis Z* and *E. gracilis var saccharophila*. It was observed that both strains produced higher amounts of paramylon when grown in a dark environment. The authors also reported that the variation in paramylon levels among species could be attributed to their light sensitivity or a light-induced reaction (Xie et al. 2022). Further studies examined the use of intricate components in the growth media to support the large-scale growing of *E. gracilis* using heterotrophic means, with the aim of producing paramylon. The study investigated other by-products of the sugar industry, including corn steep, beef extract, and yeast extract. The levels of inorganic nitrogen, phosphorus, and plant growth hormone were also compared. The greatest levels of paramylon were observed when the media contained 20 g/L of glucose and 30 g/L of maize steep solid. The study also showed that the addition of beef extract was directly associated with increased biomass concentrations (Ivušić and Šantek 2015; Wu et al. 2021). There are multiple abiotic parameters that can be enhanced to maximize β-glucan synthesis in microalgae. The primary elements that commonly influence the quantity of polysaccharides in microalgal biomass include famine and/or nutrition restriction, saline stress, light intensity, CO_2 concentration, temperature, and metabolic kinds (Naik et al. 2024).

11.5 Pharmacological and Nutritional Properties of β-glucan

A diet abundant in β-glucan possesses a beneficial impact on human health by mitigating the risk of chronic non-communicable ailments, including diabetes, hypercholesterolemia, obesity, cardiovascular illnesses, and cancer (Singh and Bhardwaj 2023). Food security, nutrition, and other elements of lifestyle are crucial aspects that significantly impact the overall health of a society. An important aspect of food safety is incorporating natural plant- and animal-based substances known as nutraceuticals, which have a significant impact on avoiding chronic non-communicable illnesses and promoting overall well-being (Dwivedi et al. 2017). The consumption of β-glucan and dietary fiber, which has been scientifically demonstrated to have a significant influence on several health issues, plays a crucial role in improving health and avoiding illnesses.

11.5.1 Anticancerous

Despite medical advancements, cancer, characterized by unregulated cell proliferation, is currently categorized as a major chronic non-communicable ailment of the twenty-first century. It is the primary cause of mortality following cardiovascular illnesses. β-Glucan is an effective phytocompound used for the prevention and treatment of several types of cancer, including pulmonary, ovarian, breast, and prostate carcinoma (Geller et al. 2019; Steimbach et al. 2021). Colorectal cancer, among the most prevalent malignant tumors in the gastrointestinal tract, is characterized by a complex progression including many stages. Multiple studies have shown the important involvement of matrix metalloproteinases and tissue inhibitors of metalloproteinases in the progression of colorectal cancer. Colorectal cancer cells possess the capacity to produce metalloproteinases, such as matrix metalloproteinase 9, an enzyme that breaks down collagen in the basement membrane and extracellular matrix. Matrix metalloproteinase 9 is important for promoting invasive growth, distant metastasis, and the formation of new blood vessels (tumor angiogenesis) (Quintero-Fabián et al. 2019; Liu et al. 2023). β-glucan has an effect on cancer cells through indirect impact caused by the immunomodulatory qualities that influence the functioning of the immune system. Indirect action refers to the process of triggering defense mechanisms, namely the activation of granulocytes (neutrophils, eosinophils), monocytes, macrophages, and natural killer cells (NK cells). β-glucan boosts the immune system by activating macrophages and cytokines, which play a role in the body's defense mechanisms (Chan et al. 2009; Han et al. 2020). Louie et al. (2010) found that combining 200 μg/mL of β-glucan with 10,000 IU/mL of IFN-α2b resulted in a 75% reduction in the growth of pancreatic cancer cell T24 in vitro. This combination yielded more favorable outcomes compared to using IFN-α2b alone. Currently, monoclonal antibody therapy that specifically targets tumor tissue has emerged as a novel approach in cancer treatment. Studies have shown that combining monoclonal antibodies with β-glucan has demonstrated more benefits in terms of tumor regression in the short-term and long-term survival, compared to using monoclonal antibodies alone. Macrophages are elements of the immune system with the distinctive capability to eliminate cancer cells by phagocytosis, making them the initial line of defense for the immune system. β-glucan exhibits immunostimulatory effects by binding to specific membrane receptors on immune cells such as macrophages, neutrophils, monocytes, NK cells, and dendritic cells. This binding influences the immune response of these cells, including the production of cytokines and the initiation of an oxygen burst (Zhong et al. 2023).

Lentinan, a variant of β-glucan, has been extensively exploited in Japan for the treatment of stomach and colorectal cancer. Lentinan, a β-glucan derived from the shiitake mushroom (*Lentinula edodes*), has gained recognition for its notable anti-cancer capabilities (Ina et al. 2013). Lentinan stimulates the activation of macrophages and dendritic cells, hence improving their capacity to display antigens to T lymphocytes. It enhances the cytotoxicity of NK cells, which play a critical role in specifically targeting and eliminating cancer cells. Lentinan stimulates the growth and function of T cells, namely cytotoxic T lymphocytes (CTLs), which have the ability to directly target and destroy cancer cells. It enhances the production of many cytokines, such as interleukin-1 (IL-1), interleukin-2 (IL-2), interferon gamma (IFN-γ), and tumor necrosis factor alpha (TNF-α). These cytokines have crucial functions in regulating the immune response and producing anti-tumor actions. Research done on cancer patients has demonstrated that β-glucan treatment effectively reduces tumor growth, halts infiltration and metastasis, and thereby enhances life expectancy. Due to its ability to modulate the immune system and its unique interactions with the immune system, it is considered a very promising option for cancer treatment (Vetvicka et al. 2021; Zhou et al. 2024).

11.5.2 Antihyperglycemic Effect

The primary cause of the positive impact on glucose metabolism is mostly ascribed to β-glucan obtained from cereal grains and microbial sources. Upon consumption, β-glucan undergoes no digestion in the gastrointestinal tract. Additionally, it possesses a significant ability to bind water and create viscous gels inside the gastrointestinal system. This leads to a delay in the emptying of the stomach, so slowing down the enzymatic degradation of starch and impeding the absorption of easily digested carbs in the intestines. This process leads to a decrease in the amount of glucose in the blood after eating, as well as the release of insulin, which may help reduce the occurrence of type 2 diabetes mellitus (Mejía et al. 2020; Murphy et al. 2023; Prabhakar et al. 2023c). The impact of β-glucan on the regulation of blood sugar levels may differ based on factors such as the amount consumed, duration of consumption, physical and chemical properties, processing techniques, and the way the foods containing β-glucan are consumed (El Khoury et al. 2012). β-Glucans have many pathways that explain their ability to reduce glucose and insulin levels. One of these processes is the capacity of β-glucan to create a thick solution that reduces the speed at which the stomach empties and prolongs the time it takes for food to pass through the intestines. This leads to a decrease in the breakdown and absorption of glucose. Many researchers have demonstrated that the intake of β-glucan may lead to an activated PI3K/Akt activity, which in turn decreases hepatic glucose production (HGP) and glycogenolysis, while simultaneously promoting glycogen synthesis and the synthesis of fatty acids for storage and later use by other organs (Huang et al. 2018; Chakraborty and Rajeswari 2022). This activity tends to decrease in individuals with diabetes. The increase in activity is facilitated by the activation of several receptors. The β-glucan stimulates several receptors such as dectin-1, CR3, lactosylceramide, scavenger receptors, and toll-like receptors. These receptors are crucial for the recognition of β-glucan polymers with (1→–6) branching on a (1→–3) chain, particularly. β-Glucans have many pathways that explain their ability to reduce glucose and insulin levels (Ariyoshi et al. 2021; Mata-Martínez et al. 2022).

11.5.3 Antihypercholesterolemic and Anti-obesity

Though improving lipid metabolism in the body, the β-glucan lowers blood cholesterol as it can function as dietary fiber and can raise the viscosity of food in the small intestine, and they have a hypocholesterolemic effect. Increased intestinal lumen viscosity increases the excretion of fat, cholesterol, and other nutrients in the feces by decreasing the absorption and binding of bile acids (Ciecierska et al. 2019). Lowering the quantity of bile acids causes the body to use more stored cholesterol for the liver's bile acid synthesis. This effect then influences the production of micelles as well as their shape and composition. β-Glucans have multiple pathways for decreasing cholesterol; therefore, including them in the diet is important for both treating and preventing cardiovascular disease (CVD) (Sima et al. 2018). On the mucosal surface of the gut, β-glucans are known to produce a pulp-like gel. This gel composition promotes bile salt production in the liver while impeding bile salt absorption. Conversely, raising bile salts causes the body to employ the cholesterol that is already in circulation and reduces blood cholesterol levels (Thandapilly et al. 2018). A healthy, balanced diet and lifestyle are the cornerstones of treating obesity, which is a condition that needs to be addressed due to its significant risk for chronic illnesses and detrimental impacts on quality of life. Dyslipidemia and increased adipose tissue are characteristics of obesity. Thus, several persistent disorders, including diabetes, metabolic syndrome, high cholesterol levels, and high blood pressure, can be brought on by these conditions and are major risk factors for cardiovascular diseases (Fruh 2017). β-glucans have been reported to decrease appetite and lose weight in humans. One well-known aspect of the metabolic syndrome is central obesity. Finding and recommending foods that naturally lower calorie intake by promoting satiety and causing satiation are possible ways to combat the obesity epidemic that is now sweeping the country. There is evidence that dietary fiber affects body weight, food intake, and satiety, although the results have not always been reliable. Numerous randomized controlled trials have demonstrated that eating a high-fiber diet or taking dietary fiber supplements can help reduce weight (Murphy et al. 2020; Mathews et al. 2023).

11.5.4 Antioxidant Activity

Oxidative stress is a significant contributor to the aging process and is closely linked to the development of various health conditions such as atherosclerosis, cardiovascular diseases, brain diseases, diabetes, and cancer. By reducing the presence of reactive oxygen species, the likelihood of these diseases occurring can be minimized. Oats and barley contain β-glucans that have a strong ability to eliminate reactive oxygen species (Bhattacharyya et al. 2014; Tan et al. 2018; Prabhakar and Banerjee 2020). β-glucan possesses antioxidant properties since it scavenges radicals to stop oxidative damage caused by free radicals. Strong antioxidant activity made glucan a potentially useful antioxidant for use in food and medicine (Khan et al. 2018b). Prior research has demonstrated that β-glucan possesses the capacity to scavenge hydroxyl radicals, which are responsible for the development of numerous diseases and the aging process. A study conducted on rats showed that orally administered β-glucan exhibited antioxidant properties by preventing oxidative stress in vital internal organs, including the liver and kidneys. The experiment revealed that β-glucan had antioxidant characteristics and effectively counteracted the detrimental effects of free radicals generated within the body (Ciecierska et al. 2019; Mirończuk-Chodakowska et al. 2021).

11.5.5 Immunomodulatory Activities

β-glucans have been identified as the most potent immunostimulants for combating viral illnesses and cancer. The immunological effectiveness of β-glucan differs depending on factors such as molecular mass, solution conformation, backbone structure, degree of branching, and the specific cell type being targeted (Zhong et al. 2023). In recent years, the significant contribution of 1,3 β-glucan derived from yeast, fungi, mushrooms, and seaweed as biological immunomodulators has been extensively demonstrated. Studies conducted in vitro, on animals, and on humans have demonstrated that 1,3 β-glucans have the ability to improve the activity and effectiveness of immune cells, hence promoting both humoral and cellular immunities. β-Glucans have been shown in laboratory tests to improve the functioning of macrophages and stimulate the antibacterial activity of mononuclear cells and neutrophils (Vetvicka and Oliveira 2014). β-glucan has been shown to enhance the immune response against microorganisms, including viruses, bacteria, parasites, and fungal pathogens, in live creatures. Multiple investigations have demonstrated that β-glucan exerts a stimulatory and regulatory influence on the human immune system, as documented in extensive relevant literature. β-glucan enhances the functional capabilities of macrophages, NK cells, and neutrophils, hence impacting the immune system (Ali et al. 2015; Bashir and Choi 2017). Moreover, β-glucan enhances the production of both cytokines and inflammatory mediators in the immune system. Furthermore, β-glucan can attach to macrophages and white blood cells through the binding system, so stimulating their activation. Moreover, the research indicates that β-glucan enhances phagocytosis, lysosomal enzyme activity, interleukin (IL)-1 production and speeds up macrophage studies. Zymosan, derived from the yeast cell wall and rich in β-glucan, stimulates the immune system of the body and thus shows immunomodulatory effect (Ali et al. 2015; Murphy et al. 2020; Ching et al. 2021).

11.5.6 Antimicrobial Activity

Vetvicka (2011) studied that the β-glucan provides protection against infections caused by both bacteria and protozoa. Studies on mice have shown that when given β-glucan, mice exhibit heightened resistance to bacterial illnesses such as *Staphylococcus aureus* and parasitic diseases such as

Leishmania braziliensis. In another study, Geller and Yan (2020) found that β-glucan possesses antiviral properties that can protect against SARS-CoV-2 infections and reduce the severity of clinical symptoms. Although there is presently no conclusive remedy for COVID-19, there are approved immunization alternatives available. Nevertheless, the pursuit of an optimal therapeutic approach devoid of any negative consequences encounters significant obstacles. Oral β-glucan is thought to have the capacity to be used as an adjuvant for the COVID-19 β-WIFE vaccine. Furthermore, there have been reports indicating that β-glucan exhibits antibacterial capabilities, particularly against illnesses caused by bacteria that are resistant to conventional antibiotics. β-glucan exhibits a defensive impact against pathogens like *Escherichia coli., Plasmodium berghei, S. aureus, C. albicans*, and *Streptococcus suis* (Şengül and Ufuk 2022).

11.6 Future Prospective β-glucan as Microbial Nutraceuticals

Exploring the contribution of microorganisms in the synthesis of nutraceuticals is a fascinating and dynamic area of study with abundant potential for future advancements. Nutraceuticals are compounds that possess both nutritional and medicinal characteristics, and they have a crucial function in enhancing health and averting illnesses. Microorganisms, including bacteria, fungi, and microalgae, have demonstrated significant promise in the synthesis of several nutraceutical substances. Microbial-based nutraceutical production may be tailored to align with specific dietary needs and individual health characteristics. The growing popularity of incorporating nutraceuticals produced by microorganisms into personalized dietary plans might lead to improved health outcomes for individuals. The bioavailability of nutraceuticals can be enhanced by microorganisms, hence improving the body's absorption of these compounds. In recent years, natural polysaccharides have been employed in the treatment of several human disorders. It has been shown that β-glucan can be beneficial in combating chronic non-communicable diseases. Because of their hypocholesterolemic and blood sugar-lowering capabilities, they might be utilized to prevent cardiovascular illnesses and diabetes and are acknowledged as promising microbial nutraceuticals. Microorganisms such as yeast, microalgae, bacteria, and mycelium provide a wide variety of sources for the extraction of β-glucan, which are not derived from cereal crops.

β-Glucan is a kind of dietary fiber that is generated spontaneously by the bonding of D-glucose monomers using β-glycoside linkages. β-glucan is commonly found in abundance in grains, fruits, fungi, algae, yeast, and bacteria. They are structural components of microbial cell walls and thus can be obtained by using various biotechnological processes and extraction methods. The advances that have been made in the production procedure and functional characteristics of β-glucan will significantly improve our capacity to assess current and future β-glucan products. By combining screening and modification techniques, it is possible to optimize the potency and cost-effectiveness of β-glucan. These will have practical uses in the fields of food, medicine, and cosmetics. In order to maximize the utilization of β-glucan, it is crucial for food manufacturers and processors to focus not only on maintaining a sufficient level of β-glucan in the raw material but also on the processing techniques and physicochemical characteristics of β-glucan. This involves minimizing the mechanical and enzymatic degradation of β-glucan in the final product and optimizing the processing conditions. Furthermore, it is crucial for food scientists to explore alternate sources of polysaccharides and determine the optimal utilization of polysaccharides to enhance their digestibility and bioavailability. Natural polymers are favored over synthetic polymers due to their lower cost, greater availability, cost-effectiveness resulting from their biodegradability, and biocompatibility with a few limited exceptions. Biotechnological processing settings, including growth conditions and pH levels, may be automated and regulated. These same mechanisms can influence the formation of β-glucan. An advantage of these procedures is that they do not require any genetic mutation or post-translational changes because the microbes already naturally produce the desired molecule that can be used as nutraceuticals.

11.7 Concluding Remarks

The research work on the nutraceutical potential of β-glucan has demonstrated its potential to be beneficial in combating chronic and non-communicable illnesses. Because of their hypocholesterolemic and blood-sugar-reducing qualities, they might be utilized to avoid cardiovascular illnesses and diabetes as well. To fully harness the advantages of β-glucan from microbial sources as nutraceuticals, it will be imperative to broaden the range of uses beyond food components. There has been a lack of extensive studies undertaken to diversify the range of products that can benefit from β-glucan as a nutraceutical component. To expedite this growth, it is necessary to develop efficient extraction techniques and get a deeper knowledge of the crucial attributes that contribute to the health-promoting benefits. In conclusion, following these methods will optimize the utilization of β-glucan to develop nutraceutical food items of exceptional sensory quality using microorganisms. Ultimately, the utilization of microorganisms in the production of nutraceuticals greatly enhances the prospects of nutrition and healthcare in the future. The advancement of our ability to utilize microorganisms in the synthesis of these crucial compounds is expected to enhance via more research, innovation, and interdisciplinary cooperation.

Contribution of Authors

Pawan Prabhakar has given the framework and designed and written the manuscript; Deepak Kumar Verma has designed, interpreted, and technically corrected the manuscript, while Mamoni Banerjee has conceptualized and provided technical suggestions and final corrections to the manuscript. All authors critically reviewed and approved the final version of the manuscript for submission.

Conflict of Research Interests

The authors declare no conflict of research interests.

References

Adefegha, S.A. (2018). Functional foods and nutraceuticals as dietary intervention in chronic diseases; novel perspectives for health promotion and disease prevention. *Journal of Dietary Supplements* 15 (6): 977–1009.

Aguilar-Uscanga, B. and Francois, J.M. (2003). A study of the yeast cell wall composition and structure in response to growth conditions and mode of cultivation. *Letters in Applied Microbiology* 37 (3): 268–274.

Ahuja, V., Bhatt, A.K., Banu, J.R. et al. (2023). Microbial exopolysaccharide composites in biomedicine and healthcare: trends and advances. *Polymers* 15 (7): 1801.

AlAli, M., Alqubaisy, M., Aljaafari, M.N. et al. (2021). Nutraceuticals: transformation of conventional foods into health promoters/disease preventers and safety considerations. *Molecules* 26 (9): 2540.

Ali, M.F., Driscoll, C.B., Walters, P.R. et al. (2015). β-glucan–activated human B lymphocytes participate in innate immune responses by releasing proinflammatory cytokines and stimulating neutrophil chemotaxis. *The Journal of Immunology* 195 (11): 5318–5326.

Alvarez-Mon, M.A., Ortega, M.A., García-Montero, C. et al. (2021). Exploring the role of nutraceuticals in major depressive disorder (MDD): rationale, state of the art and future prospects. *Pharmaceuticals* 14 (8): 821.

Ampofo, J. and Abbey, L. (2022). Microalgae: bioactive composition, health benefits, safety and prospects as potential high-value ingredients for the functional food industry. *Foods* 11 (12): 1744.

Aquinas, N., Bhat, M.R., and Selvaraj, S. (2022). A review presenting production, characterization, and applications of biopolymer curdlan in food and pharmaceutical sectors. *Polymer Bulletin* 79 (9): 6905–6927.

Ariyoshi, W., Hara, S., Koga, A. et al. (2021). Biological effects of β-glucans on osteoclastogenesis. *Molecules* 26 (7): 1982.

Babich, O., Budenkova, E., Kashirskikh, E. et al. (2022). Study of the polysaccharide production by the microalga *Vischeria punctata* in relation to cultivation conditions. *Lifestyles* 12 (10): 1614.

Bai, J., Ren, Y., Li, Y. et al. (2019). Physiological functionalities and mechanisms of β-glucans. *Trends in Food Science & Technology* 88: 57–66.

Bashir, K.M.I. and Choi, J.S. (2017). Clinical and physiological perspectives of β-glucans: the past, present, and future. *International Journal of Molecular Sciences* 18 (9): 1906.

Bedard, S., Roxborough, E., O'Neill, E., and Mangal, V. (2024). The biomolecules of *Euglena gracilis*: harnessing biology for natural solutions to future problems. *Protist* 175: 126044.

Benito-Román, Ó., Alonso, E., Palacio, L. et al. (2014). Purification and isolation of β-glucans from barley: downstream process intensification. *Chemical Engineering and Processing: Process Intensification* 84: 90–97.

Bernaerts, T.M., Gheysen, L., Kyomugasho, C. et al. (2018). Comparison of microalgal biomasses as functional food ingredients: focus on the composition of cell wall related polysaccharides. *Algal Research* 32: 150–161.

Bhagea, R., Hossen, A.M., Ruhee, D. et al. (2022). Microalgae as sources of green bioactives for health-enhancing food supplements and nutraceuticals: a review of literature. *American Journal of Biopharmacy and Pharmaceutical Sciences* 2: 10.

Bhanja, S.K., Rout, D., Patra, P. et al. (2014). Water-insoluble glucans from the edible fungus *Ramaria botrytis*. *Bioactive Carbohydrates and Dietary Fibre* 3 (2): 52–58.

Bhattacharyya, A., Chattopadhyay, R., Mitra, S., and Crowe, S.E. (2014). Oxidative stress: an essential factor in the pathogenesis of gastrointestinal mucosal diseases. *Physiological Reviews* 94 (2): 329–354.

Bisht, D., Bhatt, V., Arya, R.K.K. et al. (2024). *Plants Based Nutraceuticals: An Overview*. Wiley; Portico. https://doi.org/10.1002/9781394241576.ch1.

Bockwoldt, J.A., Fellermeier, J., Steffens, E. et al. (2021). β-Glucan production by *Levilactobacillus brevis* and *Pediococcus claussenii* for in situ enriched rye and wheat sourdough breads. *Foods* 10 (3): 547.

Bruno, G.M., Dovera, F., Ciccarone, A., and Colombo, G.L. (2022). Overview of nutraceuticals and cardiometabolic diseases following socio-economic analysis. *Endocrines* 3 (2): 255–295.

Bulmer, G.S., De Andrade, P., Field, R.A., and van Munster, J.M. (2021). Recent advances in enzymatic synthesis of β-glucan and cellulose. *Carbohydrate Research* 508: 108411.

Caseiro, C., Dias, J.N.R., de Andrade Fontes, C.M.G., and Bule, P. (2022). From cancer therapy to winemaking: the molecular structure and applications of β-glucans and β-1, 3-glucanases. *International Journal of Molecular Sciences* 23 (6): 3156.

Chakraborty, S. and Rajeswari, V.D. (2022). Biomedical aspects of β-glucan on glucose metabolism and its role on primary gene PIK3R1. *Journal of Functional Foods* 99: 105296.
Chan, G.C.F., Chan, W.K., and Sze, D.M.Y. (2009). The effects of β-glucan on human immune and cancer cells. *Journal of Hematology & Oncology* 2: 1–11.
Chandra, S., Saklani, S., Kumar, P. et al. (2022). Nutraceuticals: pharmacologically active potent dietary supplements. *BioMed Research International* 2022 (1): 2051017.
Chávez-González, M.L., Sepúlveda, L., Verma, D.K. et al. (2020). Conventional and emerging extraction processes of flavonoids. *Processes* 8 (4): 434.
Chen, H., Liu, N., He, F. et al. (2022). Specific β-glucans in chain conformations and their biological functions. *Polymer Journal* 54 (4): 427–453.
Chhetri, A., Loksztejn, A., Nguyen, H. et al. (2020). Length specificity and polymerization mechanism of (1, 3)-β-d-glucan synthase in fungal cell wall biosynthesis. *Biochemistry* 59 (5): 682–693.
Ching, J.J., Shuib, A.S., Abdul Majid, N., and Mohd Taufek, N. (2021). Immunomodulatory activity of β-glucans in fish: relationship between β-glucan administration parameters and immune response induced. *Aquaculture Research* 52 (5): 1824–1845.
Chioru, A. and Chirsanova, A. (2023). β-Glucans: characterization, extraction methods, and valorization. *Food and Nutrition Sciences* 14 (10): 963–983.
Choi, H. and Kim, S.W. (2023). Characterization of β-glucans from cereal and microbial sources and their roles in feeds for intestinal health and growth of nursery pigs. *Animals* 13 (13): 2236.
Ciecierska, A., Drywien, M., Hamulka, J., and Sadkowski, T. (2019). Nutraceutical functions of β-glucans in human nutrition. *Roczniki Państwowego Zakładu Higieny* 70 (4): 315–324.
Climova, A., Ibrahim, M.N., Salamahina, A. et al. (2021). Application of extracted β-glucan from oat for β-carotene encapsulation. *Journal of Food Science and Technology* 58: 2641–2650.
Costa, J.A.V., Lucas, B.F., Alvarenga, A.G.P. et al. (2021). Microalgae polysaccharides: an overview of production, characterization, and potential applications. *Polysaccharides* 2 (4): 759–772.
Crognale, S., Bruno, M., Fidaleo, M. et al. (2007). Production of β-glucan and related glucan-hydrolases by *Botryosphaeria rhodina*. *Journal of Applied Microbiology* 102 (3): 860–871.
Dama, A., Shpati, K., Daliu, P. et al. (2024). Targeting metabolic diseases: the role of nutraceuticals in modulating oxidative stress and inflammation. *Nutrients* 16 (4): 507.
Dhanjal, D.S., Bhardwaj, S., Sharma, R. et al. (2020). Plant fortification of the diet for anti-ageing effects: a review. *Nutrients* 12 (10): 3008.
Dineshbabu, G., Goswami, G., Kumar, R. et al. (2019). Microalgae–nutritious, sustainable aqua-and animal feed source. *Journal of Functional Foods* 62: 103545.
Du, B., Zhu, F., and Xu, B. (2014). β-Glucan extraction from bran of hull-less barley by accelerated solvent extraction combined with response surface methodology. *Journal of Cereal Science* 59 (1): 95–100.
Dwivedi, S.L., Van Bueren, E.T.L., Ceccarelli, S. et al. (2017). Diversifying food systems in the pursuit of sustainable food production and healthy diets. *Trends in Plant Science* 22 (10): 842–856.
El Khoury, D., Cuda, C., Luhovyy, B.L., and Anderson, G.H. (2012). B glucan: health benefits in obesity and metabolic syndrome. *Journal of Nutrition and Metabolism* 2012 (1): 851362.
Fazio, A., La Torre, C., Caroleo, M.C. et al. (2020). Isolation and purification of glucans from an Italian cultivar of *Ziziphus jujuba* Mill. and in vitro effect on skin repair. *Molecules* 25 (4): 968.
Fontana, C. and Widmalm, G. (2023). Primary structure of glycans by NMR spectroscopy. *Chemical Reviews* 123 (3): 1040–1102.
Fruh, S.M. (2017). Obesity: risk factors, complications, and strategies for sustainable long-term weight management. *Journal of the American Association of Nurse Practitioners* 29 (S1): S3–S14.

Garza-Juárez, A., Pérez-Carrillo, E., Arredondo-Espinoza, E.U. et al. (2023). Nutraceuticals and their contribution to preventing noncommunicable diseases. *Foods* 12 (17): 3262.

Geller, A., Shrestha, R., and Yan, J. (2019). Yeast-derived β-glucan in cancer: novel uses of a traditional therapeutic. *International Journal of Molecular Sciences* 20 (15): 3618.

Geller, A. and Yan, J. (2020). Could the induction of trained immunity by β-glucan serve as a defense against COVID-19? *Frontiers in Immunology* 11: 1782.

Gern, R.M.M., Wisbeck, E., Rampinelli, J.R. et al. (2008). Alternative medium for production of *Pleurotus ostreatus* biomass and potential antitumor polysaccharides. *Bioresource Technology* 99 (1): 76–82.

Gissibl, A., Sun, A., Care, A. et al. (2019). Bioproducts from *Euglena gracilis*: synthesis and applications. *Frontiers in Bioengineering and Biotechnology* 7: 108.

Gupta, V., Jamwal, G., Rai, G.K. et al. (2024). Biosynthesis of biomolecules from saffron as an industrial crop and their regulation, with emphasis on the chemistry, extraction methods, identification techniques, and potential applications in human health and food: a critical comprehensive review. *Biocatalysis and Agricultural Biotechnology* 59: 103260.

Han, B., Baruah, K., Cox, E. et al. (2020). Structure-functional activity relationship of β-glucans from the perspective of immunomodulation: a mini-review. *Frontiers in Immunology* 11: 658.

Huang, X., Liu, G., Guo, J., and Su, Z. (2018). The PI3K/AKT pathway in obesity and type 2 diabetes. *International Journal of Biological Sciences* 14 (11): 1483.

Ibe, C. and Munro, C.A. (2021). Fungal cell wall proteins and signaling pathways form a cytoprotective network to combat stresses. *Journal of fungi (Basel, Switzerland)* 7: 739.

Ina, K., Kataoka, T., and Ando, T. (2013). The use of lentinan for treating gastric cancer. *Anti-Cancer Agents in Medicinal Chemistry (Formerly Current Medicinal Chemistry-Anti-Cancer Agents)* 13 (5): 681–688.

Ivušić, F. and Šantek, B. (2015). Optimization of complex medium composition for heterotrophic cultivation of *Euglena gracilis* and paramylon production. *Bioprocess and Biosystems Engineering* 38: 1103–1112.

Jacobo-Velázquez, D.A., Gutiérrez-Uribe, J.A., and Antunes-Ricardo, M. (2023). Highlights of the 5th international congress of functional foods and nutraceuticals: searching for the next generation of healthy foods. *ACS Food Science & Technology* 3 (3): 361–362.

Jadhav, H.B., Sablani, S., Gogate, P. et al. (2023). Factors governing consumers buying behavior concerning nutraceutical product. *Food Science & Nutrition* 11 (9): 4988–5003.

Jaehrig, S.C., Rohn, S., Kroh, L.W. et al. (2008). Antioxidative activity of (1→3), (1→6)-β-d-glucan from *Saccharomyces cerevisiae* grown on different media. *LWT- Food Science and Technology* 41 (5): 868–877.

Jayachandran, M., Chen, J., Chung, S.S.M., and Xu, B. (2018). A critical review on the impacts of β-glucans on gut microbiota and human health. *The Journal of Nutritional Biochemistry* 61: 101–110.

de Jesus Raposo, M. F., De Morais, A. M. M. B., & De Morais, R. M. S. C. (2014). Bioactivity and applications of polysaccharides from marine microalgae. *In Polysaccharides: Bioactivity and Biotechnology*, Eds. Ramawat, K.G., Mérillon, J., Vol 1, 1683–1727, Springer Nature, Switzerland.

Kalyanasundaram, G.T., Doble, M., and Gummadi, S.N. (2012). Production and downstream processing of (1→3)-β-d-glucan from mutant strain of *Agrobacterium* sp. ATCC 31750. *AMB Express* 2: 1–10.

Kao, P.F., Wang, S.H., Hung, W.T. et al. (2012). Structural characterization and antioxidative activity of low-molecular-weights B-1, 3-glucan from the residue of extracted *Ganoderma lucidum* fruiting bodies. *BioMed Research International* 2012 (1): 673764.

Kaur, R., Sharma, M., Ji, D. et al. (2019). Structural features, modification, and functionalities of β-glucan. *Fibers* 8 (1): 1.

Khan, A.A., Gani, A., Khanday, F.A., and Masoodi, F.A. (2018b). Biological and pharmaceutical activities of mushroom β-glucan discussed as a potential functional food ingredient. *Bioactive Carbohydrates and Dietary Fibre* 16: 1–13.

Khan, M.I., Shin, J.H., and Kim, J.D. (2018a). The promising future of microalgae: current status, challenges, and optimization of a sustainable and renewable industry for biofuels, feed, and other products. *Microbial Cell Factories* 17: 1–21.

Kim, S.R., Kang, H.W., and Ro, H.S. (2013). Generation and evaluation of high β-glucan producing mutant strains of *Sparassis crispa*. *Mycobiology* 41 (3): 159–163.

Kumar, A., Rout, R.K., Rao, P.S., and Prabhakar, P. (2022). Optimization of pulsed mode sonication and in silico molecular docking study for antioxidant properties of mandarin (*Citrus reticulata* Blanco) peels. *Journal of Food Process Engineering* 46 (6): e14111.

Lante, A., Canazza, E., and Tessari, P. (2023). B-glucans of cereals: functional and technological properties. *Nutrients* 15 (9): 2124.

Lefter, R., Balyan, P., Balmus, I.-M. et al. (2024). Polysaccharides and lectins: a natural complementary approach against the SARS-CoV-2 pandemic. *Microbiology Research* 15 (2): 525–549.

Liu, J.J., Hou, Y.K., Wang, X. et al. (2024). Recent advances in the biosynthesis of fungal glucan structural diversity. *Carbohydrate Polymers* 329: 121782.

Liu, N., Zou, S., Xie, C. et al. (2023). Effect of the β-glucan from *Lentinus edodes* on colitis-associated colorectal cancer and gut microbiota. *Carbohydrate Polymers* 316: 121069.

Louie, B., Rajamahanty, S., Won, J. et al. (2010). Synergistic potentiation of interferon activity with maitake mushroom d-fraction on bladder cancer cells. *BJU International* 105 (7): 1011–1015.

Maheshwari, G., Sowrirajan, S., and Joseph, B. (2017). Extraction and isolation of β-glucan from grain sources—a review. *Journal of Food Science* 82 (7): 1535–1545.

Mahmoud, Y.A.G., El-Naggar, M.E., Abdel-Megeed, A., and El-Newehy, M. (2021). Recent advancements in microbial polysaccharides: synthesis and applications. *Polymers* 13 (23): 4136.

Makkar, R., Behl, T., Bungau, S. et al. (2020). Nutraceuticals in neurological disorders. *International Journal of Molecular Sciences* 21 (12): 4424.

Mata-Martínez, P., Bergón-Gutiérrez, M., and Del Fresno, C. (2022). Dectin-1 signaling update: new perspectives for trained immunity. *Frontiers in Immunology* 13: 812148.

Mathews, R., Shete, V., and Chu, Y. (2023). The effect of cereal B-glucan on body weight and adiposity: a review of efficacy and mechanism of action. *Critical Reviews in Food Science and Nutrition* 63 (19): 3838–3850.

Mejía, S.M.V., de Francisco, A., and Bohrer, B. (2020). A comprehensive review on cereal β-glucan: extraction, characterization, causes of degradation, and food application. *Critical Reviews in Food Science and Nutrition* 60 (21): 3693–3704.

Mirończuk-Chodakowska, I., Kujawowicz, K., and Witkowska, A.M. (2021). B-glucans from fungi: biological and health-promoting potential in the COVID-19 pandemic era. *Nutrients* 13 (11): 3960.

Moreira, J.B., Vaz, B.D.S., Cardias, B.B. et al. (2022). Microalgae polysaccharides: an alternative source for food production and sustainable agriculture. *Polysaccharides* 3 (2): 441–457.

Mouro, C., Gomes, A.P., and Gouveia, I.C. (2024). Microbial exopolysaccharides: structure, diversity, applications, and future frontiers in sustainable functional materials. *Polysaccharides* 5 (3): 241–287.

Murphy, E.J., Rezoagli, E., Collins, C. et al. (2023). Sustainable production and pharmaceutical applications of β-glucan from microbial sources. *Microbiological Research* 274: 127424.

Mykhalevych, A., Polishchuk, G., Nassar, K. et al. (2022). β-Glucan as a techno-functional ingredient in dairy and milk-based products—a review. *Molecules* 27 (19): 6313.

Nagarajan, D., Varjani, S., Lee, D.J., and Chang, J.S. (2021). Sustainable aquaculture and animal feed from microalgae–nutritive value and techno-functional components. *Renewable and Sustainable Energy Reviews* 150: 111549.

Naik, B., Mishra, R., Kumar, V. et al. (2024). Micro-algae: revolutionizing food production for a healthy and sustainable future. *Journal of Agriculture and Food Research* 15: 100939.

Nori, L.P., Sowmya, P., Madhusree, T., and Raju, K.V. (2023). Nutraceuticals: regulatory landscape and market dynamics in the Indian scenario. *Journal of Young Pharmacists* 15 (4): 595–603.

Papaspyridi, L.M., Zerva, A., and Topakas, E. (2018). Biocatalytic synthesis of fungal β-glucans. *Catalysts* 8 (7): 274.

Park, H., Ka, K.H., and Ryu, S.R. (2014). Enhancement of β-glucan content in the cultivation of cauliflower mushroom (*Sparassis latifolia*) by elicitation. *Mycobiology* 42 (1): 41–45.

Park, M.H. and Kim, M. (2018). Antioxidant and anti-inflammatory activity and cytotoxicity of ethanol extracts from Rhynchosia nulubilis cultivated with Ganoderma lucidum mycelium. *Preventive Nutrition and Food Science* 23 (4): 326–334.

Pengkumsri, N., Sivamaruthi, B.S., Sirilun, S. et al. (2016). Extraction of β-glucan from *Saccharomyces cerevisiae*: comparison of different extraction methods and *in vivo* assessment of immunomodulatory effect in mice. *Food Science and Technology (Campinas)* 37 (1): 124–130.

Philippini, R. R., Martiniano, S. E., dos Santos, J. C., da Silva, S. S., & Chandel, A. K. (2019). Fermentative production of B-glucan: properties and potential applications. In *Bioprocessing for Biomolecules Production*, Eds. Molina, G., Gupta, V., Singh, B., Gathergood, N., Vol 1, 303–320, John Wiley & Sons, NJ, USA.

Prabhakar, P. and Banerjee, M. (2020). Antidiabetic phytochemicals: s comprehensive review on opportunities and challenges in targeted therapy for herbal drug development. *International Journal of Pharmaceutical Research* 12 (2): 1673–1696.

Prabhakar, P. and Mamoni, B. (2021). Technical problems, regulatory and market challenges in bringing herbal drug into mainstream of modern medicinal practices. *Research Journal of Biotechnology* 16 (3): 212–222.

Prabhakar, P., Mukherjee, S., Kumar, A. et al. (2023a). Optimization of MAE for Carica papaya phytochemicals, and its in silico, in vitro, and ex vivo evaluation: for functional food and drug applications. *Food Bioscience* 54: 102861.

Prabhakar, P., Mukherjee, S., Kumar, A. et al. (2023b). Optimization of microwave-assisted extraction (MAE) of key phenolic compounds from pigeon pea (*Cajanus cajan* L.), their characterization, and measurement of their anti-diabetic and cytotoxic potential. *Journal of Food Measurement and Characterization* 17 (6): 5697–5720.

Prabhakar, P., Pasayat, A. K., Kumar, S., and Banerjee, M. (2023c). An in silico evaluation of dipeptidyl peptidase-4 (DPP-4) inhibiting potential of *Cajanus cajan, Coccinia grandis* and *Enhydra fluctuans. In* 2023 *International Conference on Communication, Circuits, and Systems* (IC3S) (pp. 1–6). IEEE.

Prabhakar, P., Verma, D.K., and Banerjee, M. (2024a). Advances in enzyme-assisted extraction technique for antidiabetic phytochemicals: an emerging approach for green and economical technologies. In: *Enzymatic Processes for Food Valorization* (ed. J. Jay Singh, R.P. Singh, A.K. Kaushik, et al.), 191–216. Academic Press Elsevier, New York, USA.

Prabhakar, P., Mukherjee, S., Kumar, A. et al. (2024b). In silico, in vitro and ex vivo evaluation of the antihyperglycaemic, antioxidant and cytotoxic properties of Coccinia grandis L. Leaf Extract. *Food Technology and Biotechnology* 62 (2): 188–204.

Prasad, R., Gupta, S.K., Shabnam, N. et al. (2021). Role of microalgae in global CO_2 sequestration: physiological mechanism, recent development, challenges, and future prospective. *Sustainability* 13 (23): 13061.

Puri, V., Nagpal, M., Singh, I. et al. (2022). A comprehensive review on nutraceuticals: therapy support and formulation challenges. *Nutrients* 14 (21): 4637.

Quintero-Fabián, S., Arreola, R., Becerril-Villanueva, E. et al. (2019). Role of matrix metalloproteinases in angiogenesis and cancer. *Frontiers in Oncology* 9: 1370.

Rawling, M., Schiavone, M., Apper, E. et al. (2023). Yeast cell wall extracts from *Saccharomyces cerevisiae* varying in structure and composition differentially shape the innate immunity and mucosal tissue responses of the intestine of zebrafish (*Danio rerio*). *Frontiers in Immunology* 14: 1158390.

Reitan, K.I., Øie, G., Jørgensen, H., and Wang, X. (2021). Chemical composition of selected marine microalgae, with emphasis on lipid and carbohydrate production for potential use as feed resources. *Journal of Applied Phycology* 33: 3831–3842.

Rizal, S., Kustyawati, M.E., Murhadi, and Hasanudin, U. (2021). The growth of yeast and fungi, the formation of β-glucan, and the antibacterial activities during soybean fermentation in producing tempeh. *International Journal of Food Science* 2021 (1): 6676042.

Rowland, I., Gibson, G., Heinken, A. et al. (2018). Gut microbiota functions: metabolism of nutrients and other food components. *European Journal of Nutrition* 57: 1–24.

Ruiz-Herrera, J. and Ortiz-Castellanos, L. (2019). Cell wall glucans of fungi. A review. *The Cell Surface* 5: 100022.

Russo, M.T., Rogato, A., Jaubert, M. et al. (2023). *Phaeodactylum tricornutum*: an established model species for diatom molecular research and an emerging chassis for algal synthetic biology. *Journal of Phycology* 59 (6): 1114–1122.

Saeed, A. and Zafar, S.I. (2024). *Bioactive Nutrients as Dietary Supplements and Therapeutic Agents: An Introduction to Nutraceuticals*, vol. 1. United, Kingdom: Cambridge Scholars Publishing.

Sato, H., Kobayashi, Y., Hattori, A. et al. (2012). Inhibitory effects of water-soluble low-molecular-weight β-(1, 3-1, 6) d-glucan isolated from *Aureobasidium pullulans* 1A1 strain black yeast on mast cell degranulation and passive cutaneous anaphylaxis. *Bioscience, Biotechnology, and Biochemistry* 76 (1): 84–88.

Schmidt, M. (2022). Cereal β-glucans: an underutilized health endorsing food ingredient. *Critical Reviews in Food Science and Nutrition* 62 (12): 3281–3300.

Schulze, C., Reinhardt, J., Wurster, M. et al. (2016a). A one-stage cultivation process for lipid-and carbohydrate-rich biomass of *Scenedesmus obtusiusculus* based on artificial and natural water sources. *Bioresource Technology* 218: 498–504.

Schulze, C., Wetzel, M., Reinhardt, J. et al. (2016b). Screening of microalgae for primary metabolites including β-glucans and the influence of nitrate starvation and irradiance on β-glucan production. *Journal of Applied Phycology* 28: 2719–2725.

Sen, I.K., Maji, P.K., Behera, B. et al. (2013). Structural characterization of an immunoenhancing heteroglycan of a hybrid mushroom (*pfls1h*) of *Pleurotus florida* and *Lentinus squarrosulus* (Mont.) Singer. *Carbohydrate Research* 371: 45–51.

Şengül, M. and Ufuk, S. (2022). Therapeutic and functional properties of β-glucan, and its effects on health. *Eurasian Journal of Food Science and Technology* 6 (1): 29–41.

Sharma, A., Madhu, Kondalkar, S.A. et al. (2024). Herbal Nutraceutical as Alternative Medicine. *Herbal Nutraceuticals*. Wiley; Portico. https://doi.org/10.1002/9781394241576.ch2.

Shi, L. (2016). Bioactivities, isolation and purification methods of polysaccharides from natural products: a review. *International Journal of Biological Macromolecules* 92: 37–48.

Shin, M.R., Lee, J.H., Lee, J.A. et al. (2021). Immunomodulatory and anti-inflammatory effects of *Phellinus linteus* mycelium. *BMC Complementary Medicine and Therapies* 21: 1–11.

Shivakumar, S. and Vijayendra, S.V.N. (2006). Production of exopolysaccharides by *Agrobacterium* sp. CFR-24 using coconut water–a byproduct of food industry. *Letters in Applied Microbiology* 42 (5): 477–482.

Siddik, M.A., Sørensen, M., Islam, S.M. et al. (2024). Expanded utilisation of microalgae in global aquafeeds. *Reviews in Aquaculture* 16 (1): 6–33.

Sima, P., Vannucci, L., and Vetvicka, V. (2018). β-glucans and cholesterol. *International Journal of Molecular Medicine* 41 (4): 1799–1808.

Singh, R.P. and Bhardwaj, A. (2023). β-glucans: a potential source for maintaining gut microbiota and the immune system. *Frontiers in Nutrition* 10: 1143682.

Singh, R.P., Rajarammohan, S., Thakur, R., and Hassan, M. (2020). Linear and branched β-glucans degrading enzymes from versatile *Bacteroides uniformis* JCM 13288T and their roles in cooperation with gut bacteria. *Gut Microbes* 12 (1): 1826761.

Singla, A., Gupta, O.P., Sagwal, V. et al. (2024). Β-glucan as a soluble dietary fiber source: origins, biosynthesis, extraction, purification, structural characteristics, bioavailability, biofunctional attributes, industrial utilization, and global trade. *Nutrients* 16 (6): 900.

Sofi, S.A., Singh, J., and Rafiq, S. (2017). β-glucan and functionality: a review. *Ecronicon Nutrition* 10 (2): 67–74.

Sun, A., Hasan, M.T., Hobba, G. et al. (2018). Comparative assessment of the Euglena gracilis var. saccharophila variant strain as a producer of the β-1, 3-glucan paramylon under varying light conditions. *Journal of Phycology* 54 (4): 529–538.

Stack, H.M., Kearney, N., Stanton, C. et al. (2010). Association of β-glucan endogenous production with increased stress tolerance of intestinal lactobacilli. *Applied and Environmental Microbiology* 76 (2): 500–507.

Steimbach, L., Borgmann, A.V., Gomar, G.G. et al. (2021). Fungal β-glucans as adjuvants for treating cancer patients–a systematic review of clinical trials. *Clinical Nutrition* 40 (5): 3104–3113.

Sutter, S., Thevenieau, F., Bourdillon, A., and De Coninck, J. (2017). Immunomodulatory properties of filamentous fungi cultivated through solid-state fermentation on rapeseed meal. *Applied Biochemistry and Biotechnology* 182 (3): 910–924.

Tan, B.L., Norhaizan, M.E., Liew, W.P.P., and Sulaiman Rahman, H. (2018). Antioxidant and oxidative stress: a mutual interplay in age-related diseases. *Frontiers in Pharmacology* 9: 1162.

Thandapilly, S.J., Ndou, S.P., Wang, Y. et al. (2018). Barley β-glucan increases fecal bile acid excretion and short chain fatty acid levels in mildly hypercholesterolemic individuals. *Food & Function* 9 (6): 3092–3096.

Thomas, S., Rezoagli, E., Abidin, I.Z. et al. (2022). β-glucans from yeast—immunomodulators from novel waste resources. *Applied Sciences* 12 (10): 5208.

Tian, Y., Zeng, H., Xu, Z. et al. (2012). Ultrasonic-assisted extraction and antioxidant activity of polysaccharides recovered from white button mushroom (*Agaricus bisporus*). *Carbohydrate Polymers* 88 (2): 522–529.

Ubiparip, Z., De Doncker, M., Beerens, K. et al. (2021). β-glucan phosphorylases in carbohydrate synthesis. *Applied Microbiology and Biotechnology* 105 (10): 4073–4087.

Ullah, S., Khalil, A.A., Shaukat, F., and Song, Y. (2019). Sources, extraction and biomedical properties of polysaccharides. *Foods* 8 (8): 304.

Upadhyay, S.K. and Singh, S.P. (2024). *Herbal Nutraceuticals: Products and Processes*. Wiley. https://doi.org/10.1002/9781394241576.

Upadhyay, T.K., Fatima, N., Sharma, D. et al. (2017). Preparation and characterization of β-glucan particles containing a payload of nanoembedded rifabutin for enhanced targeted delivery to macrophages. *EXCLI Journal* 16: 210.

Utama, G.L., Dio, C., Lembong, E. et al. (2020). Microorganism-based β-glucan production and their potential as antioxidant. *Systematic Reviews in Pharmacy* 11: 868–873.

Utama, G.L., Dio, C., Sulistiyo, J. et al. (2021). Evaluating comparative β-glucan production aptitude of *Saccharomyces cerevisiae, Aspergillus oryzae, Xanthomonas campestris*, and *Bacillus natto*. *Saudi Journal of Biological Sciences* 28 (12): 6765–6773.

Verma, D.K. and Thakur, M. (ed.) (2021). *Phytochemicals in Food and Health: Perspectives for Research and Technological Development*. CRC Press.

Vetvicka, V. (2011). Glucan-immunostimulant, adjuvant, potential drug. *World Journal of Clinical Oncology* 2 (2): 115.

Vetvicka, V. and Oliveira, C. (2014). β (1-3)(1-6)-d-glucans modulate immune status in pigs: potential importance for efficiency of commercial farming. *Annals of Translational Medicine* 2 (2): 16.

Vetvicka, V., Teplyakova, T.V., Shintyapina, A.B., and Korolenko, T.A. (2021). Effects of medicinal fungi-derived β-glucan on tumor progression. *Journal of Fungi* 7 (4): 250.

Veverka, M., Dubaj, T., Gallovič, J. et al. (2014). B-glucan complexes with selected nutraceuticals: synthesis, characterization, and stability. *Journal of Functional Foods* 8: 309–318.

Vishvakarma, P., Mandal, S., and Verma, A. (2023). A review on current aspects of nutraceuticals and dietary supplements. *International Journal of Pharma Professional's Research (IJPPR)* 14 (1): 78–91.

Wagener, J., Striegler, K., and Wagener, N. (2020). α-and β-1, 3-Glucan Synthesis and Remodeling. In: *The Fungal Cell Wall: An Armour and a Weapon for Human Fungal Pathogens* (ed. J.-P. Latgé), 53–82. Springer International Publishing.

Wang, X., Zhang, P., and Zhang, X. (2021). Probiotics regulate gut microbiota: an effective method to improve immunity. *Molecules* 26 (19): 6076.

Wu, M., Qin, H., Deng, J. et al. (2021). A new pilot-scale fermentation mode enhances *Euglena gracilis* biomass and paramylon (β-1, 3-glucan) production. *Journal of Cleaner Production* 321: 128996.

Xie, Y., Li, J., Qin, H. et al. (2022). Paramylon from *Euglena gracilis* prevents lipopolysaccharide-induced acute liver injury. *Frontiers in Immunology* 12: 797096.

Yang, Y.F., Li, D.W., Chen, T.T. et al. (2018). Overproduction of bioactive algal chrysolaminarin by the critical carbon flux regulator phosphoglucomutase. *Biotechnology Journal* 14 (3): 1800220.

Yi, Z., Su, Y., Brynjolfsson, S. et al. (2021). Bioactive polysaccharides and their derivatives from microalgae: biosynthesis, applications, and challenges. *Studies in Natural Products Chemistry* 71: 67–85.

Yoshimi, A., Miyazawa, K., and Abe, K. (2017). Function and biosynthesis of cell wall α-1, 3-glucan in fungi. *Journal of Fungi* 3 (4): 63.

Yu, S., Wang, J., Li, Y. et al. (2021). Structural studies of water-insoluble β-glucan from oat bran and its effect on improving lipid metabolism in mice fed high-fat diet. *Nutrients* 13 (9): 3254.

Zhong, X., Wang, G., Li, F. et al. (2023). Immunomodulatory effect and biological significance of β-glucans. *Pharmaceutics* 15 (6): 1615.

Zhou, X., He, J., Wang, L. et al. (2019). Metabolic engineering of Saccharomyces cerevisiae to improve glucan biosynthesis. *Journal of Microbiology and Biotechnology* 29 (5): 758–764.

Zhou, G., Liu, H., Yuan, Y. et al. (2024). Lentinan progress in inflammatory diseases and tumor diseases. *European Journal of Medical Research* 29 (1): 8.

Zhu, F., Du, B., and Xu, B. (2016). A critical review on production and industrial applications of β-glucans. *Food Hydrocolloids* 52: 275–288.

12

Multifaceted Role of Synbiotic Products with Nutraceutical Impact

Mariana Buranelo Egea[1], *Josemar Gonçalves de Oliveira Filho*[2], *and Ailton Cesar Lemes*[3]

[1] *Goiano Federal Institute of Education, Science and Technology, Campus Rio Verde, Goiás, Rio Verde, 75901-970, Brazil*
[2] *Brazilian Agricultural Research Corporation, Embrapa Instrumentation, São Carlos, 13561-206, São Paulo, Brazil*
[3] *School of Chemistry, Department of Biochemical Engineering, Federal University of Rio de Janeiro (UFRJ), Rio de Janeiro, 21941-909, Brazil*

12.1 Introduction

The healthy products' market has experienced significant growth in recent years, stimulated by growing awareness of the importance of a balanced diet and a healthier lifestyle (Egea et al. 2023). Therefore, consumers are increasingly interested in food options that promote well-being and meet their nutritional needs. This has stimulated the development and consumption of various products, including probiotics, prebiotics, and synbiotics (Pandey et al. 2015; Yadav et al. 2022).

Prebiotic products are nondigestible food compounds that selectively promote the growth and/or activity of beneficial microorganisms in the gastrointestinal tract, thereby improving the host's health (Davani-Davari et al. 2019). These compounds are dietary fibers, oligosaccharides, or polysaccharides that resist digestion in the small intestine and reach the large intestine, where the intestinal microbiota ferments them. The fermentation of these prebiotics results in the production of short-chain fatty acids such as acetate, propionate, and butyrate, which have beneficial effects on intestinal health, including reducing intestinal pH, stimulating the growth of beneficial bacteria, strengthening the intestinal barrier, and modulating the immune system, among other (Bamigbade et al. 2022; Neri-Numa and Pastore 2020).

Probiotics, in turn, are live microorganisms that, when administered in adequate quantities, provide health benefits to the host. These microorganisms, mainly bacteria such as *Lactobacillus* and *Bifidobacterium*, but not restricted to them, can colonize the gastrointestinal tract and interact with the existing intestinal microbiota (Ranjha et al. 2021). Through mechanisms such as competition for nutrients and binding sites, production of metabolites, and modulation of the immune system, probiotics can promote the balance of the intestinal microbiota, improve the integrity of the intestinal barrier, strengthen immune defenses, and even influence the metabolic functions of the host. These beneficial effects of probiotics have been linked to improving gastrointestinal health, preventing and treating intestinal disorders, reducing the risk of infections, and even

Microbial Nutraceuticals: Products and Processes, First Edition. Edited by Sudhir Pratap Singh and Santosh Kumar Upadhyay.

promoting benefits not restricted to the gastrointestinal tract, such as immune and mental health (George Kerry et al. 2018; Campaniello et al. 2023).

Probiotic strains' selection criteria and requirements include a series of fundamental aspects that guarantee their efficacy and safety, including their precise identification, viability and stability, resistance to gastrointestinal tract conditions, metabolic activity, safety, and clinical studies, among others. When selecting probiotic strains for probiotic products, it is essential to consider all of these criteria to ensure product and consumer quality, efficacy, and safety (Markowiak and Śliżewska 2017; Shewale et al. 2014).

Synbiotic products are formulated by combining probiotics and prebiotics in a single matrix, waiting for the beneficial effects of intestinal health. In general, it is possible to indicate that prebiotics provide the necessary substrate for the growth of probiotics, while probiotics, in turn, colonize the intestine and use these prebiotics as a source of energy, increasing their activity and effectiveness. This synergy between probiotics and prebiotics can result in better intestinal health, strengthened immune systems, and other benefits for the host (Campaniello et al. 2023; Markowiak and Śliżewska 2017).

The market for symbiotic products has grown significantly in recent years because of increased consumer awareness about the benefits for intestinal health and the search for solutions that promote the balance of the intestinal microbiota and the body. The global synbiotics market was valued at US$800 million in 2022 and is expected to reach US$1.50 billion by 2030, considering a compound annual growth rate (CAGR) of 8.20% by 2030. Furthermore, the market for synbiotic products is expected to grow as more research demonstrates the benefits of these products for human health (The Brainy Insights 2021).

Additionally, continued research into the health benefits of probiotics and prebiotics has generated more demand for products that combine these ingredients. The synbiotic products' market encompasses many products, including dairy products, dietary supplements, cereal bars, functional beverages, and intestinal healthcare products (Manassi et al. 2022; Liao et al. 2022).

Based on this, this chapter aims to present the multifaceted role of symbiotic products with nutraceutical impact, highlighting the selection criteria, types of symbiotics, mechanisms of action, and the action of symbiotics on humans.

12.2 Beneficial Effects and Selection Criteria

12.2.1 Beneficial Effects

Synbiotics are food products that combine probiotics and prebiotics in a single formulation and thus exert the prebiotic and probiotic effects recognized for this purpose (Malik et al. 2016). Prebiotic products are nondigestible food compounds that selectively promote the growth and/or activity of beneficial microorganisms in the gastrointestinal tract, ensuring additional health benefits for the host (Davani-Davari et al. 2019). Probiotics are live microorganisms that provide health benefits to the host when administered in adequate quantities. These microorganisms, mainly bacteria such as *Lactobacillus* and *Bifidobacterium*, can colonize the gastrointestinal tract and interact with the existing intestinal microbiota (Ranjha et al. 2021).

By combining probiotics and prebiotics in a synbiotic product, the beneficial effects on intestinal health are expected to be enhanced. Prebiotics provide the necessary substrate for the growth of probiotics, while probiotics, in turn, colonize the intestine and use these prebiotics as a source of energy, increasing their activity and effectiveness. This synergy between probiotics and prebiotics can result in better intestinal health, strengthened immune systems, and other benefits for the host (Campaniello et al. 2023; Markowiak and Śliżewska 2017).

Synbiotics can have many health benefits, acting on the digestive system and the immune, metabolic, mental, and nutrient absorption systems (Dahiya and Nigam 2022). These are some potential benefits of synbiotics, and research continues to explore their role in promoting overall health and well-being. The benefits of synbiotics are wide and varied, and include the following:

1) Improves gastrointestinal health: The synbiotics promote the balance of the intestinal microbiota, reducing digestive disorders such as constipation, diarrhea, and irritable bowel syndrome (Simon et al. 2021).
2) Strengthening the immune system: The body's natural defenses are strengthened, reducing the incidence of infections and allergies (Markowiak and Śliżewska 2017).
3) Regulation of metabolism: Synbiotics positively influence metabolism, helping to regulate body weight and prevent metabolic conditions such as obesity and type-2 diabetes (Hijová 2022).
4) Improved mental health: Gut health is linked to mental well-being, and synbiotics may play a role in improving mood and reducing stress and anxiety (Xiong et al. 2023).
5) Nutrient absorption: Synbiotics can improve nutrient absorption in the intestine, ensuring better food utilization and promoting adequate nutrition (Jiang et al. 2022)
6) Reduced risk of chronic diseases: Regular consumption of synbiotics has been associated with a reduced risk of chronic diseases, including cardiovascular diseases, cancer, and inflammatory diseases (Pandey et al. 2015).
7) Bone health support: Synbiotics can improve the absorption of calcium and other minerals, contributing to bone health and the prevention of osteoporosis (Whisner and Castillo 2018).
8) Improved skin health: Gut health is related to skin health, and synbiotics can help reduce problems such as acne, eczema, and psoriasis (Almeida et al. 2023).

12.2.2 Selection Criteria of Prebiotic, Probiotic, and Synbiotics

Several criteria are considered when selecting and evaluating prebiotics for foods, supplements, or pharmaceuticals intended to improve gut health and promote a balanced microbiome (Davani-Davari et al. 2019). Prebiotic selection criteria may vary depending on the specific purpose, but they generally include the aspects described below. First, prebiotics must resist digestion in the upper gastrointestinal tract to reach the colon, fermented by beneficial bacteria (Dahiya and Nigam 2022). Next, they must be fermentable, promoting the production of short-chain fatty acids, such as acetic, propionic, and butyric acids, which are beneficial for intestinal health (You et al. 2022). Additionally, prebiotics should selectively promote the growth and activity of beneficial bacteria in the gut, such as *Bifidobacterium* and *Lactobacillus*, without promoting the growth of pathogens (Ji et al. 2023). They must also have a low caloric value to avoid contributing significantly to total caloric intake (Cerdó et al. 2019). Stability during food processing, storage, and passage through the gastrointestinal tract is another important factor (Rolim 2015). The solubility and viscosity of prebiotics can affect their functionality in the gastrointestinal tract and their applicability in different food products (Holscher 2017). Furthermore, they must be safe for human consumption and not cause adverse side effects at recommended doses, in addition to having scientific evidence to support their beneficial effects on intestinal health (Davani-Davari et al. 2019). Finally, they must be compatible with the formulation of food products, ensuring that they do not negatively affect the sensory and physicochemical characteristics of the food (Valladares-Diestra et al. 2023).

As for probiotics, several essential criteria are considered to ensure their effectiveness and safety. Considering these criteria, selecting high-quality probiotics that can benefit intestinal and general

Table 12.1 Criteria for selection and use of probiotics.

Criteria	Description	References
Accurate strain identification	It is essential to know the specific strain of probiotics, as different strains can have different effects on the body.	Hill et al. (2014)
Viability and stability	Probiotics must remain viable during the manufacturing, storage, and consumption process to ensure their effectiveness.	Terpou et al. (2019)
Scientific evidence	There must be scientific evidence supporting the benefits of probiotics for human health, including clinical studies and systematic reviews	Egea et al. (2023)
Gastrointestinal survival capacity	Probiotics must be able to survive the acidic environment of the stomach and digestive enzymes to reach the intestine alive, where they can exert their beneficial effects.	Plaza-Diaz et al. (2019)
Intestinal adhesion and colonization	Probiotics must have the ability to adhere to intestinal cells and colonize the intestine to exert their beneficial effects.	Hemaiswarya et al. (2013)
Secure source	Probiotics must be derived from safe sources, preferably human or animal origin, with a history of safety and effectiveness.	Hill et al. (2014)
Formulation compatibility	Probiotics must be compatible with the final product formulation to ensure their stability and effectiveness.	Ramlucken et al. (2021)
Absence of adverse effects	Probiotics must be safe for human consumption without causing significant adverse effects at recommended doses.	Sanders et al. (2010)
Appropriate labeling	Information on the product label must be clear and accurate, including the specific strain, amount of probiotics per dose, and storage and use instructions.	Merenstein et al. (2019)
Regulatory approval	The competent regulatory authorities must approve probiotics regarding their safety and efficacy.	Žuntar et al. (2020)

health is essential. Probiotic selection criteria include accurate strain identification, viability, and stability during storage and consumption based on scientific evidence supporting their health benefits. They must survive the gastrointestinal tract, adhere to and colonize the intestine, be derived from safe sources, and be compatible with the final product formulation. Furthermore, they must be safe and have appropriate labeling, following regulatory approval. These criteria are essential to ensure the efficacy and safety of the selected probiotics. Table 12.1 presents the selection criteria and description of the importance of probiotics.

The selection criteria for synbiotics are similar to those for probiotics and prebiotics, considering the combination of these elements. Following the selection and use criteria for probiotics and prebiotics is crucial to ensure their safety and effectiveness. Adopting these criteria makes it possible to maximize the health benefits and minimize any potential risks associated with its use.

12.3 Human Synbiotic Types

Various symbiotic products are currently available for human consumption on the market. These products combine probiotics and prebiotics in a single formulation, seeking to enhance the general benefits for the consumer's health. Synbiotic products are available in different forms, such as capsules, tablets, powders, liquids, and functional foods (Kvakova et al. 2021). Synbiotics also include functional foods and beverages such as yogurts and kefir, dietary supplements in fortified infant formulas, traditional fermented foods, specific medical products, and enriched processed foods. Each category offers diverse options for consumers looking to promote gut and microbiome health conveniently and effectively (Dahiya and Nigam 2022; Linares et al. 2017).

12.3.1 Main Prebiotics for Human Consumption

The main types of prebiotics used in human foods include:

- **Fructooligosaccharides (FOS):** FOS are a type of nondigestible carbohydrate found in various plant foods such as fruits, vegetables, barley, garlic, onions, and chicory (Sabater-Molina et al. 2009). According to Costa, Vasconcelos (Costa et al. 2021), the FOS are a "mixture of oligosaccharides, consisting of glucose linked to short chain polymers of fructose units by β-1,2 glycosidic bonds". FOS are not digested in the small intestine, reaching the colon intact, where they serve as a substrate for beneficial bacteria, such as bifidobacteria, stimulating their growth in the intestine and thus promoting intestinal health and improving the microbiota balance (Singh et al. 2017). In some individuals, FOS ingestion may lead to unfavorable gastrointestinal symptoms such as bloating and flatulence (Dou et al. 2022). These microbiota microorganisms ferment FOS, producing short-chain fatty acids such as acetic, propionic, and butyric acids. These fatty acids promote a favorable intestinal environment for the growth of beneficial bacteria and inhibit pathogenic bacteria growth. This way, FOS provide a nutrient medium for beneficial bacteria, promoting their growth and activity in the intestine (Kaewarsar et al. 2023).
- **Inulin:** Inulin is a polysaccharide formed by a chain of fructose units with a terminal glucose unit (Leite et al. 2004), found mainly in chicory, onion, garlic, barley, and artichoke, among others. Inulin has a low glycemic index, low caloric value, and neutral texture and flavors, characteristics that make it a versatile ingredient widely used in food products and supplements (Qin et al. 2023; Melilli et al. 2024).
- **Galactooligosaccharides (GOS):** GOS are nondigestible carbohydrates composed of a short chain of galactose molecules, usually linked to a glucose molecule. They act as prebiotics, selectively stimulating the growth and activity of beneficial bacteria in the gut (Gonçalves et al. 2022). Like other prebiotics, GOS are resistant to digestion in the small intestine and reach the colon practically intact, where they serve as a substrate for beneficial bacteria. Furthermore, GOS can help increase the adhesion of probiotic bacteria to the intestinal mucosa, improving their colonization capacity and effectiveness in the intestine (Plaza-Diaz et al. 2019; Bruno-Barcena and Azcarate-Peril 2015).
- **Lactulose:** It is classified as a disaccharide composed of a galactose molecule and a fructose molecule, produced commercially through a lactose isomerization reaction (Gänzle 2011). Lactulose is also fermented by intestinal bacteria, producing short-chain fatty acids such as lactic and acetic acids, which promote a favorable environment for the growth of beneficial bacteria in the colon (Salminen and Salminen 1997). In addition to its prebiotic power, it acts as an osmotic laxative and has a low glycemic index and high gastrointestinal tolerance, which gives lactulose a valuable status for application in medicines, foods, and supplements for the treatment of various bodily conditions, such as constipation and, consequently, promoting gut health (Pranami et al. 2017).
- **Pectin:** It is a polysaccharide composed mainly of galacturonic acid units linked by glycosidic bonds. Pectin is widely used in the food industry as a gelling, thickening, and stabilizing agent in

products such as jellies, jams, juices, sweets, and desserts (Lara-Espinoza et al. 2018). Like other prebiotics, pectin indirectly promotes probiotics' growth in the intestine, offering a range of health benefits, from improvements in digestive and immune health to promoting mental and emotional well-being (Naeem et al. 2024; Pascale et al. 2022).

12.3.2 Main Probiotics for Human Consumption

The properties and performance of probiotics may vary in their composition, source, and application and mainly include the use of *Lactobacillus* and *Bifidobacteria*, although other bacteria and specific yeasts are also used; multispecies combinations; specific strains for health conditions; and personalized strains (Mendonça et al. 2023; Fijan 2014).

Lactobacillus is one of the best-known groups of probiotics and plays a key role in promoting intestinal health. They are facultative anaerobic Gram-positive bacteria commonly found in the human gastrointestinal tract (Dempsey and Corr 2022). Furthermore, these beneficial bacteria belong to the genus *Lactobacillus* and are commonly found in many fermented foods, probiotic supplements, and many fermented foods such as yogurt, sauerkraut, and kefir (Soemarie et al. 2021).

Among the bacteria most commonly used as probiotic products are *Lactobacillus acidophilus, Lactobacillus casei, Lactobacillus paracasei, Lactobacillus rhamnosus, Lactobacillus delbrueckii* subsp. *bulgaricus, Lactobacillus brevis, Lactobacillus johnsonii, Lactobacillus Plantarum,* and *Lactobacillus fermentum* (Fijan 2014). When administered in adequate amounts, they provide benefits to the host's health, including strengthening the immune system, preventing pathogen infections, and reducing inflammatory processes, among other benefits (Dempsey and Corr 2022).

Like other probiotics, *Lactobacillus* species are believed to act in several ways, including competing with pathogens for nutrients and space, producing organic acids that lower intestinal pH, modulating the host's immune response, and producing antimicrobial substances (Tang et al. 2023). The effects of probiotics may vary between different species and strains of *Lactobacillus*, as well as between individuals. Thus, more research is needed to fully understand their therapeutic potential and mechanisms of action (Huang et al. 2022).

Bifidobacteria, in turn, are a group of facultative anaerobic Gram-positive bacteria, meaning they can survive both in environments with and without oxygen but prefer environments with a low oxygen content (Jiang et al. 2022). Bifidobacteria are essential for intestinal health, playing several fundamental roles. They maintain the intestinal microbiota balance, competing with pathogenic bacteria and fermenting dietary fiber to produce beneficial fatty acids. Furthermore, they strengthen the intestinal barrier, stimulate mucin production, and modulate the immune response, promoting tolerance and reducing inflammation. Some strains also produce B vitamins. These combined actions contribute to healthy digestion, a strong immune system, and preventing adverse conditions in the intestine (O'Callaghan and van Sinderen 2016; Gavzy et al. 2023; Tungland 2018). Several strains of bifidobacteria are considered essential probiotics, including *Bifidobacterium infantis, Bifidobacterium adolescentis, Bifidobacterium animalis subsp animalis, Bifidobacterium animalis subsp lactis, Bifidobacterium bifidum, Bifidobacterium longum*, and *Bifidobacterium breve* (Fijan 2014; Chen et al. 2021).

Among yeasts, the use of *Saccharomyces boulardii* stands out due to its properties and advantages in administration. *S. boulardii* is a nonpathogenic yeast, recognized as safe for application in food and pharmaceuticals, and the only yeast with a probiotic status. Furthermore, *S. boulardii* has high resistance to gastric acidity and bile salts, making it capable of surviving and colonizing the small intestine and can help restore the balance of the intestinal microbiota after some disorder. This yeast acts as a regulator of the intestinal microbiota and has high antioxidant properties, a result of both yeast and metabolites produced in the fermentation process (Egea et al. 2023; McFarland 2017).

It is reported that this yeast has excellent antioxidant, antibacterial, antitumor, and anti-inflammatory capabilities due to the diversity of bioactive metabolites produced, which include 2-hydroxyisocaproic acid, gamma-aminobutyric acid, shikimic acid, P-aminobenzoic acid, tyrosol, and polylactic acid (Fu et al. 2023). Another important property compared to other probiotics, especially bacteria, is related to low resistance to antimicrobials, which contributes to and facilitates their use in the development of foods and formulations, as it guarantees their conservation and survival in the matrix (McFarland 2017).

Due to the different properties of probiotics, it is essential to consider the different properties, resistance, adequate dose of effect, safety, and parameters such as interference and/or compatibility with the food matrix in their application. Considering these factors guarantees successful application and, mainly, the adequate effect on the consumer's intestinal and bodily functions, in general.

12.3.3 Main Combinations of Probiotics and Prebiotics in Synbiotic Products

The combination of prebiotics and probiotics has already proven to be a way to enhance the benefits associated with these two components. As previously mentioned, the combination of prebiotic and probiotic components in a synbiotic product aims to make the prebiotics provide the necessary substrate for the growth of the probiotics, while the probiotics, in turn, colonize the intestine and use the prebiotics as a source of energy, increasing its activity and effectiveness. This synergy between probiotics and prebiotics can result in better intestinal health, strengthened immune systems, and other benefits for the host (Campaniello et al. 2023; Markowiak and Śliżewska 2017).

Although there are countless combinations of prebiotics and probiotics available, creating synbiotic products, some are commonly used due to their already proven beneficial effects, including the combination of inulin + *Lactobacillus*, FOS + *Bifidocaterium*, and GOS + *Lactobacillus*, among others. Different combinations of prebiotics and probiotics have already been verified and even proven in some studies. In this way, its characteristics and, mainly, its action on humans will be better presented in the item "4 Action of synbiotics in humans." In any case, in addition to their combination and verification of effectiveness, it is essential to understand the mechanism of action of these compounds and their combinations to elucidate how they promote the health of consumers of these products.

12.4 Mechanism of Action of Synbiotics

The combination of probiotics and prebiotics can result in a synergistic effect, as probiotics primarily exert their activities in the small and large intestines. In contrast, the effects of prebiotics are predominantly observed in the large intestine (Hamasalim 2016). Prebiotics are mainly employed as a selective substrate to promote the growth of a probiotic strain, favoring fermentation and passage through the intestine. Evidence in the literature indicates that, due to prebiotics, probiotic microorganisms develop an increased tolerance to environmental conditions, including factors such as oxygenation, pH, and temperature in the organism's specific intestine (Sekhon and Jairath 2010).

The concept of symbiotic action mechanisms, based on modifying the intestinal microbiota with appropriately selected probiotic microorganisms and prebiotics as their substrates, is presented in

Figure 12.1. Stimulating probiotics through prebiotics leads to regulating metabolic activity in the intestine, preserving intestinal biostructure, promoting the growth of beneficial microbiota, and inhibiting potential pathogens in the gastrointestinal tract (de Vrese and Schrezenmeir 2008). Symbiotics contribute to decreases in concentrations of undesirable metabolites and the deactivation of nitrosamines and carcinogenic substances. The use of these compounds results in a notable increase in levels of short-chain fatty acids, ketones, carbon disulfides, and methyl acetates, suggesting a potentially beneficial effect on host health (Manigandan et al. 2012).

Regarding their therapeutic efficacy, symbiotics exhibit desirable properties encompassing antibacterial, anticancer, and antiallergic attributes. Additionally, they neutralize decomposition processes in the intestine, preventing both constipation and diarrhea. Remarkably, symbiotics prove highly effective in preventing osteoporosis, reducing blood fat and sugar levels, regulating the immune system, and treating brain disorders associated with abnormal liver functions (Pandey et al. 2015).

Symbiotics have been developed to overcome potential survival challenges faced by probiotics. The decision to utilize symbiotics seems to stem from observations indicating enhanced survival of probiotic bacteria as they traverse the upper gastrointestinal tract. Achieving more effective colonization in the colon, coupled with a growth-stimulating impact on probiotics and widespread bacteria, plays a role in preserving intestinal balance and promoting overall health (Khursheed et al. 2022).

12.5 Action of Synbiotics in Humans

A search in the literature (National Library of Medicine, PubMed) for randomized controlled trials of foods with symbiotic potential carried out in the last 20 years revealed the existence of only nine articles that were briefly demonstrated in Table 12.2. Among the manuscripts found, four reported results with protocols using food in the form of drinks containing milk or dairy products (Naito et al. 2014; Bahrudin et al. 2020; Moroti et al. 2012; de Toro Martín-Consuegra et al. 2014), infant

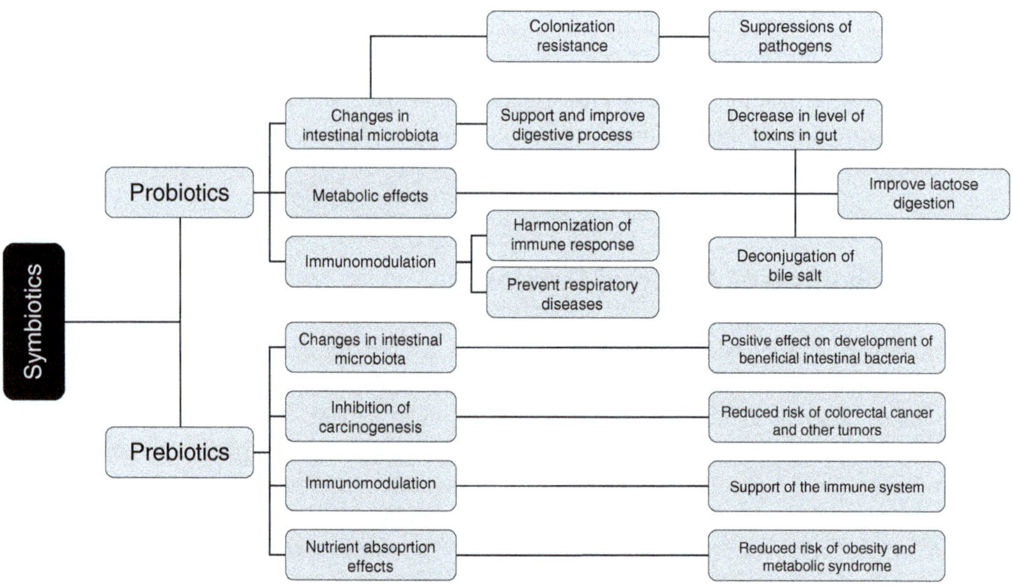

Figure 12.1 Mechanisms of action of synbiotics and their effects.

Table 12.2 The literature reports from the National Library of Medicine (PubMed) of randomized controlled trials of foods with symbiotic potential carried out in the last 20 years.

Synbiotic product	Study protocol (population × dose × duration)	Reported beneficial effects	References
Fruit juice added KC-1317 symbiotic containing microbial lysate (*Saccharomyces boulardii* lysate in a cranberry, colostrum-derived lactoferrin, fragaria, and lactose mixture)	Immunocompromised healthy elderly subjects with decreased natural killer cell activity consumed 100 mL twice daily during 9 wk.	Increased natural killer cell activity for the group treated with KC-1317.	Naito et al. (2014)
Sterilized symbiotic drink (*Lactobacillus helveticus* + 5.85 g polydextrose)	Patients fulfilling the Rome III criteria for constipation-predominant irritable bowel syndrome ($n = 163$) consumed 350 mL of the drink during a week.	Decreased median intestinal transit time (from 18 to 13 h); Same fecal pH compared to the control group (drink containing only *L. helveticus* without polydextrose). Relieved symptoms related to constipation.	Bahrudin et al. (2020)
Symbiotic shake containing 4×10^8 CFU/100 mL *Lactobacillus acidophillus*, 4×10^8 CFU/100 mL *Bifidobacterium bifidum*, and 1 g/100 mL of FOS.	Healthy female individuals ($n = 18$) with cholesterol >200 mg/ dL; triglycerides >150 mg/dL; glycemia >110 mg/dL, aged 50–65 yr, who consumed two daily doses of 100 mL symbiotic shake for 30 d.	Increase of 35.15% in HDL-C levels. Decrease of 25.84% and 37.27% in their serum levels of total cholesterol and triglycerides, respectively; 41.30% in n-HDL-C levels compared to baseline; and 38.89% of blood glucose.	Moroti et al. (2012)
Simbiotic Drink® contains partially skimmed pasteurized milk, sugar, Beneo 1.5% soluble dietary fiber (oligofructose), caramel, and lactic ferments (*Streptococcus thermophilus*, *Lactobacillus bulgaricus*, *Lactobacillus casei*, *Lactobacillus acidophilus*, and *Bifidobacterium*)	Subjects with syndrome multiorgan dysfunction patients ($n = 89$) received the drink for 7 d via enteral nutrition.	Increased fibrinogen concentration on the 5th and 7th d; and relative protein C concentration on the 5th d. Decreased biochemical parameters considered acute phase reactants and indicators of inflammation compared to the other groups; lactate concentration on the 2nd d; *Candida* colonization in mucous membranes; and D-dimers on Day 7 of intra-ICU evolution. The evolution of SOFA score of patients is same compared between groups.	de Toro Martín-Consuegra et al. (2014)

(*Continued*)

Table 12.2 (Continued)

Synbiotic product	Study protocol (population × dose × duration)	Reported beneficial effects	References
α-lactalbumin-enriched and symbiotic-supplemented infant formula	Non-breastfed term neonates ($n = 97$) consumed infant formula *ad libidum* until the babies reach 6 mo old.	Increased fecal SIgA concentration between 1 and 6 mo. Decreased crying or agitation compared to the control group (age 1 mo); and frequency of atopic dermatitis at 6 mo.	Rozé et al. (2012)
Dual potential prebiotic Metlin® or Metlos® (fructans obtained from *Agave tequilana* var Weber with β (2→1) and β (2→6) linked fructofuranosyl units) and probiotic (*Lactobacillus* GG, 10^7) incorporated to an infant formula	Healthy term babies ($n = 600$) consumed powder to the infant formula *ad libidum* until the babies reach 6 mo old.	Increased colic and greater than 10 flatulence episodes/day in the symbiotic with Metlos® group compared to the others.	López-Velázquez et al. (2013)
Symbiotic gel (Nutrihealth®) containing *Lactobacillus acidophilus* and *Bifidobacterium bifidum* (2×10^{12} CFU), 2.31 g of a prebiotic fiber (inulin), 1.5 g of omega-3 fatty acids (eicosapentaenoic and docosahexaenoic acid) and vitamins (complex B, folic acid, ascorbic acid, and vitamin E)	Patients with end-stage renal disease ($n = 18$) consumed 1 gel/d in fasting for 2 mo.	Increased *Bifidobacterium* group between the beginning and end of the study for the intervention group. Decreased *Lactobacillus* group between the beginning and end of the study for the intervention group.	Cruz-Mora et al. (2014)
Symbiotic gel (Nutrihealth®)	Hemodialysis patients ($n = 44$) consumed 1 gel/d in fasting for 2 mo.	Decreased presence of swelling in the 1st and 2nd mo of treatment in the intervention group versus baseline; prevalence and monthly episodes of vomiting, heartburn, and stomach pain; severity of gastrointestinal symptoms compared to the control group; and CRP concentration.	Viramontes-Hörner et al. (2015)

CFU, colony-forming unit; HDL-C, high-density lipoprotein cholesterol; n-HDL-C, non-high-density lipoprotein cholesterol; SOFA score, sequential organ failure assessment score; ICU, intensive care unit; SIgA, secretory immunoglobulin A.

formula (Rozé et al. 2012; López-Velázquez et al. 2013), and food in gel form (Cruz-Mora et al. 2014; Viramontes-Hörner et al. 2015). The audiences studied in *in vivo* studies are among the most varied, including healthy adult and elderly subjects, immunocompromised with constipation-predominant irritable bowel syndrome, with syndrome multiorgan dysfunction, and with end-stage renal disease or hemodialysis patients or even neonates.

Although only one of the manuscripts found reported qualification results or an abundance of microorganisms in the intestinal microbiota (Cruz-Mora et al. 2014), in general, the health benefits reported in these manuscripts are related to changes in the microbiota. However, data remain scarce or unclear due to the study's low sample population or lack of specific analysis that indicates this relationship (Bahrudin et al. 2020; Viramontes-Hörner et al. 2015).

The consumption of prebiotics in synbiotic food modulation studies appears to stimulate microbial growth in the intestinal microbiota, which can increase fecal volume and increase the speed of intestinal transit, thus reducing constipation (Bahrudin et al. 2020; Cruz-Mora et al. 2014). In newborn babies, the benefit related to improving intestinal transit is significant as they experience pain, colic, abdominal distension, pain, and agitation due to the development of gastrointestinal transit after birth (Rozé et al. 2012; López-Velázquez et al. 2013). On the other hand, an excess of it can increase bloating and abdominal distension (Bahrudin et al. 2020). Furthermore, increased beneficial microorganisms in the intestinal microbiota, such as *Bifidobacterium* spp and *Lactobacillus* spp., can increase microbial competition by inhibiting pathogenic microorganisms, increase lactose digestion, and improve nutrient absorption (Cruz-Mora et al. 2014).

Furthermore, when we refer specifically to probiotic microorganisms contained in symbiotic foods, the literature demonstrates that their potential as immunomodulators is maintained by improving the host's immunological health (Naito et al. 2014; Cruz-Mora et al. 2014; Lee et al. 2016). One of the proposed mechanisms regarding improved immunity is that an overgrowth of pathogenic microorganisms in the intestinal lumen can increase the translocation of microorganisms or liposaccharides (LPS) that can activate immunity to systemic inflammation (Egea et al. 2023; Viramontes-Hörner et al. 2015).

The administration of synbiotics was also associated with a reduction in serum cholesterol and its fractions (Bahrudin et al. 2020; Moroti et al. 2012; Cruz-Mora et al. 2014), which may be related to (i) inhibition of the cholesterol synthesis enzyme by some metabolites, mainly short-chain fatty acids and (ii) increased microbial growth can increase intestinal volume and transit by facilitating the elimination of cholesterol and bile salts through feces and inhibiting absorption back into the body through binding with these molecules (Moroti et al. 2012; Viramontes-Hörner et al. 2015).

Although the results of the manuscripts reported in the literature are promising, there are still some discrepancies and gaps that may be due to the species and strain of the microorganism used in the development of the protocol, the volunteer population selected for each study, the dose × duration of the administration protocol used in the study, and the most evident results are always present in long-term protocols.

12.6 Final Considerations

Synbiotics represent a promising approach to promoting intestinal and overall health in the human body. By combining probiotics and prebiotics, synbiotic work synergistically to improve the intestinal microbiota balance, strengthen the intestinal barrier, and modulate the immune system. These effects may contribute to various health benefits, including better digestion, a more effective immune response, and potentially a reduced risk of several gut-related conditions. However, more research is needed to understand the mechanisms of synbiotics' action and their effectiveness in different clinical contexts.

The prospects for synbiotics are centered on meeting market expectations, which continue to grow, in addition to fully elucidating the mechanisms related to the prevention of various diseases due to the consumption of synbiotics. In addition, a focus on developing personalized formulations is expected to meet each person's specific needs, the development of appropriate and resistant strains, and popularize clinical application in the treatment of various health conditions, including gastrointestinal disorders, metabolic diseases, immunological and others.

Acknowledgment

The authors acknowledge the financial support of CNPq, FAPEG, FAPERJ, CAPES, and IF Goiano.

References

Almeida, C.V., Antiga, E., and Lulli, M. (2023). Oral and topical probiotics and postbiotics in skincare and dermatological therapy: a concise review. *Microorganisms* 11 (6): 1–17.

Bahrudin, M.F., Abdul Rani, R., Tamil, A.M. et al. (2020). Effectiveness of sterilized symbiotic drink containing *Lactobacillus helveticus* comparable to probiotic alone in patients with constipation-predominant irritable bowel syndrome. *Digestive Diseases and Sciences* 65: 541–549.

Bamigbade, G.B., Subhash, A.J., Kamal-Eldin, A. et al. (2022). An updated review on prebiotics: insights on potentials of food seeds waste as source of potential prebiotics. *Molecules* 27 (18): 5947.

Bruno-Barcena, J.M. and Azcarate-Peril, M.A. (2015). Galacto-oligosaccharides and colorectal cancer: feeding our intestinal probiome. *Journal of Functional Foods* 12: 92–108.

Campaniello, D., Bevilacqua, A., Speranza, B. et al. (2023). A narrative review on the use of probiotics in several diseases. *Evidence and perspectives. Frontiers in Nutrition.* 10: 1–12.

Cerdó, T., García-Santos, J.A., and M GB, Campoy C. (2019). The role of probiotics and prebiotics in the prevention and treatment of obesity. *Nutrients* 11 (3): 1–31.

Chen, J., Chen, X., and Ho, C.L. (2021). Recent development of probiotic bifidobacteria for treating human diseases. *Frontiers in Bioengineering and Biotechnology* 9: 770248.

Costa, G.T., Vasconcelos, Q., Abreu, G.C. et al. (2021). Systematic review of the ingestion of fructooligosaccharides on the absorption of minerals and trace elements versus control groups. *Clinical Nutrition ESPEN* 41: 68–76.

Cruz-Mora, J., Martínez-Hernández, N.E., del Campo-López, F.M. et al. (2014). Effects of a symbiotic on gut microbiota in Mexican patients with end-stage renal disease. *Journal of Renal Nutrition* 24 (5): 330–335.

Dahiya, D. and Nigam, P.S. (2022). Probiotics, prebiotics, synbiotics, and fermented foods as potential biotics in nutrition improving health via microbiome-gut-brain axis. *Fermentation.* 8 (7): 303.

Davani-Davari, D., Negahdaripour, M., Karimzadeh, I. et al. (2019). Prebiotics: definition, types, sources, mechanisms, and clinical applications. *Foods (Basel, Switzerland).* 8 (3): 1–27.

de Toro Martín-Consuegra, I.L., Sanchez-Casado, M., Sánchez-Belmonte, M.J.P.-P. et al. (2014). The influence of symbiotics in multi-organ failure: randomised trial. *Medicina Clínica* 143 (4): 143–149.

de Vrese, M. and Schrezenmeir, J. (2008). Probiotics, prebiotics, and synbiotics. *Advances in Biochemical Engineering/Biotechnology* 111: 1–66.

Dempsey, E. and Corr, S.C. (2022). *Lactobacillus* spp. for gastrointestinal health: current and future perspectives. *Frontiers in Immunology* 13: 840245.

Dou, Y., Yu, X., Luo, Y. et al. (2022). Effect of fructooligosaccharides supplementation on the gut microbiota in human: a systematic review and meta-analysis. *Nutrients* 14 (16): 3298.

Egea, M.B., Oliveira Filho, J.G., and Lemes, A.C. (2023). Investigating the efficacy of *Saccharomyces boulardii* in metabolic syndrome treatment: a narrative review of what is known so far. *International Journal of Molecular Sciences* 24 (15): 12015.

Fijan, S. (2014). Microorganisms with claimed probiotic properties: an overview of recent literature. *International Journal of Environmental Research and Public Health* 11 (5): 4745–4767.

Fu, J., Liu, J., Wen, X. et al. (2023). Unique probiotic properties and bioactive metabolites of *Saccharomyces boulardii*. *Probiotics and Antimicrobial Proteins* 15 (4): 967–982.

Gänzle, M.G. (2011). Lactose and oligosaccharides/lactose: derivatives. In: *Encyclopedia of Dairy Sciences*, 2e (ed. J.W. Fuquay), 202–208. San Diego: Academic Press.

Gavzy, S.J., Kensiski, A., Lee, Z.L. et al. (2023). Bifidobacterium mechanisms of immune modulation and tolerance. *Gut Microbes* 15 (2): 2291164.

George Kerry, R., Patra, J.K., Gouda, S. et al. (2018). Benefaction of probiotics for human health: a review. *Journal of Food and Drug Analysis* 26 (3): 927–939.

Gonçalves, D.A., Teixeira, J.A., and Nobre, C. (2022). Chapter 19 - In situ enzymatic synthesis of prebiotics to improve food functionality. In: *Value-Addition in Food Products and Processing Through Enzyme Technology* (ed. M. Kuddus and C.N. Aguilar), 253–267. Academic Press.

Hamasalim, H. (2016). Synbiotic as feed additives relating to animal health and performance. *Advances in Microbiology* 06: 288–302.

Hemaiswarya, S., Raja, R., Ravikumar, R., and Carvalho, I.S. (2013). Mechanism of action of probiotics. *Brazilian Archives of Biology and Technology* 56 (1): 113–119.

Hijová, E. (2022). Synbiotic supplements in the prevention of obesity and obesity-related diseases. *Metabolites* 12 (4): 313.

Hill, C., Guarner, F., Reid, G. et al. (2014). The International Scientific Association for Probiotics and Prebiotics consensus statement on the scope and appropriate use of the term probiotic. *Nature Reviews. Gastroenterology & Hepatology* 11 (8): 506–514.

Holscher, H.D. (2017). Dietary fiber and prebiotics and the gastrointestinal microbiota. *Gut Microbes* 8 (2): 172–184.

Huang, R., Wu, F., Zhou, Q. et al. (2022). *Lactobacillus* and intestinal diseases: mechanisms of action and clinical applications. *Microbiological Research* 260: 127019.

Ji, J., Jin, W., Liu, S.J. et al. (2023). Probiotics, prebiotics, and postbiotics in health and disease. *MedComm* 4 (6): e420.

Jiang, H., Cai, M., Shen, B. et al. (2022). Synbiotics and gut microbiota: new perspectives in the treatment of type 2 diabetes mellitus. *Foods (Basel, Switzerland).* 11 (16): 1–18.

Jiang, Z., Li, M., McClements, D.J. et al. (2022). Recent advances in the design and fabrication of probiotic delivery systems to target intestinal inflammation. *Food Hydrocolloids* 125: 107438.

Kaewarsar, E., Chaiyasut, C., Lailerd, N. et al. (2023). Optimization of mixed inulin, fructooligosaccharides, and galactooligosaccharides as prebiotics for stimulation of probiotics growth and function. *Foods (Basel, Switzerland).* 12 (8): 1591.

Khursheed, R., Gulati, M., Wadhwa, S. et al. (2022). Multifaceted role of synbiotics as nutraceuticals, therapeutics and carrier for drug delivery. *Chemico-Biological Interactions* 368: 110223.

Kvakova, M., Bertkova, I., Stofilova, J., and Savidge, T.C. (2021). Co-encapsulated synbiotics and immobilized probiotics in human health and gut microbiota modulation. *Foods (Basel, Switzerland)* 10 (6): 1–28.

Lara-Espinoza, C., Carvajal-Millán, E., Balandrán-Quintana, R. et al. (2018). Pectin and pectin-based composite materials: beyond food texture. *Molecules (Basel, Switzerland)* 23 (4): 942.

Lee, M.-H., Kim, M., Kim, M. et al. (2016). Consumption of dairy yogurt with the polysaccharide rhamnogalacturonan from the peel of the Korean citrus hallabong enhances immune function and attenuates the inflammatory response. *Food & Function* 7 (6): 2833–2839.

Leite, J.T.C., Park, K.J., Ramalho, J.R.P., and Furlan, D.M. (2004). Caracterização reológica das diferentes fases de extrato de inulina de raízes de chicória, obtidas por abaixamento de temperatura. *Engenharia Agrícola* 24: 202–210.

Liao, W., Su, M., and Zhang, D. (2022). A study on the effect of symbiotic fermented milk products on human gastrointestinal health: double-blind randomized controlled clinical trial. *Food Science & Nutrition* 10 (9): 2947–2955.

Linares, D.M., Gómez, C., Renes, E. et al. (2017). Lactic acid bacteria and bifidobacteria with potential to design natural biofunctional health-promoting dairy foods. *Frontiers in Microbiology* 8: 1–11.

López-Velázquez, G., Díaz-García, L., Anzo, A. et al. (2013). Safety of a dual potential prebiotic system from Mexican agave "Metlin® and Metlos®", incorporated to an infant formula for term newborn babies: a randomized controlled trial. *Revista de Investigación Clínica* 65: 483–490.

Malik, J.K., Ahmad, A.H., Kalpana, S. et al. (2016). Chapter 57 - Synbiotics: safety and toxicity considerations. In: *Nutraceuticals* (ed. R.C. Gupta), 811–822. Boston: Academic Press.

Manassi, C.F., de Souza, S.S., Hassemer, G.S. et al. (2022). Functional meat products: trends in pro-, pre-, syn-, para- and post-biotic use. *Foodservice Research International* 154: 111035.

Manigandan, T., Mangaiyarkarasi, S.P., Hemalatha, R., and Hemalatha, V.T. (2012). Probiotics, prebiotics and synbiotics: a review. *Biomedical and Pharmacology Journal* 5: 295–304.

Markowiak, P. and Śliżewska, K. (2017). Effects of probiotics, prebiotics, and synbiotics on human health. *Nutrients* 9 (9): 1–30.

McFarland, L.V. (2017). Chapter 18 - Common organisms and probiotics: *Saccharomyces boulardii*. In: *The Microbiota in Gastrointestinal Pathophysiology* (ed. M.H. Floch, Y. Ringel, and W. Allan Walker), 145–164. Boston: Academic Press.

Melilli, M.G., Buzzanca, C., and Di Stefano, V. (2024). Quality characteristics of cereal-based foods enriched with different degree of polymerization inulin: a review. *Carbohydrate Polymers* 332: 121918.

Mendonça, A.A., Pinto-Neto, W.P., da Paixão, G.A. et al. (2023). Journey of the probiotic bacteria: survival of the fittest. *Microorganisms.* 11 (1): 95.

Merenstein, D., Guzzi, J., and Sanders, M.E. (2019). More information needed on probiotic supplement product labels. *Journal of General Internal Medicine* 34 (12): 2735–2737.

Moroti, C., Souza Magri, L.F., de Rezende, C.M. et al. (2012). Effect of the consumption of a new symbiotic shake on glycemia and cholesterol levels in elderly people with type 2 diabetes mellitus. *Lipids in Health and Disease* 11: 1–8.

Naeem, H., Hassan, H.U., Shahbaz, M. et al. (2024). Role of probiotics against human cancers, inflammatory diseases, and other complex malignancies. *Journal of Food Biochemistry* 2024: 6632209.

Naito, Y., Marotta, F., Kantah, M.K. et al. (2014). Gut-targeted immunonutrition boosting natural killer cell activity using *Saccharomyces boulardii* lysates in immuno-compromised healthy elderly subjects. *Rejuvenation Research* 17 (2): 184–187.

Neri-Numa, I.A. and Pastore, G.M. (2020). Novel insights into prebiotic properties on human health: a review. *Foodservice Research International* 131: 108973.

O'Callaghan, A. and van Sinderen, D. (2016). Bifidobacteria and their role as members of the human gut microbiota. *Frontiers in Microbiology* 7: 925.

Pandey, K.R., Naik, S.R., and Vakil, B.V. (2015). Probiotics, prebiotics and synbiotics: a review. *Journal of Food Science and Technology* 52 (12): 7577–7587.

Pascale, N., Gu, F., Larsen, N. et al. (2022). The potential of pectins to modulate the human gut microbiota evaluated by in vitro fermentation: a systematic review. *Nutrients* 14 (17): 3629.

Plaza-Diaz, J., Ruiz-Ojeda, F.J., Gil-Campos, M., and Gil, A. (2019). Mechanisms of action of probiotics. *Advances in Nutrition* 10 (suppl_1): S49–S66.

Pranami, D., Sharma, R., and Pathak, H. (2017). Lactulose: a prebiotic, laxative and detoxifying agent. *Drugs & Therapy Perspectives* 33 (5): 228–233.

Qin, Y.Q., Wang, L.Y., Yang, X.Y. et al. (2023). Inulin: properties and health benefits. *Food & Function* 14 (7): 2948–2968.

Ramlucken, U., Ramchuran, S.O., Moonsamy, G. et al. (2021). Production and stability of a multi-strain Bacillus based probiotic product for commercial use in poultry. *Biotechnology Reports* 29: e00575.

Ranjha, M.M.A.N., Shafique, B., Batool, M. et al. (2021). Nutritional and health potential of probiotics: a review. *Applied Sciences* 11 (23): 11204.

Rolim, P.M. (2015). Development of prebiotic food products and health benefits. *Food Science and Technology* 35: 3–10.

Rozé, J.-C., Barbarot, S., Butel, M.-J. et al. (2012). An α-lactalbumin-enriched and symbiotic-supplemented v. a standard infant formula: a multicentre, double-blind, randomised trial. *The British Journal of Nutrition* 107 (11): 1616–1622.

Sabater-Molina, M., Larqué, E., Torrella, F., and Zamora, S. (2009). Dietary fructooligosaccharides and potential benefits on health. *Journal of Physiology and Biochemistry* 65 (3): 315–328.

Salminen, S. and Salminen, E. (1997). Lactulose, lactic acid bacteria, intestinal microecology and mucosal protection. *Scandinavian Journal of Gastroenterology. Supplement* 222: 45–48.

Sanders, M.E., Akkermans, L.M., Haller, D. et al. (2010). Safety assessment of probiotics for human use. *Gut Microbes* 1 (3): 164–185.

Sekhon, B. and Jairath, S. (2010). Prebiotics, probiotics and synbiotics: an overview. *Journal of Pharmaceutical Education and Research.* 1: 13–36.

Shewale, R.N., Sawale, P., Khedkar, C., and Singh, A. (2014). Selection criteria for probiotics: a review. *International Journal of Probiotics and Prebiotics.* 9: 17–22.

Simon, E., Călinoiu, L.F., Mitrea, L., and Vodnar, D.C. (2021). Probiotics, prebiotics, and synbiotics: implications and beneficial effects against irritable bowel syndrome. *Nutrients* 13 (6): 1–27.

Singh, S.P., Jadaun, J.S., Narnoliya, L.K. et al. (2017). Prebiotic oligosaccharides: special focus on fructooligosaccharides, its biosynthesis and bioactivity. *Applied Biochemistry and Biotechnology* 183 (2): 613–635.

Soemarie, Y.B., Milanda, T., and Barliana, M.I. (2021). Fermented foods as probiotics: a review. *Journal of Advanced Pharmaceutical Technology & Research* 12 (4): 335–339.

Tang, H., Huang, W., and Yao, Y.F. (2023). The metabolites of lactic acid bacteria: classification, biosynthesis and modulation of gut microbiota. *Microbial cell (Graz, Austria)* 10 (3): 49–62.

Terpou, A., Papadaki, A., Lappa, I.K. et al. (2019). Probiotics in food systems: significance and emerging strategies towards improved viability and delivery of enhanced beneficial value. *Nutrients* 11 (7): 1–32.

The Brainy Insights (2021). Synbiotics market size by form (liquid, capsules, and powder), application (pharmaceutical, animal feed, food & beverages, dairy industry, and aquaculture farming), distribution channel (indirect/B2C, online retail, direct/B2B, hypermarkets/supermarkets, retail stores, convenience stores, and others), regions, global industry analysis, share, growth, trends, and forecast 2024 to 2033. https://thebrainyinsights.com/report/synbiotics-market-13103 (accessed 7 March 2023).

Tungland, B. (2018). Chapter 5 - Direct physiological effects on local GI and indirect systemic effects of prebiotic fructan treatment, and its role in disease prevention and therapy. In: *Human Microbiota in Health and Disease* (ed. B. Tungland), 155–196. Academic Press.

Valladares-Diestra, K.K., de Souza Vandenberghe, L.P., Vieira, S. et al. (2023). The potential of xylooligosaccharides as prebiotics and their sustainable production from agro-industrial by-products. *Foods (Basel, Switzerland).* 12 (14): 1–20.

Viramontes-Hörner, D., Márquez-Sandoval, F., Martín-del-Campo, F. et al. (2015). Effect of a symbiotic gel (*Lactobacillus acidophilus* + *Bifidobacterium lactis* + inulin) on presence and severity of gastrointestinal symptoms in hemodialysis patients. *Journal of Renal Nutrition* 25 (3): 284–291.

Whisner, C.M. and Castillo, L.F. (2018). Prebiotics, bone and mineral metabolism. *Calcified Tissue International* 102 (4): 443–479.

Xiong, R.G., Li, J., Cheng, J. et al. (2023). The role of gut microbiota in anxiety, depression, and other mental disorders as well as the protective effects of dietary components. *Nutrients* 15 (14): 1–20.

Yadav, M.K., Kumari, I., Singh, B. et al. (2022). Probiotics, prebiotics and synbiotics: safe options for next-generation therapeutics. *Applied Microbiology and Biotechnology* 106 (2): 505–521.

You, S., Ma, Y., Yan, B. et al. (2022). The promotion mechanism of prebiotics for probiotics: a review. *Frontiers in Nutrition* 9: 1000517.

Žuntar, I., Petric, Z., Bursać Kovačević, D., and Putnik, P. (2020). Safety of probiotics: functional fruit beverages and nutraceuticals. *Foods (Basel, Switzerland)* 9 (7): 947.

13

Postbiotic Supplements with Nutraceutical Significance

Amin Abbasi[1], *Hedayat Hosseini*[1], *Hadi Pourjafar*[2], *Leili Aghebati Maleki*[3], *Atiyeh Ghafouri Ghotbabad*[1], *Sahar Sabahi*[4], *Parvin Orojzade*[5], *and Mohammadreza Ziavand*[5]

[1] *Department of Food Science and Technology, National Nutrition and Food Technology Research Institute, Faculty of Nutrition Science and Food Technology, Shahid Beheshti University of Medical Sciences, 19395-4741, Tehran, Iran*
[2] *Dietary Supplements and Probiotic Research Center, Alborz University of Medical Sciences, 19395-4963, Karaj, Iran*
[3] *Immunology Research Center, Tabriz University of Medical Sciences, 19294-4753, Tabriz, Iran*
[4] *Department of Nutrition, School of Allied Medical Sciences, Ahvaz Jundishapur University of Medical Sciences, 19294-3342, Ahvaz, Iran*
[5] *Department of Food Sciences and Technology, Faculty of Nutrition & Food Sciences, Tabriz University of Medical Sciences, 19294-4753, Tabriz, Iran*

13.1 Introduction

Recently, there has been a significant increase in customer orientation toward assessing and selecting healthy food, as individuals are now more aware of the significance of a healthy diet. As a result, there has been a growth in the number of food goods that make functional statements. Functional foods do not have a broadly acknowledged description (Baker et al. 2022). Functional foods involve multiple groups, such as prebiotics, probiotics, synbiotics, nutraceuticals, and fermented conventional foods. Probiotics are live bacteria that offer health advantages when adequately intake, often by promoting or restoring the intestine-beneficial flora. Prebiotics are a kind of food nutrient that is broken down by gut bacteria. Prebiotics have the potential to improve human health by either replacing or functioning in tandem with probiotics, thanks to their medicinal advantages and better production and storage capabilities compared to probiotics (Damián et al. 2022). Synbiotics are a blend of probiotics and prebiotics that work together to produce health benefits that may not be achieved when taken alone. However, other terms have arisen in recent years, such as postbiotics. Research has shown that inactive cells, whether completely or broken, may also have significant impacts on the well-being of individuals. The health effects of gut microbiomes depend on their health and activity, as well as the existence of inanimate microbiome metabolites. Postbiotics, derived from probiotics, offer health advantages to the host when taken in sufficient quantities (Sabahi et al. 2023; Abbasi et al. 2022). Postbiotics are bacterial byproducts that impact the well-being of humans. The health benefits and bioactive consequences of postbiotics are still unclear or not well defined (Cuevas-González et al. 2020). Continued investigation into new technologies, methodologies, and possible applications might revise scientific comprehension and elucidate the benefits of postbiotic substances for nutrition and overall well-being (Mayorgas et al. 2021; Wegh et al. 2019).

Microbial Nutraceuticals: Products and Processes, First Edition. Edited by Sudhir Pratap Singh and Santosh Kumar Upadhyay.

Usually, the microbes that create postbiotics can be naturally present in a variety of fermented foods. Recently, investigators have shown a connection between the microbiota and immune system function. Consequently, there is an increasing interest in gastrointestinal health and nutritional items that support a healthy gut. Postbiotics are a recent subject of interest in the realm of gut health. Consuming prebiotics leads to fermentation by helpful microbes in the gastrointestinal system, which produces functional postbiotics. Postbiotics provide specific benefits associated with the intake of probiotic supplements. Scientists are now researching the specific connection between the formation of postbiotics and their effects on health, despite previous investigations highlighting the advantages of postbiotics. The latest investigations are comparing the use of pre- and probiotic meals with the use of supplements. The function of immune system and relief of digestive diseases are two capable domains. Studies indicate that consuming postbiotics may help relieve symptoms of irritable bowel-related inflammatory disorders, and decrease infections of the respiratory tract. Research conducted at preclinical scale has demonstrated that postbiotic compounds have immunomodulatory, antibacterial, anti-inflammatory, antiproliferative, and radical scavenging actions, but the precise processes are not completely known. The results suggest that these bioactive elements might enhance human health. Furthermore, integrating postbiotics as nutritious elements in food presents numerous advantages for the manufacturing and marketing of food goods. This involves the capacity to involve probiotics in some diets that are thought to be detrimental to their viability. This aids in the growth of the market for functional foods. Postbiotics are safer as well as more resilient for food and medical applications than the living microbes they originate from. Postbiotics do not require survivability for mass production or ingestion (Barros et al. 2020) (Figure 13.1). Hence, in this chapter, the biological activities of postbiotics under *in vitro* and *in vivo* conditions, clinical role in various evidenced infection prevention, applications of postbiotics for the health of newborns and premature infants, treatment of gut dysbiosis with mineral-enriched postbiotics, and utilization in the medical or pharmaceutical sectors are discussed, with an emphasis on the safety regulations and terminology challenges.

13.2 Biological Actions of Postbiotics

13.2.1 In Vitro Investigations of Bioactivities

Postbiotic therapy is not associated with undesirable effects such as inflammation throughout clinical and experimental treatments. Due to its effective doses in low quantities and recognized chemical structures, it may be suggested as a safe substitute for probiotic absorption. Due to their specialized ability to prevent and cure certain illnesses and enhance health, postbiotics are increasingly being used in the veterinary, medicinal, and food sectors. Figure 13.2 showcases the primary historical benchmarks in the biotics family, and wellness, as well as other notable milestones. The anticancer action of postbiotics is a prospective and medically significant property displayed by these metabolites. Cancer is a complex illness involving abnormal cell proliferation. Conventional treatments may lead to chemotherapy resistance, systemic damage, and cancer recurrence (Siegel et al. 2019). The role of postbiotics in the management of cancer is mostly linked to their ability to interact with the body's immune system cells, triggering certain signaling pathways that enhance the natural immune response and decrease inflammation (Francescone et al. 2014). An *in vitro* study examined exopolysaccharide substances derived from *Lactobacillus plantarum*, showing the potential for using these compounds in functional foods and as anticancer agents (Wang et al. 2014).

Research exploring the impact of metabolites revealed a correlation between dose, duration, and the effectiveness of inhibiting colon cancer, gastric cancer, and liver cancer cells. Moderate

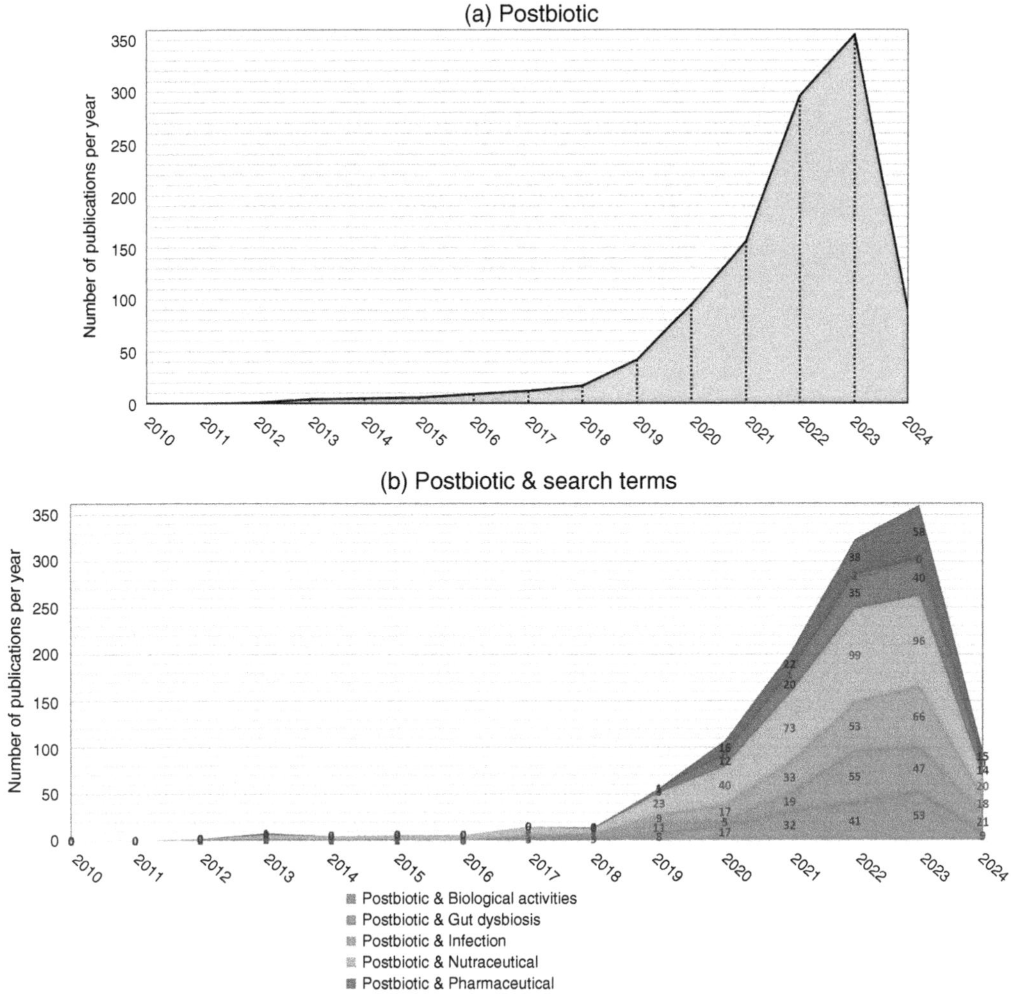

Figure 13.1 (a) Illustrates the increase in the number of papers reporting research in the field of postbiotic. Comparison between the number of publications; on top the search of postbiotic and its synonyms (nonviable microbial cells, abiotic, metabiotic, and ghost probiotics), (b) and below the search of the association between postbiotics and search terms ('biological activities'; 'gut dysbiosis'; 'infection'; 'nutraceutical'; 'pharmaceutical'; 'premature infants'; 'newborns'; and 'safety') in scientific databases.

anticancer effects were seen in liver cancer cells, with inhibition ratios of approximately 61 and 88.34% in contradiction of gastrointestinal cancer cells, respectively. The postbiotics derived from *L. plantarum* strains showed cytotoxic effects against several cancer cell lines in a manner that depended on the strain, duration, and dose, resulting in an inhibitory concentration of approximately 50% (Chuah et al. 2019). Hosseini et al. (2024), investigated the biological properties of postbiotics obtained from *Saccharomyces cerevisiae* (PSC) in controlled laboratory conditions. The generated postbiotics demonstrated high levels of flavonoid (19.87 ± 75.32 mg QE/g) and phenolic (102.46 ± 0.25 mg GAE/g) content. It showed noteworthy radical scavenging behavior ($87.34 \pm 0.56\%$) and antibacterial function against investigated pathogens in both *in vitro* and food systems (ground meat and whole milk). It also promoted the growth of probiotics in the fermentation setting and inhibited the α-glucosidase enzyme, leading to cholesterol reduction in a pH- and

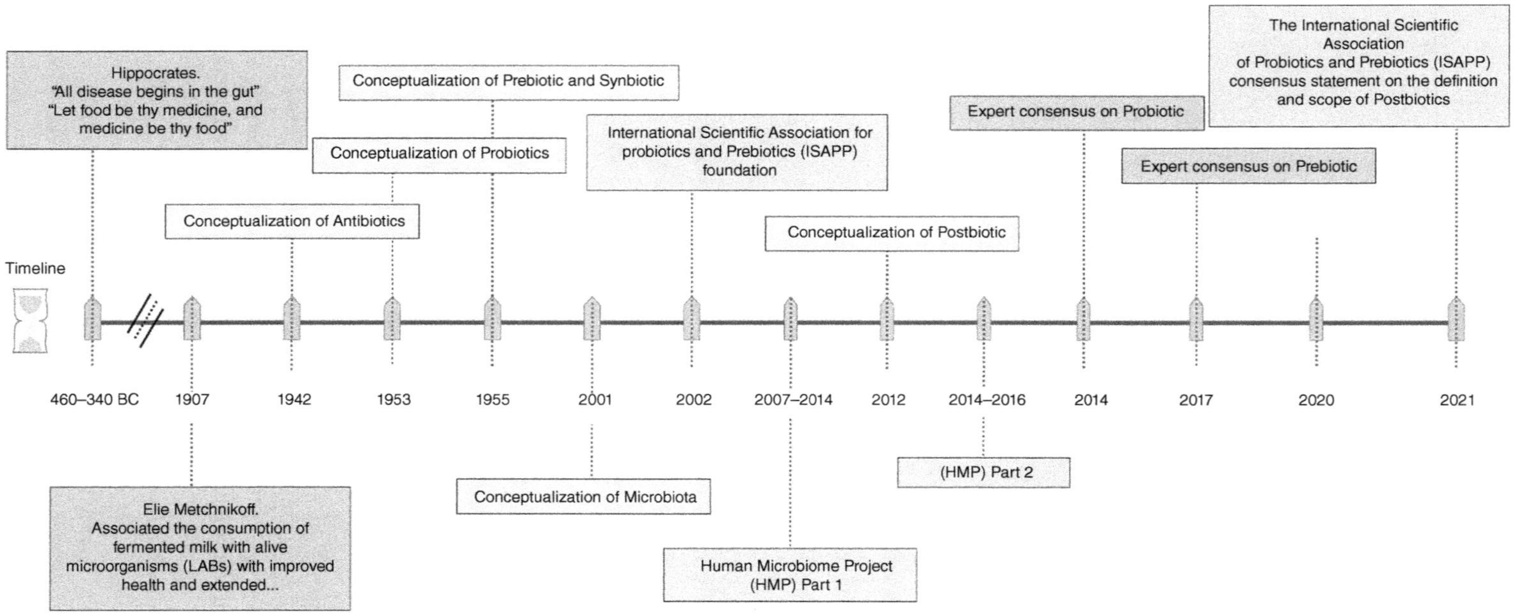

Figure 13.2 Timeline demonstrating the main historical indicators in microbiology, the biotics family and health.

concentration-dependent manner. Furthermore, it reduced cell viability with substantial IC_{50} values of 34.27 and 23.58 μg/mL for a 24- and 48-hour exposure, respectively, inhibited the initial (G0/G1) stage of dividing cells, triggered apoptosis, and augmented the expression of the phosphatase and tensin homolog (*PTEN*) gene. However, it diminished the expression of the *Bcl-XL*, *RelA*, and *IkB* genes in exposed SW480 cancer cells. The authors come to the conclusion that PSC's many health-promoting properties can be applied to medical, and food fields as new biotherapeutic styles. This will help make functional foods or extra medications that work better and faster to prevent or treat chronic or acute disorders (Hosseini et al. 2024).

No cell toxicity was confirmed in normal human breast and red blood cells, rabbits, chickens, dogs, or mouse splenocytes. In this regard, Abbasi et al. (2023), study the cytotoxic and antigenotoxic effects of cell-free supernatant (CFS) postbiotics generated from *Saccharomyces boulardii*. They use the SOS chromotest and MTT assays on HT-29 cell lines to assess these effects. In addition, additional assays relevant to biological pathways, such as cell cycle analysis, DAPI staining, and annexin V-FITC/PI staining, were conducted. The expression levels of some genes implicated in apoptosis were assessed using real-time polymerase chain reaction (PCR). The CFSs of *S. boulardii* demonstrated significant antigenotoxic effects, with a reduction in cell viability observed at IC_{50} values of 33.82, 22.68, and 27.67 μg/mL after 24, 48, and 72 hours, respectively. These CFSs also inhibited the initial phase of cell division (G0/G1), affected the nucleus of treated cells, triggered apoptosis, and raised the expression of *Caspas3* and *PTEN* genes after 48 hours. Conversely, the expression of *RelA* and *Bcl-XL* genes was reduced in treated HT-29 cells. Researchers concluded that postbiotics of *S. boulardii* derived from CFS have demonstrated notable antigenotoxic and cytotoxic effects, as well as the ability to induce apoptotic responses in HT-29 cancer cells (Abbasi et al. 2023). Additional clinical trials are needed to enhance the delivery methods, develop appropriate carriers, and evaluate the potential of postbiotics after oral administration, along with understanding the mechanisms of antiproliferative activities for their *in vivo* applications.

Furthermore, the examination of postbiotic mechanisms versus oxidative components *in vitro* is a neglected topic that needs further investigation and attention. Chang and colleagues (2021) reevaluate the importance of *L. plantarum* as a possible basis of antioxidant agents. The investigators establish a favorable relationship between the formation of organic acids and intracellular elements such as pyrrole intermediates in postbiotic preparation, leading to successful oxidative inhibitory action. A related study reveals the radical scavenging activity of intracellular substances and cellular components of *Bacillus coagulans*, and *Lactobacillus casei* strains *in vitro* (Aguilar-Toalá et al. 2020). The research showed that the cellular antioxidant activity (CAA) (units defined as quercetin equivalents) varied from 43.1 to 56.9 CAA units for *B. coagulans* and 22.7 to 42.9 CAA units for *L. casei*. Several writers corroborated reports on exploratory investigations about the efficacy and antimicrobial properties of postbiotics. More than 50% residual antibacterial activity was detected in all three LAB species, with notable decreases in the biofilm development for *Listeria monocytogenes*. The investigators link the idea of regulated pathogenic development to the postbiotic release of various elements such as organic acids, H_2O_2, and bacteriocins that inhibit microbial growth. Microbial growth prevention relies on elements such as the kind of postbiotic chemicals, contact duration, and dosage. There may be variations in the inhibitory effects across various species or strains, which might be influenced by the technique employed for eliminating them. The impact of several killing methods on the inhibitory properties of *Lactobacillus crustorum* and *Acetobacter pasteurianus* strains was studied with their ability to impede the development of *Escherichia coli* and *Streptococcus mutans* (Safari et al. 2019). *Streptococcus mutans* showed greater sensitivity to postbiotics compared to *E. coli*, with a 70% decrease in viability. The heat-killed bacteria were somewhat inefficient in reducing the number of colonies of *E. coli* compared to other means of destruction. The formaldehyde-based technique effectively reduced the viability of *S. mutans* and *E. coli* cells by over 50%. Thorough investigations should be conducted on numerous aspects that influence the effectiveness of the process due to a limited understanding of the differences in processes leading to different outcomes. The antiviral capabilities of these postbiotics are considered an additional advantage to their antibacterial characteristics. *Enterococcus hirae*, *L. plantarum*, and *Lactobacillus amylovorus* strains showed effective antiviral activity against the isolated enteroviruses (Sunmola et al. 2019). These strains exhibit antiviral activity through multiple pathways, such as producing organic acids and hydrogen peroxide, as well as competitive exclusion. Postbiotic metabolites such as bacteriocins, amino acids, vitamins, peptidoglycan-derived muropeptides, and short-chain fatty acids (SCFAs) support and improve the anti-inflammatory processes in the gastrointestinal tract of humans. Clinical experiments demonstrated that postbiotics from *Bifidobacterium longum* had anti-inflammatory properties. They effectively controlled acute inflammation and gastrointestinal disturbance by activating pathways associated with a specific immune response (Martorell et al. 2021) (Figure 13.3).

13.2.2 In Vivo Investigation of Bioactivities

Research on postbiotics demonstrates a recognized level of conformity to the attributes of prebiotic structure and surroundings. Nevertheless, the effectiveness of postbiotics in promoting their biological functions in living organisms remains unaffected. Postbiotics offer a distinct benefit over conventional preservation agents and antibiotics due to their efficacy in combating pathogenic bacteria, hence addressing the potential problem of food spoilage. The effectiveness of postbiotics in killing microorganisms is influenced by factors such as the specific type of microbe being targeted, the quantity and concentration of postbiotics used, and the characteristics of the prebiotics from which the postbiotics are derived. The existence of *Lactobacillus* spp. in rats led to the promotion of probiotics, resulting in a decrease in the infectiousness and overall pathogenicity of *E. coli*. This, in turn, prevented intestinal inflammation on a larger scale (Gupta et al. 2021). Postbiotics have been discovered to increase the concentration of cytokines that are typically regulated by pathogens' recognizing receptors, leading to immunological homeostasis toward the targeted organism. Nakamura and colleagues (2012) revealed the effectiveness of *Lactobacillus* spp. in combating *L. monocytogenes* through both laboratory experiments and investigations involving living organisms. The postbiotic direction in live tissues enabled a significant inhibition of approximately 90% of activity. The success of these populations is attributed to either artificially restricted availability of nutrients or pH drops, which indicate the existence of lactic acid bacteria. Furthermore, the antiviral properties of postbiotic metabolites are a noteworthy aspect that necessitates further contemplation. Anwar and colleagues (2021) investigated the effectiveness of postbiotics generated by cultures of *L. plantarum* in combating SARS-CoV-2 infection. These postbiotics achieve this by altering the immune system and directly fighting the virus. Plant-based probiotics effectively mitigated the infectiousness of COVID-19 via attaching to the spike glycoprotein, thereby preventing its potential binding to angiotensin 2 converting enzyme (ACE2) and subsequent infection.

Several investigators have conducted studies that focused on the defensive capabilities of postbiotic substances, particularly in relation to their ability to counteract the oxidative qualities of certain elements. The consumption of postbiotics derived from *L. plantarum* resulted in a decrease in lipid peroxidation in the blood and an increase in antioxidant capacity in both the blood and ruminal fluid. This was accompanied by an enhancement of antioxidant enzymes in the liver (Izuddin et al. 2020). The specified strain's postbiotics enhance antioxidant properties

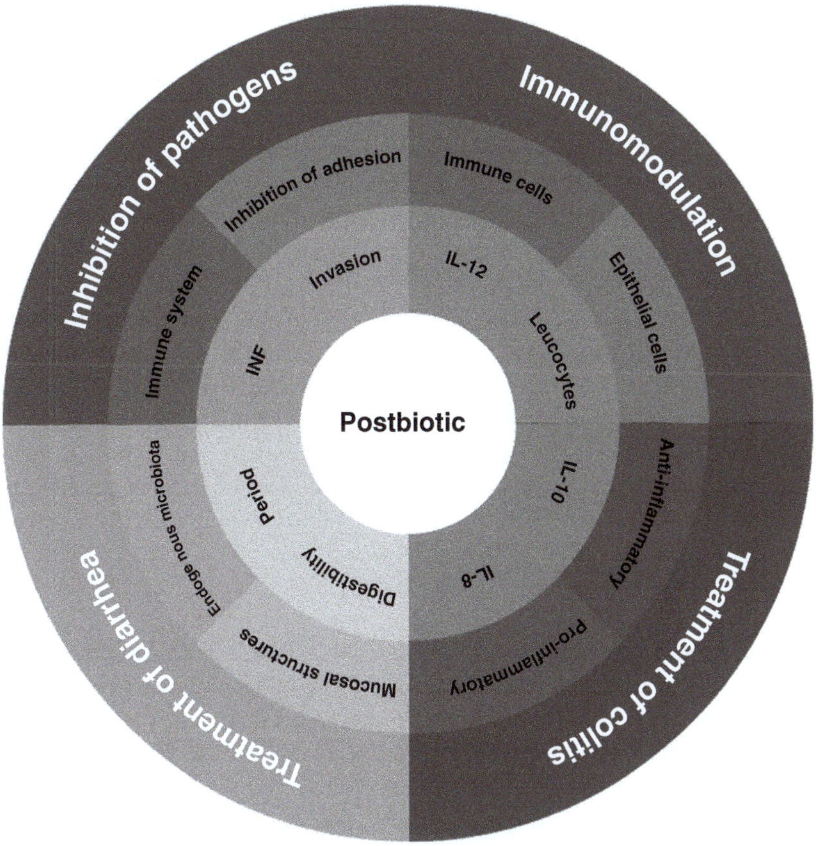

Figure 13.3 Main mechanisms of postbiotics regarding their biological role in the inhibition of pathogens, immunomodulation, treatment of colitis, and diarrhea. Inhibition of pathogens: in the *first step*, the postbiotics reduce the adherence of pathogens and promote the host's immune system; in the *second step*, they exert antibacterial and antiviral activities through reducing the invasion of pathogens and producing INF. Immunomodulation: in the *first step*, postbiotics interact with immune and epithelial cells to activate innate immunity; in the *second step*, they stimulate the secretion of IL-12 and improve immune system functioning through increasing the phagocytic activity of leucocytes. Treatment of colitis: in the *first step*, postbiotics exert significant anti-inflammatory activities; in the *second step*, they regulate the inflammatory responses in gut epithelial cells through stimulating the production of IL-10 as an anti-inflammatory cytokine and inhibiting the production of IL-8 as a pro-inflammatory cytokine. Treatment of diarrhea: in the *first step*, probiotics alter the mucosal structure by increasing the villi size and fortifying the endogenous beneficial microbiota; in the *second step*, they reduce the symptoms of diarrhea through improving digestibility and reducing the duration of the disease.

by demonstrating strong resistance to H_2O_2 and a higher level of scavenging behavior in contradiction of DDPH (1-(2,6-dimethylphenoxy)-2-(3,4-dimethoxyphenylethylamino) propane hydrochloride), and superoxide free radicals. This is attributed to the existence of metabolites such as cell-surface proteins, lipoteichoic acid, and EPS (Zhang et al. 2013). Introducing postbiotics as a dietary supplement for broiler chickens had a beneficial effect on their antioxidant qualities by raising the function of enzymes such as glutathione peroxidase and reducing the presence of heat-stress indicators. The author considers the probability of including postbiotics from *L. plantarum* as a replacement for antibiotics and as an organic source of antioxidants in feeding regimens (Humam et al. 2021). A comprehensive considerate of the transport and response modeling, pathways of signaling, and the straight impacts of ingredients on living cells is necessary for *in vivo* investigations of postbiotic metabolites. Due to the scarcity of data on these balances, the *in vivo* experimental investigation into the anticancer properties of postbiotic elements remains in its early stages. Chen and colleagues (2020) explain the efficiency of *Clostridium butyricum* metabolites in regulating and preventing the growth of intestinal tumors in mice. The probability of inhibiting the growth of tumors is linked to the effectiveness of bacteria that produce butyrate in regulating the Wnt signal and the microbiota of the gut. Additional practical instruction is required to maximize and expand the utilization possibilities of this postbiotic capability.

Postbiotics have been found to improve the immune function by modifying and fortifying the microflora population in the intestinal milieu, therefore reducing inflammatory conditions (Rad et al. 2022). Lactic acid bacteria have shown that they can regulate immune system reactions in normal situations, as demonstrated by Ménard and colleagues (2005) in mice. The study illustrates a strong association between the anti-inflammatory properties of postbiotics derived from *Streptococcus thermophilus* and *Bifidobacterium breve* in an inflammation setting, as well as their capability to trigger the immune system. The strains of *S. thermophilus* and *Lactobacillus bulgaricus* have been discovered to have beneficial effects on the health of rats by inhibiting the progression of colitis (Wasilewska et al. 2019).

13.2.2.1 Infection Prevention

Although probiotic-linked infections are not common, there is still a debate about whether probiotics can exhibit virulence factors or transfer resistance genes to pathogenic microbes (Rowan et al. 2001; Borriello et al. 2003). However, the use of postbiotics may circumvent these issues as the parent microbes become inactive (Gueimonde et al. 2013). Multiple studies assert that postbiotics have the potential to improve physiological processes (Morales-Ferré et al. 2022) and treat and prevent various illnesses such as respiratory tract enteric infections, and gastroenteritis, in addition to problems such as gut barrier disfunction (Gao et al. 2019; Malagón-Rojas et al. 2020). Certain postbiotics have the ability to compete with invasive microbes for receptors, strengthen the gut barrier integrity, alter key gene expression, or directly impact the local milieu to exhibit antimicrobic properties (Khodaii et al. 2017).

13.2.2.2 Infection of Enteric

The fundamental role of the mucosa-related immune system's primary defense, IgA, is to act as an immunological barrier against dangerous bacteria and viruses, excluding them from entering the body (Dongarrà et al. 2013). Multiple investigations have documented a decrease in the intensity of enteric infections caused by bacteria such as *E. coli* or *Salmonella enteritidis*. The decrease in question can be attributed to the higher concentrations of intestinal IgA in the lamina propria of both the large and small intestines of mice following the treatment of postbiotics (Medici et al. 2005). A further study found that an animal model that was fed a diet high in acetate demonstrated a significant enhancement in its ability to withstand infection from enterohaemorrhagic *E. coli* O157:H7. The existence is likely due to the gut barrier's capacity to effectively block hazardous toxins, therefore limiting their entry into the bloodstream (Fukuda et al. 2011). A separate investigation has shown that the nonbacterial portion (NBP) of milk that has been fermented with *Lactobacillus helveticus* has a defensive impact versus *Salmonella typhimurium* infection (Vinderola et al. 2007). The researchers determined that rats administered NBP on the 7th day after the challenge exhibited reduced levels of *Salmonella* spp. colonization in the liver, increased amounts of a particular anti-*Salmonella* S-IgA in the intestinal contents, and a decreased number of MIP-1α cells in the lining of the lamina propria. Another study evaluated the efficacy of postbiotics produced through pH-regulated fermentation in stimulating the production of secretory IgA in feces and protecting rats from infection with *Salmonella* spp. (Dunand et al. 2019). A significant increase in secretory IgA was observed in the feces of rats that were nourished the laboratory-synthesized postbiotic metabolites for a duration of 14 days when compared to the control animals (Dunand et al. 2019).

13.2.2.3 Allergic Reactions

Postbiotics are believed to have the potential to treat allergic disorders because they can help balance Th1/Th2-mediated immunological reactions and promote the development of the immune system. The current evidence (de Boer et al. 2020) supports the utility of postbiotics in minimizing asthma and asthmatic exacerbations in kids. Furthermore, the number of butyrate-generating microorganisms in the intestine was destructively associated with the intensity of atopic dermatitis signs (Nylund et al. 2015), while the consumption of oral lysate form of bacteria was associated with improved atopic dermatitis treatment results in children (Bodemer et al. 2017). In addition, postbiotics may assist individuals suffering from food allergies. A comprehensive assemblage of bacteria that produce butyrate was found to be associated with a faster resolution of cow's milk antipathy, as demonstrated by an *in vivo* study including over 200 children (Hendrickx et al. 2023) (Figure 13.4).

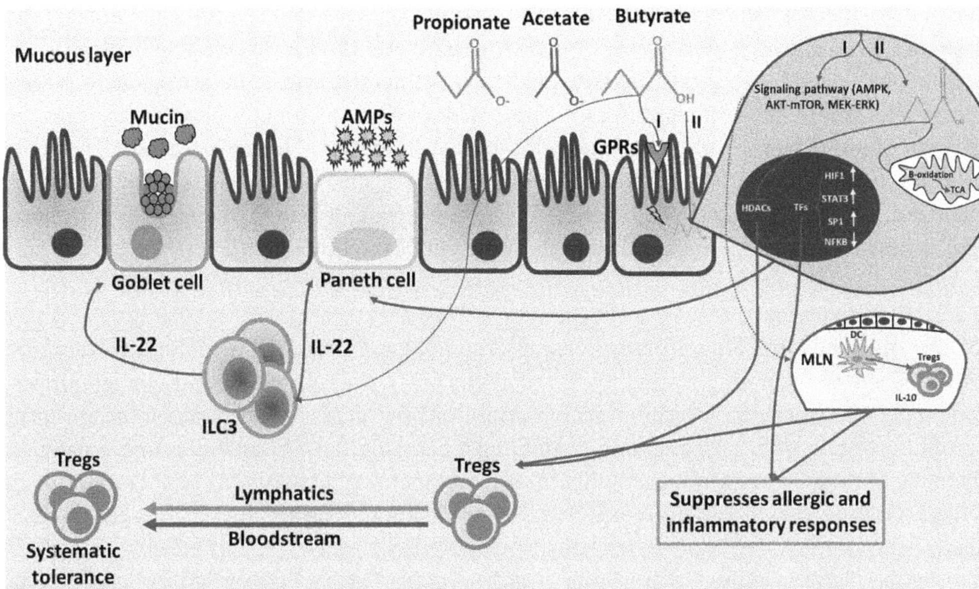

Figure 13.4 The action mechanisms of postbiotics in food allergy with a focus on the gut epithelial barrier. In healthy people, allergy-protective commensal microbes and their metabolites SCFAs preserve gut epithelial barrier integrity via driving ILC3s to generate IL-22. This response (IL-22) stimulates the generation of AMPs from Paneth cells and mucus from goblet cells and also diminishes the capability of diet allergens to obtain access to the systemic circulation. Postbiotics, particularly SCFAs, straightly stimulate Treg development and circulating Tregs develop systemic tolerance. On the other hand, in colonocytes, butyrate (one of the best-known SCFAs) is β-oxidized to acetyl-CoA and establishes the significant source of energy by log-in the TCA cycle. Besides, butyrate initiates triggering/suppression signaling pathway via GPRs and/or straightly suppresses HDACs, therefore up-regulating (HIF-1, STAT3 and SP1) or down-regulating (NF-κB) transcription factors (TFs), which lead to the improvement of gut epithelial barrier efficacy, AMPs generation, and reduce inflammation. GPRs, G-protein-coupled receptors; AMPs, antimicrobial peptides; ILC3, type 3 innate lymphoid cell; Tregs, regulatory T cells; DC, dendritic cell; MLN, mesenteric lymph node; HDACs, histone deacetylase; TFs, transcription factors; TCA, tricarboxylic acid.

13.2.2.4 Infections of Respiratory Tract

Respiratory tract infections pose a noteworthy concern for young kids in daycare or preschool, particularly during the winter. The immune system, along with the digestive and respiratory systems, are typically underdeveloped, which makes it easier for common infectious disorders to occur. Acute upper respiratory tract infections such as acute otitis media, acute rhinosinusitis, acute epiglottitis (supraglottitis), acute laryngotracheobronchitis (croup), acute pharyngitis, acute sinusitis, and common cold are a major source of illness, especially in children and older adults. Studies have also examined the effects of postbiotics on infections of the respiratory tract (Nocerino et al. 2017). Two randomized control studies were conducted to investigate the efficacy of administering heat-killed *L. paracasei* to 537 healthy Italian youngsters for a period of 90 days in avoiding infections of respiratory tract. The combined assessment of both of these investigations found that the administration of postbiotics reduced the occurrence of tracheitis, laryngitis, and pharyngitis. There was no statistically substantial difference in the incidence of otitis media or rhinitis between the postbiotic and placebo groups (Nocerino et al. 2017). The researchers noted a substantial augmentation in the synthesis of peptides related to both main arm of immunity (innate and acquired), resulting in an immunostimulatory effect. Peptides from innate immunity function as natural antibacterial elements and safeguard the host against a diverse range of infections.

Paneth cells, macrophage epithelial cells, and neutrophils create them, specifically in response to bacteria, fungi, protozoa, and viruses. Furthermore, due to their antibacterial characteristics, these peptides have been shown to regulate the activity of neutrophils, monocytes, macrophages, dendritic cells, and T cells. In addition, the fermented food product had a positive effect on the production of sIgA, an essential component of acquired immunity. Furthermore, sIgAs not only serve as the primary defense in contradiction of microorganisms in the mucosae, but they also regulate the makeup of the microbiome in the gut, enabling interaction between the immune system of the host and the commensal bacteria in the gut (Corsello et al. 2017).

13.2.2.5 Gastroenteritis

Prebiotics and postbiotics are currently being used to treat and prevent gastrointestinal diseases. Diarrhea, fever, and vomiting caused by the rotavirus (RV), can affect kids and young animals, including piglets and calves. A recent research (Morales-Ferré et al. 2022) examined the effects of daily administration with a formulation containing a combination of postbiotics and prebiotics, comparable to those present in breast milk, in an experiment of rotavirus (RV) contamination in suckling rats. The researchers determined that the combination of Lactofidus™, short-chain galactooligosaccharides (scGOS), and long-chain fructooligosaccharides (lcFOS) effectively avoided and decreased diarrhea caused by rotavirus in suckling rats. The supplementation elicited alterations in the intestinal gene expression, immunoglobulin profile, and gut flora. Both Lactofidus and scGOS/lcFOS exhibit unique mechanisms for addressing diarrhea. The postbiotic caused a drop in the acidity level (pH) in the stool and an increase in the expression of toll-like receptors. The researchers determined that the majority of the positive effects reported individually were still present when Lactofidus and scGOS/lcFOS were used together. Additionally, they detected some effects, such as a rise in the recruitment of white blood cells and lymphocytes, as well as a rise in the expression of the TLR7 and TLR9 genes. In another RCT, the efficacy of regular feedings of a fermented formula comprising non-live *S. thermophilus* and *B. breve* for a duration of 5 months was investigated in 913 otherwise healthy French babies aged between 4 and 6 months. There was no statistically significant variance in the frequency of diarrheal episodes between the treatment group with the placebo group. The study included 913 participants, and the relative risk (RR) was

1.01 with a 95% confidence interval (CI) of 0.98–1.04 (Thibault et al. 2004). Furthermore, a study was conducted to investigate the impact of administering heat-killed *L. paracasei* for a duration of 90 days on the prevention of diarrhea in a group of 537 healthy Italian children aged between 12 and 48 months (Nocerino et al. 2017). The researchers observed a notable decrease in the number of diarrheal episodes in the group that received heat-killed *L. paracasei* compared to the group that received a placebo. This finding was based on two randomized controlled trials, including a total of 537 participants. The risk ratio (RR) was calculated to be 0.51, with a 95% confidence interval (CI) ranging from 0.37 to 0.71.

13.2.2.6 Further Clinical Applications

Investigators have suggested that postbiotics may have a role in preventing and treating SARS-CoV-2 infection. This is because the makeup and metabolism of the gut microbiome may be linked to the presence of biomarkers that can anticipate the severity of COVID-19 (Guo et al. 2013). Manipulating the gut microbiota using probiotics seems to influence the inflammatory response associated with COVID-19 and directly hinder the replication of SARS-CoV-2 (Todorov et al. 2021). GM remodulation utilizing postbiotics can play a crucial role in safeguarding vulnerable individuals from acquiring SARS-CoV-2. In addition, GM remodulation has the ability to modify the biology of COVID-19, causing it to manifest in a less severe form (Spagnolello et al. 2021). Postbiotics have the potential to directly inhibit the replication of SASR-COV-2, just as probiotics.

Furthermore, a separate study indicated that the utilization of bacterial lipoteichoic acid (LTA) could potentially be efficacious in the treatment of diverse skin infections. Bacterial and viral diseases can activate the mast cell reaction in the skin through the presence of *Bifidobacteria* and *Lactobacillus* genus. The cell walls of Gram-positive bacteria contain lipoteichoic acid, which has the ability to be released into the environment spontaneously. Topical application of LTA stimulates the discharge of anti-infectious peptides such as cathelicidin and human defensin via improving non-specific defensive mechanisms. Non-pathogenic microorganisms that trigger opportunistic infections turn pathogenic when the body's immune system is weakened (Riccardi et al. 2019). *Candida* spp. is linked to opportunistic infections of yeast and can result in candidemia, cutaneous candidiasis, oral candidiasis, and vaginitis (Rossoni et al. 2020). Rossoni and colleagues (2020) conducted a study to examine the effectiveness of a crude extract and fraction 1, derived from *L. paracasei* supernatant, in fighting fungal infections. They used both *in vitro* and *in vivo* sets of probiotic cells for their investigation. The study found that *L. paracasei* and postbiotic components exhibit antifungal activity against *Candida auris* persister cells, planktonic cells, and biofilms.

Postbiotics have been studied for their potential use in treating necrotizing enterocolitis (NEC). NEC is a significant contributor to illness and death in infants. Therefore, there is a pressing requirement for preventive medications that are effective and safe. Recent therapeutic research indicates that probiotics are a highly effective treatment for lowering necrotizing enterocolitis in preterm neonates, making them the most promising solution for this life-threatening condition. Nevertheless, concerns over safety and appropriate dosing have limited the utilization of probiotic cells in preterm neonates within conventional healthcare environments. While there is limited data showing the therapeutic effectiveness of postbiotics in preventing NEC, they have the potential to serve as alternatives or complementary treatments to the administration of live probiotics. Vital bacterial elements can still have positive effects on the host despite microbes thermally damaging them. In a study conducted on the immature intestines of mice, it was discovered that heat-inactivated *Lactobacillus rhamnosus* GG (LGG) has the same therapeutic effectiveness as active/live probiotic cells in accelerating the maturation of the intestinal barrier. It is noteworthy that administering higher doses of live LGG resulted in higher mortality rates in underdeveloped neonatal mice, whereas administering heat-inactivated LGG did not have the same effect. The authors stressed the necessity of conducting additional clinical trials to determine the impact of postbiotics on the clinical consequences of prematurity, such as necrotizing enterocolitis, sepsis, or mortality.

There has been significant notice in substituting antibiotics with postbiotics in farm animals. The prohibition of antibiotics in multiple nations has led to increased popularity in seeking alternate methods to achieve antimicrobial effects against prevalent diseases in large-scale farming. The precise processes by which postbiotics exert their effects are still not completely comprehended. There is a suggestion that deactivated probiotic cells or their individual ingredients might alter the immune reactions of the host. The host immune system can be stimulated by several elements of bacteria, including the bacterial capsule, and cell-wall contents such as liposaccharides, proteins of S-layer, and peptidoglycans. Several pattern recognition receptors (PRRs) are found in the initial host response. These PRRs are proteins predominantly expressed by cells of the innate immune system and have the ability to recognize specific compounds found in bacteria, which are called pathogen-associated molecular patterns (PAMPs). The toll-like receptors (TLRs), and nucleotide-binding oligomerization domain (NOD)-like receptors (NLRs) are considered to be two types of pattern recognition receptors that have a significant influence on regulating the host's innate immune response. The many TLR subtypes have the ability to attach to different types of microbial components, such as bacterial DNA or RNA, peptides like peptidoglycans, lipoteichoic acids, lipoproteins, and flagellin (Mogensen 2009).

The NLRs, which are analogous to TLRs, mediate both the innate and adaptive immune reactions (Mogensen 2009). The NLRs have the ability to identify several types of substances, known as ligands, that come from microbial pathogens. These ligands include peptidoglycans, flagellin, and viral RNA (Kawai and Akira 2009). There is a suggestion that specific cytokines, including interferons, might trigger a response from the NLRs as well (Feerick and McKernan 2017). NLRs may facilitate the activation of both B and T cells in response to postbiotic stimuli. The presence of lipoteichoic acid in postbiotic product components may lead to a decrease in reactive oxygen species (ROS) and IL-2 synthesis while increasing the generation of IL-4, IL-6, and IL-10 (Kim et al. 2017). In situations where it is challenging or not feasible to provide probiotics, postbiotics can serve as a favorable substitute. This is because postbiotics have the advantage of not requiring the bacteria to be alive while still providing the benefits of reducing the occurrence of common infectious diseases.

13.2.3 Postbiotics in Childhood

13.2.3.1 Principal Applications of Postbiotics in Children

The study conducted by Wegh and collaborators has compiled the primary *in vivo* research conducted in the pediatric population, categorizing it based on the microbial strain utilized in the process of fermentation (Wegh et al. 2019). The favorable benefits of postbiotics on human well-being are contingent upon the specific microbial strain employed during the fermentation procedure and the type of food that undergoes fermentation. The majority of the research has conducted comparisons between the advantages of utilizing a fermented infant formula (FIF) and a regular formula. Instead of using placebos or prebiotics alone, certain research has utilized supplements, including particular deactivated bacterial strains, alongside or in the absence of their culture solution (Wegh et al. 2019).

The majority of preclinical investigations have examined the potential beneficial impact of postbiotic metabolites on a group of healthy newborns. Two investigations utilized milk from cows that had undergone fermentation with *L. paracasei*. The first established that consuming food that *L. paracasei* had fermented aided in the development of a beneficial microbiome (Berni Canani et al. 2017). Later, the authors explained how this effect resulted in an enhanced production of butyrate. The study conducted by Corsello and colleagues was a randomized controlled experiment that assessed the impact of *L. paracasei* fermented dairy product on the occurrence of typical infections in terms of clinical consequence. According to the investigators' findings, children who consumed the postbiotic experienced a decrease in the number of infections, including those affecting the respiratory and gastrointestinal systems (Corsello et al. 2017). Additional research has been carried out on newborns who are in good health and born at full term. These studies have examined the impact of different strains of *Lactobacillus*, both with and without the inclusion of prebiotics, in different proportions. The results of these investigations demonstrate the beneficial impact of such products on the functioning of the intestines. This is evident in the reduced occurrence of colic diseases, the improved quantity and consistency of stools, and the elevated levels of significant biomarkers of immunity, such as IgA (Huet et al. 2016).

Two of the most extensively studied varieties in the community of healthy infants are *S. thermophilus* and *B. breve*. Postbiotics made from a mix of these two strains, which are often used for fermenting the product, have been shown to have a positive effect on the immune system's response to immunizations, as well as on signs of immunomodulation and the intestinal infections' level of seriousness. Nevertheless, there was no discernible impact on the occurrence of these conditions. Several research studies have examined the impact of postbiotics on prevalent childhood disorders, specifically diarrhea and allergy diseases. The duration of healing in newborns with diarrhea induced by non-rotavirus was reduced by a single day after they consumed heat-inactivated *Lactobacillus acidophilus* along with their cultural media (Liévin-Le Moal et al. 2007). Nevertheless, an investigation conducted by Kaila and colleagues (1995) discovered that there were no discernible disparities in the clinical outcome of children who underwent therapy with nonviable *L. casei* compared to those who received viable units from the exact same strain.

A study conducted by Peng and Hsu (2005) demonstrated that both viable and heat-inactivated *L. paracasei* had comparable positive effects on children suffering from perennial allergic rhinitis. In their study on atopy and allergy to cow's milk, Morissett and colleagues examined the effects of the newborn formula containing heat-inactivated *S. thermophilus* and *B. breve*. They found that this formula was effective in reducing the occurrence of respiratory and digestive complications in infants who were at a high risk of developing atopy. Nevertheless, they were unable to prove any impact on the initiation of cow's milk allergy (Morisset et al. 2011). Kirjavainen and colleagues (2003) discovered that there was no distinction in the prevalence of atopic eczema and cow's milk allergy between live or heat-inactivated *L. rhamnosus* GG. Additionally, they observed a connection between heat-deactivated *L. rhamnosus* and negative intestinal indications.

Based on the aforementioned investigations, *Lactobacilli* is a bacterial strain that has been extensively researched in the generation of postbiotics. Lactobacilli are a type of bacteria that are shaped like rods and have a positive reaction to the Gram stain. There are around 237 different species of lactobacilli, and they are widely promoted and extensively researched as probiotics (Bron et al. 2013). Initial examinations of the baby microbiota have revealed the presence of several strains of *Lactobacilli* in the gastrointestinal tract from the earliest days of life (Dominguez-Bello et al. 2016). In a review conducted by Tsegay and colleagues, the different attributes of lactobacilli that elucidate their significance as postbiotics were outlined (Teame et al. 2020). Postbiotics exert their effect by interacting with the host and the proteins and chemicals present on the cell walls of different strains. These include proteins found in the cell membrane, pili, polysaccharides, peptidoglycans, and teichoic acid. Among the metabolic byproducts of *Lactobacilli* that have remarkable postbiotic activity can be identified SCFAs, conjugated linoleic acid, and bacteriocins, which have an immunoregulating and protective impact against pathogens. Another important component is the aggregation-promoting factor, which possesses a crucial role in maintaining the integrity of the intestinal epithelium. Interestingly, every strain of *Lactobacilli* possesses a distinct genetic composition that sets it apart from other strains (Bron et al. 2013). Consequently, each strain possesses distinct membrane and metabolic attributes. This is particularly important when using probiotics or postbiotics to enhance a particular condition.

13.2.3.2 Postbiotics for the Health of Newborns and Premature Infants

As previously stated, postbiotics have the potential to be a beneficial resource for the health of neonates and premature infants. Existing research substantiates the utilization of probiotic cells in these susceptible subjects to mitigate the occurrence of severe ailments such as late-onset sepsis, and necrotizing enterocolitis. Nevertheless, the potential for sepsis induced by probiotic, gastrointestinal translocation, and the spread of drug resistance presents a barrier to the utilization of probiotics in newborns, particularly those who are premature. An investigation conducted by Mosca and colleagues has surveyed the potential benefits of postbiotics in preventing necrotizing enterocolitis (LeBlanc et al. 2017). Given the complex causes of this condition, which includes imbalances in the gut bacteria, an immature intestinal barrier, and increased expression of TLR-4, it seems logical to suggest that postbiotics could directly impact many of these contributing factors. The process of postbiotics has been explained, which stimulates the production of bacteriocins. Bacteriocins are compounds that can inhibit the excessive growth of harmful bacteria. Moreover, recent studies have shown that certain postbiotics, for instance butyrate, may enhance the strength of the gut barrier integrity by stimulating the formation of the characteristic mucins found in the epithelium of the intestine (LeBlanc et al. 2017). Later, there are several postbiotics that have been found to be effective in modulating the immune system, as previously discussed in this study. Hence, it is clear that there is a necessity to carry out research focused on examining the *in vivo* tolerability and effectiveness characteristics of postbiotic metabolites in the most susceptible preterm infants.

Only a limited amount of research has examined the positive effects of postbiotics on the health of infants in a live organism. An investigation conducted by Campeotto and colleagues has made a comparison of the utilization of a preterm formula that has been undergone fermentation process with *S. thermophilus* and *B. breve* to a usual preterm formula. The study describes a potential positive effect, as seen by decreased levels of fecal calprotectin and less abdominal distention in the group that received the intervention (Campeotto et al. 2011). Postbiotics have been investigated as a potentially safer substitute for probiotics in preterm infants. Several preclinical and clinical studies have examined the possible dangers and benefits of using nonviable cells. These studies have generally found that postbiotics are well-tolerated and safe, although they have not definitively determined their effectiveness (Cuevas-González et al. 2020). Similarly, Zorzela and colleagues conducted a systematic study in an attempt to provide a more conclusive solution to the topic. In summary, Zorzela's findings suggest that there is some evidence that favors the use of postbiotics in some therapeutic scenarios. Nevertheless, the considerable diversity in study techniques, the range of microbial strains examined, and the limited sample sizes in all investigations undermine the ability to draw a more certain conclusion (Zorzela et al. 2017).

13.3 Gut Dysbiosis Therapy Based on Mineral-Enriched Postbiotics

Considering the available proof, researchers suggest that postbiotic metabolites that contain important minerals will have both the therapeutic properties of postbiotics and the added benefits of micronutrient-enriched material. The synergistic impact will amplify the advantageous health consequences of these goods, which demonstrate improved safety and many practical and technological implications. The mineral-rich postbiotic behavior may have a direct effect on the host cells because chemicals released from broken cells and lysates come into more contact with the gut epithelial cells. Minerals and postbiotic substances can indirectly affect the microbiota in the gut, which in turn affects the absorption of minerals in the gastrointestinal system. This can improve the availability of minerals, ultimately helping to maintain the balance of the host's internal environment.

Modulating gut microbiome and producing favorable properties for the human is a crucial characteristic of mineral and postbiotic function. The gut microbiome can utilize the bioactive compounds from inactivated probiotic solutions and minerals to support its development and functioning, hence impacting the overall health of the host. Biopolymers that are bonded to the cell wall, such as biosurfactants and exopolysaccharides, are generated and released by microorganisms throughout their growth. These biopolymers, together with proteins with cell-surface origin, are the primary elements of the cell envelope structures. They play a vital part in the cellular functioning of postbiotic metabolites. The teichoic acid, cell-wall-derived polysaccharides, and peptidoglycan layer are structural components of the cell wall. These materials are found in high quantities in *Lactobacillus* spp. and have been found to play a role in the dynamic postbiotic impact (Nataraj et al. 2020). In addition, a diverse array of metabolites and byproducts derived from living probiotics, such as compounds present in the cell-free supernatant (organic acids and bacteriocins), as well as soluble factors, have been discovered to have an impact on the gut microbiota (Żółkiewicz et al. 2020). Studies have demonstrated that postbiotic substances can modify the disrupted microbiota and thus produce positive impacts on the host's health (Aguilar-Toalá et al. 2021).

Iron, calcium, magnesium, zinc, and selenium are essential for the metabolic activities, cellular respiration, and DNA replication of most bacteria, including the gut microbiota. For example, certain significant bacterial species that inhabit the digestive systems of humans and animals (*Enterobacteria, Clostridia, E. coli, and Lactobacillus*) possess genes that produce selenoproteins. As a result, these bacteria are able to process and include inorganic or organic selenocompounds from the diet or supplements (Ferreira et al. 2021). Various investigations have determined that dietary selenium causes alterations in the makeup of the intestinal microbiome and the development of the digestive system (Zhai et al. 2018). Furthermore, recent studies have demonstrated that mineral-enriched probiotic cells have a substantial impact on the composition of the gut microbiome. This, in turn, leads to improved accessibility and digestion of these elements in the host as compared to inorganic types (Yang et al. 2021; Kang et al. 2020). The presence of minerals and bioactive compounds inside the carbohydrate matrix of mineral-enriched postbiotics may contribute to the enhancement of their biological function.

The bacterial strains that enhance the availability of minerals through regulation of the microbiota are primarily strains of lactic acid bacteria (LAB), specifically *Bifidobacterium* spp. and *Lactobacillus* spp. These strains are naturally existing in the gut microbiome and are also employed in the production of postbiotics (Bielik and Kolisek 2021). The LAB postbiotic preparations, which accumulate minerals as organic molecules, have the potential to create a pool of substances in the large intestine that can be utilized by the host. This has been shown in the case of selenium-enriched probiotics (Ferreira et al. 2021). Conversely, the introduction of LAB bioactive compounds by pre- or postbiotics, like SCFA, alter the intestinal ecology by reducing pH and enhancing the absorption of calcium. Additionally, these molecules also aid in the absorption of other cations, such as magnesium (Mg) and ferrous iron (Fe^{2+}) (Bielik and Kolisek 2021; Das et al. 2020). Additional metabolites, specifically reuterin and DAP (1,3-diaminopropane), which are antimicrobials created by *Lactobacillus reuteri*, as well as metabolites like propionate, have been shown to play a direct role in iron transportation in enterocytes (Zhang et al. 2019). In addition, minerals derived from enriched postbiotics may have a prebiotic property and alter the arrangement of unbalanced microbiota by affecting indigenous strains. Both mineral-enriched biomass and postbiotics have demonstrated pleiotropic qualities, which are linked to positive health impacts. Consequently, it is strongly advised to conduct extensive research and extensive clinical trials to investigate the biological activities and bioaccessibility of mineral-enriched pre- and postbiotics. This is necessary to validate the results obtained from preclinical investigations.

Certain postbiotics exhibit antibacterial and anti-adhesion properties via strengthening the gut barrier integrity, altering the immediate surroundings, or outcompeting germs for receptor binding (Żółkiewicz et al. 2020). The antibacterial and anti-biofilm properties of postbiotic compounds produced by *Lactobacillus sakei*, which were isolated from the microbiota of cow milk, have been reported to be efficacious versus pathogens responsible for mastitis (Sevin et al. 2021). Mineral insufficiency increases the vulnerability of the microbiota in the gut to dysbiosis, which in turn promotes infection with *S. typhimurium*. Additionally, the host's adaptive immune response is weakened in this situation (Zhai et al. 2018). Alternatively, using organic selenium-enriched probiotics shows promise in eliminating pathogenic bacteria linked to colon cancer and inflammatory bowel illnesses, as well as lowering oxidative stress (Mego et al. 2005).

Postbiotics, together with minerals like selenium and zinc, are associated with anti-inflammatory, anti-tumor, and antioxidant effects. The biological molecules derived from *L. paracasei* and *L. rhamnosus* GG have been found to impede the progress and advancement of gastric and colorectal cancer in laboratory models (Orlando et al. 2012). In addition, several postbiotic compounds have demonstrated anticancer activities through anti-proliferation, anti-inflammatory, and apoptosis actions (Kim et al. 2021). Short-chain fatty acids play a role in regulating oncogenes and possess a protective property for DNA transcription. Additionally, they have the potential to preserve the gut barrier integrity, which may assist in preventing or treating colorectal cancer. (Hatayama et al. 2007). In addition, SCFAs stimulate the activation of molecules that are engaged in pathways responsible for regulating the immune and inflammatory responses in mouse models (Nakkarach et al. 2021). Postbiotics have gained significant attention as a cancer adjuvant therapy due to their direct or indirect impact on host carcinogenesis (Meng et al. 2022). Various investigations have demonstrated that postbiotic compounds possess the capacity to regulate the efficacy of cancer treatment and mitigate the adverse influences of conventional treatment for cancer, thereby aiding in the recovery of cancer patients. Like short-chain fatty acids, selenium preserves DNA, and there is a suggestion that selenium retards the aging development via shielding DNA from ROS (Yildiz et al. 2019). Both postbiotic metabolites and minerals, have antioxidant properties (Mocchegiani and Malavolta 2009).

Postbiotics and vital minerals have the ability to regulate and influence adaptive and innate immunity in hosts. Postbiotics can elicit comparable immune reactions to probiotics by either stimulating the same paths or employing diverse mechanisms (Aguilar-Toalá et al. 2021). The intestinal microbiome affects cholesterol metabolism via multiple methods, such as the use of bile salt hydrolases to deconjugate glycine and taurine, as well as altering the cholesterol core via dehydroxylation, decarboxylation, and dehydrogenation processes (Hu et al. 2021). Various investigations have specified that postbiotic metabolites have the capacity to reduce cholesterol levels (Osman et al. 2021).

The primary characteristic of mineral-enriched postbiotics is their capacity to modulate the microbiota, hence enhancing the bioavailability of postbiotics and minerals. The host experiences a direct influence, which is maintained by homeostasis. This, along with the renewal of the microbial pattern in the intestines, establishes an environment of equilibrium through its linkages with the nervous system and the gut-brain axis (Rutsch et al. 2020). This bidirectional equilibrium has the impact of regulating degenerative events or other long-lasting dysfunctions, thereby necessitating *in vivo* investigations to elucidate the mechanisms governing nerve cell activity and the reaction of the gut microbiota to stress (Ceppa et al. 2020). The flexibility of this framework serves as the initial defense mechanism against external processes that impact the health of neurons by promoting the buildup of pro-inflammatory substances (Schwarz et al. 2020). Regulating the gut-brain axis response also results in the restoration of hormone production, which has a significant impact on overall health. Controlling pro-inflammatory dysbiosis to eliminate oxidative stress factors is a more appealing option compared to using traditional probiotic products, which have limited and unpredictable effects. It is therefore not advisable to administer drugs with antioxidant properties in an unregulated manner, especially when their biological effects remain unclear and do not always align with preclinical and clinical evidence (Schwarz et al. 2020).

13.4 Promising Use of Postbiotics in the Medical or Pharmaceutical Sectors

Many postbiotic compounds have garnered attention because of their distinct characteristics, including their unique chemical composition, long-term storage stability, and potential to stimulate multiple pathways that control oxidative stress, cancer, cardiovascular disease, hypertension, obesity, and inflammation (Nataraj et al. 2020). Postbiotics possess immunoregulatory properties for the host and can serve as a safer option in situations where the consumption of active probiotic cells is not recommended. A review by (Piqué et al. 2019) identified several pharmacodynamic characteristics of postbiotics that distinguish them from probiotics. These include: (Baker et al. 2022) little or no risk of transmitting bacteria from the stomach to the bloodstream in individuals with compromised immune systems; (Damián et al. 2022) absence of the potential acquisition and transmission of genes for antibiotic resistance; (Sabahi et al. 2023) more natural processes for the extraction process, standardized procedures, shipping, and storage; (Abbasi et al. 2022) potential extra advantages resulting from the loss of viability caused by cell lysis; and (Cuevas-González et al. 2020) enhanced connections between epithelial cells and metabolites discharged from disrupted cells (Piqué et al. 2019).

In modern times, companies are striving to include postbiotics in pharmaceutical product matrices because of their advantageous function in clinical circumstances. An example of a renowned postbiotic in this sector is CytoFlora®, which is produced by BioRay Inc. in Laguna Hills, CA, USA. CytoFlora consists of bacterial cell walls, specifically those isolated from *L. acidophilus* DDS-1, *L. plantarum*, and *L. casei*. CytoFlora has been documented to have significant beneficial effects, such as boosting the immune response, improving imbalances in the gut microbiota, maintaining intestinal stability, and reducing symptoms in patients with autism (Chaluvadi et al. 2015; Rad et al. 2020). Del-Immune V® is a postbiotic medicinal medication manufactured by Pure Research Products, LLC, located in Boulder, CO, USA. It is commonly recommended to consume probiotics alongside them for medicinal reasons. The Del-Immune V formulation is composed of a mixture of amino acids, muramyl peptides, and DNA fragments derived from *L. rhamnosus* V. Research has demonstrated that Del-Immune V effectively reduces the intensity of gastrointestinal disorders in individuals diagnosed with autism spectrum disorder (Rad et al. 2020). Another instance is the frequent recommendation of Zakofalk® (a combination of inulin and sodium butyrate) for the treatment of active inflammatory illnesses in the intestines (Rad et al. 2020). Hylak® Forte, a sterile liquid postbiotic drug manufactured by Ratiopharm/Merckle in Germany, consists of various biological metabolites, including organic acids, SCFAs, and other metabolites. Hylak Forte has several important health effects, including promoting the growth of beneficial microbes in the gut, regulating the pH of the gut environment, supporting healthy digestion, providing energy for cells in the intestinal lining, maintaining proper balances of water-soluble vitamins (K, B), and treating salmonellosis and gastrointestinal conditions in both adults and kids with chronic gastrointestinal conditions (Rad et al. 2020).

Besides the fact that postbiotic products are already being sold commercially, there is also a significant focus on the potential health advantages that could arise from the development of new medications in the future. A recent review examined reproductive health and the effects of postbiotics (Abbasi et al. 2022). They advocated for the utilization of postbiotic compounds as a promising instrument in a tailored medicine strategy for the restoration of vaginal microbiome eubiosis. Furthermore, it was suggested for both preventive and supplementary therapy approaches in women subjects with reproductive-linked illnesses because of their distinct characteristics in terms of technological, clinical, and financial factors (Abbasi et al. 2022). Investigators also assessed the effects of postbiotic elements in children under the age of 5 and suggested using heat-inactivated *L. acidophilus* to treat acute diarrhea, however, the evidence supporting this recommendation is limited. In addition, the authors also proposed the use of heat-inactivated *Lactobacillus paracasei* to prevent intestinal and respiratory infections, but the evidence supporting this recommendation is weak. They also emphasized the need for more studies to have a deeper understanding of this topic (Malagón-Rojas et al. 2020). Another research group conducted a study on the antigerm function of indole propionic acid, which is a product of the gut microbiome. They found that it acts as an agent against tuberculosis (Negatu et al. 2019). Through the implementation of metabolic, biochemical, genetic, and chemical rescue experiments, it was determined that indole propionic acid has the potential to mimic the physiological allosteric inhibitor of TrpE. This mimicry leads to a reduction in tryptophan production in *Mycobacterium tuberculosis*, resulting in antimycobacterial action (Negatu et al. 2019).

Animal health has also garnered attention, akin to human health, due to its potential as an invaluable instrument in the veterinary scope (Kareem et al. 2016). A study found that the addition of postbiotics from *L. plantarum* to the meals of broilers resulted in higher ultimate total weight gain, body weight, and increased height of the duodenal and ileal villus (Kareem et al. 2016). Another research found that adding *L. plantarum* postbiotics to animal diets resulted in an increase in the size of the mucosal villi and a boost in the number of beneficial intestinal microbiota. This, in turn, led to improved protein digestion and growth efficiency, as well as a decrease in the occurrence of diarrhea (Aguilar-Toalá et al. 2018). In addition, the utilization of postbiotics resulted in alterations to the arrangement of the intestinal microbiota, leading to an upsurge in the abundance of beneficial bacteria and ultimately enhancing the overall well-being of the experimental animals. Postbiotics can serve as preventive drugs, novel personalized functional foods, and germ-free dietary supplements to complement therapy for various illnesses (Aguilar-Toalá et al. 2018).

To summarize, an extensive range of postbiotics have been industrialized for use in the drug, veterinary medicine, and food sectors. These postbiotics have proven to be highly effective in preventing and treating specific diseases, improving the health of animals, and creating functional foods. Therefore, postbiotics can serve as a safe substitute and a new method of utilization in the pharmacy and/or food sectors to promote and enhance health benefits, as well as to avoid diseases and achieve therapeutic objectives. Nevertheless, the safety of food and medicinal items is largely contingent upon the safety of the substances used in their composition. Hence, further investigation is vital to establish new postbiotic metabolites and evaluate their durability and safety standards in the pharmacy and food sectors. Clinical trials are necessary to determine the optimal frequency and amount of postbiotic administration using pharmaceutical products and/or functional postbiotic meals.

13.5 Safety Regulations and Terminology Challenges

Recently, there has been an increase in consumers' interest in healthy products. The significant potential of probiotics to provide beneficial health effects has resulted in a rise in both scientific and commercial motives for using microbes as a tool to promote health. In recent years, there has been notable progress in research exploring the connections between food, the microbiota, and the human body. This has resulted in the use of advanced therapeutic approaches to administer or modify the human microbiota through microbial intervention. These new findings have resulted in the widespread use of several terms related to probiotics. However, instead of improving communication between investigators, regulatory agencies, and the food industry, these words have actually caused the term probiotics to be used incorrectly (Reid et al. 2019). The definitions of probiotics and prebiotics have lately undergone review and modification. Based on these concepts, a prebiotic refers to a substance that is specially utilized by microorganisms living in a host organism, providing a health advantage. On the other hand, a probiotic refers to active bacteria that, when administered to the host in adequate amounts, provide a health benefit. A variety of food supplements are labeled to indicate the presence of probiotics. Consumer comprehension of these terms is increasing, and numerous countries are more frequently affixing labels to them for regulatory reasons. Furthermore, there is a growing body of research indicating that non-living microorganisms and the substances they produce through fermentation or communication with dietary components have positive effects on human health (Aguilar-Toalá et al. 2018). These bioactive molecules, which cannot be classified as probiotics, prebiotics, or synbiotics, are called by several names, such as tyndallized probiotics, metabiotics, ghost probiotics, para-psychobiotics, paraprobiotics, and bacterial lysates (Lopetuso et al. 2017).

Up until now, the term "postbiotics" has been widely used. Currently, there is no widely agreed-upon definition for the phrase. Hence, it is crucial at this moment to establish a shared and precise meaning for the term "postbiotics." Providing a definition can help prevent or reduce misunderstandings of a phrase, which is a common occurrence when incorrect claims are made about certain things (De Simone 2019). A panel of specialists from the International Scientific Association for Probiotics and Prebiotics (ISAPP), who have knowledge in food science, nutrition, microbiology, microbial physiology, pediatrics, and gastroenterology, conducted a comprehensive evaluation of the definition and utilization of postbiotics in 2019 (Salminen et al. 2021). While the panel tentatively defined postbiotics as the "preparation of inanimate microorganisms and/or their components that confers a health benefit on the host," there is significant debate among experts regarding this suggested description (Aguilar-Toalá et al. 2021). A precise definition is crucial from a regulatory standpoint as it facilitates the approval of its use in food and nutritional supplements as strategies for promoting health. Moreover, the very fact that postbiotics provide biological benefits to the host without the potential risks associated with the use of active microorganisms helps reduce the legal obstacles to obtaining permission from regulatory agencies. Presently, probiotics have a promising outlook. The utilization of postbiotics in the field of food will have a positive impact on a substantial section of the population, including individuals with certain medical disorders, by potentially serving as an effective therapeutic intervention. The effectiveness of postbiotics will depend on several interconnected aspects. The ability of a postbiotic to have a positive impact on an individual's health can be influenced by many processes or routes. A postbiotic is a heterogeneous blend of several constituents. Occasionally, these pathways may bear a resemblance to the ones discovered for probiotics, and these processes may function individually or together. Hence, to ensure the successful preservation of bioactive qualities and other essential health-promoting attributes in a commercial postbiotic product, it is imperative to comprehend the pivotal molecules accountable for these advantageous effects. Postbiotics are inanimate; therefore, in order for the microorganisms to benefit from them, they must also create these bioactive substances in large enough amounts prior to inactivation. Identifying the optimal target population, dosage, and method of administration, as well as isolating the safest and most effective strains, is still a significant research gap that presents a major obstacle to the marketing of postbiotics.

To our knowledge, no regulatory bodies have established a specific postbiotic definition or framework for foods or nutritional supplements containing postbiotics. Regulatory requirements have been set for postbiotic formulations used in medical or pharmaceutical applications (Thorakkattu et al. 2022). The European Food Safety Authority (EFSA) is responsible for overseeing the assessment of food safety in Europe and regularly updating the requirements for food regulation. The European Pharmacopoeia establishes explicit standards that determine the maximum allowable quantities of live microorganisms for pharmaceuticals and medical products (Thorakkattu et al. 2022). There are no specific laws in the European Union that regulate prebiotics, probiotics, synbiotics, or postbiotics. However, certain postbiotics have been promoted or authorized as chemicals that stimulate the immune system (Yin et al. 2018). Argentina's food code in 2011 contained Articles 1389 and 1390, which specifically addressed probiotics and prebiotics, respectively. Despite the release of an infant formula by an international company in 2019 that was labeled as containing "postbiotics," the subject of postbiotics remains unresolved (Salminen et al. 2021). The Food and Drug Administration (FDA) in the United States has not specifically addressed the topic of postbiotics. The FDA is expected to regulate postbiotics depending on the specific rules that apply to the chosen regulatory category for an item in development. This is because postbiotics can be manufactured in various regulatory categories. The product must meet the criteria for the appropriate regulatory category in terms of its intended consumption, safety, and efficacy. Postbiotics are considered to be safer than live forms of probiotics. As a result, they cannot cause bacteremia or fungemia, which are risks associated with probiotic therapy (Yelin et al. 2019). Given that postbiotic foods are a somewhat recent idea, there is currently no epidemiological evidence indicating any health risks connected with their intake, as far as we know. Nevertheless, it is not appropriate to presume that postbiotics are safe only based on the safety profile of the original microbe. Lipopolysaccharides from Gram-negative bacteria, like those found in the outer membrane of bacteria that are actively growing, can cause sepsis and toxic shock when they are released when cells break down (Ginsburg 2002). Hence, it is crucial to proactively anticipate regulatory guidance for postbiotics by assessing potential risks and safety factors.

Malagón-Rojas and colleagues (2020) conducted a comprehensive assessment of probiotic-controlled clinical investigations involving 1740 children. The study specifically investigated the efficacy of postbiotics in treating and preventing common infectious diseases in kids below the age of 5. They also investigated the potential adverse effects of postbiotics. Three RCTs specifically surveyed the negative properties of postbiotics and described severe dehydration, increased nausea and abdominal pain, and child-related D-lactic acidosis (Wegh et al. 2019; Łukasik et al. 2018). Łukasik and colleagues (2018) also proposed the theoretical opportunity of subclinical gathering of D-lactate in newborns who are fed fermented infant formula. In addition, there is a scarcity of investigations that have established the detrimental adverse reactions of postbiotic compounds. Hence, it is imperative to do thorough research, encompassing both laboratory experiments and animal studies, as well as clinical trials, in order to further our comprehension of the safety concerns associated with postbiotics. Moreover, it is necessary to conduct further investigations in order to determine the correct dosage and ideal frequency of consumption, as well as to assess the health benefits associated with postbiotics.

13.6 Conclusion

Postbiotics are substances derived from microorganisms that are no longer alive and are free of any living organisms. These substances can potentially improve the beneficial effects of probiotics on the host. Consuming a diet rich in probiotic or prebiotic foods can help maintain a sufficient amount of these vital compounds in the gut, as postbiotics are generated when probiotics consume prebiotics. Postbiotics exhibit a range of bioactivities, such as antioxidant, hypocholesterolemic, antiproliferative, anti-inflammatory, anti-obesogenic, anticancer, and immunomodulatory properties. Furthermore, postbiotics show even greater potential in neonates or premature infants, as they have a safety profile that permits their use in critically ill or severely preterm babies. Nevertheless, due to the unique characteristics of each species, the considerable diversity of the substrates examined, and the wide range of potentially disruptive factors, there is still a lack of conclusive evidence about their application in therapeutic settings. As additional details on postbiotics become available, future clinical trials are expected to focus on both the composition and bioactivity of these substances.

The scientific evidence addressed in this chapter confirms the effectiveness of postbiotics; however, the exact level of efficacy remains unclear. Therefore, additional investigation is necessary to determine the therapeutic efficacy and preventive benefits of postbiotics. Additional research on postbiotics will result in novel discoveries and several advantageous impacts, hence enhancing their utilization in the food and pharmacy sectors. Further research should be undertaken to evaluate the factors, such as dietary habits, that could impact the use of these advantageous substances in functional meals. In order to have a deeper understanding of how they function within the host, it is necessary to perform further research. Furthermore, the extensive utilization of postbiotics in the food and pharmacy sectors would require their production in appropriate industrial settings and strict adherence to manufacturing standards. These endeavors will contribute to the manufacturing of safe, organic, cleanly labeled items that are free from debris, protecting human well-being. Hence, in order to achieve additional health advantages, it is imperative to thoroughly assess the characteristics of novel postbiotic constituents. In order to guarantee the profitability and long-term viability of the food and pharmacy sectors, it is imperative to take into account the enduring health advantages associated with the use of these supplements. Therefore, it is important for future investigations to focus on establishing the correlation between the health properties of postbiotic compounds and their unique mechanisms, as postbiotics are continuously developing.

References

Abbasi, A., Aghebati-Maleki, L., and Homayouni-Rad, A. (2022). The promising biological role of postbiotics derived from probiotic *Lactobacillus* species in reproductive health. *Critical Reviews in Food Science and Nutrition* 62 (32): 8829–8841.

Abbasi, A., Rad, A.H., Maleki, L.A. et al. (2023). Antigenotoxicity and cytotoxic potentials of cell-free supernatants derived from *Saccharomyces cerevisiae* var. *boulardii* on HT-29 human colon cancer cell lines. *Probiotics and Antimicrobial proteins* 15 (6): 1583–1595.

Aguilar-Toalá, J.E., Arioli, S., Behare, P. et al. (2021). Postbiotics—when simplification fails to clarify. *Nature Reviews Gastroenterology & Hepatology* 18 (11): 825–826.

Aguilar-Toalá, J., Garcia-Varela, R., Garcia, H. et al. (2018). Postbiotics: an evolving term within the functional foods field. *Trends in Food Science and Technology* 75: 105–114.

Aguilar-Toalá, J., Hall, F., Urbizo-Reyes, U. et al. (2020). In silico prediction and in vitro assessment of multifunctional properties of postbiotics obtained from two probiotic bacteria. *Probiotics and Antimicrobial Proteins* 12: 608–622.

Anwar, F., Altayb, H.N., Al-Abbasi, F.A. et al. (2021). Antiviral effects of probiotic metabolites on COVID-19. *Journal of Biomolecular Structure & Dynamics* 39 (11): 4175–4184.

Baker, M.T., Lu, P., Parrella, J.A., and Leggette, H.R. (2022). Consumer acceptance toward functional foods: a scoping review. *International Journal of Environmental Research and Public Health* 19 (3): 1217.

Barros, C.P., Guimaraes, J.T., Esmerino, E.A. et al. (2020). Paraprobiotics and postbiotics: concepts and potential applications in dairy products. *Current Opinion in Food Science* 32: 1–8.

Berni Canani, R., De Filippis, F., Nocerino, R. et al. (2017). Specific signatures of the gut microbiota and increased levels of butyrate in children treated with fermented cow's milk containing heat-killed *Lactobacillus paracasei* CBA L74. *Applied and Environmental Microbiology* 83 (19): e01206–e01217.

Bielik, V. and Kolisek, M. (2021). Bioaccessibility and bioavailability of minerals in relation to a healthy gut microbiome. *International Journal of Molecular Sciences* 22 (13): 6803.

Bodemer, C., Guillet, G., Cambazard, F. et al. (2017). Adjuvant treatment with the bacterial lysate (OM-85) improves management of atopic dermatitis: a randomized study. *PLoS ONE* 12 (3): e0161555.

Borriello, S., Hammes, W., Holzapfel, W. et al. (2003). Safety of probiotics that contain lactobacilli or bifidobacteria. *Clinical Infectious Diseases* 36 (6): 775–780.

Bron, P.A., Tomita, S., Mercenier, A., and Kleerebezem, M. (2013). Cell surface-associated compounds of probiotic lactobacilli sustain the strain-specificity dogma. *Current Opinion in Microbiology* 16 (3): 262–269.

Campeotto, F., Suau, A., Kapel, N. et al. (2011). A fermented formula in pre-term infants: clinical tolerance, gut microbiota, down-regulation of faecal calprotectin and up-regulation of faecal secretory IgA. *The British Journal of Nutrition* 105 (12): 1843–1851.

Ceppa, F.A., Izzo, L., Sardelli, L. et al. (2020). Human gut-microbiota interaction in neurodegenerative disorders and current engineered tools for its modeling. *Frontiers in Cellular and Infection Microbiology* 297.

Chaluvadi, S., Hotchkiss, A.T., and Yam, K.L. (2015). Gut microbiota: impact of probiotics, prebiotics, synbiotics, pharmabiotics, and postbiotics on human health. In: *Probiotics, Prebiotics, and Synbiotics: Bioactive Foods in Health Promotion*, 515–523. Elsevier Inc.

Chang, H.M., Foo, H.L., Loh, T.C. et al. (2021). Comparative studies of inhibitory and antioxidant activities, and organic acids compositions of postbiotics produced by probiotic *Lactiplantibacillus plantarum* strains isolated from Malaysian foods. *Frontiers in Veterinary Science* 7: 602280.

Chen, D., Jin, D., Huang, S. et al. (2020). *Clostridium butyricum*, a butyrate-producing probiotic, inhibits intestinal tumor development through modulating Wnt signaling and gut microbiota. *Cancer Letters* 469: 456–467.

Chuah, L.-O., Foo, H.L., Loh, T.C. et al. (2019). Postbiotic metabolites produced by *Lactobacillus plantarum* strains exert selective cytotoxicity effects on cancer cells. *BMC Complementary and Alternative Medicine* 19: 1–12.

Corsello, G., Carta, M., Marinello, R. et al. (2017). Preventive effect of cow's milk fermented with *Lactobacillus paracasei* CBA L74 on common infectious diseases in children: a multicenter randomized controlled trial. *Nutrients* 9 (7): 669.

Cuevas-González, P., Liceaga, A., and Aguilar-Toalá, J. (2020). Postbiotics and paraprobiotics: from concepts to applications. *Foodservice Research International* 136: 109502.

Damián, M.R., Cortes-Perez, N.G., Quintana, E.T. et al. (2022). Functional foods, nutraceuticals and probiotics: a focus on human health. *Microorganisms* 10 (5): 1065.
Das, N.K., Schwartz, A.J., Barthel, G. et al. (2020). Microbial metabolite signaling is required for systemic iron homeostasis. *Cell Metabolism* 31 (1): 115–130. e6.
de Boer, G.M., Żółkiewicz, J., Strzelec, K.P. et al. (2020). Bacterial lysate therapy for the prevention of wheezing episodes and asthma exacerbations: a systematic review and meta-analysis. *European Respiratory Review* 29 (158): 190175.
De Simone, C. (2019). The unregulated probiotic market. *Clinical Gastroenterology and Hepatology* 17 (5): 809–817.
Dominguez-Bello, M.G., De Jesus-Laboy, K.M., Shen, N. et al. (2016). Partial restoration of the microbiota of cesarean-born infants via vaginal microbial transfer. *Nature Medicine* 22 (3): 250–253.
Dongarrà, M.L., Rizzello, V., Muccio, L. et al. (2013). Mucosal immunology and probiotics. *Current Allergy and Asthma Reports* 13: 19–26.
Dunand, E., Burns, P., Binetti, A. et al. (2019). Postbiotics produced at laboratory and industrial level as potential functional food ingredients with the capacity to protect mice against Salmonella infection. *Journal of Applied Microbiology* 127 (1): 219–229.
Feerick, C.L. and McKernan, D.P. (2017). Understanding the regulation of pattern recognition receptors in inflammatory diseases–a 'Nod'in the right direction. *Immunology* 150 (3): 237–247.
Ferreira, R.L.U., Sena-Evangelista, K.C.M., De Azevedo, E.P. et al. (2021). Selenium in human health and gut microflora: bioavailability of selenocompounds and relationship with diseases. *Frontiers in Nutrition* 8: 685317.
Francescone, R., Hou, V., and Grivennikov, S.I. (2014). Microbiome, inflammation and cancer. *Cancer journal (Sudbury, Mass).* 20 (3): 181.
Fukuda, S., Toh, H., Hase, K. et al. (2011). Bifidobacteria can protect from enteropathogenic infection through production of acetate. *Nature* 469 (7331): 543–547.
Gao, J., Li, Y., Wan, Y. et al. (2019). A novel postbiotic from *Lactobacillus rhamnosus* GG with a beneficial effect on intestinal barrier function. *Frontiers in Microbiology* 10: 477.
Ginsburg, I. (2002). Role of lipoteichoic acid in infection and inflammation. *The Lancet Infectious Diseases* 2 (3): 171–179.
Gueimonde, M., Sánchez, B., de los Reyes-Gavilán, C.G., and Margolles, A. (2013). Antibiotic resistance in probiotic bacteria. *Frontiers in Microbiology* 4: 202.
Guo, Y., Pan, D., Li, H. et al. (2013). Antioxidant and immunomodulatory activity of selenium exopolysaccharide produced by *Lactococcus lactis* subsp. lactis. *Food Chemistry* 138 (1): 84–89.
Gupta, T., Kaur, H., Kapila, S., and Kapila, R. (2021). Potential probiotic *Lacticaseibacillus rhamnosus* MTCC-5897 attenuates *Escherichia coli* induced inflammatory response in intestinal cells. *Archives of Microbiology* 203 (9): 5703–5713.
Hatayama, H., Iwashita, J., Kuwajima, A., and Abe, T. (2007). The short chain fatty acid, butyrate, stimulates MUC2 mucin production in the human colon cancer cell line, LS174T. *Biochemical and Biophysical Research Communications* 356 (3): 599–603.
Hendrickx, D.M., An, R., Boeren, S. et al. (2023). Assessment of infant outgrowth of cow's milk allergy in relation to the faecal microbiome and metaproteome. *Scientific Reports* 13 (1): 12029.
Hosseini, H., Abbasi, A., Sabahi, S. et al. (2024). Assessing the potential biological activities of postbiotics derived from *Saccharomyces cerevisiae*: an in vitro study. Probiotics and antimicrobial. *Proteins* 16 (4): 1348–1364.
Hu, J., Wang, C., Huang, X. et al. (2021). Gut microbiota-mediated secondary bile acids regulate dendritic cells to attenuate autoimmune uveitis through TGR5 signaling. *Cell Reports* 36 (12): 109726.

Huet, F., Abrahamse-Berkeveld, M., Tims, S. et al. (2016). Partly fermented infant formulae with specific oligosaccharides support adequate infant growth and are well-tolerated. *Journal of Pediatric Gastroenterology and Nutrition* 63 (4): e43.

Humam, A.M., Loh, T.C., Foo, H.L. et al. (2021). Supplementation of postbiotic RI11 improves antioxidant enzyme activity, upregulated gut barrier genes, and reduced cytokine, acute phase protein, and heat shock protein 70 gene expression levels in heat-stressed broilers. *Poultry Science* 100 (3): 100908.

Izuddin, W.I., Humam, A.M., Loh, T.C. et al. (2020). Dietary postbiotic *Lactobacillus plantarum* improves serum and ruminal antioxidant activity and upregulates hepatic antioxidant enzymes and ruminal barrier function in post-weaning lambs. *Antioxidants.* 9 (3): 250.

Kaila, M., Isolauri, E., Saxelin, M. et al. (1995). Viable versus inactivated *Lactobacillus* strain GG in acute rotavirus diarrhoea. *Archives of Disease in Childhood* 72 (1): 51–53.

Kang, S., Li, R., Jin, H. et al. (2020). Effects of selenium-and zinc-enriched *Lactobacillus plantarum* SeZi on antioxidant capacities and gut microbiome in an ICR mouse model. *Antioxidants.* 9 (10): 1028.

Kareem, K.Y., Loh, T.C., Foo, H.L. et al. (2016). Effects of dietary postbiotic and inulin on growth performance, IGF1 and GHR mRNA expression, faecal microbiota and volatile fatty acids in broilers. *BMC Veterinary Research* 12: 1–10.

Kawai, T. and Akira, S. (2009). The roles of TLRs, RLRs and NLRs in pathogen recognition. *International Immunology* 21 (4): 317–337.

Khodaii, Z., Ghaderian, S.M.H., and Natanzi, M.M. (2017). Probiotic bacteria and their supernatants protect enterocyte cell lines from enteroinvasive *Escherichia coli* (EIEC) invasion. *International journal of molecular and cellular medicine.* 6 (3): 183.

Kim, K.W., Kang, S.-S., Woo, S.-J. et al. (2017). Lipoteichoic acid of probiotic *Lactobacillus plantarum* attenuates poly I: C-induced IL-8 production in porcine intestinal epithelial cells. *Frontiers in Microbiology* 8: 1827.

Kim, S., Kim, G.-H., and Cho, H. (2021). Postbiotics for cancer prevention and treatment. *The Microbiological Society of Korea.* 57 (3): 142–153.

Kirjavainen, P.V., Salminen, S.J., and Isolauri, E. (2003). Probiotic bacteria in the management of atopic disease: underscoring the importance of viability. *Journal of Pediatric Gastroenterology and Nutrition* 36 (2): 223–227.

LeBlanc, J.G., Chain, F., Martín, R. et al. (2017). Beneficial effects on host energy metabolism of short-chain fatty acids and vitamins produced by commensal and probiotic bacteria. *Microbial Cell Factories* 16 (1): 1–10.

Liévin-Le Moal, V., Sarrazin-Davila, L.E., and Servin, A.L. (2007). An experimental study and a randomized, double-blind, placebo-controlled clinical trial to evaluate the antisecretory activity of *Lactobacillus acidophilus* strain LB against nonrotavirus diarrhea. *Pediatrics* 120 (4): e795–e803.

Lopetuso, L., Graziani, C., Guarino, A. et al. (2017). Gelatin tannate and tyndallized probiotics: a novel approach for treatment of diarrhea. *European Review for Medical and Pharmacological Sciences* 21 (4): 873–883.

Łukasik, J., Salminen, S., and Szajewska, H. (2018). Rapid review shows that probiotics and fermented infant formulas do not cause d-lactic acidosis in healthy children. *Acta Paediatrica* 107 (8): 1322–1326.

Malagón-Rojas, J.N., Mantziari, A., Salminen, S., and Szajewska, H. (2020). Postbiotics for preventing and treating common infectious diseases in children: a systematic review. *Nutrients* 12 (2): 389.

Martorell, P., Alvarez, B., Llopis, S. et al. (2021). Heat-treated *Bifidobacterium longum* CECT-7347: a whole-cell postbiotic with antioxidant, anti-inflammatory, and gut-barrier protection properties. *Antioxidants* 10 (4): 536.

Mayorgas, A., Dotti, I., and Salas, A. (2021). Microbial metabolites, postbiotics, and intestinal epithelial function. *Molecular Nutrition & Food Research* 65 (5): 2000188.

Medici, M., Vinderola, C.G., Weill, R., and Perdigón, G. (2005). Effect of fermented milk containing probiotic bacteria in the prevention of an enteroinvasive *Escherichia coli* infection in mice. *The Journal of Dairy Research* 72 (2): 243–249.

Mego, M., Májek, J., Končeková, R. et al. (2005). Intramucosal bacteria in colon cancer and their elimination by probiotic strain *Enterococcus faecium* M-74 with organic selenium. *Folia Microbiologica* 50: 443–447.

Ménard, S., Laharie, D., Asensio, C. et al. (2005). *Bifidobacterium breve* and *Streptococcus thermophilus* secretion products enhance T helper 1 immune response and intestinal barrier in mice. *Experimental Biology and Medicine* 230 (10): 749–756.

Meng, Z., Ye, Z., Zhu, P. et al. (2022). New developments and opportunities of microbiota in treating breast cancers. *Frontiers in Microbiology* 13: 818793.

Mocchegiani, E. and Malavolta, M. (2009). Role of zinc and selenium in oxidative stress and immunosenescence: implications for healthy ageing and longevity. In: *Handbook on Immunosenescence: Basic Understanding and Clinical Applications*, 1367–1396. Dordrecht: Springer.

Mogensen, T.H. (2009). Pathogen recognition and inflammatory signaling in innate immune defenses. *Clinical Microbiology Reviews* 22 (2): 240–273.

Morales-Ferré, C., Azagra-Boronat, I., Massot-Cladera, M. et al. (2022). Preventive effect of a postbiotic and prebiotic mixture in a rat model of early life rotavirus induced-diarrhea. *Nutrients* 14 (6): 1163.

Morisset, M., Aubert-Jacquin, C., Soulaines, P. et al. (2011). A non-hydrolyzed, fermented milk formula reduces digestive and respiratory events in infants at high risk of allergy. *European Journal of Clinical Nutrition* 65 (2): 175–183.

Nakamura, S., Kuda, T., An, C. et al. (2012). Inhibitory effects of Leuconostoc mesenteroides 1RM3 isolated from narezushi, a fermented fish with rice, on *Listeria monocytogenes* infection to Caco-2 cells and A/J mice. *Anaerobe* 18 (1): 19–24.

Nakkarach, A., Foo, H.L., Song, A.A.-L. et al. (2021). Anti-cancer and anti-inflammatory effects elicited by short chain fatty acids produced by *Escherichia coli* isolated from healthy human gut microbiota. *Microbial Cell Factories* 20: 1–17.

Nataraj, B.H., Ali, S.A., Behare, P.V., and Yadav, H. (2020). Postbiotics-parabiotics: the new horizons in microbial biotherapy and functional foods. *Microbial Cell Factories* 19 (1): 1–22.

Negatu, D.A., Yamada, Y., Xi, Y. et al. (2019). Gut microbiota metabolite indole propionic acid targets tryptophan biosynthesis in *Mycobacterium tuberculosis*. *MBio* 10 (2): e02781.

Nocerino, R., Paparo, L., Terrin, G. et al. (2017). Cow's milk and rice fermented with *Lactobacillus paracasei* CBA L74 prevent infectious diseases in children: a randomized controlled trial. *Clinical Nutrition* 36 (1): 118–125.

Nylund, L., Nermes, M., Isolauri, E. et al. (2015). Severity of atopic disease inversely correlates with intestinal microbiota diversity and butyrate-producing bacteria. *Allergy* 70 (2): 241–244.

Orlando, A., Refolo, M., Messa, C. et al. (2012). Antiproliferative and proapoptotic effects of viable or heat-killed *Lactobacillus paracasei* IMPC2. 1 and *Lactobacillus rhamnosus* GG in HGC-27 gastric and DLD-1 colon cell lines. *Nutrition & Cancer* 64 (7): 1103–1111.

Osman, A., El-Gazzar, N., Almanaa, T.N. et al. (2021). Lipolytic postbiotic from *Lactobacillus paracasei* manages metabolic syndrome in albino wistar rats. *Molecules* 26 (2): 472.

Peng, G.C. and Hsu, C.H. (2005). The efficacy and safety of heat-killed *Lactobacillus paracasei* for treatment of perennial allergic rhinitis induced by house-dust mite. *Pediatric Allergy and Immunology* 16 (5): 433–438.

Piqué, N., Berlanga, M., and Miñana-Galbis, D. (2019). Health benefits of heat-killed (Tyndallized) probiotics: an overview. *International Journal of Molecular Sciences* 20 (10): 2534.

Rad, A.H., Abbasi, A., Kafil, H.S., and Ganbarov, K. (2020). Potential pharmaceutical and food applications of postbiotics: a review. *Current Pharmaceutical Biotechnology* 21 (15): 1576–1587.

Rad, A.H., Hosseini, S., and Pourjafar, H. (2022). Postbiotics as dynamic biological molecules for antimicrobial activity: a mini-review. *Biointerface Research in Applied Chemistry* 12 (5): 6543–6556.

Reid, G., Gadir, A.A., and Dhir, R. (2019). Probiotics: reiterating what they are and what they are not. *Frontiers in Microbiology* 10: 424.

Riccardi, N., Rotulo, G.A., and Castagnola, E. (2019). Definition of opportunistic infections in immunocompromised children on the basis of etiologies and clinical features: a summary for practical purposes. *Current Pediatric Reviews* 15 (4): 197–206.

Rossoni, R.D., De Barros, P.P., Mendonça, I.C. et al. (2020). The postbiotic activity of *Lactobacillus paracasei* 28.4 against *Candida auris*. *Frontiers in Cellular and Infection Microbiology* 10: 397.

Rowan, N.J., Deans, K., Anderson, J.G. et al. (2001). Putative virulence factor expression by clinical and food isolates of *Bacillus* spp. after growth in reconstituted infant milk formulae. *Applied and Environmental Microbiology* 67 (9): 3873–3881.

Rutsch, A., Kantsjö, J.B., and Ronchi, F. (2020). The gut-brain axis: how microbiota and host inflammasome influence brain physiology and pathology. *Frontiers in Immunology* 11: 604179.

Sabahi, S., Homayouni Rad, A., Aghebati-Maleki, L. et al. (2023). Postbiotics as the new frontier in food and pharmaceutical research. *Critical Reviews in Food Science and Nutrition* 63 (26): 8375–8402.

Safari, M.S., Keyhanfar, M., and Shafiei, R. (2019). Investigating the antibacterial effects of some *Lactobacillus*, *Bifidobacterium* and *Acetobacter* strains killed by different methods on *Streptococcus mutans* and Escherichia coli. *Molecular Biology Research Communications* 8 (3): 103.

Salminen, S., Collado, M.C., Endo, A. et al. (2021). The International Scientific Association of Probiotics and Prebiotics (ISAPP) consensus statement on the definition and scope of postbiotics. *Nature Reviews Gastroenterology & Hepatology* 18 (9): 649–667.

Schwarz, S., Lehmbecker, A., Tongtako, W. et al. (2020). Neurotrophic effects of GM1 ganglioside, NGF, and FGF2 on canine dorsal root ganglia neurons in vitro. *Scientific Reports* 10 (1): 5380.

Sevin, S., Karaca, B., Haliscelik, O. et al. (2021). Postbiotics secreted by *Lactobacillus sakei* EIR/CM-1 isolated from cow milk microbiota, display antibacterial and antibiofilm activity against ruminant mastitis-causing pathogens. *Italian Journal of Animal Science* 20 (1): 1302–1316.

Siegel, R.L., Miller, K.D., and Jemal, A. (2019). Cancer statistics, 2019. *CA: A Cancer Journal for Clinicians* 69 (1): 7–34.

Spagnolello, O., Pinacchio, C., Santinelli, L. et al. (2021). Targeting microbiome: an alternative strategy for fighting SARS-CoV-2 infection. *Chemotherapy* 66 (1-2): 24–32.

Sunmola, A.A., Ogbole, O.O., Faleye TO et al. (2019). Antiviral potentials of *Lactobacillus plantarum*, *Lactobacillus amylovorus*, and *Enterococcus hirae* against selected Enterovirus. *Folia Microbiologica* 64: 257–264.

Teame T, Wang, A., Xie, M. et al. (2020). Paraprobiotics and postbiotics of probiotic Lactobacilli, their positive effects on the host and action mechanisms: a review. *Frontiers in Nutrition* 7: 570344.

Thibault, H., Aubert-Jacquin, C., and Goulet, O. (2004). Effects of long-term consumption of a fermented infant formula (with *Bifidobacterium breve* c50 and *Streptococcus thermophilus* 065) on acute diarrhea in healthy infants. *Journal of Pediatric Gastroenterology and Nutrition* 39 (2): 147–152.

Thorakkattu, P., Khanashyam, A.C., Shah, K. et al. (2022). Postbiotics: current trends in food and pharmaceutical industry. *Food* 11 (19): 3094.

Todorov, S.D., Tagg, J.R., and Ivanova, I.V. (2021). Could probiotics and postbiotics function as "silver bullet" in the post-COVID-19 era? *Probiotics and Antimicrobial Proteins* 13: 1499–1507.

Vinderola, G., Matar, C., and Perdigón, G. (2007). Milk fermented by *Lactobacillus helveticus* R389 and its non-bacterial fraction confer enhanced protection against *Salmonella enteritidis* serovar *Typhimurium* infection in mice. *Immunobiology* 212 (2): 107–118.

Wang, K., Li, W., Rui, X. et al. (2014). Characterization of a novel exopolysaccharide with antitumor activity from *Lactobacillus plantarum* 70810. *International Journal of Biological Macromolecules* 63: 133–139.

Wasilewska, E., Zlotkowska, D., and Wroblewska, B. (2019). Yogurt starter cultures of *Streptococcus thermophilus* and *Lactobacillus bulgaricus* ameliorate symptoms and modulate the immune response in a mouse model of dextran sulfate sodium-induced colitis. *Journal of Dairy Science* 102 (1): 37–53.

Wegh, C.A., Geerlings, S.Y., Knol, J. et al. (2019). Postbiotics and their potential applications in early life nutrition and beyond. *International Journal of Molecular Sciences* 20 (19): 4673.

Yang, J., Wang, J., Huang, K. et al. (2021). Selenium-enriched *Bacillus subtilis* yb-1 14246 improved growth and immunity of broiler chickens through modified ileal bacterial composition. *Scientific Reports* 11 (1): 21690.

Yelin, I., Flett, K.B., Merakou, C. et al. (2019). Genomic and epidemiological evidence of bacterial transmission from probiotic capsule to blood in ICU patients. *Nature Medicine* 25 (11): 1728–1732.

Yildiz, A., Kaya, Y., and Tanriverdi, O. (2019). Effect of the interaction between selenium and zinc on DNA repair in association with cancer prevention. *Journal of Cancer Prevention.* 24 (3): 146.

Yin, J., Xu, B., Zeng, X., and Shen, K. (2018). Broncho-Vaxom in pediatric recurrent respiratory tract infections: a systematic review and meta-analysis. *International Immunopharmacology* 54: 198–209.

Zhai, Q., Cen, S., Li, P. et al. (2018). Effects of dietary selenium supplementation on intestinal barrier and immune responses associated with its modulation of gut microbiota. *Environmental Science & Technology Letters* 5 (12): 724–730.

Zhang, L., Liu, C., Li, D. et al. (2013). Antioxidant activity of an exopolysaccharide isolated from *Lactobacillus plantarum* C88. *International Journal of Biological Macromolecules* 54: 270–275.

Zhang, Z., Tang, H., Chen, P. et al. (2019). Demystifying the manipulation of host immunity, metabolism, and extraintestinal tumors by the gut microbiome. *Signal Transduction and Targeted Therapy* 4 (1): 41.

Żółkiewicz, J., Marzec, A., Ruszczyński, M., and Feleszko, W. (2020). Postbiotics—a step beyond pre-and probiotics. *Nutrients* 12 (8): 2189.

Zorzela, L., Ardestani, S., McFarland, L., and Vohra, S. (2017). Is there a role for modified probiotics as beneficial microbes: a systematic review of the literature. *Beneficial Microbes* 8 (5): 739–754.

14

Innovative Approaches for the Microbial Production of Carotenoids

Rajni Kumari[1], Monika[1], V Vivekanand[2], and Nidhi Pareek[1]

[1] *Microbial Catalysis and Process Engineering Laboratory, Department of Microbiology, School of Life Sciences, Central University of Rajasthan Bandarsindri, Kishangarh, Ajmer, Rajasthan, 305817, India*
[2] *Centre for Energy and Environment, Malaviya National Institute of Technology, Jaipur, Rajasthan, 302017, India*

14.1 Introduction to Microbial Carotenoid Production

Natural products have been increasingly useful in the quest to improve quality of life. They can be used as industrial supplements, tastes, and scents, among other things. Carotenoids, a prominent subgroup within the realm of natural products, hold significant importance (Maoka 2020). Defined as a subset of terpenoids typically comprising 40 carbon atoms and 8 isoprene units, carotenoids exhibit spectrum shades of orange, red, and yellow owing to their extended polyene chain conjugation (Barford 2022). This unique chemical architecture not only imbues them with antioxidant properties and provitamin A activity but also renders them effective in shielding against UV radiation (Pérez-Gálvez et al. 2020). Throughout different organisms, such as fungi and bacteria, carotenoids serve as potent photoprotective and stress-alleviating agents (Afroz Toma et al. 2023). Furthermore, new studies demonstrate their critical role in preserving animal health by highlighting their involvement in vital biological processes like the regulation of genes and apoptosis in mammalian cells. Due to their biological activities, carotenoids find extensive applications as food and feed additives, cosmetics, and nutraceuticals (Foong et al. 2021). The international market for carotenoids reached \$1.5 billion in 2018 and is projected for expanding to \$2.0 billion by 2022 (Bogacz-Radomska et al. 2020). Despite the escalating demand for carotenoids, their limited supply persists due to inefficient production methods. While natural plant-derived carotenoids are esteemed for their purity, they often entail high production costs and susceptibility to seasonal variations. In contrast, microbial-derived carotenoids are gaining preference owing to their inherent efficiency, cost-effectiveness, and suitability for repetitive-flow industrial production (Mussagy et al. 2021).

This chapter explores innovative strategies aimed at enhancing the microbial production of carotenoids. Understanding the factors influencing microbial synthesis is pivotal in optimizing production efficiency. Environmental parameters such as temperature, pH, strain selection, substrate availability, and oxygen levels profoundly impact carotenoid biosynthesis in microbial hosts (Igreja et al. 2021). Additionally, genetic factors, including strain selection, metabolic pathways, and regulatory mechanisms, play a principal role in modulating carotenoid yields

Microbial Nutraceuticals: Products and Processes, First Edition. Edited by Sudhir Pratap Singh and Santosh Kumar Upadhyay.

(Liu et al. 2020). To address these challenges and augment carotenoid production, various approaches have been devised. Metabolic engineering stands at the forefront, enabling the manipulation of microbial metabolism to bolster carotenoid biosynthesis (Torres-Haro et al. 2021). Techniques such as gene expression modulation and knockout strategies empower researchers to fine-tune metabolic pathways for enhanced carotenoid accumulation. Pathway engineering further optimizes substrate utilization and flux toward carotenoid biosynthesis, amplifying production yields. Furthermore, advanced fermentation strategies, including fed-batch and continuous processes, offer precise control over nutrient supply and waste removal, fostering optimal conditions for carotenoid production. Consortia engineering, harnessing the synergistic interactions among microbial species, presents a novel approach to amplify carotenoid synthesis through mutualistic relationships. The advent of CRISPR-Cas genome editing has revolutionized microbial engineering, facilitating targeted modifications in the microbial genome to enhance carotenoid production. By precisely altering regulatory elements and metabolic networks, CRISPR-Cas technology enables unprecedented control over carotenoid biosynthesis (Awasthi et al. 2022). The implementation of CRISPR-Cas9 technology for the production of microbial carotenoids has emerged as a transformative approach in biotechnology. Precisely, CRISPR-Cas9 allows for the optimization of metabolic pathway, improving yields and reducing production costs (Mo et al. 2020). Moreover, stress induction strategies and directed evolution techniques offer alternative avenues for augmenting carotenoid production. By subjecting microbial strains to selective pressures or evolutionary forces, researchers can iteratively improve carotenoid yields, harnessing the adaptability of microbial hosts to optimize production (Giani & Martínez-Espinosa 2020). The primary aim of this chapter is to elucidate pivotal plan of action and recent advancements in optimizing these modules to maximize carotenoid production, shedding light on the transformative potential of microbial biotechnology in meeting the growing demand for natural pigments with diverse industrial applications. Natural products have been increasingly useful in the quest to improve quality of life. They can be used as industrial supplements, tastes, and scents, among other things. Carotenoids, a prominent subgroup within the realm of natural products, hold significant importance (Maoka 2020). Defined as a subset of terpenoids typically comprising 40 carbon atoms and 8 isoprene units, carotenoids exhibit a spectrum shades of orange, red, and yellow owing to their extended polyene chain conjugation (Barford 2022). This unique chemical architecture not only imbues them with antioxidant properties and provitamin A activity but also renders them effective in shielding against UV radiation (Pérez-Gálvez et al. 2020). Throughout different organisms, such as fungi and bacteria, carotenoids serve as potent photoprotective and stress-alleviating agents (Afroz Toma et al. 2023). Furthermore, new studies demonstrate their critical role in preserving animal health by highlighting their involvement in vital biological processes like the regulation of genes and apoptosis in mammalian cells. Due to their biological activities, carotenoids find extensive applications as food and feed additives, cosmetics, and nutraceuticals (Foong et al. 2021). The international market for carotenoids reached \$1.5 billion in 2018 and is projected for expanding to \$2.0 billion by 2022 (Bogacz-Radomska et al. 2020). Despite the escalating demand for carotenoids, their limited supply persists due to inefficient production methods. While natural plant-derived carotenoids are esteemed for their purity, they often entail high production costs and susceptibility to seasonal variations. In contrast, microbial-derived carotenoids are gaining preference owing to their inherent efficiency, cost-effectiveness, and suitability for repetitive-flow industrial production (Mussagy et al. 2021).

This chapter explores innovative strategies aimed at enhancing the microbial production of carotenoids. Understanding the factors influencing microbial synthesis is pivotal in optimizing production efficiency. Environmental parameters such as temperature, pH, strain selection, substrate availability, and oxygen levels profoundly impact carotenoid biosynthesis in microbial hosts (Igreja et al. 2021). Additionally, genetic factors, including strain selection, metabolic pathways, and regulatory mechanisms, play a principal role in modulating carotenoid yields (Liu et al. 2020). To address these challenges and augment carotenoid production, various approaches have been devised. Metabolic engineering stands at the forefront, enabling the manipulation of microbial metabolism to bolster carotenoid biosynthesis (Torres-Haro et al. 2021). Techniques such as gene expression modulation and knockout strategies empower researchers to fine-tune metabolic pathways for enhanced carotenoid accumulation. Pathway engineering further optimizes substrate utilization and flux toward carotenoid biosynthesis, amplifying production yields. Furthermore, advanced fermentation strategies, including fed-batch and continuous processes, offer precise control over nutrient supply and waste removal, fostering optimal conditions for carotenoid production. Consortia engineering, harnessing the synergistic interactions among microbial species, presents a novel approach to amplify carotenoid synthesis through mutualistic relationships. The advent of CRISPR-Cas genome editing has revolutionized microbial engineering, facilitating targeted modifications in the microbial genome to enhance carotenoid production. By precisely altering regulatory elements and metabolic networks, CRISPR-Cas technology enables unprecedented control over carotenoid biosynthesis (Awasthi et al. 2022). The implementation of CRISPR-Cas9 technology for the production of microbial carotenoids has emerged as a transformative approach in biotechnology. Precisely, CRISPR-Cas9 allows for the optimization of metabolic pathway, improving yields and reducing production costs (Mo et al. 2020). Moreover, stress induction strategies and directed evolution techniques offer alternative avenues for augmenting carotenoid production. By subjecting microbial strains to selective pressures or evolutionary forces, researchers can iteratively improve carotenoid yields, harnessing the adaptability of microbial hosts to optimize production (Giani and Martínez-Espinosa 2020). The primary aim of this chapter is to elucidate a pivotal plan of action and recent advancements in optimizing these modules to maximize carotenoid production, shedding light on the transformative potential of microbial biotechnology in meeting the growing demand for natural pigments with diverse industrial applications.

14.2 Carotenoids: A Structure-based Approach to Biosynthesis

Carotenoids are known to be naturally produced by all photosynthetic organisms; in addition, certain non-photosynthetic organisms, like archaea, bacteria, and fungi, possess a composite carotenogenic metabolism. They are classified on the basis of number of carbon atoms contained in their structure, namely carotenoids C30, C40, C45, and C50. However, this literature focuses more on C40 carotenoids since they are the most plentiful in nature. These C40 carotenoids are synthesized by an extensive range of organisms including eukaryotes, archaea, and bacteria, exhibiting diverse chemical structures with various terminal groups (Yabuzaki 2017). On the other hand, C30 and C50 carotenoids are synthesized by archaea and bacteria, containing only six and ten C5 isoprenoid units, respectively. Notably, bacteria alone are capable of producing the C45 carotenoids, which consist of nine isoprenoid units. The presence of functional groups and cyclic end groups, together with changes in the length and organization of the polyene chain, contribute to the structural diversity of carotenoids. These structural variations contribute to differences in their optical properties, chemical reactivity, and biological functions. For example, carotenoids with keto groups, such as astaxanthin, exhibit stronger antioxidant activity compared to their hydroxylated counterparts (Nair et al. 2023).

Carotenoids undergo a complex biosynthetic pathway that involves multiple enzymatic reactions and regulatory mechanisms. To produce carotenoids, the process starts with the condensation of dimethylallyl pyrophosphate (DMAPP) and isopentenyl pyrophosphate (IPP), derived from either the mevalonate (MVA) pathway initiating through acetoacetyl-CoA or the methylerythritol phosphate (MEP) pathway using 1-deoxy-D-xylulose-5-phosphate, depending on the organism. These precursors are then further elongated to form geranyl pyrophosphate (GPP), farnesyl diphosphate (FPP), and geranyl geranyl pyrophosphate (GGPP) and cyclized to form phytoene, a colorless carotenoid, by using phytoene synthase. Lycopene, the first colored carotenoid in the biosynthetic route, is produced when phytoene passes through desaturation processes facilitated by phytoene desaturase enzymes (Zhang et al. 2023). Subsequent enzymatic steps involve the modification of lycopene and its derivatives through processes such as cyclization, oxidation, and hydroxylation, leading to the production of a diverse array of carotenoids zeaxanthin, astaxanthin, lutein, etc., with distinct chemical structures and functions (Ashraf et al. 2022). The general carotenoid biosynthesis pathway in microorganisms is shown in Figure 14.1. Valuable insights have been gained into the molecular mechanisms that govern carotenoid biosynthesis and

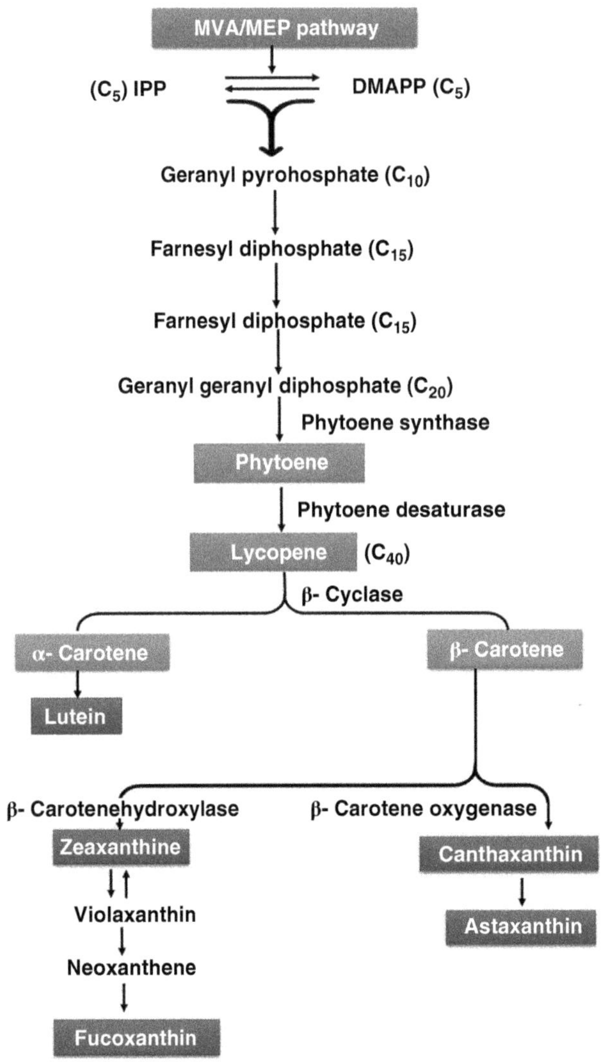

Figure 14.1 Systematic representation of carotenoid biosynthesis pathway in microorganisms.

regulation, shedding light on the functions of key enzymes and regulatory factors in these processes. Notably, the discovery and characterization of new enzymes like carotenoid cleavage di-oxygenases (CCDs) and cytochrome P450 mono-oxygenases have broadened our knowledge of carotenoid metabolism and its control. Furthermore, technological developments in synthetic biology and genetic engineering have empowered the manipulation of carotenoid biosynthetic pathways for various biotechnological aims, such as producing high-value carotenoids in microbial hosts and augmenting carotenoid levels in crops for nutritional benefits (Madhavan et al. 2023). Overall, the biosynthesis of carotenoids represents a dynamic and intricate process that is regulated by multiple factors, including environmental cues and developmental signals. Continued research into the molecular mechanisms underlying carotenoid biosynthesis promises to uncover new insights into the metabolic engineering of carotenoid pathways for agricultural, nutritional, and industrial applications.

14.3 Microbial Sources of Carotenoid Production

Microbial carotenoid synthesis represents a sophisticated interplay between microorganisms and biotechnological advancements, captivating the scientific community and industry alike (Liu et al. 2021). The utilization of diverse microorganisms, encompassing bacteria, yeast, fungi, and microalgae, underscores the versatility of nature's biochemical machinery harnessed for the manufacture of highly sought-after pigments such as astaxanthin, lycopene, canthaxanthin, zeaxanthin, lutein, and β-carotene as shown in Table 14.1.

Table 14.1 Carotenoids derived from microorganisms.

Microbe	Strain	Carotenoid	Aggregate yield (dry cell weight)	References
Bacteria	*Rhodococcus maris*	Canthaxanthin	13.4 mg/g DCW	Gharibzahedi et al. (2013)
	Escherichia coli	β-carotene	10.6 mg/g DCW	Scaife et al. (2012)
	Escherichia coli	Zeaxanthin	12.0 mg/g DCW	Li et al. (2015a)
	Algibacter sp.	Zeaxanthin	11.4 mg/g DCW	Mohamed et al. (2013)
Fungi	*Blakeslea trispora*	β-carotene	78.0 mg/g DCW	Nanou et al. (2012)
	Aspergillus carbonarius	Canthaxanthin	206.0 mg/g DCW	Krupa et al. (2010)
	Yarrowia lipolytica	Astaxanthin	6.0 mg/g DCW	Tramontin et al. (2019)
Algae	*Coelastrella striolata*	Astaxanthin	1.5 mg/g DCW	Abe et al. (2007)
	Chlorella sp.	Lutein	2.3 mg/g DCW	Minhas et al. (2016b)
	Chlorella zofingiensis	Lutein	13.8 mg/g DCW	Huang et al. 2018)
	Chlamydomonas reinhardtii	Zeaxanthin	7.3 mg/g DCW	Song et al. 2020)
Yeast	*Xanthophyllomyces dendrorhous*	Astaxanthin	4.1 mg/g DCW	Mircheska et al. 2020)
	Pichia pastoris	Lycopene	9.3 mg/g DCW	Zhang et al. (2020a)
	Saccharomyces cerevisiae	Lycopene	3.3 mg/g DCW	Shi et al. (2019)

Among bacteria, species such as *Pantoea*, *Rhodobacter*, and *Paracoccus* exhibit natural proficiency in carotenoid biosynthesis. *Pantoea* strains, for instance, are recognized for their synthesis of astaxanthin, a carotenoid renowned for its antioxidant properties (Choi et al. 2021a). *Rhodobacter* species contribute to the pool with the production of spirilloxanthin, a carotenoid that holds promise for diverse applications (Nannan et al. 2023). Additionally, *Paracoccus* carotenoids, including canthaxanthin and astaxanthin, showcase the diversity within bacterial carotenoid biosynthesis (Hayashi et al. 2021; Ye et al. 2006).

In the realm of yeast, the naturally occurring strains of *Rhodosporidium, Phaffia,* and *Xanthophyllomyces* demonstrate remarkable carotenogenic capabilities. *Rhodosporidium* strains are known for synthesizing β-carotene (Sereti et al. 2024), while *Phaffia* yeast stands out for its impressive production of astaxanthin, a red pigment of increasing commercial interest (Luna-Flores et al. 2022). *Xanthophyllomyces* strains contribute to the diversity with their ability to produce various xanthophylls, expanding the repertoire of naturally occurring yeast-derived carotenoids (Rodríguez-Sáiz et al. 2010).

Fungi, being eukaryotic microorganisms, engage in complex biosynthetic pathways to generate an array of carotenoids with diverse chemical structures and functionalities. A literature by Sajid and Akbar (2018) highlights the significance of carotenoids in fungi, emphasizing their roles in photo-protection, anti-oxidation, and adaptation to environmental stressors. Fungal carotenoid biosynthesis typically occurs within specialized cellular organelles, such as the mitochondria or plastids, and involves a series of enzymatic reactions. Notably, certain fungal genera, like *Neurospora* and *Fusarium*, have been extensively studied for their carotenoid-producing capabilities (Avalos et al. 2012).

Microalgae being photosynthetic microorganisms naturally biosynthesize carotenoids as a part of their cellular processes. *Haematococcus pluvialis* is a notable example, excelling in the production of astaxanthin, particularly when subjected to stress conditions (Mularczyk et al. 2020). An appealing source for sustainable pigment production is *Dunaliella salina*, which grows well in salty conditions and is known for its natural β-carotene synthesis (Pourkarimi et al. 2020).

These microorganisms, with their intrinsic capacity for microbial carotenoid synthesis, represent a rich reservoir for biotechnological applications. The exploration and understanding of their natural pathways not only offer insights into the fundamental aspects of carotenogenesis but also lay the foundation for harnessing their capabilities in environmentally friendly and economically viable industrial processes (Paniagua-Michel et al. 2012). The diverse array of carotenoids naturally produced by these microorganisms contributes to the ever-expanding palette of pigments available for applications in food, pharmaceuticals, and cosmetics (Poonia and Pandey 2022).

14.4 Factors Affecting Microbial Production of Carotenoid

Microbial carotenoid production is primarily a biosynthetic process. Enhancing carotenoid biosynthesis effectively boosts the accumulation of these pigments in microbial cells. Recent research has identified diverse environmental and cultural stimuli that can elevate microbial yield and augment carotenoid accumulation, particularly in microalgae and fungi of significant commercial interest. Controlling temperature and light, in addition to adding metal ions, salts, pH, strain selection, and co-cultivation, are crucial factors in promoting carotenoid production. (Liu et al. 2021).

Current studies have highlighted the significance of light intensity, wavelength, and photoperiod in modulating carotenoid production. For instance, studies have shown that optimal light conditions boost the production of essential enzymes involved in the carotenoid biosynthetic pathway, resulting in a buildup of carotenoids (Wu et al. 2020). Additionally, specific wavelengths of light, such as blue and red light, have been found to stimulate carotenoid synthesis in certain microbial strains by activating light-sensitive receptors and transcription factors (Sun et al. 2019). Another factor is temperature, which plays a pivotal role in enzyme concentrations, thereby regulating carotenoid levels in microorganisms. In microorganisms such as *Rhodotorula glutinis* and *Rhodotorula vulgaris*, temperature variations lead to distinct alterations in carotenoid levels. For instance, at lower temperatures, *R. glutinis* demonstrates elevated β-carotene content, while *R. vulgaris* exhibits increased biomass at higher temperatures. This differential response is attributed to the thermodynamic regulation of carotenoid-producing enzymes (Zhang et al. 2014). Specifically, producing torularhodin, a key carotenoid, is inhibited at lower temperatures due to reduced enzyme activity compared to β-carotene synthase (Tang et al. 2019). Consequently, fluctuations in temperature play a crucial role in regulating microbial carotenoid production by modulating enzyme kinetics and metabolic pathways, thereby influencing the overall composition and abundance of carotenoids in microbial populations.

Meanwhile, the interplay between pH, metals, and salt concentrations profoundly influences microbial carotenoid production. pH serves as a crucial determinant, as it directly affects enzyme activities, substrate solubility, and microbial physiology. Different microbial species exhibit distinct pH optima for carotenoid biosynthesis, with deviations from the optimal pH often resulting in decreased yields (Kot et al. 2017). In the process of carotenoid synthesis, metals serve a dual purpose as inhibitors and necessary cofactors. In order to speed up the process of carotenoid biosynthesis, enzymes need essential metals including iron, magnesium, and zinc, which act as cofactors. On the other hand, heavy metals and non-essential metals can impair enzyme performance and metabolic pathways, leading to reduced carotenoid production (Jannel et al. 2020; Locascio et al. 2019). And majorly, microbial carotenoid production is influenced by the specific strain of microorganisms and their interactions within co-cultivation systems (Fernandes et al. 2018; Rojas-Villalta et al. 2023). Different strains have varying genetic makeup and metabolic capabilities affecting carotenoid biosynthesis. For example, *R. glutinis* and *R. vulgaris* show strain-specific responses to environmental factors, impacting carotenoid production (Choi et al. 2021b). Co-cultivation adds complexity by promoting cooperative or competitive interactions among strains, altering nutrient utilization and metabolic pathways, thus influencing carotenoid production (Rojas-Villalta et al. 2023). Synergistic interactions between strains can enhance carotenoid biosynthesis, while competitive interactions may limit resources, affecting yields (Zhang et al. 2019b). Understanding strain selection and co-cultivation dynamics is crucial for optimizing carotenoid production in biotechnological applications.

14.5 Approaches for Enhancement of Carotenoid Production

Enhancement of microbial carotenoid biosynthesis employs multifaceted strategies, drawing upon various advanced techniques and approaches. Pathway engineering lies at the forefront, involving the modification and optimization of carotenoid biosynthesis pathways within microbial hosts to enhance precursor availability and flux toward target compounds. Metabolic engineering complements this by manipulating cellular metabolism to redirect carbon and energy flux toward carotenoid biosynthesis (Madhavan et al. 2023). Furthermore, carbon source optimization is crucial, with the selection and optimization of substrates influencing carotenoid production efficiency. Nutrient optimization and process control variables including oxygen levels, temperature, and pH are meticulously managed to create conducive environments for carotenoid accumulation. Consortia engineering leverages the synergistic interactions between multiple microbial species to enhance carotenoid production through co-cultivation or engineered consortia with complementary metabolic capabilities (González-González and de-Bashan 2021). By using

CRISPR-Cas genetic editing, precise changes can be made to microbial genomes, allowing for the targeted improvement of regulatory elements and pathways involved in carotenoid production (Ciurkot 2021). Stress induction strategies induce mild stress conditions, such as nutrient limitation or environmental stressors, to stimulate carotenoid accumulation as protective responses in microbial cells (Minhas et al. 2016a). Direct evolution techniques harness the power of natural selection to iteratively improve carotenoid-producing strains through adaptive evolution under selective pressure. Gene expression modulation, including upregulation or knockout of key genes involved in carotenoid biosynthesis or regulatory pathways, fine-tunes cellular metabolism for enhanced production (Su et al. 2023). Fermentation optimization techniques, such as adjusting agitation, aeration, and duration, are employed to maximize carotenoid productivity. Fed-batch and continuous fermentation strategies optimize cultivation conditions and nutrient supply to sustain high cell densities and metabolic activity, thereby maximizing carotenoid yields over extended cultivation periods (Larroude et al. 2018). Integrating these diverse approaches offers a comprehensive toolkit for engineering microbial hosts and cultivation processes to achieve enhanced carotenoid production for various industrial applications (Table 14.2).

Table 14.2 Microbial carotenoid production.

Approaches	Microorganisms used	Carotenoid	Total yield (dry cell weight)	References
Metabolic pathway engineering	*Escherichia coli*	Astaxanthin	7.12 mg/g DCW	Park et al. (2018)
	Escherichia coli	Astaxanthin	6.98 mg/g DCW	Park et al. (2018)
	Saccharomyces cerevisiae	β-carotene	16.2 ± 1.4 mg/g DCW	Olson (2014)
Overexpression of genes	*Yarrowia lipolytica*	β-carotene	2.2 mg/g DCW	Gao et al. (2014)
	Chlamydomonas reinhardtii	β-carotene	22.8 mg/g DCW	Rathod et al. (2020)
	Saccharomyces cerevisiae	β-carotene	5.9 mg/g DCW	Li et al. (2017)
Fed-batch fermentation	*Xanthophyllomyces dendrorhous*	Astaxanthin	7.44 mg/L	Villegas-Méndez et al. (2021)
	Streptomyces avermitilis	Lycopene	82 mg/g DCW	Bai et al. (2015)
CRISPR-Cas gene editing	*Escherichia coli*	β-carotene	2.0 g/L	Li et al. (2015a,b)
	Escherichia coli	β-carotene	1900 mg/g DCW	Jin et al. (2017)
Pathway engineering	*Bacillus subtilis*	Carotenoids	9.03 ± 0.16 mg/g DCW	Xue et al. (2015)
	Bacillus subtilis	Carotenoids	10.65 ± 0.13 mg/g DCW	Xue et al. (2015)
Consortia engineering	*Paracoccus haeundaensis* and lactic acid bacteria	Astaxanthin	821.09 ± 30.90 mg/g DCW	Ju et al. (2023)
Direct evolution	*Saccharomyces cerevisiae*	Lycopene	96.661 ± 1.481 mg/g DCW	Xie et al. (2015)
	Saccharomyces cerevisiae	Astaxanthin	1.87 mg/g DCW	Ding et al. (2023)
Stress induction	*Scenedesmus almeriensis*	Lutein	3.1 g/L	Sánchez et al. (2008)

14.5.1 Metabolic Pathway Engineering

The goal of this strategy is to improve the synthesis of target substances by manipulating metabolic pathways within microbes. In the case of carotenoids, metabolic engineering focuses on redirecting metabolic flux toward precursor molecules and enzymes involved in carotenoid biosynthesis. This can be achieved through the optimization of cellular metabolic networks. By redesigning the metabolic architecture of microbial hosts, metabolic engineering enables the efficient conversion of available nutrients into carotenoids, leading to increased yields and improved productivity (Guo et al. 2023). One primary target in metabolic pathway engineering is the enhancement of precursor supply. By increasing the pool of these precursors such as IPP and DMAPP, researchers aim to boost the substrate availability for carotenoid biosynthesis. Su et al. (2020) used metabolic engineering technique on *Saccharomyces cerevisiae* which further induced carotenoid production. Modifying the expression levels of enzymes within the process for the biosynthesis of carotenoids is another key strategy. Improved flow through the system, brought about by overexpression of important enzymes like phytoene synthase and phytoene desaturase, increases carotenoid synthesis. Ajikumar et al. (2008) found that by overexpressing 1-deoxy-D-xylose 5-phosphate synthase (dxs), a small change in the first stage of the non-mevalonate pathway leads to a twofold to threefold increase in lycopene synthesis. It can include targeting rate-limiting steps by overexpressing enzymes like phytoene synthase or phytoene desaturase to increase precursor conversion rates. Introducing heterologous pathways from organisms with high carotenoid production capabilities can also boost production by accessing novel enzymes (Su et al. 2023). Fine-tuning regulatory elements, such as transcription factors, allows for precise control over pathway flux. Modular pathway engineering breaks down the pathway into smaller modules for individual optimization before reassembling them in the host organism (López et al. 2020). These approaches collectively enable scientists to overcome metabolic bottlenecks, optimize enzyme activities, and fine-tune regulatory mechanisms, resulting in increased microbial carotenoid production for various applications. Several metabolically engineered microorganisms were used to produce various carotenoids with optimal yield, as detailed in Table 14.3.

Table 14.3 Production of microbial carotenoids from metabolic engineered microorganisms.

Strain	Carotenoid	Yield	Engineering strategy	References
Saccharomyces cerevisiae	Astaxanthin	218 mg/L	Analyzing genome evolution by ARTP	Jin et al. (2018)
	Lycopene	56 mg/g	Enhancing the acetyl-CoA reservoir and fine-tuning the lycopene synthesis pathway through genome manipulation for improved production	Chen et al. (2016)
Blakeslea trispora	Lycopene	256 mg/L	Refinement of fermentation through inhibition of lycopene cyclase for enhanced efficiency	Mantzouridou and Naziri (2017)
	β-carotene	704.1 mg/L	Precision in oxygen transfer rate management	Mantzouridou et al. (2005)

(*Continued*)

Table 14.3 (Continued)

Strain	Carotenoid	Yield	Engineering strategy	References
Escherichia coli	Lycopene	0.5 g/g	Control of gene expression in the lycopene biosynthetic pathway	Coussement et al. 2017)
	β-carotene	2.1 g/L	Modifying the MEP pathway to enhance the supply of IPP and DMAPP and redirecting carbon flux through the central pathway	Zhao et al. (2013)
Yarrowia lipolytica	β-carotene	6.5 g/L	Fine-tuning promoter gene pairs in the heterologous *crt* pathway for enhanced efficiency	Larroude et al. (2018)
	β-carotene	4 g/L	Sequential incorporation of replicated pathway genes for iterative enhancement	Gao et al. (2017)
Xanthophyllomyces dendrorhous	Zeaxanthin	0.5 mg/g	Genetic alteration of astaxanthin synthase and amplification of β-carotene hydrolase expression	Pollmann et al. (2017)
Rhodobacter glutamicum	Lycopene	10 mg/g	Substitution of *crtl*, enhancement of MEP pathway, and inhibition of PPP pathway	Su et al. (2018)
Corynebacterium glutamicum	β-carotene	7 mg/L	Deletion of *crtR* and incorporation of *crt* pathway genes	Henke et al. (2018)

14.5.2 Gene Overexpressing and Knockout

The overexpression of essential genes involved in the biosynthesis of carotenoids stands out as a critical strategy. Gene expression involves the activation of specific genes responsible for carotenoid biosynthesis, typically achieved through genetic engineering to enhance production efficiency. Knockout techniques, on the other hand, involve the targeted disruption of genes that may compete for metabolic resources or inhibit carotenoid synthesis. By eliminating these barriers, microbial hosts can allocate more resources toward carotenoid production, thereby increasing yields. Through a combination of gene expression and knockout strategies, researchers can optimize microbial strains for enhanced carotenoid production, contributing to the production of microbial carotenoids in a sustainable and effective manner (Watcharawipas and Runguphan 2023). Zhang et al. (2020b) demonstrated the efficiency of this method by focusing on the phytoene synthase gene. Overexpression of this gene led to a substantial increase in carotenoid production in the microbial host. For instance, overexpression of pivotal genes involved in carotenogenesis, such as *crtB* and *crtI*, has been demonstrated to significantly raise the yields of carotenoids. In a study by Alper et al. (2005), the overexpression of *crtE, crtB*, and *crtI* in *Escherichia coli* resulted in a nearly 50-fold increase in β-carotene synthesis. Similarly, in yeast systems, overexpression of carotenogenic genes has proven effective in enhancing carotenoid accumulation. On the other hand, gene silencing, achieved through RNA interference (RNAi) or CRISPR interference (CRISPRi), can be strategically employed to downregulate competing pathways or negative regulators, further amplifying carotenoid production. In a recent example, CRISPRi was used to silence a competitive pathway in *Yarrowia lipolytica*, causing the production of β-carotene to surge significantly, as reported by Liu et al. (2022). These studies underscore the pivotal role of gene overexpression and silencing strategies in tailoring microbial hosts for optimal carotenoid biosynthesis, providing valuable insights for the development of industrially relevant strains (Alper et al. 2005; Liu et al. 2022).

14.5.3 Fed-Batch and Continuous Fermentation

Fed-batch and continuous fermentation techniques provide efficient means to sustain high cell densities and metabolic activity, resulting in improved carotenoid yields in microbial fermentation processes. In fed-batch fermentation, nutrients are continuously added to the culture during the fermentation process to sustain cell growth and productivity. This approach allows for better control over nutrient concentrations and prevents substrate depletion, enabling microbial cells to maintain high metabolic activity and carotenoid production rates over extended periods (Gharibzahedi et al. 2014). A recent research of Thanapimmetha et al. (2017) reported that fed-batch fermentation by *Rhodococcus opacus* is induced to overproduce carotenoids under suitable conditions. Khumrangsee et al. (2022) reported that fed-batch fermentation by *Aurantiochytrium* sp. FIKU018 with stepwise aeration induced carotenoid production. Continuous fermentation, on the other hand, involves the continuous removal of culture broth while fresh medium is continuously added to the bioreactor. A study by Malisorn and Suntornsuk (2009) showed that continuous culture of *R. glutinis* DM28 produced more β-carotene as compared to the batch mode. Continuous fermentation offers advantages such as higher volumetric productivity and reduced process variability compared to batch fermentation, leading to enhanced carotenoid yields (Mapelli-Brahm et al. 2020). Additionally, monitoring and controlling the accumulation of toxic by-products or inhibitory metabolites during fermentation are crucial for maximizing carotenoid production.

14.5.4 Consortia Engineering

Consortia engineering leverages the synergistic interactions among different microbial species so that carotenoids can be produced more effectively. This approach involves the assembly of microbial consortia consisting of multiple species with complementary metabolic capabilities. Each member of the consortium contributes unique functionalities that collectively improve carotenoid production. For instance, certain microbes may specialize in producing precursor molecules required for carotenoid synthesis, while others may possess enzymes that efficiently convert these precursors into carotenoids (Dahake and Dhoble 2022). Consortia engineering for enhanced microbial carotenoid production has gained significant attention in recent years, as evidenced by numerous research articles. These studies highlight the effectiveness of consortia of different microbial strains to leverage synergistic interactions for improved carotenoid yields. Szotkowski et al. (2021) found that a combination of yeast *Rhodotorula kratochvilovae* and several microalgae with complementary metabolic capabilities increased the production of carotenoids such as lycopene, torularhodin, and β-carotene. Cultivating *Chlorella vulgaris* and *Chlorella sorokiniana* with *Azospirillum brasilense* leads to the enhancement in the yield of lutein, violaxanthin, and β-carotene along with other algal pigments (Subasankari et al. 2020). An essential aspect of engineering synthetic microbial consortia is ensuring a balanced distribution of each microbial population, crucial for fostering successful collaboration toward achieving the desired goal (Johns et al. 2016). Research by Wang et al. (2018) revealed the possibilities of consortium engineering to optimize carotenoid production using *E. coli* and yeasts. Through systematic selection and optimization of microbial consortia, the researchers demonstrated improved carotenoid productivity by exploiting metabolic cross-feeding and cooperative interactions among different microbial strains. This approach not only increased carotenoid yields but also enhanced the stability and robustness of the production process (Kalra et al. 2021).

14.5.5 CRISPR-Cas Genome Editing

Gene editing techniques have become indispensable tools for enhancing microbial carotenoid production, offering precise control over the genetic makeup of microorganisms to optimize their biosynthetic capabilities (Srivastava et al. 2022). CRISPR-Cas offers a formidable tool to drive the development of microbial platforms for sustainable and scalable production of carotenoids, which hold significant promise in several sectors, including the cosmetics, food, and pharmaceutical industries (Bhagwat et al. 2022). CRISPR-Cas systems enable the precise manipulation of microbial genomes, such that metabolic pathways can be optimized and carotenoid yields can be enhanced (Cho et al. 2018). Additionally, CRISPR-Cas9 offers the potential to engineer microbial strains with desirable traits such as improved tolerance to environmental stresses, further augmenting carotenoid production. By precisely editing certain regions of the genome, scientists are able to optimize metabolic pathways and engineer new microbial strains with desirable characteristics like increased tolerance, stability, and yield (Song et al. 2020). In pioneering work by Hu et al. (2021) *D. salina* is transformed by CRISPR-Cas9 which results in the enhancement of microbial carotenoid production. A highly efficient CRISPR-iCas9 genetic manipulation tool, incorporating D147Y and P411T mutants, was developed for *Y. lipolytica*. This tool was then employed to integrate key genes involved in carotene synthesis, resulting in a significant improvement in the desired product output (Chen et al. 2023). Meanwhile, Zhao et al. (2022) have done whole-genome mutagenesis by developing CRIPR-Cas random mutation (CARM) technique which was employed to enhance the production capacity of *S. cerevisiae* for β-carotene. Gene repression through Cas9/ CRISPRi has been done by Göttl et al. (2021) which facilitate the pinpointing of carotenoid-inducing genes for metabolic engineering in *Corynebacterium glutamicum*, leading to the surpluses of the carotenoids. By precisely editing promoter regions or transcription factor binding sites, researchers have successfully adjusted the levels of expression for critical genes, leading to increased yields of carotenoids (Li et al. 2022).

14.5.6 Stress Induction

Stress induction involves subjecting microbial cells to various stress conditions such as nutrient limitation, osmotic stress, or UV radiation to trigger increased carotenoid production as a protective mechanism. When microorganisms face unfavorable environmental conditions, they often respond by activating stress response pathways, which can potentially result in the buildup of secondary metabolites like carotenoids that act as antioxidants and protect against oxidative damage. Research conducted by Sandesh Kamath et al. (2008) found that culturing *H. pluvialis* in mutagenic circumstances such as UV radiations, 1-methyl 3-nitro 1-nitrosoguanidine (NTG), and ethyl methane sulfonate (EMS) increased the levels of carotenoids. *Phaffia rhodozyma*, under stress conditions in the presence of TiO_2, induces the production of β-carotene, astaxanthin, and lycopene (Zhang et al. 2019a). Several research studies have already used this strategy to great effect, with the goal of increasing carotenoid yields for use in food, feed, and pharmaceutical manufacturing. It is easy to implement and does not cost a fortune.

14.5.7 Directed Evolution

Directed evolution is the process of using evolutionary concepts to enhance the carotenoid production capacity of microbial strains via repeated rounds of mutation and selection. This technique mimics natural selection processes in microbial populations to drive desired traits, such as increased carotenoid production, over successive generations. Through successive rounds of mutation and selection, microbial strains with desired traits, such as increased carotenoid production or improved substrate utilization, are enriched. Directed evolution harnesses natural evolutionary principles to drive the adaptation of microbial populations toward desired phenotypes over time (Tominaga et al. 2021). The introduction of metabolically modified *E. coli* by Wang et al. (2000) led to a twofold increase in lycopene synthesis. A study by Xie et al. (2015) demonstrates that the expression of various phytoene synthase genes in *S. cerevisiae* results in the overproduction of lycopene. Directed evolution can be applied to improve various aspects of carotenoid production, such as enzyme efficiency, substrate specificity, and regulatory control. Researchers can create microbial strains that are ideal for particular industrial uses by methodically screening and selecting strains that have enhanced carotenoid production capabilities (Zhou et al. 2017).

14.6 Fermentation Processes, Bioreactor Design, and Downstream Processing

Fermentation techniques and bioreactor design significantly influence microbial carotenoid production, with careful consideration of parameters such as agitation, aeration, and substrate availability critical for optimizing yields. Integration of scalable downstream processing methods and innovative bioreactor designs further enhances the efficiency and sustainability of carotenoid production processes, paving the way for broader applications in food, feed, pharmaceuticals, and nutraceutical industries (Leong et al. 2022). The quantity and quality of the final product are affected by fermentation processes and bioreactor design, which are key components of the synthesis of microbial carotenoids. Several factors such as agitation, aeration, pH control, substrate concentration, and temperature play crucial roles in optimizing carotenoid production within the bioreactor. The choice of fermentation strategy whether batch, fed-batch, or continuous also impacts the productivity and stability of carotenoid production over time (Igreja et al. 2021). Furthermore, bioreactor design features such as vessel geometry, impeller configuration, and gas sparging methods affect mass transfer rates, substrate utilization, and overall microbial growth kinetics, thereby influencing carotenoid production efficiency (Khumrangsee et al. 2022). The extraction and purification of carotenoids from the fermentation broth are a key downstream processing step, and bioreactor design issues also apply at this stage. The scale-up of fermentation processes necessitates scalable downstream processing techniques that can handle large volumes of broth while maintaining product purity and recovery efficiency (Basiony et al. 2022). Integration of in situ extraction methods within the bioreactor or downstream processing unit can streamline the recovery process, minimizing product losses and reducing processing time. Innovative bioreactor designs, such as immobilized cell systems or membrane bioreactors, offer advantages in terms of enhanced productivity, ease of operation, and reduced contamination risks, which can further optimize carotenoid production processes (Zhang et al. 2024). Additionally, advances in monitoring and control technologies allow for real-time optimization of fermentation parameters, leading to improved process robustness and consistency in carotenoid production.

14.7 Applications of Microbial Carotenoids

Microbial carotenoids, pigments produced by microorganisms such as bacteria, fungi, and algae, have found diverse applications across various industries due to their unique properties. In the food and beverage sector, these carotenoids serve as natural colorants, enhancing the visual appeal of products without the inclusion of synthetic additives. In addition, the antioxidant characteristics of these foods help keep them fresh for longer (Poorniammal and Prabhu, 2024). Microbial carotenoids show promise as nutraceuticals in the pharmaceutical industry; they promote human health as vitamin A precursors and demonstrate anti-inflammatory and immune-modulatory properties (Papapostolou et al. 2023). Furthermore, their role in aquaculture has gained attention, as these pigments enhance the pigmentation of fish and shrimp, improving market value. In the cosmetics industry, microbial carotenoids are utilized for their skin-friendly coloring properties (García-Chavarría and Lara-Flores 2013). Their versatility and sustainable production make microbial carotenoids promising candidates for addressing contemporary challenges in diverse fields, in keeping with the increasing need for sustainable and environmentally friendly remedies.

14.7.1 Nutraceutical and Pharmaceutical Applications

Microbial carotenoids have gained significant attention in both nutraceutical and pharmaceutical industries due to their diverse applications. These natural pigments, synthesized by microorganisms, exhibit a broad spectrum of colors ranging from yellow to red and possess valuable properties that make them attractive for various health-related applications.

In the field of nutraceuticals, microbial carotenoids are used as powerful antioxidants to protect cells from oxidative stress by scavenging free radicals in keeping with the increasing need for sustainable and environmentally friendly remedies. Astaxanthin, a carotenoid produced by the yeast *P. rhodozyma* and the microalga *H. pluvialis*, stands out for its exceptional antioxidant capacity. Studies have shown its potential in supporting cardiovascular health, reducing inflammation, and enhancing skin health (Fassett and Coombes 2012). Furthermore, microbial carotenoids contribute to eye health, with zeaxanthin and lutein playing crucial roles in preventing disease affecting the eyes caused by aging. These carotenoids are usually synthesized by microorganisms such as bacteria and microalgae. Incorporating them into nutraceutical formulations can provide consumers with natural alternatives for maintaining optimal vision.

In the pharmaceutical realm, microbial carotenoids exhibit promising applications in disease prevention and treatment. For instance, β-carotene, produced by fungi and certain bacteria, has been investigated for its function in strengthening the defenses against chronic diseases and lowering their prevalence. Additionally, microbial carotenoids demonstrate antimicrobial properties, making them potential candidates for developing novel antibiotics and antifungal agents (Lopes et al. 2020). The anti-inflammatory properties of some microbial carotenoids, including canthaxanthin and astaxanthin, have implications for pharmaceutical interventions in inflammatory conditions. These carotenoids may modulate immune responses and aid in the management of inflammatory bowel disorders and rheumatoid arthritis (Al-Okbi 2014).

Several studies have explored the potential anticancer effects of microbial carotenoids. Bacterial and fungal carotenoids, such as lycopene, have been studied for their potential to suppress the proliferation of specific cancer cells. While more research is needed, these findings suggest a potential path for the research and creation of drug therapies (Mircheska et al. 2020).

14.7.2 Food Coloring and Cosmetic Industry

Microbial carotenoids have become pivotal components in both the food coloring and cosmetic industries, offering natural alternatives to synthetic additives. The carotenoids astaxanthin, β-carotene, and others that come from microbes like bacteria and microalgae are used as natural hues in food goods, and they are both brilliant and stable. These compounds provide an appealing solution for consumers seeking clean-label options, given their resistance to various processing conditions and extended shelf life in food formulations (Bhosale 2004; Del Campo et al. 2004). In the cosmetic industry, microbial carotenoids contribute to the trend of incorporating natural colorants into skincare and makeup products. Astaxanthin, known for its vibrant red-orange hue, stands out as a sought-after ingredient due to its antioxidant properties and potential anti-aging effects on the skin (Ambati et al. 2014). The UV-absorbing capacity of astaxanthin further positions it as a natural choice for sun protection in skincare formulations, emphasizing its multifaceted role in promoting both color and health in the cosmetic industry (Tominaga et al. 2017). This dual application of microbial carotenoids underscores their versatility and sustainable appeal in meeting the evolving preferences of consumers in both the food and cosmetic sectors.

14.7.3 Antioxidant Properties and Health Benefits

Microbial carotenoids, such as astaxanthin, xanthophyll, β-carotene, lutein, and lycopene, have garnered attention for their remarkable antioxidant capabilities and associated health benefits. These natural pigments, synthesized by microorganisms like bacteria and microalgae, serve as potent scavengers of free radicals, thereby mitigating oxidative stress and cellular damage. Astaxanthin, in particular, stands out because of its powerful antioxidant capabilities, which effectively shield cells and tissues from harmful oxidative stress (Fassett and Coombes 2012). The antioxidant potential of microbial carotenoids extends to supporting cardiovascular health by reducing inflammation, enhancing lipid profiles, and improving endothelial function (Fassett and Coombes 2012). Furthermore, these carotenoids contribute to keeping mental processes running smoothly and reducing the risk of age-related degenerative diseases (Johnson et al. 2013). β-Carotene, another prominent microbial carotenoid, exhibits immune-enhancing effects, suggesting the part it plays in lowering the likelihood of developing chronic diseases and bolstering the immune system generally (Britton 2020). Overall, the antioxidant properties of microbial carotenoids underscore their significance in promoting health and preventing various oxidative stress-related conditions, positioning them as valuable components in the pursuit of holistic well-being.

14.8 Challenges and Future Perspectives

Microbial carotenoids, renowned for their diverse applications in the food sector, cosmetics, and, pharmaceuticals, are at the forefront of biotechnological advancements. Nevertheless, there are significant obstacles to overcome in the manufacture of these pigments. One key obstacle is the optimization of fermentation conditions to enhance microbial growth and carotenoid synthesis, a task complicated by the sensitivity of microorganisms to environmental changes. Additionally, achieving high yields in a cost-effective manner requires navigating intricate metabolic pathways and genetic engineering hurdles. The sustainability of production processes, considering the reliance on substrates like glucose or agro-industrial by-products, poses another challenge. As the industry strives for more eco-friendly practices, addressing this issue becomes imperative.

Looking ahead, the future perspective in microbial carotenoid production is characterized by a promising blend of innovation and sustainability. Advances in synthetic biology and metabolic engineering hold potential for tailoring microorganisms to yield higher quantities of carotenoids efficiently. The exploration of non-conventional microbial strains, coupled with the development of robust bioprocessing strategies, could open new avenues. Moreover, the integration of biorefinery concepts, aiming to extract multiple value-added compounds from microbial cultures, could enhance the overall economic viability of carotenoid production. Embracing circular economy principles and exploring alternative carbon sources, such as waste streams, align with the broader goal of establishing a greener and more resilient microbial carotenoid industry. In summary, while challenges persist, the future of microbial carotenoid production is bright, driven by technological innovation and a commitment to sustainable practices.

14.9 Conclusion

The exploration of innovative approaches for the microbial production of carotenoids has yielded promising advancements in the field of biotechnology. Through a comprehensive review of various strategies, including metabolic engineering, synthetic biology, and fermentation optimization, this chapter has provided valuable insights into enhancing carotenoid production in microbial systems. The integration of cutting-edge technologies and the manipulation of metabolic pathways showcase the potential for developing robust and efficient microbial platforms for carotenoid biosynthesis. As researchers continue to refine these approaches, the prospect of sustainable and cost-effective production of carotenoids for diverse applications, ranging from food and feed industries to pharmaceuticals, becomes increasingly attainable. Researchers and engineers can use this chapter as a springboard for further work in understanding how to maximize the use of microbial systems to produce these important chemicals.

References

Abe, K., Hattori, H., and Hirano, M. (2007). Accumulation and antioxidant activity of secondary carotenoids in the aerial microalga *Coelastrella striolata* var. multistriata. *Food Chemistry* 100 (2): 656–661.

Afroz Toma, M., Rahman, M.H., Rahman, M.S. et al. (2023). Fungal pigments: carotenoids, riboflavin, and polyketides with diverse applications. *Journal of Fungi* 9 (4): 454.

Ajikumar, P.K., Tyo, K., Carlsen, S. et al. (2008). Terpenoids: opportunities for biosynthesis of natural product drugs using engineered microorganisms. *Molecular Pharmaceutics* 5 (2): 167–190.

Al-Okbi, S.Y. (2014). Nutraceuticals of anti-inflammatory activity as complementary therapy for rheumatoid arthritis. *Toxicology and Industrial Health* 30 (8): 738–749.

Alper, H., Miyaoku, K., and Stephanopoulos, G. (2005). Construction of lycopene-overproducing *E. coli* strains by combining systematic and combinatorial gene knockout targets. *Nature Biotechnology* 23 (5): 612–616.

Ambati, R.R., Phang, S.-M., Ravi, S., and Aswathanarayana, R.G. (2014). Astaxanthin: sources, extraction, stability, biological activities and its commercial applications—a review. *Marine Drugs* 12 (1): 128–152.

Ashraf, W., Latif, A., Lianfu, Z. et al. (2022). Technological advancement in the processing of lycopene: a review. *Food Reviews International* 38 (5): 857–883.

Avalos, J., Prado-Cabrero, A., and Estrada, A.F. (2012). Neurosporaxanthin production by Neurospora and Fusarium. In: *Microbial Carotenoids from Fungi: Methods and Protocols*, 263–274. Humana Press.

Awasthi, P., Khan, S., Lakhani, H. et al. (2022). Transgene-free genome editing supports the role of carotenoid cleavage dioxygenase 4 as a negative regulator of β-carotene in banana. *Journal of Experimental Botany* 73 (11): 3401–3416.

Bai, C., Zhang, Y., Zhao, X. et al. (2015). Exploiting a precise design of universal synthetic modular regulatory elements to unlock the microbial natural products in Streptomyces. *Proceedings of the National Academy of Sciences* 112 (39): 12181–12186.

Barford, W. (2022). Theory of the dark state of polyenes and carotenoids. *Physical Review B* 106 (3): 35201.

Basiony, M., Ouyang, L., Wang, D. et al. (2022). Optimization of microbial cell factories for astaxanthin production: biosynthesis and regulations, engineering strategies and fermentation optimization strategies. *Synthetic and Systems Biotechnology* 7 (2): 689–704.

Bhagwat, A.C., Patil, A.M., and Saroj, S.D. (2022). CRISPR/Cas 9-based editing in the production of bioactive molecules. *Molecular Biotechnology* 1–7.

Bhosale, P. (2004). Environmental and cultural stimulants in the production of carotenoids from microorganisms. *Applied Microbiology and Biotechnology* 63: 351–361.

Bogacz-Radomska, L., Harasym, J., and Piwowar, A. (2020). Commercialization aspects of carotenoids. In: *Carotenoids: Properties, Processing and Applications*, 327–357. Elsevier.

Britton, G. (2020). Carotenoid research: history and new perspectives for chemistry in biological systems. *Biochimica et Biophysica Acta (BBA)-Molecular and Cell Biology of Lipids* 1865 (11): 158699.

Chen, Y., Xiao, W., Wang, Y. et al. (2016). Lycopene overproduction in *Saccharomyces cerevisiae* through combining pathway engineering with host engineering. *Microbial Cell Factories* 15 (1): 1–13.

Chen, Q.H., Qian, Y.D., Niu, Y.J. et al. (2023). Characterization of an efficient CRISPR-iCas9 system in *Yarrowia lipolytica* for the biosynthesis of carotenoids. *Applied Microbiology and Biotechnology* 107 (20): 6299–6313.

Cho, S., Shin, J., and Cho, B.-K. (2018). Applications of CRISPR/Cas system to bacterial metabolic engineering. *International Journal of Molecular Sciences* 19 (4): 1089.

Choi, O., Kang, B., Lee, Y. et al. (2021a). Pantoea ananatis carotenoid production confers toxoflavin tolerance and is regulated by Hfq-controlled quorum sensing. *MicrobiologyOpen* 10 (1): e1143.

Choi, S.S., Seo, Y.B., Nam, S.-W., and Kim, G.-D. (2021b). Enhanced production of astaxanthin by co-culture of *Paracoccus haeundaensis* and lactic acid bacteria. *Frontiers in Marine Science* 7: 597553.

Ciurkot, K. (2021). *Expanding CRISPR-based toolbox for enhanced production of β-carotene in S. cerevisiae*. Staats-und Universitätsbibliothek Hamburg Carl von Ossietzky. https://ediss.sub.uni-hamburg.de/handle/ediss/9567.

Coussement, P., Bauwens, D., Maertens, J., and De Mey, M. (2017). Direct combinatorial pathway optimization. *ACS Synthetic Biology* 6 (2): 224–232.

Dahake, A.P. and Dhoble, A.S. (2022). Microbial consortium engineering for the improvement of biochemicals production. In: *Current Developments in Biotechnology and Bioengineering*, 201–233. Elsevier.

Del Campo, J.A., Rodriguez, H., Moreno, J. et al. (2004). Accumulation of astaxanthin and lutein in *Chlorella zofingiensis* (Chlorophyta). *Applied Microbiology and Biotechnology* 64: 848–854.

Ding, Y.-W., Lu, C.-Z., Zheng, Y. et al. (2023). Directed evolution of the fusion enzyme for improving astaxanthin biosynthesis in *Saccharomyces cerevisiae*. *Synthetic and Systems Biotechnology* 8 (1): 46–53. https://doi.org/10.1016/j.synbio.2022.10.005.

Fassett, R.G. and Coombes, J.S. (2012). Astaxanthin in cardiovascular health and disease. *Molecules* 17 (2): 2030–2048.

Fernandes, A.S., Paixão, S.M., Silva, T.P. et al. (2018). Influence of culture conditions towards optimal carotenoid production by *Gordonia alkanivorans* strain 1B. *Bioprocess and Biosystems Engineering* 41: 143–155.

Foong, L.C., Loh, C.W.L., Ng, H.S., and Lan, J.C.-W. (2021). Recent development in the production strategies of microbial carotenoids. *World Journal of Microbiology and Biotechnology* 37: 1–11.

Gao, S., Han, L., Zhu, L. et al. (2014). One-step integration of multiple genes into the oleaginous yeast *Yarrowia lipolytica*. *Biotechnology Letters* 36: 2523–2528.

Gao, S., Tong, Y., Zhu, L. et al. (2017). Iterative integration of multiple-copy pathway genes in *Yarrowia lipolytica* for heterologous β-carotene production. *Metabolic Engineering* 41: 192–201.

García-Chavarría, M. and Lara-Flores, M. (2013). The use of carotenoid in aquaculture. *Research Journal of Fisheries and Hydrobiology* 8 (2): 38–49.

Gharibzahedi, S.M.T., Razavi, S.H., and Mousavi, S.M. (2013). Microbial canthaxanthin: perspectives on biochemistry and biotechnological production. *Engineering in Life Sciences* 13 (4): 408–417.

Gharibzahedi, S.M.T., Razavi, S.H., and Mousavi, M. (2014). Feeding strategies for the improved biosynthesis of canthaxanthin from enzymatic hydrolyzed molasses in the fed-batch fermentation of *Dietzia natronolimnaea* HS-1. *Bioresource Technology* 154: 51–58.

Giani, M. and Martínez-Espinosa, R.M. (2020). Carotenoids as a protection mechanism against oxidative stress in *Haloferax mediterranei*. *Antioxidants* 9 (11): 1060.

González-González, L.M. and de-Bashan, L. E. (2021). Toward the enhancement of microalgal metabolite production through microalgae–bacteria consortia. *Biology* 10 (4): 282.

Göttl, V.L., Schmitt, I., Braun, K. et al. (2021). CRISPRi-library-guided target identification for engineering carotenoid production by *Corynebacterium glutamicum*. *Microorganisms* 9 (4): 670.

Guo, Q., Peng, Q.-Q., Li, Y.-W. et al. (2023). Advances in the metabolic engineering of *Saccharomyces cerevisiae* and *Yarrowia lipolytica* for the production of β-carotene. *Critical Reviews in Biotechnology* 1–15.

Hayashi, M., Ishibashi, T., Kuwahara, D., and Hirasawa, K. (2021). Commercial production of astaxanthin with *Paracoccus carotinifaciens*. In: *Carotenoids: Biosynthetic and Biofunctional Approaches*, 11–20. Springer.

Henke, N.A., Wiebe, D., Pérez-García, F. et al. (2018). Coproduction of cell-bound and secreted value-added compounds: simultaneous production of carotenoids and amino acids by *Corynebacterium glutamicum*. *Bioresource Technology* 247: 744–752.

Hu, L., Feng, S., Liang, G. et al. (2021). CRISPR/Cas9-induced β-carotene hydroxylase mutation in *Dunaliella salina* CCAP19/18. *AMB Express* 11 (1): 83.

Huang, W., Lin, Y., He, M. et al. (2018). Induced high-yield production of zeaxanthin, lutein, and β-carotene by a mutant of *Chlorella zofingiensis*. *Journal of Agricultural and Food Chemistry* 66 (4): 891–897.

Igreja, W.S., de Andrade Maia, F., Lopes, A.S., and Chisté, R.C. (2021). Biotechnological production of carotenoids using low cost-substrates is influenced by cultivation parameters: a review. *International Journal of Molecular Sciences* 22 (16): 8819.

Jannel, S., Caro, Y., Bermudes, M., and Petit, T. (2020). Novel insights into the biotechnological production of *Haematococcus pluvialis*-derived astaxanthin: advances and key challenges to allow its industrial use as novel food ingredient. *Journal of Marine Science and Engineering* 8 (10): 789.

Jin, Y.F., Han, L., Zhang, S.S. et al. (2017). Construction of β-carotene high-yielding strains through optimization of lycopene cyclase. *Chinese Journal of Bioengineering* 33: 1814–1826.

Jin, J., Wang, Y., Yao, M. et al. (2018). Astaxanthin overproduction in yeast by strain engineering and new gene target uncovering. *Biotechnology for Biofuels* 11: 1–15.

Johns, N.I., Blazejewski, T., Gomes, A.L.C., and Wang, H.H. (2016). Principles for designing synthetic microbial communities. *Current Opinion in Microbiology* 31: 146–153.

Johnson, E.J., Vishwanathan, R., Johnson, M.A. et al. (2013). Relationship between serum and brain carotenoids,-tocopherol, and retinol concentrations and cognitive performance in the oldest old from the Georgia Centenarian Study. *Journal of Aging Research* 2013: 951786.

Ju, M., Zhang, J., Mai, T. et al. (2023). Co-culture of *Rhodotorula mucilaginosa* and *Monascus purpureus* increased the yield of carotenoids and *Monascus* pigments. *LWT* 114949.

Kalra, R., Gaur, S., and Goel, M. (2021). Microalgae bioremediation: a perspective towards wastewater treatment along with industrial carotenoids production. *Journal of Water Process Engineering* 40: 101794.

Khumrangsee, K., Charoenrat, T., Praiboon, J., and Chittapun, S. (2022). Development of a fed-batch fermentation with stepwise aeration to enhance docosahexaenoic acid and carotenoid content in *Aurantiochytrium* sp. FIKU018. *Journal of Applied Phycology* 34 (3): 1243–1253.

Kot, A.M., Błażejak, S., Kurcz, A. et al. (2017). Effect of initial pH of medium with potato wastewater and glycerol on protein, lipid and carotenoid biosynthesis by *Rhodotorula glutinis*. *Electronic Journal of Biotechnology* 27: 25–31. https://doi.org/10.1016/j.ejbt.2017.01.007.

Krupa, D., Nakkeeran, E., Kumaresan, N. et al. (2010). Extraction, purification and concentration of partially saturated canthaxanthin from *Aspergillus carbonarius*. *Bioresource Technology* 101 (19): 7598–7604.

Larroude, M., Celinska, E., Back, A. et al. (2018). A synthetic biology approach to transform *Yarrowia lipolytica* into a competitive biotechnological producer of β-carotene. *Biotechnology and Bioengineering* 115 (2): 464–472.

Leong, Y.K., Chen, C.-Y., Varjani, S., and Chang, J.-S. (2022). Producing fucoxanthin from algae–recent advances in cultivation strategies and downstream processing. *Bioresource Technology* 344: 126170.

Li, X.-R., Tian, G.-Q., Shen, H.-J., and Liu, J.-Z. (2015a). Metabolic engineering of *Escherichia coli* to produce zeaxanthin. *Journal of Industrial Microbiology and Biotechnology* 42 (4): 627–636.

Li, Y., Lin, Z., Huang, C. et al. (2015b). Metabolic engineering of *Escherichia coli* using CRISPR–Cas9 meditated genome editing. *Metabolic Engineering* 31: 13–21.

Li, J., Shen, J., Sun, Z. et al. (2017). Discovery of several novel targets that enhance β-carotene production in *Saccharomyces cerevisiae*. *Frontiers in Microbiology* 8: 1116.

Li, C., Jiang, T., Li, M. et al. (2022). Fine-tuning gene expression for improved biosynthesis of natural products: from transcriptional to post-translational regulation. *Biotechnology Advances* 54: 107853.

Liu, H., Zhang, C., Zhang, X. et al. (2020). A novel carotenoids-producing marine bacterium from noble scallop *Chlamys nobilis* and antioxidant activities of its carotenoid compositions. *Food Chemistry* 320: 126629.

Liu, C., Hu, B., Cheng, Y. et al. (2021). Carotenoids from fungi and microalgae: a review on their recent production, extraction, and developments. *Bioresource Technology* 337: 125398.

Liu, M., Zhang, J., Liu, X. et al. (2022). Rapid gene target tracking for enhancing β-carotene production using flow cytometry-based high-throughput screening in *Yarrowia lipolytica*. *Applied Environmental Microbiology* 88 (19): e0114922.

Locascio, A., Andrés-Colás, N., Mulet, J.M., and Yenush, L. (2019). *Saccharomyces cerevisiae* as a tool to investigate plant potassium and sodium transporters. *International Journal of Molecular Sciences* 20 (9): 2133.

Lopes, G., Clarinha, D., and Vasconcelos, V. (2020). Carotenoids from cyanobacteria: a biotechnological approach for the topical treatment of psoriasis. *Microorganisms* 8 (2): 302.

López, J., Bustos, D., Camilo, C. et al. (2020). Engineering *Saccharomyces cerevisiae* for the overproduction of β-ionone and its precursor β-carotene. *Frontiers in Bioengineering and Biotechnology* 8: 578793.

Luna-Flores, C.H., Wang, A., von Hellens, J., and Speight, R.E. (2022). Towards commercial levels of astaxanthin production in *Phaffia rhodozyma*. *Journal of Biotechnology* 350: 42–54.

Madhavan, A., Arun, K.B., Alex, D. et al. (2023). Microbial production of nutraceuticals: metabolic engineering interventions in phenolic compounds, poly unsaturated fatty acids and carotenoids synthesis. *Journal of Food Science and Technology* 60 (8): 2092–2104.

Malisorn, C. and Suntornsuk, W. (2009). Improved β-carotene production of *Rhodotorula glutinis* in fermented radish brine by continuous cultivation. *Biochemical Engineering Journal* 43 (1): 27–32.

Mantzouridou, F.T. and Naziri, E. (2017). Scale translation from shaken to diffused bubble aerated systems for lycopene production by *Blakeslea trispora* under stimulated conditions. *Applied Microbiology and Biotechnology* 101: 1845–1856.

Mantzouridou, F., Roukas, T., and Achatz, B. (2005). Effect of oxygen transfer rate on β-carotene production from synthetic medium by *Blakeslea trispora* in shake flask culture. *Enzyme and Microbial Technology* 37 (7): 687–694.

Maoka, T. (2020). Carotenoids as natural functional pigments. *Journal of Natural Medicines* 74 (1): 1–16.

Mapelli-Brahm, P., Barba, F.J., Remize, F. et al. (2020). The impact of fermentation processes on the production, retention and bioavailability of carotenoids: an overview. *Trends in Food Science & Technology* 99: 389–401.

Minhas, A.K., Hodgson, P., Barrow, C.J., and Adholeya, A. (2016a). A review on the assessment of stress conditions for simultaneous production of microalgal lipids and carotenoids. *Frontiers in Microbiology* 7: 546.

Minhas, A.K., Hodgson, P., Barrow, C.J. et al. (2016b). The isolation and identification of new microalgal strains producing oil and carotenoid simultaneously with biofuel potential. *Bioresource Technology* 211: 556–565.

Mircheska, V.B., Karakashova, L., Ristovska, N. et al. (2020). Lycopene and its antioxidant properties in the prevention of chronic illnesses and cancer. *Journal of Hygienic Engineering and Design* 31: 69–72.

Mo, X.-H., Zhang, H., Wang, T.-M. et al. (2020). Establishment of CRISPR interference in *Methylorubrum extorquens* and application of rapidly mining a new phytoene desaturase involved in carotenoid biosynthesis. *Applied Microbiology and Biotechnology* 104: 4515–4532.

Mohamed, I., Mearns, A. S., Fraser, K., & Hodgson, R. (2013). Biological production of zeaxanthin and carotenoid biosynthesis control. Google Patents, filed 5 May 2006 and issued 13 June 2012.

Mularczyk, M., Michalak, I., and Marycz, K. (2020). Astaxanthin and other nutrients from *Haematococcus pluvialis*—multifunctional applications. *Marine Drugs* 18 (9): 459.

Mussagy, C.U., Khan, S., and Kot, A.M. (2021). Current developments on the application of microbial carotenoids as an alternative to synthetic pigments. *Critical Reviews in Food Science and Nutrition* 62 (25): 6932–6946.

Nair, A., Ahirwar, A., Singh, S. et al. (2023). Astaxanthin as a king of ketocarotenoids: structure, synthesis, accumulation. Bioavailability and antioxidant properties. *Marine Drugs* 21 (3): 176.

Nannan, L.I., Siyu, P., Hai, Y.A.N., and Chunhua, Y.I.N. (2023). Production and antioxidant activities of carotenoids from *Rhodopseudomonas palustris*. *Chemical Industry and Engineering Progress* 42 (11): 5891.

Nanou, K., Roukas, T., and Papadakis, E. (2012). Improved production of carotenes from synthetic medium by *Blakeslea trispora* in a bubble column reactor. *Biochemical Engineering Journal* 67: 203–207.

Olson, M.L. (2014). Metabolic engineering of *S. cerevisiae* for carotenoid production optimization. *Biology Engineering* 4828106.

Paniagua-Michel, J., Olmos-Soto, J., and Ruiz, M.A. (2012). Pathways of carotenoid biosynthesis in bacteria and microalgae. In: *Microbial Carotenoids from Bacteria and Microalgae: Methods and Protocols*, 1–12. Humana Press.

Papapostolou, H., Kachrimanidou, V., Alexandri, M. et al. (2023). Natural carotenoids: recent advances on separation from microbial biomass and methods of analysis. *Antioxidants* 12 (5): 1030.

Park, S.Y., Binkley, R.M., Kim, W.J. et al. (2018). Metabolic engineering of *Escherichia coli* for high-level astaxanthin production with high productivity. *Metabolic Engineering* 49: 105–115. https://doi.org/10.1016/j.ymben.2018.08.002.

Pérez-Gálvez, A., Viera, I., and Roca, M. (2020). Carotenoids and chlorophylls as antioxidants. *Antioxidants* 9 (6): 505.

Pollmann, H., Breitenbach, J., and Sandmann, G. (2017). Engineering of the carotenoid pathway in *Xanthophyllomyces dendrorhous* leading to the synthesis of zeaxanthin. *Applied Microbiology and Biotechnology* 101: 103–111.

Poonia, A. and Pandey, S. (2022). Production of microbial pigments from whey and their applications: a review. *Nutrition & Food Science* 53 (2): 265–284.

Poorniammal, R. and Prabhu, S. (2024). Microbial pigments as antioxidants in the food and beverage industries. In: *Microbial Pigments*, 173–183. CRC Press.

Pourkarimi, S., Hallajisani, A., Alizadehdakhel, A. et al. (2020). Factors affecting production of beta-carotene from *Dunaliella salina* microalgae. *Biocatalysis and Agricultural Biotechnology* 29: 101771.

Rathod, J.P., Vira, C., Lali, A.M., and Prakash, G. (2020). Metabolic engineering of *Chlamydomonas reinhardtii* for enhanced β-carotene and lutein production. *Applied Biochemistry and Biotechnology* 190: 1457–1469.

Rodríguez-Sáiz, M., de la Fuente, J.L., and Barredo, J.L. (2010). *Xanthophyllomyces dendrorhous* for the industrial production of astaxanthin. *Applied Microbiology and Biotechnology* 88 (3): 645–658.

Rojas-Villalta, D., Gómez-Espinoza, O., Murillo-Vega, F. et al. (2023). Insights into co-cultivation of photosynthetic microorganisms for novel molecule discovery and enhanced production of specialized metabolites. *Fermentation* 9 (11): 941.

Sajid, S. and Akbar, N. (2018). 1. Applications of fungal pigments in biotechnology. *Pure and Applied Biology (PAB)* 7 (3): 922–930.

Sánchez, J.F., Fernández, J.M., Acién, F.G. et al. (2008). Influence of culture conditions on the productivity and lutein content of the new strain *Scenedesmus almeriensis*. *Process Biochemistry* 43 (4): 398–405. https://doi.org/10.1016/j.procbio.2008.01.004.

Sandesh Kamath, B., Vidhyavathi, R., Sarada, R., and Ravishankar, G.A. (2008). Enhancement of carotenoids by mutation and stress induced carotenogenic genes in *Haematococcus pluvialis* mutants. *Bioresource Technology* 99 (18): 8667–8673. https://doi.org/10.1016/j.biortech.2008.04.013.

Scaife, M.A., Prince, C.A., Norman, A., and Armenta, R.E. (2012). Progress toward an *Escherichia coli* canthaxanthin bioprocess. *Process Biochemistry* 47 (12): 2500–2509.

Sereti, F., Alexandri, M., Papadaki, A. et al. (2024). Natural lycopene and β-carotene synthesis by *Rhodosporidium kratochvilovae* yeasts: sustainable production, chemical characterization and antioxidative properties. *Food Bioscience* 57: 103425.

Shi, B., Ma, T., Ye, Z. et al. (2019). Systematic metabolic engineering of *Saccharomyces cerevisiae* for lycopene overproduction. *Journal of Agricultural and Food Chemistry* 67 (40): 11148–11157.

Song, I., Kim, J., Baek, K. et al. (2020). The generation of metabolic changes for the production of high-purity zeaxanthin mediated by CRISPR-Cas9 in *Chlamydomonas reinhardtii*. *Microbial Cell Factories* 19: 1–9.

Srivastava, A., Kalwani, M., Chakdar, H. et al. (2022). Biosynthesis and biotechnological interventions for commercial production of microalgal pigments: a review. *Bioresource Technology* 352: 127071.

Su, A., Chi, S., Li, Y. et al. (2018). Metabolic redesign of *Rhodobacter sphaeroides* for lycopene production. *Journal of Agricultural and Food Chemistry* 66 (23): 5879–5885.

Su, B., Song, D., and Zhu, H. (2020). Metabolic engineering of *Saccharomyces cerevisiae* for enhanced carotenoid production from xylose-glucose mixtures. *Frontiers in Bioengineering and Biotechnology* 8: 435.

Su, B., Deng, M.-R., and Zhu, H. (2023). Advances in the discovery and engineering of gene targets for carotenoid biosynthesis in recombinant strains. *Biomolecules* 13 (12): 1747.

Subasankari, K., Thanappan, V., Jeyapragash, D. et al. (2020). Growth promoting studies on co-culturing *Nannochloropsis oceanica* with *Halomonas aquamarina* actively enhance the algal biomass and lipid production. *Biocatalysis and Agricultural Biotechnology* 29: 101790. https://doi.org/10.1016/j.bcab.2020.101790.

Sun, D., Zhang, Z., Zhang, Y. et al. (2019). Light induces carotenoids accumulation in a heterotrophic docosahexaenoic acid producing microalga *Crypthecodinium* sp. SUN. *Bioresource Technology* 276: 177–182.

Szotkowski, M., Holub, J., Šimanský, S. et al. (2021). Bioreactor co-cultivation of high lipid and carotenoid producing yeast *Rhodotorula kratochvilovae* and several microalgae under stress. *Microorganisms* 9 (6): 1160.

Tang, W., Wang, Y., Zhang, J. et al. (2019). Biosynthetic pathway of carotenoids in Rhodotorula and strategies for enhanced their production. *Journal of Microbiology and Biotechnology* 29 (4): 507–517.

Thanapimmetha, A., Suwaleerat, T., Saisriyoot, M. et al. (2017). Production of carotenoids and lipids by *Rhodococcus opacus* PD630 in batch and fed-batch culture. *Bioprocess and Biosystems Engineering* 40: 133–143.

Tominaga, K., Hongo, N., Fujishita, M. et al. (2017). Protective effects of astaxanthin on skin deterioration. *Journal of Clinical Biochemistry and Nutrition* 61 (1): 33–39.

Tominaga, M., Nozaki, K., Umeno, D. et al. (2021). Robust and flexible platform for directed evolution of yeast genetic switches. *Nature Communications* 12 (1): 1846.

Torres-Haro, A., Verdín, J., Kirchmayr, M.R., and Arellano-Plaza, M. (2021). Metabolic engineering for high yield synthesis of astaxanthin in *Xanthophyllomyces dendrorhous*. *Microbial Cell Factories* 20 (1): 1–17.

Tramontin, L.R.R., Kildegaard, K.R., Sudarsan, S., and Borodina, I. (2019). Enhancement of astaxanthin biosynthesis in oleaginous yeast *Yarrowia lipolytica* via microalgal pathway. *Microorganisms* 7 (10): 472.

Villegas-Méndez, M.Á., Papadaki, A., Pateraki, C. et al. (2021). Fed-batch bioprocess development for astaxanthin production by *Xanthophyllomyces dendrorhous* based on the utilization of *Prosopis* sp. pods extract. *Biochemical Engineering Journal* 166: 107844. https://doi.org/10.1016/j.bej.2020.107844.

Wang, C., Oh, M., and Liao, J.C. (2000). Directed evolution of metabolically engineered *Escherichiacoli* for carotenoid production. *Biotechnology Progress* 16 (6): 922–926.

Wang, C., Liwei, M., Park, J.-B. et al. (2018). Microbial platform for terpenoid production: *Escherichia coli* and yeast. *Frontiers in Microbiology* 9: 2460.

Watcharawipas, A. and Runguphan, W. (2023). Red yeasts and their carotenogenic enzymes for microbial carotenoid production. *FEMS Yeast Research* 23: foac063.

Wu, M., Zhu, R., Lu, J. et al. (2020). Effects of different abiotic stresses on carotenoid and fatty acid metabolism in the green microalga *Dunaliella salina* Y6. *Annals of Microbiology* 70 (1): 1–9.

Xie, W., Lv, X., Ye, L. et al. (2015). Construction of lycopene-overproducing *Saccharomyces cerevisiae* by combining directed evolution and metabolic engineering. *Metabolic Engineering* 30: 69–78.

Xue, D., Abdallah, I.I., de Haan, I.E.M. et al. (2015). Enhanced C 30 carotenoid production in *Bacillus subtilis* by systematic overexpression of MEP pathway genes. *Applied Microbiology and Biotechnology* 99: 5907–5915.

Yabuzaki, J. (2017). Carotenoids database: structures, chemical fingerprints and distribution among organisms. *Database* 2017 (1): bax004.

Ye, R.W., Stead, K.J., Yao, H., and He, H. (2006). Mutational and functional analysis of the β-carotene ketolase involved in the production of canthaxanthin and astaxanthin. *Applied and Environmental Microbiology* 72 (9): 5829–5837.

Zhang, Z., Ji, H., Gong, G. et al. (2014). Synergistic effects of oleaginous yeast *Rhodotorula glutinis* and microalga *Chlorella vulgaris* for enhancement of biomass and lipid yields. *Bioresource Technology* 164: 93–99. https://doi.org/10.1016/j.biortech.2014.04.039.

Zhang, J., Li, Q.-R., Zhang, M.-H. et al. (2019a). Enhancement of carotenoid biosynthesis in *Phaffia rhodozyma* PR106 under stress conditions. *Bioscience, Biotechnology, and Biochemistry* 83 (12): 2375–2385.

Zhang, Z., Pang, Z., Xu, S. et al. (2019b). Improved carotenoid productivity and COD removal efficiency by co-culture of *Rhodotorula glutinis* and *Chlorella vulgaris* using starch wastewaters as raw material. *Applied Biochemistry and Biotechnology* 189: 193–205.

Zhang, C., Chen, X., and Too, H.-P. (2020a). Microbial astaxanthin biosynthesis: recent achievements, challenges, and commercialization outlook. *Applied Microbiology and Biotechnology* 104 (13): 5725–5737.

Zhang, X., Wang, D., Duan, Y. et al. (2020b). Production of lycopene by metabolically engineered *Pichia pastoris*. *Bioscience, Biotechnology, and Biochemistry* 84 (3): 463–470.

Zhang, Q.-W., Kong, C.-L., and Tao, Y.-S. (2023). Fate of carotenoids in yeasts: synthesis and cleavage. *Critical Reviews in Food Science and Nutrition* 63 (25): 7638–7652.

Zhang, Z.-X., Xu, L.-W., Xu, Y.-S. et al. (2024). Integration of genetic engineering and multi-factor fermentation optimization for co-production of carotenoid and DHA in *Schizochytrium* sp. *Bioresource Technology* 394: 130250.

Zhao, J., Li, Q., Sun, T. et al. (2013). Engineering central metabolic modules of *Escherichia coli* for improving β-carotene production. *Metabolic Engineering* 17: 42–50.

Zhao, M., Gao, M., Xiong, L. et al. (2022). CRISPR-Cas assisted shotgun mutagenesis method for evolutionary genome engineering. *ACS Synthetic Biology* 11 (5): 1958–1970. https://doi.org/10.1021/acssynbio.2c00112.

Zhou, P., Xie, W., Li, A. et al. (2017). Alleviation of metabolic bottleneck by combinatorial engineering enhanced astaxanthin synthesis in *Saccharomyces cerevisiae*. *Enzyme and Microbial Technology* 100: 28–36.

15

Exploring the Chemistry and Sources of Microbial 1,2-Propanediol [Propylene glycol] with a Focus on Biosynthesis, Extraction, and Identification for Nutraceutical Significance and Human Health

Alaa Kareem Niamah[1], Shayma Thyab Gddoa Al-Sahlany[1], Deepak Kumar Verma[2], Amit Kumar Singh[3], Manish Kumar Singh[4], Rakesh Mohan Shukla[5], Smita Singh[6], Ami R. Patel[7], Gemilang Lara Utama[8,9], Mónica L. Chávez González[10], José Sandoval-Cortés[10], Prem Prakash Srivastav[2], and Cristobal Noe Aguilar[10]

[1] *Department of Food Science, College of Agriculture, University of Basrah, Basra City, 61004, Iraq*
[2] *Agricultural and Food Engineering Department, Indian Institute of Technology Kharagpur, Kharagpur, 721302, West Bengal, India*
[3] *Department of Post Harvest Technology, College of Horticulture, Banda University of Agriculture and Technology, Banda, 210001, Uttar Pradesh, India*
[4] *Department of Vegetable Science, College of Horticulture, Banda University of Agriculture and Technology, Banda, 210001, Uttar Pradesh, India*
[5] *Processing and Food Engineering Division, College of Agricultural Engineering and Technology (CoAE&T), Sher-e-Kashmir University of Agricultural Sciences and Technology of Kashmir, Srinagar, 190025, Jammu and Kashmir, India*
[6] *Department of Allied Health Sciences, Chitkara School of Health Sciences, Chitkara University, Shambu Kalan, Punjab, 140401, India*
[7] *Division of Dairy Microbiology, Mansinhbhai Institute of Dairy and Food Technology-MIDFT, Kamdhenu University, Mehsana, 384002, Gujarat, India*
[8] *Faculty of Agro-Industrial Technology, Universitas Padjadjaran, Sumedang, West Java, 45363, Indonesia*
[9] *Center for Environment and Sustainability Science, Universitas Padjadjaran, Bandung, Jawa Barat, 40132, Indonesia*
[10] *Bioprocesses Research Group, Food Research Department, School of Chemistry, Universidad Autonoma de Coahuila, Unidad Saltillo, 25280, Coahuila, Mexico*

15.1 Introduction

During the early twentieth century, microbes were the primary source of bulk chemicals. These chemicals were produced by fermenting biomass such as corn and potatoes (Detroy 2018). During this time period, there was a significant change in understanding the process of "cracking" petroleum into simpler hydrocarbons. This discovery led to chemists becoming the dominant figures in this field (Bull et al. 2000). This transition marked the beginning of complex and diverse methodologies developed by chemists to convert these basic units into larger quantities of chemicals and specialized products on a smaller scale. During this period, there was a significant shift in chemical synthesis techniques and strategies. The approach changed from relying on microbes to using petroleum-derived methods (Voit and Lederer 2009). As a result, it is crucial to revive fermentation as an economical approach to manufacture various bulk chemicals, including diols (such as succinic acid, lactic acid, 1,2-propanediol, and 1,3-propanediol), pharmaceuticals (medications, vitamins), and biofuels (ethanol, butanol, and hydrogen). The fermentation process utilized to generate these compounds underscores their sustainable characteristics, while also ensuring industrial and environmental safety (Jiang et al. 2014; Verma et al. 2022; Niamah et al. 2023).

Microbial Nutraceuticals: Products and Processes, First Edition. Edited by Sudhir Pratap Singh and Santosh Kumar Upadhyay.

1,2-propanediol (PDO), also known as propylene glycol, is a colorless liquid with a thick consistency. It has a faint smell and a slightly sweet flavor. The compound has a chemical formula of $CH_3CH(OH)CH_2OH$, which classifies it as a diol because it contains two alcohol groups (Chakraborty et al. 2020; Główka and Krawczyk 2023). To distinguish it from the isomer propane-1,3-diol, the compound is sometimes called alpha(α)-PDO, while the other is known as beta(β)-PDO (Główka and Krawczyk 2023). Chirality is observed in PDO. Industrial practices often utilize the racemate. The *S*-isomer is produced using biotechnological pathways, as described by various scientific reports (Altaras and Cameron 1999; Siebert and Wendisch 2015; Pyne et al. 2016; Sun et al. 2021; Tao et al. 2021; Jo et al. 2023). The substance shows compatibility with a diverse range of solvents including water (H_2O), acetone, and chloroform. In general, glycols are known for their non-irritating nature and low volatility (Chakraborty et al. 2020).

According to Restrepo et al. (2021), the projected value of the global PDO market is set to increase to US$ 4.3 billion in FY 2022, compared to US$ 4.13 billion in 2021. Lyondell and Dow, the top manufacturers, collectively produce 60% of the global output, i.e., 35 and 25%, respectively. The significant rise in usage can be credited to its diverse array of practical uses, as documented in various studies (Oliveira et al. 2019; Restrepo et al. 2021; Główka and Krawczyk 2023). These applications encompass a wide range of industries, such as the production of unsaturated polyester resins, incorporation into nutritional products, formulation of non-ionic detergents, inclusion in cosmetics, utilization in fluids, functioning as an anti-freezing agent, and serving as a de-icing agent. Due to the need for petrochemicals and its expensive manufacturing process, chemically produced PDO faced challenges in finding its application in the industrial sector (Fiume et al. 2012; Sabra et al. 2016; Oliveira et al. 2019; Restrepo et al. 2021; Główka and Krawczyk 2023). Throughout history, scientists have made significant strides in developing methods and products that are environment-friendly and sustainable, impacting various aspects of life. Given the circumstances, it is crucial to establish that the biologically produced PDO is a cost-effective alternative to the chemically synthesized product currently in use (Marinas et al. 2015; Scioli et al. 2020).

PDO is manufactured on a large scale, primarily for polymer production (Restrepo et al. 2021; Główka and Krawczyk 2023). In the European Union, it is given the E-number E1520 for use in the food industry, while for cosmetics and pharmacology, it is referred to as E490 (Fiume et al. 2012; Younes et al. 2018; Kathiravan 2021). In addition, PDO can be found in PDO alginate, which is also known as E405. Considered to be generally regarded as safe (GRAS) for consumption by the US Food and Drug Administration, PDO is approved for certain applications as an indirect additive in food (Saxena and Saxena 2015; Younes et al. 2018; Kathiravan 2021). It is widely accepted and commonly used as a carrier in different pharmaceutical formulations for topical, oral, and certain intravenous applications in both the United States and Europe (Kathiravan 2021; Scott et al. 2024).

PDO is a widely used component in various food products, such as prepackaged items like coffee, creamers, breakfast foods, snacks, and fast food (Saxena et al. 2010; Younes et al. 2018; Główka and Krawczyk 2023). It can be found in a variety of bread products, canned beans, dairy items, condiments, and certain frozen vegetables. Moreover, this compound is present in artificial flavor extracts, food coloring products, and specific medications like coated pills, gelcaps, liquids, chewables, and lozenges. However, non-coated pills generally do not contain PDO (Jacob et al. 2018; Tao et al. 2021).

Given the critical conversation that has taken place in the various paragraphs above, we have recognized multiple research gaps and possibilities. This has led us to conceive and define the effective aims in the following manner, which we have described in this chapter: (i) to define microbial PDO chemically and structurally; (ii) to address various sources of microbial PDO; (iii) to discuss the methods for synthesizing PDO, including chemical and microbial production; (iv) to describe the biochemical pathways for PDO synthesis; (v) to emphasize sample preparation, isolation, and extraction methods along with identification and characterization techniques; (vi) to highlight the nutraceutical value of PDO to human health; and (vii) to address the potential future research areas for researchers, scientists, and industrial people who are looking forward to novel discovery.

15.2 Structure and Chemistry of Microbial 1,2-Propanediol

Microbial PDO can be produced by a range of microorganisms, such as bacteria and yeast, as demonstrated in studies by Saxena and Saxena (2015), Zielińska et al. (2017), and Tao et al. (2021). It is widely utilized in the food and pharmaceutical sectors as a food additive and a constituent of different medications (Główka and Krawczyk 2023). The production of PDO through microbial processes entails the fermentation of sugars, such as glucose or glycerol, by microorganisms. This has been extensively studied by researchers (Bennett and San 2001; Saxena and Saxena 2015; Tao et al. 2021). During fermentation, microorganisms utilize enzymes to break down sugars into PDO and other by-products, including ethanol and carbon dioxide.

The PDO is a diol with a chemical formula of C_3H8O_2. It is composed of three carbon atoms and has two hydroxyl groups (—OH) attached to a central carbon atom (Bennett and San 2001). There are two forms of PDO, namely (*R*)-PDO and (*S*)-PDO, as shown in Figure 15.1. This compound exhibits a diol structure that possesses a pronounced affinity for water and a significantly elevated boiling point (Bennett and San 2001). PDO is a transparent, scentless liquid with a slightly thick consistency. It can easily mix with water and various organic solvents.

Generally, the chemistry of microbial PDO involves understanding its molecular structure, synthesis pathways, and the biochemical processes that facilitate its production in microorganisms which mean to say that it encompasses its molecular structure, physical and chemical properties, and the biochemical pathways involved in its microbial synthesis.

The study of microbial PDO chemistry focuses on analyzing the molecular properties and reactions associated with this compound, which is synthesized by microorganisms (Wang et al. 2023). This involves gaining knowledge about the structure, composition, and processes of synthesis and metabolism by microbial cells. Additionally, it involves examining the physical and chemical properties of the compound, including its solubility, reactivity, and stability. Table 15.1 provides information on the physical and chemical characteristics of PDO. Understanding this information is essential for unlocking the possibilities of microbial PDO in a range of industrial uses, including biofuels, bioplastics, and pharmaceuticals (Klemm et al. 2005). Through a comprehensive grasp of the chemistry behind microbial PDO, researchers have the ability to enhance production

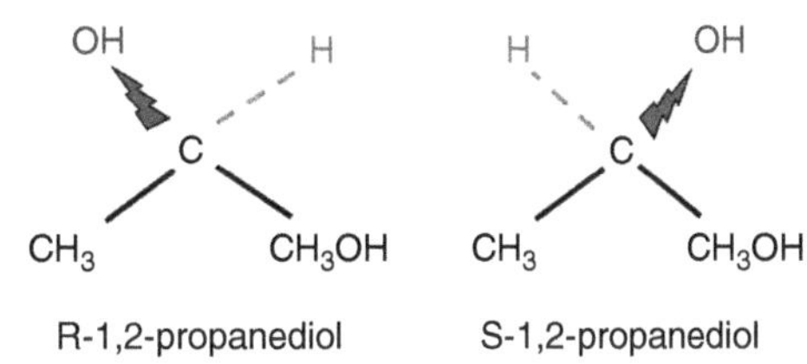

Figure 15.1 Chemical structure of 1,2-propanediol (propylene glycol).

Table 15.1 Detailed information on the chemical and physical characteristics of 1,2-propanediol.

Property	Remarks
Molecular formula	The molecular formula of the PDO is $C_3H_8O_2$
Molecular weight	The molecular weight of PDO is 76.09 g/mol
Boiling point	The PDO has a boiling point of 187°C
Melting point	The PDO has a melting point of −60°C
Density	At a temperature of 20°C, the density of PDO is measured to be 1.036 g/cm^3
Specific rotation(α)	The PDO has a specific rotation of n20/D 1.432
Flash point	The PDO has a flash point of 107.22°C
Odor	The PDO has no discernible odor
Water solubility	The PDO has a water solubility that is classified as miscible
Solubility	The PDO has the ability to dissolve in various solvents such as water, acetone, ethyl acetate, chloroform, and ether. This PDO can dissolve in a range of essential oils, but it is not compatible with petroleum ether, paraffin, or grease
Vapor density	The PDO has a vapor density of 2.62 when compared to air
Vapor pressure	The vapor pressure of the PDO at 20°C is 0.08 mm Hg
Appearance	The PDO has a viscous liquid appearance
Specific gravity	The specific gravity of the PDO is 1.038 (20/20°C)1.036~1.040
Color	The PDO color is measured using the APHA scale which is equal to or less than 10
pKa	The pKa value of the PDO is 14.49 ± 0.20
pH	The pH value of the PDO falls within the range of 6–8
Storage condition	The recommended storage condition for the PDO is between 5°C and 30°C
Explosive limit	The PDO has an explosive limit between 2.4% and 17.4%(V)
Refractive index (RI)	The PDO has a RI of 1.432, calculated as n20/D
Toxicity	The toxicity of the PDO is measured by its LD_{50} values in rabbits. When administered orally, the LD_{50} is between 19,400 and 36,000 mg/kg, while the LD_{50} for dermal exposure is 20,800 mg/kg
Hydrophilic nature	PDO exhibits high solubility in water and has the ability to form hydrogen bonds, due to its hydroxyl groups
Sensitivity/ hygroscopicity	The PDO is highly sensitive to moisture due to its hygroscopic nature. It has the ability to take in water from the surrounding atmosphere, making it valuable for situations where keeping moisture is important
Reactivity	Its reactivity with acids, bases, and other alcohols is due to the presence of hydroxyl groups, which enable a range of chemical transformations

Sources: Modified from ChemBK (2024), Cheméo (2024), ChemSpider (2024), and NCBI (2024).

processes, develop catalysts that are more effective, and enhance the overall yield and quality of the compound. The knowledge and utilization of the beneficial functions of such microorganisms can pave the way for the creation of eco-friendly technologies that harness microbial PDO as a valuable asset in the bio-based economy (González-Figueredo et al. 2019). The study of microbial PDO involves examining its molecular properties and reactions, such as its structure, synthesis, metabolism, and potential applications (Klemm et al. 2005). Gaining a comprehensive understanding of the chemistry behind microbial PDO is crucial in order to maximize its potential in various industrial applications. It enables the optimization of production processes, refinement of catalysts, and augmentation of overall yield and quality (González-Figueredo et al. 2019).

15.3 Sources and Synthesis of 1,2-Propanediol

15.3.1 Sources of Microbial 1,2-Propanediol

PDO can be produced by microbial sources through fermentation (Tao et al. 2021). Some common microbial sources of PDO include bacteria and yeast. Microbial sources of PDO include *Escherichia coli*, *Klebsiella pneumoniae*, *Saccharomyces cerevisiae*, *Lactobacillus* spp., and *Clostridium thermosaccharolyticum* have been discussed in detail in Table 15.2. These microbial sources convert different carbon sources (such as glycerol, glucose, or lactate) into PDO as a metabolic end product. Certain species of bacteria like *K. pneumoniae*, *Citrobacter freundii*, and *Limosilactobacillus reuteri* have the ability to produce PDO through fermentation of different carbon sources (Sun et al. 2021; Jo et al. 2023). Moreover, some strains of yeast, such as *Candida stellata* and *Candida. tropicalis*, have been found to produce PDO during fermentation processes. Furthermore, genetically engineered strains of *E. coli* have been developed to produce PDO as a metabolite.

15.3.2 Synthesis of 1,2-Propanediol

The current dominant method for producing PDO involves petrochemically derived routes, but it is known to have high power consumption (Tao et al. 2021). In addition, it is possible to produce PDO from renewable resources by utilizing microbes (Tao et al. 2021). The microbial biosynthesis of PDO shows great potential as it can incorporate important biochemical pathways into fermentation processes for production (Saxena et al. 2010; Tao et al. 2021). Certain microbial sources have the ability to produce PDO. One example is bacteria that are present in the human intestinal microbiota. These bacteria can produce PDO as a significant end product through the anaerobic breakdown of sugars like fucose and rhamnose (Saxena et al. 2010). In addition, it has been reported that specific strains of *C. thermobutyricum* are capable of producing PDO while breaking down cellulose (Dank et al. 2021). It is important to point out that although microbial production of PDO holds promise, commercial production through microbial biosynthesis is currently constrained by low efficiency.

15.3.2.1 Chemical Production

Due to increasing environmental regulations and the fluctuating price of crude oil, leading industry players are increasingly shifting toward the use of renewable raw materials for the production of PDO. Despite the emergence of innovative technologies, further modifications are necessary to enhance energy efficiency and sustainability in production (Główka and Krawczyk 2023). There are three main methods for producing PDO: the traditional conversion of propylene to propylene oxide and hydrolysis, the hydrogenolysis of glycerol and sugar derivatives, and biotechnological approaches that involve microorganisms, enzymatic catalysis, or fermentation of sugars and glycerol (Figure 15.2). These chemical techniques have the potential for producing PDO, but they require significant investment or produce waste streams that contain harmful contaminants for the environment. Recently, there have been some intriguing advancements in biological techniques that offer a cost-effective solution for reducing high-quality commercial PDO derived from renewable resources. Microbes that thrive in anaerobic and facultative anaerobic environments are fascinating organisms because they produce a primary metabolic by-product known as PDO (Saxena et al. 2010). PDO is sẏnthesized through the hydration of propylene oxide, which is derived from non-renewable petrochemicals. The production of propylene oxide involves complex chemical processes, such as the chlorohydrin process or the hydroperoxide process. These processes require the use of potentially hazardous chemicals. In today's context, the production of

Table 15.2 Some key microbial sources of 1,2-propanediol along with their natural habitat and biosynthesis pathway.

Microbial species	Natural habitat	Biosynthesis pathway	References
Escherichia coli	*E. coli* is a frequently encountered bacterium that is typically present in the intestinal tracts of warm-blooded creatures	*E. coli* is a highly researched microorganism that is commonly employed in the production of PDO. By utilizing metabolic engineering techniques, it is possible to alter the glycolytic pathway in order to increase the production of PDO from glucose and other sugars. The procedure entails the conversion of dihydroxyacetone phosphate (DHAP) into methylglyoxal, followed by its reduction to PDO	Jain et al. (2015); Aziz et al. (2017); Niu et al. (2019)
Klebsiella pneumoniae	*K. pneumoniae* is commonly present in various environments such as soil and water, and it is also a natural inhabitant of the human mouth, skin, and intestines	The bacterium has the ability to produce PDO through the fermentation process of sugars like glucose and glycerol. The DHAP pathway is used to convert sugars to PDO by utilizing intermediates such as methylglyoxal and hydroxyacetone	Sun et al. (2021) Jo et al. (2023)
Saccharomyces cerevisiae	*S. cerevisiae*, commonly known as baker's yeast, can be found naturally on the skins of fruits and other sugary surfaces	Using genetic engineering techniques, it is possible to modify *S. cerevisiae* to produce PDO from glucose and other sugars. Through the manipulation of genetic material and optimizing metabolic pathways, yeast has the ability to transform these substances into PDO	Islam et al. (2017)
Lactobacillus spp.	*Lactobacillus* spp. can be found in a wide range of environments, such as fermented foods like yogurt and sauerkraut, as well as in the natural gut flora	Some strains of *Lactobacillus* have the ability to produce a compound known as PDO from sugars such as glucose and xylose. They commonly utilize the methylglyoxal pathway to convert sugars into PDO, with intermediates like methylglyoxal	Sabra et al. (2016); Tao et al. (2021)
Clostridium thermosaccharolyticum	*C. thermosaccharolyticum* is a thermophilic bacterium that thrives in high-temperature environments, like compost heaps and hot springs	*C. thermosaccharolyticum* has the ability to produce PDO from glucose and other carbohydrates using the methylglyoxal pathway. The bacterium's remarkable capacity to operate in extreme temperatures renders it highly suitable for various industrial applications	Tao et al. (2021)

Figure 15.2 Illustration depicting various techniques for synthesizing 1,2-propanediol using diverse raw materials.

enantiomerically pure PDO can be achieved through three main methods (Lete et al. 2024). Several methods can be employed to achieve the desired results, such as the catalytic hydrogenation of lactic acid esters, the bioreduction of acetol, and the resolution of racemic PDO. It is noteworthy that companies in the agribusiness sector, such as Archer Daniels Midland, have developed a new chemical process that relies on glycerol as a starting material. Glycerol is used as the starting material for the conversion to PDO. This conversion process can be aided by either heterogeneous or homogeneous catalysts in the reaction medium. Another method for producing propylene oxide that has garnered significant interest recently is the catalytic oxidation of propylene using hydrogen peroxide. It was produced and marketed independently in South Korea (Amada et al. 2010; Wang et al. 2015).

15.3.2.2 Microbial Production

There is currently a wealth of knowledge regarding the fermentation of carbohydrates to PDO by a diverse range of microorganisms (Table 15.3). This PDO has been demonstrated to be produced by both yeasts and bacteria. In 1954, Enebo reported that *C. thermobutyricum* was capable of producing PDO (Enebo 1954). According to Turner and Roberton (1979), *Bacteroides ruminicola* was identified as the source of this PDO ten years later. The metabolism and growth yields of *B. ruminicola* were studied when grown on xylose, arabinose, and rhamnose. It was noticed that rhamnose was primarily fermented to PDO.

It has been reported that *Salmonella typhimurium* has the ability to produce PDO from de-oxy sugars (Saxena et al. 2010). When using methyl pentose, rhamnose, or fucose as substrates, an equal amount of PDO was detected. However, in anaerobic cultures, PDO did not undergo any additional metabolic processes. It was noted that there was a gradual decrease in *S. typhimurium* cultures under similar conditions. Through a thorough examination, it was discovered that when grown on rhamnose, *S. typhimurium* released a significant amount of PDO into the surrounding medium. After the sugar source was used up, the level of PDO reached its highest point and then slowly decreased over time, even though the culture was kept in a stable environment. It was noted that the decrease in PDO concentration did not occur when the cells were isolated from the medium using centrifugation. This suggested that the activity of the cells played a vital role in the

Table 15.3 Comprehensive information from important studies related to the microbial synthesis of 1,2-propanediol using various strains.

Organisms		Substrates	Yield	References
Bacteria	*Escherichia coli*	10 g glucose	0.25 (g/L)	Altaras and Cameron (1999)
	Enterobacter ludwigii	10 g glucose	0.56 (g/g)	Karayannis et al. (2023)
	Klebsiella pneumoniae	30 g glycerol	1.01 (g/L)	Jo et al. (2023)
	Corynebacterium glutamicum	20 g glucose	0.131 (mol/mol)	Siebert and Wendisch (2015)
	Clostridium phytofermentans	10 g fucose	0.41 (g/g)	Petit et al. (2013)
	Bacteroides ruminicola	10 g rhamnose	0.08 (g/g)	Niu et al. (2019)
Yeasts	*Saccharomyces cerevisiae*	10 g glycerol	2.19 (g/L)	Jung et al. (2011)
	Candida polymorpha	9.35 g rhamnose	1.83 (g/g)	Suzuki and Onishi (1968)
	Pichia robertsii	9.45 g rhamnose	0.45 (g/g)	

metabolism and elimination of the PDO compound. This finding emphasizes the reliance of PDO elimination on the existence and effectiveness of the bacterial cells in the medium, indicating an intricate interaction between cellular processes and chemical dynamics in the environment (Saxena et al. 2010).

Several strains of *Lactobacillus*, including *Levilactobacillus brevis* and *Lentilactobacillus buchneri*, have been found to produce PDO by breaking down lactic acid and using glycerol as an electron acceptor (Oude Elferink et al. 2001; Krooneman et al. 2002; Huang et al. 2021). The lactobacilli mentioned here have shown an impressive capability to decompose lactic acid in environments without oxygen, without relying on an external electron acceptor to aid the process (Oude Elferink et al. 2001). During this biochemical transformation, it was interesting to observe that every mole of lactic acid underwent a systematic conversion into approximately 0.5 M of acetic acid. Additionally, an equivalent amount of PDO was produced, along with small traces of ethanol. In addition, the acidity of the environment was considered crucial for stimulating and improving the natural ability of these microbial cells to break down lactic acid. The complex metabolic pathways and enzymatic mechanisms involved in this fascinating biochemical process highlight the significance of comprehending the intricate interplay between different microbial species and their metabolic capabilities (Krooneman et al. 2002; Mitrík et al. 2023). Thus, it is necessary to conduct additional research to gain a more comprehensive understanding of the molecular mechanisms behind these phenomena and to investigate their potential uses in biotechnology and industry.

Studies have found that different industrial strains of *S. cerevisiae* contain low levels of this PDO, suggesting its presence in various biological contexts (Islam et al. 2017). The production of PDO from glucose using genetic and metabolic manipulation techniques in *S. cerevisiae* highlights the exciting possibilities for biotechnological progress in this area. The strain mentioned in the study demonstrated its ability to produce approximately 0.24 g/L of PDO from glucose, highlighting the effectiveness of the engineered biological system in generating this valuable compound (Islam et al. 2017). The results of this study emphasized the potential of manipulating microorganisms to support the environment-friendly production of chemicals that can be used in various industries.

There are notable differences in the substrates, fermentation conditions, and reaction mechanisms used by various categories of microorganisms (Bennett and San 2001; Tao et al. 2021; Główka and Krawczyk 2023). It is interesting to observe that bacteria require strictly anaerobic environments to produce PDO, which is in stark contrast to fungi that need aerobic conditions for the same process (Cameron and Cooney 1986; Oude Elferink et al. 2001; Zielińska et al. (2017); Tao et al. 2021). Different substrate metabolisms have an impact on the stereochemistry of PDO that is naturally synthesized by bacteria. For example, although fucose and rhamnose can produce *S*-PDO, glucose and xylose can produce the *R*-isomer (Cameron and Cooney 1986; Niu and Guo 2015; Zhu et al. 2016; Niu et al. 2019; Tao et al. 2021). A significant number of microbial genera are recognized for their ability to produce PDO primarily by utilizing different sugars such as fucose, rhamnose, and glucose (Sabra et al. 2016; Tao et al. 2021; Hlynsdóttir 2022). The discovery of PDO formation as a by-product of cellulose degradation in *C. thermobutyricum* was a major breakthrough (Cameron and Cooney 1986). Several bacteria have been found to produce *S*-PDO from fucose or rhamnose, highlighting the wide range of microbial pathways involved in PDO biosynthesis (Cameron and Cooney 1986; Zhu et al. 2016; Aziz et al. 2017; Dank et al. 2021; Tao et al. 2021; Hlynsdóttir 2022; Główka and Krawczyk 2023).

A bacterial strain, *K. pneumoniae* GEM167ΔadhE/pBR-PDO, was modified to halt ethanol production and instead utilize glycerol as a carbon source to convert lactic acid into PDO. This modification was achieved through engineering techniques described in studies by (González-Figueredo et al. 2019; Sun et al. (2021) and Jo et al. (2023)). An impressive increase in PDO production by 2.24-fold to 731.0 ± 24.7 mg/L at 48 hours was noted when the agitation speed was raised from 200 to 400 rpm (Jo et al. 2023). In addition to the increased production of PDO, there was also an observed rise in the formation of 2,3-butanediol, which is a by-product. Various methods were developed to impede the production of 2,3-butanediol, which included controlling pH levels and making genetic modifications. Keeping the pH level at 7.0 led to a significant increase in PDO production to 1016.5 ± 37.3 mg/L after 48 hours. Unfortunately, the genetic engineering approach did not achieve the intended result. The plasmid's stability within this strain remained consistently at 100% for a duration of 72 hours, as reported by Jo et al. (2023).

15.3.2.3 Deployed Biochemical Pathways for Synthesis

The pathways in biotechnology demonstrate variations depending on the specific microorganisms used in the processes. These variations have been extensively studied by researchers (Niu et al. 2019; Sun et al. 2021; Tao et al. 2021; Hlynsdóttir 2022; Jo et al. 2023). One of the pathways involves a crucial step where a sugar is phosphorylated, resulting in the subsequent splitting of the sugar into dihydroxyacetone phosphate and lactic aldehyde. These intermediate compounds play a significant role in the process. Afterward, the lactic aldehyde goes through a reduction process to produce PDO, which is a crucial product in this pathway (Figure 15.3). Another pathway, which relies on glucose, uses methylglyoxal as an intermediate compound. This highlights the wide range of routes available in biotechnological processes. Within these pathways, by-products like dihydroxyacetone phosphate and lactic aldehyde are produced, demonstrating the intricate nature of the biochemical reactions that it performs. The complex nature of these pathways highlights the significance of comprehending the distinct functions of each compound and intermediate to enhance the production of desired end products. Additional investigation and examination are required to fully understand the complexities of these biotechnological pathways and to improve their effectiveness and output (David et al. 2018; Hlynsdóttir 2022). Nevertheless, the large-scale production of PDO through biotechnology has proven to be a complex task. This is mainly due to the intricate metabolic pathways of microorganisms, the challenging growing conditions, and the less-than-optimal yields achieved.

Therefore, we can conclude based on the available evidence-based scientific reports that there has been a clear decrease in scientific publications and patents focused on the bioproduction of PDO using native microorganisms, suggesting a lack of progress in research in this field. In

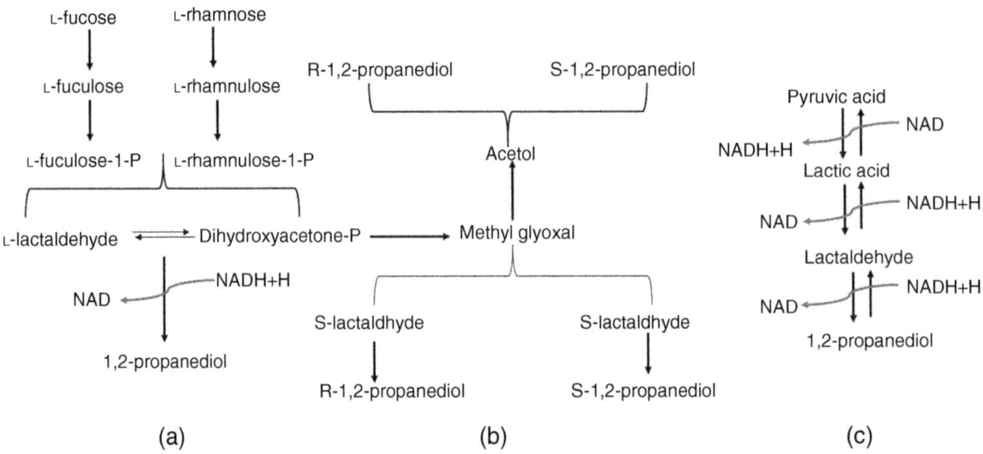

Figure 15.3 A schematic representation of the several metabolic pathways used to produce 1,2-propanediol, (a) fucose and rhamnose metabolic pathways, (b) methylglyoxl metabolic pathways, (c) lactic acid metabolic pathways (Zielińska et al. 2017; Hlynsdóttir 2022).

addition, the current state of technology has not yet achieved a level of economic competitiveness in comparison with traditional manufacturing processes. This emphasizes the importance of further advancements in biotechnological methods for PDO production.

15.4 Extraction, Identification, and Characterization Process

The process of extracting, identifying, and characterizing microbial PDO involves a series of important steps, as outlined in various studies (Yagi and Yamada 1969; Bennett and San 2001; Krooneman et al. 2002; Zielińska et al. 2017). This section provides detailed instructions for carrying out the sample preparation, isolation, and extraction process, as well as the identification and characterization process using analytical instruments. The extraction process is designed to separate PDO from the fermentation broth. In general, solvent extraction is used to extract PDO from the aqueous phase. A suitable solvent, such as ethyl acetate or butanol, is employed for this purpose. The selection of a solvent plays a critical role (Chávez-González et al. 2020; Verma and Srivastav 2020a, b; Tripathy et al. 2021, 2022), as it needs to have a high level of selectivity toward PDO while having minimal solubility in water. The sample preparation and extraction method for microbial PDO are detailed in Table 15.4, providing a clear step-by-step procedure for conducting the experiment. Additionally, the table highlights important points to consider during the experiments.

Following the isolation and extraction process, the identification and quantification process occurs. Confirming the presence and measuring the concentration of PDO are crucial to identify it in the extract. Gas chromatography (GC) combined with mass spectrometry (MS) is frequently employed for this purpose. GC separates various compounds by analyzing their volatility and polarity, while MS offers insights into the molecular structure of these separated compounds (Verma and Srivastav 2020a, b; Tripathy et al. 2021, 2022). Through the comparison of the acquired mass spectra with established standards, it is possible to identify and measure PDO. The process of conducting the identification and quantification method of microbial PDO using analytical instruments is clearly outlined in Table 15.5. This table provides step-by-step procedures for carrying out the experiment, as well as important considerations to keep in mind during the experiments.

Table 15.4 Detailed instructions on how to carry out the sample preparation, isolation, and extraction process of microbial 1,2-propanediol.

(1) *Materials and equipment*		
Microbial culture	• *Escherichia coli* and *Saccharomyces cerevisiae* are microorganisms that produce PDO • Media for culture (such as lysogeny broth, minimal media containing glucose or glycerol)	
Reagents	• Distilled water organic solvents (e.g., ethanol, acetone) • Buffer solutions (e.g., phosphate buffer) • Salts (e.g., sodium chloride) • Activated charcoal (optional for decolorization)	
Fermentation equipment	• Bioreactor or fermentation flasks • Incubator or shaker • pH meter • Temperature control system	
Extraction and purification equipment	• Centrifuge filtration setup (e.g., vacuum filtration) • Liquid–liquid extraction apparatus • Rotary evaporator or distillation setup • Chromatography setup (e.g., ion-exchange or adsorption columns)	
(2) *Step-by-step procedures for conducting the sample preparation and extraction process*		
Step 1: Microbial cultivation	*Preparation of inoculum*	• Add a small amount of sterile culture medium and introduce the microbial strain • Allow the inoculum culture to incubate overnight at the recommended temperature (e.g., 37°C for *E. coli* or 30°C for *S. cerevisiae*) with gentle shaking
	Fermentation	• Transfer the overnight culture to a larger volume of fermentation medium in a bioreactor or large flasks • Optimize fermentation conditions (temperature, pH, aeration, and agitation) to achieve the highest possible PDO production • Monitor the fermentation process by measuring the growth (OD600) and product formation (e.g., using high-performance liquid chromatography (HPLC))
Step 2: Sample preparation	*Harvesting cells*	• Once the fermentation process is finished, the microbial cells can be gathered through the utilization of centrifugation • Adjust the centrifuge to a speed range of 4000–6000 rpm and allow it to operate for a duration of 10–20 min • This method will efficiently isolate the cells from the culture broth
	Cell lysis (if necessary)	• Cells with intracellular PDO should be broken down using mechanical disruption, sonication, or enzymatic digestion (such as lysozyme for *E. coli*) • It is recommended to centrifuge the lysate to remove any cell debris
Step 3: Extraction of PDO	*Liquid–liquid extraction*	• Add an organic solvent, like ethanol or acetone, to the supernatant or cell lysate for PDO extraction • Thoroughly mix the solution and allow the distinct layers to separate • If necessary, can be used as separatory funnel • Collect the organic phase that contains PDO
	Aqueous extraction	• Alternatively, if the product is primarily in the aqueous phase, proceed directly to the purification steps

(*Continued*)

Table 15.4 (Continued)

Step 4: Purification	*Distillation*	• Use a rotary evaporator or a distillation setup to remove the organic solvent and increase the concentration of the PDO • Retrieve the solution that has been distilled to obtain purified PDO
	Chromatography (optional)	• To attain a greater level of purity, researchers can utilize chromatography techniques such as ion-exchange or adsorption chromatography • Apply the concentrated extract onto the chromatography column and utilize an appropriate solvent or buffer to elute it • Collect and combine the fractions that contain PDO
	Decolorization (optional)	• If needed, might employ activated charcoal to remove any color impurities from the extract. Mix the extract with activated charcoal, ensuring a thorough combination, and then separate the charcoal using filtration
Step 5: Final processing	*Drying and concentration*	• To achieve a solid form, it is essential to dry the purified product through techniques like freeze-drying or evaporation • Ensure that the final product is free from any remaining solvents or impurities
	Quality control	• Analyze the purity and concentration of the final PDO by using techniques such as HPLC, gas chromatography (GC), or nuclear magnetic resonance (NMR) spectroscopy • Make sure there are no impurities and confirm that the product is appropriate for its intended use
(3) *Important points to consider*		
Sterility	• It is crucial to maintain sterility in all equipment and reagents to avoid any possible contamination	
Optimization	• Optimize fermentation conditions (pH, temperature, nutrient concentration) specific to the microbial strain used	
Safety	• Ensure safety by following all required protocols when working with microorganisms and organic solvents	

Sources: Prepared from Tabah et al. (2016), Zhu et al. (2016), Mahato et al. (2017), Islam et al. (2017), and Yang et al. (2018).

Once the identification and quantification are complete, the characterization process begins (Prabhakar et al. 2023). Understanding microbial PDO requires analyzing its physicochemical properties, including purity, enantiomeric composition, and isotopic composition. Nuclear magnetic resonance (NMR) spectroscopy is an incredibly valuable technique for analyzing molecular structures (Mahato et al. 2017). It offers a wealth of information, allowing for a thorough understanding of the number and organization of atoms and functional groups within a molecule. Through the utilization of advanced techniques like gas chromatography–isotope ratio mass spectrometry (GC-IRMS), isotopic analysis can unveil the isotopic composition of PDO. This analysis offers valuable insights into the metabolic pathways that are involved in the biosynthesis of PDO. Enantiomeric purity can be determined by utilizing chiral chromatography techniques, such as high-performance liquid chromatography (HPLC) with a chiral stationary phase.

Table 15.5 Detailed instructions on how to carry out the identification and characterization process of microbial 1,2-propanediol using advanced analytical instruments.

(1) *Materials and equipment*		
Analytical instruments		• High-performance liquid chromatography (HPLC) • Gas chromatography–mass spectrometry (GC-MS) • Nuclear magnetic resonance (NMR) • Spectroscopy infrared spectroscopy • Ultraviolet-visible (UV-Vis) spectroscopy
Reagents		• Standard PDO for calibration • Solvents (e.g., water, methanol, ethanol) • Deuterated solvents for NMR, e.g., water (D_2O) and chloroform ($CDCl_3$)
(2) *Step-by-step procedures for conducting the identification and characterization process using deployed analytical instruments*		
Step 1: Initial identification by HPLC	*Sample preparation*	• To prepare the microbial PDO sample, it is necessary to dilute it in a suitable solvent, such as water or methanol • Use a 0.45-μm filter to effectively eliminate any particulates from the sample
	Analysis	• Set up the HPLC system using an appropriate column, such as a C18 reversed-phase column • Choose a suitable mobile phase, like a combination of water and methanol (70 : 30 v/v) • Place the sample into the HPLC system and adjust the detection wavelength, usually around 210 nm • To confirm the presence of the sample, it is necessary to compare its retention time with that of a standard PDO
Step 2: Confirmation by GC-MS	*Sample preparation*	• If needed, the sample can be derivatized to enhance its volatility and detectability • Popular derivatization agents include silylation reagents such as BSTFA (*N,O*-bis(trimethylsilyl)trifluoroacetamide)
	Analysis	• Set up the GC-MS system using an appropriate column, such as a DB-5 or HP-5 column • Introduce the derivatized sample into the GC-MS system • Examine the mass spectrum of the eluted compound and make a comparison with the mass spectrum of a standard PDO
Step 3: Structural characterization by NMR spectroscopy	*Sample preparation*	• To ensure accurate results, the purified PDO sample should be dissolved in a deuterated solvent like D2O or CDC13
	Analysis	• Utilize 1H-NMR and 13C-NMR spectroscopy techniques to ascertain the chemical composition • Examine the spectra to determine the distinct peaks that correspond to the hydrogen and carbon atoms found in PDO • Examine the NMR spectra alongside standard reference spectra for PDO in order to validate the structure

(*Continued*)

Table 15.5 (Continued)

Step 4: Functional group analysis by infrared spectroscopy	*Sample preparation*	• Prepare a thin layer of the PDO sample on a substrate that is transparent to infrared light, or utilize an attenuated total reflectance (ATR) accessory
	Analysis	• Remember to document the infrared spectrum of the sample • Understand the distinctive absorption bands associated with hydroxyl (—OH) groups and carbon–hydrogen (C—H) bonds • Examine the infrared spectrum alongside standard reference spectra to verify the existence of functional groups commonly found in PDO
Step 5: Purity assessment by UV-Vis spectroscopy	*Sample preparation*	• Prepare the PDO sample by diluting it in a suitable solvent, such as water or ethanol
	Analysis	• Make sure the record UV-vis spectrum of the sample is within the appropriate wavelength range, which is typically 190–400 nm • Identify and quantify any impurities by analyzing their absorbance spectra in comparison to pure PDO
(3) *Quality control and validation*		
Standard curve calibration	• Prepare standard solutions of known concentrations of PDO • Establish calibration curves for HPLC, GC-MS, and UV-vis spectroscopy in order to measure the concentration of PDO in samples	
Repeatability and reproducibility	• Conduct multiple analyses to ensure the consistency and reliability of the results • Confirm the methods by examining established standards and comparing the outcomes with certified reference materials	
(4) *Documentation and reporting*		
Data compilation	• Gather all the analytical data, which include chromatograms, spectra, and quantitative results • Include comprehensive information about the experimental conditions and parameters employed in each analytical technique	
Results interpretation	• Interpret the data to verify the identity, purity, and structure of microbial PDO • Examine any detected impurities and their potential effects on the overall quality of the final product	

Sources: Prepared from Jain et al. (2015), Pyne et al. (2016), Mahato et al. (2017) and Dolejš et al. (2019).

15.5 Nutraceutical Importance and Human Health

PDO has attracted significant interest in the nutraceutical industry because of its promising health benefits and wide range of applications. It has garnered significant interest as a possible nutraceutical because of its numerous health-promoting properties, which are described in Table 15.6. This section delves into the importance of microbial PDO in terms of nutrition and its impact on human health. PDO serves as an important cosolvent and food emulsifier, with extensive use in industries like food essence, cosmetics, and medicine (Fiume et al. 2012; Younes et al. 2018; Oliveira

Table 15.6 Importance of microbial 1,2-propanediol in nutraceuticals and its effects on human health.

Nutraceuticals and biological properties	Impacts on human health	References
Anti-inflammatory effects	• Studies have demonstrated that PDO possesses anti-inflammatory properties • It reduces the production of pro-inflammatory cytokines, such as TNF-α, IL-1β, and IL-6, which play a role in chronic inflammation	Younes et al. (2018); Bajaj et al. (2024)
Antioxidant properties	• PDO exhibits strong antioxidant activity, scavenging free radicals and protecting cells from oxidative damage • It has been shown to inhibit lipid peroxidation, a major mechanism of oxidative stress in the body	Younes et al. (2018); Kao et al. (2021)
Cardiovascular benefits	• PDO has been associated with improved cardiovascular health • It has been shown to reduce blood pressure, improve blood cholesterol levels, and prevent atherosclerosis	Younes et al. (2018); Zeng et al. (2021)
Gut health	• PDO has prebiotic properties and supports the growth of beneficial gut bacteria • It promotes the production of short-chain fatty acids, which enhance gut barrier function and reduce inflammation	Younes et al. (2018); Cheng et al. (2020); Bajaj et al. (2024)
Immune modulation	• Some studies indicate that PDO may modulate immune responses • It has been found to enhance the production of certain cytokines and reduce the severity of immune disorders	Younes et al. (2018); Bushmeleva et al. (2021)
Metabolic regulation	• Some studies suggest that PDO may play a regulatory role in metabolism • It may enhance glucose utilization and suppress appetite, contributing to weight management	Morshed et al. (1991); Yakan et al. (2023)
Neuroprotective effects	• PDO may have neuroprotective properties • It has been shown to reduce oxidative stress and inflammation in the brain, potentially benefiting conditions such as Alzheimer's and Parkinson's diseases	Sood et al. (2007)
Nutritional value and absorption	• PDO is a water-soluble alcohol with a slightly sweet taste • It is metabolized to pyruvate in the body, contributing to energy production • Unlike other short-chain alcohols such as ethanol, PDO does not exhibit toxic effects on the liver and other organs	Morshed et al. (1991); Fowles et al. (2013); Younes et al. (2018)

et al. 2019; Karayannis et al. 2023). Nevertheless, the detection of its presence is strictly prohibited in pure milk or foods intended for specific dietary purposes, as specified in the list of food additives in Regulation (EC) No. 1334/2008. In June 2022, an incident emerged when PDO (at a concentration of 0.3 g/kg) was found in pure milk. This scandal was linked to the excessive use of flavoring, and it has raised concerns about consumer trust in the dairy sector and the regulation of dairy product markets. Although PDO is commonly considered safe, there have been reports suggesting potential toxicity. The use of PDO in dissolving substances such as phenytoin, multivitamins, IV nitroglycerine, etomidate, enoximone, and topical sulfadiazine silver cream has been associated with adverse effects including seizure, coma, and agitation (Wilson et al. 2005). Research has shown that PDO is absorbed more quickly from the gastrointestinal tract compared to subcutaneous tissues and inhalation (LaKind et al. 1999).

Previous study has also highlighted that exposure to low-dose PDO can result in hyperosmolarity, metabolic acidosis, hemolysis, neurotoxicity, and reduced renal clearing, as indicated by Lim et al. (2015). In the primary culture of human proximal tubule cells, the use of PDO at concentrations ranging from 10 to 50 mmol/L for periods of 1–5 days caused minimal harm to cells in the proliferating phase. The observed renal dysfunction in both human exposure cases and animal studies could potentially be attributed to the cytotoxicity. In addition, studies on human blood mononuclear cells have shown that PDO can effectively reduce the activity of natural killer cells in laboratory settings, as previously stated in a study by Zar et al. (2007). Overconsumption of PDO can lead to the accumulation of lactic acid in the bloodstream and kidneys, causing symptoms of poisoning like impaired respiration, lowered heart rate, and loss of consciousness. The detrimental impacts of these phenomena have been thoroughly recorded in numerous studies encompassing both animal and human subjects (Pillai et al. 2014; Fan and Jia 2023). It is of utmost importance to thoroughly study the fundamental processes through which PDO affects the quality of food and the potential hazards it may present to human well-being. This will enable us to improve the safety and regulations surrounding food items. Research in this field is crucial to guarantee the safety and quality of food products for consumers.

A research report provides a comprehensive understanding of the ecological consequences that result from a trophic interaction between gut symbionts (Cheng et al. 2020). The research was carried out by conducting experiments in gnotobiotic mice using isogenic mutants of bacterial strains that produce (*Bifidobacterium breve*) and utilize (*L. reuteri*) (Cheng et al. 2020). By shedding light on a potential selective pressure that has played a role in shaping the evolution of this species over time, the insights that were gained from this study represent a significant contribution to our knowledge regarding how cross-feeding influences the competitive fitness of *L. reuteri* within the environment of the vertebrate gut. These results are highly relevant because they provide a foundation for future microbial strategies that aim to modify gut ecosystems. These strategies can be tailored to individual patients by considering their microbiota's resource-providing capabilities or by using bacterial strain mixtures to establish trophic interactions (Cheng et al. 2020).

It is therefore possible to draw the conclusion that microbial PDO is a potential nutraceutical that possesses many health-promoting qualities. This conclusion can be reached on the basis of the facts that have been critically reviewed in the preceding paragraphs as well as in Table 15.6. As a result of its antioxidant, anti-inflammatory, cardiovascular, metabolic, gut health, neuroprotective, and immune-modulating properties, it is an extremely valuable contribution to the expanding area of functional foods. It is necessary to do more study to completely investigate the possible advantages of PDO and its uses in the field of human food and health.

15.6 Prospective Future and Research Opportunities

There are immense and diverse possibilities for future prospects and research in the field of microbial PDO. These opportunities range from fundamental biochemical studies to practical biotechnological advancements. Through tackling these obstacles and harnessing cutting-edge technologies, scientists can tap into the complete capabilities of microbial PDO as an environment-friendly and adaptable chemical foundation with a wide range of uses in various fields such as industry, agriculture, and healthcare. Exploring the field of microbial PDO opens up exciting possibilities for future research and provides ample opportunities for innovation and discovery (Table 15.7).

Table 15.7 Promising avenues for future research and numerous opportunities for exploration and innovation in studying microbial 1,2-propanediol.

Areas of study		Research opportunities
Characterization of novel strains and metabolic pathways	Microbial diversity studies	Exploring microbial biodiversity can lead to the discovery of new strains that have the ability to efficiently produce PDO in various environmental habitats
	Metabolic engineering approaches	Utilizing systems biology and omics technologies can allow for the exploration of metabolic networks and the identification of crucial enzymes and regulatory elements to enhance pathway optimization.
Enhanced production strains and pathways	Genetic engineering advances	Enhanced production capabilities of microbial strains can be achieved by employing precise genetic modifications, such as the utilization of cutting-edge gene editing techniques like CRISPR-Cas
	Pathway optimization	Metabolic pathways have the potential to be modified in order to improve the efficiency of substrate utilization and increase the yield of desired products. The main objective is to minimize the production of unwanted by-products and maximize the synthesis of PDO
Elucidation of biochemical mechanisms and regulatory networks	Enzyme kinetics and mechanistic studies	Studying the biochemical kinetics and mechanisms of enzymes involved in PDO synthesis can provide valuable insights for developing effective engineering strategies
	Transcriptional regulation	Decoding the transcriptional regulatory networks involved in PDO biosynthesis can lead to the optimization of gene expression and metabolic flux distribution
Integration of advanced biotechnologies and computational tools	Synthetic biology tools	Utilizing synthetic biology techniques, it is possible to develop specific microbial chassis strains that are specifically engineered to produce PDO with enhanced characteristics
	High-throughput screening methods	Advanced screening platforms can be established to quickly assess microbial strains and culture conditions, expediting efforts to enhance strain performance
	Computational modeling and simulation	Utilizing computational modeling and simulation techniques can allow for the prediction of metabolic fluxes, optimization of bioprocess parameters, and guidance of strain engineering strategies
Sustainable and eco-friendly production processes	Bioprocess optimization	Exploring novel bioprocessing strategies may help minimize resource consumption, energy requirements, and waste generation in the production of PDO
	Utilization of renewable resources	One potential opportunity for investigation is the utilization of renewable carbon sources like lignocellulosic biomass, industrial waste streams, and carbon dioxide to facilitate sustainable microbial fermentation
Sustainable development and industrial scale-up	Life cycle assessment	Comprehensive life cycle assessments can be performed to evaluate the environmental impact, sustainability, and economic feasibility of microbial PDO production processes
	Scale-up strategies	Addressing the challenges related to scale-up, process integration, and downstream processing is crucial for facilitating the industrial-scale production of microbial PDO
Development of novel applications and products	Functional foods and nutraceuticals	Exploring the potential of incorporating microbial PDO into functional foods, dietary supplements, and nutraceutical formulations can improve stability, bioavailability, and health benefits
	Biomedical and pharmaceutical applications	Exploring the potential therapeutic uses of microbial PDO is a fascinating area of research that has implications in drug delivery systems, tissue engineering, and regenerative medicine

Since the beginning of time, the manufacturing of PDO has been mostly dependent on synthetic processes. These methods begin with petrochemical feedstock and catalysis, and they involve a wide variety of chemical intermediates. However, there is an increasing chance that microbiological pathways might develop as a viable method for producing PDO from sustainable feedstocks. This would involve the utilization of technologies that would integrate intermediates that are favorable to the environment. This trend toward microbial manufacturing methods offers new paths for the synthesis of PDO, which may offer solutions that are more sustainable and favorable to the environment than conventional petrochemical-based approaches. Furthermore, there have been substantial scientific developments in the field of producing genetically modified strains, as well as improvements in the methodologies that are used to analyze various substances. These advancements have been made possible by these advancements. In addition to these improvements, there has been an emphasis placed on theoretical assessments of metabolic pathways to promote the production of compounds with a high value-added. It is vital to take into consideration the dual strategy of employing optimal hosts and optimized expression, in addition to having an awareness of the enzymatic features of pathway enzymes, to guarantee the continuous evolution of these production procedures. Furthermore, the strategy of removing rival routes and strengthening preferred pathways by means of genetic alterations and manipulation of the cultural environment will play a significant part in the progression of such processes. In addition, there is a significant amount of potential in the incorporation of biological processes into downstream processing, which looks to be not only practical but also economically advantageous. Several technological breakthroughs are necessary for the effective commercialization of products or chemicals. One of these advancements is the combination of fermentation production technology with metabolic pathway engineering. With the goal of increasing production, enhancing product purity, diversifying product variety, and expanding market reach, this approach is continually changing. In order to simplify metabolic engineering, which is a relatively new technology that has the potential to dramatically affect the development of bio-based economies, a comprehensive understanding of biochemical pathways and fermentation processes as well as fermentation methods will be required. The catalyst that serves as the focal point of that specific process was developed as a result of a metabolic engineering endeavor that was probably unrivaled in its scale. It is important to note that this initiative resulted in the invention of the catalyst.

In the realm of industrial biotechnology, there is a vision of a future where microorganisms play a vital role in producing a diverse range of products, such as medicine, fuels, and materials. This would lead to significant improvements in their overall quality. This vision is based on the concept that, in contrast to conventional chemicals and fossil fuels, these resources are more sustainable and can be replenished on an ongoing basis. As a result, it can be deduced that renewable resource-focused industrial biotechnology is expected to gradually replace the dominant roles currently held by petroleum and chemistry in various sectors.

15.7 Concluding Remarks

The compound discussed in this chapter, specifically 1,2-propanediol (PDO), has significant connections to a wide range of food products and pharmaceutical preparations. Because of its dependence on synthetic sources, the chemical production of PDO is costly and harmful to the environment. As a result, the natural production of PDO using appropriate microorganisms is a very promising, cost-effective, and environmentally friendly alternative. Several substrates have been investigated for the cost-effective production of PDO, including glycerol, glucose, cellulose, and its derivatives. Throughout the fermentation process, microorganisms employ various enzymes to decompose sugars into PDO and other by-products, such as ethanol and carbon dioxide. Yeasts have demonstrated a greater capacity for PDO production compared to bacteria among various microorganisms. The production of PDO from natural strains is currently quite low, which highlights the need to identify more efficient microorganisms from different environmental habitats that can produce PDO. The results of utilizing genetically modified strains or employing metabolic engineering seem to be a more intriguing method for PDO production. However, biotechnological methods for producing PDO have not yet been implemented on an industrial scale due to various factors. The challenges in this field stem from the intricate growth conditions required to cultivate microorganisms, limited knowledge about their complex metabolism, and the resulting low yields. It is crucial to address existing obstacles, tackle challenges, and prioritize the advancement of commercially viable, environmentally friendly, and economically feasible methods for PDO production. Therefore, future research efforts should focus on strategically planning and conducting concrete studies to better comprehend the kinetics of enzymes involved in this process. Additionally, the utilization of bioinformatic tools in combination with the systematic application of genetic engineering, metabolomics, or other relevant biotechnological tools should be explored.

References

Altaras, N.E. and Cameron, D.C. (1999). Metabolic engineering of a 1, 2-propanediol pathway in *Escherichia coli*. *Applied and Environmental Microbiology* 65 (3): 1180–1185.

Amada, Y., Koso, S., Nakagawa, Y., and Tomishige, K. (2010). Hydrogenolysis of 1, 2-propanediol for the production of biopropanols from glycerol. *ChemSusChem* 3 (6): 728–736.

Aziz, R.K., Monk, J.M., Andrews, K.A. et al. (2017). The aldehyde dehydrogenase, AldA, is essential for L-1, 2-propanediol utilization in laboratory-evolved *Escherichia coli*. *Microbiological Research* 194: 47–52.

Bajaj, A., Markandey, M., Kedia, S., and Ahuja, V. (2024). Gut bacteriome in inflammatory bowel disease: an update on recent advances. *Indian Journal of Gastroenterology* 43: 1–9.

Bennett, G.N. and San, K.Y. (2001). Microbial formation, biotechnological production and applications of 1, 2-propanediol. *Applied Microbiology and Biotechnology* 55: 1–9.

Bull, A.T., Ward, A.C., and Goodfellow, M. (2000). Search and discovery strategies for biotechnology: the paradigm shift. *Microbiology and Molecular Biology Reviews* 64 (3): 573–606.

Bushmeleva, K., Vyshtakalyuk, A., Terenzhev, D. et al. (2021). Radical scavenging actions and immunomodulatory activity of *Aronia melanocarpa* propylene glycol extracts. *Plants* 10 (11): 2458.

Cameron, D.C. and Cooney, C.L. (1986). A novel fermentation: the production of R (–)–1, 2–propanediol and acetol by *Clostridium thermosaccharolyticum*. *Bio/Technology* 4 (7): 651–654.

Chakraborty, N., Juglan, K.C., and Kumar, H. (2020). Study of thermodynamic and acoustic properties of niacin in aqueous hexylene glycol and propylene glycol at different temperatures. *ACS Omega* 5 (50): 32357–32365.

Chávez-González, M.L., Sepúlveda, L., Verma, D.K. et al. (2020). Conventional and emerging extraction processes of flavonoids. *Processes* 8: 434. https://doi.org/10.3390/pr8040434.

ChemBK (2024). 1,2-propanediol. https://www.chembk.com/en/chem/1,2-Propanediol (accessed 11 June 2024).

Cheméo (2024). Chemical properties of 1,2-propanediol, 2-acetate (CAS 6214-01-3). https://www.chemeo.com/cid/86-719-9/1-2-Propanediol-2-acetate (accessed 11 June 2024).

ChemSpider (2024). (±)-1,2-propanediol. https://www.chemspider.com/Chemical-Structure.13835224.html (accessed 11 June 2024).

Cheng, C.C., Duar, R.M., Lin, X. et al. (2020). Ecological importance of cross-feeding of the intermediate metabolite 1, 2-propanediol between bacterial gut symbionts. *Applied and Environmental Microbiology* 86 (11): e00190–e00120.

Dank, A., Zeng, Z., Boeren, S. et al. (2021). Bacterial microcompartment-dependent 1, 2-propanediol utilization of *Propionibacterium freudenreichii*. *Frontiers in Microbiology* 12: 679827.

David, C., Schmid, A., Adrian, L. et al. (2018). Production of 1, 2-propanediol in photoautotrophic *Synechocystis* is linked to glycogen turn-over. *Biotechnology and Bioengineering* 115 (2): 300–311.

Detroy, R.W. (2018). Bioconversion of agricultural biomass to organic chemicals. In: *Organic Chemicals from Biomass* (ed. B.I.S. Goldstein), 19–43. CRC press.

Dolejš, I., Líšková, M., Krasňan, V. et al. (2019). Production of 1,3-propanediol from pure and crude glycerol using immobilized *Clostridium butyricum*. *Catalysts* 9 (4): 317. https://doi.org/10.3390/catal9040317.

Enebo, L. (1954). *Studies in Cellulose Decomposition by an Anaerobic Thermophilic Bacterium and Two Associated Non-cellulolytic Species*. Stockholm, Sweden: KTH Royal Institute of Technology.

Fan, Z. and Jia, W. (2023). Ambient 1, 2-propanediol exposure accelerates the degradation of lipids and amino acids in milk via allosteric effects and affects the utilization of nutrients containing amide bond. *Food Research International* 170: 112965.

Fiume, M.M., Bergfeld, W.F., Belsito, D.V. et al. (2012). Safety assessment of propylene glycol, tripropylene glycol, and PPGs as used in cosmetics. *International Journal of Toxicology* 31 (5_suppl): 245S–260S.

Fowles, J.R., Banton, M.I., and Pottenger, L.H. (2013). A toxicological review of the propylene glycols. *Critical Reviews in Toxicology* 43 (4): 363–390.

Główka, M. and Krawczyk, T. (2023). New trends and perspectives in production of 1, 2-propanediol. *ACS Sustainable Chemistry & Engineering* 11 (19): 7274–7287.

González-Figueredo, C., Alejandro Flores-Estrella, R., and O.A. Rojas-Rejón (2019). Fermentation: Metabolism, Kinetic Models, and Bioprocessing. Current Topics in Biochemical Engineering. IntechOpen.

Hlynsdóttir, S. H. (2022). Microbial production of enantiomerically pure 1, 2-propanediol. Doctoral dissertation, Department of Human Resources Business and Science, University of Akureyri, Iceland.

Huang, Z., Wang, M., Ke, W., and Guo, X. (2021). Screening of high 1, 2-propanediol production by *Lactobacillus buchneri* strains and their effects on fermentation characteristics and aerobic stability of whole-plant corn silage. *Agriculture* 11 (7): 590.

Islam, Z.U., Klein, M., Aßkamp, M.R. et al. (2017). A modular metabolic engineering approach for the production of 1, 2-propanediol from glycerol by *Saccharomyces cerevisiae*. *Metabolic Engineering* 44: 223–235.

Jacob, S.E., Scheman, A., and McGowan, M.A. (2018). Propylene glycol. *Dermatitis* 29 (1): 3–5.

Jain, R., Sun, X., Yuan, Q., and Yan, Y. (2015). Systematically engineering *Escherichia coli* for enhanced production of 1, 2-propanediol and 1-propanol. *ACS Synthetic Biology* 4 (6): 746–756.

Jiang, Y., Liu, W., Zou, H. et al. (2014). Microbial production of short chain diols. *Microbial Cell Factories* 13: 1–17.

Jo, M.H., Ju, J.H., Heo, S.Y. et al. (2023). Production of 1, 2-propanediol from glycerol in *Klebsiella pneumoniae* GEM167 with flux enhancement of the oxidative pathway. *Biotechnology for Biofuels and Bioproducts* 16 (1): 18.

Jung, J.Y., Yun, H.S., Lee, J.W., and Oh, M.K. (2011). Production of 1, 2-propanediol from glycerol in *Saccharomyces cerevisiae*. *Journal of Microbiology and Biotechnology* 21 (8): 846–853.
Kao, T.T., Wang, M.C., Chen, Y.H. et al. (2021). Propylene glycol improves stability of the anti-inflammatory compounds in *Scutellaria baicalensis* extract. *Processes* 9 (5): 894.
Karayannis, D., Vasilakis, G., Charisteidis, I. et al. (2023). Screening of new industrially important bacterial strains for 1, 3-propanediol, 2, 3-butanediol and ethanol production through biodiesel-derived glycerol fermentations. *Microorganisms* 11 (6): 1424.
Kathiravan, G. (2021). Formulation of novel nano transdermal using effective combination of acyclovir and omeprazole for enhanced anti-viral activity. (Doctoral dissertation, Karpagam College of Pharmacy, Coimbatore).
Klemm, D., Heublein, B., Fink, H.P., and Bohn, A. (2005). Cellulose: fascinating biopolymer and sustainable raw material. *Angewandte Chemie International Edition* 44 (22): 3358–3393.
Krooneman, J., Faber, F., Alderkamp, A.C. et al. (2002). *Lactobacillus diolivorans* sp. nov., a 1, 2-propanediol-degrading bacterium isolated from aerobically stable maize silage. *International Journal of Systematic and Evolutionary Microbiology* 52 (2): 639–646.
LaKind, J.S., McKenna, E.A., Hubner, R.P., and Tardiff, R.G. (1999). A review of the comparative mammalian toxicity of ethylene glycol and propylene glycol. *Critical reviews in Toxicology* 29 (4): 331–365.
Lete, A., Raso, R., García, L. et al. (2024). Synthesis of ketones from glycerol and 1, 2-propanediol using copper and nickel catalysts: unraveling the impact of reaction phase and active metal. *Fuel* 371: 132001.
Lim, E.S., Zhou, Y., Zhao, G. et al. (2015). Early life dynamics of the human gut virome and bacterial microbiome in infants. *Nature Medicine* 21 (10): 1228–1234.
Mahato, D.K., Verma, D.K., Billoria, S. et al. (2017). Applications of nuclear magnetic resonance in food processing and packaging management. In: *Developing Technologies in Food Science* (ed. M. Meghwal and M.R. Goyal), 109–142. Apple Academic Press.
Marinas, A., Bruijnincx, P., Ftouni, J. et al. (2015). Sustainability metrics for a fossil-and renewable-based route for 1, 2-propanediol production: a comparison. *Catalysis Today* 239: 31–37.
Mitrík, T., Mitrík, A., and Kořínek, D. (2023). 1, 2-propanediol in maize silages with nutrients measurement using NIRS technology. *Acta fytotechnica et zootechnica: ISSN 1336-9245* 26 (2): 138–146.
Morshed, K.M., L'Helgoualch, A., Nagpaul, J.P. et al. (1991). The role of propylene glycol metabolism in lactatemia in the rabbit. *Biochemical Medicine and Metabolic Biology* 46 (2): 145–151.
NCBI (National Center for Biotechnology Information) (2024). S-1,2-propanediol. https://pubchem.ncbi.nlm.nih.gov/compound/S-1_2-Propanediol (accessed 11 June 2024).
Niamah, A.K., Al-Fekaiki, D.F., Thyab Gddoa Al-Sahlany, S. et al. (2023). Investigating the effect of addition of probiotic microorganisms (bacteria or yeast) to yoghurt on the viability and volatile aromatic profiles. *Journal of Food Measurement and Characterization* 17 (5): 5463–5473.
Niu, W. and Guo, J. (2015). Stereospecific microbial conversion of lactic acid into 1, 2-propanediol. *ACS Synthetic Biology* 4 (4): 378–382.
Niu, W., Kramer, L., Mueller, J. et al. (2019). Metabolic engineering of *Escherichia coli* for the de novo stereospecific biosynthesis of 1, 2-propanediol through lactic acid. *Metabolic Engineering Communications* 8: e00082.
Oliveira, G.N., Barbosa, N.C., Araújo, F.C. et al. (2019). Conversion of glycerine into 1, 2-propanediol for industrial applications. In: *Jatropha, Challenges for a New Energy Crop: Volume 3: A Sustainable Multipurpose Crop* (ed. B. Bahadur, N. Carels, and S. Mulpuri), 383–414. Springer.
Oude Elferink, S.J., Krooneman, J., Gottschal, J.C. et al. (2001). Anaerobic conversion of lactic acid to acetic acid and 1, 2-propanediol by *Lactobacillus buchneri*. *Applied and Environmental Microbiology* 67 (1): 125–132.

Petit, E., LaTouf, W.G., Coppi, M.V. et al. (2013). Involvement of a bacterial microcompartment in the metabolism of fucose and rhamnose by *Clostridium phytofermentans*. *PLoS One* 8 (1): e54337.

Pillai, U., Hothi, J.C., and Bhat, Z.Y. (2014). Severe propylene glycol toxicity secondary to use of anti-epileptics. *American Journal of Therapeutics* 21 (4): e106–e109.

Prabhakar, P., Mukherjee, S., Kumar, A. et al. (2023). Optimization of microwave-assisted extraction (MAE) of key phenolic compounds from pigeon pea (*Cajanus cajan* L.), their characterization, and measurement of their anti-diabetic and cytotoxic potential. *Journal of Food Measurement and Characterization* 17: 1–24.

Pyne, M.E., Sokolenko, S., Liu, X. et al. (2016). Disruption of the reductive 1, 3-propanediol pathway triggers production of 1, 2-propanediol for sustained glycerol fermentation by *Clostridium pasteurianum*. *Applied and Environmental Microbiology* 82 (17): 5375–5388.

Restrepo, J.B., Paternina-Arboleda, C.D., and Bula, A.J. (2021). 1, 2—propanediol production from glycerol derived from biodiesel's production: technical and economic study. *Energies* 14 (16): 5081.

Sabra, W., Groeger, C., and Zeng, A.P. (2016). Microbial cell factories for diol production. In: *Bioreactor Engineering Research and Industrial Applications I: Cell Factories* (ed. J.-J. Zhong, J. Bao, and Q. Ye), 165–197. Springer Berlin Heidelberg.

Saxena, R.K., Anand, P., Saran, S. et al. (2010). Microbial production and applications of 1, 2-propanediol. *Indian Journal of Microbiology* 50: 2–11.

Saxena, S. and Saxena, S. (2015). Microbes in production of commodity chemicals: ethanol, acrylamide, citric acid, adipic acid, 1, 2-propanediol and penicillin. In: *Applied Microbiology* (ed. S. Saxena), 71–81. Springer.

Scioli, G., Tonucci, L., Di Profio, P. et al. (2020). New green route to obtain (bio)-propene through 1, 2-propanediol deoxydehydration. *Sustainable Chemistry and Pharmacy* 17: 100273.

Scott, L.N., Fiume, M., Bergfeld, W.F. et al. (2024). Safety assessment of alkane diols as used in cosmetics. *International Journal of Toxicology* 43: 70S–131S. 10.1177/10915818231224234.

Siebert, D. and Wendisch, V.F. (2015). Metabolic pathway engineering for production of 1, 2-propanediol and 1-propanol by *Corynebacterium glutamicum*. *Biotechnology for Biofuels* 8: 1–13.

Sood, R., Taheri, S., Estrada, E.Y., and Rosenberg, G.A. (2007). Quantitative evaluation of the effect of propylene glycol on BBB permeability. *Journal of Magnetic Resonance Imaging: An Official Journal of the International Society for Magnetic Resonance in Medicine* 25 (1): 39–47.

Sun, S., Shu, L., Lu, X. et al. (2021). 1, 2-Propanediol production from glycerol via an endogenous pathway of *Klebsiella pneumoniae*. *Applied Microbiology and Biotechnology* 105: 9003–9016.

Suzuki, T. and Onishi, H. (1968). Aerobic dissimilation of L-rhamnose and the production of L-rhamnonic acid and 1, 2-propanediol by yeasts. *Agricultural and Biological Chemistry* 32 (7): 888–893.

Tabah, B., Varvak, A., Pulidindi, I.N. et al. (2016). Production of 1, 3-propanediol from glycerol via fermentation by *Saccharomyces cerevisiae*. *Green Chemistry* 18 (17): 4657–4666.

Tao, Y.M., Bu, C.Y., Zou, L.H. et al. (2021). A comprehensive review on microbial production of 1, 2-propanediol: micro-organisms, metabolic pathways, and metabolic engineering. *Biotechnology for Biofuels* 14: 1–16.

Tripathy, S., Verma, D.K., Thakur, M. et al. (2022). Recent trends in extraction, identification and quantification methods of *Centella asiatica* phytochemicals with potential applications in food industry and therapeutic relevance: a review. *Food Bioscience* 49: 101864. https://doi.org/10.1016/j.fbio.2022.101864.

Tripathy, S., Verma, D.K., Thakur, M. et al. (2021). Curcumin extraction, isolation, quantification and its application in functional foods: a review with a focus on immune enhancement activities and COVID-19. *Frontiers in Nutrition* 675: https://doi.org/10.3389/fnut.2021.747956.

Turner, K.W. and Roberton, A.M. (1979). Xylose, arabinose, and rhamnose fermentation by *Bacteroides ruminicola*. *Applied and Environmental Microbiology* 38 (1): 7–12.

Verma, D.K., Al-Sahlany, S.T.G., Niamah, A.K. et al. (2022). Recent trends in microbial flavour compounds: a review on chemistry, synthesis mechanism and their application in food. *Saudi Journal of Biological Sciences* 29 (3): 1565–1576.

Verma, D.K. and Srivastav, P.P. (2020a). A paradigm of volatile aroma compounds in rice and their product with extraction and identification methods: a comprehensive review. *Food Research International* 130: 1–33. https://doi.org/10.1016/j.foodres.2019.108924.

Verma, D.K. and Srivastav, P.P. (2020b). Extraction, identification and quantification methods of rice aroma compounds with emphasis on 2-acetyl-1-pyrroline (2-AP) and its relation with rice quality: a comprehensive review. *Food Reviews International* 38: 111–162. https://doi.org/10.1080/87559129.2020.1720231.

Voit, B.I. and Lederer, A. (2009). Hyperbranched and highly branched polymer architectures synthetic strategies and major characterization aspects. *Chemical Reviews* 109 (11): 5924–5973.

Wang, Y., Wan, Z., Zhu, Y. et al. (2023). Enhanced 1, 3-propanediol production with high yield from glycerol through a novel Klebsiella–Shewanella coculture. *Biotechnology for Biofuels and Bioproducts* 16 (1): 50.

Wang, Y., Zhou, J., and Guo, X. (2015). Catalytic hydrogenolysis of glycerol to propanediols: a review. *RSC Advances* 5 (91): 74611–74628.

Wilson, K.C., Reardon, C., Theodore, A.C., and Farber, H.W. (2005). Propylene glycol toxicity: a severe iatrogenic illness in ICU patients receiving IV benzodiazepines: a case series and prospective, observational pilot study. *Chest* 128 (3): 1674–1681.

Yagi, O. and Yamada, K. (1969). Studies on the utilization of petrochemicals by microorganisms: part I. Production of lactic acid from 1, 2-propanediol by *Arthrobacter oxydans*. *Agricultural and Biological Chemistry* 33 (11): 1587–1593.

Yakan, A., Özkan, H., Kaya, U. et al. (2023). Effects of propylene glycol used at different doses in Akkaraman lambs rations on metabolism-related parameters and liver gene and protein expression during different feeding periods. *Animal Science Journal* 94 (1): e13886.

Yang, B., Liang, S., Liu, H. et al. (2018). Metabolic engineering of *Escherichia coli* for 1, 3-propanediol biosynthesis from glycerol. *Bioresource Technology* 267: 599–607.

Younes, M., Aggett, P., Aguilar, F. et al. (2018). Re-evaluation of propane-1, 2-diol (E 1520) as a food additive. *EFSA Journal* 16 (4): e05235.

Zar, T., Graeber, C., and Perazella, M.A. (2007). Reviews: recognition, treatment, and prevention of propylene glycol toxicity. In: *Seminars in Dialysis* (ed. S. Agarwal) (Vol. 20, No. 3), 217–219. Oxford, UK: Blackwell Publishing Ltd.

Zeng, G., Zhang, Q., Wang, X., and Wu, K.H. (2021). Association between blood ethylene oxide levels and the risk of cardiovascular diseases in the general population. *Environmental Science and Pollution Research* 28 (45): 64921–64928.

Zhu, L., Guan, X., Xie, N. et al. (2016). Fermentative production of enantiomerically pure S-1, 2-propanediol from glucose by engineered *E. coli* strain. *Applied Microbiology and Biotechnology* 100: 1241–1251.

Zielińska, K., Fabiszewska, A., Świątek, M., and Szymanowska-Powałowska, D. (2017). Evaluation of the ability to metabolize 1, 2-propanediol by heterofermentative bacteria of the genus Lactobacillus. *Electronic Journal of Biotechnology* 26: 60–63.

16

Innovations in the Production of Multivitamins in Microbial Factories

Nupur[1], *Mohit Kumar*[1], *Aditi Singh*[1], *Neeraj Agarwal*[2], *Narendra Kumar*[3], *and Santosh Kumar Mishra*[1]

[1] *Department of Life Sciences, Sharda School of Bioscience & Technology, Sharda University, Greater Noida, 201310, Uttar Pradesh, India*
[2] *Department of Biotechnology, Noida Institute of Engineering & Technology, Greater Noida, 201310, Uttar Pradesh, India*
[3] *Department of Biotechnology and Bioengineering, Institute of Advanced Research, Gandhinagar, 382426, Gujarat, India*

16.1 Introduction

Vitamins and minerals are two different types of micronutrients that are required in trace amounts for survival and to maintain a healthy body. Humans as well as animals cannot synthesize vitamins on their own; thus, in order to maintain optimal health, they must either take supplements or eat enough through food. However, plants and microbes can create them naturally. Vitamins are being introduced and widely employed as therapeutic agents, food and feed additives, and health aids because of their significance as growth factors. Nowadays, the bulk of vitamins are produced artificially in factories and utilized as fortifiers in food, medicine, and cosmetics (Acevedo-Rocha et al. 2019). The process of chemical synthesis for vitamin production has various downsides. These processes entail costly waste disposal costs and high energy consumption. Chemical synthesis processes are commonly employed to create a diverse range of goods for industrial and consumer applications. While these procedures can be extremely effective, they frequently need the use of high temperatures or pressurized reactors, implying that they may be energy-intensive. Furthermore, the chemicals utilized in these processes may be non-renewable or hazardous, raising issues about the environment, product safety, and public health. It is critical that these chemical synthesis technologies be utilized responsibly, with suitable safety precautions in place, to avoid any potential harm to the environment or human health. Furthermore, customers are becoming aware of the safety of food additives (Wang et al. 2021). This sparked a great interest in the scientific community and industries in using biotechnological methods to replace chemical synthesis. Though algal and microbiological processes already exist, new methods for the environmentally friendly synthesis of vitamins are being developed at a rapid pace. Numerous techniques, including genetic engineering, mutation and screening, medium optimization, and biocatalyst conversion, have been employed to enhance vitamin output (Averianova et al. 2020). Therefore, in order to improve the production of multivitamins will require (i) knowledge of the role that microbes play in the process, (ii) the use of genetic engineering tools and techniques for enhanced synthesis, (iii) fermentation process optimization, (iv) novel approaches to extraction and purification, (v) and advancements in encapsulation and delivery systems.

Microbial Nutraceuticals: Products and Processes, First Edition. Edited by Sudhir Pratap Singh and Santosh Kumar Upadhyay.

16.1.1 Overview and Classification of Multivitamins

Multivitamins can be defined as supplements that contain a combination of multiple vitamins and minerals in varying quantities. They are typically formulated to meet the recommended daily allowances (RDA). Multivitamins may include vitamins as well as minerals like calcium, magnesium, zinc, and iron. The word vitamin derives from Latin term "vita" meaning "life," which denotes its essential roles for good health and well-being of humans (Aguilera-Méndez et al. 2021). They are classified as on the basis of solubility. Those vitamins which dissolve in dietary fat are classified as fat-soluble vitamins that include vitamins A, D, E, and K. This classification is due to their ability to be stored in our body's fatty tissues and liver for later use. However, those vitamins which are readily dissolved in water are classified under water-soluble vitamins and are not stored by the body. It includes vitamin C and B complexes (Wang et al. 2021; Surana et al. 2023). The scientific names, sources, functions, and deficiency of vitamins are summarized in Table 16.1.

16.1.2 Definition and Need of Microbial Factories

Microbial factories are cellular bio-factories that use microorganisms such as yeast, fungi, bacteria, and microalgae to generate compounds useful in many applications. Microbial factories offer significant alternative means of production (Singh and Upadhyay 2021, 2023). The remarkable biotechnology tools and techniques offer absolutely new and very advantageous alternatives for large-scale production. In fact, they possess a number of advantages as compared to traditional chemical synthesis processes and the lengthy process of extracting target chemicals from plants or animals (Gohil et al. 2021; Upadhyay and Singh 2021, 2023, 2024). Microbial factories use biological systems to create valuable compounds, which are then purified by downstream techniques such as filtering and chromatography. The benefit of this strategy is that the purifying process can be completed in less time and relatively occupying less area, resulting in high-quality goods with minimum waste. This makes microbial factories a popular alternative for manufacturing a wide range of products, including pharmaceuticals, biofuels, and chemicals (Du et al. 2011; Montaño López et al. 2022; Singh et al. 2023). The compelling need to create effective alternative ways of manufacturing has become a vital issue, particularly in today's world when limited resources are expected and consumption for these valuable compounds continues to rise. The need to transit from our traditional dependency on finite fossil fuels has become undeniable. Furthermore, it has become more essential to aim for developing beneficial and sustainable conditions for rural communities, eventually leading to their development and economic success (Ko et al. 2020; Su et al. 2020). In essence, the ultimate goal is to create an environment that encourages and maintains the use of renewable resources while also advancing the ideals of sustainable development. To properly accomplish this, three interconnected and distinct difficulties must be tackled with steadfast drive and resolution. These challenges are the (i) optimization and enhancement of production efficiency, (ii) the development and implementation of innovative and environmentally conscious bioprocessing techniques, and (iii) the establishment of robust and reliable systems for the commercialization of products derived from microbial factories.

16.2 Microbial Factories for Multivitamin Production

Microbial factories are essential for the manufacturing of multivitamins. These factories use a variety of microbes to manufacture essential vitamins, offering a more effective and ecological approach than traditional approaches. Multivitamin synthesis can be optimized and scaled up to

Table 16.1 Summary of different vitamins with their function and deficiency.

Vitamins	Scientific name	Sources	Function	Deficiency	References
Vitamin B1	Thiamine	Meat, legumes, whole grains, nuts, beef, pork	Metabolism of glucose, proteins, and lipids	Wet beriberi, dry beriberi, or Wernicke–Korsakoff syndrome	Hanna et al. (2022)
Vitamin B2	Riboflavin	Milk, organ meats, e.g., calf liver, egg, fish, nuts, certain fruits and legumes, wild rice, mushrooms, dark green leafy vegetables, yeast, beer, cheese, and dietary products	To prevent cataracts, lower homocysteine levels, evidence for migraine headache prophylaxis	Sore tongue; a scaly rash on the scrotum or vulva; Ariboflavinosis causes cheilitis	Kennedy (2016); Suwannasom et al. (2020)
Vitamin B3	Niacin	In animal and plant-based foods, including soy, nuts, seeds, legumes, and grains. Many grains, such as bread and cereals	It is a precursor for NAD^+ and $NADP^+$ coenzymes. Required for DNA repair and cholesterol synthesis.	Pellagra is characterized by dementia, diarrhea, and dermatitis. It also includes memory loss, depression, disorientation, headaches, apathy, fatigue, vomit, a swollen mouth, and a scaly rash on sun-exposed skin	Hanna et al. (2022)
Vitamin B5	Pantothenic acid	Eggs, milk, vegetables, beef, chicken, and whole grains	For coenzyme A (CoA): break down fatty acids and acyl carrier protein	Fatigue, abdominal cramps, headache, muscle cramps, paresthesia, nausea, and impaired muscle coordination	Kennedy (2016)
Vitamin B6	Pyridoxine	Beef liver, potatoes, organ meats, starchy vegetables, and fruit (except citrus), fish	Metabolism of protein. Involved in immune function and hemoglobin formation	Peripheral neuropathy with seborrheic dermatitis, glossitis, and cheilosis	Hanna et al. (2022); Sriram et al. (2012)
Vitamin B7	Biotin	Biotin is found naturally in organ meats, eggs, fish, seeds, soybeans, and nuts but is also available through supplementation	Biotin plays an essential role in gene regulation, cell signaling, and replication. It catalyzes the metabolism of fatty acids, glucose, and amino acids	Biotin deficiency is associated with hair thinning, a scaly rash around the eyes, nose, mouth, and perineum, nail changes, skin infections, and neurologic symptoms such as ataxia, seizures, depression, lethargy, and paresthesia	Hanna et al. (2022); Kennedy (2016)
Vitamin B9	Folate	Folate is present in plenty of foods, with the highest levels in dark green leafy vegetables, dairy products, meat, poultry, grains, nuts, Brussel sprouts, beans	Nucleic acid synthesis and red blood cell production	Megaloblastic anemia, oral ulcerations. Maternal low folate levels during pregnancy increase the chance of congenital birth defects, including fetal neural tube defects, and congenital heart defects, in addition to low birth weight, preterm labor, and delayed fetal growth	Kennedy (2016); Hanna et al. (2022)

(Continued)

Table 16.1 (Continued)

Vitamins	Scientific name	Sources	Function	Deficiency	References
Vitamin B12	Cyanocobalamin	Cobalamin is found in animal products and fortified foods.	Red blood cell production, neurologic function, and myelin synthesis. It serves as a cofactor in DNA and RNA synthesis as well as hormone, protein, and lipid synthesis and metabolism	Deficiency may present as megaloblastic anemia, fatigue, low appetite, and neuropsychiatric symptoms. If not treated, neuropsychiatric illness and irreversible neurologic damage occur	Hanna et al. (2022)
Vitamin C	Ascorbic acid	Citrus fruits, tomatoes, potatoes, red kiwifruit, strawberries, Brussel sprouts, broccoli and cantaloupe, green peppers	Antioxidant, role in immune function, prevention of cancer; treatment of cancer; heart diseases, common cold, age-related macular degeneration (AMD), and cataracts	Acute vitamin C deficiency leads to scurvy. Malaise, and inflammation of the gums. As vitamin C deficiency progresses, collagen synthesis becomes impaired and connective tissues become weakened, causing petechiae, ecchymoses, purpura, joint pain, poor wound healing, hyperkeratosis, and corkscrew hairs	Bendich and Langseth (1995); Camarena and Wang (2016)
Vitamin A	Retinoic acid	Dairy products, leafy green vegetables, eggs, orange and yellow vegetables, tomato and its products, liver, fish, fruits, and some vegetable oils	Vitamin A for normal vision, the immune system, reproduction, and growth, development, functioning of heart, lungs	Xerophthalmia. Chronic vitamin A deficiency associated with respiratory diseases, abnormal lung development; anemia, increased severity, and mortality risk of infections	Underwood and Arthur (1996); Wolf (1984)
Vitamin D	Calcitriol	Fatty fish, fish liver oil, and egg yolk, cod liver oil, Salmon, Swordfish Tuna fish	Help the body absorb and retain calcium and phosphorus hence for the development of bone, reduce cancer cell growth, help control infections, and reduce inflammation	Osteomalacia, heart disease, asthma, hay fever, rickets, weak bones	Cashman (2020)
Vitamin E	Tocopherol	Vegetable oils, green leafy vegetables, nuts, soybean, canola, corn, and other vegetable oils and seeds	Antioxidant enhances immune function and prevents clots from forming in heart arteries	Ataxia, skeletal myopathy, retinopathy, peripheral neuropathy, and reduce the functioning of the immune response, poor transmission of nerve impulses, muscle weakness, and retinal degeneration that leads to blindness and other eye-related diseases, cardiovascular disease, and reduction in cognitive learning	Kowdley et al. (1992); Yuan et al. (2020)
Vitamin K	Phylloquinone	Geen leafy vegetables kale, spinach, broccoli, sprouts, cabbage, lettuces, soybean, and canola oil	Vitamin K for blood clotting and the building of bones, liver, brain, heart, and pancreas	Bleeding, hemorrhaging, osteopenia, or osteoporosis	Suttie (1992); Weber (2001)

meet increasing worldwide demand by leveraging the metabolic capabilities of microorganisms such as bacteria and yeast. Microbial factories provide a viable approach for producing high-quality multivitamins while resolving the limitations associated with chemical synthesis and traditional manufacturing procedures (Ko et al. 2020; Rusu et al. 2023).

16.2.1 Role of Microbial Factories in Vitamin Synthesis

Microbial factories use biotechnology and genetic engineering to improve the ability of microorganisms like bacteria, yeast, and filamentous fungi to biosynthesize vitamins. The few key innovations in this field include metabolic engineering, directed evolution, fermentation optimization, and synthetic biology (Ko et al. 2020). Scientists can manipulate the genetic code of certain microorganisms to enhance their natural ability to produce vitamins. Microorganisms can be genetically modified to improve vitamin yield and simultaneously reduce production costs. The directed evolution strategy includes intentionally advancing the evolutionary process by selecting mutations that maximize bacteria's vitamin production. It frequently produces more efficient enzyme catalysts for vitamin production. Optimizing fermentation processes such as temperature, pH, aeration, and nutrient delivery can be customized to further maximize microbial production, resulting in increased vitamin production (Singh et al. 2017). Another major strategy is the development of novel biosynthetic pathways utilizing synthetic biology tools, which enables the production of vitamins that are difficult to synthesize chemically or are not naturally generated in large numbers by microbes. Advances in bioreactor design and process engineering have enabled industrial-scale microbial vitamin production, ensuring uniformity and quality control on a massive scale. Modern biotechnology is being used to improve the yields and efficiency of vitamin B production (Acevedo-Rocha et al. 2019). Furthermore, advancement in nanotechnology strives at enhancing agricultural bioproducts, which potentially include microbially manufactured vitamins. Together, these advancements pave the path for more sustainable, efficient, and cost-effective methods of generating multivitamins in microbial factories, potentially leading to increased accessibility and new forms of supplements and fortified foods (Malini et al. 2022).

16.2.2 Advantages of Using Microbial Factories

Using microbial factories to manufacture multivitamins has various advantages. The most significant advantage of these factories is that they provide a cost-effective and scalable manufacturing process, enabling large-scale production to fulfill the increasing demand for multivitamins. Large-scale fermenters can generate considerable amounts of vitamins in relatively modest facilities compared to the land and resources required for traditional agriculture-based manufacturing (Singh et al. 2017). Secondly, microbial factories may manufacture multivitamins that have relatively pure product quality and dependability as compared to chemical synthesis. Furthermore, the utilization of microbial factories permits the production of vitamins that would be otherwise difficult or impractical to extract and purify using traditional methods. Microorganisms are more efficient in producing vitamins than plants or animals, reducing the time and cost from production to market. Microbial technologies reduce production costs by allowing the use of waste materials as substrates, lowering raw material costs while also handling trash in an efficient manner. Advances in genetic engineering tools and techniques can help to increase yields and efficiency. Further, microbial manufacturing is frequently more sustainable than traditional chemical synthesis or extraction from natural sources since it can use renewable resources and work under regulated circumstances with little environmental impact (Adrio and Demain 2014). Microbial methods are highly controllable, resulting in consistent product purity and batch quality. This is critical for meeting the strict regulatory requirements for dietary supplements. Some vitamins, like vitamin B12, are difficult to manufacture chemically, but microorganisms may be designed to create them naturally, making microbial production the preferred approach for such complicated vitamins and chemical products (Fang et al. 2017). Furthermore, microbial production can be customized to develop multivitamin formulations or improve certain qualities, such as bioavailability. Microbial production may generate vitamins without using animal products, making it an important resource for the vegetarian and vegan sectors, where demand for animal-free supplements is increasing. Unlike crop-based production systems, microbial vitamin manufacturing does not require significant tracts of arable land, which contributes to the preservation of natural ecosystems. These advantages explain why there is a growing interest in using microbial platforms to produce multivitamins and other nutritional supplements. The adaptability and effectiveness of microbial factories make them a significant asset in the multivitamin sector (Acevedo-Rocha et al. 2019; Gohil et al. 2021; Wang et al. 2021).

16.2.3 Types of Microorganisms Used in Multivitamin Production

There are four types of microbes capable of producing multivitamins. These include yeast and filamentous fungus, bacteria being the most frequent. Each of these microbes has unique advantages and disadvantages, but they all require similar nutritional inputs. As a result, the simultaneous activity of microorganisms in the same production fermenter might generate a complicated set of interactions. Various types of microorganisms are utilized in multivitamin production within microbial factories. Bacteria, such as *Escherichia coli* and *Bacillus subtilis* (Su et al. 2020), are commonly used due to their well-characterized genetic systems and ease of manipulation. Yeasts, such as *Saccharomyces cerevisiae* and *Pichia pastoris*, are also used for their robustness and capacity to generate complex vitamins. Furthermore, filamentous fungi such as *Aspergillus oryzae* and *Trichoderma reesei* have been used for their ability to produce certain vitamins. Each species of microbe has distinct benefits and qualities that can be used to maximize multivitamin production in microbial factories. Here are some common types of microorganisms used in the production of various vitamins as summarized in Table 16.2. Each of these microorganisms brings different metabolic pathways, which can be naturally occurring or genetically engineered to optimize the production of specific vitamins. Advances in biotechnology have broadened the use of these microorganisms, enabling more efficient and cost-effective ways to produce multivitamins on an industrial scale (Acevedo-Rocha et al. 2019; Averianova et al. 2020; Gohil et al. 2021; Rusu et al. 2023).

16.3 Innovations in Multivitamin Production

Innovations in multivitamin production have revolutionized the way these essential nutrients are synthesized. This section explores some of the key advancements in the field, ranging from genetic engineering techniques to optimization of fermentation processes, novel approaches for vitamin extraction and purification, and advances in encapsulation and delivery systems. These innovations have greatly enhanced the efficiency, quality, and sustainability of multivitamin production, enabling the development of more effective and accessible nutritional supplements.

16.3.1 Genetic Engineering Techniques for Enhanced Vitamin Synthesis

Genetic engineering techniques have paved the way for enhanced vitamin synthesis in microbial factories. By manipulating the genetic makeup of microorganisms, scientists can optimize their ability to produce specific vitamins. This involves introducing or modifying genes responsible for the synthesis of target vitamins, enabling greater production yields and improving overall

Table 16.2 Summary of microbes used for the production of vitamins.

Vitamins	Microorganism	Production method	Advantages	References
Vitamin A	*Saccharomyces cerevisiae*	Production of retinol is done efficiently from xylose by introducing human *RDH12* and NADH oxidase into *S. cerevisiae*	Genetic engineered for the production of vitamin A	Lee et al. (2022)
	Escherichia coli	Recombinant protein is usually produced by refolding inclusion bodies in *E. coli play role in the production of vitamin A*	*E. coli* cell is genetically engineered for the production of vitamin A	Wysocka-Kapcinska et al. (2010)
Vitamin B1	*Bacillus subtilis TH95*	Mutation of gene encoding thiamine pyrophosphate kinase activity (*thiN*) and thiamine-related transport protein (*ykoJ*)	*B. subtilis* may be genetically engineered, and researchers can change its genetic composition to produce more vitamin B1	Schyns et al. (2005)
	E. coli	TPP biosensor (plasmid Ptpp_Bios); overexpression of native *thiFSGH; thiC; thiE*; and *thiD;* genetic–metabolic coupling	Genetically modified, fast growth rate	Cardinale et al. (2017)
	Aspergillus oryzae	Overexpression of *thiP, thiA,* and *nmtA*	*A. oryzae* can be engineered or genetic altered to maximize its potential	Tokui et al. (2011)
Vitamin B2	*B. subtilis*	Decrease the activity of flavinase RibCF activity; overexpression of riboflavin biosynthetic genes; improved the de *novo* purine synthesis and pentose supply	*B. subtilis* genetically and metabolically engineered for the production of vitamin B6	Abbas and Sibirny (2011)
	Ashbya gossypii	Introduced the *icl* gene; overexpression of *gly1, prs2, prs4,* and *prs3* genes: knocked out *vma4, shm2,* and *bas 1* genes	The overexpression of structural and regulatory genes helps in the synthesis of riboflavin	Schwechheimer et al. (2016); Averianova et al. (2020)
	Candida famata	Conventional mutagenesis by overexpression of *sef1* and *imh3*	*Candida famata* can genetically manipulate and able to tolerate harsh climate	Dmytruk et al. (2011)
Vitamin B3	*Yeast*	Knock out NR importer Nrt1 in the NR non-salvaging genotype *nrkl, urhl, pmpl* (strain PABO38).	Genetic manipulation and fermentation capability	Belenky et al. (2011)
	E. coli	Expressing *Rhodochrous* nitrile hydratase	The total production of vitamin B3 can be increased by metabolic engineering	Wang et al. (2017)

(*Continued*)

Table 16.2 (Continued)

Vitamins	Microorganism	Production method	Advantages	References
Vitamin B5	*Corynebacterium glutamicum*	Deletion of *ilvA* gene and overexpression of *ilvBNCD* and *panBC* genes	*Corynebacterium glutamicum* may develop by effectively using a variety of carbon sources. Because of its metabolic flexibility, fermentation may be carried out using a variety of feedstocks, which might lower manufacturing costs	Leonardi and Jackowski (2007)
	B. subtilis	Overexpression of *ilvBHCD* and *panBNCD*; overexpression of *SerA* and *GlyA* of the enzymes of the glycine cleavage cycle	*B. subtilis* grows well and can be grown on a range of low-cost carbon sources	Hohmann et al. (2017)
Vitamin B6	*E. coli*	Overexpression of native *Epd*, *PdxJ*, and *Dxs* enzymes	Fast growth rate, genetically engineered	Liu et al. (2023)
	Sinorhizobium meliloti IFO14782	Overexpression of *E. coli*, *Epd*, and native *PdxJ* enzyme	Genetically engineered for the production of biotin	Commichau et al. (2014)
	B. subtilis	Overexpression of *E. coli*, PdxA, and *S. meliloti* PdxJ enzymes	Genetically manipulated and shows fast growth rate	Commichau et al. (2014)
Vitamin B7	*Agrobacterium/ Rhizobium* HK4	Overexpression of a strong biotin operon from *E. coli*; use of a powerful artificial tac promoter and introduction of a modified RBS in front of BioB	*Agrobacterium/Rhizobium* HK4 can be metabolically engineered	Streit and Entcheva (2003)
	E. coli	Overexpression of native biotin operon from a high-copy number plasmid	Genetically modified, fast growth rate	Ifuku et al. (1995)
	B. subtilis	Overexpression of native biotin operon and selection on S-2-aminoethyl-L-cysteine	*B. subtilis* is metabolically versatile	Van Arsdell et al. (2005)
Vitamin B9	*Ashbya gossypii* (ATCC 10895)	Overexpression of FOL genes and deletion of *AgMEY7*; deletion of *AgADE12* and *AgRIB1* at the same time	*Ashbya gossypii* are considered sustainable sources for vitamin production as they are possible to be grown using sustainable resources	Li et al. (2008)
Vitamin B12	*S. meliloti (MC5-20)*	High-throughput screening of mutants using riboswitch ARTP irradiation was used to induce random mutations; deletion of *cobl*; overexpression of *hemE*.	The *S. meliloti* bluB mutant is unable to grow in minimal media and fails to establish a symbiosis with alfalfa, and these defects can be rescued by the addition of vitamin B12	Cai et al. (2018); Campbell et al. (2006)

	Pseudomonas denitrificans	Random mutagenesis and genetic engineering; overexpression of *cobF-cobM* gene cluster and *cogA* and *cobE* genes; optimize the best PH range; optimize promoters	*P. denitrificans* is highly productive over anaerobic vitamin B12-producing bacteria; aerobic *P. denitrificans* is commonly utilized for industrial and commercial vitamin B12 fermentation	Li et al. (2008); Xia et al. (2015)
	E. coli	Heterologously expressed the hemO, hemB, hemC, and hemD genes; optimizing of fermentation conditions	Genetically and metabolically engineered	Fang et al. (2018)
	Propionibacterium shermanii	Overexpressing of biosynthetic genes	They have the capacity to grow and synthesize metabolites on substrates containing different industrial waste products, which considerably elevates the economic profitability of biotechnological processes	Marie Sych et al. (2016); Piwowarek et al. (2018)
Vitamin C	*S. cerevisiae and Zygosaccharomyces bailii*	Overexpressing the endogenous D-arabinono-1, 4-lactone oxidase and l-galactose dehydrogenase (overexpression of *lgdh* and *alo1*)	*S. cerevisiae and Zygosaccharomyces bailii* can be metabolically engineered for the production of vitamin C	Sauer et al. (2004)
	Ketogulonicigenium vulgare DSM 4025	Oxidation and lactonization	Metabolically engineered for the production of *K. vulgare*	Sugisawa et al. (2005)
	Xanthomonas campestris 2286	Lactonization under oxidative stress; direct synthesis of glucose (carbohydrate source) induced by free radicals (HOCL treatment)	*X. campestris* is not a natural producer of vitamin C, and it is genetically engineered	Rao et al. (2003)
	Gluconobacter oxydans and k. vulgare and Bacillus endophyticus	Cell–cell interaction; one-step 2-KGA fermentation	*Gluconobacter oxydans* ability is to oxidize various sugars and alcohols which help in production of vitamin C *K. vulgare* plays intermediate role in the industrial synthesis of vitamin C The ability of *Bacillus endophyticus* to synthesize gluconic acid—a necessary precursor for the production of vitamin C	SUGISAWA et al. (2005)
Vitamin D	*S. cerevisiae*	UV-B irradiation	*S. cerevisiae* was treated with UV-B irradiation to induce alteration of yeast sterol to ergocalciferol for using as food supplement	Mcdonnell et al. (1989)

(*Continued*)

Table 16.2 (Continued)

Vitamins	Microorganism	Production method	Advantages	References
Vitamin E	*Euglena gracilis*	Fermentative production from glucose, tocopherol synthesis pathway	Although *Euglena gracilis* does not synthesize vitamin E on its own, it has been investigated for the possibility of accumulating intermediate or precursor molecules that are part of the route leading to the production of tocopherol	Takeyama et al. (1997)
	E. coli	Recombinant tocochromanol biosynthesis, the genes encoding hydroxyphenylpyruvate dioxygenase (*hpd*), geranylgeranyl pyrophosphate synthase (*crtE*), geranylgeranyl pyrophosphate reductase (*ggh*), homogentisate phytyltransferase (*hpt*), and tocopherol-cyclase (*cyc*) were cloned in a stepwise fashion and expressed in *E. coli*	Genetically engineered or modified	Albermann et al. (2008)
Vitamin K	*B. subtilis*	Fed-batch fermentation	*B. subtilis* is a GRAS (generally recognized as safe) host microorganism widely used for the production of food ingredients	Kang et al. (2022)
	Lactic acid bacteria	The menaquinone was extracted from lyophilized cells from the growth medium and from soymilk culture using a mixture of chloroform and methanol	LAB strains may be developed as a probiotic strain producing vitamin K in human gut	

efficiency. Genetic engineering offers the potential to enhance the nutritional profile of multivitamins, ensuring a more comprehensive and diverse range of essential vitamins in the final product.

16.3.2 Optimization of Fermentation Processes

In addition to genetic alterations, culture medium optimization and modifications to the bioprocessing environment have also been used to increase vitamins production using microbes. The optimization of fermentation processes has significantly contributed to the production of multivitamins in microbial factories. Through careful control and modification of various factors such as temperature, pH, nutrient availability, and fermentation duration, scientists can maximize vitamin production. Culture optimization refers to the process of improving and enhancing the nutrient-rich medium composition that is used in laboratory settings to cultivate and support a variety of cells, microbes, or tissues. But the way that microorganisms make vitamins is greatly influenced by the environment. It is therefore also a goal to create an environment that offers the best circumstances for the viability, development, and productivity of the organisms or cells being cultivated. In this procedure, a number of variables (including pH, temperature, nutritional concentrations, carbon sources, amino acids, vitamins, and growth factors) are meticulously adjusted and fine-tuned to support the intended cellular or microbial development (Khattab et al. 2016). This optimization ensures higher yields, shorter production cycles, and improved overall efficiency. By fine-tuning the fermentation conditions, microbial factories can now produce multivitamins in a more cost-effective and environmentally sustainable manner. A variety of environmental elements, including carbon sources, phosphorus, and nitrogen, have an impact on the fermentation process that produces secondary metabolite. Optimizing culture media helps the researchers to enlarge the productivity of their cultures, efficiency of the biotechnological processes can be maximized, and the production of desired products such as pharmaceuticals, enzymes, or biofuels can be enhanced. It is a fascinating area where science and creativity come together to develop the best possible environment that can be achieved for the organisms being cultured. This optimization is necessary because the composition of the culture media copiously influences the growth rate, yield, and quality of the cultured organisms or cells (Elibol 2004). Vitamin synthesis was significantly impacted by the media composition, which also had an impact on the pH stability of cultures. A feeding method employing betaine as a methyl donor and glucose as a carbon source was devised to better manage pH. When applied to 120 m^3 bioreactor cultures, this feeding strategy was found to be helpful for vitamin synthesis (Wang et al. 2010) The exploration for optimization of culture conditions is usually performed using non-statistics, using only single-variable approach and statistical experimental framework approaches. As the former one is exceedingly strenuous and resource draining than the statistical methods, therefore, techniques of data-driven experimental design that particularly include Plackett–Burman design (PBD) and response surface methodology (RSM) are extensively utilized to select the pertinent variable and acquire the pinnacle levels, sequentially. As compared to the conventional practice of optimization of single factor, the implementation of these data-driven modeling techniques of experimental optimization of media can lead to the enhanced production of the products, diminished process fluctuation, reduced time, and all-encompassing costs. To select the influential feature among the constituents of an intricate medium, many researchers have applied the Plackett–Burman design (Korayem et al. 2015).

16.3.3 Novel Approaches for Vitamin Extraction and Purification

Novel approaches for vitamin extraction and purification have emerged as valuable innovations in multivitamin production. These techniques aim to effectively and selectively isolate vitamins from the microbial cultures, ensuring high purity and quality of the final product. Various methods, such as membrane filtration, liquid–liquid extraction, and chromatography, have been developed to achieve efficient and selective separation of vitamins. These advancements enable the production of highly concentrated and bioavailable multivitamins, enhancing their efficacy and ease of consumption.

16.3.4 Advances in Encapsulation and Delivery Systems

Advances in encapsulation and delivery systems have played a crucial role in improving the effectiveness and stability of multivitamins. Encapsulation techniques involve the protection of vitamins within microspheres, nanoparticles, or coatings to prevent degradation and enhance their bioavailability. Additionally, innovative delivery systems such as time-release capsules, liposomes, and nanocarriers have been developed to ensure controlled and targeted release of the vitamins in the body. These advances enhance the absorption and utilization of multivitamins, maximizing their health benefits for consumers.

16.4 Future Prospects

As the field of microbial factories for multivitamin manufacture advances, new applications and challenges emerge. Multivitamins derived from microbial factories show promise for usage in animal feed, medications, and food fortification, among other applications. Microbial factories can produce multivitamins that could be utilized for animal feed, medications, and food fortification, among other uses. However, regulatory considerations and safety concerns must be addressed to ensure that these items are of high quality and safe for use. Simultaneously, the sustainability and scalability of microbial vitamin production must be carefully assessed to ensure long-term viability and low environmental effect.

References

Abbas, C.A. and Sibirny, A.A. (2011). Genetic control of biosynthesis and transport of riboflavin and flavin nucleotides and construction of robust biotechnological producers. *Microbiology and Molecular Biology Reviews* 75 (2): 321–360. https://doi.org/10.1128/MMBR.00030-10.

Acevedo-Rocha, C.G., Gronenberg, L.S., Mack, M. et al. (2019). Microbial cell factories for the sustainable manufacturing of B vitamins. *Current Opinion in Biotechnology* 56: 18–29. https://doi.org/10.1016/J.COPBIO.2018.07.006.

Adrio, J.L. and Demain, A.L. (2014). Microbial enzymes: tools for biotechnological processes. *Biomolecules* 4 (1): 117. https://doi.org/10.3390/BIOM4010117.

Aguilera-Méndez, A., Boone-Villa, D., Nieto-Aguilar, R. et al. (2021). Role of vitamins in the metabolic syndrome and cardiovascular disease. *Pflügers Archiv - European Journal of Physiology* 474 (1): 117–140. https://doi.org/10.1007/S00424-021-02619-X.

Albermann, C., Ghanegaonkar, S., Lemuth, K. et al. (2008). Biosynthesis of the vitamin E compound δ-tocotrienol in recombinant *Escherichia coli* cells. *ChemBioChem* 9 (15): 2524–2533. https://doi.org/10.1002/cbic.200800242.

Averianova, L.A., Balabanova, L.A., Son, O.M. et al. (2020). Production of vitamin B2 (riboflavin) by microorganisms: an overview. *Frontiers in Bioengineering and Biotechnology* 8: 570828. https://doi.org/10.3389/FBIOE.2020.570828.

Belenky, P., Stebbins, R., Bogan, K.L. et al. (2011). Nrt1 and Tna1-independent export of NAD^+ precursor vitamins promotes NAD^+ homeostasis and allows engineering of vitamin production. *PLoS ONE* 6 (5): e19710. https://doi.org/10.1371/journal.pone.0019710.

Bendich, A. and Langseth, L. (1995). The health effects of vitamin C supplementation: a review. *Journal of the American College of Nutrition* 14 (2): 124–136. https://doi.org/10.1080/07315724.1995.10718484.

Cai, Y., Xia, M., Dong, H. et al. (2018). Engineering a vitamin B12 high-throughput screening system by riboswitch sensor in *Sinorhizobium meliloti*. *BMC Biotechnology* 18 (1): 27. https://doi.org/10.1186/s12896-018-0441-2.

Camarena, V. and Wang, G. (2016). The epigenetic role of vitamin C in health and disease. *Cellular and Molecular Life Sciences* 73 (8): 1645–1658. https://doi.org/10.1007/s00018-016-2145-x.

Campbell, G.R.O., Taga, M.E., Mistry, K. et al. (2006). *Sinorhizobium meliloti bluB* is necessary for production of 5,6-dimethylbenzimidazole, the lower ligand of B_{12}. *Proceedings of the National Academy of Sciences* 103 (12): 4634–4639. https://doi.org/10.1073/pnas.0509384103.

Cardinale, S., Tueros, F.G., and Sommer, M.O.A. (2017). Genetic-metabolic coupling for targeted metabolic engineering. *Cell Reports* 20 (5): 1029–1037. https://doi.org/10.1016/j.celrep.2017.07.015.

Cashman, K.D. (2020). Vitamin D deficiency: defining, prevalence, causes, and strategies of addressing. *Calcified Tissue International* 106 (1): 14–29. https://doi.org/10.1007/s00223-019-00559-4.

Commichau, F.M., Alzinger, A., Sande, R. et al. (2014). Overexpression of a non-native deoxyxylulose-dependent vitamin B6 pathway in *Bacillus subtilis* for the production of pyridoxine. *Metabolic Engineering* 25: 38–49. https://doi.org/10.1016/j.ymben.2014.06.007.

Dmytruk, K.V., Yatsyshyn, V.Y., Sybirna, N.O. et al. (2011). Metabolic engineering and classic selection of the yeast Candida famata (Candida flareri) for construction of strains with enhanced riboflavin production. *Metabolic Engineering* 13 (1): 82–88. https://doi.org/10.1016/j.ymben.2010.10.005.

Du, J., Shao, Z., and Zhao, H. (2011). Engineering microbial factories for synthesis of value-added products. *Journal of Industrial Microbiology & Biotechnology* 38 (8): 873. https://doi.org/10.1007/S10295-011-0970-3.

Elibol, M. (2004). Optimization of medium composition for actinorhodin production by *Streptomyces coelicolor* A3(2) with response surface methodology. *Process Biochemistry* 39 (9): 1057–1062. https://doi.org/10.1016/S0032-9592(03)00232-2.

Fang, H., Kang, J., and Zhang, D. (2017). Microbial production of vitamin B12: a review and future perspectives. *Microbial Cell Factories* 16 (1): 1–14. https://doi.org/10.1186/S12934-017-0631-Y/METRICS.

Fang, H., Li, D., Kang, J. et al. (2018). Metabolic engineering of *Escherichia coli* for de novo biosynthesis of vitamin B12. *Nature Communications* 9 (1): 4917. https://doi.org/10.1038/s41467-018-07412-6.

Gohil, N., Bhattacharjee, G., and Singh, V. (2021). An introduction to microbial cell factories for production of biomolecules. *Microbial Cell Factories Engineering for Production of Biomolecules* 1–19. https://doi.org/10.1016/B978-0-12-821477-0.00021-0.

Hanna, M., Jaqua, E., Nguyen, V., and Clay, J. (2022). B vitamins: functions and uses in medicine. *The Permanente Journal* 26 (2): 89–97. https://doi.org/10.7812/TPP/21.204.

Hohmann, H., van Dijl, J.M., Krishnappa, L., and Prágai, Z. (2017). Host organisms: *Bacillus subtilis*. In: *Industrial Biotechnology: Microorganisms* (ed. C. Wittmann and J.C. Liao), 221–297. Wiley https://doi.org/10.1002/9783527807796.ch7.

Ifuku, O., Koga, N., Haze, S. et al. (1995). Molecular analysis of growth inhibition caused by overexpression of the biotin operon in *Escherichia coli*. *Bioscience, Biotechnology, and Biochemistry* 59 (2): 184–189. https://doi.org/10.1271/bbb.59.184.

Kang, M.-J., Baek, K.-R., Lee, Y.-R. et al. (2022). Production of vitamin K by wild-type and engineered microorganisms. *Microorganisms* 10 (3): 554. https://doi.org/10.3390/microorganisms10030554.

Kennedy, D. (2016). B vitamins and the brain: mechanisms, dose and efficacy—a review. *Nutrients* 8 (2): 68. https://doi.org/10.3390/nu8020068.

Khattab, A.I., Babiker, E.H., and Saeed, H.A. (2016). Streptomyces: isolation, optimization of culture conditions and extraction of secondary metabolites. *International Current Pharmaceutical Journal* 5 (3): 27–32. https://doi.org/10.3329/ICPJ.V5I3.26695.

Ko, Y.S., Kim, J.W., Lee, J.A. et al. (2020). Tools and strategies of systems metabolic engineering for the development of microbial cell factories for chemical production. *Chemical Society Reviews* 49 (14): 4615–4636. https://doi.org/10.1039/D0CS00155D.

Korayem, A.S., Abdelhafez, A.A., Zaki, M.M., and Saleh, E.A. (2015). Optimization of biosurfactant production by *Streptomyces* isolated from Egyptian arid soil using Plackett–Burman design. *Annals of Agricultural Sciences* 60 (2): 209–217. https://doi.org/10.1016/J.AOAS.2015.09.001.

Kowdley, K.V., Mason, J.B., Meydani, S.N. et al. (1992). Vitamin E deficiency and impaired cellular immunity related to intestinal fat malabsorption. *Gastroenterology* 102 (6): 2139–2142. https://doi.org/10.1016/0016-5085(92)90344-X.

Lee, Y., Kim, C., Sun, L. et al. (2022). Selective production of retinol by engineered *Saccharomyces cerevisiae* through the expression of retinol dehydrogenase. *Biotechnology and Bioengineering* 119 (2): 399–410. https://doi.org/10.1002/bit.28004.

Leonardi, R. and Jackowski, S. (2007). Biosynthesis of pantothenic acid and coenzyme A. *EcoSal Plus* 2 (2): https://doi.org/10.1128/ecosalplus.3.6.3.4.

Li, K.-T., Liu, D.-H., Chu, J. et al. (2008). An effective and simplified pH-stat control strategy for the industrial fermentation of vitamin B_{12} by *Pseudomonas denitrificans*. *Bioprocess and Biosystems Engineering* 31 (6): 605–610. https://doi.org/10.1007/s00449-008-0209-5.

Liu, L., Li, J., Gai, Y. et al. (2023). Protein engineering and iterative multimodule optimization for vitamin B_6 production in *Escherichia coli*. *Nature Communications* 14 (1): 5304. https://doi.org/10.1038/s41467-023-40928-0.

Malini, S., Raj, K., Madhumathy, S. et al. (2022). Bioinspired advances in nanomaterials for sustainable agriculture. *Journal of Nanomaterials* 2022: https://doi.org/10.1155/2022/8926133.

Marie Sych, J., Lacroix, C., and Stevens, M.J.A. (2016). Vitamin B_{12}– physiology, production and application. In: *Industrial Biotechnology of Vitamins, Biopigments, and Antioxidants* (ed. E.J. Vandamme and J.L. Revuelta), 129–159. Wiley https://doi.org/10.1002/9783527681754.ch6.

Mcdonnell, D.P., Pike, J.W., Drutz, D.J. et al. (1989). Reconstitution of the vitamin D-responsive osteocalcin transcription unit in *Saccharomyces cerevisiae*. *Molecular and Cellular Biology* 9 (8): 3517–3523. https://doi.org/10.1128/mcb.9.8.3517-3523.1989.

Montaño López, J., Duran, L., and Avalos, J.L. (2022). Physiological limitations and opportunities in microbial metabolic engineering. *Nat Rev Microbiol* 20: 35–48. https://doi.org/10.1038/s41579-021-00600-0.

Piwowarek, K., Lipińska, E., Hać-Szymańczuk, E. et al. (2018). *Propionibacterium* spp.—source of propionic acid, vitamin B12, and other metabolites important for the industry. *Applied Microbiology and Biotechnology* 102 (2): 515–538. https://doi.org/10.1007/s00253-017-8616-7.

Rao, Y.M., Suresh, A.K., and Suraishkumar, G.K. (2003). Free radical aspects of *Xanthomonas campestris* cultivation with liquid phase oxygen supply strategy. *Process Biochemistry* 38 (9): 1301–1310. https://doi.org/10.1016/S0032-9592(02)00328-X.

Rusu, A.V., Trif, M., and Rocha, J.M. (2023). Microbial Secondary Metabolites via Fermentation Approaches for Dietary Supplementation Formulations. *Molecules* 2023, 28: 6020. https://doi.org/10.3390/MOLECULES28166020.

Sauer, M., Branduardi, P., Valli, M., and Porro, D. (2004). Production of l-ascorbic acid by metabolically engineered *Saccharomyces cerevisiae* and *Zygosaccharomyces bailii*. *Applied and Environmental Microbiology* 70 (10): 6086–6091. https://doi.org/10.1128/AEM.70.10.6086-6091.2004.

Schwechheimer, S.K., Park, E.Y., Revuelta, J.L. et al. (2016). Biotechnology of riboflavin. *Applied Microbiology and Biotechnology* 100 (5): 2107–2119. https://doi.org/10.1007/s00253-015-7256-z.

Schyns, G., Potot, S., Geng, Y. et al. (2005). Isolation and characterization of new thiamine-deregulated mutants of *Bacillus subtilis*. *Journal of Bacteriology* 187 (23): 8127–8136. https://doi.org/10.1128/JB.187.23.8127-8136.2005.

Singh, A.K., Upadhyay, S.K., and Singh, S.P. (2023). *Microbial Bioreactors: An Introduction*. https://doi.org/10.1002/9781119874096.ch1.

Singh, S.P. and Upadhyay, S.K. (2021). *Bioprospecting of Microorganism-Based Industrial Molecules*. Wiley. https://doi.org/10.1002/9781119717317.

Singh, S.P. and Upadhyay, S.K. (2023). *Microbial Bioreactors for Industrial Molecules*. Wiley. https://doi.org/10.1002/9781119874096.

Singh, V., Haque, S., Niwas, R. et al. (2017). Strategies for fermentation medium optimization: an in-depth review. *Frontiers in Microbiology* 7 (JAN): 227613. https://doi.org/10.3389/FMICB.2016.02087/BIBTEX.

Sriram, K., Manzanares, W., and Joseph, K. (2012). Thiamine in nutrition therapy. *Nutrition in Clinical Practice* 27 (1): 41–50. https://doi.org/10.1177/0884533611426149.

Streit, W.R. and Entcheva, P. (2003). Biotin in microbes, the genes involved in its biosynthesis, its biochemical role and perspectives for biotechnological production. *Applied Microbiology and Biotechnology* 61 (1): 21–31. https://doi.org/10.1007/s00253-002-1186-2.

Su, Y., Liu, C., Fang, H., and Zhang, D. (2020). *Bacillus subtilis*: a universal cell factory for industry, agriculture, biomaterials and medicine. *Microbial Cell Factories* 19 (1): 1–12. https://doi.org/10.1186/S12934-020-01436-8.

Sugisawa, T., Miyazaki, T., and Hoshino, T. (2005). Microbial production of l-ascorbic acid from d-sorbitol, l-sorbose, l-gulose, and l-sorbosone by *Ketogulonicigenium vulgare* DSM 4025. *Bioscience, Biotechnology, and Biochemistry* 69 (3): 659–662. https://doi.org/10.1271/bbb.69.659.

Surana, K.R., Ahire, E.D., Patil, S.J. et al. (2023). Introduction to nutraceutical vitamins. In: *Vitamins as Nutraceuticals: Recent Advances and Applications* (ed. E.D. Ahire, R.K. Keservani, K.R. Surana, et al.), 1–34. Wiley https://doi.org/10.1002/9781394175543.CH1.

Suttie, J.W. (1992). Vitamin K and human nutrition. *Journal of the American Dietetic Association* 92 (5): 585–590. https://doi.org/10.1016/S0002-8223(21)00681-7.

Suwannasom, N., Kao, I., Pruß, A. et al. (2020). Riboflavin: the health benefits of a forgotten natural vitamin. *International Journal of Molecular Sciences* 21 (3): 950. https://doi.org/10.3390/ijms21030950.

Takeyama, H., Kanamaru, A., Yoshino, Y. et al. (1997). Production of antioxidant vitamins, β-carotene, vitamin C, and vitamin E, by two-step culture of *Euglena gracilis* Z. *Biotechnology and Bioengineering* 53 (2): 185–190. https://doi.org/10.1002/(SICI)1097-0290(19970120)53:2<185::AID-BIT8>3.0.CO;2-K.

Tokui, M., Kubodera, T., Gomi, K. et al. (2011). Construction of a thiamine pyrophosphate high-producing strain of Aspergillus oryzae by overexpression of three genes involved in thiamine biosynthesis. *Journal of Bioscience and Bioengineering* 111 (4): 388–390. https://doi.org/10.1016/j.jbiosc.2010.12.011.

Underwood, B.A. and Arthur, P. (1996). The contribution of vitamin A to public health. *The FASEB Journal* 10 (9): 1040–1048. https://doi.org/10.1096/fasebj.10.9.8801165.

Upadhyay, S.K. and Singh, S.P. (2021). *Bioprospecting of Plant Biodiversity for Industrial Molecules*. Wiley. https://doi.org/10.1002/9781119718017.

Upadhyay, S.K. and Singh, S.P. (2023). *Plants as Bioreactors for Industrial Molecules*. Wiley. https://doi.org/10.1002/9781119875116.

Upadhyay, S.K. and Singh, S.P. (2024). *Herbal Nutraceuticals: Products and Processes*. Wiley. https://doi.org/10.1002/9781394241576.

Van Arsdell, S.W., Perkins, J.B., Yocum, R.R. et al. (2005). Removing a bottleneck in the *Bacillus subtilis* biotin pathway: BioA utilizes lysine rather than S-adenosylmethionine as the amino donor in the KAPA-to-DAPA reaction. *Biotechnology and Bioengineering* 91 (1): 75–83. https://doi.org/10.1002/bit.20488.

Wang, Y., Liu, L., Jin, Z., and Zhang, D. (2021). Microbial cell factories for green production of vitamins. *Frontiers in Bioengineering and Biotechnology* 9: 661562. https://doi.org/10.3389/FBIOE.2021.661562/BIBTEX.

Wang, Z.J., Wang, H.Y., Li, Y.L. et al. (2010). Improved vitamin B12 production by step-wise reduction of oxygen uptake rate under dissolved oxygen limiting level during fermentation process. *Bioresource Technology* 101 (8): 2845–2852. https://doi.org/10.1016/J.BIORTECH.2009.10.048.

Wang, Z., Liu, Z., Cui, W., and Zhou, Z. (2017). Establishment of bioprocess for synthesis of nicotinamide by recombinant *Escherichia coli* expressing high-molecular-mass nitrile hydratase. *Applied Biochemistry and Biotechnology* 182 (4): 1458–1466. https://doi.org/10.1007/s12010-017-2410-y.

Weber, P. (2001). Vitamin K and bone health. *Nutrition* 17 (10): 880–887. https://doi.org/10.1016/S0899-9007(01)00709-2.

Wolf, G. (1984). Multiple functions of vitamin A. *Physiological Reviews* 64 (3): 873–937. https://doi.org/10.1152/physrev.1984.64.3.873.

Wysocka-Kapcinska, M., Campos-Sandoval, J.A., Pal, A., and Findlay, J.B.C. (2010). Expression and characterization of recombinant human retinol-binding protein in Pichia pastoris. *Protein Expression and Purification* 71 (1): 28–32. https://doi.org/10.1016/j.pep.2010.01.015.

Xia, W., Chen, W., Peng, W., and Li, K. (2015). Industrial vitamin B12 production by *Pseudomonas denitrificans* using maltose syrup and corn steep liquor as the cost-effective fermentation substrates. *Bioprocess and Biosystems Engineering* 38 (6): 1065–1073. https://doi.org/10.1007/s00449-014-1348-5.

Yuan, P., Cui, S., Liu, Y. et al. (2020). Metabolic engineering for the production of fat-soluble vitamins: advances and perspectives. *Applied Microbiology and Biotechnology* 104 (3): 935–951. https://doi.org/10.1007/s00253-019-10157-x.

17

An Overview of GABA Production by Microorganisms

Hend Altaib[1], *Mahmoud A. M. El-Nouby*[2,3], *and Yassien Badr*[4]

[1] *Department of Research and Development, Bacterial Vaccines Section, Middle East for Veterinary Vaccines (MEVAC) Company, El-Salhya El-Gdeda, 44813, EL-Sharkya, Egypt*
[2] *Department of Pesticide Chemistry and Technology, Faculty of Agriculture, Alexandria University, Alexandria, 21545, El-Shatby, Egypt*
[3] *Unit of Analysis and Scientific Services (Central Lab), Faculty of Agriculture, Alexandria University, Alexandria, 21545, El-Shatby, Egypt*
[4] *Department of Infectious Diseases and Epidemics, Faculty of Veterinary Medicine, Damanhour University, Damanhour, El-Beheira, 22511, Egypt*

17.1 Introduction

Gamma-aminobutyric acid (GABA) is a four-carbon non-protein amino acid which is widely distributed among prokaryotes as well as eukaryotes. In animals, GABA acts as the primary inhibitory neurotransmitter (Ochoa-de la Paz et al. 2021; Koh et al. 2023). In plants, it is critical for growth and development (Khan et al. 2021). In microorganisms, it is important for growth and acid tolerance (Otaru et al. 2021). Additionally, GABA serves as a communication molecule between bacteria and their host as well as among bacteria themselves (Quillin et al. 2021).

Microorganisms have emerged as efficient producers of GABA, utilizing their enzymatic pathways to convert glutamate into GABA through the action of glutamate decarboxylase (GAD) (Cui et al. 2020). This process is not only of biological interest but also of commercial significance, as it offers a sustainable and eco-friendly alternative to chemical synthesis. Several bacterial species can synthesize GABA. For example, members of the genus *Bacillus* have been frequently reported to produce GABA, with the presence of GAD genes and a glutamate/GABA antiporter being key determinants of this capability (Sun et al. 2021). *Corynebacterium glutamicum* is another notable bacterium, recognized for its endogenous production of L-glutamate, which is a precursor for GABA synthesis (Son et al. 2022).

Moreover, certain strains of *Streptomyces*, such as *Streptomyces bacillaris* and *Streptomyces cinereus*, have been observed to increase GABA content in fermented tea, highlighting the diverse applications of microbial GABA production (Jeng et al. 2007). Human intestinal microbiota, including genera like *Bacteroides*, *Bifidobacteria*, and various lactic acid bacteria (LAB), also contribute to GABA levels in the gut through the metabolism of dietary monosodium glutamate (MSG) (Otaru et al. 2021; Tamés et al. 2024).

Several methods were investigated for GABA production including chemical synthesis, plant enrichment, enzymatic process, and microbial production. Microbial production stands out from

Microbial Nutraceuticals: Products and Processes, First Edition. Edited by Sudhir Pratap Singh and Santosh Kumar Upadhyay.

other methods for its use of renewable substrates, reduced pollutant output, mild reaction conditions, and compliance with green standards in the food and pharma sectors.

The current chapter emphasizes the significance of GABA and its diverse applications. It explores the various GABA-producing microbial strains and the metabolic pathways they utilize. A key focus is placed on innovative strategies for enhancing GABA production, such as fermentation process optimization, whole-cell conversion, and metabolic engineering. Additionally, GABA producer model organisms were discussed, aiming to guide future research for more effective microbial production of GABA.

17.2 Chemical Structure and Biosynthesis of GABA

GABA, a non-protein amino acid with the molecular formula of $C_4H_9NO_2$ and a molecular weight of 103.12, is broadly synthesized by a variety of organisms. A summary of the sources and physical and chemical properties of GABA is summarized in Table 17.1. The GABA shunt is a crucial metabolic pathway involved in the production of GABA. This pathway is conserved in prokaryotes and eukaryotes. It comprises three enzymatic reactions and is often termed a "shunt" as it circumvents two stages of the tricarboxylic acid (TCA) cycle. The three enzymatic reactions were catalyzed by cytosolic enzyme GAD (E.C. 4.1.1.15) and mitochondrial enzymes GABA transaminase (GABA-T; E.C. 2.6.1.19) and succinate semialdehyde dehydrogenase (SSADH; E.C. 1.2.2.16) (Rashmi et al. 2018; Sarasa et al. 2020). The initial step in the GABA shunt is the production of glutamate from alpha-ketoglutarate through a transamination reaction catalyzed by glutamate dehydrogenase. The next step is the irreversible decarboxylation of glutamate to GABA, catalyzed by GAD, which consumes a proton and releases CO2 (Figure 17.1). GAD is present in various organisms across all kingdoms of life and requires PLP as a cofactor. In humans, two isoforms of GAD, GAD67 and GAD65, are expressed. GAD67 synthesizes GABA in the brain, while GAD65 is the major GAD isoform in pancreatic cells (Kim et al. 1993; Ueno 2000). The third enzyme involved in the GABA shunt is GABA transaminase (GABA-T), which catabolizes GABA to produce succinate semialdehyde (SSA). The final step is the conversion of SSA to succinate by succinate semialdehyde dehydrogenase (SSADH) allowing it to enter the TCA cycle. Interestingly, the GABA shunt operates as a closed-loop system that generates and preserves GABA. The transamination of α-ketoglutarate, resulting from glucose metabolism in the Krebs cycle, by GABA-T results in l-glutamic acid, indicating the initiation of the GABA shunt (Figure 17.1).

17.3 Physiological and Biological Functions of GABA for Microorganisms and Carrier Hosts

The physiological and biological functions of GABA extend beyond its role as a neurotransmitter (Mazzoli and Pessione 2016). The synthesis of GABA in a variety of organisms, including microorganisms, leads to a multitude of applications and potential therapeutic benefits. The roles of GABA in different host systems are summarized in Figure 17.2.

17.3.1 Role in Microorganisms

GABA is crucial in bacterial metabolism, especially in the Krebs cycle, with its role varying across different bacterial strains. In bacteria, GABA's primary function is to aid in acid tolerance, a key characteristic of potential probiotic strains (Ayyash et al. 2021). This is facilitated by glutamate-antiporter reactions where glutamate is introduced into the cell and undergoes

Table 17.1 γ-Amino butyric acid, source, physical and chemical properties.

Item	Description	References
Sources	Naturally found in plants and microorganisms; also synthesized in the brain	Sarasa et al. (2020)
Preparation	Typically extracted from fermented products or produced synthetically from glutamate decarboxylation	Sarasa et al. (2020)
Biological properties	Acts as the main inhibitory neurotransmitter in the central nervous system	Liu, Li, Liu, Ko, and Kim (2022)
Physical and chemical properties		
Structure	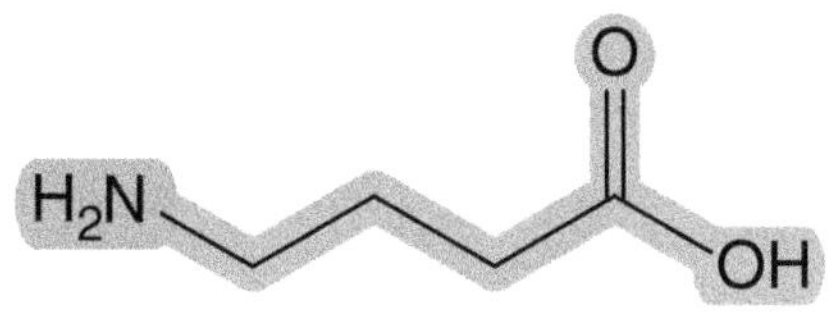	Han et al. (2023)
Chemical formula	$C_4H_9NO_2$	
Form	White, powdery solid; savory, meat-like aroma	Singh et al. (2023)
Molecular weight	103.12 g/mol	Heli et al. (2022)
Composition	Gamma-amino group, carboxyl group, and hydrocarbon chain	
Reactivity	**C_3-NH_2 (primary amine):** Covalent bonding (imine formation), semi-covalent bonding amide formation, and non-covalent bonding (ionic and hydrogen bonding) **Carboxyl group (COOH):** Covalent bonding (amide formation), non-covalent ionic bonding **Hydrocarbon chain (CH_2-CH_2-):** Non-covalent bonding (Van der Waals forces). Coordinate covalent bonding (potential for metal complexation)	
Type	Non-proteinogenic amino acid; functions as a neurotransmitter in the central nervous system	
Solubility	Slightly soluble in water (1.2 µg/mL); soluble in many non-polar solvents	Machado (2018)
LogP	-3.17	Oh et al. (2008)
pKa	10.6	
Viscosity	Low viscosity aqueous solutions	
Crystallinity	Forms crystalline structures at low temperatures	
Humidity	Hygroscopic; susceptible to moisture absorption, affecting stability	

cytoplasmic decarboxylation, consuming an intracellular proton. The resulting GABA is then expelled from the cell, leading to an increase in both intracellular and extracellular pH due to the removal of hydrogen ions (Wu et al. 2017). Certain bacteria, such as *Escherichia coli*, *Shigella*, and LAB like *Lactobacillus lactis*, have genes for glutamate decarboxylase (Sarasa et al. 2020). The activation of this gene enables these bacteria to survive in acidic pH conditions (Feehily and Karatzas 2013). An increase in GAD activity is vital for survival in acidic environments and helps the bacterium withstand the low-pH environment of fermented foods, gastric juice, and volatile

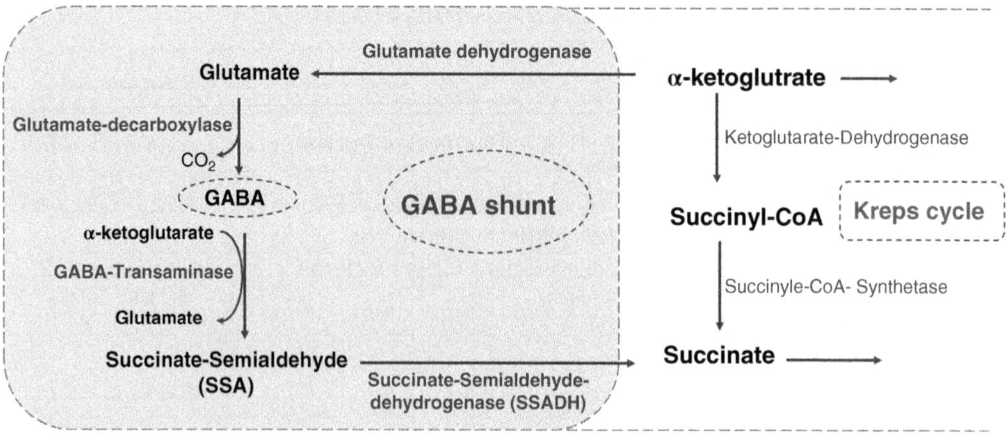

Figure 17.1 GABA shunt. It begins with the transamination of ketoglutarate from the tricarboxylic acid (TCA) cycle, leading to the formation of L-glutamate catalyzed by mitochondrial glutamate dehydrogenase. Subsequently, GABA is metabolized to succinate semialdehyde (SSA) and further to succinate. This succinate can re-enter the tricarboxylic acid (TCA) cycle. The enzyme succinate semialdehyde dehydrogenase (SSADH) catalyzes the conversion of SSA to succinate.

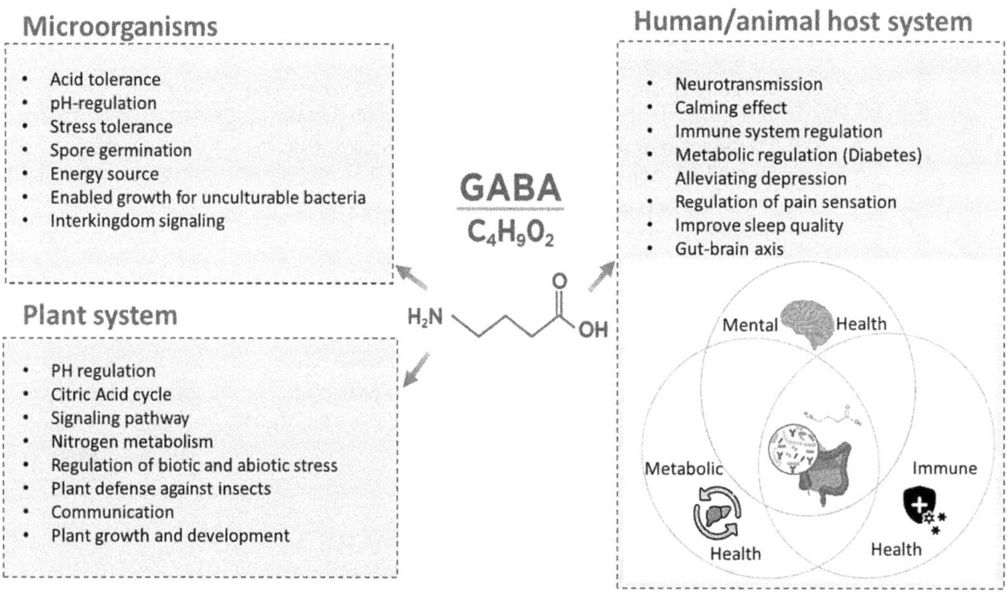

Figure 17.2 Roles of GABA in various host systems, including microorganisms, plants, and human/animal hosts. Microbial GABA contributes to mental, metabolic, and immune health in humans and animals.

fatty acids in the gastrointestinal tract (De Biase and Pennacchietti 2012). GABA is produced by Bacteroides and is crucial for their acid stress tolerance. The production process is pH-regulated and uses glutamate and glutamine as precursors. The GAD system, including the GAD gene, protects Bacteroides against acid stress, reducing cell death and maintaining metabolic activity. The GAD system includes highly conserved genes that encode a GAD, a glutaminase, a glutamate/

GABA antiporter, and a potassium channel. This system might be the only one in Bacteroides that is dependent on amino acids for acid tolerance (Otaru et al. 2021).

GABA also influences virulence. For example, it has been found to affect the virulence of various strains of *Pseudomonas* (Dagorn et al. 2013; Tumewu et al. 2020). GABA can regulate the cytotoxicity of *Pseudomonas aeruginosa* (Dagorn et al. 2013). In *Agrobacterium tumefaciens*, plant GABA stimulates the deactivation of N-3-oxo-octanoyl-l-homoserine lactone (3-oxo-C8-HSL) and modulates quorum sensing (Wang et al. 2006; Crépin et al. 2012). In *Bacillus megaterium*, spore germination relies on the decarboxylation of endogenous l-glutamic acid and the production of gamma-aminobutyric acid (Wang et al. 2006). In *P. aeruginosa*, GABA controls intracellular polyamine levels by acting as an inducer for the enzymatic pathway (Foerster 1971). In yeast, specifically the *Kluyveromyces marxianus* strain, GABA is involved in both catabolism and anabolism. This makes the yeast a useful platform for GABA production (Perpetuini et al. 2020). In fungi, GABA has a crucial role in metabolism, pathogenicity, and sporulation. For example, in the wheat pathogen *Stagonospora nodorum*, it has been discovered that GABA metabolism is necessary for full pathogenicity (Mead et al. 2013). Mutants that lack the Sdh1 gene, which encodes for succinic semialdehyde dehydrogenase, struggle to grow when GABA is their nitrogen source and are more vulnerable to reactive oxygen stress. Additionally, adding GABA to the minimal media of *S. nodorum* leads to limited growth but promotes sporulation (Mead et al. 2013). Moreover, GABA plays a part in nitrogen acquisitions (Ansari et al. 2021). During the infection of tomato by *Cladosporium fulvum*, the fungus can manipulate nitrogen acquisition to its advantage through the metabolism of GABA (Bolton and Thomma 2008; Ansari et al. 2021).

17.3.2 Role of Microbial GABA for the Carrier Host

Microbial GABA plays a substantial role in the host organism, particularly in the context of the gut–brain axis (Sarasa et al. 2020). GABA, the primary inhibitory neurotransmitter in the mammalian central nervous system, is essential in decreasing neuronal excitability across the nervous system. This could potentially explain the influence of the gut microbiome on mood (Mazzoli and Pessione 2016). In recent years, the significant neuro-active potential of metabolites, including GABA, synthesized by gut microbes has seen a growing interest between scientists. Metagenomic studies underscore the significance of microbiota-derived GABA in modulating mental health (Zhu et al. 2020). Furthermore, certain gut bacterial species have been found to produce (and consume) GABA (Altaib et al. 2021). Animal model studies have shown that introducing a bacterial species capable of producing GABA can lead to alterations in GABA receptors in the brain and sensory neuron activity (Morais et al. 2021). Interestingly, microbial-produced GABA has been reported to be effective in alleviating clinical depression (Quillin et al. 2021). This suggests the potential for utilizing probiotics that produce neuroactive molecules as therapeutic agents for treating neuro-gastroenteric and/or psychiatric disorders.

17.3.3 Role of GABA in Plants

Plants utilize GABA primarily for growth and development through the GABA shunt. GABA, a key metabolite, plays a crucial role in both primary and secondary pathways, serving as a significant link in nitrogen metabolism and the creation of amino acids (Ramos-Ruiz et al. 2019). The GABA shunt, through which GABA is metabolized, provides carbon skeletons and energy for subsequent biosynthetic pathways. Aside from its metabolic role, GABA is involved in signaling or regulatory mechanisms, indirectly impacting plant growth and development during the course of the crop cycle (Yuan et al. 2023). GABA's response to stressful conditions is another key aspect of its role. It accumulates rapidly in response to abiotic stresses, helping plants cope with adverse environmental conditions (Li et al. 2021). Moreover, it contributes to biotic stress responses through various mechanisms, such as constraining the spread of necrotrophic fungi like Botrytis (Hatmi et al. 2015; Janse van Rensburg and Van den Ende 2020).

Interestingly, GABA also exhibits activity against insects, which could be due to a direct inhibitory effect or the induction of downstream defense reactions (Ramos-Ruiz et al. 2019). This adds another layer to GABA's role in plant defense mechanisms. GABA's dual function as both a metabolite and a component of signaling pathways equips plants to handle a variety of conditions (Shelp et al. 2021). The capacity of GABA to enhance plant vigor through external application highlights its importance in plant physiology (Hayat et al. 2023). Hence, GABA's function in plants is not just crucial but also highly adaptable.

17.4 Applications for GABA

GABA is widely distributed in nature and plays a crucial role in the biological systems of animals, plants, and microorganisms. It has several applications in various fields including the medical and industrial sectors (Figure 17.3).

In the medical sector, GABA is often used in the medical field due to its various therapeutic effects. It has been found to have anti-anxiety properties, making it a popular choice for anxiety remedies (Savage et al. 2018). It can also provide relief for chronic pain associated with inflammation and conditions such as premenstrual syndrome (Tiranini and Nappi 2022). GABA has also been used to decrease blood pressure and accelerate metabolic function (Shimada et al. 2009; Hou et al. 2024). Furthermore, it can mitigate pain or discomfort from injuries, increase tolerance to exercise, burn fat, and increase the growth of thin muscle mass (Lee et al. 2010; Breus 2019). GABA has been used extensively in pharmaceuticals and nutritional supplements due to its calming effect on the brain (Ngo and Vo 2019). It is often used in supplements to reduce stress, fatigue, anxiety, and insomnia (Alex and Srivastava 2019). Barbiturates, anesthetics, benzodiazepines, antidepressants, and anti-seizure medications are some of the medications that target GABA (Mathur and Malik 2020; Kaczmarski et al. 2023).

Regarding the food industry, there has been a notable increase in the use of GABA in this field recently. A notable example of this is the commercial sale of a clustered regularly interspaced short palindromic repeats (CRISPR)-engineered tomato in Japan, which was genetically modified to contain a higher concentration of GABA compared to its non-genetically modified counterparts (Nonaka et al. 2017; Camerlengo et al. 2022). It is important to note that naturally occurring GABA is predominantly found in fermented food products, examples of which include kimchi, miso, and tempeh (Hashimoto et al. 2023; Lee and Kim 2023; Lee et al. 2023).

In livestock, GABA is integral to the health and productivity of livestock. In birds like chickens, GABA supplementation can mitigate stress effects and boost growth, especially when given through water. The quantity and period of GABA supplementation can profoundly affect the growth and immune responses in chickens (Ncho et al. 2021). In cattle, combining GABA with ascorbic acid can lessen the stress associated with transportation (Su et al. 2023). In pigs, dietary GABA boosts growth performance, improves nutrient digestion, and reduces aggressive tendencies (Litta et al. 2023). In sheep, the inclusion of rumen-protected GABA in their diet leads to an increase in dry matter consumption (Wang et al. 2023). Nevertheless, further research is required

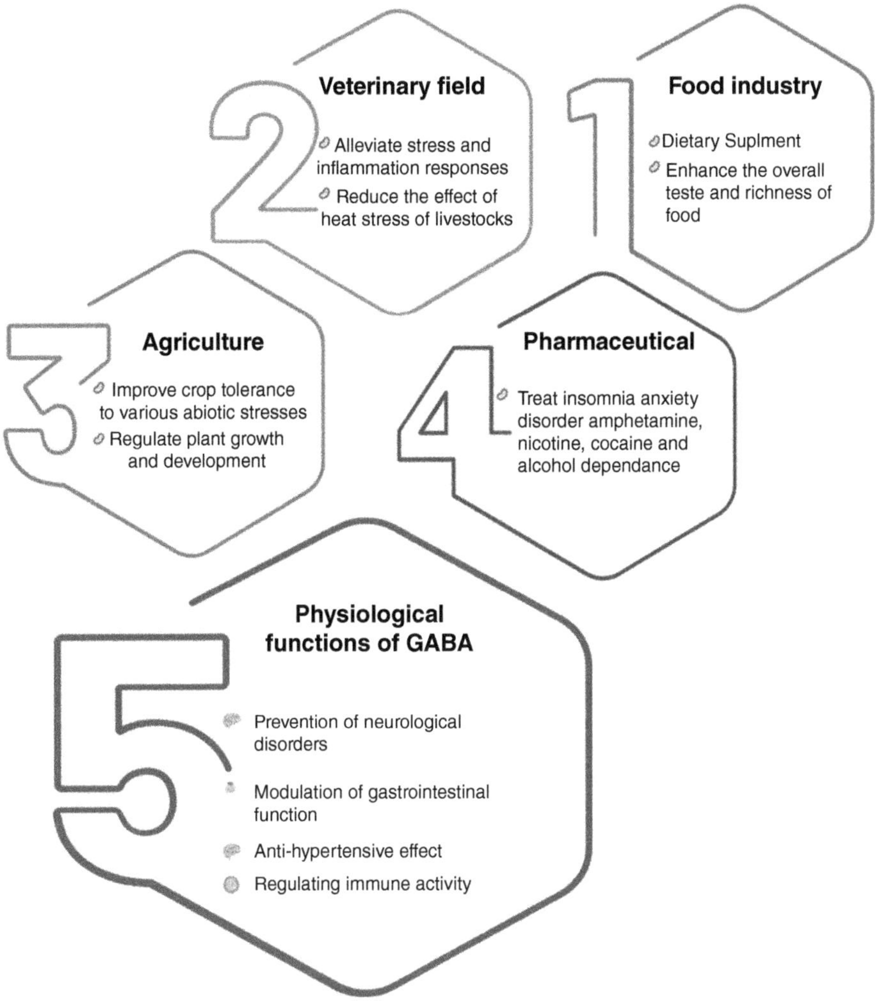

Figure 17.3 Overview of GABA applications in food, veterinary, agricultural, and pharmaceutical fields, along with its physiological roles.

to fully comprehend the impacts of GABA on livestock and explore its potential uses in animal husbandry.

In agriculture, GABA is involved in numerous metabolic pathways and confers a degree of protection to plants against abiotic stressors. GABA bolsters plant resilience to such adversities by modulating intracellular pH and ion flux, triggering antioxidant mechanisms, and mitigating reactive oxygen species. In legumes, GABA protects against a variety of abiotic stresses by augmenting leaf turgidity, amplifying osmolytes, and mitigating oxidative damage via the stimulation of antioxidants (Sita and Kumar 2020; Yuan et al. 2023). It also assumes a vital role in plant stress tolerance associated with the regulation of the TCA cycle, nitrogen sequestration, cytoplasmic pH, antioxidant defense, and osmotic potential. For example, exogenous GABA significantly lowered the salt damage index in tomato plants subjected to salt stress while also increasing plant growth, chlorophyll content, and weight gain (Ullah et al. 2023).

Despite these applications, research is ongoing to increase the endogenous GABA pool and its extraction in a cost-effective way to encounter the rising demand due to its various physiological roles on human health (Heli et al. 2022). This includes developing GABA-enriched plant types and dietary products, as well as providing insights into the efficient manufacture of GABA via synthetic biology technologies.

17.5 Critical Parameters for Enhanced Microbial GABA Production from Microorganisms

GABA production from microorganisms is affected by diverse optimization strategy factors, including fermentation conditions, cultivation temperature, pH, substrate concentration, cofactors, specific nutrient addition, and enzymatic hydrolysis-assisted cultivation (Figure 17.4). The optimization of these factors is crucial for increasing GABA yield. Different strains respond to unique optimization methods, indicating the need for tailored approaches based on microbial characteristics. Several fermentation strategies have been developed to enhance GABA productivity through microorganisms. These strategies aim to reduce production costs and broaden the applications of GABA. Recent advancements have led to increased GABA production through various methods such as optimizing fermentation processes, employing whole-cell bioconversion and physiological approaches, and introducing new co-culture systems (Table 17.2).

17.5.1 Optimizing Fermentation Process (Type of Fermentation–Substrate–pH)

The modulation of GABA synthesis by microorganisms is intricately influenced by multiple factors associated with the fermentation process. Key determinants in this regard include pH levels, temperature, fermentation duration, and the inclusion of additives in the culture medium, such as glutamic acid and pyridoxal-5-phosphate (PLP). The unique characteristics of glutamate decarboxylases (GADs) observed in different microorganisms contribute to variations in optimal fermentation conditions. Achieving precision in adjusting these conditions to yield optimal outcomes requires an in-depth exploration of the biochemical properties inherent to the GAD enzyme

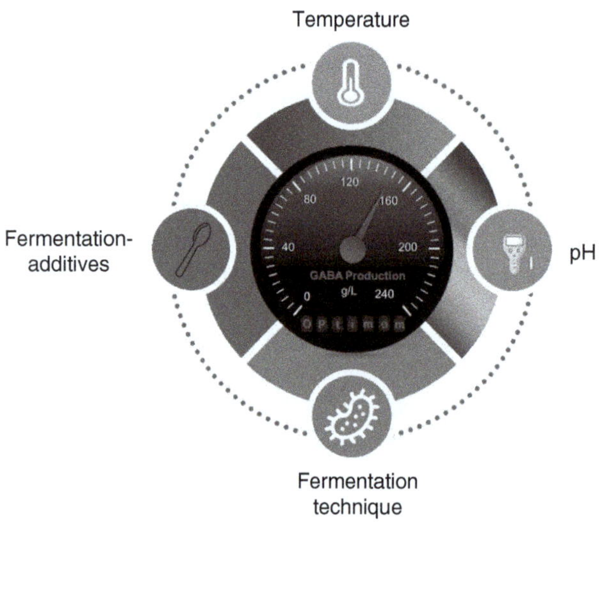

Figure 17.4 Diagram illustrating key parameters for optimizing microbial GABA production.

Table 17.2 Selected microbial systems and their optimized fermentation conditions for GABA production.

Strain	Fermentation condition	GABA yield	References
Lactobacillus fermentum 4–17	29.27% dairy sludge, 24.77% molasses, and 10.49% soybean meal in 48 h fermentation	359.45 mg/L	Falah, Vasiee et al. (2022)
Tetragenococcus halophilus	pH 8, MSG 5 g/L, and NaCl 20%	653.101 mg/L	Zamanpour, Bakhtiari Jami et al. (2024)
Escherichia coli	1.0 mol/L L-Glu L-MSG (4 : 1), 37°C,	206.2 g/L	Chang et al. (2024)
Lactococcus lactis 311	Temperature 35.4°C, pH 4.5, and MSG 89 mM	0.395 mg/mL	Edalatian Dovom, Habibi Najafi et al. (2023)
L. lactis 491	Temperature 30°C, pH 4.6, and MSG 147.4 mM	0.179 mg/mL	
Lactobacillus brevis	3% MSG	16.935 g/L	Phuengjayaem et al. (2023)
L. brevis	1g sucrose at 27°C for 28 h	352 mg/L	Mohd Lazim et al. (2023)
Bifidobacterium adolescentis	Production of recombinant *Bifidobacterium* and optimization of culture medium with pH 6.0	42,000 mg/L	Altaib et al. (2022a)
L. lactis L-571 or L-572	Temperature 37°C and 109 CFU/mL and 3 g/L of glutamate and 100 μM of PLP	1153.1 ± 13.5 mg/L	Santos-Espinosa et al. (2020)
Lplantarum plantarum Taj-Apis362	Glutamic acid 497.97 mM, culture temperature of 36°C, initial pH 5.31, and incubation time of 60 h	11.15 mM	Tajabadi et al. (2015)
L. brevis RK03	GM broth recipe of 1% glucose; 2.5% yeast extract; 2 ppm each of $CaCO_3$, MnSO4, and Tween 80; and 10 μM pyridoxal phosphate (PLP) 88 h	62,523 mg/L	Wu et al. (2018)
L. brevis CRL 1942	MSG 53 mM PLP, pH 4.5, 30°C 96 h	50 mM	Villegas et al. (2016)
L. brevis GABA 057	534 mM MSG; iso-malto-oligosaccharide to alginate beads, 34°C, pH 4.2, 48 h	223 mM	Geun-Eog and Tae-Ryeon (2006)
Streptococcus salivarius subsp. *thermophilus* Y2	PLP 0.02 mmol/L, peptone, beef extract, MSG, ammonium citrate, 40–45°C, pH 4.5–5.0 48 h	7,984.75 ± 293.33 mg/L	Yang et al. (2008)
Streptococcus thermophilus, co-culture with *Lactobacillus rhamnosus*	1 g/L skim milk powder, or yeast extract or soy protein hydrolysate, 10 g/L MSG at 37 and 42°C	8.3 g/L	Han, Liao et al. (2020)

(Continued)

Table 17.2 (Continued)

Strain	Fermentation condition	GABA yield	References
L. brevis	MRS, 1% MSG, 0.05% cysteine. 37°C PLP (some synthesized Vit-B6, needed no PLP supplement)	675 mg/L	Yunes, Poluektova et al. (2016)
L. plantarum		300 mg/L	
Bifidobacterium sp.		2500–6000 mg/L	
L. plantarum FNCC 260	MSG, MRS medium 0.1 mM PLP, 37°C, pH 4.1, and 108 h	945.3 mg/L	Yogeswara, Kittibunchakul et al. (2020)
Rhizopus oligosporus	Quinoa, soybeans pH 5.5, PLP, MSG-0.5% 50 h	540 mg/kg	Hur, Nguyen et al. (2018)
Corynebacterium glutamicum	gadB1 and gadB2 from *L. brevis* and urea supplementation	30.18 ± 1.33 g/L 36 h	Shi et al. (2013)
S. thermophilus APC151	14% (w/v) skim milk; 2.25 mg/mL MSG; 42°C for 48 h	2 mg/mL	Linares, O'Callaghan et al. (2016)
R44	GAD enzyme; 497.973 mM glutamic acid, 36°C, pH 5.31	11.09 mM	Tajabadi et al. (2015)
L. plantarum EJ2014	100 g/L yeast extract, 10 g/L glucose, 2.25% MSG	19,800 mg/L	Park, Kim et al. (2021)
L. fermentum YS2,	GAD molecular cloning, expression, immobilization	5,150 mg/L	Lin, Li et al. (2017)
L. brevis cloned into pMG36e	Glucose yeast extract peptone medium, engineering of bacterial cells for physiology orientation	43,650 mg/L	Lyu et al. (2017)
L. lactis subsp. *Lactis* CV56	50°C, pH 4.7 genetic engineering of *E. coli* cells	25,610 mg/L	Lyu, Zhao et al. (2018)
Lactobacillus delbrueckii subsp.	MSG 1.3 mg/mL, NaCl 2%, amount of PLP:MSG 0.72 : 1.3, pH 6.5, 33°C		Hudec et al. (2015)
L. plantarum DSM19463	Grape, 18.4 mM/L glutamate, pH 6.0, 30°C, fermentation 72 h.	4.83 mM	Di Cagno, Mazzacane et al. (2010)
L. brevis NCL912	30–35°C, 5.0 and 250–500 mM glutamate	1,005.81mM	Li, Qiu et al. (2010)
L. brevis RK03	1% glucose; 2.5% yeast extract; 2 ppm, $CaCO_3$, $MnSO_4$, Tween 80; 10 µM PLP, 650 mM MSG, and 88 h	62,523 mg/L,	Wu, Hsueh et al. (2018)
L. brevis NCL912	295 g/L glutamic acid, 25 g/L each glucose and yeast extract; 25 mg/L $MnSO_4{\cdot}H_2O$; 2 g/L Tween-80; 32°C, 48 h	20,580 mg/L	Wang et al. (2018)
L. brevis CRL 2013	Fermentation, add 267mM MSG, 1% (w/v) yeast extract or trypsin	25,200 mg/L	Cataldo, Villena et al. (2020)

L. plantarum LMG6907	Enzymatic hydrolysis-assisted fermentation	859 mg/L	Le, Parmentier et al. (2021)
L. plantarum EJ2014	10 g/L dextrose, 100 g/L yeast extract, and 22.5 g/L (w/v) MSG	19,800 mg/L	Park, Kim et al. (2021)
L. plantarum BC114	20 g/L MSG in MRS	3,500 mg/L	Zhang, Zeng et al. (2017)
L. brevis CRL2013	267mM MSG in MRS broth containing 1% glucose and 1% fructose	26,600 mg/L	Cataldo et al. (2020)
L. rhamnosus (GG)	2.27% MSG, 1.44% galactose, and 0.20% PLP	1,120 mg/L	Song and Yu (2018)
L. brevis	pH ¼ 4.65, temperature: 34.09°C, time 96 h, and GA concentration of 650 mmol	19,960 mg/L	Sharafi and Nateghi (2020)
L. brevis HYE1	4.01% (w/v) tryptone, 2.14% (w/v) maltose, 2.38% (w/v) MSG, and an initial pH 4.7	3,376.8 mg/L	Lim, Cha et al. (2017)
L. plantarum	Wheat bran, rice bran, and corn bran	1,300 mg/L	Sharma, Sharma et al. (2021)
L. plantarum K16	25 g/L of glucose, MSG, an initial pH of 5.5, and 1.2% inoculum	90,000 mg/L	Diez-Gutiérrez, Vicente et al. (2022)
Lactobacillus futsaii CS3	MRS broth containing 20 mg/mL MSG for 120 h	8,000 mg/L	Sanchart, Rattanaporn et al. (2016)
S. thermophilus	20 g/L of MSG, culture temperature 38.8°C, and culture time 120 h	9,660 mg/L	Sanchart, Rattanaporn et al. (2016)
L. delbrueckii ssp. bulgaricus	Cysteine (50 mg/L)		Xiao and Shah (2022)
L. brevis RK03	GM fermentation broth containing 500mM MSG fixed on the PM ceramic porous material, with 25 cycles of fermentation at 30°C for 1200 h	55,000 mg/L	Hsueh, Yang et al. (2021)
L. fermentum	1% MSG concentration and 1.19% glucose concentration with 48 h at 37°C	5,340 mg/L	Rayavarapu et al. (2019)
L. sp. Makhdzir Naser-1 (GQ451633)	506.99 (mM) of glutamic acid, 37.11°C, pH ¼ 5.26, and 53 h	773.9 mg/L	Gharehyakheh, Elhami Rad et al. (2019)
Bacillus subtilis HA and *L. plantarum* EJ2014	Glucose (3% for the first fermentation, 1.5% for the second fermentation), MSG (5%), and skim milk (5%)	16,080 mg/L	Yoon, Garcia et al. (2018)
B. subtilis HA and *L. plantarum* EJ2014	Co-fermentation of *Dendropanax morbifera* extract by *B. subtilis* and *L. plantarum* by adding 5% MSG, 3% glucose, and 2.5% skim milk	12,800 mg/L	Yoon, Garcia et al. (2018)
L. brevis F064A	pH of 5.26, 2.71% (w/v) of glucose, and 2.96% (w/v) of MSG	4,530 mg/L	Kanklai, Thongwai et al. (2022)

(*Continued*)

Table 17.2 (Continued)

Strain	Fermentation condition	GABA yield	References
Enterococcus malodoratus SJC25	Pasteurized sweet whey richened in glutamic acid	3,000 mg/L	Cunha, Coelho et al. (2022)
L. sp. Makhdzir Naser-1 (GQ451633)	Temperature 36°C, cherry juice concentration (Wt.%) 13, glutamic acid concentration, mM 360, time (h) 38.7, pH 4.4	3,818 mg/L	Gharehyakheh (2021)
L. plantarum SPS109	MRS containing 5 mg/mL MSG	1,157.01 mg/L	Jitpakdee, Kantachote et al. (2021)
L. brevis F064A	Added to MSG, 2% (w/v), and incubated at 37°C for 48 h	63,310 mg/L	Kanklai, Somwong et al. (2020)
B. subtilis ATCC 6051	Temperature 30°C, pH 8.0, 11.481 g potato starch/L, 60 g peptone/L, 5 g NaCl/L, 2.5 g K_2HPO_4/L, and sodium L-glutamate 11.825 g/L	19,740 mg/L	Wang, Huang et al. (2019)
L. brevis	5% soy protein isolate, 3% inulin, and 96 h fermenting time at 37°C	1,473.44 mg/L	Zareie, Tabatabaei Yazdi et al. (2019)
L. plantarum GB01-21	Two-stage fermentation strategy using *C. glutamicum* G01, *L. plantarum* GB01-21, and cassava	80,500 mg/L	Yang et al. (2015)
L. plantarum KCTC 3103	5 g/L glucose, 10 g/L yeast extract, 100 g/L rice bran extract, and 1.0 g/L ascorbic acid	670 mg/L	Kim et al. (2021)
L. brevis TCCC 13007	50 g/L 48 h cultivated wet resting. cells, 0.1mM PLP in glutamate-containing 0.6M citrate buffer (pH 4.5), and performed at 45°C and 180 rpm	201,180 mg/L	Shi, Chang et al. (2017)
L. brevis A3	14.96% whey powder, 4.95% MSG at a temperature of 37°C, and fermentation for 48 h	553.5 mg/L	Alizadeh Behbahani, Jooyandeh et al. (2020)
L. lactis NCDO 2118	Supplemented with both arginine and glutamate	1,547.9 mg/L	Laroute et al. (2016)
L. plantarum	MRS, pH 5–6, 48 h, and 4% (w/v)	15,740 mg/L	Tanamool, Hongsachart et al. (2019)
L. plantarum HU-C2W	The litchi juice fermented by *L. plantarum* HU-C2W at 37°C for 40 h	1,340 mg/L	Wang, Wang et al. (2021)

L. plantarum FBT215	The one-factor-at-a-time strategy was applied	1,812.160 mg/L	Kim, Lee et al. (2022)
L. lactis and *L. rhamnosus*	5 mmol of monosodium glutamate (up to 249.31 mg/L) and co-culture	395.84612.55mg/L	Galli, Venturi et al. (2022)
L. brevis KCL010	Monosodium glutamate (MSG) (3%, w/v) and PLP (30 μM) hydrolytic conditions were a slurry content of 8% (w/v), 180 mM H_2SO_4, and 121°C for 30 min. Es using Celluclast 1.5 L at 8 U/mL produced 2.7 g/L glucose after 120 h	2,500 mg/L	Kim, Kim et al. (2023)
L. futsaii CS3	3.48% (w/v) cane sugar, 3.84% (w/v) tuna condensate, and 10.77% (w/v) MSG at an initial pH of 5, with a temperature of 37°C	20.63 g/L	Thongruck and Maneerat (2023)
L. brevis F109-MD3	MRS broth containing 10% monosodium glutamate for 72 h	520 mmol	Liu, Li et al. (2022)
L. brevis	6% (w/v) L-glutamic acid, 4% (w/v) maltose, 2% (w/v) yeast extract, 1% (w/v) NaCl, 1% (w/v) $CaCl_2$, 2 g Tween 80/L, and 0.02 mM pyridoxal 5′-phosphate at initial pH 5.25 and 37°C	44.4 g/L	Binh, Ju et al. (2014)
L. brevis BJ-20	Yeast extract at 3% (w/v) with 5 brix STE	76.8 ± 3.4 μg/mL	Kim et al. (2018)
L. *brevis* FPA 3709	Cooked black soybean milk supplemented with 1% MSG, 1% brown sugar, and 0.1% peptone (without adjusting the initial pH level) was fermented (37°C, 48 h)	(5.42 ± 0.39 mg/mL)	Ko, Lin et al. (2013)
L. plantarum NDC75017	L-MSG at 80 mM, PLP at 18 μM, and a culture temperature of 36°C	314.56 mg/100 g	Shan et al. (2015)
C. glutamicum	50 g/L glucose and 0.1 mM PLP, 30°C with 10% (v/v) seed culture	12.37 g/L	Takahashi, Shirakawa et al. (2012)
L. plantarum K154	3% monosodium glutamate (MSG) and 0.25% yeast extract, pH of 4.14	10 mg/mL	Kwon, Garcia et al. (2016)
Lactobacillus sakei B2-16	4% sucrose, 1% yeast extract, and 12% MSG	223 mM	Verni, Rizzello et al. (2019)

within the fermenting microorganisms. This critical examination serves as a guiding framework for delineating the most effective parameters conducive to GABA production by microorganisms (Altaib et al. 2022a).

17.5.1.1 The Effect of Media Additives and Fermentation–Substrate

The nutritional makeup of the growth medium, encompassing carbon and nitrogen sources, exerts a substantial impact on the production of GABA. The inclusion of glutamate and PLP as coenzymes for GAD significantly enhances GABA yields, although the precise concentrations and timings of their addition are critical parameters. Various microorganisms have demonstrated heightened GABA production upon the introduction of glutamate, and the supplementation of sulfate ions or alternative substrates, such as whole meal–wheat sourdough, has proven effective in enhancing GABA synthesis. Achieving optimal GABA production necessitates meticulous optimization of both media additives and substrate concentrations to ensure maximum efficiency. The differential responses observed in GABA production at varying pH levels among distinct microorganisms emphasize the complexity of this phenomenon. The optimal synthesis of GABA in *Lactobacillus plantarum* NDC75017 is achieved through the combined presence of L-monosodium glutamate (L-MSG) and PLP (Shan et al. 2015). A similar enhancement in GABA production was observed by Takahashi et al. (2012), who reported that the supplementation of PLP positively influenced GABA synthesis in *C. glutamicum* GAD. Additionally, the utilization of PLP in the bioconversion of glutamate to GABA employs barley or wheat bran as substrates (Jin et al. 2013). Huang et al. (2016) demonstrated that the increase in pyridoxine, a precursor of PLP, significantly improved the activity of recombinant GAD, an enzyme pivotal in GABA production.

GABA production through fermentation is a complex process that involves the use of various substrates. Wang et al. (2018) found that L-glutamic acid was a superior substrate for GABA production by *Lactobacillus brevis* NCL912, while using monosodium glutamate as a substrate for GABA production by *Lactobacillus rhamnosus* YS9 and *Streptococcus salivarius* subsp. has been reported (Lin 2013). *Thermophilus* Y2 used glucose as the main carbon source for GABA production by recombinant *C. glutamicum* expressing GAD from *Escherichia coli* (*E. coli*) (Yang et al. 2008). These studies collectively demonstrate the diversity of substrates that can be used for GABA production through fermentation.

Co-culturing of LAB has been found to enhance GABA production, potentially through increased acid resistance. The addition of arginine and malate has been shown to improve GABA production in *Lactococcus lactis* (Laroute et al. 2016). Furthermore, the use of *L. plantarum* NDC75017 in yogurt production has been found to increase GABA yield, with optimal conditions including specific concentrations of substrate and coenzyme, as well as a particular culture temperature (Shan 2015).

17.5.1.2 The Effect of pH

The pH of the culture media is a crucial factor to maintain high GAD activity (Mountfort and Pybus 1992; Choi et al. 2015; Xu et al. 2017; Wang et al. 2024). The pH of the culture media affects both bacterial growth and GAD activity; therefore, the growth condition and pH must be fine-tuned to achieve the optimum condition for GABA production. To achieve that, a two-step batch fermentation is often used (Yang et al. 2015; Kim et al. 2021; Son et al. 2022; Pannerchelvan et al. 2023). In *L. brevis* TCCC13007, two-step biotransformation of monosodium glutamate was explored. The process resulted in approximately 38 g/L of GABA after 66 hours (hr) of fermentation at a conversion rate of 98.6% (Pannerchelvan et al. 2023). Decarboxylation of glutamate results in stoichiometric GABA release and proton consumption. This proton consumption results in an increase in the pH of the reaction medium. Paradoxically, this rise in pH leads to an alkalization of the cytosol in the reaction environment (Crawford et al. 1994; Xu et al. 2017; Cataldo et al. 2020). The altered pH, in turn, affects the enzyme's active site and overall structure. To address this challenge and maintain optimal enzyme function, strategies must be employed to regulate the pH of the reaction medium. Some approaches include the addition of acidic substances, such as H_2SO_4, to counterbalance the alkalizing effect of the reaction environment. This pH regulation helps sustain the enzymatic activity of GAD and ensures the efficient synthesis of GABA. The paramount factor in maximizing GABA production during microbial fermentation is the pH of the fermentation broth, which intricately affects GAD activity, the principal player in GABA synthesis (Kook and Cho 2013). The pH exhibits various specific subtleties, with the microorganism displaying a distinct preference for a specific pH range. *L. plantarum* thrives at a pH of 6.0, while *Lactobacillus paracasei* excels at pH 5.0. Deviating from this tailored trend can significantly hinder GABA production. Shin et al. (2014) highlight the critical role of acidic pH in enhancing GAD enzyme activity, thereby influencing its catalytic efficiency. Le Vo et al. (2013) contribute another layer to this narrative, disclosing that disrupting GAD's cooperative behavior extends its active pH range, enabling it to function even at neutral pH. This pH reliance is not arbitrary; Cotter et al. (2005) uncovered the association between the presence of GadD1 glutamate decarboxylase in *Listeria monocytogenes* strains and their ability to thrive in low-pH environments, highlighting the intimate connection between GAD and acidic conditions. De Biase and Pennacchietti (2012) expand on this, proposing that the glutamate decarboxylase-dependent acid resistance system, including GAD, acts as a protective mechanism for orally acquired bacteria navigating the acidic environment of the host stomach. Clearly, the pH of the fermentation broth is more than simply a background to the outstanding performance of GABA production. It emerges as a pivotal variable, directing the activity of GAD and ultimately dictating the magnitude of GABA synthesized. Proficiency in comprehending and mastering this pH dependence is imperative for unlocking the full potential of microbial fermentation for GABA production (Sarasa et al. 2020; Khongkool et al. 2024).

17.5.1.3 Temperature Influence

The temperature at which incubation occurs plays a pivotal role in the biocatalytic production of GABA. This is due to its impact on key elements of the enzymatic reaction, encompassing the activity and stability of the biocatalyst, as well as the thermodynamic equilibrium governing the process (Luo et al. 2021; Zhang et al. 2022; Mikulski et al. 2023; Son et al. 2023). As elucidated by Fan et al. (2018), there exists a demonstrable relationship between the enhancement of GABA production and the augmentation of the thermal stability of the biocatalyst, particularly the enzyme GAD (Laroute et al. 2021). Tung et al. (2011) further emphasize the significance of temperature in enzyme-catalyzed reactions, featuring its pronounced impact, especially at subsaturating substrate concentrations. This highlights the importance of considering the thermodynamic equilibrium when delineating the intricate processes involved in GABA production. Huang et al. (2007b) and Chunlong et al. (2013) both highlighted the importance of optimizing the reaction temperature for GABA production and the thermal stability of immobilized cells and achieving increased GABA production through temperature and pH adjustments. These studies collectively underline the critical role of incubation temperature in GABA production. Various microorganisms display distinct temperature preferences when it comes to optimal GABA production, as exemplified by instances such as *L. brevis* NCL912, which attains its maximum growth potential at 35°C. In a broader context, temperature conditions falling within the range of 25°C–40°C are generally conducive to achieving elevated GABA yields; however, GABA yield decreased with further increase in temperature (Huang et al. 2007a; Chunlong et al. 2013). Shan et al. (2015) and Tajabadi et al. (2015) both found that a culture temperature of 36°C was optimal for GABA production in yogurt and De Man–Rogosa–Sharpe (MRS) broth, respectively. Similarly, Santos-Espinosa et al. (2020) and Lu et al. (2009) reported that a temperature of 37°C and 31.8°C, respectively, led to the highest GABA yield in fermented milk and *L. lactis*. Hudec et al. (2015) also highlighted the importance of temperature, with an optimal temperature of 33°C for GABA production by bacteria from bananas. The impact of temperature influenced the kinetics and thermodynamics of GABA production by *Enterococcus faecium* CFR 3003 (Divyashri and Prapulla 2015).

17.5.1.4 Cultivation Time Impact

The cultivation time significantly influences GABA production, particularly in microbial fermentation processes. Typically, GABA synthesis is time-dependent, with an observed increase in its concentration as the cultivation time progresses. There exists an optimal cultivation time at which GABA production reaches its peak, and beyond this point, factors such as nutrient depletion or changes in environmental conditions may impact GABA accumulation (Lu et al. 2009; Wu et al. 2018; Sarasa et al. 2020; Zhang et al. 2022; Pannerchelvan et al. 2023; Wang et al. 2024). The impact of cultivation time on GABA production varies among microbial strains, with some exhibiting faster kinetics than others. The availability of nutrients and control over environmental factors during the cultivation period play crucial roles in optimizing GABA synthesis. Some fermentation processes employ cyclic or fed-batch strategies to extend cultivation time and enhance GABA production. Monitoring GABA levels at different time points is essential for understanding the kinetics of synthesis and optimizing cultivation time to achieve desired GABA yields efficiently. Sassi et al. (2022) found that the GABA yield of *Tetragenococcus halophilus* strain Kwong Bee Chun (KBC) increased significantly after five days of cultivation. The highest GABA production by *Aspergillus oryzae* NSK in submerged-liquid fermentation occurred at 216 hours after cultivation (Wan-Mohtar et al. 2019). These findings are consistent with the review by Dhakal et al. (2012), which highlighted fermentation time as a key factor in GABA production. Additionally, Lu et al. (2009) stressed the significance of time in optimizing fermentative factors for GABA enrichment by *L. lactis*. *L. plantarum* DSM19463 and *L. paracasei* NFRI 7415 required 72 hours and 144 hours, respectively, to achieve maximum GABA production. Nutrient composition, including carbon and nitrogen sources, strongly influences GABA production. Glutamate and PLP as GAD coenzymes enhance GABA yield, but their concentrations and additional times are critical. The incubation time significantly influenced GABA production by *L. plantarum* Taj-Apis362, with the optimal time being 60 hours (Tajabadi et al. 2015). Similarly, Tung et al. (2011) and Wan-Mohtar et al. (2019) both identified extended incubation as a key factor in maximizing GABA production by *L. plantarum* NTU 102 and *L. brevis* RK03, respectively. The importance of time in GABA production by *L. brevis* CRL 1942, with the highest yield achieved after 48 hours of incubation, was shown (Villegas et al. 2016).

17.5.2 Methodology of Design of Experiments (DOE)

In the optimization of GABA production, experimental design plays a crucial role, aiming to identify the optimal conditions for achieving the best performance (Figure 17.5). The initial step in designing any experiment involves defining the factors or independent variables and specifying the response of interest. Factors are experimental variables that can be altered independently of each other, representing the potential causes influencing the studied phenomenon. These factors encompass parameters such as pH, temperature, and substrate concentrations (e.g., MSG). Factors can be explored at different values or levels, each representing conditions under which experiments are conducted (Zareian et al. 2013; Altaib et al. 2022a).

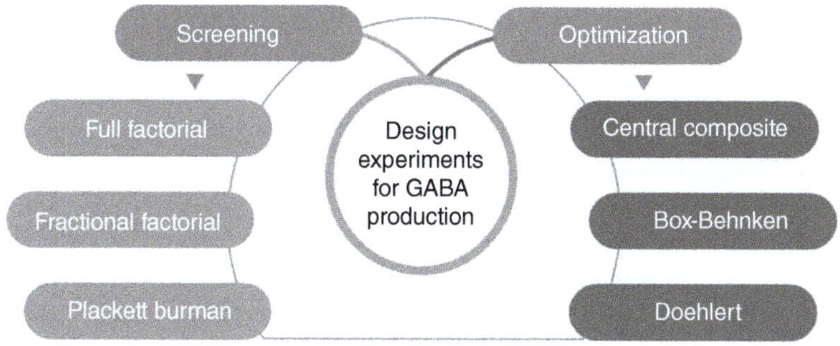

Figure 17.5 Strategies for screening and optimizing GABA production.

In the development of DOE for optimizing GABA production, two significant groups are considered: screening design and optimization design. These designs are instrumental in systematically exploring the various factors and their interactions to enhance the efficiency of GABA production processes (Suwanmanon and Hsieh 2014).

Screening designs are instrumental in the optimization of GABA production processes, enabling the evaluation of numerous factors with a limited number of experiments. Given the complexity of factors that can influence the GABA production system and the challenge of identifying and controlling each variable's small contributions, screening designs become crucial. Screening designs can be classified into full factorial, fractional factorial, and Plackett–Burman design (Figure 17.5) (Huang et al. 2007b; Suwanmanon and Hsieh 2014; Dikshit and Tallapragada 2015; Chang et al. 2018; Song et al. 2021).

After the screening stage identifies statistically significant factors, the subsequent step involves optimization, where these factors are further examined to determine the optimum conditions for GABA production. To streamline this process and reduce the number of experiments involving multiple factors, multivariate statistical techniques are often employed. One widely used multivariate technique for the optimization of GABA production is the response surface methodology (RSM) (Rayavarapu et al. 2019; Song et al. 2021). RSM, grounded in factorial designs, has demonstrated significant applicability in determining experimental conditions for various processes. RSM entails a suite of mathematical and statistical techniques that involve fitting a polynomial equation to experimental data. This approach aims to describe the behavior of a set of observations, enabling the derivation of statistical predictions. Notable response surface designs commonly employed in GABA production optimization include the central composite, Box–Behnken, and Doehlert designs (Figure 17.5) (Al-Quraan and Al-Omari 2017; Sharafi and Nateghi 2020; Rayavarapu et al. 2021; Sharma et al. 2021; Song et al. 2021; Mohd Lazim et al. 2023). These designs facilitate a systematic and efficient exploration of the experimental space, leading to the identification of optimal conditions for maximizing GABA production.

17.5.3 Genetic Engineering

Genetic engineering is a versatile strategy for enhancing GABA bioconversion and increasing yields by manipulating metabolic pathways. Overexpression of key enzymes, such as GAD, in model strains like *E. coli*, *C.glutamicum*, and LAB has been shown to significantly improve GABA production (Altaib et al. 2022a; Pannerchelvan et al. 2023; Chang et al. 2024). Additionally, the overproduction of other genes in the GABA synthesis pathway, such as *gadC* and *gadR*, can further enhance GABA productivity. Inactivation of competing mechanisms, such as the GABA aminotransferase gene *gabT*, can also enhance GABA synthesis (Huang et al. 2022; Pannerchelvan et al. 2023). Furthermore, the use of recombinant strains that can provide adequate precursor L-glutamate, such as those developed for GABA production by *E. coli* (Hu et al. 2022), and the engineering of *GADs* with improved pH stability and adaptability can further boost GABA production (Hu et al. 2022; Son et al. 2022). In a recombinant *C. glutamicum*, co-expressing two *GAD* genes (gadB1 and gadB2) significantly increases GABA production, reaching 18.66 grams per liter (g/L) after 84 hours of fermentation was achieved (Shi et al. 2013). Genetic engineers are also exploring the modulation of genes associated with GABA synthesis, such as the glutamate: GABA antiporter gene *gadC* and the regulator gene *gadR*. *E. coli* strains engineered with co-overexpression of gad and *gadC* experienced a 2.5-fold increase in GABA yield. Similarly, a genetically modified *C. glutamicum* strain incorporating the *gadB2, gadC*, and *gadR* complex from *L. brevis* achieved a high GABA yield (Cha et al. 2023; Pannerchelvan et al. 2023; Rehman et al. 2023; Yao et al. 2023). Efforts to enhance GABA biosynthesis efficiency include inactivating competing pathways, such as the GABA aminotransferase gene *gabT*. Co-overexpression of *gadB* and *gadC* in a *gabT* mutant strain increased the final GABA concentration (Le Vo et al. 2012). Ensuring an ample supply of precursor L-glutamate is crucial for GABA production. Ingenious approaches involve developing strains that overproduce L-glutamate, enhancing the substrate pool for GABA synthesis. For instance, a recombinant *E. coli* strain produced GABA directly using glucose through the introduction of a synthetic protein scaffold (Pham et al. 2015).

C. glutamicum, known for its robust amino acid-producing capabilities, has been utilized for GABA production using endogenous L-glutamate (Sheng et al. 2021). Through genetic manipulation, bacterial strains achieved satisfactory GABA yields without exogenous L-glutamate (Altaib et al. 2022a; Rehman et al. 2023). Optimizing the pH for *GAD* activity, typically ranging from 4.0 to 5.0, is essential. Mutants of *GAD GadB1* from *L. brevis* Lb85 were engineered through direct evolution, exhibiting superior activity and efficacy at a near-neutral pH of 6.0 (Shi et al. 2014). Genetic engineering offers diverse strategies for enhancing GABA production, from precise pathway modulation to optimizing precursor availability. These advancements hold promise for applications in various industries (Choi et al. 2015; Altaib et al. 2022a; Pannerchelvan et al. 2023; Chang et al. 2024).

17.5.4 Physiology-oriented Engineering

Several studies have delved into the application of physiology-oriented engineering strategies to amplify GABA production within LAB. Successful creation of a FoF1-ATPase-deficient strain with heightened GAD activity was attempted. This innovative approach resulted in a noteworthy 1.22-fold increase in GABA production, showcasing the efficacy of targeting specific physiological pathways to enhance metabolite yields (Lyu et al. 2017; Pannerchelvan et al. 2023). Building on this foundation, the potential of LAB in GABA production has been comprehensively reviewed (Cui et al. 2020 and Pannerchelvan et al. 2023). Their analyses encompassed insights into the GABA biosynthesis pathway and explored various methodologies, including fermentation optimization, co-cultural techniques, and genetic engineering (Zhang et al. 2023). By focusing on these aspects, both reviews provided valuable perspectives on utilizing physiological interventions to boost GABA production. This holistic approach recognizes the multifaceted nature of LAB physiology and its impact on GABA biosynthesis. In parallel, the current landscape of process engineering strategies is aimed at enhancing GABA production (Hayat et al. 2023). The focus extended beyond traditional approaches, encompassing fermentation optimization techniques and whole-cell conversion methodologies applied to diverse feedstocks by employing different host strains. By encapsulating a wide array of strategies, Lyu et al.'s (2017) work presented a comprehensive panorama of the evolving field of GABA production process engineering. Collectively, tremendous potential is embedded in physiology-oriented engineering strategies for advancing GABA production within LAB. The success stories and comprehensive reviews not only affirm the significance of manipulating physiological factors but also inspire future research avenues. As the exploration of LAB physiology continues, the integration of innovative engineering strategies promises to further unlock the full biotechnological potential of GABA production, paving the way for enhanced applications in various industries (Lyu et al. 2017; Song et al. 2021; Son et al. 2022; Hayat et al. 2023).

17.5.5 Co-culture Engineering

Several studies have delved into the promising field of co-cultural engineering for GABA production, particularly involving LAB and *Corynebacteria*. This approach aims to optimize fermentation conditions and increase the activity of GAD in order to maximize GABA synthesis, offering a multifaceted strategy for improving yields.

The effectiveness of co-culturing *L. brevis* GABA 100 with *Bifidobacterium bifidum* BGN4 results in enhanced GABA production during the fermentation of *Gastrodia elata* Bl (Kim et al. 2018). This early study showcased the potential synergy between different bacterial strains in elevating GABA levels. Lyu et al. (2017) took co-culture engineering a step further by adopting a physiology-oriented strategy to improve GABA production in *L. brevis*. The significance of addressing the physiological aspects of LAB is to enhance GABA biosynthesis, offering insights into the intricate interplay between microbial physiology and metabolite production. The cooperative potential of *Streptococcus thermophilus* IFO13957 and *Lactobacillus delbrueckii* subsp. *bulgaricus* IAM1120 in co-culture has led to increased GABA levels (Watanabe et al. 2011). An extensive overview of the several co-cultural techniques available for GABA synthesis by LAB was conducted (Cui et al. 2020). Sharma et al.'s (2021) work accomplished the co-production of GABA and lactic acid using *L. plantarum* LP-9, demonstrating a twofold increase in GABA production under optimized conditions. Phuengjayaem et al. (2023) contributed to the optimization of GABA production by lactic acid bacteria from plant materials, achieving a significant maximum production of 32.48 g/L. A novel avenue where co-culture engineering might contribute to increased GABA yields by harnessing the growth-promoting capabilities of specific bacteria. Exploring another facet, Sultana et al. (2024) investigated the antifungal potential of mannopyranoside derivatives. While the primary focus is on antifungal properties, the study implies potential applications in co-cultural engineering for GABA production. This suggests that the antimicrobial properties of certain derivatives may contribute to a more controlled and favorable environment for GABA-producing microorganisms.

17.6 Models of Engineered GABA Producer Organisms

17.6.1 *Corynebacterium*

C. glutamicum is an ideal candidate for GABA production owing to its established role as a robust industrial microbial cell factory for the commercial synthesis of glutamate from glucose (Shimizu and Hirasawa 2007). Moreover, it is a Gram-positive nonpathogenic bacterium, which is generally regarded as a safe bacterium (Heider and Wendisch 2015). GABA production using one-step fermentation from a renewable carbon source such as glucose represents a cost-effective way for GABA production. The one-step fermentation process was effectively utilized with engineered *C. glutamicum*. The strains demonstrated proficiency in generating a substantial quantity of intracellular L-glutamate, a crucial precursor for GABA. This capability in glutamate production ensures a reliable source for the synthesis of GABA during the fermentation process (Zhang et al. 2014).

In the preliminary phases of GABA biosynthesis utilizing a recombinant strain of *C. glutamicum*, the heterologous expression of gadBCR genes derived from *L. brevis* was employed. These genes, encompassing glutamate decarboxylase, L-glutamate/GABA antiporter, and a transcriptional regulator, constituted a strategic approach. The outcome of this methodology yielded a GABA production level of 2.15 ± 0.16 g/L, initiated from an initial glucose concentration of 160 g/L (Shi and Li 2011).

In order to boost GABA production within *C. glutamicum*, a strategic approach involved the elimination of GABA uptake systems to elevate the intracellular concentration of glutamate. This attempt encompassed the identification, characterization, and subsequent deletion of the GABA-specific transporter (*Gab*PCg), encoded by the ncgl0464 gene. Concurrently, the introduction of the *gad*B1 gene from *L. brevis* was executed. These modifications collectively resulted in a significant enhancement of GABA production in the recombinant *C. glutamicum* ATCC 13032 strain, reaching a level of 25.6 ± 2.3 g/L from an initial glucose concentration of 50 g/L (Zhao et al. 2012).

Continued advancements were realized by targeting the deletion of the *pkn*G gene, encoding serine/threonine protein kinase G, coupled with the introduction of *E. coli* GAD into the recombinant *C. glutamicum*. This strategic modification culminated in a heightened GABA production, yielding 31 g/L from an initial glucose concentration of 100 g/L (Okai et al. 2014). In a recent development, a genetically modified variant of *C. glutamicum* ATCC 13032 was engineered to express the *E. coli gad*B mutant gene (Glu89Gln/Δ452-466) under the potent synthetic H36 promoter. This specific GAD mutant demonstrated enhanced activity over an expanded pH range of 4–7. Notably, the recombinant strain exhibited improved GABA production, achieving a higher yield of 39 g/L in a fed-batch culture maintained at pH 6 (Choi et al. 2015). In another investigation, GABA production in *C. glutamicum* CGY700 was increased by implementing a CO_2 anaplerotic reaction and the upregulation of citrate synthase. The engineered strain demonstrated a significant ability to produce GABA. Shake flask cultures yielded GABA at a concentration of approximately 41 g/L, while fed-batch fermentation achieved even higher production, reaching around 58 g/L. This translates to a yield of roughly 0.3 g of GABA per gram of glucose consumed (Yao et al. 2022). These studies collectively emphasize the potential of *C. glutamicum* as an efficient and sustainable platform for GABA production.

17.6.2 LAB

LAB have been intricately engineered to serve as robust models for the proficient production of GABA. Genetic engineering techniques have enabled the targeted modification of specific LAB strains to enhance the activity of GAD, the enzyme essential for catalyzing the conversion of glutamic acid to GABA (Cui et al. 2020). This genetic manipulation has resulted in the development of LAB strains exhibiting heightened GABA production capabilities, thereby holding substantial promise for the creation of functional foods and supplements imbued with the health benefits of GABA.

A notable example is the utilization of *L. plantarum* in a two-step fermentation process optimized for GABA production. The conditions of a modified synthetic medium were fine-tuned to elevate GABA yields significantly, highlighting the potential of this strain for biotechnological applications (Lee and Lee 2014; Park et al. 2019).

The potential of using xylose, a biomass-derived substrate, for direct GABA production has been demonstrated, showcasing the versatility of LAB in utilizing alternative carbon sources (Iorizzo et al. 2023). Researchers developed a high-efficiency fermentation process for GABA production using *L. brevis* (*Lb. brevis*) CE701. Xylose was identified as the optimal carbon source, leading to a significant increase in GABA production (40.35 g/L) and cell density (OD600 of 8.64) in shake flasks compared to glucose as carbon source. Studies on *Lb. brevis* CE701's metabolic pathway revealed that xylose utilization activates the xyl operon. This activation likely leads to increased adenosine triphosphate (ATP) and organic acid production, which can in turn promote the growth of the bacteria and its synthesis of GABA (Cha et al. 2023).

By strategically inactivating competing pathways, studies have successfully increased the final GABA concentration, further highlighting the versatility and adaptability of LAB in optimizing GABA production (Cui et al. 2020).

While these strategies have showcased the significant potential of LAB as vectors for GABA production, further research and optimization efforts are essential to attain high efficiency and productivity. The integration of multiple strategies and the development of innovative tools and methods for GABA biosynthesis and bioprocessing are crucial focal points for future research in this domain.

17.6.3 *Bifidobacterium*

Bifidobacteria, which are gram-positive, probiotic, and commonly considered safe bacteria, are amenable to genetic modification techniques such as transformation, gene deletion, and heterologous gene expression (He et al. 2023). This establishes *Bifidobacterium* as a potential cell factory platform. Nevertheless, the limited exploration in this domain can be attributed to challenges associated with the highly anaerobic nature of these bacteria (De Vries and Stouthamer 1969; Kawasaki et al. 2006).

Among *Bifidobacterium* species, *Bifidobacterium adolescentis* stands out as particularly oxygen-sensitive, showcasing strain-specific GABA production. *B. adolescentis* 4-2 strain, isolated from a healthy adult, can generate GABA at concentrations around 14 mM. The presence of *gad*B and *gad*C genes in this strain implies it can produce GAD and the glutamate GABA antiporter (Altaib et al. 2022b).

Recombinant *Bifidobacterium* strains were generated using plasmids harboring *gad*BC driven by both the original *gad*B (ori) and gap promoters. Fermentation parameters like substrate concentration, initial culture pH, and cofactor addition were optimized to enhance GABA production. The choice of promoter region influenced the optimum initial cultivation pH, with the ori promoter induced under acidic conditions (pH 5.2 : 4.4), while the constitutive gap promoter exhibited improved GABA production at pH 6.0 (Altaib et al. 2022a).

In fed-batch fermentation, a substantial GABA production of approximately 415 mM was achieved, with a conversion ratio of glutamate to GABA ranging from 92% to 100%. This research paves the way for utilizing *Bifidobacterium* as a platform for large-scale production of GABA.

Furthermore, a recent genomic analysis of 1022 strains of *Bifidobacteria* revealed the conservation of genes linked to GABA metabolism, with *B. adolescentis* PRL2019 identified as the most efficient GABA-producing strain. This strain demonstrated the ability to convert more than 65% of glutamate into GABA and exhibited the highest expression of *gad*B and *gad*C genes among the analyzed bifidobacteria (Duranti et al. 2020).

Approximately 80% of *B. adolescentis* strains, including notable producers like *B. adolescentis* PRL2019 and *B. adolescentis* HD17T2H, have been found to produce GABA, showcasing increased production in living models (Duranti et al. 2020). By assessing GABA levels in over 80 bifidobacterial strains, researchers identified specific strains, particularly two *B. adolescentis* strains, capable of stimulating GABA production in rats, denoting the potential for therapeutic applications.

17.6.4 *E. coli*

E. coli has been effectively engineered as a model to produce GABA. Initial attempts for GABA production in *E. coli* were initiated in 1999, marking a significant milestone in the pursuit of bioactive compound synthesis. Researchers cloned and expressed GAD from an eukaryotic source, specifically *Saccharomyces cerevisiae*, within *E. coli* strains (Park et al. 2007). This initial recombinant strain exhibited the ability to produce 0.5 g/L of GABA from 10 g/L of glutamate in batch culture, laying the foundation for subsequent advancements in GABA bioproduction.

Following this pioneering work, numerous successive attempts have been made to enhance GABA production in *E. coli*, driven by the diverse physiological and psychological functions of GABA. Strategies have evolved to include co-expression of GAD and GABA transporter genes to facilitate GABA export and prevent its degradation (Ke et al. 2016). Additionally, efforts have focused on co-localizing GAD, glutamate synthase, and GABA transporter on synthetic scaffolds to redirect TCA cycle flux toward the GABA pathway (Dung Pham et al. 2016).

Further refinements have targeted engineering glutamate supply, pH, and cofactor balance, as well as transport and secretion systems within *E. coli* (Ye et al. 2023). Varied carbon sources, such as glycerol and acetate, have been explored to enhance glutamate availability for GABA synthesis. Notably, the co-utilization of glucose and xylose, derived from lignocellulosic biomass, has been investigated to enable GABA production from renewable resources (Choi et al. 2015).

Recently, genetic circuits were designed for feedback control of GABA biosynthesis in *E. coli* Nissle (EcN). EcN is engineered to regulate the expression of GAD. Researchers were able to alter and tune the kinetics of GABA production by utilizing a library of repressor logic gates in the feedback circuit. These loops sense the intracellular GABA concentration and adjust GAD expression accordingly. This technology could be used to create designable smart microorganisms that function as living therapies (Lebovich et al. 2024).

These comprehensive approaches have yielded diverse outcomes, with reported GABA production levels ranging from 0.4 to 245 g/L, reflecting the influence of strain selection, substrate utilization, and culture conditions (Xiong et al. 2017). The collective efforts underscore the dynamic landscape of GABA bioproduction in *E. coli* and its potential for sustainable synthesis of this bioactive compound with broad application prospects.

17.7 Conclusion

The microbial production of GABA represents a promising approach for the sustainable synthesis of this valuable compound with diverse applications in the food, pharmaceutical, and agricultural sectors. By harnessing the metabolic capabilities of microorganisms, researchers can unlock new opportunities for GABA production and pave the way for innovative solutions to address health, nutritional, and environmental challenges. Continued research and development in this field are warranted for realizing the full potential of microbial GABA production and its beneficial impacts on society.

Abbreviations

Abbreviation	Item
ATP	Adenosine triphosphate
B. adolescentis	*Bifidobacterium adolescentis*
C. glutamicum	*Corynebacterium glutamicum*
CRISPR	Clustered regularly interspaced short palindromic repeats
DOE	Design of experiments
E. coli	*Escherichia coli*
GABA	γ-amino butyric acid
GABA-T	GABA transaminase
GAD	Glutamate decarboxylase
g/L	Grams per liter
hr	Hours
LAB	Lactic acid bacteria
MSG	Monosodium glutamate
MRS	De Man–Rogosa–Sharpe
SSA	Succinate semialdehyde
SSADH	Succinate semialdehyde dehydrogenase
TCA	Tricarboxylic acid cycle
PLP	Pyridoxal-5-phosphate
3-oxo-C8-HSL	N-3-oxo-octanoyl-l-homoserine lactone

References

Alex, A. and Srivastava, A. (2019). Nutraceuticals for calming and stress. In: *Nutraceuticals in Veterinary Medicine* (ed. A. Srivastava, R. Lall, and R.C. Gupta), 417–425. Springer International Publishing.

Al-Quraan, N. and Al-Omari, H. (2017). GABA accumulation and oxidative damage responses to salt, osmotic and H_2O_2 treatments in two lentil (Lens culinaris Medik) accessions. *Plant Biosystems-An International Journal Dealing with all Aspects of Plant Biology* 151: 148–157.

Alizadeh, B.B., Jooyandeh, H., Falah, F., and Vasiee, A. (2020). Gamma-aminobutyric acid production by *Lactobacillus brevis* A3: optimization of production, antioxidant potential, cell toxicity, and antimicrobial activity. *Food Science & Nutrition* 8: 5330–5339.

Altaib, H., Kozakai, T., Badr, Y. et al. (2022a). Cell factory for γ-aminobutyric acid (GABA) production using *Bifidobacterium adolescentis*. *Microbial Cell Factories* 21: 1–13.

Altaib, H., Nakamura, K., Abe, M. et al. (2021). Differences in the concentration of the fecal neurotransmitters GABA and glutamate are associated with microbial composition among healthy human subjects. *Microorganisms* 9: 378.

Altaib, H., Niwa, R., Abe, M., and Suzuki, T. (2022b). Draft genome sequence of *Bifidobacterium adolescentis* 4-2, isolated from healthy human feces. *Microbiology Resource Announcements* 11: e00846–e00821.

Ansari, M.I., Jalil, S.U., Ansari, S.A., and Hasanuzzaman, M. (2021). GABA shunt: a key-player in mitigation of ROS during stress. *Plant Growth Regulation* 94: 131–149.

Ayyash, M.M., Abdalla, A.K., AlKalbani, N.S. et al. (2021). Invited review: characterization of new probiotics from dairy and nondairy products—insights into acid tolerance, bile metabolism and tolerance, and adhesion capability. *Journal of Dairy Science* 104: 8363–8379.

Binh, T.T.T., Ju, W.-T., Jung, W.-J., and Park, R.-D. (2014). Optimization of γ-amino butyric acid production in a newly isolated *Lactobacillus brevis*. *Biotechnology Letters* 36: 93–98.

Bolton, M.D. and Thomma, B.P. (2008). The complexity of nitrogen metabolism and nitrogen-regulated gene expression in plant pathogenic fungi. *Physiological and Molecular Plant Pathology* 72: 104–110.

Breus, M.J., 2019. 3 amazing benefits of GABA. Psychology Today.

Camerlengo, F., Frittelli, A., and Pagliarello, R. (2022). CRISPR towards a sustainable agriculture. *Encyclopedia* 2: 538–558.

Cataldo, P.G., Villegas, J.M., de Giori, G.S. et al. (2020). Enhancement of γ-aminobutyric acid (GABA) production by *Lactobacillus brevis* CRL 2013 based on carbohydrate fermentation. *International Journal of Food Microbiology* 333: 108792.

Cataldo, P.G., Villena, J., Elean, M. et al. (2020b). Immunomodulatory properties of a γ-aminobutyric acid-enriched strawberry juice produced by *Levilactobacillus brevis* CRL 2013. *Frontiers in Microbiology* 3176.

Cha, X., Ding, J., Ba, W. et al. (2023). High production of γ-aminobutyric acid by activating the *xyl* operon of *Lactobacillus brevis*. *ACS Omega* 8: 8101–8109.

Chang, C.-Y., Ma, S.-X., Zhang, J., Gao, Q. 2018. Medium optimization for γ-aminobutyric acid production by response surface methodology. In: *Advances in Applied Biotechnology: Proceedings of the 3rd International Conference on Applied Biotechnology (ICAB2016)*, November 25–27, 2016, Tianjin, China 3, 403-412.

Chang, F., Wang, Y., Zhang, J. et al. (2024). Efficient production of γ-aminobutyric acid using engineered *Escherichia coli* whole-cell catalyst. *Enzyme and Microbial Technology* 174: 110379.

Choi, J.W., Yim, S.S., Lee, S.H. et al. (2015). Enhanced production of gamma-aminobutyrate (GABA) in recombinant *Corynebacterium glutamicum* by expressing glutamate decarboxylase active in expanded pH range. *Microbial Cell Factories* 14: 1–11.

Chunlong, P., Huang, J., Sheng, H. et al. (2013). A two-stage pH and temperature control with substrate feeding strategy for production of gamma-aminobutyric acid by lactobacillus brevis CGMCC 1306. *Chinese Journal of Chemical Engineering* 21: 1190–1194.

Cotter, P.D., Ryan, S., Gahan, C.G., and Hill, C. (2005). Presence of GadD1 glutamate decarboxylase in selected *Listeria monocytogenes* strains is associated with an ability to grow at low pH. *Applied and Environmental Microbiology* 71: 2832–2839.

Crawford, L.A., Bown, A.W., Breitkreuz, K.E., and Guinel, F.C. (1994). The synthesis of [gamma]-aminobutyric acid in response to treatments reducing cytosolic pH. *Plant Physiology* 104: 865–871.

Crépin, A., Barbey, C., Beury-Cirou, A. et al. (2012). Quorum sensing signaling molecules produced by reference and emerging soft-rot bacteria (*Dickeya* and *Pectobacterium* spp.). *PLoS One* 7: e35176.

Cui, Y., Miao, K., Niyaphorn, S., and Qu, X. (2020). Production of gamma-aminobutyric acid from lactic acid bacteria: a systematic review. *International Journal of Molecular Sciences* 21: 995.

Cunha, D.S., Coelho, M.C., Ribeiro, S.C., and Silva, C.C. (2022). Application of *Enterococcus malodoratus* SJC25 for the manufacture of whey-based beverage naturally enriched with GABA. *Foods* 11: 447.

Dagorn, A., Hillion, M., Chapalain, A. et al. (2013). Gamma-aminobutyric acid acts as a specific virulence regulator in *Pseudomonas aeruginosa*. *Microbiology* 159: 339–351.

De Biase, D. and Pennacchietti, E. (2012). Glutamate decarboxylase-dependent acid resistance in orally acquired bacteria: function, distribution and biomedical implications of the gadBC operon. *Molecular Microbiology* 86: 770–786.

De Vries, W. and Stouthamer, A. (1969). Factors determining the degree of anaerobiosis of *Bifidobacterium* strains. *Archiv für Mikrobiologie* 65: 275–287.

Dhakal, R., Bajpai, V.K., and Baek, K.-H. (2012). Production of GABA (γ-aminobutyric acid) by microorganisms: a review. *Brazilian Journal of Microbiology* 43: 1230–1241.

Di Cagno, R., Mazzacane, F., Rizzello, C.G. et al. (2010). Synthesis of γ-aminobutyric acid (GABA) by *Lactobacillus plantarum* DSM19463: functional grape must beverage and dermatological applications. *Applied Microbiology and Biotechnology* 86: 731–741.

Diez-Gutiérrez, L., Vicente, L.S., Sáenz, J. et al. (2022). Biosynthesis of gamma-aminobutyric acid by *Lactiplantibacillus plantarum* K16 as an alternative to revalue agri-food by-products. *Scientific Reports* 12: 18904.

Dikshit, R. and Tallapragada, P. (2015). Screening and optimization of γ-aminobutyric acid production from *Monascus sanguineus* under solid-state fermentation. *Frontiers in Life Science* 8: 172–181.

Divyashri, G. and Prapulla, S.G. (2015). An insight into kinetics and thermodynamics of gamma-aminobutyric acid production by *Enterococcus faecium* CFR 3003 in batch fermentation. *Annals of Microbiology* 65: 1109–1118.

Dung Pham, V., Somasundaram, S., Lee, S.H. et al. (2016). Efficient production of gamma-aminobutyric acid using *Escherichia coli* by co-localization of glutamate synthase, glutamate decarboxylase, and GABA transporter. *Journal of Industrial Microbiology and Biotechnology* 43: 79–86.

Duranti, S., Ruiz, L., Lugli, G.A. et al. (2020). Bifidobacterium adolescentis as a key member of the human gut microbiota in the production of GABA. *Scientific Reports* 10: 14112.

Edalatian Dovom, M.R., Habibi Najafi, M.B., Rahnama, V.P. et al. (2023). Screening of lactic acid bacteria strains isolated from Iranian traditional dairy products for GABA production and optimization by response surface methodology. *Scientific Reports* 13: 440.

Falah, F., Vasiee, A., Tabatabaei-Yazdi, F. et al. (2022). Optimization of γ-aminobutyric acid (GABA) production by *Lactobacillus spp.* from agro-food waste. *Biomass Conversion and Biorefinery* 1–13.

Fan, L.-Q., Li, M.-W., Qiu, Y.-J. et al. (2018). Increasing thermal stability of glutamate decarboxylase from *Escherichia. coli* by site-directed saturation mutagenesis and its application in GABA production. *Journal of Biotechnology* 278: 1–9.

Feehily, C. and Karatzas, K. (2013). Role of glutamate metabolism in bacterial responses towards acid and other stresses. *Journal of Applied Microbiology* 114: 11–24.

Foerster, H.F. (1971). γ-Aminobutyric acid as a required germinant for mutant spores of *Bacillus megaterium*. *Journal of Bacteriology* 108: 817–823.

Galli, V., Venturi, M., Mari, E. et al. (2022). Gamma-aminobutyric acid (GABA) production in fermented milk by lactic acid bacteria isolated from spontaneous raw milk fermentation. *International Dairy Journal* 127: 105284.

Geun-Eog, J. and Tae-Ryeon, H. (2006). Improvement of $\gamma-aminobutyric $ acid (GABA) production using cell entrapment of *Lactobacillus brevis* GABA 057. *Journal of Microbiology and Biotechnology* 16: 562–568.

Gharehyakheh, S. (2021). Gamma aminobutyric acid (GABA) production using *Lactobacillus sp.* Makhdzir Naser-1 (GQ451633) in the cherry-kefir beverage. *Journal of Food Processing and Preservation* 45: e15521.

Gharehyakheh, S., Elhami Rad, A.H., Nateghi, L., and Varmira, K. (2019). Production of GABA-enriched honey syrup using *Lactobacillus bacteria* isolated from honey bee stomach. *Journal of Food Processing and Preservation* 43: e14054.

Han, M., W-y, L., S-m, W. et al. (2020). Use of *Streptococcus thermophilus* for the in situ production of γ-aminobutyric acid-enriched fermented milk. *Journal of Dairy Science* 103: 98–105.

Han, J., Zhao, X., Zhao, X. et al. (2023). Microbial-derived γ-aminobutyric acid: synthesis, purification, physiological function, and applications. *Journal of Agricultural and Food Chemistry* 71: 14931–14946.

Hashimoto, Y., Hamaguchi, M., and Fukui, M. (2023). Fermented soybean foods and diabetes. *Journal of Diabetes Investigation* 14: 1329–1340.

Hatmi, S., Gruau, C., Trotel-Aziz, P. et al. (2015). Drought stress tolerance in grapevine involves activation of polyamine oxidation contributing to improved immune response and low susceptibility to *Botrytis cinerea*. *Journal of Experimental Botany* 66: 775–787.

Hayat, F., Khan, U., Li, J. et al. (2023). γ Aminobutyric acid (GABA): a key player in alleviating abiotic stress resistance in horticultural crops: current insights and future directions. *Horticulturae* 9: 647.

He, B.-L., Xiong, Y., Hu, T.-G. et al. (2023). *Bifidobacterium* spp. as functional foods: a review of current status, challenges, and strategies. *Critical Reviews in Food Science and Nutrition* 63: 8048–8065.

Heider, S.A. and Wendisch, V.F. (2015). Engineering microbial cell factories: metabolic engineering of *Corynebacterium glutamicum* with a focus on non-natural products. *Biotechnology Journal* 10: 1170–1184.

Heli, Z., Hongyu, C., Dapeng, B. et al. (2022). Recent advances of γ-aminobutyric acid: physiological and immunity function, enrichment, and metabolic pathway. *Frontiers in Nutrition* 9: 1076223.

Hou, D., Tang, J., Feng, Q. et al. (2024). Gamma-aminobutyric acid (GABA): a comprehensive review of dietary sources, enrichment technologies, processing effects, health benefits, and its applications. *Critical Reviews in Food Science and Nutrition* 64 (24): 8852–8874.

Hu, M., Li, M., Miao, M., and Zhang, T. (2022). Engineering *Escherichia coli* for the high-titer biosynthesis of lacto-N-tetraose. *Journal of Agricultural and Food Chemistry* 70: 8704–8712.

Huang, H., He, Y., Cui, A. et al. (2022). Genome-wide identification of GAD family genes suggests GhGAD6 functionally respond to Cd2+ stress in cotton. *Frontiers in Genetics* 13: 965058.

Huang, J., Mei, L.-H., Wu, H., and Lin, D.-Q. (2007a). Biosynthesis of γ-aminobutyric acid (GABA) using immobilized whole cells of *Lactobacillus brevis*. *World Journal of Microbiology and Biotechnology* 23: 865–871.

Huang, J., Mei, L.H., and Xia, J. (2007b). Application of artificial neural network coupling particle swarm optimization algorithm to biocatalytic production of GABA. *Biotechnology and Bioengineering* 96: 924–931.

Huang, Y., Su, L., and Wu, J. (2016). Pyridoxine supplementation improves the activity of recombinant glutamate decarboxylase and the enzymatic production of gama-aminobutyric acid. *PLoS One* 11: e0157466.

Hudec, J., Kobida, Ľ., Čanigová, M. et al. (2015). Production of γ-aminobutyric acid by microorganisms from different food sources. *Journal of the Science of Food and Agriculture* 95: 1190–1198.

Hur, J., Nguyen, T.T.H., Park, N. et al. (2018). Characterization of quinoa (Chenopodium quinoa) fermented by *Rhizopus oligosporus* and its bioactive properties. *Amb Express* 8: 1–8.

Iorizzo, M., Paventi, G., and Di Martino, C. (2023). Biosynthesis of gamma-aminobutyric acid (GABA) by *Lactiplantibacillus plantarum* in fermented food production. *Current Issues in Molecular Biology* 46: 200–220.

Janse van Rensburg, H.C. and Van den Ende, W. (2020). Priming with γ-aminobutyric acid against *Botrytis cinerea* reshuffles metabolism and reactive oxygen species: dissecting signalling and metabolism. *Antioxidants* 9: 1174.

Jeng, K.-C., Chen, C.-S., Fang, Y.-P. et al. (2007). Effect of microbial fermentation on content of statin, GABA, and polyphenols in Pu-Erh tea. *Journal of Agricultural and Food Chemistry* 55: 8787–8792.

Jin, W.J., Kim, M.J., and Kim, K.S. (2013). Utilization of barley or wheat bran to bioconvert glutamate to γ-aminobutyric acid (GABA). *Journal of Food Science* 78: C1376–C1382.

Jitpakdee, J., Kantachote, D., Kanzaki, H., and Nitoda, T. (2021). Selected probiotic lactic acid bacteria isolated from fermented foods for functional milk production: Lower cholesterol with more beneficial compounds. *LWT* 135: 110061.

Kaczmarski, P., Sochal, M., Strzelecki, D. et al. (2023). Influence of glutamatergic and GABAergic neurotransmission on obstructive sleep apnea. *Frontiers in Neuroscience* 17: 1213971.

Kanklai J, Somwong T.C., Rungsirivanich P., Thongwai N., 2020. Screening of GABA-producing lactic acid bacteria from Thai fermented foods and probiotic potential of *Levilactobacillus brevis* F064A for GABA-fermented mulberry juice production. *Microorganisms* 9, 33.

Kanklai, J., Thongwai, N., Rungsirivanich, P. et al. (2022). Optimization of GABA production and partial gad gene identification of potential probiotic *Levilactobacillus brevis* F064A isolated from Thai fermented sausage. *Chiang Mai Journal of Science* 49: 1013–1027.

Kawasaki, S., Mimura, T., Satoh, T. et al. (2006). Response of the microaerophilic *Bifidobacterium* species, *B. boum* and *B. thermophilum*, to oxygen. *Applied and Environmental Microbiology* 72: 6854–6858.

Ke, C., Yang, X., Rao, H. et al. (2016). Whole-cell conversion of l-glutamic acid into gamma-aminobutyric acid by metabolically engineered Escherichia coli. *Springerplus* 5: 591.

Khan, M.I.R., Jalil, S.U., Chopra, P. et al. (2021). Role of GABA in plant growth, development and senescence. *Plant Gene* 26: 100283.

Khongkool, K., Prakit, B., Chaiyod, R. et al. (2024). Gamma-aminobutyric acid (GABA) producing *Lactobacillus plantarum* TSUB-17 and probiotic properties for using as probiotics additive in swine feed. *Trends in Sciences* 21: 7198–7198.

Kim, D.-H., Dasagrandhi, C., Park, S.-K. et al. (2018). Optimization of gamma-aminobutyric acid production using sea tangle extract by lactic acid bacterial fermentation. *Lwt* 90: 636–642.

Kim, J., Lee, M.-H., Kim, M.-S. et al. (2022). Probiotic properties and optimization of gamma-aminobutyric acid production by Lactiplantibacillus plantarum FBT215. *Journal of Microbiology and Biotechnology* 32 (6): 783–791.

Kim, N.Y., Kim, J.M., Son, J.-Y., and Ra, C.H. (2023). Synbiotic Fermentation of Undaria pinnatifida and *Lactobacillus brevis* to produce prebiotics and probiotics. *Applied Biochemistry and Biotechnology* 1–13.

Kim, N.Y., Kim, S.-K., and Ra, C.H. (2021). Evaluation of gamma-aminobutyric acid (GABA) production by *Lactobacillus plantarum* using two-step fermentation. *Bioprocess and Biosystems Engineering* 44: 2099–2108.

Ko C.Y., Lin H.-TV., Tsai GJ., 2013. Gamma-aminobutyric acid production in black soybean milk by *Lactobacillus brevis* FPA 3709 and the antidepressant effect of the fermented product on a forced swimming rat model. *Process Biochemistry* 48, 559–568.

Kim, J., Richter, W., Aanstoot, H.-J. et al. (1993). Differential expression of GAD65 and GAD67 in human, rat, and mouse pancreatic islets. *Diabetes* 42: 1799–1808.

Koh, W., Kwak, H., Cheong, E., and Lee, C.J. (2023). GABA tone regulation and its cognitive functions in the brain. *Nature Reviews Neuroscience* 24: 523–539.

Kook, M.-C. and Cho, S.-C. (2013). Production of GABA (gamma amino butyric acid) by lactic acid bacteria. *Korean Journal for Food Science of Animal Resources* 33: 377–389.

Kwon S.-Y., Garcia CV., Song Y.-C., Lee S-P., 2016. GABA-enriched water dropwort produced by co-fermentation with *Leuconostoc mesenteroides* SM and *Lactobacillus plantarum* K154. *LWT* 73, 233–238.

Laroute, V., Mazzoli, R., Loubière, P. et al. (2021). Environmental conditions affecting GABA production in *Lactococcus lactis* NCDO 2118. *Microorganisms* 2021, 9: 122. (s Note: MDPI stays neu-tral with regard to jurisdictional clai-ms in . . .).

Laroute, V., Yasaro, C., Narin, W. et al. (2016). GABA production in *Lactococcus lactis* is enhanced by arginine and co-addition of malate. *Frontiers in Microbiology* 7: 1050.

Le, P.H., Parmentier, N., Le, T.T., and Raes, K. (2021). Evaluation of using a combination of enzymatic hydrolysis and lactic acid fermentation for γ-aminobutyric acid production from soymilk. *LWT* 142: 111044.

Le Vo, T.D., Kim, T.W., and Hong, S.H. (2012). Effects of glutamate decarboxylase and gamma-aminobutyric acid (GABA) transporter on the bioconversion of GABA in engineered *Escherichia coli*. *Bioprocess and Biosystems Engineering* 35: 645–650.

Le Vo, T.D., Ko, J.-S., Park, S.J. et al. (2013). Efficient gamma-aminobutyric acid bioconversion by employing synthetic complex between glutamate decarboxylase and glutamate/GABA antiporter in engineered *Escherichia coli*. *Journal of Industrial Microbiology and Biotechnology* 40: 927–933.

Lebovich, M., Lora, M.A., Gracia-David, J., and Andrews, L.B. (2024). Genetic circuits for feedback control of gamma-aminobutyric acid biosynthesis in probiotic *Escherichia coli* Nissle 1917. *Metabolites* 14: 44.

Lee, H., Chang, M.-J., and Kim, S.-H. (2010). Effects of poly-γ-glutamic acid on serum and brain concentrations of glutamate and GABA in diet-induced obese rats. *Nutrition Research and Practice* 4: 23–29.

Lee, J.-S. and Kim, K.-S. (2023). Optimization of culture conditions for and assessment of kimchi-originated lactic acid bacterial isolates toward their extracellular GABA-producing ability. *Emirates Journal of Food and Agriculture* 11: 1033–1040.

Lee, E.-J. and Lee, S.-P. (2014). Novel bioconversion of sodium glutamate to γ-amino butyric acid by co-culture of *Lactobacillus plantarum* K154 in *Ceriporia lacerata* culture broth. *Food Science and Biotechnology* 23: 1997–2005.

Lee, X., Tan, J., and Cheng, L. (2023). Gamma aminobutyric acid (GABA) enrichment in plant-based food–a mini review. *Food Reviews International* 39: 5864–5885.

Li, H., Qiu, T., Huang, G., and Cao, Y. (2010). Production of gamma-aminobutyric acid by *Lactobacillus brevis* NCL912 using fed-batch fermentation. *Microbial Cell Factories* 9: 1–7.

Li, L., Dou, N., Zhang, H., and Wu, C. (2021). The versatile GABA in plants. *Plant Signaling & Behavior* 16: 1862565.

Lim, H.S., Cha, I.-T., Roh, S.W. et al. (2017). Enhanced production of gamma-aminobutyric acid by optimizing culture conditions of *Lactobacillus brevis* HYE1 isolated from kimchi, a Korean fermented food. *Journal of Microbiology and Biotechnology* 27: 450–459.

Lin, Q. (2013). Submerged fermentation of *Lactobacillus rhamnosus* YS9 for γ-aminobutyric acid (GABA) production. *Brazilian Journal of Microbiology* 44: 183–187.

Lin, Q., Li, D., and Qin, H. (2017). Molecular cloning, expression, and immobilization of glutamate decarboxylase from *Lactobacillus fermentum* YS2. *Electronic Journal of Biotechnology* 27: 8–13.

Linares D.M., O'Callaghan T.F., O'Connor PM., Ross R.P., Stanton C., 2016. *Streptococcus thermophilus* APC151 strain is suitable for the manufacture of naturally GABA-enriched bioactive yogurt. *Frontiers in Microbiology* 7, 1876.

Litta, G., López-Bote, C., Hernández, J.M. et al. (2023). *Optimum Vitamin Nutrition for More Sustainable Swine Farming*, vol. 2. 5m Books Ltd.

Liu, W., Li, H., Liu, L. et al. (2022). Screening of gamma-aminobutyric acid-producing lactic acid bacteria and the characteristic of glutamate decarboxylase from *Levilactobacillus brevis* F109-MD3 isolated from kimchi. *Journal of Applied Microbiology* 132: 1967–1977.

Lu, X., Xie, C., and Gu, Z. (2009). Optimisation of fermentative parameters for GABA enrichment by *Lactococcus lactis*. *Czech Journal of Food Sciences* 27: 433–442.
Luo, H., Liu, Z., Xie, F. et al. (2021). Microbial production of gamma-aminobutyric acid: applications, state-of-the-art achievements, and future perspectives. *Critical Reviews in Biotechnology* 41: 491–512.
Lyu, C.-J., Zhao, W.-R., Hu, S. et al. (2017). Physiology-oriented engineering strategy to improve gamma-aminobutyrate production in *Lactobacillus brevis*. *Journal of Agricultural and Food Chemistry* 65: 858–866.
Lyu, C., Zhao, W., Peng, C. et al. (2018). Exploring the contributions of two glutamate decarboxylase isozymes in *Lactobacillus brevis* to acid resistance and γ-aminobutyric acid production. *Microbial Cell Factories* 17: 1–14.
Machado, P.A.P. (2018). *Development of a Method Based on HPLC-DAD for the Detection and Quantification of Glutamate and Gamma-Aminobutyric Acid*. Instituto Politecnico do Porto (Portugal).
Mathur, M. and Malik, M.T. (2020). Benzodiazepines and barbiturates. In: *Sedation and Analgesia for the Pediatric Intensivist: A Clinical Guide* (ed. J.W. Berkenbosch and P.P. Kamat), 401–410. Springer.
Mazzoli, R. and Pessione, E. (2016). The neuro-endocrinological role of microbial glutamate and GABA signaling. *Frontiers in Microbiology* 7: 1934.
Mead, O., Thynne, E., Winterberg, B., and Solomon, P.S. (2013). Characterising the role of GABA and its metabolism in the wheat pathogen *Stagonospora nodorum*. *PLoS One* 8: e78368.
Mikulski, T., Górecka, M., Smorawiński, J. et al. (2023). The effect of diminished metabolic acidosis on thermoregulatory response during exercise. *Biology of Sport* 41: 287–293.
Mohd Lazim, M., Mohamad, S., Abdul Rahman, I. et al. (2023). Optimization of process parameters for GABA and L-DOPA production on fermented coconut drink. *Food Research* 6: 102–108.
Morais, L.H., Schreiber, H.L. IV, and Mazmanian, S.K. (2021). The gut microbiota–brain axis in behaviour and brain disorders. *Nature Reviews Microbiology* 19: 241–255.
Mountfort, D.O. and Pybus, V. (1992). Effect of pH, temperature and salinity on the production of gamma aminobutyric acid (GABA) from amines by marine bacteria. *FEMS Microbiology Letters* 101: 237–244.
Ncho, C.M., Jeong, C., Gupta, V., and Goel, A. (2021). The effect of gamma-aminobutyric acid supplementation on growth performances, immune responses, and blood parameters of chickens reared under stressful environment: a meta-analysis. *Environmental Science and Pollution Research International* 28: 45019–45028.
Ngo, D.-H. and Vo, T.S. (2019). An updated review on pharmaceutical properties of gamma-aminobutyric acid. *Molecules* 24: 2678.
Nonaka, S., Arai, C., Takayama, M. et al. (2017). Efficient increase of γ-aminobutyric acid (GABA) content in tomato fruits by targeted mutagenesis. *Scientific Reports* 7: 7057.
Ochoa-de la Paz, L.D., Gulias-Cañizo, R., Ruíz-Leyja, E.D. et al. (2021). The role of GABA neurotransmitter in the human central nervous system, physiology, and pathophysiology. *Revista mexicana de neurociencia* 22: 67–76.
Oh, S.-H., Kim, H.-J., Kim, Y.-H. et al. (2008). Changes in some physico-chemical properties and γ-aminobutyric acid content of Kimchi during fermentation and storage. *Preventive Nutrition and Food Science* 13: 219–224.
Okai, N., Takahashi, C., Hatada, K. et al. (2014). Disruption of pknG enhances production of gamma-aminobutyric acid by *Corynebacterium glutamicum* expressing glutamate decarboxylase. *AMB Express* 4: 1–8.
Otaru, N., Ye, K., Mujezinovic, D. et al. (2021). GABA production by human intestinal *Bacteroides* spp.: prevalence, regulation, and role in acid stress tolerance. *Frontiers in Microbiology* 12: 656895.

Pannerchelvan, S., Rios-Solis, L., Wong, F.W.F. et al. (2023). Strategies for improvement of gamma-aminobutyric acid (GABA) biosynthesis via lactic acid bacteria (LAB) fermentation. *Food & Function* 14 (9): 3929–3948.

Park, E.-J., Garcia, C.V., Youn, S.-J. et al. (2019). Fortification of γ-aminobutyric acid and bioactive compounds in *Cucurbita moschata* by novel two-step fermentation using *Bacillus subtilis* and *Lactobacillus plantarum*. *Lwt* 102: 22–29.

Park, J.H., Lee, K.H., Kim, T.Y., and Lee, S.Y. (2007). Metabolic engineering of *Escherichia coli* for the production of L-valine based on transcriptome analysis and in silico gene knockout simulation. *Proceedings of the National Academy of Sciences* 104: 7797–7802.

Perpetuini, G., Tittarelli, F., Battistelli, N. et al. (2020). γ-aminobutyric acid production by *Kluyveromyces marxianus* strains. *Journal of Applied Microbiology* 129: 1609–1619.

Pham, V.D., Lee, S.H., Park, S.J., and Hong, S.H. (2015). Production of gamma-aminobutyric acid from glucose by introduction of synthetic scaffolds between isocitrate dehydrogenase, glutamate synthase and glutamate decarboxylase in recombinant *Escherichia coli*. *Journal of Biotechnology* 207: 52–57.

Phuengjayaem, S., Pakdeeto, A., Kingkaew, E. et al. (2023). Genome sequences and functional analysis of *Levilactobacillus brevis* LSF9-1 and *Pediococcus acidilactici* LSF1-1 from fermented fish cake (*Som-fak*) with gamma-aminobutyric acid (GABA) production. *Functional & Integrative Genomics* 23: 158.

Quillin, S.J., Tran, P., and Prindle, A. (2021). Potential roles for gamma-aminobutyric acid signaling in bacterial communities. *Bioelectricity* 3: 120–125.

Ramos-Ruiz, R., Martinez, F., and Knauf-Beiter, G. (2019). The effects of GABA in plants. *Cogent Food & Agriculture* 5: 1670553.

Rashmi, D., Zanan, R., John, S. et al. (2018). γ-aminobutyric acid (GABA): biosynthesis, role, commercial production, and applications. *Studies in Natural Products Chemistry* 57: 413–452.

Rayavarapu, B., Tallapragada, P., and Subbaiah, U. (2021). Optimization and comparison of γ-aminobutyric acid (GABA) production by LAB in soymilk using RSM and ANN models. *Beni-Suef University Journal of Basic and Applied Sciences* 10: 1–15.

Rayavarapu, B., Tallapragada, P., and Usha, M. (2019). Statistical optimization of γ-aminobutyric acid production by response surface methodology and artificial neural network models using *Lactobacillus fermentum* isolated from palm wine. *Biocatalysis and Agricultural Biotechnology* 22: 101362.

Rehman, A., Di Benedetto, G., Bird, J.K., Dabene, V., Vadakumchery, L., May, A., Schyns, G., Sybesma, W., Mak, T.N., 2023. Selection, identification and optimization of lactic acid bacteria with high γ-aminobutyric acid production. Scientific Reports 13, 13663.

Sanchart, C., Rattanaporn, O., Haltrich, D. et al. (2016). Technological and safety properties of newly isolated GABA-producing *Lactobacillus futsaii* strains. *Journal of Applied Microbiology* 121: 734–745.

Santos-Espinosa, A., Beltrán-Barrientos, L.M., Reyes-Díaz, R. et al. (2020). Gamma-aminobutyric acid (GABA) production in milk fermented by specific wild lactic acid bacteria strains isolated from artisanal Mexican cheeses. *Annals of Microbiology* 70: 1–11.

Sarasa, S.B., Mahendran, R., Muthusamy, G. et al. (2020). A brief review on the non-protein amino acid, gamma-amino butyric acid (GABA): its production and role in microbes. *Current Microbiology* 77: 534–544.

Sassi, S., Ilham, Z., Jamaludin, N.S. et al. (2022). Critical optimized conditions for gamma-aminobutyric acid (GABA)-producing tetragenococcus halophilus strain KBC from a commercial soy sauce moromi in batch fermentation. *Fermentation* 8: 409.

Savage, K., Firth, J., Stough, C., and Sarris, J. (2018). GABA-modulating phytomedicines for anxiety: a systematic review of preclinical and clinical evidence. *Phytotherapy Research* 32: 3–18.

Shan, Y., Man, C., Han, X. et al. (2015). Evaluation of improved γ-aminobutyric acid production in yogurt using *Lactobacillus plantarum* NDC75017. *Journal of Dairy Science* 98: 2138–2149.

Sharafi, S. and Nateghi, L. (2020). Optimization of gamma-aminobutyric acid production by probiotic bacteria through response surface methodology. *Iranian Journal of Microbiology* 12: 584.

Sharma, P., Sharma, A., Singh, J. et al. (2021). Co-production of gamma amino butyric acid (GABA) and lactic acid using *Lactobacillus plantarum* LP-9 from agro-residues. *Environmental Technology & Innovation* 23: 101650.

Shelp, B., Aghdam, M., and Flaherty, E. (2021). γ-aminobutyrate (GABA) regulated plant defense: mechanisms and opportunities. *Plants* 2021, 10: 1939. (s Note: MDPI stays neutral with regard to jurisdictional claims in published . . .).

Sheng, Q., Wu, X.Y., Xu, X. et al. (2021). Production of l-glutamate family amino acids in *Corynebacterium glutamicum*: physiological mechanism, genetic modulation, and prospects. *Synthetic and Systems Biotechnology* 6: 302–325.

Shi, F., Jiang, J., Li, Y. et al. (2013). Enhancement of γ-aminobutyric acid production in recombinant *Corynebacterium glutamicum* by co-expressing two glutamate decarboxylase genes from *Lactobacillus brevis*. *Journal of Industrial Microbiology and Biotechnology* 40: 1285–1296.

Shi, F. and Li, Y. (2011). Synthesis of γ-aminobutyric acid by expressing *Lactobacillus brevis*-derived glutamate decarboxylase in the *Corynebacterium glutamicum* strain ATCC 13032. *Biotechnology Letters* 33: 2469–2474.

Shi, F., Xie, Y., Jiang, J. et al. (2014). Directed evolution and mutagenesis of glutamate decarboxylase from *Lactobacillus brevis* Lb85 to broaden the range of its activity toward a near-neutral pH. *Enzyme and Microbial Technology* 61–62: 35–43.

Shi, X., Chang, C., Ma, S. et al. (2017). Efficient bioconversion of l-glutamate to γ-aminobutyric acid by *Lactobacillus brevis* resting cells. *Journal of Industrial Microbiology and Biotechnology* 44: 697–704.

Shimada, M., Hasegawa, T., Nishimura, C. et al. (2009). Anti-hypertensive effect of gamma-aminobutyric acid (GABA)-rich Chlorella on high-normal blood pressure and borderline hypertension in placebo-controlled double blind study. *Clinical and experimental hypertension (New York, N.Y.: 1993)* 31: 342–354.

Shimizu, H. and Hirasawa, T. (2007). Production of glutamate and glutamate-related amino acids: molecular mechanism analysis and metabolic engineering. In: *Amino Acid Biosynthesis~ Pathways, Regulation and Metabolic Engineering* (ed. V.F. Wendisch), 1–38. Physica-Verlag.

Shin, S.-M., Kim, H., Joo, Y. et al. (2014). Characterization of glutamate decarboxylase from *Lactobacillus plantarum* and its C-terminal function for the pH dependence of activity. *Journal of Agricultural and Food Chemistry* 62: 12186–12193.

Singh, U., Tiwari, P., Kelkar, S. et al. (2023). Edible mushrooms: a sustainable novel ingredient for meat analogs. *eFood* 4: e122.

Sita, K. and Kumar, V. (2020). Role of gamma amino butyric acid (GABA) against abiotic stress tolerance in legumes: a review. *Plant Physiology Reports* 25: 654–663.

Son, J., Baritugo, K.-A., Sohn, Y.J. et al. (2022). Production of γ-aminobutyrate (GABA) in recombinant *Corynebacterium glutamicum* by expression of glutamate decarboxylase active at neutral pH. *ACS Omega* 7: 29106–29115.

Son, J., Sohn, Y.J., Baritugo, K.-A. et al. (2023). Recent advances in microbial production of diamines, aminocarboxylic acids, and diacids as potential platform chemicals and bio-based polyamides monomers. *Biotechnology Advances* 62: 108070.

Song, C., Zhu, L., Shao, Y., and Chen, F. (2021). Enhancement of GABA content in Hongqu wine by optimisation of fermentation conditions using response surface methodology. *Czech Journal of Food Sciences* 39: 297–304.

Song, H.Y. and Yu, R.C. (2018). Optimization of culture conditions for gamma-aminobutyric acid production in fermented adzuki bean milk. *Journal of Food and Drug Analysis* 26: 74–81.

Su, Y., Cheng, Z., Liu, W. et al. (2023). Effects of rumen-protective γ-aminobutyric acid additive on lactation performance and serum biochemistry in heat-stressed cows. *Frontiers in Veterinary Science* 10: 1228155.

Sultana, S., Hossain, M., Islam, M., and Kawsar, S. (2024). Antifungal potential of mannopyranoside derivatives through computational and molecular docking studies against *Candida albicans* 1IYL and 1AI9 proteins. *Current Chemistry Letters* 13: 1–14.

Sun, L., Bai, Y., Zhang, X. et al. (2021). Characterization of three glutamate decarboxylases from *Bacillus* spp. for efficient γ-aminobutyric acid production. *Microbial Cell Factories* 20: 1–12.

Suwanmanon, K. and Hsieh, P.-C. (2014). Isolating *Bacillus subtilis* and optimizing its fermentative medium for GABA and nattokinase production. *CyTA Journal of Food* 12: 282–290.

Tajabadi, N., Ebrahimpour, A., Baradaran, A. et al. (2015). Optimization of γ-aminobutyric acid production by *Lactobacillus plantarum* Taj-Apis362 from honeybees. *Molecules* 20: 6654–6669.

Takahashi, C., Shirakawa, J., Tsuchidate, T. et al. (2012). Robust production of gamma-amino butyric acid using recombinant *Corynebacterium glutamicum* expressing glutamate decarboxylase from *Escherichia coli*. *Enzyme and Microbial Technology* 51: 171–176.

Tanamool, V., Hongsachart, P., and Soemphol, W. (2019). Screening and characterisation of gamma-aminobutyric acid (GABA) producing lactic acid bacteria isolated from Thai fermented fish (Plaa-som) in Nong Khai and its application in Thai fermented vegetables (Som-pak). *Food Science and Technology* 40: 483–490.

Thongruck, K. and Maneerat, S. (2023). Enhanced Production of Gamma-Aminobutyric Acid (GABA) from *Lactobacillus futsaii* CS3 Using Agri-Food Industries By-Products Under Batch and Fed-Batch Fermentation. *Indian Journal of Microbiology* 63: 467–482.

Tamés, H., Sabater, C., Royo, F. et al. (2024). Mouse intestinal microbiome modulation by oral administration of a GABA-producing *Bifidobacterium adolescentis* strain. *Microbiology Spectrum* 12: e02580–e02523.

Tiranini, L. and Nappi, R.E. (2022). Recent advances in understanding/management of premenstrual dysphoric disorder/premenstrual syndrome. *Faculty Reviews* 11: 11.

Tumewu, S.A., Matsui, H., Yamamoto, M. et al. (2020). Requirement of γ-aminobutyric acid chemotaxis for virulence of *Pseudomonas syringae* pv. *tabaci* 6605. *Microbes and Environments* 35: ME20114.

Tung, Y.T., Lee, B.H., Liu, C.F., and Pan, T.M. (2011). Optimization of culture condition for ACEI and GABA production by lactic acid bacteria. *Journal of Food Science* 76: M585–M591.

Ueno, H. (2000). Enzymatic and structural aspects on glutamate decarboxylase. *Journal of Molecular Catalysis B: Enzymatic* 10: 67–79.

Ullah, A., Ali, I., Noor, J. et al. (2023). Exogenous γ-aminobutyric acid (GABA) mitigated salinity-induced impairments in mungbean plants by regulating their nitrogen metabolism and antioxidant potential. *Frontiers in Plant Science* 13: 1081188.

Verni, M., Rizzello, C.G., and Coda, R. (2019). Fermentation biotechnology applied to cereal industry by-products: Nutritional and functional insights. *Frontiers in Nutrition* 6: 42.

Villegas, J.M., Brown, L., de Giori, G.S., and Hebert, E.M. (2016). Optimization of batch culture conditions for GABA production by *Lactobacillus brevis* CRL 1942, isolated from quinoa sourdough. *LWT- Food Science and Technology* 67: 22–26.

Wang, Q., Liu, X., Fu, J. et al. (2018). Substrate sustained release-based high efficacy biosynthesis of GABA by *Lactobacillus brevis* NCL912. *Microbial Cell Factories* 17: 1–8.

Wang, X., Wang, Y., Nan, B. et al. (2024). Optimization of Fermentation for gamma-aminobutyric acid (Gaba) production by *Lactobacillus plantarum* Lp3 and the development of fermented soymilk. *LWT 195*: 115841.

Wang, Y.-L., Zhang, Z.-H., Wang, W.-K. et al. (2023). The effect of γ-aminobutyric acid addition on *in vitro* ruminal fermentation characteristics and methane production of diets differing in forage-to-concentrate ratio. *Fermentation* 9: 105.

Wang, C., Zhang, H.B., Wang, L.H., and Zhang, L.H. (2006). Succinic semialdehyde couples stress response to quorum-sensing signal decay in *Agrobacterium tumefaciens. Molecular Microbiology* 62: 45–56.

Wang, D., Wang, Y., Lan, H. et al. (2021). Enhanced production of γ-aminobutyric acid in litchi juice fermented by *Lactobacillus plantarum* HU-C2W. *Food Bioscience* 42: 101155.

Wang, H., Huang, J., Sun, L. et al. (2019). An efficient process for co-production of γ-aminobutyric acid and probiotic *Bacillus subtilis* cells. *Food Science and Biotechnology* 28: 155–163.

Wan-Mohtar, W.A.A.Q.I., Ab Kadir, S., Halim-Lim, S.A. et al. (2019). Vital parameters for high gamma-aminobutyric acid (GABA) production by an industrial soy sauce koji *Aspergillus oryzae* NSK in submerged-liquid fermentation. *Food Science and Biotechnology* 28: 1747–1757.

Watanabe, Y., Hayakawa, K., and Ueno, H. (2011). Effects of co-culturing LAB on GABA production. *Journal of Biological Macromolecules* 11: 3–13.

Wu, C.-H., Hsueh, Y.-H., Kuo, J.-M., and Liu, S.-J. (2018). Characterization of a potential probiotic *Lactobacillus brevis* RK03 and efficient production of γ-aminobutyric acid in batch fermentation. *International Journal of Molecular Sciences* 19: 143.

Wu, Q., Tun, H.M., Law, Y.-S. et al. (2017). Common distribution of gad operon in *Lactobacillus brevis* and its GadA contributes to efficient GABA synthesis toward cytosolic near-neutral pH. *Frontiers in Microbiology* 8: 206.

Xiao T., Shah NP., 2022. Role of cysteine in the improvement of γ-aminobutyric acid production by nonproteolytic *Levilactobacillus brevis* in coculture with Streptococcus thermophilus. *Journal of Dairy Science* 105, 3883–3895.

Xiong, Q., Xu, Z., Xu, L. et al. (2017). Efficient production of γ-GABA using recombinant *E. coli* expressing glutamate decarboxylase (GAD) derived from eukaryote *Saccharomyces cerevisiae. Applied Biochemistry and Biotechnology* 183: 1390–1400.

Xu, N., Wei, L., and Liu, J. (2017). Biotechnological advances and perspectives of gamma-aminobutyric acid production. *World Journal of Microbiology and Biotechnology* 33: 1–11.

Yang, S.-Y., Lü, F.-X., Lu, Z.-X. et al. (2008). Production of γ-aminobutyric acid by *Streptococcus salivarius* subsp. *thermophilus* Y2 under submerged fermentation. *Amino Acids* 34: 473–478.

Yang, T., Rao, Z., Kimani, B.G. et al. (2015). Two-step production of gamma-aminobutyric acid from cassava powder using *Corynebacterium glutamicum* and *Lactobacillus plantarum. Journal of Industrial Microbiology and Biotechnology* 42: 1157–1165.

Yao, C., Liu, Y., Hu, X., and Wang, X. (2022). Improve gamma-aminobutyric acid production in *Corynebacterium glutamicum* by optimizing the metabolic flux. *Systems Microbiology and Biomanufacturing* 2 (9): 1–12.

Yao, C., Shi, F., and Wang, X. (2023). Chromosomal editing of *Corynebacterium glutamicum* ATCC 13032 to produce gamma-aminobutyric acid. *Biotechnology and Applied Biochemistry* 70: 7–21.

Ye, D.Y., Moon, J.H., and Jung, G.Y. (2023). Recent progress in metabolic engineering of *Escherichia coli* for the production of various C4 and C5-dicarboxylic acids. *Journal of Agricultural and Food Chemistry* 71: 10916–10931.

Yogeswara, I.B.A., Kittibunchakul, S., Rahayu, E.S. et al. (2020). Microbial production and enzymatic biosynthesis of γ-aminobutyric acid (GABA) using *Lactobacillus plantarum* FNCC 260 isolated from Indonesian fermented foods. *Processes* 9: 22.

Yoon, W.-K., Garcia, C.V., Kim, C.-S., and Lee, S.-P. (2018). Fortification of mucilage and GABA in Hovenia dulcis extract by co-fermentation with Bacillus subtilis HA and *Lactobacillus plantarum* EJ2014. *Food Science and Technology Research* 24: 265–271.

Yuan, D., Wu, X., Gong, B. et al. (2023). GABA metabolism, transport and their roles and mechanisms in the regulation of abiotic stress (hypoxia, salt, drought) resistance in plants. *Metabolites* 13: 347.

Yunes, R., Poluektova, E., Dyachkova, M. et al. (2016). GABA production and structure of gadB/gadC genes in *Lactobacillus* and *Bifidobacterium* strains from human microbiota. *Anaerobe* 42: 197–204.

Zamanpour, S., Bakhtiari, J.P., and Afshari, A. (2024). Gamma-aminobutyric acid enrichment optimized methods in probiotics foods: a systematic review. *Nutrition & Food Science* 54: 131–150.

Zareian, M., Ebrahimpour, A., Mohamed, A.K.S., and Saari, N. (2013). Modeling of glutamic acid production by *Lactobacillus plantarum* MNZ. *Electronic Journal of Biotechnology* 16: 12–12.

Zareie, Z., Tabatabaei, Y.F., and Mortazavi, S.A. (2019). Optimization of gamma-aminobutyric acid production in a model system containing soy protein and inulin by *Lactobacillus brevis* fermentation. *Journal of Food Measurement and Characterization* 13: 2626–2636.

Zhang, Q., Zeng, L., Tan, X. et al. (2017). An efficient γ-aminobutyric acid (GABA) producing and nitrite reducing ability of *Lactobacillus plantarum* BC114 isolated from Chinese Paocai. *Food Science and Technology Research* 23: 749–755.

Zhang, R., Yang, T., Rao, Z. et al. (2014). Efficient one-step preparation of γ-aminobutyric acid from glucose without an exogenous cofactor by the designed *Corynebacterium glutamicum*. *Green Chemistry* 16: 4190–4197.

Zhang, L., Yue, Y., Wang, X. et al. (2022). Optimization of fermentation for γ-aminobutyric acid (GABA) production by yeast *Kluyveromyces marxianus* C21 in okara (soybean residue). *Bioprocess and Biosystems Engineering* 45: 1111–1123.

Zhang, Y., Zhu, M., Lu, W. et al. (2023). Optimizing *Levilactobacillus brevis* NPS-QW 145 fermentation for gamma-aminobutyric acid (GABA) production in soybean sprout yogurt-like product. *Food* 12: 977.

Zhao, Z., Ding, J.-Y., Ma, W.-H. et al. (2012). Identification and characterization of γ-aminobutyric acid uptake system GabP Cg (NCgl0464) in *Corynebacterium glutamicum*. *Applied and Environmental Microbiology* 78: 2596–2601.

Zhu, F., Ju, Y., Wang, W. et al. (2020). Metagenome-wide association of gut microbiome features for schizophrenia. *Nature Communications* 11: 1612.

18

Promising GRAS Strains for Production of Nutraceuticals

Sanjay Kala[1], Shashank Singh[2], Chayanika Kala[3], and Anurag Singh[1]

[1] *Department of General Surgery, G.S.V.M. Medical College, Kanpur, 208002, Uttar Pradesh, India*
[2] *Multi-Disciplinary Research Unit, G.S.V.M. Medical College, Kanpur, 208002, Uttar Pradesh, India*
[3] *Department of Pathology, G.S.V.M. Medical College, Kanpur, 208002, Uttar Pradesh, India*

18.1 Introduction

18.1.1 Nutraceuticals

Nutraceuticals are food-based products that offer extra health advantages that go beyond the fundamental nutritional content present in food. They can be used for therapeutic or wellness reasons such as illness prevention and treatment (DeFelice 1995). The term "nutraceutical," which combines the words "nutrition" and "pharmaceutical," was first used in 1989 by Stephen L. DeFelice. Later, "a product isolated or purified from foods that is generally sold in medicinal forms not usually associated with food" was added to the definition of nutraceutical (Pandey et al. 2010). Unlike functional foods and medications, nutraceuticals have lately become more focused on the functional and structural aspects of diverse bioactive substances that have long-term medical or physiological benefits in addition to solely nutritional or direct pharmacological effects. Natural food sources, such as vegetables, fruits, spices, herbs, and livestock products, are frequently the source of nutraceuticals. Additionally, they can be purified and isolated into a variety of forms, like liquids, capsules, tablet forms, and powders. They can be consumed as prebiotics and probiotics (*Lactobacillus*, *bifidobacterial*, and indigestible carbohydrate), functional foods (fortified grains, yogurt with introduced probiotics, and omega-3 enriched eggs), therapeutic meals (meals prepared under medical guidance for certain illnesses or medical state), and herbal supplements (ginseng, echinacea, and turmeric). Because of their ability to promote wellness and cure ailments, applications are generally increasingly being used. This is particularly relevant when it comes to preventing illnesses linked to aging, such as oxidative anxiety, depressive disorder, inflammatory disorders, osteoporosis, gastrointestinal disorders, cardiovascular conditions, cancer, and diabetes (Jain and Ramawat 2013). The market for nutraceuticals has grown significantly as a result of ongoing increases in consumer demand and curiosity toward preserving human wellness via food. A recent analysis by Global Information Inc. estimates that the U.S. nutraceutical industry was worth approximately $75.9 billion, whereas the international market had been anticipated to be worth USD 171.8 billion in 2014 and reach USD 241.1 billion by 2019 (Jain and Ramawat 2013). Moreover,

Microbial Nutraceuticals: Products and Processes, First Edition. Edited by Sudhir Pratap Singh and Santosh Kumar Upadhyay.

the marketplace is expected to expand rapidly at a compound annual growth rate (CAGR) of higher than 4.5% from 2023 to 2032 due to the increasing incidence of metabolic ailments around the globe. In 2022, the market was valued at greater than USD 423 billion.

18.1.2 Generally Regarded As Safe (GRAS) Strains

Based on empirical data or an extended record of widespread usage in food, these are microbial strains that the U.S. Food and Drug Administration (FDA) has deemed harmless for the intended application in food items. With this classification, the FDA's preliminary market authorization is not anymore necessary regarding these strains, which are deemed safe for intake. All of these strains must have passed thorough safety assessments by professionals who conclude that they are safe for usage in the intended environments (e.g., lactic acid bacteria, yeasts, and Bacillus species). A prolonged tradition of secure usage in food prior to 1958, when the designation of GRAS was created, might serve as the basis for the safety determination, or it can be based on scientific techniques, such as published investigations and professional opinions. The beverage and food industry makes extensive use of GRAS strains, especially when producing foods that are fermented, probiotics, enzymes, and other additives for food. Additionally, they are utilized as biological preservation agents in nutritional supplements, preventing the growth of spoilage organisms and extending the possible lifespan of food. Any GRAS classification is limited to a strain's specific applications, such as fermentation, and does not apply to other purposes, such as usage as a medicinal agent.

A comprehensive and ongoing study is necessary to guarantee that a strain's GRAS certification is still genuine. Anyone who is curious about the specifics of a given GRAS determination can view the FDA's public inventory of GRAS notices. When it comes to food safety along with innovation, GRAS strains are essential. They provide producers with certainty that their microorganisms can be used safely in a variety of contexts. By easing the production of fresh products while upholding strict safety requirements, the GRAS accreditation aids in streamlining the certification procedure.

We shall discuss the GRAS strains and how they are utilized in this chapter, as they are commonly employed in the manufacturing of nutritional supplements.

18.1.2.1 *Lactobacillus* Strains

One of the beneficial bacterial species most frequently researched and utilized in nutraceuticals is *Lactobacillus* species. These microorganisms, which belong to the lactic acid bacteria (LAB) group, frequently appear in food that has been fermented and inside the human gastrointestinal system, wherein their presence is essential for good wellness and proper gut function. LAB can be categorized into two groups based on their end products of carbohydrate fermentation: homofermentative strains, which convert glucose into lactic acid as the primary end product, and heterofermentative strains, which convert glucose into lactic acid, ethanol or acetic acid, and carbon dioxide (CO_2) as end products of fermentation (Mayo et al. 2010). The following *Lactobacillus* strains are frequently utilized in the manufacturing of nutraceuticals:

a) ***Lactobacillus rhamnosus*:** Initially, this strain was classified as a subspecies of *Lactobacillus casei*. However, via genomic research, it was discovered to be a distinct species within the *L. casei* clade, which also includes *Lactobacillus paracasei* and *Lactobacillus zeae* (Wuyts et al. 2017; Liu and Gu 2020). *Lactobacillus rhamnosus* is a Gram-positive bacterium that is short, rod-shaped, non-spore-forming and often forms chains. It is a homofermentative facultative anaerobe, meaning it can produce lactic acid as its main metabolic product and can survive in both oxygen-rich and oxygen-poor environments. Certain strains of *L. rhamnosus* bacteria are utilized in the production of probiotic preparations such as capsules and powders, as well as functional foods like yogurts and fortified beverages. These uses have been linked to enhanced gut health, increased immune function, and a decreased likelihood of gastrointestinal infections.

b) ***Lactobacillus acidophilus***: The strain was initially obtained from the human gastrointestinal tract in 1900 by Ernst Moro and was originally named *Bacillus acidophilus*. *Lactobacillus acidophilus* is a bacterium that has a rod-shaped structure, is Gram-positive, and is capable of fermenting sugars without producing any by-products other than lactic acid. It was initially discovered in 1900 from the feces of infants. This species is frequently encountered in people, particularly in the gastrointestinal system and oral cavity, as well as in certain specialized fermented foods such fermented milk or yogurt. The species exhibits optimal growth at a temperature of 37°C and is most commonly found in environments with low pH levels, specifically below 5.0. Specific strains of *L. acidophilus* exhibit potent probiotic properties and are commonly employed in the mass manufacturing of dairy products such as yogurts and other fermented items.

c) ***Lactobacillus plantarum***: This particular strain was obtained from human saliva and is a commonly found member of the *Lactiplantibacillus* genus. This strain is a Gram-positive bacterium that has a rod-shaped morphology with rounded ends. It can be found as individual cells or in pairs or short chains (Landete et al. 2010; Huang et al. 2021). *Lactobacillus plantarum* is predominantly present in fermented food items like sauerkraut, kimchi, pickles, and functional beverages such as probiotic drinks and kombucha (Kleerebezem et al. 2003). Within the human body, *L. plantarum* is recognized for its capacity to endure in the acidic conditions of the stomach and aid in maintaining a healthy balance of gut microbiota. It demonstrates potent antioxidant effects, aids in reducing bloating, and enhances symptoms of irritable bowel syndrome (IBS). This strain boosts immunological function by augmenting the body's defense mechanisms through the synthesis of bacteriocins.

d) ***Lactobacillus fermentum***: *L. fermentum* is a type of bacteria that is Gram-positive, meaning it retains a stain used in microbiology, and has a shape that can be either rod-like or spherical. It is capable of fermenting different substances and can survive in environments with or without oxygen. This bacteria is commonly found in fermented cereals, plant materials undergoing fermentation, dairy products, manure, sewage, feces, and the human vagina (Cvetković et al. 2019). This strain is linked to improved intestinal health and enhances the immune system. It is also used in the commercial manufacturing of industrial enzymes such as X-prolyl dipeptidyl aminopeptidase, L-arabinose isomerase, and glutamate decarboxylase (Dowarah et al. 2018). This strain also has antioxidant characteristics while undergoing fermentation and promotes the growth and regulation of other gut bacteria (Zhao et al. 2019; Sampaio et al. 2022).

18.1.2.2 *Bifidobacterium* Strains

Henri Tissier was the pioneering individual to separate *Bifidobacterium* from the excrement of breastfed infants and officially classified it as *Bacillus bifidus* (Tissier 1899; Tissier 1900). In 1924, Orla-Jensen suggested the creation of the genus *Bifidobacterium* to include *B. bifidus*. Later, Poupard et al. (Poupard et al. 1973) further categorized it as a distinct taxonomic group and identified it as the genus *Bifidobacterium*, which comprises 11 species. *Bifidobacterium* belongs to the phylum *Actinobacteria*, which is one of the largest phyla and consists of 5 subclasses, 6 orders, and 14 suborders. These bacteria exhibit pleomorphic shapes, are Gram-positive, lack catalase activity, are stationary, and do not form spores. Here are several *bifidobacterial* strains that are frequently employed in the manufacturing of nutraceuticals:

a) ***Bifidobacterium bifidum***: *B. bifidum* is a type of bacterium that is Gram-positive, catalase-negative, γ-hemolytic, and anaerobic. It is not capable of movement and does not produce spores (Hoover 2014). The bacterium exhibits a rod-shaped morphology and can be observed residing in clusters, pairs, or even in isolation. *B. bifidum* is predominantly located in the colon, lower small intestine, breast milk, and frequently in the female reproductive tract (Palmer et al. 2007). *Bifidobacterium bifidum* is mostly utilized in functional foods, such as yogurts, kefir, and other fermented dairy products, to improve digestive and metabolic well-being. It is a vital ingredient in infant formulas that imitate the composition of breast milk and aid in the development of the gut microbiota in newborns, hence promoting digestion and immunity.
b) ***Bifidobacterium longum***: These bacteria are Gram-positive, catalase-negative, γ-hemolytic, and have a rod-shaped morphology. They can be found in the human gastrointestinal tract. The *Bifidobacterium longum* strain is a microaerotolerant anaerobe and is known to be one of the first organisms to inhabit the gastrointestinal tract of babies (Schell et al. 2002). This strain is also present in the gastrointestinal tract and breast milk. *B. longum* has been utilized in symbiotic formulations in the form of capsules and powder. *B. longum* is present in non-dairy products such as probiotic-rich soy and coconut yogurts, which are suitable for individuals with lactose intolerance. It has also been utilized in newborn formulas to promote gastrointestinal health. Individuals with particular medical disorders, such as IBS, inflammatory bowel disease (IBD), or compromised immune systems, frequently incorporate *B. longum* into their regimen to obtain unique health advantages.
c) ***Bifidobacterium animalis***: This bacterium is Gram-positive, anaerobic, rod-shaped, catalase-negative, γ-hemolytic bacteria that mostly colonize the human gastrointestinal tract. *B. lactis* maintains equilibrium in the gut microbiota by inhibiting the proliferation of pathogenic bacteria and facilitating the proliferation and maturation of advantageous bacteria. *B. lactis* has been widely utilized in the production of newborn formulas, dietary supplements, and fermented milk products. This has been achievable thanks to its advantageous technological attributes, including its fermentation activity, high air tolerance, stability, and susceptibility to acidic and bile conditions, even in freeze-dried products. This strain possesses efficacy in both preventing and treating constipation by stimulating regular bowel movements and aiding in the relief of symptoms like bloating, gas, and indigestion. As a result, it proves advantageous for persons suffering from IBS or other gastrointestinal disorders.

18.1.2.3 *Saccharomyces* Species

These strains are classified within the genus of yeasts, belonging to the family *Saccharomycetaceae*, which is part of the phylum Ascomycota and the kingdom Fungi. Several species within this genus play a vital role in the preparation of food and beverages, commonly referred to as brewer's yeast, baker's yeast, and sourdough starter. These fungi are single-celled and obtain nutrients from decaying organic matter. An impressive characteristic of *Saccharomyces* members is their capacity to enzymatically transform sugar into carbon dioxide and alcohol. Yeasts are employed to undergo fermentation of sugars in the production of baked foods, beers, wines, and distilled spirits. The following strains of *Saccharomyces* have been used in the synthesis of nutraceuticals:

a) ***Saccharomyces boulardii***: The tropical yeast was initially discovered in 1923 by French scientist Henri Boulard (Malgoire et al. 2005), who isolated it from the peels of lychee and mangosteen fruits. *S. boulardii* is a well-known yeast that is commonly used in the manufacturing of nutraceuticals due to its advantageous probiotic characteristics. This non-pathogenic yeast strain is well-known for its capacity to promote gastrointestinal well-being by rebalancing the gut microbiota and avoiding or relieving different digestive illnesses, such as antibiotic-associated diarrhea, traveler's diarrhea, and IBS. *S. boulardii* functions by improving the integrity of the intestinal barrier, regulating immunological responses, and exerting anti-inflammatory actions. The use of it in dietary supplements and functional meals aids in preserving a robust digestive system and enhancing general immunological function, rendering it an important constituent in the nutraceutical business.

b) ***Saccharomyces cerevisiae***: *S. cerevisiae*, sometimes known as brewer's yeast or baker's yeast, is a unicellular fungus belonging to the yeast species. Since ancient times, this species has been widely utilized in winemaking, baking, and brewing. According to research, it is thought that this substance was first discovered from the outer layer of grapes (Feldmann 2010). *S. cerevisiae* is a highly adaptable yeast that is often employed in the development of nutraceuticals because of its abundant nutritional content and positive effects on health. This yeast is a highly concentrated provider of B vitamins, proteins, and key minerals, which play a vital role in sustaining energy metabolism, immunological function, and overall health. *S. cerevisiae* is used in supplements to enhance gut health due to its nutritional content and the presence of helpful chemicals that support a balanced microbiome and facilitate digestion. The role of nutraceuticals is to improve overall well-being by using fortified foods and dietary supplements that leverage its inherent health-enhancing characteristics.

18.1.2.4 *Bacillus* Species

These bacteria are Gram-positive, have a rod-shaped structure, and belong to the phylum *Bacillota*. There are now 266 recognized species of these bacteria. Bacillus species can exhibit either obligate aerobism, relying on oxygen, or facultative anaerobism, enabling survival in the absence of oxygen. Certain species of Bacillus have the ability to generate enormous quantities of enzymes, which find use in several industries. For instance, α-amylase produced by these bacteria is utilized in the process of starch hydrolysis, while the protease subtilisin is employed in the manufacturing of detergents. *Bacillus subtilis* is an invaluable model organism for bacterial research. Certain species of Bacillus have the ability to produce and release lipopeptides, specifically surfactins and mycosubtilins (Favaro et al. 2016; Nigris et al. 2018; Paul et al. 2021). Here are some of the Bacillus strains that have been used in the production of nutraceuticals:

a) ***Bacillus coagulans***: The species was initially isolated and characterized in 1915 by B.W. Hammer at the Iowa Agricultural Experiment Station (Hammer 1915). *B. coagulans* is a rod-shaped bacterium that is Gram-positive, catalase-positive, forms spores, is capable of movement, and can survive with or without oxygen. *B. coagulans* is a probiotic bacteria that is gaining popularity in the manufacturing of nutraceuticals. It is well-known for its ability to withstand harsh conditions and its positive effects on health. *B. coagulans*, unlike most probiotics, produces spores that are resistant to heat. This allows it to withstand extreme circumstances, such as high temperatures and acidity, both during processing and in the gastrointestinal system. This strain has the ability to enhance digestive health by re-establishing and sustaining a harmonious equilibrium of gut microbiota, enhancing the process of digestion, and relieving symptoms associated with IBS and other gastrointestinal illnesses. Moreover, scientific studies have demonstrated that *B. coagulans* has the ability to boost immune function, bolster the body's innate defenses, and promote general well-being. Due to its stability and efficacy, it is highly valued as an ingredient in nutritional supplements and functional meals. It serves as a dependable supply of beneficial bacteria that promotes digestive health and general vigor.

b) ***Bacillus subtilis***: This strain is a rod-shaped bacterium that is Gram-positive and catalase-positive. It may be found in soil, as well as in the gastrointestinal system of ruminants, humans, and marine sponges (Errington and Aart 2020; Rahman et al. 2020; Paul and Rahman 2022; European Bioinformatics Institute, n.d.). *B. subtilis* has traditionally been categorized as an obligate aerobe, yet there is evidence suggesting that it can also function as a facultative anaerobe. *B. subtilis* is a resilient probiotic bacteria that is extensively used in the production of nutraceuticals because of its capacity to flourish in various settings and provide substantial health advantages. This bacteria, which forms spores, is highly esteemed for its capacity to withstand harsh conditions like intense heat, acidic surroundings, and extended storage periods. Consequently, it is an excellent choice for incorporation into dietary supplements and functional meals. *B. subtilis* promotes digestive health by fostering a harmonious gut microbiota, augmenting nutritional assimilation, and synthesizing enzymes that facilitate the decomposition of intricate carbohydrates, proteins, and lipids. Additionally, it enhances immunological health by promoting the synthesis of antimicrobial peptides and bolstering the body's innate protective systems. The extensive use of this substance in nutraceuticals is due to its adaptability and effectiveness in promoting gastrointestinal health and overall well-being. It is specifically used to enhance digestion, increase the body's defense, and improve general health.

18.1.2.5 *Streptococcus* Species

This species exhibits Gram-positive characteristics, lacks motility and spore formation, and consists of catalase-negative cocci that are found in pairs or chains (Anon, n.d.). Ancient cultures may lose their Gram-positive nature over time. The majority of streptococci are capable of surviving in both aerobic and anaerobic conditions, whereas a few can only survive in anaerobic conditions. The majority necessitate the use of enriched medium, specifically blood agar. The organisms are categorized into three groups based on the kind of haemolysis observed on blood agar: β-hemolytic (characterized by clear and full lysis of red blood cells), α-hemolytic (characterized by partial and green haemolysis), and γ-hemolytic (characterized by no haemolysis). The subsequent species of streptococci have been extensively employed for human applications in the production of nutraceuticals:

a) ***Streptococcus thermophilus***: This strain belongs to the Gram-positive group, namely the viridans group, and is a facultative fermentative anaerobe (Tannock 2005; Roger et al. 2011). *S. thermophilus* is a prominent probiotic bacterium extensively utilized in the production of nutraceuticals, specifically in the dairy industry for the manufacturing of yogurts and symbiotic formulations. It has a vital function in the fermentation process of yogurt and cheese, where it not only improves the texture and taste but also provides important health advantages. This bacterium possesses the capability to enzymatically degrade lactose, hence enhancing the digestibility of dairy products, particularly for persons afflicted with lactose intolerance. In addition, *S. thermophilus* enhances gut health by fostering a harmonious microbiota and facilitating the synthesis of vital nutrients such as B vitamins. Additionally, it aids in bolstering the immune system by promoting the synthesis of advantageous substances such exopolysaccharides, which possess anti-inflammatory and immuno-modulating characteristics. *S. thermophilus* is highly stable and effective, making it a great ingredient in functional foods and dietary supplements. It provides digestive and immunological health advantages that improve overall well-being.
b) ***Streptococcus salivarius***: These bacteria are spherical in shape, belong to the Gram-positive group, and may survive with or without oxygen. They produce lactic acid and do not produce catalase or oxidase enzymes (Roger et al. 2011). *S. salivarius* establishes itself in the oral cavity and upper respiratory tract of humans, typically forming chains, within a few hours of birth.

As a result, subsequent encounters with the bacteria are generally not harmful. *S. salivarius* is a probiotic bacteria that is becoming increasingly recognized in the manufacturing of nutraceuticals due to its important function in enhancing oral and upper respiratory health. *S. salivarius*, which is naturally present in the human oral cavity, plays a crucial role in maintaining a harmonious oral microbiome (Santacroce et al. 2023). This, in turn, inhibits the proliferation of detrimental bacteria that can lead to dental caries, halitosis, and throat infections. Strains such as *S. salivarius* K12 and M18 are highly esteemed for their capacity to generate bacteriocins, which are innate antimicrobial peptides that impede the growth of harmful bacteria. *S. salivarius* is commonly used in oral probiotics, lozenges, and dietary supplements to promote oral hygiene and decrease the likelihood of disorders including gingivitis, pharyngitis, and halitosis. The significance of its function in regulating the immune response and enhancing dental health highlights its crucial role in the creation of nutraceuticals that target oral and respiratory protection.

18.1.2.6 *Enterococcus faecium*

This strain is characterized by being Gram-positive, exhibiting γ-hemolysis, having a coccal or spherical shape, and appearing in chains or pairs (Ryan et al. 2004). These organisms are facultative anaerobes, meaning they can thrive in a variety of circumstances including basic, acidic, isotonic, and hypertonic environments, and at temperatures between 10 and 45°C. *E. faecium* is a probiotic bacterium utilized in the development of nutraceuticals due to its potential advantages in promoting gastrointestinal health and boosting the immune system (Im et al. 2023). While certain strains of *E. faecium* can cause disease, specific non-pathogenic strains are deliberately chosen for inclusion in nutritional supplements and functional meals. These advantageous strains are recognized for their capacity to endure the challenging conditions of the gastrointestinal tract, where they aid in preserving a well-balanced gut microbiota by impeding detrimental bacteria and fostering the proliferation of advantageous ones. *E. faecium* plays a role in the synthesis of short-chain fatty acids, which are essential for maintaining a healthy gut, and in regulating the immune response (Nogal et al. 2021). This makes it beneficial for treating digestive diseases and improving overall gut function. The incorporation of this substance in nutraceuticals is highly esteemed for its ability to enhance digestive health, prevent infections, and bolster the immune system.

18.1.2.7 *Propionibacterium freudenreichii*

These microorganisms are Gram-positive, non-motile, non-spore-producing, anaerobic to facultative anaerobes (Van Niel 1928). They have a pleomorphic rod shape and tend to cluster together, resembling Chinese characters. *P. freudenreichii* is a beneficial bacteria that is extensively utilized in the creation of nutraceuticals, mostly due to its ability to enhance gut health and provide protection. This bacterium is primarily recognized for its utilization in the dairy sector, namely in the fermentation procedure for the creation of Swiss-type cheeses like Emmental. It accomplishes this by generating the distinctive holes and nutty taste that are characteristic of these cheeses. In addition to its involvement in cheese production, *P. freudenreichii* is highly esteemed for its probiotic characteristics, specifically its capacity to generate short-chain fatty acids such as propionate (Cousin et al. 2010). These fatty acids are essential for promoting gastrointestinal well-being and diminishing inflammation. In addition, it synthesizes vitamin B12, a vital component that aids in neurological functioning and energy metabolism. Moreover, *P. freudenreichii* has been researched for its capacity to augment immunological function and safeguard against gastrointestinal illnesses, rendering it a beneficial constituent in functional meals and dietary supplements designed to promote digestive health and overall nutritional status.

18.1.2.8 *Clostridium butyricum*

These bacteria are rod-shaped, anaerobic, spore-forming bacilli that belong to the Gram-positive group. It is infrequently documented as a disease-causing organism in humans and is commonly utilized as a beneficial bacteria in Japan, Korea, and China (Seki et al. 2003). The utilization of *C. butyricum* in nutraceutical production has been growing due to its capacity to promote gut health and enhance overall well-being. This bacterium is specifically recognized for its synthesis of butyrate, a type of short-chain fatty acid that has a crucial function in preserving intestinal well-being. Butyrate functions as the main source of energy for colonocytes, which are cells found in the colon (Cazzaniga et al. 2024). It also plays a role in regulating the integrity of the gut barrier and has anti-inflammatory qualities that can lower the chances of developing gastrointestinal illnesses, such as IBD. *C. butyricum* is also efficacious in restoring equilibrium to the gut microbiome through the inhibition of detrimental pathogens and the stimulation of the proliferation of advantageous bacteria. The spore-forming capabilities of this organism enables it to withstand the acidic conditions of the stomach and remain intact as it travels to the intestines (Koopman et al. 2022). This quality makes it a dependable and efficient ingredient in dietary supplements and functional meals that aim to enhance digestive health and boost immune function.

18.1.2.9 *Leuconostoc mesenteroides*

These bacteria are Gram-positive, non-motile, facultative anaerobes, and non-sporogenous. They have a spherical shape. The organism often assumes a lenticular coccoid morphology, occurring in pairs and chains (Özcan et al. 2019). However, it can alternatively adopt a short rod shape with rounded ends, arranged in elongated chains. *L. mesenteroides* is commonly acknowledged in the field of dairy fermentation for its multifaceted functions, including the synthesis of dextran, gas, and taste compounds. It is known to promote gastrointestinal health and contribute to overall wellness. *L. mesenteroides* is commonly employed in the fermentation process of vegetables, such as sauerkraut and kimchi. It is highly regarded for its capacity to generate lactic acid, which effectively reduces pH levels and establishes an environment that hampers the proliferation of detrimental pathogens. Moreover, this bacterium synthesizes exopolysaccharides, which exhibit prebiotic properties by stimulating the proliferation of advantageous gut microbiota (Miyamoto et al. 2023). Furthermore, these chemicals play a significant role in enhancing the texture and overall quality of fermented foods, hence increasing their palatability and nutritional value. *L. mesenteroides* is incorporated into probiotic supplements and functional foods in the field of nutraceuticals to promote digestive health, boost immunological function, and enhance nutrient absorption. Due to its role in fermentation and its beneficial effects on health, it is a valuable ingredient in products designed to enhance overall well-being via natural methods.

18.1.2.10 *Escherichia coli* Nissle

This bacterium is a Gram-negative, rod-shaped organism, generally around 1–3 μm in length, and was isolated from the feces of a German soldier in 1917 by the German researcher Alfred Nissle. The bacteria exhibit motility owing to its existence with peritrichous flagella, which are arranged circumferentially along its surface, facilitating movement. Similar to other *E. coli* strains, it possesses fimbriae (pili) that facilitate adherence to surfaces. *E. coli Nissle* is utilized as a probiotic strain in nutraceutical production owing to its well-documented health benefits, especially in enhancing gut health. The non-pathogenic strain of *E. coli* is deemed safe and beneficial for repairing and maintaining a balanced gut microbiome (Park et al. 2021). *E. coli* Nissle has demonstrated the ability to suppress the proliferation of pathogenic bacteria by competing for nutrients and adhesion sites in the intestinal epithelium, in addition to synthesizing antimicrobial compounds (Hu et al. 2024). It is commonly utilized in nutraceuticals designed to treat and prevent gastrointestinal problems, including IBS, IBD, and diarrhea. Furthermore, it enhances intestinal barrier function and regulates the immune system, aiding in the reduction in inflammation and fostering general digestive health (Darwish et al. 2022). Its stability, safety profile, and resilience in the adverse conditions of the gastrointestinal tract render it a significant element in probiotic supplements and functional foods.

18.1.2.11 *Torulaspora delbrueckii*

This yeast exhibits an oval to ellipsoidal cell shape, generally ranging from 2 to 5 micrometers in diameter (Ramírez and Velázquez 2018). *T. delbrueckii* produces smooth, creamy to whitish colonies like to other yeasts and possesses a unique yeast cell wall comprised of glucans and mannoproteins. It thrives under both aerobic and anaerobic environments, contingent upon oxygen supply. This yeast is garnering attention in nutraceutical production due to its distinctive fermentation capabilities and health-enhancing attributes, and it is extensively utilized in the food and beverage sectors, particularly in winemaking, brewing, and baking. *T. delbrueckii* improves the flavor, fragrance, and texture of fermented items. Its reduced generation of unwanted by-products, such as acetic acid, in comparison with other yeasts like *S. cerevisiae*, renders it especially advantageous for the manufacturing of high-quality, health-oriented meals. Moreover, *T. delbrueckii* has demonstrated the capacity to produce bioactive molecules, such as polysaccharides and antioxidants, which may enhance gut health, immunological function, and overall well-being (Zhao et al. 2021).

18.1.2.12 *Corynebacterium glutamicum*

Bacteria in this group are Gram-positive, pleomorphic, non-branching, catalase-positive, asporogenous bacilli with irregular morphology. They are omnipresent in the environment and can be extracted from the skin and mucous membranes of animals, plants, dirt, the exterior of smear-ripened cheeses, and freshwater and sea. *Corynebacterium glutamicum* is extensively employed in nutraceutical manufacturing due to its proficiency in synthesizing amino acids, notably L-glutamate and L-lysine, which are vital for human health, as well as its metabolic adaptability and safety. *C. glutamicum* is essential in the industrial biosynthesis of these amino acids, utilized in dietary supplements to enhance muscle function, immunological health, and general nutrition. L-glutamate is essential for gastrointestinal health and neurotransmitter activity. This bacterium has been investigated for the biosynthesis of important chemicals, including vitamins and carotenoids, in addition to amino acid production, which include antioxidant and immune-enhancing effects (Wolf et al. 2021). The GRAS designation of *C. glutamicum* guarantees its safety for use in food and nutraceutical applications, establishing it as a fundamental component in the manufacture of vital nutrients that enhance health and well-being in diverse dietary supplements and functional meals.

18.1.2.13 *Yarrowia lipolytica*

It is an ascomycete with biotechnological potential, capable of forming yeast cells or hyphae and pseudohyphae in response to environmental circumstances (Bellou et al. 2014). *Y. lipolytica* exhibits dimorphic growth, indicating its ability to adopt two distinct phenotypes. The typical cellular morphology is round and spherical; however, under stressful conditions such as temperature fluctuations, pH variations, mechanical stress, or osmotic pressure, the organism can transition to a filamentous or hyphal growth form (Ruiz-Herrera and Sentandreu 2002; Nicaud 2012). *Y. lipolytica* is gaining prominence in nutraceutical production for its capacity to synthesize diverse useful chemicals, such as lipids, organic acids, and bioactive substances. It is especially recognized for its significant ability to store lipids, including omega-3 and omega-6 fatty acids, which are vital for cardiovascular health and the regulation of inflammation. This yeast can synthesize essential nutraceutical substances such as carotenoids, polyunsaturated fatty acids, and antioxidant molecules, which offer considerable health benefits (Gu et al. 2023). Moreover, *Y. lipolytica* is utilized in the synthesis of bioactive peptides and proteins, hence improving the nutritional profiles of dietary supplements and functional meals (Jach and Malm 2022). Its metabolic versatility and capacity to utilize diverse substrates, including waste products, render it a compelling choice for sustainable nutraceutical production. *Y. lipolytica*, with GRAS status, is a promising bacterium for the development of health-enhancing products that promote cardiovascular health, immunological function, and overall well-being.

18.2 Conclusion

In conclusion, several GRAS strains, including *Lactobacillus, Bifidobacteria, Saccharomyces, Bacillus*, *Streptococcus*, *Clostridium butyricum*, and *Leuconostoc mesenteroides*, hold significant promise for the production of nutraceuticals due to their diverse health benefits and robust safety attributes. These microorganisms contribute to various aspects of human health, from supporting immune and digestive functions to enhancing nutrient absorption and producing bioactive compounds. Their stability, resilience, and effectiveness make them ideal candidates for inclusion in dietary supplements, functional foods, and other health-promoting products. The increasing interest in the microbiome's influence on human health necessitates the enhancement of studies incorporating the artificial gut for illustration, thus facilitating the execution of several research related to microbial interactions, generation, digestion, and nutrient absorption, including secondary metabolites generated by these strains. It offers a regulated environment to evaluate the behavior of probiotics and other microbial strains in the gastrointestinal tract. Furthermore, animal studies are necessary to assess the synergistic effects of these GRAS strains; unfortunately, there is a paucity of available research.

As research continues to explore the full potential of these and other GRAS strains, their role in nutraceutical production is likely to expand, offering natural, effective solutions for improving overall wellness and addressing specific health concerns. The continued development of these probiotic and fermentative strains will play a key role in advancing the nutraceutical industry and enhancing public health.

Acknowledgment

The authors are thankful to MRU Lab G.S.V.M. Medical College, funded by the Department of Health Research (DHR), New Delhi, for providing the facility.

Conflict of Interest

The authors declare no conflict of interest.

References

Anon. (n.d.). Streptococcus. *Online Etymology Dictionary*. Retrieved 25 July 2018.

Bellou, S., Makri, A., Triantaphyllidou, I.-E. et al. (2014). Morphological and metabolic shifts of *Yarrowia lipolytica* induced by alteration of the dissolved oxygen concentration in the growth environment. *Microbiology* 160 (4): 807–817. https://doi.org/10.1099/mic.0.074302-0.

Cazzaniga, M., Cardinali, M., Di Pierro, F. et al. (2024). the role of short-chain fatty acids, particularly butyrate, in oncological immunotherapy with checkpoint inhibitors: the effectiveness of complementary treatment with *Clostridium butyricum* 588. *Microorganisms* 12 (6): 1235. https://doi.org/10.3390/microorganisms12061235.

Cousin, F.J., Mater, D.D.G., Foligne, B., and Jan, G. (2010). Dairy propionibacteria as human probiotics: a review of recent evidence. *Dairy Science and Technology* 91: 1–26.

Cvetković, D., Ranitović, A., Savić, D. et al. (2019). Survival of wild strains of lactobacilli during Kombucha fermentation and their contribution to functional characteristics of beverage. *Polish Journal Of Food And Nutrition Sciences* 69 (4): 407–415.

Darwish, M.S., Qiu, L., Taher, M.A. et al. (2022). Health benefits of postbiotics produced by *E. coli* Nissle 1917 in functional yogurt enriched with cape gooseberry (*Physalis peruviana* L.). *Fermentation* 8 (3): 128. https://doi.org/10.3390/fermentation8030128.

DeFelice, S.L. (1995). The nutraceutical revolution: its impact on food industry R&D. *Trends in Food Science & Technology* 6 (2): 59–61.

Dowarah, R., Verma, A.K., Agarwal, N. et al. (2018). Selection and characterization of probiotic lactic acid bacteria and its impact on growth, nutrient digestibility, health and antioxidant status in weaned piglets. *PLoS One* 13 (3): e0192978.

Errington, J. and Aart, L.T. (2020). Microbe profile: *Bacillus subtilis*: model organism for cellular development, and industrial workhorse. *Microbiology* 166 (5): 425–427.

European Bioinformatics Institute. (n.d.). Bacteria Genomes – Streptococcus Thermophilus. European Bioinformatics Institute. Archived from the original on 19 February 2013.

Favaro, G., Bogialli, S., Di Gangi, I.M. et al. (2016). Characterization of lipopeptides produced by *Bacillus licheniformis* using liquid chromatography with accurate tandem mass spectrometry. *Rapid Communications in Mass Spectrometry* 30 (20): 2237–2252.

Feldmann, H. (2010). *Yeast*. John Wiley & Sons.

Gu, Y., Lu, X., Liu, T. et al. (2023). Engineering the oleaginous yeast *Yarrowia lipolytica* to produce nutraceuticals: from metabolic design to industrial applications. *Food Bioengineering* 2 (3): 187–199. https://doi.org/10.1002/fbe2.12062.

Hammer, B.W. (1915). Bacteriological studies on the coagulation of evaporated milk. *Research Bulletin (Iowa Agriculture and Home Economics Experiment Station)* 2 (19): 1.

Hoover, D.G. (2014). *Bifidobacterium*. In: *Encyclopedia of Food Microbiology*, 2e (ed. C.A. Batt and M.L. Tortorello), 216–222. Oxford: Academic Press.

Hu, M., Zhang, T., Miao, M. et al. (2024). Expectations for employing *Escherichia coli* Nissle 1917 in food science and nutrition. *Critical Reviews in Food Science and Nutrition* 8: 1–9. https://doi.org/10.1080/10408398.2023.2301416.

Huang, Z., Zhou, X., Stanton, C. et al. (2021). Comparative genomics and specific functional characteristics analysis of *Lactobacillus acidophilus*. *Microorganisms* 9 (9): 1992.

Im, E.J., Lee, H.H.-Y., Kim, M., and Kim, M.-K. (2023). Evaluation of enterococcal probiotic usage and review of potential health benefits, safety, and risk of antibiotic-resistant strain emergence. *Antibiotics* 12 (8): 1327. https://doi.org/10.3390/antibiotics12081327.

Jach, M.E. and Malm, A. (2022). *Yarrowia lipolytica* as an alternative and valuable source of nutritional and bioactive compounds for humans. *Molecules* 27 (7): 2300. https://doi.org/10.3390/molecules27072300.

Jain, N. and Ramawat, K.G. (2013). *Nutraceuticals and Antioxidants in Prevention of Diseases*, 2559–2580. *Springer eBooks.*

Kleerebezem, M., Boekhorst, J., Van Kranenburg, R. et al. (2003). Complete genome sequence of *Lactobacillus plantarum* WCFS1. *Proceedings of the National Academy of Sciences* 100 (4): 1990–1995.

Koopman, N., Remijas, L., Seppen, J. et al. (2022). Mechanisms and applications of bacterial sporulation and germination in the intestine. *International Journal of Molecular Sciences* 23 (6): 3405. https://doi.org/10.3390/ijms23063405.

Landete, J.M., Rodríguez, H., Curiel, J.A. et al. (2010). *Degradation of Phenolic Compounds Found in Olive Products by Lactobacillus plantarum Strains*, 387–396. *Elsevier eBooks.*

Liu, D.D. and Gu, C.T. (2020). Proposal to reclassify *Lactobacillus zhaodongensis, Lactobacillus zeae, Lactobacillus argentoratensis* and *Lactobacillus buchneri* subsp. *silagei* as *Lacticaseibacillus zhaodongensis* comb. nov., *Lacticaseibacillus zeae* comb. nov., *Lactiplantibacillus argentoratensis* comb. nov. and *Lentilactobacillus buchneri* subsp. *silagei* comb. nov., respectively and *Apilactobacillus kosoi* as a later heterotypic synonym of *Apilactobacillus micheneri. International Journal of Systematic and Evolutionary Microbiology* 70 (12): 6414–6417.

Malgoire, J.Y., Bertout, S., Renaud, F. et al. (2005). Typing of *Saccharomyces cerevisiae* clinical strains by using microsatellite sequence polymorphism. *Journal of Clinical Microbiology* 43 (3): 1133–1137.

Mayo, B., Aleksandrzak-Piekarczyk, T., Fernández, M. et al. (2010). Updates in the metabolism of lactic acid bacteria. In: *Biotechnology of Lactic Acid Bacteria: Novel Applications* (ed. F. Mozzi, R.R. Raya, and G.M. Vignolo), 3–33. Iowa: John Wiley & Sons.

Miyamoto, J., Shimizu, H., Hisa, K. et al. (2023). Host metabolic benefits of prebiotic exopolysaccharides produced by *Leuconostoc mesenteroides. Gut Microbes* 15 (1).

Nicaud, J. (2012). *Yarrowia lipolytica. Yeast* 29 (10): 409–418. https://doi.org/10.1002/yea.2921.

Nigris, S., Baldan, E., Tondello, A. et al. (2018). Biocontrol traits of *Bacillus licheniformis* GL174, a culturable endophyte of *Vitis vinifera* cv. Glera. *BMC Microbiology* 18 (1): 133.

Nogal, A., Valdes, A.M., and Menni, C. (2021). The role of short-chain fatty acids in the interplay between gut microbiota and diet in cardio-metabolic health. *Gut Microbes* 13 (1): https://doi.org/10.1080/19490976.2021.1897212.

Özcan, E., Selvi, S.S., Nikerel, E. et al. (2019). A genome-scale metabolic network of the aroma bacterium *Leuconostoc mesenteroides* subsp. *cremoris. Applied Microbiology and Biotechnology* 103 (7): 3153–3165.

Palmer, C., Bik, E.M., DiGiulio, D.B. et al. (2007). Development of the human infant intestinal microbiota. *PLoS Biology* 5 (7): e177.

Pandey, M., Verma, R.K., and Saraf, S.A. (2010). Nutraceuticals: new era of medicine and health. *Asian Journal of Pharmaceutical and Clinical Research* 3: 11–15.

Park, J., Kim, D.H., Kim, S. et al. (2021). Anti-inflammatory properties of *Escherichia coli* Nissle 1917 in a murine colitis model. *Intestinal Research* 19 (4): 478–481. https://doi.org/10.5217/ir.2021.00121.

Paul, S.I. and Rahman, M.M. (2022). Draft genome sequence of *Bacillus subtilis* YBS29, a potential fish probiotic that prevents motile *Aeromonas* septicemia in *Labeo rohita. Microbiology Resource Announcements* 11 (10): e0091522.

Paul, S.I., Rahman, M.M., Salam, M.A. et al. (2021). Identification of marine sponge-associated bacteria of the Saint Martin's island of the Bay of Bengal emphasizing on the prevention of motile *Aeromonas* septicemia in *Labeo rohita. Aquaculture* 545: 737156.

Poupard, J.A., Husain, I., and Norris, R.F. (1973). Biology of the bifidobacteria. *Bacteriological Reviews* 37 (2): 136–165.

Rahman, M.M., Paul, S.I., Akter, T. et al. (2020). Whole-genome sequence of *Bacillus subtilis* WS1A, a promising fish probiotic strain isolated from marine sponge of the Bay of Bengal. *Microbiology Resource Announcements* 9 (39).

Ramírez, M. and Velázquez, R. (2018). The yeast *Torulaspora delbrueckii*: an interesting but difficult-to-use tool for winemaking. *Fermentation* 4 (4): 94. https://doi.org/10.3390/fermentation4040094.

Roger, P., Delettre, J., Bouix, M., and Béal, C. (2011). Characterization of *Streptococcus salivarius* growth and maintenance in artificial saliva. *Journal of Applied Microbiology* 111 (3): 631–641.

Roger, P., Delettre, J., Bouix, M., and Béal, C. (2011). Characterization of *Streptococcus salivarius* growth and maintenance in artificial saliva: Growth of S. salivarius in saliva. *Journal of Applied Microbiology* 111: 631–641.

Ruiz-Herrera, J. and Sentandreu, R. (2002). Different effectors of dimorphism in *Yarrowia lipolytica*. *Archives of Microbiology* 178 (6): 477–483. https://doi.org/10.1007/s00203-002-0478-3.

Ryan, K.J., Ray, C.G., and Sherris, J.C. (2004). *Sherris Medical Microbiology*, 4the, 294–295. McGraw Hill.

Sampaio, K.B., De Brito Alves, J.L., Nascimento, Y.M.D. et al. (2022). Nutraceutical formulations combining *Limosilactobacillus fermentum*, quercetin, and or resveratrol with beneficial impacts on the abundance of intestinal bacterial populations, metabolite production, and antioxidant capacity during colonic fermentation. *Food Research International* 161: 111800.

Santacroce, L., Passarelli, P.C., Azzolino, D. et al. (2023). Oral microbiota in human health and disease: a perspective. *Experimental Biology and Medicine* 248 (15): 1288–1301.

Schell, M.A., Karmirantzou, M., Snel, B. et al. (2002). The genome sequence of *Bifidobacterium longum* reflects its adaptation to the human gastrointestinal tract. *Proceedings of the National Academy of Sciences* 99 (22): 14422–14427.

Seki, H., Shiohara, M., Matsumura, T. et al. (2003). Prevention of antibiotic-associated diarrhea in children by *Clostridium butyricum* MIYAIRI. *Pediatrics International* 45 (1): 86–90.

Tannock, G.W. (2005). *Probiotics and Prebiotics*, 43. Caister Academic Press.

Tissier, H. (1899). Le bacterium coli et la reaction chromophile d'Escherich. *Critical Review of Social Biology* 51: 943–945.

Tissier, H. (1900). Recherches sur la flore intestinale des nourrissons (état normal et pathologique). M.D. thesis. University of Paris, Paris, France.

Van Niel, C.B. (1928). *The Propionic Acid Bacteria*. Holland: Uitgeverszaak and Boissevain and Co.

Wolf, S., Becker, J., Tsuge, Y. et al. (2021). Advances in metabolic engineering of *Corynebacterium glutamicum* to produce high-value active ingredients for food, feed, human health, and well-being. *Essays in Biochemistry* 65 (2): 197–212. https://doi.org/10.1042/EBC20200134.

Wuyts, S., Wittouck, S., De Boeck, I. et al. (2017). Large-scale phylogenomics of the *Lactobacillus casei* group highlights taxonomic inconsistencies and reveals novel clade-associated features. *mSystems* 2 (4): e00061–e00017.

Zhao, Y.-S., Eweys, A.S., Zhang, J.-Y. et al. (2021). Fermentation affects the antioxidant activity of plant-based food material through the release and production of bioactive components. *Antioxidants* 10 (12): 2004. https://doi.org/10.3390/antiox10122004.

Zhao, Y., Hong, K., Zhao, J. et al. (2019). *Lactobacillus* fermentum and its potential immunomodulatory properties. *Journal of Functional Foods* 56: 21–32.

19

Microalgae: A Sustainable Source for Next-Generation Nutraceuticals

Neha Goel and Poonam Choudhary

Biomanufacturing Process Lab, Department of Biosciences and Bioengineering, Indian Institute of Technology Roorkee, Roorkee, 247667, Uttarakhand, India

19.1 Introduction

In the modern era, diseases are outbreaking globally. According to the Global Health Estimates of the World Health Organization (WHO), the top 10 global causes of death as reported in 2019 include heart diseases, stroke, chronic diseases, lower respiratory infections, neonatal conditions, lung cancers, Alzheimer's disease, diarrheal diseases, diabetes, and kidney diseases (World Health Organization 2019). All these accounted for 55.4 million deaths worldwide. Poor nutrition and lifestyle choices have a big impact on the risk of chronic diseases. Malnutrition, which includes both under- and overnutrition, disturbs off the body's metabolic equilibrium, erodes immunity, and makes people more vulnerable to illness. On the other hand, a healthy diet promotes optimum health and lowers the chance of illness. Leveraging health breakthroughs for better diet and lifestyles requires immediate action. During 460–377 B.C., Hippocrates first stated "let food be thy medicine and medicine be thy food" (Nwosu and Ubaoji 2020). But due to lack of awareness, it could not achieve much momentum. Then, after 3000 years, the term "nutraceuticals" was first coined in 1989 by Stephen L. DeFelice, founder and chairman of Foundation of Innovation Medicine (Andlauer and Fürst 2002). The term "nutraceuticals" is composed of two words—"nutrient" that denotes the nourishing food component and "pharmaceuticals" that refers to a medical drug (Meštrović 2022; Upadhyay and Singh 2024). Combining "nutrient" and "pharmaceutical," it describes food-derived products offering health benefits beyond basic nutrition. Nutraceuticals represent a modern evolution of Hippocrates' vision, leveraging food for proactive health and well-being. Over the past few years, these products have been available in supermarkets, health food shops, and pharmaceutical shops. Precisely, nutraceuticals can be categorized as the dietary supplements, functional foods, medicinal foods, and pharmaceutical products that help in claiming the physiological benefits (Chaudhari and Fanion 2008; Sharma et al. 2024; Bisht et al. 2024). Recently, nutraceuticals have gained importance and have attracted substantial interests due to their fascinating properties (Bagchi 2014). They are being used in improving health, increasing life expectancy by preventing life-threatening diseases, and maintaining the functions and integrity of human body (Das et al. 2012). Initial studies exploring nutraceuticals for COVID-19, a major global health concern, fuel consumer interest in their therapeutic potential, driving market expansion

Microbial Nutraceuticals: Products and Processes, First Edition. Edited by Sudhir Pratap Singh and Santosh Kumar Upadhyay.

(Talukdar et al. 2020). In 2019, the global nutraceuticals market size was USD 247 billion, whereas the compound annual growth rate (CAGR) is expected to rise 8% in upcoming years and the global market should reach USD 336 billion (India's Nutraceutical Market Growth 2019). The primary factors leading to the expansion of nutraceuticals market globally include preventive health care, lifestyle-related disorders, and the demand of consumers for health-promoting diet and products. The major sources of nutraceuticals include natural food, micro-organisms (algae, fungi, yeasts), plants, and animals (Srivastava et al. 2021; Upadhyay and Singh 2021, 2023, 2024; Singh and Upadhyay 2021, 2023). While seaweeds have long been dietary staples, the hidden gems of the marine world—microalgae—are gaining momentum in the exciting realm of nutraceuticals (Guiry 2012). It has been reported that microalgae contribute 40% of the oxygen in the atmosphere (Nicoletti 2016). These tiny powerhouses not only contribute significantly to Earth's oxygen but also boast over 40,000 identified species, each with unique potential (Singh and Dhar 2011). What truly sets them apart is their ease and affordability of cultivation, making them a sustainable and readily available source of valuable compounds (Choudhary et al. 2016). This, coupled with their renewable nature, has ignited research into their ability to produce high-value nutraceutical products, positioning them as potential game-changers in the future of health and wellness (Mehariya et al. 2021).

This chapter delves into the world of microalgal nutraceuticals, exploring their untapped potential and current market trends. It unveils the diverse array of nutraceutical products derivable from microalgae, highlighting their biological significance and health benefits. Moving beyond theory, the chapter dives deep into the practicalities of large-scale microalgal nutraceutical production. It elucidates the intricacies of bioprocessing and downstream processing techniques crucial for efficient product recovery. Furthermore, it navigates the intricate pathways of commercialization, outlining the challenges and strategies for bringing microalgal nutraceuticals to market. Finally, the chapter examines the economic viability and market demand for these innovative products, providing valuable insights for stakeholders.

19.2 Microalgae: A Source for Nutraceutical Products

Microalgae are a diverse group of aquatic organisms having the ability to conduct photosynthesis due to the presence of chlorophyll. Algae can thrive in any habitat ranging from freshwater lakes to the salty oceans (Choudhary et al. 2016). They also endure wide temperature ranges, carbon dioxide concentration, acidic and alkaline environments, etc. (Choudhary et al. 2020). These species can be single cells or even multicellular and complex. Algae can be classified into various categories depending on the life history and reproductive structures, morphological characteristics of cells, chemistry and physical features of cells, assimilatory products of photosynthesis, and nature and properties of the obtained products (Niklas et al. 1978). The major categories of algae include *Rhodophycophyta* (the red algae), *Xanthophycophyta* (yellow-green algae), *Chrysophycophyta* (the golden algae), *Phaeophycophyta* (the brown algae), *Baccillariophycophyta* (diatoms), *Euglenophycophyta* (Euglenoids), *Chlorophycophyta* (the green algae), *Cryptophycophyta* (the Cryptomonads), and *Pyrrophycophyta* (the Dinoflagellates) as illustrated in Figure 19.1.

Algae exist in a wide range of sizes and shapes, which can be simply aggregation of single identical cells or may be quite complex that subsequently superficially resembles higher plants in structure. The cells may be prokaryotic or eukaryotic, and the cell wall also has a variety of characteristics (Douglas et al. 2003). Some algae have thin and rigid cell wall, whereas the cell walls of diatoms are impregnated with silica making them thicker and more complex (Figure 19.1).

Algal species may be surrounded by a flexible, gelatinous outer matrix that is secreted through the cell wall. With the aging of cells, the outer matrix often becomes pigmented and stratified. The

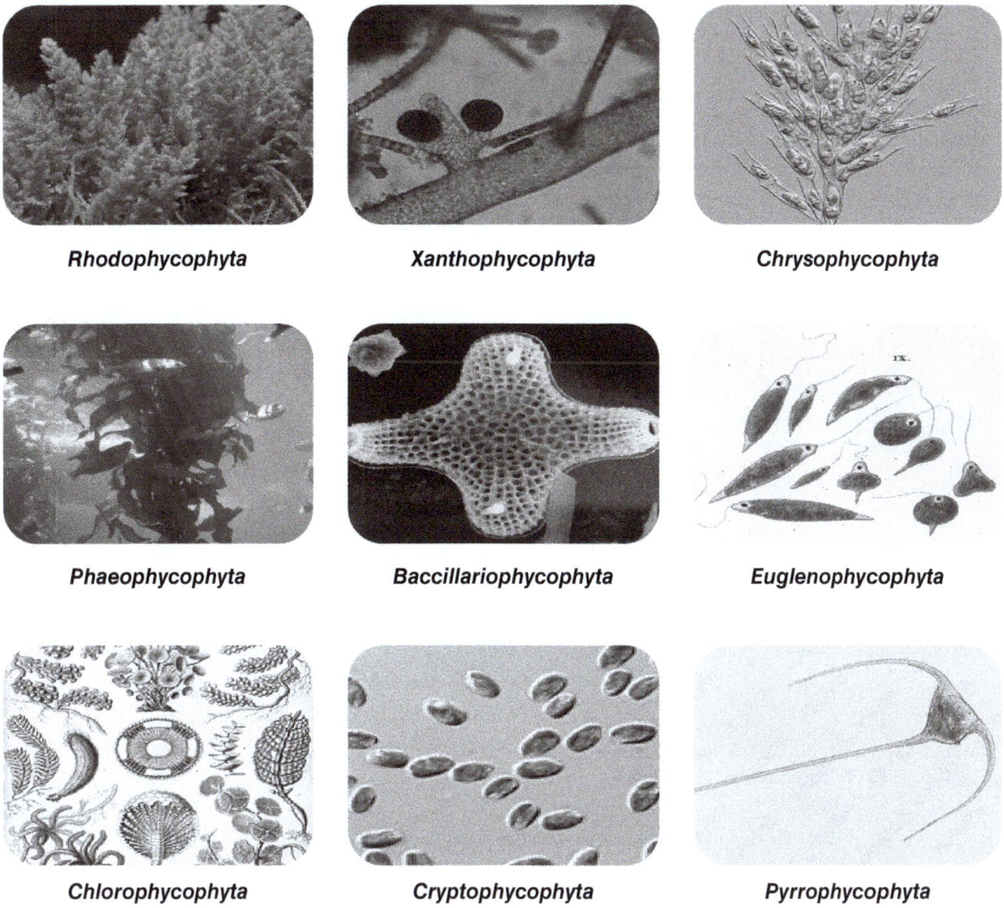

Figure 19.1 Classification of algae. *Source:* damedias/Adobe Stock, Keisotyo/Wikimeadia Commons/CC BY SA 4.0., Frank Fox/Wikimedia Commons/CC BY SA 3.0., Stef/Flickr/CC BY SA 2.0, Photo by Karen Wetmore, CG Ehrenberg/Wikimedia commons/Public domain, Pixabay/Public domain, CSIRO/Wikimedia Commons/CC BY 3.0, Ekky/Adobe Stock.

cell membrane or the periplast of motile algae is flexible in nature. Chloroplasts are the pigment-containing membrane-bounded organelles. They consist of chlorophyll and other pigments synthesized by algae (Bajpai 2019). Thylakoids are flattened membrane vesicles found within the chloroplast matrix. The algal phylogeny exclusively depends on the ultrastructure of chloroplasts and the chemistry of pigments found within the chloroplasts.

This section explains about the microalgae and its biological importance followed by the categories and the significance of nutraceutical products obtained from microalgae.

19.2.1 Microalgae and Its Biological Importance

Microalgae being a source of bioactive compounds have versatile applications in human health. Commonly derived microalgal biomolecules categorized as proteins, carbohydrates, lipids and fatty acids, vitamins, and their applications as nutraceutical products have been discussed in Table 19.1. Within the microalgal cell, carbohydrates are present as cellulose or starch and are

Table 19.1 Microalgal-derived biomolecules and their nutraceutical applications.

Microalgal-derived biomolecules	Microalgal sources	Nutraceutical applications	References
Carbohydrates	*Chlorella* sp., *Dunaliella* sp., *Porphyridium purpureum, Chlorella pyrenoidosa, Chlorella ellipsoidea*	• Immunostimulant activity • Anti-microbial effects	de Souza et al. (2019); Hamed (2016); Li et al. (2020)
Proteins and amino acids	*Scenedesmus* sp., *Chlorella* sp., *Porphyridium aerugineum, Tetraselmis chuii, Tetraselmis* sp.,*Spirulina sp., Botryococcus braunii, Nannochloropsis granulate*	• As human feed and animal feed • Anti-inflammatory activity • Anti-oxidant activity • Reduce allergy symptoms • Protective effect against cellular damage	Schwenzfeier et al. (2011); Tibbetts et al. (2015c); Sathasivam et al. (2019); Tibbetts et al. (2015c)
Lipids and fatty acids	*Chlorella* sp., *Nannochloropsis* sp., *Neochloris oleobundas,Hematococcus* sp., *Spirulina, Chlorella reinhardtii*	• Maintain cellular and tissue metabolism • Prevents cardiovascular diseases • Balances cholesterol level • Anti-microbial activity • Dietary supplements	Hamed (2016); Ördög et al. (2016); Sathasivam et al. (2019); Sun et al. (2019)
Vitamins	Vitamin B—*Chlorella* sp. *Tetraselmis* sp., *Dunaliella tertiolecta* Vitamin C—*Chlorella* sp., *Dunaliella* sp., *Tetraselmis* sp. Vitamin E—*Tetraselmis sp., Tetraselmis suecica, Chlamydomonas reinhardtiiDunaliella tertiolecta* Vitamin K—*Chlamydomonas reinhardtii, Dunaliella salina, Chlorella* sp., *Nannochloropsis oculate, Desmodesmus asymmetricus, Tetraselmis suecica*	• Anticarcinogenic activities • Reduces depression • Protects skin • Help in RBC movement in body • Prevents weakness • Anticarcinogenic • Biosynthesis of neurotransmitter • Food additive and supplement • Antioxidant activities • Anti-diabetic properties • Prevents age-related diseases	Andrade et al. (2018); Koyande et al. (2019); Sathasivam et al. (2019); Simes et al. (2020)
Chlorophyll	Chlorophyll a *(all algae)* Chlorophyll b *Euglenophycophyta and Chlorophycophyta* Chlorophyll c *Xanthophycophyta, Bacillariophycophyta, Chrysophycophyta, Pyrrophycophyta, Phaeophycophyta, Cryptophycophyta* Chlorophyll d *Rhodophycophyta* Chlorophyll e *Xanthophycophyta (Triboneara* and *zoospores* of *Vaucheria)*	• Abundant in vital minerals like magnesium, potassium, iron, calcium, and essential fatty acids, as well as vitamins E, A, C, K, and β-carotene • Aid in reducing inflammation, increasing glucose tolerance, and decreasing weight gain—all of which can help prevent obesity. • Anti-diabetic activity • Can be used as natural colorants and food additives	Aiamla-or et al. (2010); Pelczar et al. (1993); Wunderlich et al. (2020); Zen et al. (2020)

Table 19.1 (Continued)

Microalgal-derived biomolecules	Microalgal sources	Nutraceutical applications	References
Carotenoids	Astaxanthin—*Haematococcus pluvialis,Chlorococcum* sp., *Chlorella zofingiensis, Scenedesmus* sp. β-Carotene—*Chlorella sp., Dunaliella tertiolecta, Dunaliella salina* Lutein—*Haematococcus pluvialis,Chlorella zofingiensis, Tetraselmis* sp. *CTP4, Chlorella zofingiensis,Scenedesmus* sp., *Tetraselmis* sp.,*Chlorococcum* sp., *Chlorella sorokiniana*	• Prevents cardiovascular diseases • Anti-inflammatory activities • Neuroprotective activities • Anticarcinogenic activities • Improve immune response • Used as food colorants • Work in synthesis of Vitamin A • Food supplement • Food colorant • Prevents cataract • Nutrition source for infants • Prevents neurodegenerative diseases • Antioxidant properties	Cordero et al. (2011); Hamed (2016); Kim et al. (2020); Koyande et al. (2019); Lin et al. (2015); Low et al. (2020); Ma and Lin (2010); Ma et al. (2020a); Marino et al. (2020); Mehariya et al. (2019); Molino et al. (2020); Ochoa Becerra et al. (2020); Sathasivam and Ki (2018); Schüler et al. (2020)
Phycobilins	2 types: Phycocyanin and phycoerythrin present only in *Rhodophycophyta* and *Cryptophycophyta*	• Used as food supplements • Anti-cancer activity • Anti-oxidant properties • Neuroprotective activity • Antiviral activity • Immunomodulatory activity	Cuellar-Bermudez et al. (2015); Eriksen (2008); Gantar et al. (2012); Liu et al. (2000); Manirafasha et al. (2016); Mysliwa-Kurdziel and Solymosi (2016); Pelczar et al. (1993)

vital biomolecules. Due to their exceptional antimicrobial, antiviral, and immunostimulant effects, microalgal carbohydrates are being investigated for their potential as nutraceuticals (Table 19.1). It has been reported that the microalgal biomass constitutes 10–52% carbohydrates (Hossain and Mahlia 2019). All species contain different amounts of carbohydrates. With a productivity rate of 15.47 mg/L/day, *Scenedesmus parvus* generated 12,640 mg/L of biomass (K. M. Tan et al. 2020b). Applications for amino acids obtained from microalgae are numerous in the domains of medicine, nutrition, and health (Table 19.1). The cells of *Porphyridium aerugineum* contains tryptophan, arginine, lysine, and leucine. While *Nannochloropsis granulate* has leucine in its cells, other microalgae species including *Tetraselmis chuii* and *Botrycoccus braunii* include arginine (Tibbetts et al. 2015a). Compared to *Dunaliella sp.*and *Chlorella* sp., which have biomass containing roughly 57% and 53.3% of crude protein, respectively, *Scenedesmus* sp. cells have 33–44% protein (Tibbetts et al. 2015b). The high-protein content has been found in *Scenedesmus obliquus* and *Scenedesmus almeriensis* with the percentage extent of 50–55% and 41.7%, respectively (Tomás-Almenar et al. 2018). Lipids, another significant biomolecule obtained from microalgae, are utilized as nutraceuticals and have a wide range of uses (Table 19.1). They support tissue and cell metabolism in addition to being employed as dietary supplements. The microalgae have a lipid content between 30% and 70% of the dry weight of the cell biomass (Sun et al. 2019). The lipid content accumulated in *Chlorella* sp., *Neochloris oleoabundans, Botryococcus Braunii, Dunaliella* sp., and *Nannochloropsis* sp. ranges from 12% to 65.8% (Hossain and Mahlia 2019). Microalgae *Botryococcus terribilis, Botryococcus braunii, Chlorella vulgaris, Chlorella* sp., *Desmodesmus* sp., and *Desmodesmus brasiliensis* produce 37.0 ± 5.3%, 30.2 ± 5.1%, 21.9 ± 3.6%, 21.8 ± 0.7%, 19.4 ± 1.3%, and 14.5±1.7% lipids, respectively. Certain microalgal species, like *Picochlorum* sp. SBL2, *Nannochloris* sp. SBL1, *Nannochloris* sp. SBL4, and *Desmochloris* sp. SBL3, have comparatively greater biomass productivity rates, making them appropriate to produce microalgal-based bioactive compounds (Pereira et al. 2013). Microalgae is a potential source of vitamins having the capability to produce provitamin A, vitamin B, vitamin C, and vitamin K (Hamed 2016; Sathasivam et al. 2019). The synthesis and production of vitamins depend on the growth phase of microalgae and the harvesting conditions. Vitamins C and E are produced at the highest rate in the exponential growth phase, whereas provitamin A and vitamin B2 are found in less amount in the stationary phase (Levasseur et al. 2020).

The major sources of the synthesis of vitamin E are *Chlamydomonas reinhardtii, Tetraselmis tetrathele, Scenedesmus* sp., *Chlamydomonas nivalis, Haematococcus pluvialis, Tetraselmis striata, Desmodesmus* sp., *Botryococcus braunii, Chlorococcum novaeangliae,* and *Chlorella* sp. (Mudimu et al. 2017). The metabolic pathways can also be tuned to increase production. Since microalgae may be synthesized to produce high-value-added compounds (HVACs), they are utilized in the food industry and in the treatment of diseases. Microalgae rely on three main classes of pigments for their photosynthetic pathways and cell growth: chlorophylls (0.5–1.0% of dry weight), carotenoids (typically 0.1–0.2% of dry weight or as high as 14% within some species), and phycobilins (8% of dry weight) (Sun et al. 2023). When compared to other naturally occurring pigment sources like fruits, plants, or animals, microalgae-derived pigments have demonstrated exceptional physiological activities (including antibacterial and antioxidant properties) and an extensive spectrum of health benefits (Sun et al. 2023). Each of them is described in Table 19.1.

Chlorophyll exists in five distinct forms, classified as a, b, c, d, and e. Chlorophyll a is present in all forms of algae, whereas the other types of chlorophyll are present in specific categories of algae only (Table 19.1). Secondly, carotenoids are the secondary metabolites and essential pigments found in the microalgae (Sathasivam and Ki 2018). Carotenoids can be divided into two main categories: carotenes and xanthophylls. Carotenes are linear, unsaturated hydrocarbons, while xanthophylls are often derived from carotenes and contain oxygen atoms. The primary class includes α-carotene, β-carotene, and lutein, whereas the secondary class is composed of astaxanthin and canthaxanthin. Carotenes and xanthophylls have been found predominantly in all forms of algae. Higher amounts of β-carotene have been found in the cells of *Chlorella* sp., *Dunaliella salina,* and *Dunaliella tertiolecta* (Zhu et al. 2020). β-Carotene finds diverse applications across the food, nutraceutical, and pharmaceutical sectors (Sathasivam and Ki 2018). Lutein, a yellow-colored carotenoid, is produced by *Muriellopsis* sp., *Tetraselmis* sp., *Haematococcus pluvialis, Chlamydomonas* sp. JSC4, *Chlorella sorokiniana, Chlorococcum* sp., and *Chlorella zofingiensis* (Cordero et al. 2011).

The production of lutein can be improved by genetic modifications in the bio-carotenogenesis pathway. It has wide applications as supplements in feed, food industries, and human health (Low et al. 2020). Astaxanthin, a red-colored xanthophyll, is found in the cytoplasm of *Chlorococcum* sp., *Chromochloris zofingiensis, Scenedesmus* sp., and *Haematococcus pluvialis,* (Sathasivam et al. 2019). Lastly, phycobilins are the pigment–protein complexes. Their types, sources, and applications are summarized in Table 19.1.

Molecular engineering techniques can be used to increase the synthesis of a variety of bioactive chemicals, allowing the targeted compound to be generated in an optimal quantity to meet market demands. A summary of numerous experiments using genetic transformation to escalate microalgae's synthesis of bioactive compounds is summarized in Table 19.2. To produce high-value compounds, gene silencing, knock-in/knock-out, and overexpression are among the strategies being used (Saini et al. 2020). Recent research indicates that numerous other developments, including Clustered Regularly Interspaced Short Palindromic Repeats-CRISPR-associated protein 9 (CRISPR-Cas9), have been applied to gene editing (Fajardo et al. 2020). It has been reported that some species of microalgae such as *Chromochloris zofingiensis, Chlamydomonas* sp., *Dunaliella* sp., *Chlorella* sp. have been engineered genetically to produce bioactive compounds for nutraceutical applications (Saini et al. 2020).

Table 19.2 Genetic engineering strategies for improving productivity of bioactive compounds from microalgae.

Microalgal species	Genetic modifications	Productivity	References
Chlamydomonas reinhardtii	Plastid-targeted 2-lysophosphatidic acid acyltransferase (*CrLPAAT-1*) for overexpression of LPAAT enzyme	Lipid productivity increased by 20%	Yamaoka et al. (2016)
Phaeodactylum tricornutum	Upregulation of diacylglycerol acyltransferases-2 (*DGAT-2*) genes	The lipid productivity enhanced by 35%	Niu et al. (2013)
Phaeodactulum tricornutum	Knocking down pyruvate carboxylase enzyme using antisense cDNA construct	Enhanced 82% of lipid productivity	Ma et al. (2014)
Chlorella ellipsoidea	Genetic modification in the lipid biosynthesis pathway	Lipid productivity rose from 46.4% to 52.9%	Zhang et al. (2014)
Haematococcus pluvialis	Genetic engineering within the phytoene synthase (*psy*) gene	Enhanced the production of astaxanthin	Steinbrenner and Sandmann (2006)
Haematococcus pluvialis	Endogenous nuclear phytoene desaturase (*pds*) gene expression in the chloroplast	67% higher astaxanthin	Galarza et al. (2018)
Dunaliella viridis	Insertion of two genes, β-carotenoid ketolase (*bkt*) and β-carotene hydroxylase (*CRTR-B*), involved in the production of astaxanthin	Higher amounts of astaxanthin were produced	Lin et al. (2019)
Chlamydomonas reinhardtii	Overexpression of glycerol-3-phosphate acyltransferase (GPAT)	The lipid productivity increased by 50%	Iskandarov et al. (2016)
Nannochloropsis oceanica	Overexpression of diacylglycerol acyltransferases-1 (*DGAT-1*)	Enhanced the lipid productivity by 2.4-fold	Wei et al. (2017)
Chlamydomonas reinhardtii	Overexpression of diacylglycerol acyltransferases-2 (*DGAT-2*)	Increase in lipid productivity	Iwai et al. (2014)
	Knock-out of lipid-oxidizing enzyme	Lipid accumulation was enhanced	Liang and Jiang (2013)
	Upregulation of *psy* gene and downregulation of (Lycopene epsilon cyclase) *Lcye* gene	Enhancement of carotenoid production	Saini et al. (2020)

19.2.2 Nutraceuticals from Microalgae: Types and Significance

From past, many algal species especially the red and brown algae have been used extensively as food. *Porphyra*, one of the species of the red algae cultivated since 1570, is used as food in Japan (Hansen 1981; Baweja et al. 2016). It is commonly called as nori, and after toasting, it is sprinkled over rice or in soup or used to make popular snack called sushi. The nori industry relies mainly on *Porphyra tenera* and *Porphyra yezoensis*. Apart from this, red algae such as *Chondrus, Eucheuma, Acanthopeltis, and Nemalion* are eaten as vegetables (Pelczar et al. 1993). Some are even used in soup or to prepare sweetened jellies. The red algae have been the most significant food source in China for thousands of years. China has also been cultivating the brown algae *Laminaria japonica, Porphyra,* and *Gracilaria,* which are increasingly important in Chinese diet. In Europe and North America, red algae food usage is centered mainly around three genera, viz. *Porphyra, Chrondus,* and *Palmaria*. *Porphyra* has been used extensively in British Isles, with the minors of southern Wales being the major consumers (Pelczar et al. 1993). Growing awareness of the value of algae as a food source has led to a rise in interest in employing microalgae or smaller varieties of algae. Current biotechnology research offers valuable insights into microalgae that could be investigated as a novel nutritional supplement (Kumari et al. 2023).

Some of the bioactive compounds derived from microalgae and their applications are summarized in Figure 19.2 (Kumari et al. 2023). Although they can be found in both natural and artificial environments, microalgae need particular conditions in order to produce significant amounts of biomass (Nicoletti 2016). When cultivated in the right conditions, microalgae produce a nutrient-rich medium that contains every amino acid required for growth. The chemical makeup of microalgae is intricate. To put it another way, it is a combination of vitamins, minerals, and primary and secondary products that provide a wide range of benefits for people. The properties include anti-aging, antioxidant, and nutritional qualities in addition to their preventive effects. Genetic engineering and metabolic modification of microalgal strains have been investigated to address short- and long-term needs for nutritional supplements (Guihéneuf et al. 2016; Vaz et al. 2016). Technology is essential to the conversion of raw resources into a wide range of finished goods (Spolaore et al. 2006). Microalgae have been shown to have several health benefits, including the ability to prevent disease, enhance diet items, and support the rapidly growing nutraceutical industry (Bishop and Zubeck 2012). Depending upon the functionality, the nutraceuticals obtained from microalgae can be majorly categorized into as dietary supplements, functional foods and medicinal foods, and pharmaceuticals as depicted in Figure 19.3.

19.2.2.1 Dietary Supplements

Demand for dietary supplements is always rising in order to feed the world's population. The value of the global market was estimated at USD 140.1 billion in 2018, and by 2026, it is projected to reach USD 216.3 billion (Laamanen et al. 2021). Microalgae are now a potentially valuable source of health-promoting substances such phycobiliproteins, carotenoids, and omega-3 fatty acids (Manirafasha et al. 2016;). But the use of microalgae in food is not new; *Chlorella* sp. has been cultivated in Japan as a protein supplement since the 1960s. In recent times, the growth of *Chlorella* and *Spirulina* sp. has extended to numerous other nations, yielding a diverse array of products such as carotenoids and omega-3 fatty acids (Camacho et al. 2019).

There are two primary approaches to look at microalgae as a dietary supplement: first, as a whole cell that generate large quantities of primary metabolites such as lipids, proteins, and carbohydrates resemble popular foods like grains, nuts, and vegetable crops (Draaisma et al. 2013; Shao et al. 2018); second, by removing particular secondary metabolites—like carotenoids and omega-3

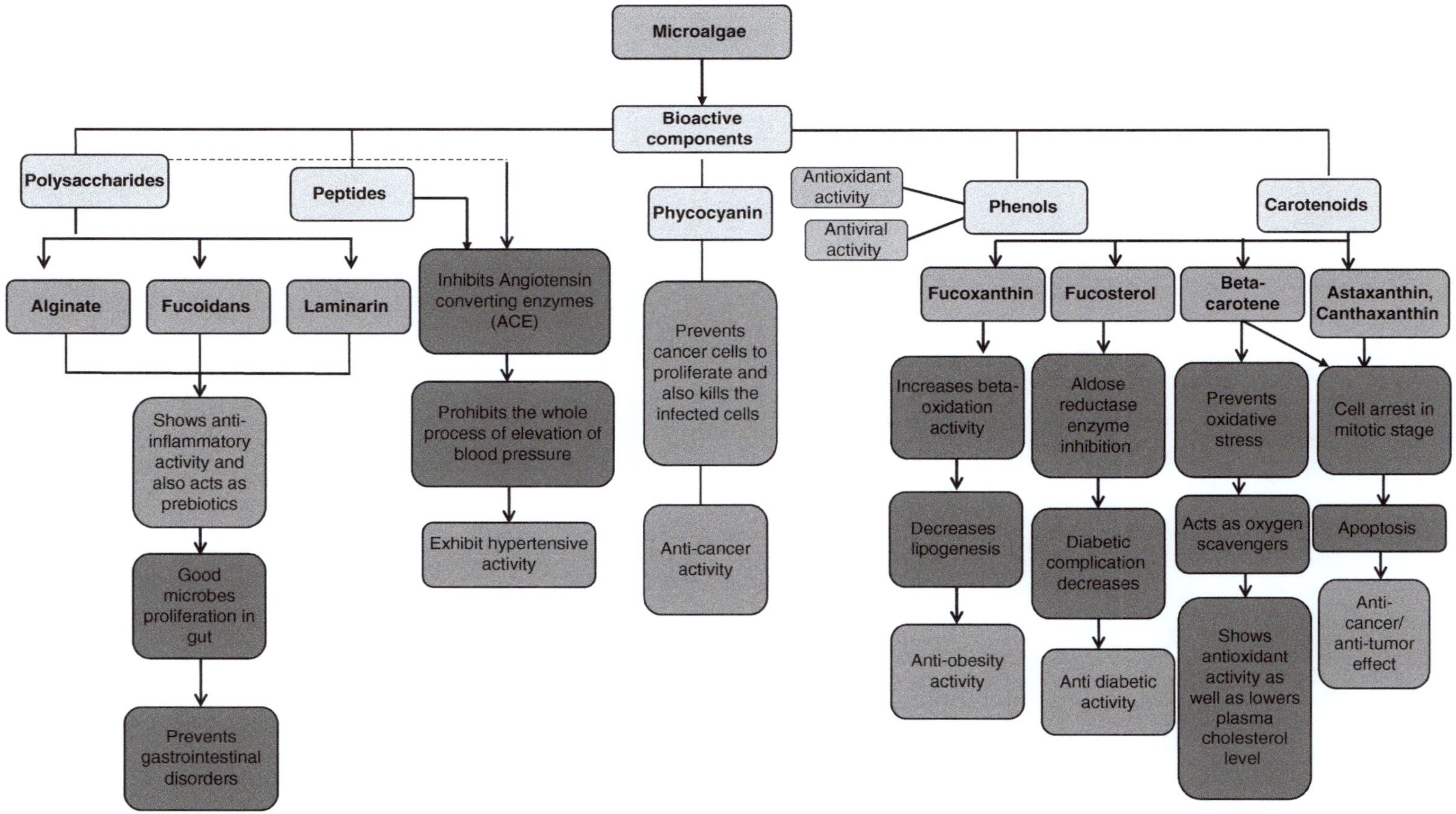

Figure 19.2 Microalgae, their bioactive compounds, and applications.

Figure 19.3 Classification of nutraceuticals based on functions. *Source:* (Left) U.S. Food and Drug Administration/Public domain, (Right-Top) Keith Weller/Wikimedia Commons/Public domain, (Right-Middle) Jirakit/Adobe Stock Photos and (Right-Bottom) New Africa/Adobe Stock Photos.

fatty acids from the cell (El- Baz et al. 2020). The ingestion of these bioactive molecules, often referred to as secondary metabolites, has been linked to favorable health outcomes in both humans and animals. Among these are qualities that are anti-inflammatory, antibacterial, anticancer, and anti-obesity (Senhorinho et al. 2019). Probiotics, prebiotics, and functional food are just a few of the suggested industrial uses for the diverse array of bioactive chemicals found in microalgae. Carotenoids, such as astaxanthin, β-carotene, fucoxanthin, and lutein, and fatty acids, such as docosahexaenoic acid (DHA) and eicosapentaenoic acid (EPA), are high-value prospective dietary supplement items derived from microalgae (Vazhappilly and Chen 1998; Olaizola 2000; Gong and Bassi 2016; Panis and Carreon 2016; Úbeda et al. 2017; Khoo et al. 2021). Numerous undiscovered microalgae chemicals also possess pharmacologic properties, such as antibacterial and anticancer properties (Senhorinho et al. 2015, 2019; Tang et al. 2020). The omega-3 fatty acids DHA and EPA are the most widely available nutritional supplements on the market and are mostly found in fish oil-based supplements. Fish cannot, however, produce these substances on their own or obtain omega-3 fatty acids by eating algae (Ward and Singh 2005). Thus, it is possible to avoid a number of problems associated with fish oil tablets such as indigestion, toxin accumulation, fishy after-taste, taste, poor oxidative stability, bloating, and vegetarian concerns by directly consuming microalgae (Chua et al. 2020; Katiyar and Arora 2020). While dietary supplements may be used for specific health conditions, functional and medicinal foods play a much larger role in the nutraceutical industry, alongside pharmaceutical products.

19.2.2.2 Functional and Medicinal Foods

Functional foods possess exceptional properties and are therefore critical in nutraceutical industries. Microalgal carbohydrates are used as food supplement because of their ease in digestion. Reports indicate this may enhance human health by stimulating immune function (Sathasivam et al. 2019). Amino acids and proteins derived from microalgae are used as food source and assist in fulfillment of daily protein diet. It has been reported that omega-3 fatty acids from microalgae prevent the cardiovascular diseases (Leray et al. 2001), reduce the concentration of lipids in serum, and have anti-thrombotic activity (Bucher et al. 2002). Omega-3 oils are also being used in improving health including pre-mature infant health (Carlson 1999), depressive disorders (Stoll et al. 1999), asthma (Hodge et al. 1996), and diabetes (Connor 2000). The marketed nutraceutical products are majorly based on *Spirulina* and *Chlorella* that are prokaryotic (Nicoletti 2016). *Spirulina*, a cyanobacterium, is globally known for its potential nutritional value (Nicoletti 2016). It possesses low purine concentration, which makes it edible and allows it to pose minimal risk of the buildup of uric acids in the body (Pittman et al. 2011). It is rich in vitamins, proteins, fatty acids, minerals, β-carotene, and antioxidants, therefore facilitating its commercial production as human food supplement from past decades (Campanella et al. 2002; Juárez-Oropeza et al. 2009; Capelli and Cysewski 2010). *Arthrospira platensis* is classified as a single-celled protein and is considered edible with high food value (FAO Fisheries 2008). *Chlorella* consists of proteins, amino acids, vitamins, minerals, and pigments and therefore is considered as an ideal food (Shoaf and Lium 1976; Chia et al. 2013). *Chlorella* sp., *Dunaliella salina,* and *Haematococcus pluvialis* are frequently explored as nutraceutical products (Hamed 2016). Besides, many medicinal foods are being produced for improving health and treating diseases.

Other products include the prebiotics and probiotics. Prebiotics are often described as non-digestible food ingredients that are beneficial for hosts as they aid in stimulating the growth of a limited bacteria in colon and thereby improving the health of the host (Davani-Davari et al. 2019). The prebiotic and probiotic roles of microalgae are considered to mutually complement each other (Oleskin and Boyang 2022). Probiotics are generally referred as the live organisms which when consumed or applied on the body have health benefits. Microalgae help in inhibiting the growth of many pathogenic organisms such as *Bacillus subtilis, Proteus vulgaris,* and *Bacillus pumulis* (Hamed et al. 2015; Camacho et al. 2019). It has been reported that *Chlorella ellipsoidea* and *Chlorella pyrenoidosa* help in the suppression of *Candida albicans* and *Listeria monocytogenes* (Hamed et al. 2015; Camacho et al. 2019). Therefore, microalgae can be explored for new therapeutics. Microalgae as probiotics facilitate in eliminating the metabolites and the toxins that are harmful for organisms (Oleskin and Boyang 2022). On the top of it, they can detoxify the harmful organic compounds that might get accumulated in the food and drinking water or the products released from industries. Probiotics constitute various antioxidants, protective compounds, and the bioactive substances. These biomolecules aid in regulation of host gene expression, elimination of carcinogenic, and other toxic compounds (Oleskin and Boyang 2022). *Dunaliella salina* has 10–13% β-carotene that helps humans and mice to protect against atherosclerosis (Oleskin and Boyang 2022). Omega-3 acids produced by *Isochrysis galbana* and *Porphyridium purpureum* assist in lowering down the blood cholesterol level further normalizing the blood pressure (Oleskin and Boyang 2022). *Chlorella vulgaris* tablets have been used in the treatment of nonalcoholic fatty acid liver disease (Ebrahimi-Mameghani et al. 2014). Phycocyanins obtained from *Arthrospira* have anti-inflammatory activities (Camacho et al. 2019; Khavari et al. 2021). Poly-unsaturated fatty acids are even used to treat arthritis (Camacho et al. 2019). Anti-diabetic effects have been found in *Arthrospira* that contribute to the production of phycocyanin and astaxanthin (Camacho et al. 2019; Eseberri et al. 2020). It has been reported that N-phycocyanin exerts an inhibitory effect on various cancer cell lines such as lung cancer (A549 and NSCLC) (Hao et al. 2018, 2019), leukemia (E562) (Liu et al. 2000), and liver cancer (HepG2) (Prabakaran et al. 2020). *Limnothrix* sp. are a source of phycocyanins that potentiates the effect of topotecan (an antitumor drug) on prostate cancer cell line (Gantar et al. 2012). The lipids produced from *Skeletonema marinoi* have the potential to induce apoptosis of hematological cancer (U-937) and colon cancer (HCT-116) (Remize et al. 2021). Fucoidan, a sulfated polysaccharide obtained from *Sargassum henslowianum, Fucus vesiculosus, Coccophora longdorfii,* and *Cladosiphon fucoidan,* aids in the apoptosis of melanoma, colon, lymphoma, lung carcinoma, and breast cancer cell lines (Oleskin and Boyang 2022). *Dunaliella tertiolecta* synthesize violaxanthin which has been reported to have anticarcinogenic activity (Oleskin and Boyang 2022). The microalgal probiotics are essential as they improve the metabolism of body and assist in treating the metabolic syndromes such as obesity. When the biomass powder of microalgal species such as *Chlorella* sp. and *Arthrospira platensis* is added to food products such as cookies and breads, it assists in declining the lipid and cholesterol levels (Camacho et al. 2019). Several microalgae such as *Spirulina* are being used to treat duodenal ulcers, whereas

Chlorella contributes to the regeneration of gut wall tissue (Oleskin and Boyang 2022). *Arthrospira* has been found to have neuroprotective properties and therefore facilitate the normal functioning of the brain. Certain microalgae such as *Ochrophyta, Chlorophyta, Rhodophyta,* and *Charophyta* can synthesize significant amounts of histamine, dopamine, acetylcholine, serotonin, and other neurotransmitters that have potential in improving the mental health and treating the brain-related disorders such as Parkinson's disease (Oleskin and Boyang 2022). To sum up, microalgae are explored as prebiotics and probiotics as they aid in treating various diseases and thereby promoting health.

19.2.2.3 Pharmaceuticals

Microalgae can produce pharmaceutical products (Figure 19.4) such as monoclonal antibodies, antibiotics, neurotoxic compounds, subunit vaccines, and hormones (Khavari et al. 2021). It has been reported that these compounds cannot be synthesized easily by chemical methods (Rizwan et al. 2018). Microalgae are demonstrated to be a potential host to produce therapeutic and recombinant proteins as they exhibit unique properties exhibiting similarity in post-translational modifications as present in the mammalian cells (Ramana et al. 2017). *Chlamydomonas reinhardtii* is regarded as the popular algae explored in pharmaceutical biotechnology. Some of the common recombinant proteins produced in this microalga comprise vaccines, viral protein 28, antibodies, and erythropoietin (Yan et al. 2016). Other species involved in the production of recombinant protein products include *Dunaliella, Chlorella,* and *Scenedesmus* (Geng et al. 2003; Ryan Georgianna et al. 2012).

Microalgae are currently being explored as vaccine transporters. Recombinant proteins are expressed in microalgae and find applications in pharmaceutical industries. It was reported that gene of human papillomavirus (HPV)16 E7 protein was inserted into genome of *Chlamydomonas reinhardtii* that aid in the expression of E7GGG protein to produce therapeutic vaccines (Demurtas et al. 2013). Some of the pharmaceutical products used as subunit vaccines are mentioned in

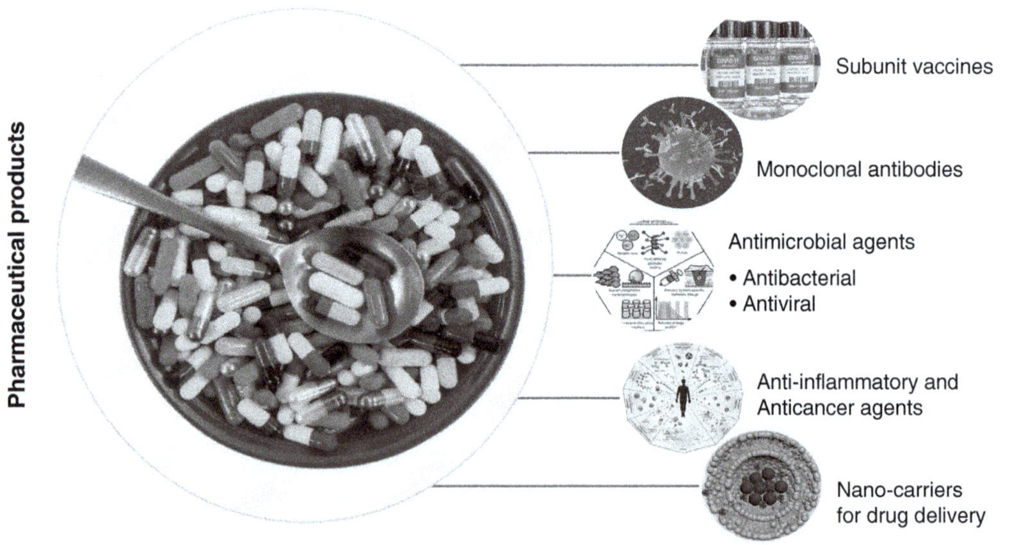

Figure 19.4 Categories of pharmaceutical products produced from microalgae. *Source:* (Left) trekandphoto /Adobe Stock Photos, (Right-Top) Markus Mainka/Adobe Stock Photos and (Right-second from top) Steven/ Adobe Stock Photos.

Table 19.3. Owing to the benefits and cost effectiveness, microalgae are utilized for producing monoclonal antibodies. The organisms being eukaryotic are preferred over bacterial cells due to the resemblance of post-translational modifications with human recombinant proteins. Microalgal extracts also possess antibacterial, antiviral, and anticarcinogenic properties. It has been reported that *Prymnesium parvum* and *Ochromonas* sp. produce toxins having the pharmaceutical applications (Rizwan et al. 2018). Nowadays, significant challenge occurring in the treatment of infectious diseases is the antibiotic resistance. Therefore, novel antibiotics are required to treat diseases. It has been found that mixture of fatty acids derived from *Chlorella* has inhibitory effects on the bacteria (Khavari et al. 2021). In addition, free fatty acids can also inhibit and kill growth of certain bacteria. The bioactive compounds extracted from microalgae exhibit antibiofilm activities, thereby playing preeminent role in treating infectious diseases (Soha 2012; López and Soto 2019). Biomolecules such as carbohydrates, lipids, and phycobiliproteins have apoptotic and antiproliferative effects on various types of cancers (Deniz et al. 2017). Some of the anticarcinogenic products extracted from microalgae are mentioned in Table 19.3.

Most predominantly, microalgae can be used for treating coronavirus disease 2019 (COVID-19) which is a major issue in the entire world responsible for many deaths (Talukdar et al. 2020). The major cause of death in COVID-19 patients is acute respiratory distress syndrome (ARDS). It results from the cytokine storm syndrome where the level of pro-inflammatory cytokines such as IL-1, IL-6, and TNF-α and chemokines such as CXCL9, CXCL10, CCL2, and CCL3 have been found to be increased, thereby resulting in hyperactivity of immune system and acute lung injury (ALI) (Yang et al. 2020). Astaxanthin, produced from *Haematococcus pluvialis,* assists in preventing ARDS and ALI by attenuating the cytokine storm and therefore has been administered for treating COVID-19 patients (Talukdar et al. 2020). *Spirulina* also has therapeutic applications in treating the viral diseases (Singh et al. 2020). It possesses anti-inflammatory and antiviral activities as mentioned in Table 19.3. Therefore, it can be an adjuvant therapy for COVID-19 patients as well (Singh et al. 2020). Another important application of microalgae is in drug delivery for treating the patients with genetic disorders (Khavari et al. 2021). Conventional drugs possess many limitations such as poor solubility and high toxicity. Also, the chemically synthesized nanoparticles require a higher energy and toxins. The synthesis process is also time-consuming and costly. Therefore, these must be overcome by advanced drug delivery systems as they aid in carrying the drugs to specific tissues of the body. The bioactive compounds produced from microalgae can be converted to nanoparticles (Shankar et al. 2016). The silica nanoparticles when modified using cationic reagents are then used for the delivery of gene (Ezzati et al. 2011). The major sources of silica nanoparticles include live diatoms and diatomaceous earth (Maher et al. 2018). Many drugs such as gentamicin and indomethacin can be delivered orally by passing them through porous silica diatoms (Aw et al. 2011). Diatomite silica nanoparticles are also regarded as the superior nanocarriers and have been explored in cancer therapy (Rigby et al. 2008). Precisely, microalgal species are the effective candidates for the pharmaceutical industries.

To recapitulate, the microalgae biomass is explored to extract and formulate many beneficial compounds including oil, tablets, proteins, carotenoids, vitamins, minerals, and fatty acids, which are further used in functional food such as the desserts, dietary products, oil derivatives, milk products, and pasta; medicinal food (prebiotics and probiotics); and pharmaceutical industries as discussed. Some of the commercial products produced from microalgae and their applications in nutraceutical industries are summarized in Table 19.4. Understanding the significance of nutraceutical products in the daily lifestyle, these compounds play a key role in the market across the world. Owing to their necessity in biotechnological sectors, the production of these compounds is necessary which is discussed in the upcoming section.

Table 19.3 Applications of microalgae-derived pharmaceutical products in medical biotechnology.

Source	Category	Pharmaceutical product	Applications	References
Chlamydomonas reinhardtii	Subunit vaccine	E2 protein D2-CTB (*Chlorella* toxin B) fusion protein E7 oncoprotein	• Treating swine fever virus • Oral vaccine against *Staphylococcus aureus* • Applied in HPV vaccine	Specht and Mayfield (2014); Yan et al. (2016)
Engineered *Phaeodactylum tricornutum*	Monoclonal antibodies	IgG antibody	• Works against nucleoprotein of Marburg virus	Fu et al. (2016); Hempel et al. (2017)
Coccomyxa sp.	Antiviral	Monogalactogyl diacylglycerol	• Antiviral effect by preventing virus to attach on host cell	Hayashi et al. (2019)
Chlorella vulgaris and *Dunaliella salina*	Antibacterial	Extracts	• Prevent dental caries by inhibiting biofilm formation on teeth by *Streptococcus mutans* and *Lactobacillus* sp.	López and Soto (2019); Soha (2012)
Chlorococcum humicola	Antibacterial	Chlorophyll II and β-carotene	• Antibacterial effects against *Salmonella typhimurium, Bacillus subtilis, Vibrio cholera, Staphylococcus aureus*	Khavari et al. (2021); Soha (2012)
Sargassum henslowianam, Coccophora longsdorfii, Fucus vesiculosus, Cladosiphon fucoidan	Anticancer	Fucoidan	• Inhibition of metastasis and angiogenesis in melanoma, lung carcinoma, breast cancer, human lymphoma cell line, promyeloid leukemia, human colon cancer	Deniz et al. (2017); Khavari et al. (2021)
Porphyra haitanensis, Spirulina platensis, and *Arthronema africanum*	Anticancer	Phycocyanin	• Inhibits growth of human hepatocellular carcinoma, leukemia cells, colon cancer, lung cancer	Deniz et al. (2017); Khavari et al. (2021)
Nostoc ellipsosporum	Antiviral	Cyanovirin-N	• Effective against Ebola virus, influenza, and HIV	Khavari et al. (2021)
Spirulina	Antiviral	Calcium spirulan	• Antiviral activity against influenza, mumps, and HIV	Singh et al. (2020)
Spirulina	Anti-inflammatory	Phycocyanin	• Inhibitor of NADPH oxidase	Singh et al. (2020)

Table 19.4 Commercial products from microalgae and their applications in nutraceutical industries.

Products	Source	Applications in nutraceutical industries	References
Agar	*Gelidium, Gracilaria*	• Valuable material due to its stabilizing properties • Improved bioavailability and biodegradability • Used in manufacturing of processed cheese, puddings, mayonnaise, jellies, canned goods, and the baking products • Used as a drug carrier • Anti-diabetic, anti-tumor, skin-moisturizing, immunomodulatory, anti-carcinogenic, and antioxidative properties	Park et al. (2020); Shahruzzaman et al. (2019)
Carrageenan	*Chondrus, Eucheuma, Gigartina*	• Increases the texture quality of cottage cheese • Used as stabilizer in production of sausages, meat processing, and patties • Used to maintain viscosity of dairy products • Used as an emulsifier in foods and milk products • Used as an agent in ulcer therapy	Dong et al. (2021); Prasetyaningrum et al. (2019); Udo et al. (2023)
Alginic Acid	*Macrocystis, Agarum, Ascophyllum, Laminaria,* and *Fucus*	• Excellent water solubility • Biocompatible and biodegradable • Provide smooth consistency and eliminate the ice crystal formation • Used in bakery products especially for frostings • Alginate material has been used extensively by the dentists to make the teeth impressions	Abka-khajouei et al. (2022); Bi et al. (2022); Guo et al. (2020)
Milk	*Chlorella* sp.,*Spirulina* sp.	• Improved flavor and mouthfeel • Prevents anemia	Shalaby (2013)
Wheat crackers	*Phaeodactylum tricornutum*	• Reduction in hardness on adding the algae • Increase in total dietary fiber by 16–26% • Lighter and thinner	Batista et al. (2019)
Edible *Spirulina*	*Spirulina*	• Enhanced the production of antibodies • Improves immunity • Fights of chronic illness • Cardiovascular positive effects • Reduce cholesterol • Anticarcinogenic	Capelli and Cysewski (2010); Díaz del Castillo and Garcia (1928); Hirahashi et al. (2002); Juárez-Oropeza et al. (2009); Ku et al. (2013); Mišurcová et al. (2012); Mosulishvili et al. (2002); Nyenje and Ndip (2013)

(Continued)

Table 19.4 (Continued)

Products	Source	Applications in nutraceutical industries	References
Cheese	*Arthrospira platensis, Chlorella* sp.	• Improved texture • Anti-cancer activity • Reduce risk of hypertension, anemia, diabetes, neurosis, infant malnutrition	Camacho et al. (2019)
Vegetarian mayonnaise	*Chlorella vulgaris,Haematococcus pluvialis*	• Improved color and stability • Anti-oxidant activity	Crampon et al. (2017); Gouveia et al. (2006)
Gluten free bread	*Tetraselmis chuii, Chlorella sorokiniana*	• High crumb brightness • Reduction in green color • Lipid content increased by 62.86%	Diprat et al. (2020); Lafarga et al. (2022)
Cookies and biscuits	*Arthrospira platensis,Phaeodactylum tricornutum,Tetraselmis suecica*	• Improved color, stability, and texture • Anti-oxidant activity	Batista et al. (2017); Singh et al. (2015)
Yoghurt	*Arthrospira platensis*	• Improved texture and viscosity • Anti-cancer • Anti-oxidant • Anti-inflammatory	Golmakani et al. (2019)
Miso	*Dunaliella* sp.,*Spirulina* sp.	• Slightly seaweed taste • Anti-oxidant activity	Kamel El-Baz et al. (2017)
Bread and cookies	*Arthrospira platensis, Chlorella* sp.	• Improved flavor, texture, and appearance • Reduce fats and cholesterol	Camacho et al. (2019)
Desserts	*Arthrospira maxima,Chlorella protothecoides,*	• Improved color and stability • Anti-oxidative properties	Gouveia et al. (2007); Gouveia et al. (2008)
Soybean oil	*Chlorella vulgaris*	• Improved color and stability • Anti-oxidant activity	Camacho et al. (2019)

19.3 Bioprocess Development of Nutraceutical Products

A variety of strategies have been adopted to enhance the production of bioactive compounds derived from microalgae. Every bioprocess involves the two-step process: firstly, the upstream process that deals with the development of inoculum; secondly, the downstream process that deals with the purification, clarification, and the development of final product. Here, the generation of bioactive chemicals in the algal host systems has been explained, and then, the methods for downstream processing to produce the finished products have been discussed.

19.3.1 Bioprocessing of Microalgal Nutraceuticals

The master key for producing the bioactive compounds discussed previously is the selection of the microalgal species and the bioprocessing technologies involved in the production of these compounds. Media optimization is one of the key factors playing a vital role in biomolecule productivity (Aziz et al. 2020). Biotic and abiotic factors form the basis of media optimization and are therefore considered as the major aspects determining the production of bioactive compounds derived from microalgae. Abiotic factors including the pH, temperature, light, incubation time, external radiations, nutrient stress have been known to play a remarkable role in biomolecule productivity.

The effect of carbon supplementation for synthesis of bioactive compounds has been studied in *Dunaliella bardawil* and *Haematococcus pluvialis* where it has been found that the productivity of lipids and astaxanthin enhanced in these species, respectively (Srinivasan et al. 2015; Wen et al. 2020). Four different species of *Chlorella* were induced with the nitrogen stress that finally increased the productivity of carbohydrates, lipids, and carotenoids, respectively (Kim et al. 2014; Ördög et al. 2012, 2016; Minyuk et al. 2020). The effect of temperature and light intensity was also studied in some species including *Tetraselmis sp. CTP4* (Schüler et al. 2020). Lutein productivity incremented by 1.5-fold in high-light intensity and production of β-carotene increased by threefold under low-light intensity. *Tetraselmis chui* and *Nannochloropsis gaditana* were supplied with low-frequency light that finally helped in escalating the total chlorophyll content (Lima et al. 2021). Another factor affecting the productivity of compounds is the nitrogen stress and CO_2 supplementation induced together in *Testradesmus obliquus* PF3, *Scenedesmus* sp, *Botryococcus braunii,* and *Chlorella vulgaris* (Yoo et al. 2010; Jiang et al. 2011; Ma et al. 2020b). The lipid and carbohydrate production increased significantly upon the induction of the stress conditions. *Picochlorum* sp. was used to study the effect of nitrogen, and it was observed that the biomass and fat content escalated in this species (Dahmen-Ben Moussa et al. 2019). *Nannochloropsis oculata* and *Tetraselmis suecica* were selected for inducing the nitrogen, phosphate, and $FeCl_3$ stress where the lipid content was found to increase significantly (Usman Shah and Abdullah 2018). *Porphyridium* sp. was induced with the Fe^{+3} and Co^{+2} wherein the total content of phycobiliproteins was incremented (Priatni et al. 2018).

Lutein production was studied under the effect of different CO_2 concentration in *Scenedesmus almeriensis*. The results showed production of 3.7 g/L biomass with 5.71 mg/g of lutein under 3.0% v/v CO_2 concentration. Increasing the light intensity and continuous supply of 2.5% CO_2 aided in attaining higher amounts of lutein (Molino et al. 2020). In addition to this, *Scenedesmus obliquus* FSP-3 was also explored for lutein production, where 8 mM calcium nitrate was used as nitrogen source during the time of cultivation and 5.52 mg/g lutein production was observed (Ho et al. 2014). Besides, *Chlorella minutissima* MCC-27 was cultivated in 2 L airlift photo-bioreactor that helped in attaining 3.4 mg/L/day of lutein productivity in batch mode which was further improved to 4.32.

mg/L/day by optimizing light intensity, gas flow rate, and CO_2 concentration (Dineshkumar et al. 2015). pH also plays a significant role in the production of bioactive compounds. *Haematococcus lacustris* was explored for synthesis of astaxanthin in the presence of pH shock. It was reported that pH shock treatment significantly incremented the astaxanthin content by 39 % as compared to the control (Han et al. 2020). pH and temperature were changed in *Arthrospira* sp., and it was found that the concentration of phycobiliproteins increased at a significant rate (H. T. Tan et al. 2020a). In addition, various phytohormones and inducers like indole-3-acetic acid, gibberellic acid, salicylic acid, zeatin, and auxins when supplemented also enhanced the production of specific compounds. The supplementation of phytohormones was done in *Scenedesmus obliquus and Monoraphidium* sp. QLY-1 (Salama et al. 2014; Li et al. 2019). The total lipids incremented as compared to the control (without stress). In *Spirulina* sp. LEB 18, it was investigated that by addition of crude glycerol with concentration of 2.5 g/L, biomass and protein productivity were enhanced to 430 mg/L/day and 75.7% w/w, respectively (de Morais et al. 2022).

Not only limited to this, protein was produced from *Chlorella vulgaris* FSP-E and *Chlorella sorokiniana* under optimum conditions where 5% v/v CO_2 was supplemented. Along with it, nitrogen sources of 12 mM of $NaNO_3$ and NH_4Cl were provided, respectively, and ammonium iron (III) citrate was added at concentration of 12 mg/L. In semi-batch cultivation conditions, the protein productivity increased by 4.86 and 2.77-fold to 793.3 and 812.8 mg/L in both species, respectively (Lai et al. 2019). Optimization of different parameters plays a remarkable role for large-scale production of different bioactive compounds produced from microalgae. Once the compounds are synthesized in a cell, they need to be extracted from the host cell, which is further done by the downstream processing technique.

19.3.2 Downstream Processing Techniques for Product Recovery

The successful utilization of microalgae and cyanobacteria requires an intricate downstream processing strategy to harvest, extract, and purify the desired products efficiently. Various stages involved in the downstream processing of any product are illustrated in Figure 19.5. Owing to the exceptional benefits and fascinating properties of microalgal nutraceuticals, the upcoming section highlights the extraction methods that are utilized in the industries to extract various bioactive compounds from microalgae.

19.3.2.1 Cell Recovery

It is necessary to remove water from the microalgae using a variety of techniques, including centrifugation, filtration, electrically powered procedures, and flotation, to effectively harvest

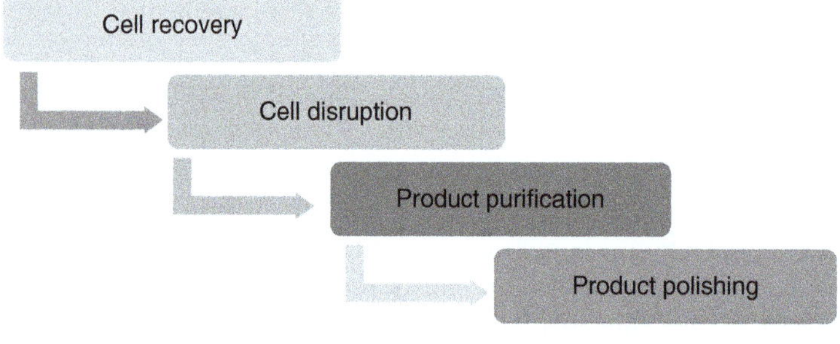

Figure 19.5 Stages in downstream processing.

the algae (Rahman et al. 2022). The two main processes in the creation of microalgal products are cultivation and harvesting. Sedimentation, flotation, centrifugation, and filtering are examples of traditional harvesting techniques (Khanra et al. 2018). Because flocculation requires less energy and is therefore energy-efficient, it is typically advised as a pre-treatment. Pre-treatment is typically performed to break down the algae's cell wall, increasing the effectiveness of the biomolecule extraction process. Nevertheless, there is no a single culture and harvesting technique that works for all species of microalgae. The type of microalgae and the intended product or products are taken into consideration while choosing an effective culture and harvesting technique.

19.3.2.2 Cell Disruption

Following cultivation and harvesting, the most often utilized pre-treatment techniques are drying, particle size reduction, and cell disruptions; these are chosen depending on the extraction process and final product (Khanra et al. 2018). In order to maximize the extraction efficiency of biomolecules, pre-treatment typically involves breaking down the algal cell wall. Large-scale operations have found success with mechanical pre-treatment techniques like steam explosion, microwave radiation, and sonication. Freeze-thaw, chemical lysis, enzymatic treatment, high-pressure homogenization (French press), bead milling/ball milling, and other techniques are also available for disrupting cells (Günerken et al. 2015; Masarin et al. 2016). There are two types of cell disruption techniques: mechanical and non-mechanical. The bead mill, microwave, ultrasonication, and high-pressure homogenization are examples of mechanical processes. Conversely, non-mechanical techniques involve breaking microalgal cells using acids, alkalis, enzymes, or osmotic shocks. In their analytical work, Al hattab and Ghaly (2015) noted that the energy usage in bead milling can vary (35–810 W/kg of biomass) because different microalgal species have different cell wall structures and because of the operating environment. The energy required for ultrasonic pre-treatment varies depending on the species and can range from 1.7 to 119 W/g of biomass. Guldhe et al. (2014) observed in a different investigation that *Scenedesmus* sp. consumed 119 W/g of biomass. As per Adam et al. (2012), *Nannochloropsis* sp. exhibited an energy consumption of 16.66 W/g of biomass in a separate investigation. According to two separate investigations by Chemat et al. (2011) and Wang and Weller (2006), the ultrasound-aided extraction of biomolecules from microalgae is more cost-effective than the freeze-drying approach and may be conducted at low temperatures to limit thermal denaturation. Steam explosion is another type of disruption of cells to facilitate the easy removal of intracellular components. It operates by subjecting the biomass to high temperatures (160–260 °C) and vapor pressures (1.03–3.45 MPa), which are subsequently reduced to room temperature by depressurization, ultimately causing the cell to rupture.

In addition to cell disruption, other factors that affect extraction efficiency include the kind of organic solvent used, the length of time the process during the extraction method, and the length of time the microalgae are stored (Halim et al. 2010). The choice of extraction process depends on the microalgal species, the specific target product, and the intended application. Additionally, a combination of methods or a sequential extraction approach may be employed to maximize the yield of multiple products from microalgae (Table 19.5). For instance, hexane, chloroform, or other organic solvents are used to extract lipids from microalgae. However, other biomolecules such as polysaccharides isolated from algae for nutraceuticals include agar, carrageenan, and alginates. The mentioned products are extracted by various methods such as alkali extraction, drum drying, KCl process, and drum process (Khanra et al. 2018). To improve the alkali extraction process, the particle size of the algal biomass must be reduced. Next, the algal biomass is rehydrated for 1–12 hours, depending on the species, using a formaldehyde solution (0.1%) to soften the biomass

Table 19.5 Extraction of bioactive compounds by different techniques.

Biomolecules	Microalgae species	Extraction method	Recovery	References
Lipids	*Nannochloropsis* sp.	• Extract with chloroform/methanol in 2:1 v/v for 1 h and stir at 37 °C	90% of total biomass (w/w)	Zuorro et al. (2016)
	Auxenochlorella protothecoides	• Pulsed electric field followed by 100% ethanol	220 mg/g	Qu et al. (2018)
	Nannochloropsis salina	• Dilute acid treatment followed by hexane	88.4% at CSF of 1.9	Eing et al. (2013)
	Nannochloropsis sp.	• Alkaline treatment followed by extraction with 30 ml chloroform.	83.5% of lipid yield at CSF' between (−0.9 and −0.3)	Lee and Han (2015)
	Chlorella vulgaris	• Bead milling followed by supercritical–CO_2 extraction method at 60 °C, with a conc. of 5% ethanol	67% of lipid at 1 : 13 solid: water (W/V)	Wu et al. (2017)
	Chlorella sp. KR-1	• Using 2 mM $K_2S_2O_8$ and 5 ml chloroform	95% lipid extraction	Safi et al. (2014)
				Seo et al. (2015)
Carbohydrates	*Chlorella vulgaris*	• Extraction by high-pressure homogenization	41.9% (w/w)	Carullo et al. (2018)
	Tetraselmis suecica	• Extracted by bead milling	99% carbohydrate of total biomass slurry	Postma et al. (2017)
	Chlorella pyrenoidosa	• Hydrothermal treatment	80.5% of total carbohydrate content	Xiao et al. (2019)
	Mixed culture	• Acid treatment using dilute H_2SO_4.	94% total reducing sugar content of pretreatment liquid	Shokrkar et al. (2017)
	Hindakia tetrachotoma ME03	• Alkaline treatment using NaOH	80% of total carbohydrate recovered in hydrolysates at CSF' 2.4	Onay (2019)
Proteins	*Chlorella vulgaris*	• Alkaline treatment	42% protein yield at CSF' −1.8	Mendez et al. (2013)
	Tetraselmis. suecica	• Bead milling	99% of total biomass slurry	Postma et al. (2017)
	Chlorella vulgaris	• Extraction using high-pressure homogenization	54.1% of total protein	Carullo et al. (2018)
	Chlorella sorokiniana	• By hydrothermal steam explosion Hydrothermal lysis at 200 °C for 60 min	75.9% of total protein determined in pretreatment liquid	Xiaojian et al. (2017)
		• Extraction by dilute acid treatment using sulfuric acid (5%w/w)	55.7% of total protein content	Lorente et al. (2018)

Pigments	*Desmodesmus sp. F51*	• Extraction of carotenoids using high-pressure homogenization followed by treatment with ethyl ether	7.9 mg/g biomass	Xie et al. (2016)
	Haematococcus pluvialis	• Cell disruption by hydrothermal treatment followed by extraction of astaxanthin using acetone	100% of total astaxanthin present	Cheng et al. (2017)
	Haematococcus pluvialis	• Pulsed electric field followed by extracting astaxanthin by using 99.5% methanol	96% of astaxanthin present	Martínez et al. (2019)
	Haematococcus pluvialis	• Treatment with dilute acid followed by extracting astaxanthin using acetone	80% of total astaxanthin present	Liu et al. (2018)
	Chromochloris zofingiensis	• Extraction of lutein by using ball mill grinding followed by solvent extraction using acetone	100% of total biomass	Araya et al. (2014)
	Scenedesmus almeriensis	• Alkaline treatment followed by solvent extraction using hexane to extract lutein	99% of lutein yield at CSF' between (−2 and 0)	Cerón et al. (2008)

CSF: Combined severity factor for hydrothermal pretreatment with three parameters—time, temperature, and pH.
CSF': Modified combined severity factor for hydrothermal pretreatment with three parameters—time, temperature, and pH
Source: Adapted from Nitsos et al. (2020).

and prevent alginate coloring (Khanra et al. 2018). This technique is well-established and has been used on a variety of algae, including *Sargassum* sp. and *Laminaria digitata* (Hernández-Carmona et al. 2013). The process of extracting agar involves heating algal biomass to boiling point in water. The careful addition of acid to maintain the pH between 6.3 and 6.5 is essential for effective extraction (Hernández-Carmona et al. 2013). Furthermore, the application of pressure improves the extraction efficiency and results in a higher yield of agar with a shorter processing time. The optimum conditions for extraction vary with the kind of algae. Under hot conditions, the filtration helps in removing the residual biomass followed by the cooling of hot filtrate to allow the formation of gel. Occasionally, the gel is bleached (typically with sodium hypochlorite) to remove color and enhance quality. Lastly, water is eliminated using a freeze-thaw procedure or by applying pressure to squeeze it out.

Spirulina and *Chlorella*, two unicellular microalgal species, have gained attention as important sources of protein for mass-scale production. The nutritional benefits of algae and their usage as dietary supplements have been well demonstrated by several review studies (Wells et al. 2017). The most crucial step in protein extraction is the elimination of all other interacting substances, such as nucleic acids, carbohydrates, lipids, and pigments (Kose et al. 2017). A common method for protein precipitation involves using acetone with trichloroacetic acid (TCA) for the recovery of tiny amounts of proteins used for proteome research (Slocombe et al. 2013). In one investigation, the protein was obtained from dinoflagellate *Alexandrium tamarense* cells using ultrasonic disrupter followed by treatment of 10% TCA/acetone (w/v) and 80% acetone (v/v) (Wang et al. 2009). According to the study report, using this procedure would enable the greatest amount of protein (about 6.73 μg/μl) to be extracted from 1×10^6 cells (Lim and Teo 2015). The protein yield obtained from the solvent extraction method employed by Schwenzfeier et al. (2011) to extract protein from *Tetraselmis* sp. was 64% (w/w). Before developing a manufacturing process based on proteins, a significant problem in industrial-scale operations is the scalability of protein extraction. The ability to successfully disrupt cells to gain access to the molecule is necessary for the efficient extraction of protein from cells. One method for disrupting cells for small-scale cultures is sonication. By applying various sonication waves (0.02–4.3 MHz) to algal cell suspensions of three different species (*Chaetoceros gracilis, Chaetoceros calcitrans*, and *Nannochloropsis* sp.), it was demonstrated that cell disruption efficiency varied according to the species. The maximum efficiency is attained by *Chaetoceros gracilis* at 2.2 MHz, *Chaetoceros calcitrans* at 3.3 MHz, and *Nannochloropsis* sp. at 4.3 MHz (Kurokawa et al. 2016). Different extraction methods reported for extracting the bioactive compounds from microalgae are summarized in Table 19.5.

The pigment extraction process typically entails subcritical dimethyl ether extraction, organic solvent extraction, supercritical fluid extraction, and supercritical CO_2 extraction (Khanra et al. 2018). Chlorophyll is extracted using organic solvents such as acetone, methanol, ethanol, and diethyl formamide. In a study on algae (*Selenastrum capricornuturn*), Sartory and Grobbelaar (1984) found that ethanol and methanol performed better at extracting chlorophyll a than acetone and acetone with dimethyl sulfoxide (DMSO). A combination of dimethyl formamide and mechanical homogenization treatment (Schumann et al. 2005), demonstrated the effective extraction of chlorophyll a from *Stichococcus* sp. and *Chlorella* sp. Since the nutraceutical industries are the primary consumers of chlorophyll, its quality has to be regulated strictly. As a result, achieving a solvent-free pure extract using supercritical CO_2 extraction is a straightforward process (Halim et al. 2010). Supercritical CO_2 extraction of chlorophyll from microalgae is mostly dependent on temperature and operating pressure. Using supercritical CO_2, Macías-Sánchez et al. (2007) conducted several investigations on the extraction of chlorophyll from microalgae *Synechococcus* sp. The findings indicated 0.715 μg of chlorophyll a per mg dry weight of biomass was extracted when operated at 60°C and 500 bar. Goto et al. (2015) have demonstrated extraction of 390 μg/g of fucoxanthin from algae *Undaria Pinnatifida* using 286 g dimethyl ether (DME). Numerous investigations on the supercritical fluid carotenoid extraction method indicate that, while running at 40–80°C and 20–40 MPa pressures, the extraction process improves the selectivity of the carotenoid molecules.

19.3.2.3 Product Purification

Another significant difficulty is to find a cost-effective and sustainable way to separate the desired products from the grown microalgal broth for industrial use. The purification of extracted products from different species of microalgae involves a series of steps aimed at isolating the target compounds with high purity. The type of extracted products, their chemical and physical characteristics, and the intended use influence the separation and purification methods chosen. To obtain the required degree of purity for each target product from various microalgal species, a well-optimized sequence consisting of a combination of these techniques is frequently used.

Diverse protein components from algae are regularly purified in laboratories using a variety of chromatography techniques, including size exclusion, affinity, ion-exchange, and hydrophobic interaction chromatography (Bleakley and Hayes 2017). Using hydrophobic interaction chromatography and ion-exchange chromatography, Wang et al. (2014) were able to isolate a monomeric protein from the microalgae *Betaphycus gelatinus* with a 3.7% yield of the cell lysate. In addition to the various chromatography techniques, membrane-based filtration techniques (such as cross-flow and ultrafiltration filtration) are also widely used for protein purification from microalgae (Henderson et al. 2008; Zhang et al. 2010).

Di Sanzo et al. (2018) performed supercritical CO_2 extraction at 50 °C and 550 bars of pressure to obtain lutein (52.3%), astaxanthin (98.6%), and fatty acids (93.2%) from *Haematococcus pluvialis* biomass at 65 °C and 550 bar of pressure. The recovery of proteins, carbohydrates, and pigments from *Parachlorella kessleri* biomass can be enhanced by the combination of high-pressure homogenization and ultrasonication (Zhang et al. 2019). Martínez et al. (2019) extracted 18.3 mg/g of dry weight of astaxanthin from the Nordic biomass of *Haematococcus pluvialis* using pulsed electric field permeabilization. Halim et al. (2016) used mechanical and chemical extraction techniques to fractionate and separate concentrated microalgae biomass for bioactive compound production and extraction. Nonetheless, the amalgamation of bead milling and the enzymatic hydrolysis technique enhanced the separation of proteins (68%), carbohydrates (74%), and lipids (88%) in the liquid phase and solid phase, respectively, from the *Chlorella vulgaris* biomass (Alavijeh et al. 2020). Algae like *Chlorella vulgaris*, *Arthrospira platensis*, *Nostoc* sp., *Dunaliella salina*, and *Pleurochrysis carterae* are known to accumulate useful components and build biomass quickly. The study of Andreeva et al. (2021) showed purification process for protein concentrate derived from the biomass of microscopic algae cell cultures involved ultrafiltration (active acidity of 7.0, process pressure of 0.2 MPa, and membrane with pore diameter of 50.0 kDa) and high-performance liquid chromatography (HPLC) in a gradient of sodium chloride concentration.

Many patent applications have been made in relation to the downstream processing of different bioactive substances that are extracted from microalgae. A patent had been awarded for the novel column chromatography method of separating chlorophyll a and chlorophyll b from plants, including algae (Anderson and Melvin 1963). Many brown algae, including *Phaeodactykum* sp., *Naviculla* sp., *Isochrysis* sp., and *Amphora* sp., have fucoxanthin, a pigment (xanthophyll) in their chloroplasts. Dr. Oran Ayalon patented the fucoxanthin extraction method, which states that supercritical fluid (CO_2) and solvent extraction were used to extract 1.5% (w/w) from collected algae (Ayalon 2016).

Valuable molecules called carotenoids and chlorophyll are utilized as ingredients in food and nutritional supplements. A patent covering the process of producing lipids, carotenoids, and chlorophyll from algal biomass has been awarded to Heliae Development Limited Liability Company (LLC), a platform technology firm located in the United States (Kale 2012). Czartoski et al. (2010) filed for patent on a technique of fractionating algal biomass by digesting algae in a renewable way and recovering cells and macromolecules from a non-polar solvent.

19.3.2.4 Product Polishing

The creation of resilient polishing processes in downstream processing is essential to resolving possible problems with installation, adaptability, and scale-up in the biologic product manufacturing process. The biologic product is ready for the final product packing with a biopolishing finish. Any traces of impurities and pollutants from the target product are eliminated from the feedstock during this biopolishing process. Typical polishing techniques in downstream include crystallization, desiccation, lyophilization, and spray drying. End users, such as nutraceutical corporations and contract research organizations, are using downstream processing techniques to develop and generate products in bulk due to the increase in the demand (Future Market Insights 2024).

19.4 Economics and Market Demand of Nutraceuticals from Microalgae

Today, industrial-scale bioprocessing plays a major role in the production of food, drink, pharmaceuticals, and nutraceuticals. These manufacturing processes gradually leverage their economies of scale and scope, which are so dependent on titers, productivity, and product yields to boost their profitability. The yearly dry matter production of microalgae for commercial purposes is estimated to be 5000 tons (Raja et al. 2008). In the Asia–Pacific area, there are about 110 commercial microalgae producers with capacities ranging from 3 to 500 tons/year. Asia accounts for almost nine-tenths of the world's algae farming. Among these, China, Taiwan, and India are home to a sizable population of commercial microalgae growers. *Crypthecodinium, Spirulina, Botryococcus, Chlorella, Haematococcus, Isochrysis, Dunaliella, Porphyridium, Schizochytrium, Phaeodactylum, Nannochloris, Chaetoceros, Nitzschia, Skeletonema,* and *Tetraselmis* are among the very few species of microalgae that are important commercially. Nonetheless, microalgae products are widely utilized in aquaculture, human nutrition, and nutraceuticals as a high-protein supplement (Del Campo et al. 2007). Algal biomass is sold for a range of prices on the market, with reported prices for human use being 100 €/kg and for animal and fish feed being 5–20 €/kg. According to Wijffels and Barbosa (2010), microalgal products are 40% lipids, 50% proteins, and 10% carbohydrates. Of these, 25% are sold for 2 € kg to the food and chemical industry, the remainder for 0.50 €/kg for biodiesel, 20% for food at 5 €/kg for soluble proteins, and the remaining 80% for feed at 0.75 €/kg. Figure 19.6 depicts the various countries that are manufacturing microalgal products globally (Kumari et al. 2023).

The economic and market demands of microalgae-derived nutraceuticals are mainly regulated by cultivation charges, extraction, and purification methods, followed by market demand, supply, and competition. Carrageenan, agar, and alginate are considered "generally regarded as safe (GRAS) substances" for human consumption by the European Food Safety Authority (EFSA) and Food and Drug Administration (FDA, USA). Carrageenan, agar, and alginate are the most exploited and industrially used hydrocolloids in the current market. Although agar possesses the highest retail

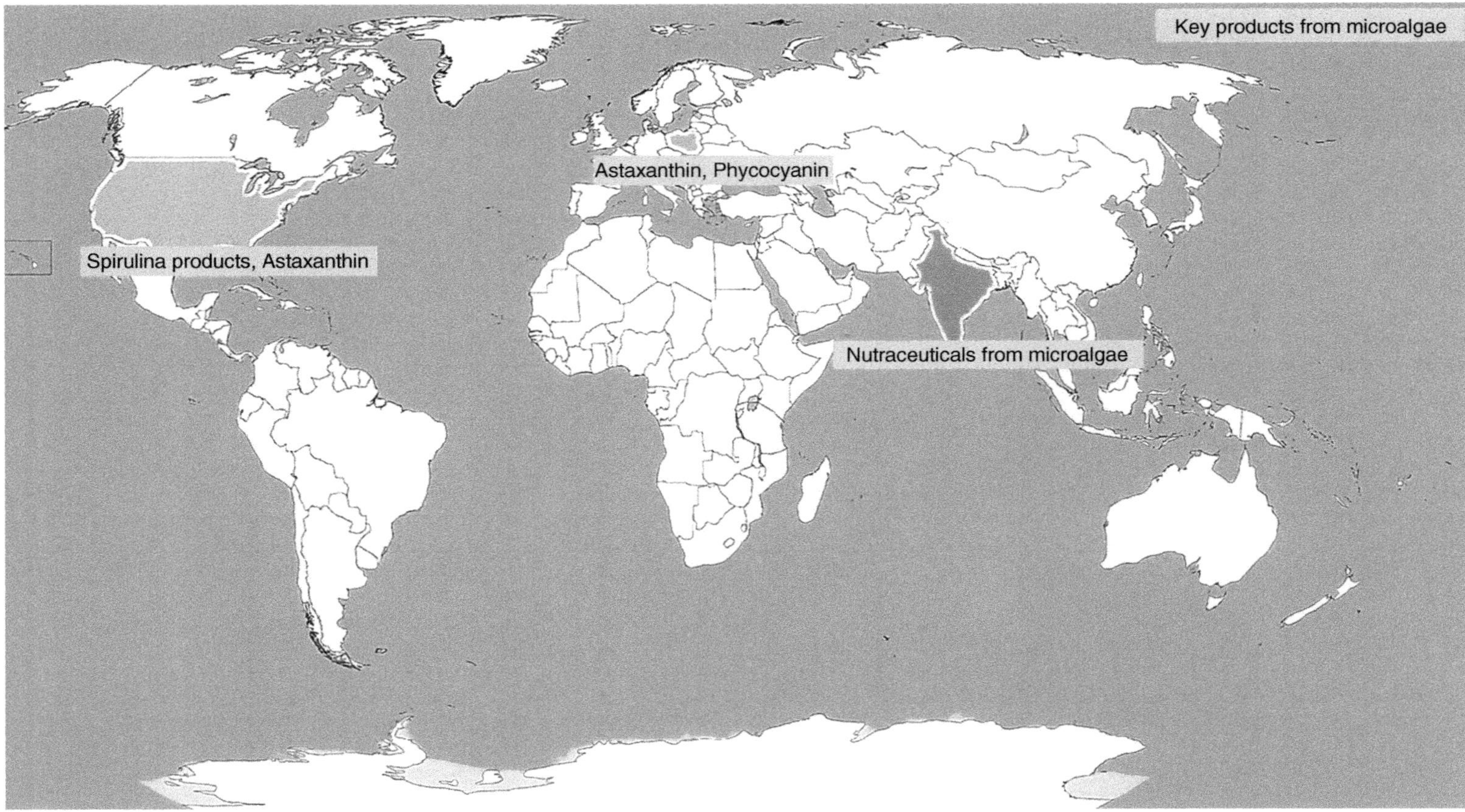

Figure 19.6 Countries manufacturing microalgal products having commercial value globally. *Source:* Petr Dlouhý/Wikimedia Commons.

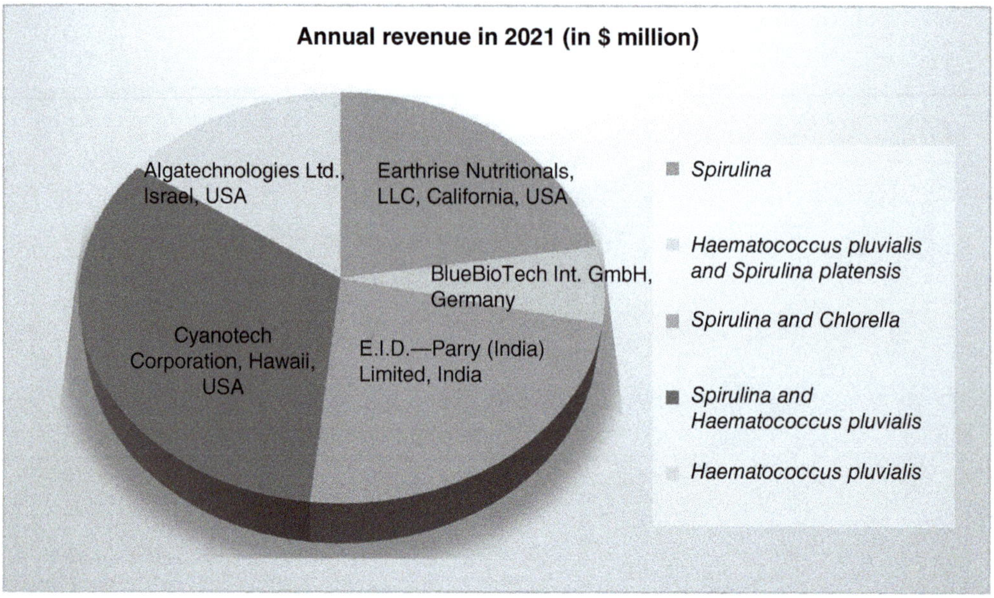

Figure 19.7 Annual revenue generated by major companies for production of microalgae-based nutraceutical products.

price per kg (18 USD/kg), carrageenan leads the total commercial production (60,000 tons/year), accounting for USD 626 million per year. On the other hand, the global market of fucoidan is continuously growing as it reached USD 33 million in 2020 and is estimated to achieve USD 42 million by 2027 with a CAGR of 3.4 % during 2021–2027. In the same line, the fucoxanthin global market is in continuous expansion with a CAGR of 3.8 % during 2021–2027 (Garcia-Perez et al. 2023).

Recent studies show that algae-based functional food products have high opportunities for market share. Figure 19.7 depicts the annual revenue generated by major manufacturers in 2021 that share major market shareholding for the microalgae-based nutraceutical products.

By the end of 2022, the downstream processing market is projected to have grown from its 2021 estimate of US$ 22.0 billion to US$ 24.9 billion. According to recent study of Future Market Insights, 2023, the algae protein market produced total revenue of US$ 750 million in 2022, and moreover by 2023, the industry is expected to rise at a healthy CAGR of 5.5 %, for attaining market revenue of US$ 1351.57 million (Bhatnagar et al. 2024). Furthermore, the market for downstream processing goods is expected to reach a worth of US$ 72.3 billion by 2032 due to increased adoption among biopharmaceutical businesses and contract research organizations. This growth in demand is expected to occur at a CAGR of 11.3% between 2022 and 2032 (Future Market Insights 2024). Hence, algae-based products play a remarkable role in the economics and subsequently the market value.

19.5 Conclusion

Microalgae offer a treasure trove of bioactive compounds with a promising future in nutraceuticals. Research has highlighted their potential anti-inflammatory, anti-cancer, and cardiovascular benefits, making them attractive to both the food and pharmaceutical industries. Collaboration between scientists, industry experts, and regulatory bodies is critical for navigating the complex landscape of microalgal nutraceuticals. Further research is needed to explore a deeper understanding of the long-term effects, ideal dosages, and potential interactions of microalgal nutraceuticals with human health. Ensuring sustainable microalgae farming practices that minimize environmental impact remains a priority. However, significant hurdles remain before large-scale commercialization can be achieved. Here are the key challenges to address:

- **Scaling Up Production:** Current methods for harvesting microalgal biomass are often expensive and energy-intensive. Efficient, cost-effective techniques are needed for large-scale production.
- **Optimizing Downstream Processing:** Selecting the most appropriate extraction method is crucial for obtaining high-quality bioactive compounds. Research is ongoing to develop efficient and scalable extraction processes.
- **Ensuring Product Stability and Quality:** Maintaining product stability and addressing taste and odor issues are essential for consumer acceptance.

Despite the challenges, microalgae nutraceuticals hold immense potential for the future of healthcare. Their sustainable production methods and potential health benefits make them a compelling area for continued research and development. This field has the potential to revolutionize functional foods and pharmaceuticals, offering innovative and eco-friendly solutions for preventive and therapeutic healthcare.

References

Abka-khajouei, R., Tounsi, L., Shahabi, N. et al. (2022). Structures, properties and applications of alginates. *Marine Drugs* 2022, 20 (6): 364. https://doi.org/10.3390/MD20060364.

Adam, F., Abert-Vian, M., Peltier, G., and Chemat, F. (2012). "Solvent-free" ultrasound-assisted extraction of lipids from fresh microalgae cells: a green, clean and scalable process. *Bioresource Technology* 114: 457–465. https://doi.org/10.1016/J.BIORTECH.2012.02.096.

Aiamla-or, S., Kaewsuksaeng, S., Shigyo, M., and Yamauchi, N. (2010). Impact of UV-B irradiation on chlorophyll degradation and chlorophyll-degrading enzyme activities in stored broccoli (*Brassica oleracea* L. Italica Group) florets. *Food Chemistry* 120 (3): 645–651. https://doi.org/10.1016/j.foodchem.2009.10.056.

Al hattab, M. and Ghaly, A. (2015). Microalgae oil extraction pre-treatment methods: critical review and comparative analysis. *Journal of Fundamentals of Renewable Energy and Applications* 5 (4): https://doi.org/10.4172/2090-4541.1000172.

Alavijeh, R.S., Karimi, K., Wijffels, R.H. et al. (2020). Combined bead milling and enzymatic hydrolysis for efficient fractionation of lipids, proteins, and carbohydrates of *Chlorella vulgaris* microalgae. *Bioresource Technology* 309: https://doi.org/10.1016/j.biortech.2020.123321.

Anderson and Melvin. (1963). *Extraction and purification of chlorophyll - Patent US-3258467-A - PubChem*. https://pubchem.ncbi.nlm.nih.gov/patent/US-3258467-A

Andlauer, W. and Fürst, P.F. (2002). Nutraceuticals: a piece of history, present status and outlook. *Food Research International* 35 (2–3): 171–176. www.elsevier.com/locate/foodres.

Andrade, L.D.O., Awasthi, R., Dua, K., and Andreoli Pinto, T.D.J. (2018). Matrix-assisted laser desorption ionization–time of flight mass spectrometry for identification of bacteria isolated from pharmaceutical clean rooms. *Interventional Medicine and Applied Science* 10 (1): 45–53. https://doi.org/10.1556/1646.9.2017.40.

Andreeva, A., Budenkova, E., Babich, O. et al. (2021). Production, purification, and study of the amino acid composition of microalgae proteins. *Molecules* 26 (9): https://doi.org/10.3390/molecules 26092767.

Araya, B., Gouveia, L., Nobre, B. et al. (2014). Evaluation of the simultaneous production of lutein and lipids using a vertical alveolar panel bioreactor for three *Chlorella* species. *Algal Research* 6 (PB): 218–222. https://doi.org/10.1016/j.algal.2014.06.003.

Aw, M.S., Simovic, S., Addai-Mensah, J., and Losic, D. (2011). Silica microcapsules from diatoms as new carrier for delivery of therapeutics. *Nanomedicine (London, England)* 6 (7): 1159–1173. https://doi.org/10.2217/NNM.11.29.

Ayalon, O. (2016). *AU2016314624A1 - Improved process for producing fucoxanthin and/or polysaccharides from microalgae - Google Patents*. https://patents.google.com/patent/AU2016314624A1/en

Aziz, M.M.A., Kassim, K.A., Shokravi, Z. et al. (2020). Two-stage cultivation strategy for simultaneous increases in growth rate and lipid content of microalgae: a review. *Renewable and Sustainable Energy Reviews* 119: 109621. https://doi.org/10.1016/j.rser.2019.109621.

Bagchi, D. (2014). *Nutraceutical and Functional Food Regulations in the United States and Around the World*, 2ee, 1–549. Academic Press https://doi.org/10.1016/C2012-0-00335-3.

Bajpai, P. (2019). Characteristics of Algae. In: *Third Generation Biofuels*, 11–15. Springer https://doi.org/10.1007/978-981-13-2378-2_3.

Batista, A.P., Niccolai, A., Fradinho, P. et al. (2017). Microalgae biomass as an alternative ingredient in cookies: sensory, physical and chemical properties, antioxidant activity and in vitro digestibility. *Algal Research* 26: 161–171. https://doi.org/10.1016/J.ALGAL.2017.07.017.

Batista, A.P., Niccolai, A., Bursic, I. et al. (2019). Microalgae as functional ingredients in savory food products: application to wheat crackers. *Foods* 8 (12): 611. https://doi.org/10.3390/FOODS8120611.

Baweja, P., Kumar, S., Sahoo, D., and Levine, I. (2016). Biology of seaweeds. In: *Seaweed in Health and Disease Prevention* (ed. J. Fleurence and I.A. Levine), 41–106. Academic Press https://doi.org/10.1016/B978-0-12-802772-1.00003-8.

Bhatnagar, P., Gururani, P., Parveen, A. et al. (2024). Algae: a promising and sustainable protein-rich food ingredient for bakery and dairy products. *Food Chemistry* 441: 138322. https://doi.org/10.1016/j.foodchem.2023.138322.

Bi, D., Yang, X., Yao, L. et al. (2022). Potential food and nutraceutical applications of alginate: a review. *Marine Drugs* 20 (9): https://doi.org/10.3390/MD20090564.

Bishop, W.M. and Zubeck, H.M. (2012). Evaluation of microalgae for use as nutraceuticals and nutritional supplements. *Journal of Nutrition & Food Sciences* 02 (05): https://doi.org/10.4172/2155-9600.1000147.

Bisht, D., Bhatt, V., Arya, R.K.K. et al. (2024). Plants Based Nutraceuticals: An Overview. *Herbal Nutraceuticals*. Wiley; Portico. https://doi.org/10.1002/9781394241576.ch1.

Bleakley, S. and Hayes, M. (2017). Algal proteins: extraction, application, and challenges concerning production. *Foods* 6 (5): 33. https://doi.org/10.3390/FOODS6050033.

Bucher, H.C., Hengstler, P., Schindler, C., and Meier, G. (2002). N-3 polyunsaturated fatty acids in coronary heart disease: a meta-analysis of randomized controlled trials. *American Journal of Medicine* 112 (4): 298–304. https://doi.org/10.1016/S0002-9343(01)01114-7.

Camacho, F., Macedo, A., and Malcata, F. (2019). Potential industrial applications and commercialization of microalgae in the functional food and feed industries: a short review. *Marine Drugs* 17 (6): 312. https://doi.org/10.3390/md17060312.

Campanella, L., Russo, M.V., and Avino, P. (2002). by Società Chimica Italiana). Free and total amino acid composition in blue-green algae. *Annali di Chimica* 92 (4): 343–352.
Capelli, B. and Cysewski, G.R. (2010). Potential health benefits of *Spirulina* microalgae*. *Nutrafoods* 9 (2): 19–26. https://doi.org/10.1007/BF03223332.
Carlson, S.E. (1999). Long-chain polyunsaturated fatty acids and development of human infants. *Acta Paediatrica (Oslo, Norway: 1992). Supplement* 88 (430): 72–77. https://doi.org/10.1111/J.1651-2227.1999.TB01304.X.
Carullo, D., Abera, B.D., Casazza, A.A. et al. (2018). Effect of pulsed electric fields and high pressure homogenization on the aqueous extraction of intracellular compounds from the microalgae *Chlorella vulgaris*. *Algal Research* 31: 60–69. https://doi.org/10.1016/j.algal.2018.01.017.
Cerón, M.C., Campos, I., Sánchez, J.F. et al. (2008). Recovery of lutein from microalgae biomass: development of a process for *Scenedesmus almeriensis* biomass. *Journal of Agricultural and Food Chemistry* 56 (24): 11761–11766. https://doi.org/10.1021/jf8025875.
Chaudhari, R. and Fanion, M. (2008). Technical aspects of micronutrient addition to foods. In: *Food Fortification and Supplementation: Technological, Safety and Regulatory Aspects* (ed. P.B. Ottaway), 108–125. Woodhead Publishing Ltd https://doi.org/10.1533/9781845694265.1.108.
Chemat, F., Zill-E-Huma, and Khan, M.K. (2011). Applications of ultrasound in food technology: processing, preservation and extraction. *Ultrasonics Sonochemistry* 18 (4): 813–835. https://doi.org/10.1016/J.ULTSONCH.2010.11.023.
Cheng, X., Riordon, J., Nguyen, B. et al. (2017). Hydrothermal disruption of algae cells for astaxanthin extraction. *Green Chemistry* 19 (1): 106–111. https://doi.org/10.1039/c6gc02746f.
Chia, M.A., Lombardi, A.T., Da, M., and Melão, G.G. (2013). Growth and biochemical composition of *Chlorella vulgaris* in different growth media. *Anais da Academia Brasileira de Ciências* 85 (4): https://doi.org/10.1590/0001-3765201393312.
Choudhary, P., Prajapati, S.K., and Malik, A. (2016). Screening native microalgal consortia for biomass production and nutrient removal from rural wastewaters for bioenergy applications. *Ecological Engineering* 91: 221–230. https://doi.org/10.1016/j.ecoleng.2015.11.056.
Choudhary, P., Assemany, P.P., Naaz, F. et al. (2020). A review of biochemical and thermochemical energy conversion routes of wastewater grown algal biomass. *Science of the Total Environment* 726 (271): 137961. https://doi.org/10.1016/j.scitotenv.2020.137961.
Chua, E.T., Shekh, A.Y., Eltanahy, E. et al. (2020). Effective harvesting of *Nannochloropsis* microalgae using mushroom chitosan: a pilot-scale study. *Frontiers in Bioengineering and Biotechnology* 8: 771. https://doi.org/10.3389/fbioe.2020.00771.
Connor, W.E. (2000). Importance of n-3 fatty acids in health and disease. *The American Journal of Clinical Nutrition* 71 (1 Suppl): https://doi.org/10.1093/AJCN/71.1.171S.
Cordero, B.F., Obraztsova, I., Couso, I. et al. (2011). Enhancement of lutein production in *Chlorella sorokiniana* (Chorophyta) by improvement of culture conditions and random mutagenesis. *Marine Drugs* 9 (9): 1607–1624. https://doi.org/10.3390/md9091607.
Crampon, C., Nikitine, C., Zaier, M. et al. (2017). Oil extraction from enriched *Spirulina platensis* microalgae using supercritical carbon dioxide. *The Journal of Supercritical Fluids* 119: 289–296. https://doi.org/10.1016/J.SUPFLU.2016.10.006.
Cuellar-Bermudez, S.P., Aguilar-Hernandez, I., Cardenas-Chavez, D.L. et al. (2015). Extraction and purification of high-value metabolites from microalgae: essential lipids, astaxanthin and phycobiliproteins. In: *Microbial Biotechnology* 8(2), 190–209. John Wiley & Sons Ltd and Society for Applied Microbiology https://doi.org/10.1111/1751-7915.12167.

Czartoski, T. J., Perkins, R., Villanueva, J. L., & Richards, G. (2010). *WO2010104922A1 - Algae biomass fractionation - Google Patents.* https://patents.google.com/patent/WO2010104922A1/ja

Dahmen-Ben Moussa, I., Chtourou, H., Hassairi, I. et al. (2019). The effect of switching environmental conditions on content and structure of lipid produced by a wild strain *Picochlorum* sp. *Renewable Energy* 134: 406–415. https://doi.org/10.1016/J.RENENE.2018.11.050.

Das, L., Bhaumik, E., Raychaudhuri, U., and Chakraborty, R. (2012). Role of nutraceuticals in human health. *Journal of Food Science and Technology* 49 (2): 173. https://doi.org/10.1007/S13197-011-0269-4.

Davani-Davari, D., Negahdaripour, M., Karimzadeh, I. et al. (2019). Prebiotics: definition, types, sources, mechanisms, and clinical applications. *Foods* 8 (3): https://doi.org/10.3390/FOODS8030092.

Del Campo, J.A., García-González, M., and Guerrero, M.G. (2007). Outdoor cultivation of microalgae for carotenoid production: current state and perspectives. *Applied Microbiology and Biotechnology* 74 (6): 1163–1174. https://doi.org/10.1007/s00253-007-0844-9.

Demurtas, O.C., Massa, S., Ferrante, P. et al. (2013). A *Chlamydomonas*-derived Human Papillomavirus 16 E7 vaccine induces specific tumor protection. *PLoS One* 8 (4): https://doi.org/10.1371/JOURNAL.PONE.0061473.

Deniz, I., Vaquero, M.-G., and Imamoglu, E. (2017). Chapter 19 Trends in red biotechnology: microalgae for pharmaceutical applications. In: *Microalgae-Based Biofuels and Bioproducts*, 1ee (ed. R. Muñoz and C. Gonzalez-Fernandez), 432–464. Elsevier.

Di Sanzo, G., Mehariya, S., Martino, M. et al. (2018). Supercritical carbon dioxide extraction of astaxanthin, lutein, and fatty acids from *Haematococcus pluvialis* microalgae. *Marine Drugs* 16 (9): https://doi.org/10.3390/md16090334.

Díaz del Castillo, B., and Garcia, G. (1928). The discovery and conquest of Mexico, 1517-1521. https://www.routledge.com/The-Discovery-and-Conquest-of-Mexico-1517-1521/Garcia-Castillo/p/book/9780415847087

Dineshkumar, R., Dash, S.K., and Sen, R. (2015). Process integration for microalgal lutein and biodiesel production with concomitant flue gas CO_2 sequestration: a biorefinery model for healthcare, energy and environment. *RSC Advances* 5 (90): 73381–73394. https://doi.org/10.1039/c5ra09306f.

Diprat, A.B., Silveira Thys, R.C., Rodrigues, E., and Rech, R. (2020). Chlorella sorokiniana: a new alternative source of carotenoids and proteins for gluten-free bread. *LWT* 134: 109974. https://doi.org/10.1016/J.LWT.2020.109974.

Dong, Y., Wei, Z., and Xue, C. (2021). Recent advances in carrageenan-based delivery systems for bioactive ingredients: a review. *Trends in Food Science & Technology* 112: 348–361. https://doi.org/10.1016/J.TIFS.2021.04.012.

Douglas, S.E., Raven, J.A., and Larkum, A.W.D. (2003). The Algae and their General Characteristics. In: *Photosynthesis in Algae*, 1–10. Springer https://doi.org/10.1007/978-94-007-1038-2_1.

Draaisma, R.B., Wijffels, R.H., Slegers, P.M. et al. (2013). Food commodities from microalgae. *Current Opinion in Biotechnology* 24 (2): 169–177. https://doi.org/10.1016/j.copbio.2012.09.012.

Ebrahimi-Mameghani, M., Aliashrafi, S., Javadzadeh, Y., and AsghariJafarabadi, M. (2014). The effect of Chlorella vulgaris supplementation on liver enzymes, serum glucose and lipid profile in patients with non-alcoholic fatty liver disease. *Health Promotion Perspectives* 4 (1): 107. https://doi.org/10.5681/HPP.2014.014.

Eing, C., Goettel, M., Straessner, R. et al. (2013). Pulsed electric field treatment of microalgae - benefits for microalgae biomass processing. *IEEE Transactions on Plasma Science* 41 (10): 2901–2907. https://doi.org/10.1109/TPS.2013.2274805.

El-Baz, F.K., Aly, H.F., and Abd-Alla, H.I. (2020). The ameliorating effect of carotenoid rich fraction extracted from *Dunaliella salina* microalga against inflammation- associated cardiac dysfunction in obese rats. *Toxicology Reports* 7: 118–124. https://doi.org/10.1016/J.TOXREP.2019.12.008.

Eriksen, N.T. (2008). Production of phycocyanin - a pigment with applications in biology, biotechnology, foods and medicine. *Applied Microbiology and Biotechnology* 80 (1): 1–14. https://doi.org/10.1007/s00253-008-1542-y.

Eseberri, I., Gómez-Zorita, S., Trepiana, J. et al. (2020). Anti-obesity effects of microalgae. *International Journal of Molecular Sciences* 21 (1): 41. https://doi.org/10.3390/IJMS21010041.

Ezzati, J., Dolatabadi, N., and De La Guardia, M. (2011). Applications of diatoms and silica nanotechnology in biosensing, drug and gene delivery, and formation of complex metal nanostructures. *TrAC Trends in Analytical Chemistry* 30 (9): 1538–1548. https://doi.org/10.1016/j.trac.2011.04.015.

Fajardo, C., De Donato, M., Carrasco, R. et al. (2020). Advances and challenges in genetic engineering of microalgae. *Reviews in Aquaculture* 12 (1): 365–381. https://doi.org/10.1111/RAQ.12322.

FAO Fisheries and Aquaculture. (2008). *A REVIEW ON CULTURE, PRODUCTION AND USE OF SPIRULINA AS FOOD FOR HUMANS AND FEEDS FOR DOMESTIC ANIMALS AND FISH.* http://www.fao.org

Fu, W., Chaiboonchoe, A., Khraiwesh, B. et al. (2016). Algal cell factories: approaches, applications, and potentials. *Marine Drugs* 14 (12): https://doi.org/10.3390/MD14120225.

Future Market Insights. (2023). .:: *Investor Portal - Nivesh Bandhu* ::. https://www.foodprocessingindia.gov.in/newsletter/emailer/two

Future Market Insights. (2024). *Downstream Processing Market Trends, Demand & Forecast | FMI.* https://www.futuremarketinsights.com/reports/downstream-processing-market

Galarza, J.I., Gimpel, J.A., Rojas, V. et al. (2018). Over-accumulation of astaxanthin in *Haematococcus pluvialis* through chloroplast genetic engineering. *Algal Research* 31: 291–297. https://doi.org/10.1016/J.ALGAL.2018.02.024.

Gantar, M., Dhandayuthapani, S., and Rathinavelu, A. (2012). Phycocyanin induces apoptosis and enhances the effect of topotecan on prostate cell line LNCaP. *Journal of Medicinal Food* 15 (12): 1091–1095. https://doi.org/10.1089/JMF.2012.0123.

Garcia-Perez, P., Cassani, L., Garcia-Oliveira, P. et al. (2023). Algal nutraceuticals: a perspective on metabolic diversity, current food applications, and prospects in the field of metabolomics. *Food Chemistry* 409: 135295. https://doi.org/10.1016/j.foodchem.2022.135295.

Geng, D., Wang, Y., Wang, P. et al. (2003). Stable expression of hepatitis B surface antigen gene in *Dunaliella salina* (Chlorophyta). *Journal of Applied Phycology* 15: 451–456.

Golmakani, M.T., Soleimanian-Zad, S., Alavi, N. et al. (2019). Effect of *Spirulina* (*Arthrospira platensis*) powder on probiotic bacteriologically acidified feta-type cheese. *Journal of Applied Phycology* 31 (2): 1085–1094. https://doi.org/10.1007/S10811-018-1611-2/FIGURES/4.

Gong, M. and Bassi, A. (2016). Carotenoids from microalgae: a review of recent developments. *Biotechnology Advances* 34 (8): 1396–1412. https://doi.org/10.1016/j.biotechadv.2016.10.005.

Goto, M., Kanda, H., Wahyudiono, and Machmudah, S. (2015). Extraction of carotenoids and lipids from algae by supercritical CO_2 and subcritical dimethyl ether. *Journal of Supercritical Fluids* 96: 245–251. https://doi.org/10.1016/j.supflu.2014.10.003.

Gouveia, L., Raymundo, A., Batista, A.P. et al. (2006). *Chlorella vulgaris* and *Haematococcus pluvialis* biomass as colouring and antioxidant in food emulsions. *European Food Research and Technology* 222 (3–4): 362–367. https://doi.org/10.1007/S00217-005-0105-Z/FIGURES/6.

Gouveia, L., Batista, A.P., Miranda, A. et al. (2007). *Chlorella vulgaris* biomass used as colouring source in traditional butter cookies. *Innovative Food Science & Emerging Technologies* 8 (3): 433–436. https://doi.org/10.1016/J.IFSET.2007.03.026.

Gouveia, L., Coutinho, C., Mendonça, E. et al. (2008). Functional biscuits with PUFA-ω3 from *Isochrysis galbana*. *Journal of the Science of Food and Agriculture* 88 (5): 891–896. https://doi.org/10.1002/JSFA.3166.

Guihéneuf, F., Khan, A., and Tran, L.S.P. (2016). Genetic engineering: a promising tool to engender physiological, biochemical, and molecular stress resilience in green microalgae. *Frontiers in Plant Science* 7 (MAR2016): https://doi.org/10.3389/FPLS.2016.00400.

Guiry, M.D. (2012). How many species of algae are there? *Journal of Phycology* 48 (5): 1057–1063. https://doi.org/10.1111/J.1529-8817.2012.01222.X.

Guldhe, A., Singh, B., Rawat, I. et al. (2014). Efficacy of drying and cell disruption techniques on lipid recovery from microalgae for biodiesel production. *Fuel* 128: 46–52. https://doi.org/10.1016/J.FUEL.2014.02.059.

Günerken, E., D'Hondt, E., Eppink, M.H.M. et al. (2015). Cell disruption for microalgae biorefineries. *Biotechnology Advances* 33 (2): 243–260. https://doi.org/10.1016/J.BIOTECHADV.2015.01.008.

Guo, X., Wang, Y., Qin, Y. et al. (2020). Structures, properties and application of alginic acid: a review. *International Journal of Biological Macromolecules* 162: 618–628. https://doi.org/10.1016/J.IJBIOMAC.2020.06.180.

Halim, R., Hosikian, A., Lim, S., and Danquah, M.K. (2010). Chlorophyll extraction from microalgae: a review on the process engineering aspects. *International Journal of Chemical Engineering* 2010: https://doi.org/10.1155/2010/391632.

Halim, R., Webley, P.A., and Martin, G.J.O. (2016). The CIDES process: fractionation of concentrated microalgal paste for co-production of biofuel, nutraceuticals, and high-grade protein feed. *Algal Research* 19: 299–306. https://doi.org/10.1016/j.algal.2015.09.018.

Hamed, I. (2016). The evolution and versatility of microalgal biotechnology: a review. *Comprehensive Reviews in Food Science and Food Safety* 15 (6): 1104–1123. https://doi.org/10.1111/1541-4337.12227.

Hamed, I., Özogul, F., Özogul, Y., and Regenstein, J.M. (2015). Marine bioactive compounds and their health benefits: a review. *Comprehensive Reviews in Food Science and Food Safety* 14 (4): 446–465. https://doi.org/10.1111/1541-4337.12136.

Han, S.I., Chang, S.H., Lee, C. et al. (2020). Astaxanthin biosynthesis promotion with pH shock in the green microalga, *Haematococcus lacustris*. *Bioresource Technology* 314: 123725. https://doi.org/10.1016/j.biortech.2020.123725.

Hansen, J.E., Packard, J.E., and Doyle, W.T. (1981). *Mariculture of Red Seaweeds. A California Sea Grant College Program Publication.*

Hao, S., Yan, Y., Li, S. et al. (2018). The in vitro anti-tumor activity of phycocyanin against non-small cell lung cancer cells. *Marine Drugs* 16 (6): https://doi.org/10.3390/MD16060178.

Hao, S., Li, S., Wang, J. et al. (2019). Phycocyanin exerts anti-proliferative effects through down-regulating TIRAP/NF-κB activity in human non-small cell lung cancer cells. *Cells* 8 (6): https://doi.org/10.3390/CELLS8060588.

Hayashi, K., Lee, J.B., Atsumi, K. et al. (2019). *In vitro* and *in vivo* anti-herpes simplex virus activity of monogalactosyl diacylglyceride from *Coccomyxa sp. KJ* (IPOD FERM BP-22254), a green microalga. *PLoS One* 14 (7): https://doi.org/10.1371/JOURNAL.PONE.0219305.

Hempel, F., Maurer, M., Brockmann, B. et al. (2017). From hybridomas to a robust microalgal-based production platform: molecular design of a diatom secreting monoclonal antibodies directed against the Marburg virus nucleoprotein. *Microbial Cell Factories* 16 (1): https://doi.org/10.1186/S12934-017-0745-2.

Henderson, R.K., Baker, A., Parsons, S.A., and Jefferson, B. (2008). Characterisation of algogenic organic matter extracted from cyanobacteria, green algae and diatoms. *Water Research* 42 (13): 3435–3445. https://doi.org/10.1016/j.watres.2007.10.032.

Hernández-Carmona, G., Freile-Pelegrín, Y., and Hernández-Garibay, E. (2013). Conventional and alternative technologies for the extraction of algal polysaccharides. In: *Functional Ingredients from Algae for Foods and Nutraceuticals*, 475–516. Elsevier Ltd. Woodhead Publishing Series in Food Science, Technology and Nutrition. https://doi.org/10.1533/9780857098689.3.475.

Hirahashi, T., Matsumoto, M., Hazeki, K. et al. (2002). Activation of the human innate immune system by Spirulina: augmentation of interferon production and NK cytotoxicity by oral administration of hot water extract of *Spirulina platensis. International Immunopharmacology* 2 (4): 423–434. https://doi.org/10.1016/S1567-5769(01)00166-7.

Ho, S.H., Chan, M.C., Liu, C.C. et al. (2014). Enhancing lutein productivity of an indigenous microalga *Scenedesmus obliquus* FSP-3 using light-related strategies. *Bioresource Technology* 152: 275–282. https://doi.org/10.1016/j.biortech.2013.11.031.

Hodge, L., Salome, C.M., Peat, J.K. et al. (1996). Consumption of oily fish and childhood asthma risk. *The Medical Journal of Australia* 164 (3): 137–140. https://doi.org/10.5694/J.1326-5377.1996.TB122010.X.

Hossain, N. and Mahlia, T.M.I. (2019). Progress in physicochemical parameters of microalgae cultivation for biofuel production. *Critical Reviews in Biotechnology* 39 (6): 835–859. https://doi.org/10.1080/07388551.2019.1624945.

India's Nutraceutical Market Growth (2019). https://foodprocessingindia.gov.in/newsletter/emailer/two (Retrieved 7 December 2023).

Iskandarov, U., Sitnik, S., Shtaida, N. et al. (2016). Cloning and characterization of a GPAT-like gene from the microalga *Lobosphaera incisa* (Trebouxiophyceae): overexpression in *Chlamydomonas reinhardtii* enhances TAG production. *Journal of Applied Phycology* 28 (2): 907–919. https://doi.org/10.1007/S10811-015-0634-1.

Iwai, M., Ikeda, K., Shimojima, M., and Ohta, H. (2014). Enhancement of extraplastidic oil synthesis in *Chlamydomonas reinhardtii* using a type-2 diacylglycerol acyltransferase with a phosphorus starvation-inducible promoter. *Plant Biotechnology Journal* 12 (6): 808–819. https://doi.org/10.1111/PBI.12210.

Jiang, L., Luo, S., Fan, X. et al. (2011). Biomass and lipid production of marine microalgae using municipal wastewater and high concentration of CO_2. *Applied Energy* 88 (10): 3336–3341. https://doi.org/10.1016/j.apenergy.2011.03.043.

Juárez-Oropeza, M.A., Mascher, D., Torres-Durán, P.V. et al. (2009). Effects of dietary *Spirulina* on vascular reactivity. *Journal of Medicinal Food* 12 (1): 15–20. https://doi.org/10.1089/JMF.2007.0713.

Kale, A. (2012). *US8152870B2 - Methods of and systems for producing biofuels - Google Patents.* https://patents.google.com/patent/US8152870B2/en

Kamel El-Baz, F., Abdo, S.M., and Hussien, A. (2017). Microalgae *Dunaliella salina* for use as food supplement to improve pasta quality. *International Journal of Pharmaceutical Sciences Review and Research* 46 (2): 45–51. https://www.researchgate.net/publication/363712123.

Katiyar, R. and Arora, A. (2020). Health promoting functional lipids from microalgae pool: a review. *Algal Research* 46: 101800. https://doi.org/10.1016/j.algal.2020.101800.

Khanra, S., Mondal, M., Halder, G. et al. (2018). Downstream processing of microalgae for pigments, protein and carbohydrate in industrial application: a review. *Food and Bioproducts Processing* 1 (1): 60–84. https://doi.org/10.1016/j.fbp.2018.02.002.

Khavari, F., Saidijam, M., Taheri, M., and Nouri, F. (2021). Microalgae: therapeutic potentials and applications. *Molecular Biology Reports* 48 (5): 4757–4765. https://doi.org/10.1007/S11033-021-06422-W.

Khoo, K.S., Chong, Y.M., Chang, W.S. et al. (2021). Permeabilization of *Chlorella sorokiniana* and extraction of lutein by distillable CO_2-based alkyl carbamate ionic liquids. *Separation and Purification Technology* 256: 117471. https://doi.org/10.1016/j.seppur.2020.117471.

Kim, K.H., Choi, I.S., Kim, H.M. et al. (2014). Bioethanol production from the nutrient stress-induced microalga *Chlorella vulgaris* by enzymatic hydrolysis and immobilized yeast fermentation. *Bioresource Technology* 153: 47–54. https://doi.org/10.1016/j.biortech.2013.11.059.

Kim, Y.E., Matter, I.A., Lee, N. et al. (2020). Enhancement of astaxanthin production by *Haematococcus pluvialis* using magnesium aminoclay nanoparticles. *Bioresource Technology* 307: 123270. https://doi.org/10.1016/J.BIORTECH.2020.123270.

Kose, A., Ozen, M.O., Elibol, M., and Oncel, S.S. (2017). Investigation of in vitro digestibility of dietary microalga *Chlorella vulgaris* and cyanobacterium *Spirulina platensis* as a nutritional supplement. *3 Biotech* 7 (3): 170. https://doi.org/10.1007/s13205-017-0832-4.

Koyande, A.K., Chew, K.W., Rambabu, K. et al. (2019). Microalgae: a potential alternative to health supplementation for humans. *Food Science and Human Wellness* 8 (1): 16–24. https://doi.org/10.1016/J.FSHW.2019.03.001.

Ku, C.S., Pham, T.X., Park, Y. et al. (2013). Edible blue-green algae reduce the production of pro-inflammatory cytokines by inhibiting NF-κB pathway in macrophages and splenocytes. *Biochimica et Biophysica Acta* 1830 (4): 2981–2988. https://doi.org/10.1016/J.BBAGEN.2013.01.018.

Kumari, A., Garima, and Bharadvaja, N. (2023). A comprehensive review on algal nutraceuticals as prospective therapeutic agent for different diseases. In: *3 Biotech* 13(2), 44. Springer Science and Business Media Deutschland GmbH https://doi.org/10.1007/s13205-022-03454-2.

Kurokawa, M., King, P.M., Wu, X. et al. (2016). Effect of sonication frequency on the disruption of algae. *Ultrasonics Sonochemistry* 31: 157–162. https://doi.org/10.1016/j.ultsonch.2015.12.011.

Laamanen, C.A., Desjardins, S.M., Senhorinho, G.N.A., and Scott, J.A. (2021). Harvesting microalgae for health beneficial dietary supplements. *Algal Research* 54: 102189. https://doi.org/10.1016/J.ALGAL.2021.102189.

Lafarga, T., Gabriel, F., Fernández, A. et al. (2022). Improving the nutritional, structural, and sensory properties of gluten-free bread with different species of microalgae. *Foods* 11 (3): 397. https://doi.org/10.3390/FOODS11030397.

Lai, Y.C., Chang, C.H., Chen, C.Y. et al. (2019). Towards protein production and application by using *Chlorella* species as circular economy. *Bioresource Technology* 289: 121625. https://doi.org/10.1016/j.biortech.2019.121625.

Lee, I. and Han, J.I. (2015). Hydrothermal-acid treatment for effectual extraction of eicosapentaenoic acid (EPA)-abundant lipids from *Nannochloropsis salina*. *Bioresource Technology* 191: 1–6. https://doi.org/10.1016/j.biortech.2015.04.124.

Leray, C., Wiesel, M.L., Freund, M. et al. (2001). Long-chain n-3 fatty acids specifically affect rat coagulation factors dependent on vitamin K: relation to peroxidative stress. *Arteriosclerosis, Thrombosis, and Vascular Biology* 21 (3): 459–465. https://doi.org/10.1161/01.ATV.21.3.459.

Levasseur, W., Perré, P., and Pozzobon, V. (2020). A review of high value-added molecules production by microalgae in light of the classification. *Biotechnology Advances* 41: https://doi.org/10.1016/J.BIOTECHADV.2020.107545.

Li, X., Li, X., Han, B. et al. (2019). Improvement in lipid production in *Monoraphidium* sp. QLY-1 by combining fulvic acid treatment and salinity stress. *Bioresource Technology* 294: 122179. https://doi.org/10.1016/j.biortech.2019.122179.

Li, S., Ji, L., Chen, C. et al. (2020). Efficient accumulation of high-value bioactive substances by carbon to nitrogen ratio regulation in marine microalgae *Porphyridium purpureum*. *Bioresource Technology* 309: https://doi.org/10.1016/J.BIORTECH.2020.123362.

Liang, M.H. and Jiang, J.G. (2013). Advancing oleaginous microorganisms to produce lipid via metabolic engineering technology. *Progress in Lipid Research* 52 (4): 395–408. https://doi.org/10.1016/J.PLIPRES.2013.05.002.

Lim, H.T. and Teo, S.S. (2015). Comparison of protein extraction protocols for proteomic analysis of red algae, *Eucheuma cottonii*. *Pertanika Journal of Tropical Agricultural Science* 38 (2): 279–293.

Lima, S., Schulze, P.S.C., Schüler, L.M. et al. (2021). Flashing light emitting diodes (LEDs) induce proteins, polyunsaturated fatty acids and pigments in three microalgae. *Journal of Biotechnology* 325: 15–24. https://doi.org/10.1016/j.jbiotec.2020.11.019.

Lin, J.H., Lee, D.J., and Chang, J.S. (2015). Lutein production from biomass: marigold flowers versus microalgae. *Bioresource Technology* 184: 421–428. https://doi.org/10.1016/J.BIORTECH.2014.09.099.

Lin, B., Cui, Y., Yan, M. et al. (2019). Construction of astaxanthin metabolic pathway in the green microalga *Dunaliella viridis*. *Algal Research* 44: https://doi.org/10.1016/J.ALGAL.2019.101697.

Liu, Y., Xu, L., Cheng, N. et al. (2000). Inhibitory effect of phycocyanin from *Spirulina platensis* on the growth of human leukemia K562 cells. *Journal of Applied Phycology* 12 (2): 125–130. https://doi.org/10.1023/A:1008132210772/METRICS.

Liu, Z.W., Zeng, X.A., Cheng, J.H. et al. (2018). The efficiency and comparison of novel techniques for cell wall disruption in astaxanthin extraction from *Haematococcus pluvialis*. *International Journal of Food Science and Technology* 53 (9): 2212–2219. https://doi.org/10.1111/ijfs.13810.

López, Y. and Soto, S.M. (2019). The usefulness of microalgae compounds for preventing biofilm infections. *Antibiotics (Basel, Switzerland)* 9 (1): https://doi.org/10.3390/ANTIBIOTICS9010009.

Lorente, E., Hapońska, M., Clavero, E. et al. (2018). Steam explosion and vibrating membrane filtration to improve the processing cost of microalgae cell disruption and fractionation. *Processes* 6 (4): https://doi.org/10.3390/pr6040028.

Low, K.L., Idris, A., and Mohd Yusof, N. (2020). Novel protocol optimized for microalgae lutein used as food additives. *Food Chemistry* 307: https://doi.org/10.1016/J.FOODCHEM.2019.125631.

Ma, Y.H., Wang, X., Niu, Y.F. et al. (2014). Antisense knockdown of pyruvate dehydrogenase kinase promotes the neutral lipid accumulation in the diatom *Phaeodactylum tricornutum*. *Microbial Cell Factories* 13 (1): 1–9. https://doi.org/10.1186/S12934-014-0100-9/FIGURES/4.

Ma, R., Zhao, X., Ho, S.H. et al. (2020a). Co-production of lutein and fatty acid in microalga *Chlamydomonas* sp. JSC4 in response to different temperatures with gene expression profiles. *Algal Research* 47: https://doi.org/10.1016/J.ALGAL.2020.101821.

Ma, S., Yu, Y., Cui, H. et al. (2020b). Unsterilized sewage treatment and carbohydrate accumulation in *Tetradesmus obliquus* PF3 with CO_2 supplementation. *Algal Research* 45: 101741. https://doi.org/10.1016/j.algal.2019.101741.

Macías-Sánchez, M.D., Mantell, C., Rodríguez, M. et al. (2007). Supercritical fluid extraction of carotenoids and chlorophyll a from *Synechococcus* sp. *Journal of Supercritical Fluids* 39 (3): 323–329. https://doi.org/10.1016/j.supflu.2006.03.008.

Maher, S., Kumeria, T., Aw, M.S., and Losic, D. (2018). Diatom silica for biomedical applications: recent progress and advances. *Advanced Healthcare Materials* 7 (19): https://doi.org/10.1002/ADHM.201800552.

Manirafasha, E., Ndikubwimana, T., Zeng, X. et al. (2016). Phycobiliprotein: potential microalgae derived pharmaceutical and biological reagent. In: *Biochemical Engineering Journal*, vol. Vol. 109, 282–296. Elsevier B.V. https://doi.org/10.1016/j.bej.2016.01.025.

Marino, T., Iovine, A., Casella, P. et al. (2020). From *Haematococcus pluvialis* microalgae a powerful antioxidant for cosmetic applications. *Chemical Engineering Transactions* 79: 271–276. https://doi.org/10.3303/CET2079046.

Martínez, J.M., Gojkovic, Z., Ferro, L. et al. (2019). Use of pulsed electric field permeabilization to extract astaxanthin from the Nordic microalga *Haematococcus pluvialis*. *Bioresource Technology* 289: https://doi.org/10.1016/j.biortech.2019.121694.

Masarin, F., Cedeno, F.R.P., Chavez, E.G.S. et al. (2016). Chemical analysis and biorefinery of red algae *Kappaphycus alvarezii* for efficient production of glucose from residue of carrageenan extraction process. *Biotechnology for Biofuels* 9 (1): https://doi.org/10.1186/S13068-016-0535-9.

Mehariya, S., Iovine, A., Di Sanzo, G. et al. (2019). Supercritical fluid extraction of lutein from *Scenedesmus almeriensis*. *Molecules* 24 (7): 1324. https://doi.org/10.3390/MOLECULES24071324.

Mehariya, S., Kumar Goswami, R., Karthikeysan, O.P. et al. (2021). Microalgae for high-value products: a way towards green nutraceutical and pharmaceutical compounds. *Chemosphere* 280: 130553. https://doi.org/10.1016/j.chemosphere.2021.130553.

Mendez, L., Mahdy, A., Timmers, R.A. et al. (2013). Enhancing methane production of *Chlorella vulgaris* via thermochemical pretreatments. *Bioresource Technology* 149: 136–141. https://doi.org/10.1016/j.biortech.2013.08.136.

Meštrović T. (2022). What are Nutraceuticals? https://www.news-medical.net/health/What-are-Nutraceuticals.aspx (accessed 14 December 2023)

Minyuk, G., Sidorov, R., and Solovchenko, A. (2020). Effect of nitrogen source on the growth, lipid, and valuable carotenoid production in the green microalga *Chromochloris zofingiensis*. *Journal of Applied Phycology* 32 (2): 923–935. https://doi.org/10.1007/s10811-020-02060-0.

Mišurcová, L., Škrovánková, S., Samek, D. et al. (2012). Health benefits of algal polysaccharides in human nutrition. *Advances in Food and Nutrition Research* 66: 75–145. https://doi.org/10.1016/B978-0-12-394597-6.00003-3.

Molino, G., Montalbano, G., Pontremoli, C. et al. (2020). Imaging techniques for the assessment of the bone osteoporosis-induced variations with particular focus on micro-ct potential. *Applied Sciences (Switzerland)* 10 (24): 1–27. https://doi.org/10.3390/APP10248939.

de Morais, E.G., Nunes, I.L., Druzian, J.I. et al. (2022). Increase in biomass productivity and protein content of *Spirulina* sp. LEB 18 (Arthrospira) cultivated with crude glycerol. *Biomass Conversion and Biorefinery* 12 (3): 597–605. https://doi.org/10.1007/S13399-020-00934-4/FIGURES/3.

Mosulishvili, L.M., Kirkesali, E.I., Belokobylsky, A.I. et al. (2002). Experimental substantiation of the possibility of developing selenium- and iodine-containing pharmaceuticals based on blue-green algae *Spirulina platensis*. *Journal of Pharmaceutical and Biomedical Analysis* 30 (1): 87–97. https://doi.org/10.1016/S0731-7085(02)00199-1.

Mudimu, O., Koopmann, I.K., Rybalka, N. et al. (2017). Screening of microalgae and cyanobacteria strains for α-tocopherol content at different growth phases and the influence of nitrate reduction on α-tocopherol production. *Journal of Applied Phycology* 29 (6): 2867–2875. https://doi.org/10.1007/S10811-017-1188-1.

Mysliwa-Kurdziel, B. and Solymosi, K. (2016). Phycobilins and phycobiliproteins used in food industry and medicine. *Mini-Reviews in Medicinal Chemistry* 17 (13).

Nicoletti, M. (2016). Microalgae nutraceuticals. *Foods* 5 (3): 1–13. https://doi.org/10.3390/FOODS5030054.

Niklas, K.J., Bold, H.C., and Wynne, M.J. (1978). Introduction to the algae: structure and reproduction. *Brittonia* 30 (4): 482. https://doi.org/10.2307/2806355.

Nitsos, C., Filali, R., Taidi, B., and Lemaire, J. (2020). Current and novel approaches to downstream processing of microalgae: a review. In: *Biotechnology Advances*, vol. 45. Elsevier Inc. https://doi.org/10.1016/j.biotechadv.2020.107650.

Niu, Y.F., Zhang, M.H., Li, D.W. et al. (2013). Improvement of neutral lipid and polyunsaturated fatty acid biosynthesis by overexpressing a type 2 diacylglycerol acyltransferase in marine diatom *Phaeodactylum tricornutum*. *Marine Drugs* 11 (11): 4558–4569. https://doi.org/10.3390/MD11114558.

Nwosu, O.K. and Ubaoji, K.I. (2020). Nutraceuticals: history, classification and market demand. In: *Functional Foods and Nutraceuticals* (ed. C. Egbuna and G. Dable-Tupas), 13–22. Springer https://doi.org/10.1007/978-3-030-42319-3_2.

Nyenje, M.E. and Ndip, N.R. (2013). The challenges of foodborne pathogens and antimicrobial chemotherapy: a global perspective. *African Journal of Microbiology Research* 7 (14): 1158–1172. https://doi.org/10.5897/AJMRX12.014.

Ochoa Becerra, M., Mojica Contreras, L., Hsieh Lo, M. et al. (2020). Lutein as a functional food ingredient: stability and bioavailability. *Journal of Functional Foods* 66: 103771. https://doi.org/10.1016/J.JFF.2019.103771.

Olaizola, M. (2000). Commercial production of astaxanthin from *Haematococcus pluvialis* using 25,000-liter outdoor photobioreactors. *Journal of Applied Phycology* 12 (3): 499–506. https://doi.org/10.1023/a:1008159127672.

Oleskin, A.V. and Boyang, C. (2022). Microalgae in terms of biomedical technology: probiotics, prebiotics, and metabiotics. *Applied Biochemistry and Microbiology* 58 (6): 813. https://doi.org/10.1134/S0003683822060126.

Onay, M. (2019). Bioethanol production via different saccharification strategies from *H. tetrachotoma* ME03 grown at various concentrations of municipal wastewater in a flat-photobioreactor. *Fuel* 239: 1315–1323. https://doi.org/10.1016/j.fuel.2018.11.126.

Ördög, V., Stirk, W.A., Bálint, P. et al. (2012). Changes in lipid, protein and pigment concentrations in nitrogen-stressed *Chlorella* minutissima cultures. *Journal of Applied Phycology* 24 (4): 907–914. https://doi.org/10.1007/s10811-011-9711-2.

Ördög, V., Stirk, W.A., Bálint, P. et al. (2016). Effect of temperature and nitrogen concentration on lipid productivity and fatty acid composition in three *Chlorella* strains. *Algal Research* 16: 141–149. https://doi.org/10.1016/J.ALGAL.2016.03.001.

Panis, G. and Carreon, J.R. (2016). Commercial astaxanthin production derived by green alga *Haematococcus pluvialis*: a microalgae process model and a techno-economic assessment all through production line. *Algal Research* 18: 175–190. https://doi.org/10.1016/j.algal.2016.06.007.

Park, S.H., Lee, C.R., and Hong, S.K. (2020). Implications of agar and agarase in industrial applications of sustainable marine biomass. *Applied Microbiology and Biotechnology* 104 (7): 2815–2832. https://doi.org/10.1007/S00253-020-10412-6.

Pelczar, Chan E.C.S, & Krieg N.R. (1993). *The World of Bacteria IV: Gram-Positive Filamentous Bacteria of Complex. Morphology*. Part Five Microorganisms- Fungi, Algae, Protozoa, and Viruses. Chapter 18- Algae.

Pereira, H., Barreira, L., Custódio, L. et al. (2013). Isolation and fatty acid profile of selected microalgae strains from the red sea for biofuel production. *Energies* 6 (6): 2773–2783. https://doi.org/10.3390/en6062773.

Pittman, J.K., Dean, A.P., and Osundeko, O. (2011). The potential of sustainable algal biofuel production using wastewater resources. *Bioresource Technology* 102 (1): 17–25. https://doi.org/10.1016/J.BIORTECH.2010.06.035.

Postma, P.R., Suarez-Garcia, E., Safi, C. et al. (2017). Energy efficient bead milling of microalgae: effect of bead size on disintegration and release of proteins and carbohydrates. *Bioresource Technology* 224: 670–679. https://doi.org/10.1016/j.biortech.2016.11.071.

Prabakaran, G., Sampathkumar, P., Kavisri, M., and Moovendhan, M. (2020). Extraction and characterization of phycocyanin from *Spirulina platensis* and evaluation of its anticancer, antidiabetic and antiinflammatory effect. *International Journal of Biological Macromolecules* 153: 256–263. https://doi.org/10.1016/J.IJBIOMAC.2020.03.009.

Prasetyaningrum, A., Praptyana, I.R., and Nurfiningsih, & Ratnawati. (2019). Carrageenan: nutraceutical and functional food as future food. *IOP Conference Series: Earth and Environmental Science* 292 (1): 012068. https://doi.org/10.1088/1755-1315/292/1/012068.

Priatni, S., Ratnaningrum, D., Warya, S., and Audina, E. (2018). Phycobiliproteins production and heavy metals reduction ability of *Porphyridium* sp. *IOP Conference Series: Earth and Environmental Science* 160 (1): 012006. https://doi.org/10.1088/1755-1315/160/1/012006.

Qu, Z., Zeng, J., Zhang, Y. et al. (2018). Hydrothermal cell disruption of *Nannochloropsis* sp. and its influence on lipid extraction. *Algal Research* 35: 407–415. https://doi.org/10.1016/j.algal.2018.09.015.

Rahman, M.M., Hosano, N., and Hosano, H. (2022). Recovering microalgal bioresources: a review of cell disruption methods and extraction technologies. *Molecules* 27 (9): https://doi.org/10.3390/MOLECULES27092786.

Raja, R., Hemaiswarya, S., Kumar, N.A. et al. (2008). A perspective on the biotechnological potential of microalgae. *Critical Reviews in Microbiology* 34 (2): 77–88. https://doi.org/10.1080/10408410802086783.

Ramana, K.V., Xavier, J.R., and Sharma, R.K. (2017). Recent trends in pharmaceutical biotechnology. *Pharmaceutical Biotechnology: Current Research* 1 (1): https://www.imedpub.com/articles/recent-trends-in-pharmaceutical-biotechnology.php?aid=18187.

Remize, M., Brunel, Y., Silva, J.L. et al. (2021). Microalgae n-3 PUFAs production and use in food and feed industries. *Marine Drugs* 19 (2): https://doi.org/10.3390/MD19020113.

Rigby, S., Fairhead, M., and van der Walle, C. (2008). Engineering silica particles as oral drug delivery vehicles. *Current Pharmaceutical Design* 14 (18): 1821–1831. https://doi.org/10.2174/138161208784746671.

Rizwan, M., Mujtaba, G., Memon, S.A. et al. (2018). Exploring the potential of microalgae for new biotechnology applications and beyond: a review. *Renewable and Sustainable Energy Reviews* 92: 394–404. https://doi.org/10.1016/j.rser.2018.04.034.

Ryan Georgianna, D., Hannon, M.J., Marcuschi, M. et al. (2012). Production of recombinant enzymes in the marine alga *Dunaliella tertiolecta*. *Algal Research* 2: 2–9. https://doi.org/10.1016/j.algal.2012.10.004.

Safi, C., Camy, S., Frances, C. et al. (2014). Extraction of lipids and pigments of *Chlorella vulgaris* by supercritical carbon dioxide: influence of bead milling on extraction performance. *Journal of Applied Phycology* 26 (4): 1711–1718. https://doi.org/10.1007/s10811-013-0212-3.

Saini, D.K., Chakdar, H., Pabbi, S., and Shukla, P. (2020). Enhancing production of microalgal biopigments through metabolic and genetic engineering. *Critical Reviews in Food Science and Nutrition* 60 (3): 391–405. https://doi.org/10.1080/10408398.2018.1533518.

Salama, E.S., Kabra, A.N., Ji, M.K. et al. (2014). Enhancement of microalgae growth and fatty acid content under the influence of phytohormones. *Bioresource Technology* 172: 97–103. https://doi.org/10.1016/j.biortech.2014.09.002.

Sartory, D.P. and Grobbelaar, J.U. (1984). Extraction of chlorophyll a from freshwater phytoplankton for spectrophotometric analysis. *Hydrobiologia* 114 (3): 177–187. https://doi.org/10.1007/BF00031869.

Sathasivam, R. and Ki, J.S. (2018). A review of the biological activities of microalgal carotenoids and their potential use in healthcare and cosmetic industries. *Marine Drugs* 16 (1): 26. https://doi.org/10.3390/MD16010026.

Sathasivam, R., Radhakrishnan, R., Hashem, A., and Abd_Allah, E.F. (2019). Microalgae metabolites: a rich source for food and medicine. *Saudi Journal of Biological Sciences* 26 (4): 709–722. https://doi.org/10.1016/J.SJBS.2017.11.003.

Schüler, L.M., Santos, T., Pereira, H. et al. (2020). Improved production of lutein and β-carotene by thermal and light intensity upshifts in the marine microalga *Tetraselmis* sp. CTP4. *Algal Research* 45: https://doi.org/10.1016/J.ALGAL.2019.101732.

Schumann, R., Häubner, N., Klausch, S., and Karsten, U. (2005). Chlorophyll extraction methods for the quantification of green microalgae colonizing building facades. *International Biodeterioration and Biodegradation* 55 (3): 213–222. https://doi.org/10.1016/j.ibiod.2004.12.002.

Schwenzfeier, A., Wierenga, P.A., and Gruppen, H. (2011). Isolation and characterization of soluble protein from the green microalgae *Tetraselmis* sp. *Bioresource Technology* 102 (19): 9121–9127. https://doi.org/10.1016/J.BIORTECH.2011.07.046.

Senhorinho, G.N.A., Ross, G.M., and Scott, J.A. (2015). Cyanobacteria and eukaryotic microalgae as potential sources of antibiotics. *Phycologia* 54 (3): 271–282. https://doi.org/10.2216/14-092.1.

Senhorinho, G.N.A., Lannér, C., and Scott, J.A. (2019). Effect of green microalgal extracts exhibiting antibacterial activity on viability of human malignant and non-malignant cells. *Phycological Research* 67 (2): 145–151. https://doi.org/10.1111/pre.12362.

Seo, Y.H., Sung, M., Oh, Y.K., and Han, J.I. (2015). Lipid extraction and esterification for microalgae-based biodiesel production using pyrite (FeS2). *Bioresource Technology* 191: 420–425. https://doi.org/10.1016/j.biortech.2015.02.083.

Shahruzzaman, M., Biswas, S., Sakib, M.N. et al. (2019). Pharmaceutical applications of agar-agar. In: *Natural Polymers for Pharmaceutical Applications* (ed. A.K. Nayak, M.S. Hasnain, and D. Pal), 71–86. Apple Academic Press https://doi.org/10.1201/9780429328299-3.

Shalaby, S. (2013). Quality of novel healthy processed cheese analogue enhanced with marine microalgae *Chlorella vulgaris* biomass. *World Applied Sciences Journal* 23 (7): 914–925. https://doi.org/10.5829/idosi.wasj.2013.23.07.13122.

Shankar, P.D., Shobana, S., Karuppusamy, I. et al. (2016). A review on the biosynthesis of metallic nanoparticles (gold and silver) using bio-components of microalgae: formation mechanism and applications. *Enzyme and Microbial Technology* 95: 28–44. https://doi.org/10.1016/J.ENZMICTEC.2016.10.015.

Shao, W., Zhang, J., Wang, K. et al. (2018). Cocamidopropyl betaine-assisted foam separation of freshwater microalgae *Desmodesmus brasiliensis*. *Biochemical Engineering Journal* 140: 38–46. https://doi.org/10.1016/j.bej.2018.09.006.

Sharma, A., Madhu, Kondalkar, S.A. et al. (2024). Herbal Nutraceutical as Alternative Medicine. *Herbal Nutraceuticals*. Wiley; Portico. https://doi.org/10.1002/9781394241576.ch2.

Shoaf, W.T. and Lium, B.W. (1976). Improved extraction of chlorophyll a and b from algae using dimethyl sulfoxide. *Limnology and Oceanography* 21 (6): 926–928. https://doi.org/10.4319/LO.1976.21.6.0926.

Shokrkar, H., Ebrahimi, S., and Zamani, M. (2017). Bioethanol production from acidic and enzymatic hydrolysates of mixed microalgae culture. *Fuel* 200: 380–386. https://doi.org/10.1016/j.fuel.2017.03.090.

Simes, D.C., Viegas, C.S.B., Araújo, N., and Marreiros, C. (2020). Vitamin K as a diet supplement with impact in human health: current evidence in age-related diseases. *Nutrients* 12 (1): 138. https://doi.org/10.3390/NU12010138.

Singh, N.K. and Dhar, D.W. (2011). Phylogenetic relatedness among *Spirulina* and related cyanobacterial genera. *World Journal of Microbiology and Biotechnology* 27 (4): 941–951. https://doi.org/10.1007/S11274-010-0537-X/FIGURES/2.
Singh, P., Singh, R., Jha, A. et al. (2015). Optimization of a process for high fibre and high protein biscuit. *Journal of Food Science and Technology* 52 (3): 1394–1403. https://doi.org/10.1007/S13197-013-1139-Z.
Singh, S.D., Singh, S., Dwivedi, V. et al. (2020). Therapeutic and Nutritional Potential of Spirulina in Combating COVID-19 Infection. *AIJR Preprints* 49: 2–8. https://doi.org/10.13140/RG.2.2.16527.41127.
Singh, S.P. and Upadhyay, S.K. (2021). *Bioprospecting of Microorganism-Based Industrial Molecules*. Wiley. https://doi.org/10.1002/9781119717317.
Singh, S.P. and Upadhyay, S.K. (2023). *Microbial Bioreactors for Industrial Molecules*. Wiley. https://doi.org/10.1002/9781119874096.
Slocombe, S.P., Ross, M., Thomas, N. et al. (2013). A rapid and general method for measurement of protein in micro-algal biomass. *Bioresource Technology* 129: 51–57. https://doi.org/10.1016/J.BIORTECH.2012.10.163.
Soha S. M. Mostafa. (2012). *Microalgal Biotechnology: Prospects and Applications*.
de Souza, M.P., Hoeltz, M., Gressler, P.D. et al. (2019). Potential of microalgal bioproducts: general perspectives and main challenges. *Waste and Biomass Valorization* 10 (8): 2139–2156. https://doi.org/10.1007/S12649-018-0253-6.
Specht, E.A. and Mayfield, S.P. (2014). Algae-based oral recombinant vaccines. *Frontiers in Microbiology* 5 (FEB): https://doi.org/10.3389/FMICB.2014.00060.
Spolaore, P., Joannis-Cassan, C., Duran, E., and Isambert, A. (2006). Commercial applications of microalgae. *Journal of Bioscience and Bioengineering* 101 (2): 87–96. https://doi.org/10.1263/JBB.101.87.
Srinivasan, R., Kumar, V.A., Kumar, D. et al. (2015). Effect of dissolved inorganic carbon on β-carotene and fatty acid production in *Dunaliella* sp. *Applied Biochemistry and Biotechnology* 175 (6): 2895–2906. https://doi.org/10.1007/s12010-014-1461-6.
Srivastava, R.K., Nedungadi, S.V., Alharthi, M.A. et al. (2021). Nutraceutical products based on polysaccharides: sources, properties and applications. In: *Food, Medical, and Environmental Applications of Polysaccharides*, 531–554. Elsevier https://doi.org/10.1016/B978-0-12-819239-9.00021-X.
Steinbrenner, J. and Sandmann, G. (2006). Transformation of the green alga *Haematococcus pluvialis* with a phytoene desaturase for accelerated astaxanthin biosynthesis. *Applied and Environmental Microbiology* 72 (12): 7477–7484. https://doi.org/10.1128/AEM.01461-06.
Stoll, A.L., Severus, W.E., Freeman, M.P. et al. (1999). Omega 3 fatty acids in bipolar disorder: a preliminary double-blind, placebo-controlled trial. *Archives of General Psychiatry* 56 (5): 407–412. https://doi.org/10.1001/ARCHPSYC.56.5.407.
Sun, X.M., Ren, L.J., Zhao, Q.Y. et al. (2019). Enhancement of lipid accumulation in microalgae by metabolic engineering. *Biochimica et Biophysica Acta - Molecular and Cell Biology of Lipids* 1864 (4): 552–566. https://doi.org/10.1016/J.BBALIP.2018.10.004.
Sun, H., Wang, Y., He, Y. et al. (2023). Microalgae-derived pigments for the food industry. *Marine Drugs* 21 (2): https://doi.org/10.3390/MD21020082.
Talukdar, J., Dasgupta, S., Nagle, V., & Bhadra, B. (2020). COVID-19: potential of microalgae derived natural astaxanthin as adjunctive supplement in alleviating cytokine storm. *SSRN Electronic Journal*. https://doi.org/10.2139/SSRN.3579738
Tan, H.T., Khong, N.M.H., Khaw, Y.S. et al. (2020a). Optimization of the freezing-thawing method for extracting phycobiliproteins from *Arthrospira* sp. *Molecules* 25 (17): 3894. https://doi.org/10.3390/molecules25173894.

Tan, K.M., Kassim, M.A., Ng, Z.J., and Lalung, J. (2020b). Isolation and characterization of novel acidophilic microalgae from abandoned mining site area for carbohydrate biosynthesis and its kinetic growth study in photobioreactor. *IOP Conference Series: Materials Science and Engineering* 716 (1): 012011. https://doi.org/10.1088/1757-899X/716/1/012011.

Tang, D.Y.Y., Khoo, K.S., Chew, K.W. et al. (2020). Potential utilization of bioproducts from microalgae for the quality enhancement of natural products. *Bioresource Technology* 304: 122997. https://doi.org/10.1016/j.biortech.2020.122997.

Tibbetts, S.M., Milley, J.E., and Lall, S.P. (2015a). Chemical composition and nutritional properties of freshwater and marine microalgal biomass cultured in photobioreactors. *Journal of Applied Phycology* 27 (3): 1109–1119. https://doi.org/10.1007/S10811-014-0428-X.

Tibbetts, S.M., Melanson, R.J., Park, K.C. et al. (2015b). Nutritional evaluation of whole and lipid-extracted biomass of the microalga *Scenedesmus* sp. AMDD isolated in Saskatchewan, Canada for animal feeds: proximate, amino acid, fatty acid, carotenoid and elemental composition. *Current Biotechnology* 4: 530–546.

Tibbetts, S.M., Whitney, C.G., MacPherson, M.J. et al. (2015c). Biochemical characterization of microalgal biomass from freshwater species isolated in Alberta, Canada for animal feed applications. *Algal Research* 11: 435–447. https://doi.org/10.1016/j.algal.2014.11.011.

Tomás-Almenar, C., Larrán, A.M., de Mercado, E. et al. (2018). *Scenedesmus almeriensis* from an integrated system waste-nutrient, as sustainable protein source for feed to rainbow trout (*Oncorhynchus mykiss*). *Aquaculture* 497: 422–430. https://doi.org/10.1016/j.aquaculture.2018.08.011.

Úbeda, B., Gálvez, J.Á., Michel, M., and Bartual, A. (2017). Microalgae cultivation in urban wastewater: *Coelastrum* cf. *pseudomicroporum* as a novel carotenoid source and a potential microalgae harvesting tool. *Bioresource Technology* 228: 210–217. https://doi.org/10.1016/j.biortech.2016.12.095.

Udo, T., Mummaleti, G., Mohan, A. et al. (2023). Current and emerging applications of carrageenan in the food industry. *Food Research International* 173: 113369. https://doi.org/10.1016/J.FOODRES.2023.113369.

Upadhyay, S.K. and Singh, S.P. (2021). *Bioprospecting of Plant Biodiversity for Industrial Molecules*. Wiley. https://doi.org/10.1002/9781119718017.

Upadhyay, S.K. and Singh, S.P. (2023). *Plants as Bioreactors for Industrial Molecules*. Wiley. https://doi.org/10.1002/9781119875116.

Upadhyay, S.K. and Singh, S.P. (2024). *Herbal Nutraceuticals: Products and Processes*. Wiley. https://doi.org/10.1002/9781394241576.

Usman Shah, S.M. and Abdullah, M.A. (2018). Effects of macro/micronutrients on green and brown microalgal cell growth and fatty acids in photobioreactor and open-tank systems. *Biocatalysis and Agricultural Biotechnology* 14: 10–17. https://doi.org/10.1016/J.BCAB.2018.01.011.

Vaz, B.d.S., Moreira, J.B., Morais, M.G.d., and Costa, J.A.V. (2016). Microalgae as a new source of bioactive compounds in food supplements. *Current Opinion in Food Science* 7: 73–77. https://doi.org/10.1016/J.COFS.2015.12.006.

Vazhappilly, R. and Chen, F. (1998). Eicosapentaenoic acid and docosahexaenoic acid production potential of microalgae and their heterotrophic growth. *JAOCS, Journal of the American Oil Chemists' Society* 75 (3): 393–397. https://doi.org/10.1007/s11746-998-0057-0.

Wang, L. and Weller, C.L. (2006). Recent advances in extraction of nutraceuticals from plants. *Trends in Food Science and Technology* 17 (6): 300–312. https://doi.org/10.1016/J.TIFS.2005.12.004.

Wang, D.Z., Lin, L., Chan, L.L., and Hong, H.S. (2009). Comparative studies of four protein preparation methods for proteomic study of the dinoflagellate *Alexandrium* sp. using two-dimensional electrophoresis. *Harmful Algae* 8 (5): 685–691. https://doi.org/10.1016/j.hal.2009.01.001.

Wang, A., Islam, M.N., Qin, X. et al. (2014). Purification, identification, and characterization of d-galactose-6-sulfurylase from marine algae (*Betaphycus gelatinus*). *Carbohydrate Research* 388 (1): 94–99. https://doi.org/10.1016/J.CARRES.2013.12.010.

Ward, O.P. and Singh, A. (2005). Omega-3/6 fatty acids: alternative sources of production. *Process Biochemistry* 40 (12): 3627–3652. https://doi.org/10.1016/j.procbio.2005.02.020.

Wei, H., Shi, Y., Ma, X. et al. (2017). A type-I diacylglycerol acyltransferase modulates triacylglycerol biosynthesis and fatty acid composition in the oleaginous microalga, *Nannochloropsis oceanica*. *Biotechnology for Biofuels* 10 (1): 1–18. https://doi.org/10.1186/S13068-017-0858-1/FIGURES/10.

Wells, M.L., Potin, P., Craigie, J.S. et al. (2017). Algae as nutritional and functional food sources: revisiting our understanding. *Journal of Applied Phycology* 29 (2): 949–982. https://doi.org/10.1007/s10811-016-0974-5.

Wen, X., Wang, Z., Ding, Y. et al. (2020). Enhancing the production of astaxanthin by mixotrophic cultivation of *Haematococcus pluvialis* in open raceway ponds. *Aquaculture International* 28 (2): 625–638. https://doi.org/10.1007/s10499-019-00483-2.

Wijffels, R.H. and Barbosa, M.J. (2010). An outlook on microalgal biofuels. *Science* 329 (5993): 796–799. https://doi.org/10.1126/science.1189003.

World Health Organization (2019). *WHO: Global health estimates*. https://www.who.int/data/gho/data/themes/mortality-and-global-health-estimates/ghe-leading-causes-of-death (Retrieved 6 December 2023).

Wu, C., Xiao, Y., Lin, W. et al. (2017). Aqueous enzymatic process for cell wall degradation and lipid extraction from *Nannochloropsis* sp. *Bioresource Technology* 223: 312–316. https://doi.org/10.1016/j.biortech.2016.10.063.

Wunderlich, A.L.M., Azevedo, S.C.S.F., Yamada, L.A. et al. (2020). Chlorophyll treatment combined with photostimulation increases glycolysis and decreases oxidative stress in the liver of type 1 diabetic rats. *Brazilian Journal of Medical and Biological Research* 53 (1): https://doi.org/10.1590/1414-431x20198389.

Xiao, C., Liao, Q., Fu, Q. et al. (2019). A solar-driven continuous hydrothermal pretreatment system for biomethane production from microalgae biomass. *Applied Energy* 236: 1011–1018. https://doi.org/10.1016/j.apenergy.2018.12.014.

Xiaojian, D., Yun, H., Qiang, L. et al. (2017). Medium-low temperature hydrothermal hydrolysis kinetic characteristics of concentrated wet microalgae biomass. *International Journal of Agricultural and Biological Engineering* 10 (1): 154–162. https://doi.org/10.3965/j.ijabe.20171001.2699.

Xie, Y., Ho, S.H., Chen, C.N.N. et al. (2016). Disruption of thermo-tolerant *Desmodesmus* sp. F51 in high pressure homogenization as a prelude to carotenoids extraction. *Biochemical Engineering Journal* 109: 243–251. https://doi.org/10.1016/j.bej.2016.01.003.

Yamaoka, Y., Achard, D., Jang, S. et al. (2016). Identification of a *Chlamydomonas* plastidial 2-lysophosphatidic acid acyltransferase and its use to engineer microalgae with increased oil content. *Plant Biotechnology Journal* 14 (11): 2158–2167. https://doi.org/10.1111/PBI.12572.

Yan, N., Fan, C., Chen, Y., and Hu, Z. (2016). The potential for microalgae as bioreactors to produce pharmaceuticals. *International Journal of Molecular Sciences* 17 (6): https://doi.org/10.3390/IJMS17060962.

Yang, X., Yu, Y., Xu, J. et al. (2020). Clinical course and outcomes of critically ill patients with SARS-CoV-2 pneumonia in Wuhan, China: a single-centered, retrospective, observational study. *The Lancet Respiratory Medicine* 8 (5): 475–481. https://doi.org/10.1016/S2213-2600(20)30079-5.

Yoo, C., Jun, S.Y., Lee, J.Y. et al. (2010). Selection of microalgae for lipid production under high levels carbon dioxide. *Bioresource Technology* 101 (1): S71–S74. https://doi.org/10.1016/J.BIORTECH.2009.03.030.

Zen, C.K., Tiepo, C.B.V., da Silva, R.V. et al. (2020). Development of functional pasta with microencapsulated *Spirulina*: technological and sensorial effects. *Journal of the Science of Food and Agriculture* 100 (5): 2018–2026. https://doi.org/10.1002/jsfa.10219.

Zhang, X., Hu, Q., Sommerfeld, M. et al. (2010). Harvesting algal biomass for biofuels using ultrafiltration membranes. *Bioresource Technology* 101 (14): 5297–5304. https://doi.org/10.1016/j.biortech.2010.02.007.

Zhang, J., Hao, Q., Bai, L. et al. (2014). Overexpression of the soybean transcription factor GmDof4 significantly enhances the lipid content of *Chlorella ellipsoidea*. *Biotechnology for Biofuels* 7 (1): 1–16. https://doi.org/10.1186/S13068-014-0128-4/FIGURES/5.

Zhang, R., Grimi, N., Marchal, L., and Vorobiev, E. (2019). Application of high-voltage electrical discharges and high-pressure homogenization for recovery of intracellular compounds from microalgae *Parachlorella kessleri*. *Bioprocess and Biosystems Engineering* 42 (1): 29–36. https://doi.org/10.1007/s00449-018-2010-4.

Zhu, Q.L., Zheng, J.L., and Liu, J. (2020). Transcription activation of β-carotene biosynthetic genes at the initial stage of stresses as an indicator of the increased β-carotene accumulation in isolated *Dunaliella salina* strain GY-H13. *Aquatic Toxicology* 222: https://doi.org/10.1016/J.AQUATOX.2020.105472.

Zuorro, A., Maffei, G., and Lavecchia, R. (2016). Optimization of enzyme-assisted lipid extraction from *Nannochloropsis* microalgae. *Journal of the Taiwan Institute of Chemical Engineers* 67: 106–114. https://doi.org/10.1016/j.jtice.2016.08.016.

Index

Microbial Nutraceuticals: Products and Processes, First Edition. Edited by Sudhir Pratap Singh and Santosh Kumar Upadhyay.